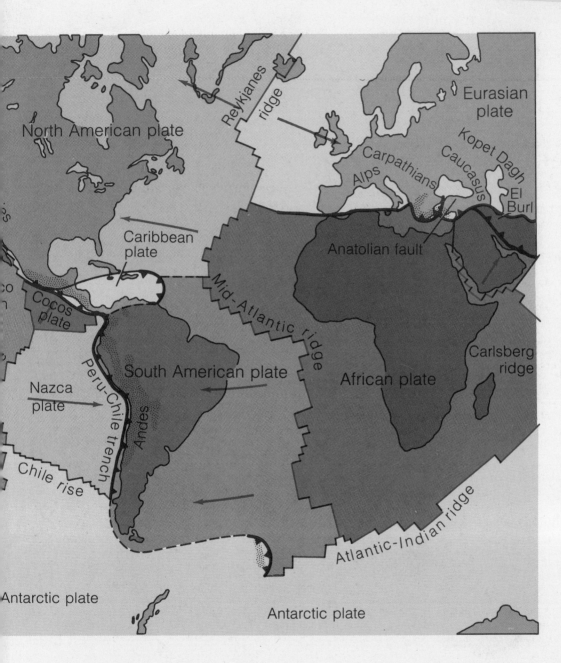

North American plate

Reykjanes ridge

Eurasian plate

Kopet Dagh

Caucasus

Alps Carpathians El Burl

Caribbean plate

Anatolian fault

Cocos plate

Mid-Atlantic ridge

Carlsberg ridge

South American plate

African plate

Nazca plate

Peru-Chile trench

Andes

Chile rise

Atlantic-Indian ridge

Antarctic plate

Antarctic plate

After "Plate Tectonics" by J. F. Dewey.

# Earth

Snake River, the Teton Range, Wyoming. [Photo by Ansel Adams.]

# Earth

Frank Press
MASSACHUSETTS INSTITUTE OF TECHNOLOGY

Raymond Siever
HARVARD UNIVERSITY

W. H. Freeman and Company
San Francisco

# A Series of Books in Geology

EDITORS: James Gilluly and A. O. Woodford

Library of Congress Cataloging in Publication Data

Press, Frank.
   Earth.

    Includes bibliographies.
    1.  Physical geology.  I.  Siever, Raymond, joint
author.
QE28.2.P7          550        73-21594
ISBN 0–7167–0261–4

Printed in the United States of America

9  8  7  6  5  4  3  2

*To Billie and Doris*

# Contents

# Preface

This book is an introduction to the study of the Earth as it is carried on by geologists of our day. We hope that it is as up to date as a current meeting of the Geological Society of America and, at the same time, as understandable as today's newspaper. It is intended for beginning students who have had no previous college science courses and who may not necessarily intend to specialize in geology. For this reason, we have deliberately emphasized a broad view, one that stresses concepts and shows by many examples what is meant by "scientific method." Through the use of analogies to familiar processes, the use of "kitchen physics" and "kitchen chemistry" in explanations, and the use of many diagrammatic illustrations, we try to show the evidential bases of geological theories and the strong dependence of geology on the basic scientific disciplines of physics and chemistry. The organic world plays an important role in many geological processes, and thus we introduce some notions of biological processes where appropriate. Yet first and foremost, the book is about geological processes.

As much as possible, we have tried to impart something about what motivates contemporary geologists and about the methods they use, the old ones as well as the new. We want the student to share some of the excitement and exhilaration triggered by the many recent discoveries that have greatly increased our understanding of how this planet works and by the beginnings that have been made in obtaining direct knowledge of how our nearby sister planets compare. Both of us are active in research, but we are also teachers, and believe that the gap between what is new and what is taught to beginners should be narrowed. We took it as a challenge to see if we could put down in a book what we do in our classes — integrate in a natural way the newest discoveries of plate tectonics, marine geology, geochemistry, geophysics, and lunar and Martian geology into the traditional discussions of such topics as geomorphology, sedimentation, petrology, volcanism, and structural geology.

We have not introduced the very new at the expense of eliminating the essential material, both traditional and modern, that a good course in geology should cover. We introduce the minimum vocabulary necessary to allow free discussion of the concepts involved; this is enough to serve as an introduction to any more-specialized courses that may follow. The coverage of subject matter is sufficiently complete that the book could, if used in its entirety, serve as a foundation on which to build a career in earth science. Nevertheless, our major aim is to reach the many students whose course in geology may be their sole college exposure to science and the study of the Earth. Ultimately they will participate as citizens in governmental policy decisions that pertain to geological questions.

Geology is a subject of growing importance. The decrease in available energy and materials and the increasing difficulties we encounter in managing our environment have heightened the strategic position of our discipline. More than at any time in the past century, geologists are today being asked to explain how oil is distributed in the Earth, why we have no domestic supply of some valuable metals, and many other and varied questions. For this reason, we have shown, in many places in the book, how *knowledge* of the science is linked to its *uses*, both in detailed practical ways and in the making of policy decisions. As geology has become the focus of more attention, it has aroused the curiosity of young people about nature in general. Enrollment in introductory geology courses has increased almost fourfold since 1960. Today's young people travel more than ever before, and as they see the diversity of the Earth, they want to understand what they see.

Geology is in a golden age as measured by the impact of recent discoveries and the profound insights that have emerged from them. For the first time in the history of the discipline, an all-encompassing synthesis of much of geological knowledge has been advanced. It is our hope that this book will reach many students and introduce them to geology with what we believe is a new approach, and impart something of the intellectual excitement, the growing relevance to societal problems, and the esthetics of the subject. If we succeed in this, the large investment we have made in time and energy will all have been worthwhile.

*Note on organization.* We have made each chapter as self-sufficient as possible, so that chapters may be skipped in short courses or be taken up out of sequence according to the instructor's individual taste. (Professor Roger Thomas, in his instructor's guide, has several suggestions along these lines.) We have designed *repetition, review,* and *alternative restatement* into the book with a definite purpose — to enhance learning and to increase flexibility in the way the book is used.

The extensive use of line drawings and photographs, plus the use of boxed information and other aids to learning, has resulted in a greater number of pages than we originally planned for, but it is our feeling that these aids will be of considerable value in simplifying otherwise difficult material and will serve to motivate the student. Explanatory sketches, diagrams, and photographs are used in places as substitutes for equations, and help to make up for lack of prior knowledge of other sciences. We also use illustrations as alternative statements of concepts presented in the text and as summaries of material covered earlier. To avoid interrupting the flow of the main text, occasional bits of parenthetical material appear in smaller type in the margins. These brief notes serve as slight amplifications of the text, as interesting sidelights, and as brief comments on some of the extraordinary personalities who have been part of the quest to understand the Earth. In a few places we have used boxes to expand in a more detailed way some materials of the text. These boxes are for the student who wants to understand more deeply some of the background of the subject. They are not necessary for understanding the text; some are pitched at a slightly higher level than the rest of the book.

The introductions at the beginning of the three parts of the book and the brief abstracts at the beginnings of chapters are designed to forecast in a general way the nature of the subject matter, how it fits together, and how it relates to other chapters. Summaries in list form appear at the ends of chapters to serve as systematic reviews of major conclusions. Questions were devised to help the student test his comprehension of the materials either by essay or by solving concrete problems. The bibliographies include *Scientific American* Offprints on closely related subjects, elementary or slightly advanced paperbacks, government reports on specialized topics, and a few readily available technical articles from the geological literature.

This book is divided into three parts. Each part consists of chapters grouped together according to their relations to the major concepts of the Earth's dynamics. Part I groups topics relating to the Earth as an evolving planet and how we study it and its materials. In the first chapter we give a capsule history of the Earth and the first glimpse of the general theory about how it operates. A brief outline of plate tectonics is presented to serve as a guide to succeeding chapters, where ramifications and implications of the theory are discussed with reference to the entire range of geological subjects considered in the book. The second chapter explores time in geology, the relation of process to history, and emphasizes how field observations form the central basis of our knowledge of the geological cycle. The third chapter, complementary to the second, shows how experimentation and instrumental observations work in conjunction with field information to deepen our understanding. The last chapter of Part I is concerned with the prime source of information on the Earth: rocks and minerals. The major concepts of mineralogy and petrology are introduced and linked to a brief but systematic discussion of the major rock-forming minerals and the three major rock groups.

Part II covers those aspects of the Earth that are dominated by the external solar heat machine, all of the surface processes that result from the Sun's radiant energy impinging on the surface of the planet, its atmosphere, and oceans. Erosion, transportation, and deposition of chemically altered and physically fragmented rocks, and the resulting sculpture of the surface are discussed in relation to tectonics and the dynamics of the atmosphere and oceans. Part II concludes with a chapter on the interactions of the biological world and Earth's inorganic materials, and how man as a geological agent has been profoundly changing the surface environment.

Part III explores the consequences of the internal heat machine of the Earth, and how it drives major movements of the interior and determines the structure of the whole planet. Internal heat, volcanism, and the kinds of igneous and metamorphic rocks that are produced by thermal processes are the subjects of the first group of chapters. The structure of the interior as deduced from seismology, gravity, and magnetism is then explored in preparation for a detailed systematic explanation of plate tectonics. It is only after this that we come to structural geology, which then can be treated in the context of the large-scale motions of lithospheric plates. Following this is a chapter on what we know of the nature and evolution of the other planets in the solar system, with major emphasis on lunar exploration. The book concludes with a chapter on earth materials as resources for Man, including an extended treatment of energy reserves and the central importance of energy costs in the recovery of all other resources.

*Frank Press*
*Raymond Siever*

*December 1973*

# Acknowledgments

James Gilluly (formerly of the U.S. Geological Survey) and A. O. Woodford (Pomona College) read the entire manuscript. They were our most useful critics as measured by the amount of rewriting that followed their reviews. Individual chapters or various portions of the manuscript were reviewed in detail by Robert E. Boyer (The University of Texas at Austin), William F. Brace (Massachusetts Institute of Technology), Charles W. Burnham (Harvard University), Robert H. Dott, Jr. (University of Wisconsin), Ronald J. Gibbs (Northwestern University), Stephen J. Gould (Harvard University), Thomas R. McGetchin (Massachusetts Institute of Technology), Ronald L. Shreve (University of California, Los Angeles), John B. Southard (Massachusetts Institute of Technology), and Seiya Uyeda (Tokyo University). Robert S. Fiske (U.S. Geological Survey), John Haller (Harvard University), Warren B. Hamilton (U.S. Geological Survey), Edwin D. McKee (U.S. Geological Survey), Paul E. Potter (University of Cincinnati), and the staff of the U.S. Geological Survey Photographic Library were particularly helpful in supplying photographs. Peter Bird helped us greatly in compiling the glossary and index. The staff of W. H. Freeman and Company showed interest and commitment beyond the call of duty. Richard Johnson and Gunder Hefta taught us much about literary style and how to communicate complex notions simply, yet accurately. John Staples not only sparked our efforts with encouragement and enthusiasm but served us admirably as a source of ideas, advice, criticism, and help in moments of difficulty. The attractive and functional book design is due to the good work of Robert Ishi, and for efficiently coordinating the book-manufacturing process we thank Batyah Janowski.

In order to ensure that the level and scope of the book be appropriate for the intended audience, we sent chapters to a number of well-known teachers of introductory courses. Their responses to our questionnaire and their individual comments proved invaluable in setting the proper tone, style, level, and content. We are indebted to the following colleagues for this special contribution: Randall S. Babcock (Western Washington State College), Walker H. Baker (Onondaga Community College), Mervin J. Bartholomew (Virginia Polytechnic Institute and

State University), David Bickel (Minot State College), Shawn Biehler (University of California, Riverside), Ruth B. Boeckerman (Fullerton Junior College), Roger F. Boneham (Indiana University at Kokomo), Graeme Bonham-Carter (University of Rochester), R. H. Bruns (The University of Arizona), John A. Burger (Beloit College), John D. Cooper (California State University, Fullerton), Charles M. Davis (The American University), John G. Dennis (California State University, Long Beach), L. D. deYampert (Broward Community College), Hugh W. Dresser (Montana College of Mineral Science and Technology), J. Mark Erickson (St. Lawrence University), William J. Frazier (Madison College), Ansel M. Gooding (Indiana University East), Gary O. G. Greiner (Southampton College), N. Timothy Hall (Foothill College), David V. Harris (Colorado State University), Arthur W. Hayes (Campbell College), Frank Hensley (Otero Junior College), I. M. Jamieson (Riverside City College), Roger L. Kaesler (The University of Kansas), Stephen J. Kridelbaugh (University of Oregon), Leonard Kubicek (University of Northern Colorado), Edwin E. Larson (University of Colorado), Robert D. Lawrence (Oregon State University), Jere H. Lipps (University of California, Davis), John D. Longshore (California State University, Humboldt), Lawrence Lundgren (University of Rochester), Michael J. McLane (The University of Wisconsin, Marathon County Campus), Marshall E. Maddock (California State University, San Jose), Stanley A. Mertzman (Franklin and Marshall College), Barry B. Miller (Union College), Donald W. Newberg (Colgate University), Won C. Park (Boston University), Susan Patch (New York University), Donald D. Runnells (University of Colorado), Dewey D. Sanderson (Marshall University), Melvin C. Schroeder (Texas A&M University), Robert E. Stevenson (University of South Dakota), Richard L. Threet (California State University, San Diego), Russell O. Utgard (The Ohio State University), John A. Vargas, Jr. (Pennsylvania State University, DuBois Campus), Janet K. Warter (California State University, Long Beach), and John T. Whetten (University of Washington).

# Part I

# THE EARTH AS AN
# HISTORICALLY EVOLVED BODY
# AND HOW WE STUDY IT

This planet on which we live has undergone constant change throughout a long history. To understand what the Earth is and how it works today we link direct observation of processes operating at the surface with indirect measurement of forces working in the interior. The fullest knowledge comes from deducing how the planet evolved—from its beginnings to its present state.

Formed almost five billion years ago from a mass of dust rotating around the infant Sun, the Earth grew into a medium-sized planet whose history has been dominated by two driving machines. The first is the internal heat produced by radioactivity inside the Earth. The second is the external heat supplied to the surface by the Sun. The internal heat melts rocks, makes volcanoes, and thrusts mountains upward. The external heat drives the atmosphere and the oceans and causes the erosion of mountains and the reduction of rock to sediment. New methods of studying how the internal and external forces drive the Earth have generated a wealth of new information and raised many exciting new questions. In the past decade, geologists have gradually developed a new unifying theory that relates all of the dynamic Earth processes to the motions of large plates that constitute the outer shell of the planet. Called the theory of plate tectonics, it offers the most comprehensive model that geologists have ever had for explaining how the Earth works.

In the first part of this book we survey the ways in which we study this planet and what we have learned of its origin. These first chapters are a capsule of the book; in them we preview many subjects—particularly the nature of time and the materials of Earth—before they are explored in detail in later chapters.

# 1

# History of the Earth and Solar System

*An introduction to the Earth, beginning with a cloud of dust and gas from which the solar system formed some 5 billion years ago, to the birth of the Earth about 4.7 billion years ago, to the planet we know today, with its hospitable atmosphere and rich resources, a planet still active inside—as evidenced by earthquakes, volcanoes, ocean basins that open and close, and continents that drift apart.*

## THE UNIQUENESS OF PLANET EARTH

"Civilization exists by geological consent, subject to change without notice," said philosopher-historian Will Durant, reminding us of the remarkable circumstances that make this planet congenial to life as we know it. The Earth, after all, is a very special place— and not just because we humans inhabit it. More than a million life forms have developed on this unique spot in the solar system. *Homo sapiens,* the one species with the power of reason, is a rather recent arrival. In the study of geology, we explore not only the Earth as it exists today; we also seek answers to how it was formed, what it was like when first born, how it evolved to the planet of today, and, perhaps most exciting of all, what made it capable of supporting life.

No one knows precisely when the composition and state of the Earth's primitive atmosphere were just right for life to begin and evolve. We do know, however, that the large organic molecules that apparently preceded the evolution of the earliest forms of

It is quite probable that the conditions that would allow life to flourish anywhere in the Universe would not differ much from those that have allowed life to evolve on Earth. This observation led astronomers to propose that other planets, situated about as far from their suns as we are from ours, might also have life.

life could not have formed if the primitive atmosphere contained as much oxygen as the one we now enjoy. Chemists tell us the oxygen would have destroyed them. The Earth's atmosphere and magnetic field acted as a shield against some of the biologically damaging radiation from space, just as they do today. Meteors, unbraked by the cushion of gases, would have bombarded Earth, leaving a crater-pocked, desolate surface that could never be softened by erosion. The atmosphere that exists today not only filters out the greatest part of destructive ultraviolet radiation, but together with the ocean, it stores and redistributes solar energy, thus moderating climate. Without atmosphere and oceans, there would be much more extreme temperature differences between day and night, summer and winter, and equator and pole.

The list of fortuitous conditions so propitious to life is long. Life as we know it is possible over a very narrow temperature interval — essentially within the limits set by the freezing and boiling points of water. This interval is perhaps 1 or 2 percent of the range between a temperature of absolute zero and the surface temperature of the Sun. How fortunate that Earth formed where it did in the Solar System, neither too far from the Sun nor too close to it! And Earth's size was just about right — not so small as to lose its atmosphere because its gravity was too small to keep it from escaping into space, and not so large that its gravity would hold on to too much atmosphere, including harmful gases.

We will see that the Earth's interior is a gigantic but delicately balanced heat engine fueled by radioactivity, which has much to do with how the surface evolved. Were it running more slowly, geological activity would have proceeded at a slower pace. The continents might not have evolved to their present form, and volcanoes might not have spewed out the water and gases that became the oceans and atmosphere. Iron might not have melted and sunk to form the liquid core, and the magnetic field would never have developed. The Earth would then have evolved as a cratered, dead planet similar to the Moon. Another scenario can be imagined: if there had been more radioactive fuel and a faster-running engine, volcanic gas and dust would have blotted out the Sun, the atmosphere would have been oppressively dense, and the surface would have been racked by daily earthquakes and volcanic explosions. Perhaps the Earth had such a fast-running era early in its history.

## ASPECTS OF GEOLOGY

Although geology has ranked as a modern scholarly discipline for only two centuries, man has been curious about the Earth and its origins from the beginnings of pre-history. Stories of creation are invariably found in the sagas and folktales of early civilizations.

**Figure 1-1**
The Earth, geologically active, with atmosphere and ocean and conditions
hospitable to life; the Moon, geologically inactive, without atmosphere or water,
a lifeless body. [Composite made from National Aeronautics and Space
Administration photographs.]

In doing so, perhaps the ancients derived some feeling of security;
perhaps they satisfied their own curiosity by reciting creation
myths to their young, as if to put behind them the primeval chaos
of an unknown creation. But a common theme was the creation,
the bringing into existence, as contrasted with the idea of always
having been here.

**Nature's Threat** The human need to understand and to be
able to explain nature in order to gain protection from her vagaries
survives to this day as a major motivation toward the study of
geology. Modern man seeks safeguards against nature's threats,
which come in the form of earthquakes, landslides, volcanic eruptions, floods, and the destructive sea waves known as tsunamis.
Perhaps even more dangerous are man-made catastrophes, for
our unique species has gained the power to trigger earthquakes,
foul the atmosphere and oceans, and alter climates to the point of
either initiating ice ages or melting the polar ice caps and flooding
coastal cities. If man can do these things, so can he eventually
predict or control most natural disasters.

Nature appreciation

Environment

Natural hazards

Natural resources

Scientific

**Figure 1-2**
Different aspects of geology — nature
appreciation, environmental protection,
hazard reduction, natural resources,
scientific research.

**Economic Geology**  Were it not for the accessibility and diversity of minerals in the Earth's crust, man's cultural level would not have progressed beyond the Stone Age. The discovery of the Earth's mineral wealth is the geologist's task. If we are to extend to the many the affluence enjoyed by the few, let alone maintain our own present standard of living, new mineral deposits must be found. Prospectors have long since found the obvious deposits of iron, copper, tin, uranium, oil, and other important minerals. It is a challenge to geologists to re-explore the world, using new tools and techniques to ferret out undiscovered deposits. New and pressing concerns of geologists are conservation and environmental protection. How can we most efficiently exploit nature's wealth without waste and without devastating the landscape? Somehow we must find answers to these questions.

**Scholarly Geology**  Geology also has its "pure" aspects for those who work at it, in that it is interesting of and for itself. How a planet is born — the course of its evolution — and how it works today are only partially answered questions. Geologists are motivated to find answers because, like all scientists, they have unbounded curiosity and perhaps even a sense of uneasiness when important natural phenomena remain unexplained. Geologists will be hammering at outcrops, making geologic maps, exploring the sea floor, and scrutinizing Moon rocks as long as mountain-making, continental drift, sea-floor spreading, earthquakes, and other planetary features remain incompletely explained.

**Geology for the Poets**  Most students who enroll in geology courses have no intention of becoming professional geologists. They elect geology for many reasons. Perhaps they hope to heighten their appreciation of nature by gaining insight into her many ways. Perhaps the current crisis of the environment has induced them to learn more about a key environmental science. Perhaps Norman Mailer expressed their motivation when he wrote of the moon flight of the Apollo astronauts: "yes, we might have to go out into space until the mystery of new discovery would force us to regard the world once again as poets."

## ORIGIN OF THE SYSTEM OF PLANETS

Let us start at the beginning with the first and most difficult problem: How did the **system of planets** originate? This question has attracted the attention of many of the great philosophers and scientists of the past two centuries. Yet it is a rare geological or astronomical congress that does not witness a fresh debate triggered by the latest experimental data or the newest theoretical advance pertaining to this question. Perhaps this chapter should have been published in loose leaf form to facilitate frequent updating.

**Explain the Observations** If anyone wishes to enter the lists with a pet hypothesis, the procedure is simple enough. On the basis of logical reasoning, develop a mechanism for the formation of the planetary system—a mechanism that is self-consistent and which can explain the mass and size distribution of the planets, the peculiarities of their orbits, the relative abundances of elements in the planets and the sun. Useful hints can be gleaned from studies of other stars.

Whatever did happen, beginning about 4.7 billion years ago at the start of the process, resulted in several amazing regularities and curious groupings in the solar system:

1. The planets all revolve around the Sun in the same direction, in **elliptical**, but almost circular, **orbits** that lie in nearly the same plane; most of their moons also revolve in the same direction.

We will show in Chapter 2 how certain radioactive elements, uranium, potassium, and rubidium, serve as clocks that enable us to chronicle major events in the solar system, such as its beginning 4.7 billion years ago.

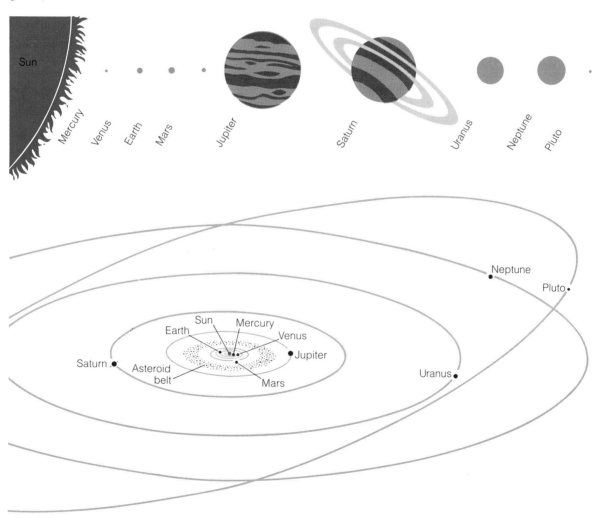

**Figure 1-3**
The Solar System. Diagrammatic representations of the sun and planets.

2. The planets, except for Uranus, *rotate* in the same direction as their revolution around the Sun — that is, counterclockwise as one looks from the North pole to the South pole of the Earth.

3. The distance of each planet from the Sun is roughly twice as far as that of the next planet closer to the Sun (an ordering known as the **Titius-Bode rule**).

4. Although the Sun makes up about 99.9 percent of the mass of the solar system, 99 percent of the **angular momentum** is concentrated in the large planets (see Fig. 1-4 for an explanation of angular momentum).

5. The planets form two groups: the so-called **terrestrial planets**, Mercury, Venus, Earth and Mars, which form an inner group of small, dense bodies (densities about 4 to 5.5 times that of water); and the giant planets, Jupiter, Saturn, Uranus and Neptune, which are an outer group of large bodies with low densities (between 0.7 and 1.7 times that of water). In some respects — for example, their high gas content and low density — the giant planets are more like the Sun than like the terrestrial planets.

From chemical analysis of Earth rocks, Moon rocks, and meteorites that reach the Earth from interplanetary space, we surmise that the terrestrial planets are mostly (±90 percent) made up of four elements: iron, oxygen, silicon and magnesium. Spectroscopic studies of the sun show it to be composed almost entirely (99 percent) of hydrogen and helium. Presumably the high abundances of hydrogen and helium are features of the giant planets also.

**The Nebular Hypotheses**   There has been no dearth of theories of creation over the centuries. The modern approach to the problem, however, began in 1755 when the German philosopher *Immanuel Kant* hypothesized a primeval, slowly rotating cloud of gas, now called a nebula, which in some unspecified fashion condensed into a number of discrete, globular bodies. In this way Kant neatly explained the consistency of revolution and rotation directions, in that the rotation of the parent nebula is preserved in the rotation of the Sun, the revolution of the planets about the Sun, and the rotation of the planets about their axes — all in the same direction (Fig. 1-5).

The great French mathematician *Laplace* proposed essentially the same theory in 1796 — surprisingly, without the mathematical formulation he was capable of providing. Historians of science will have to resolve the questions of whether Laplace knew of Kant's work and why he chose not to subject his own *nebular hypothesis* to mathematical examination, for had he done so he might have discovered some serious flaws.

According to Kant and Laplace, the original mass of gas cooled and began to contract. As it did, the rotational speed increased (a consequence of the **law of conservation of angular momentum**, explained in Figure 1-6) to the point where successive

Light is emitted or absorbed in a characteristic way by different elements in gaseous form, as they incandesce. Analysis of light into color components (more accurately, its spectral components) reveals the composition of its source. Thus the yellow color produced when common salt (sodium chloride, NaCl) is burned (vaporized) in a natural gas flame reveals the presence of the element Na.

Kant's hypothesis, published anonymously, carried an impressive title: *Universal Natural History and Theory of the Heavens, or an Essay on the Constitution and Mechanical Origin of the Whole Universe, Treated according to Newtonian Principles.* The publisher went bankrupt and the stock was seized by the creditors, so that very few copies reached public hands.

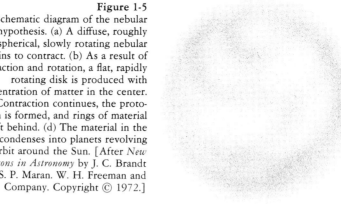

a

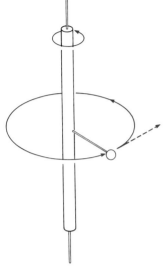

b

**Figure 1-4**
Angular momentum, illustrated by the example of a heavy steel ball fastened to the central shaft by a very light rod. Turning the shaft causes the ball to rotate around it. The ball's angular momentum is defined as the product of the mass of the ball, its velocity, and its distance from the rotation axis. [From *New Horizons in Astronomy* by J. C. Brandt and S. P. Maran. W. H. Freeman Company. Copyright © 1972.]

c

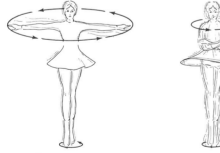

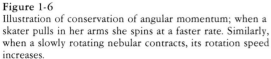

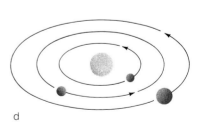

d

**Figure 1-6**
Illustration of conservation of angular momentum; when a skater pulls in her arms she spins at a faster rate. Similarly, when a slowly rotating nebular contracts, its rotation speed increases.

rings of gaseous material were spun off from the central mass by centrifugal force. In the final stages the rings condensed into planets.

Not so, according to the great British physicists *James Clerk Maxwell,* and *Sir James Jeans,* who showed about one hundred years later that there was not enough mass in the rings to provide the gravitational attraction for condensation into individual planets. The coup de grâce was delivered at the close of the nineteenth century, when astronomer *F. R. Moulton* of Chicago showed that the nebular hypothesis violated item 4 above—namely, that the planets have most of the angular momentum. The law of conservation of angular momentum requires each part of a rotating, condensing nebula to hold onto its angular momentum; the Sun, which collected most of the mass, should have gathered up most of the angular momentum of the system. Simply stated, the Sun doesn't rotate fast enough; it should have spun faster, just as the skater in Figure 1-6 spins faster when she pulls in her arms.

**Collision Hypotheses** Wanted—a theory, now that Kant and Laplace were in disfavor. Geologist *T. C. Chamberlin* collaborated on one theory with Moulton, his astronomer fellow professor at Chicago, and some years later Sir James Jeans and *Sir Harold Jeffreys,* a mathematician-astronomer-geophysicist in England, came forward with a revival of an early proposal (1749) of *Count Buffon* of France—the **collision hypothesis.** Stated simply, this hypothesis holds that material was torn from the pre-existing Sun by the gravitational attraction of a passing star. Giant tongues of matter were pulled out. According to Chamberlin and Moulton, these broke into small chunks, or **planetesimals,** which went flying as cold bodies into orbits around the sun in the plane of the passing star. By collision and gravitational attraction, the larger planetesimals swept up the smaller pieces, and those that survived became the planets. Unfortunately, the several versions of collision theories have fatal weaknesses. According to astronomers, much of the material ejected from the Sun would have come from the interior and would have been so hot, perhaps 1,000,000°C, that the gases would have been dispersed throughout space with explosive violence rather than condensed into planets. Although more angular momentum would have been imparted to the planets by a passing star than by the rotation of a nebula, the amount is still less than that observed. Finally, the vastness of space makes the probability of such a close approach of two stars extremely small.

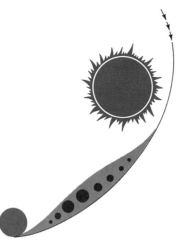

**Figure 1-7**
Schematic representation of the formation of planets according to the collision hypothesis. A passing star pulls giant tongues of matter from the Sun, and these develop into the planets and satellites.

**Recent Theories** The thinking of the past few decades has been influenced by the discovery that space is not as empty as had been thought. Astronomical observations have detected, both in interstellar space and in nebulae, rarefied matter consisting of about 99 percent gas and 1 percent dust. The gases are mostly hydrogen and helium; the dust-size particles have compositions similar to those of terrestrial materials, such as silicon compounds, iron oxides, ice crystals, and a host of other small molecules and compounds, including organic ones. It is not surprising, then, that recent theories tend to be neo-Laplacian in that they revive the idea of a primordial, rotating cloud of gas and dust whose shape and internal motions were determined by gravitational forces and the forces of rotation. At some moment gravitational attraction became the dominant factor, contraction began, and the rotation speeded up (again, conservation of angular momentum). The cloud tended to flatten into a disc; matter began to drift toward the center, accumulating into the **proto-Sun**. The proto-Sun collapsed under its own gravitation; it became dense and opaque as the material was compressed. Its internal temperature rose to about 1,000,000°C at which point "nuclear burning," or fusion, began. More precisely, the Sun began to shine with the initiation of the **thermonuclear reaction** (now unfortunately duplicated on Earth by H-bombs), in which hydrogen nuclei combine under intense pressure to form helium nuclei, releasing a vast amount of energy.

What about the disk of planetary material enveloping the primitive Sun? How did it form into planets? How did the planets pick up the necessary angular momentum, and what about the segregation of the outer, large, sunlike planets from the small, dense, inner planets? There is very little agreement among the experts about the answers to these questions.

*Harold Urey,* an American chemist and Nobel Prize winner, *W. A. Fowler,* an American physicist, and *Sir Fred Hoyle,* an English astronomer, have used arguments drawn from nuclear physics and chemistry to try to find some answers. From their work as well as that of others, a model emerges in which the Sun did acquire most of the angular momentum, as the law of conservation of angular momentum requires, but afterwards lost it to the planetary disk by a process of **magnetic coupling**. See Figure 1-8 for a familiar example of magnetic coupling. Stars are known to have magnetic fields that extend into surrounding space. The more rapidly rotating Sun dragged the less rapidly rotating disk, the linkage being a rotating solar magnetic field interacting with gases whose particles acted as miniature magnets. In taking up the Sun's angular momentum—that is, in taking on a faster rotation—most of the gaseous components in the planetary disk were flung to the far reaches of the solar system, where they condensed into the great planets. Following the outflow of the gases, the remaining particles consisted mostly of iron, silicon, and magnesium oxides —the major raw materials of rock. These formed into small, cold planetesimals that ultimately grew into the terrestrial planets by collision and gravitational attraction. How long did all of this take? The experts say about 100 million years, or less. The Titius-Bode rule about the mathematically regular spacing of the planets still awaits a satisfactory explanation. As one of the specialists recently

11

Exciting recent discoveries of such biologically important compounds as ammonia, methane, formic acid, and formaldehyde in interstellar space and of amino acids in meteorites have profound implications for the origin of life on Earth and elsewhere in the universe. Their presence in space shows that the formation of the simple organic compounds, the building blocks of life, is an ordinary accomplishment of galactic evolution.

**Figure 1-8**
The rotating magnet beneath a toy skating rink can impart a rotation to the magnets (attached to the form of skaters) above by "magnetic coupling." Similarly, the magnetic field of the Sun, rotating with the Sun, could have imparted a rotation to the surrounding gases whose particles acted as miniature magnets. In this way the more rapidly rotating Sun could have given up its angular momentum to the planetary disk, the Sun slowing down and the planetary disk spinning up as a result.

## Box 1-1 More Details on the Formation of the Planets

For the reader who wishes more examples of the kinds of reasoning used, we expand on how it is inferred that most of the gases moved to the outer part of the solar system before the planetesimals formed and grew into the terrestrial planets. Gases differ from liquids and solids in that the molecules move freely in irregular zig-zag paths as a result of repeated collisions. Physics tells us that the hotter the gas, the more rapid the movements and also that light molecules move faster than heavy ones. Unless the gas is contained by walls or held by the gravity of a planet, the rapidly moving molecules would escape. It is easy to see how a planet's atmosphere could lose light gases like hydrogen and helium. The molecules move too fast for gravity to hold them very long. But the slower-moving molecules of heavy gases like krypton and xenon should have been held selectively by the planet's gravitational field. The fact that they are deficient on Earth argues that the growth of planetesimals into planets took place in a region previously swept clear of both light and heavy gases.

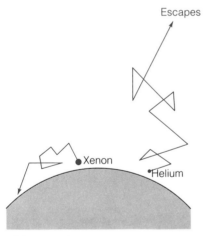

Schematic illustration of how Earth's gravity should have held on to heavy gas atoms like xenon and krypton, whereas light gases like hydrogen and helium would have escaped into space. The deficiency of both light and heavy gases suggests that planet Earth originated in a region previously swept clear of almost all gases.

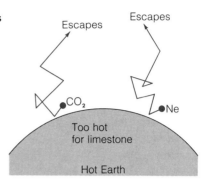

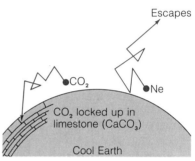

Schematic diagram showing why the rarity of such inert gases as neon and the abundance of such chemically reactive gases as water and carbon dioxide indicate that the Earth was probably cool in its beginning days.

Taking a chemical approach, we also infer that the Earth was probably cool when it first formed, rather than hot, because neon, the gas used in electric signs, is extremely rare in the atmosphere, but comparatively abundant in the stars. Somehow, most neon escaped from the Earth. Water molecules, however, weigh nearly as much as neon and for that reason should have escaped along with neon, yet there are oceans of water on Earth today. Water differs from neon chemically in that it reacts under "cool" conditions to form compounds that do not easily evaporate, whereas neon is chemically inert and remains a gas. Thus a cool primitive Earth would selectively retain water (and other chemically reactive gases like carbon dioxide and ammonia) by locking them up as compounds in rock, whereas inert gases like neon would gradually escape. If the Earth were initially hotter than a few hundred degrees, both the chemically reactive and inert gases would be lacking today.

wrote us, "nothing works in all details, and so there is still some fun left in the old problem."

The preceding should be taken for what it is—a hypothesis, a possible model. Perhaps some of these notions come close to what actually happened: we will only know after much additional work is carried out, some of it already underway. Nebulae at different stages of development are now being studied not only with the familiar optical telescope but also with special devices that magnify x-rays and radio waves. These invisible but detectable waves provide additional information about what goes on in the remote sections of the universe. The first planetary probes have already been returning data on the composition of the atmospheres and crusts of Mars, and Venus, and the Moon. Mathematicians are developing new tools to improve computer reproduction of the motions of rotating gas-dust clouds and growing planets. All of this activity should in time give us better answers about how it all started.

We have dwelt on the question of the origin of the solar system for several reasons. The evolutionary course followed by a planet is set by its initial state. The current state of Earth, some 4.7 billion years later, is reasonably well known to us. These two times in the course of **planetary evolution**—the beginning and now—are important constraints in developing models of how the Earth has changed throughout its history. The growth of ideas on the origin of the planets is an interesting story in the history of science. It illustrates how successive hypotheses are advanced, rejected, resurrected, and modified in the light of new observational data and theoretical concepts. Typical of all of science, it is especially pertinent to modern geology, where a revolution in thought and concept is currently underway.

## EARTH AS AN EVOLVING PLANET

This section serves as an introduction and preview to all that pertains, in this book, to the Earth's large-scale evolution. We will sketch the transformation of the Earth from an initially homogeneous body to a differentiated planet—that is, one in which the interior is divided into layers or zones, each differing chemically and mineralogically. We will see how the process of differentiation is indirectly related to the formation of the atmosphere, oceans, continents, mountains, volcanoes, and magnetic field. The general sequence of events rather than details will be emphasized, for no other reason than that many of the details have yet to be worked out by future generations of Earth scientists.

**Initial State—A Homogenized Conglomeration** Think of the Earth as it was before the beginning of the geologic record—that is, before the formation of the oldest rocks now known, which

A 4000-kilogram (5-ton) planetesimal impacting with a velocity of 30 km/sec (19 miles/sec) delivers as much energy as a 1-kiloton nuclear explosion.

are nearly 4 billion years old. This period covers the first billion years in the history of the Earth, the events of which we now attempt to reconstruct, even though direct evidence in the form of rocks dating back to these early times is lacking. The stage was set for the accumulation of the planet by the gathering up or accretion of planetesimals about 4.7 billion years ago. The new planet was probably an unsorted conglomeration, mostly of silicon compounds, iron and magnesium oxides, and smaller amounts of all of the natural chemical elements. Although the planetesimals were relatively cold, the growing planet soon began heating up because of three different effects.

**The Initial Temperature** Each infalling planetesimal carried much energy of motion—energy that was converted to heat upon impact. Although much of the heat was radiated back into space, a significant fraction was retained by the growing planet. Just how much is uncertain, since it depended on the mass, velocity, and temperature of the planetesimals and on the rate of accretion. At a high rate of accretion, the heated impact zone would have been covered by material that arrived later before the energy could radiate back into space, and the "buried" heat would have raised the temperature of the interior.

Compression is another effect that leads to a temperature rise. A common example is the heating of the barrel of an air pump, like those we use to inflate bicycle tires. The air is compressed quickly as we push down, too quickly for the heat to be diffused away, and the barrel heats up. The inner parts of the planet were squeezed under the growing weight of the accumulating outer parts. The energy expended in compressing the interior was converted to heat, which was locally retained. The heat did not flow out because heat moves, or is conducted, very slowly in the rocks. As a result, the heat accumulated, and the temperature increased inside the Earth. Most geophysicists who have calculated the magnitude of heating think that accretion and compression resulted in an average internal temperature of about 1000°C for the newly organized planet.

**Heat from Radioactive Disintegration** The heavy elements uranium and thorium, and the small fraction of potassium atoms that are heavier than ordinary potassium, are not very plentiful on Earth. Their occurrence is measured in a few parts per million (one gram in a thousand kilograms of rock). Yet these elements have had a profound effect on the evolution of the Earth because of their **radioactivity**. The atoms of these elements spontaneously disintegrate by emitting atomic particles (helium nuclei and electrons), and are thus transformed into different elements. The emitted particles are the important part of this discussion. As they are absorbed by the surrounding matter, their energy of motion is transformed into heat. At first thought, the heat generated may seem inconsequential: about 20 calories of heat is emitted by a cubic centimeter of granite in the course of a million years. This does not seem impressive when we realize that it would take some 500 million years to brew a cup of coffee using the radioactive heat released by a cubic centimeter of granite (assuming, of course, that the heat would be retained and not flow out). What *is* impressive is that the temperature would rise to the melting point of granite in several billion years — even sooner if the temperature increase due to accretion and compression were also included. Thus the disintegration of the radioactive elements provides a heat source that has persisted for billions of years. Radioactively generated heat, its outward flow slowed by the low thermal conductivity of rocks, caused the newly formed Earth to warm up. The process of planetary development was thus initiated — energized by radioactivity.

## HEATING UP, OVERTURN, AND FORMATION OF A DIFFERENTIATED EARTH

**The Earth Heats Up**   One-hundred fifty years ago, the French mathematician *Fourier* derived his heat-conduction equation, which is still used to calculate the flow of heat from one part of a body to another, together with the corresponding temperature changes. To start the computation, one needs to know the temperature at the boundaries (e.g., the Earth's surface), the beginning temperature, and the location and strength of heat sources in the body. So far as the Earth is concerned, the last two factors are uncertain. Nevertheless, one can make reasonable guesses about the starting temperature and radioactivity of the Earth and compute models of how temperature may have changed inside in the years following the "birth" of the planet.

Almost everyone who has done this finds the same important feature. Figure 1-10 is a graph of one of the computational results. The curves illustrate how the internal temperature increased in the years following the planet's formation. The computer verifies what we might have guessed — that radioactive heat was generated more rapidly than it could flow away (remember, rocks are poor heat conductors), and that the interior gradually warmed up. Figure 1-10 also shows the temperature at which metallic iron would melt within the Earth. The melting point increases with depth in the Earth because of the increasing pressure. The key feature of the model is that it indicates that at depths of 400 to 800 kilometers (250 to 500 miles), the temperature would have risen to the melting point of iron about one billion years after the Earth was formed. Other models show that this state of affairs may have been reached only a few hundred million years after Earth's formation.

(1) Accretion

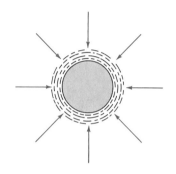

(2) Gravity compresses the original Earth into smaller volume

(3) Disintegration of radioactive elements and heat flowing away from points where particles are absorbed

**Figure 1-9**
Diagrammatic representation of three mechanisms that would cause the early Earth to heat up. (1) Accretion, in which impacting bodies bombard the Earth and their energy of motion is converted to heat. (2) Gravitational compression of the earth into a smaller volume, causing its interior to heat up. (3) The disintegration of radioactive elements releases particles and radiation which are absorbed by the surrounding rock, heating it up.

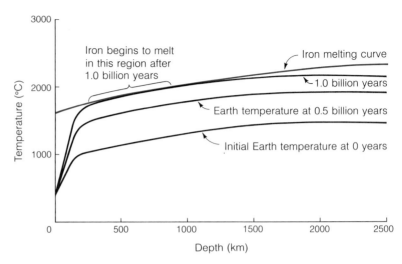

Figure 1-10
Temperature in the Earth's interior at different times in its history, according to
a calculation by T. C. Hanks and D. L. Anderson. The lowest curve shows the
initial temperature due to accretion and compression at 0 years. After 500 million
years radioactivity caused the Earth to warm up to the temperature shown by the
next curve. After 1 billion years the interior heated up and reached the melting
point of iron at depths between 400 and 800 km, and iron began to melt in this
region.

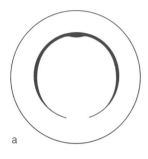

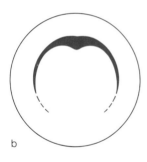

Figure 1-11
(a) The melting of iron
leads to the formation of
a heavy liquid layer.
According to W. Elsasser,
drops begin to develop in
later stages (b, c) and sink
toward the center.

**The Iron Catastrophe**  Iron is heavier than the other common
elements of the Earth, and when the iron in a layer began to melt,
large drops would probably have formed and fallen toward the
center, displacing the lighter materials that were there. Figure
1-11 shows how this may have occurred.

Iron is an abundant element; we will see that it accounts for
about one-third of the mass of the Earth. The melting and sinking
of iron to form a **liquid core** at the center was therefore an event
of catastrophic proportions. Iron "falling" toward the center would
release a huge amount of gravitational energy, which would eventually have been converted to heat. The process is basically the
same as using the gravitational energy in a waterfall to turn turbines and generate electrical energy. The additional heat released
during the formation of the iron core would produce an average
temperature rise of some 2000°C, causing a large fraction of the
Earth to melt.

**Planetary Differentiation**  We have seen that the Earth probably underwent a profound reorganization after it warmed to the
temperature at which iron melts. Approximately one-third of the
primitive planet's material sank to the center, and in the process
a large part of the body was converted to a partially molten state.
The molten material, which was lighter than the parent material
from which it separated, floated upward to cool and form a primitive crust. Core formation was the beginning stage of the **differ-**

entiation of the Earth, wherein it was converted from a homogeneous body, with roughly the same kind of material at all depths, to a zoned or layered or shelled body with a dense **iron core**, a **surficial crust** composed of lighter materials with lower melting points, and between them the remaining mantle. *Differentiation is perhaps the most significant event in the history of the Earth. It led to the formation of a crust and eventually the continents. Differentiation probably initiated the escape of gases from the interior, which eventually led to the formation of the atmosphere and oceans.* It was as if our planet gave a "big burp" during this violent upheaval.

But what of the other planets: did they go through the same early history? This question must remain unanswered, perhaps for a few more years, by which time we should know whether other terrestrial planets have iron cores and how much radioactivity is contained in their rocks. But lack of direct information is no bar to speculation, for on the basis of what they know so far, geophysicists have made computer studies of the thermal regimes of Mars and Venus, using reasonable guesses about conditions on these planets. They find that Mars is just enough smaller than the Earth so that it may not yet have warmed up to the melting point of iron. Without the triggering mechanism of an iron catastrophe, differentiation may not have been initiated or it may just be beginning. The computer models show that Venus, close to Earth in size, has probably melted its iron and is well advanced in the differentiation process. Is there support for these results? Support may lie in the fact that Mars has a thin atmosphere of carbon dioxide, which measures less than 1 percent the density and pressure of Earth's atmosphere. In marked contrast, Venus has a thick atmosphere, consisting mostly of carbon dioxide, which exerts a pressure of 100 Earth atmospheres. The atmospheric compositions of these planets are important because of what they will tell us about the planets' interiors and of their stage of evolution.

The giant planets will remain a puzzle for a long time. They are chemically so distinct from the terrestrial planets, and are so much larger, that if they have undergone differentiation the process must have followed an entirely different course. Perhaps most puzzling of all of these giants is Jupiter: through some unexplained internal process, that planet radiates two to three times as much energy as it receives from the Sun! Perhaps Jupiter, with its 12 moons, is akin to a small solar system whose sun never quite made it to the stage of shining.

**The Earth Reborn**   We take up the story of the evolution of the Earth following the catastrophic formation of the core. Early continents, if they existed, would have been engulfed and resorbed. In a sense the Earth was reborn without leaving a trace of its early history. According to the best estimates, all of this took place sometime between 3.7 and 4.5 billion years ago. We prefer to

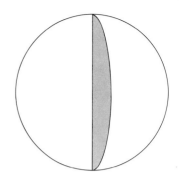

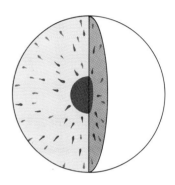

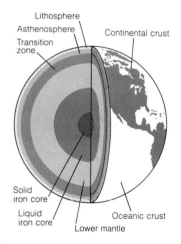

Lithosphere
Asthenosphere
Transition zone
Continental crust
Solid iron core
Liquid iron core
Oceanic crust
Lower mantle

**Figure 1-12**
The early Earth (top) was probably a homogeneous mixture, with no continents or oceans. In the process of differentiation, iron sank to the center, and light material floated upward to form a crust (middle). As a result, the Earth is a zoned planet (bottom), with a dense iron core, a surficial crust of light rock, and between them, a residual mantle.

date the event at around 4 billion years ago, near the beginning of geological time, when the oldest rocks we can now find were first formed.

**Convective Overturn**  We have mentioned that the flow of heat by conduction to the surface took place so slowly in the Earth that most of the heat accumulated within the Earth, causing the internal temperature to rise. But when the interior became so hot that it became molten, and material could move, a more efficient mechanism took over the transfer of internal heat to the surface — namely, overturn, or **convection**. Convection takes place in liquids and gases that are hotter at the bottom than at the top. The hotter material expands, becomes lighter than the material above it, and floats upward, carrying its heat to the surface, where it cools and sinks again. A schematic model of convection currents in the Earth is shown in Figure 1-13.

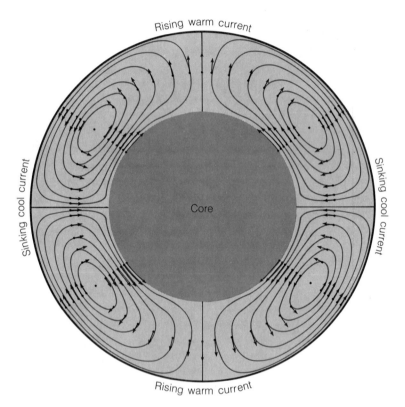

**Figure 1-13**
Highly schematic model of convection in the Earth's mantle, computed by C. L. Pekeris in 1935. Although the model is mathematically exact, it is too simple and regular to represent the actual Earth. Nevertheless, the warm rising currents, the cool sinking currents, and the horizontal flow at the tops and bottoms of the convection cells are features common to the recent, more realistic models, which were influenced by this early work. [After "The Origin of the Ocean Ridges" by E. Orowan. Copyright © 1969 by Scientific American, Inc. All rights reserved.]

Once convection began in the Earth, heat was dissipated rapidly, and the planet quickly cooled. The mantle solidified, but the underlying iron core did not; it remains molten, for even 4 billion years is too short a time for it to have cooled off. Convective overturn also produced a chemically zoned Earth.

Although convection is most common in heated liquids and gases, surprisingly enough, it can also occur in solids. Under certain conditions, heated rock will slowly move upward, because its density is reduced as it expands. We will see that convective flow in the Earth's solid interior has been proposed as the driving force for such large-scale geological processes as sea-floor spreading and continental drift.

**Chemical Zonation** The eight most abundant elements on Earth are listed in Figure 1-14. How we know this will become evident in later chapters. Together these elements account for more than 99 percent of the weight of the Earth. About 90 percent of the Earth is made of the four elements iron, oxygen, silicon, and magnesium. Compare the abundance of elements in the crust with the values for the Earth as a whole. Because most of the

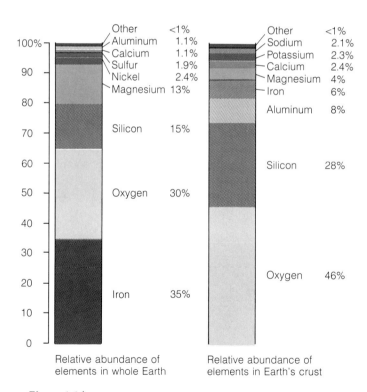

**Figure 1-14**
Relative abundance of elements in the whole Earth and in the Earth's crust. Differentiation has resulted in a light crust depleted in iron and enriched in oxygen, silicon, aluminum, calcium, potassium, and sodium.

iron sank to the core, that element drops to fourth-place rank in the crust. On the other hand, silicon, aluminum, calcium, potassium, and sodium have moved upward with a manyfold increase in concentration in the crust.

It is interesting and significant that differentiation did not lead to a vertical arrangement of elements based entirely on their relative weights, as we might have supposed. The reason is that various elements formed compounds, and it was the chemical and physical properties of those compounds—properties such as melting points, chemical affinities, and densities—that governed the distribution of elements, rather than the properties of the elements themselves. For example, certain silicates of calcium, sodium, potassium, and aluminum—notably the *feldspars* ($CaAl_2Si_2O_8$, $NaAlSi_3O_8$, $KAlSi_3O_8$)—are easily melted. They begin to melt at temperatures as low as 700 to 1000°C, and when molten they are relatively light. One might speculate that compounds like these would melt early, rise to the surface by convection, and accumulate as crust. The details are obscure, but it should be no surprise that feldspars are the most common minerals found in the Earth's crust.

The mantle, situated between the crust and the core, became the reservoir for magnesium-iron silicates, which melt less easily and are heavier than feldspars. From the abundances shown in Figure 1-13, we would expect that the dominant constituents are combinations of Fe, Mg, Si, and O. We should therefore not be surprised when we see later why most geophysicists and petrologists believe that the principal minerals in the mantle are olivine ($Mg_2SiO_4$— $Fe_2SiO_4$) and pyroxene ($MgSiO_3$—$FeSiO_3$).

Heavy elements like gold and platinum have little chemical attraction to oxygen and silicon, and most of these important metals probably sank into the Earth's core. But other heavy elements, like uranium and thorium, have strong tendencies to form oxides and silicates, which are light and could rise to accumulate in the crust. Gravity, however, is only secondarily responsible for the zonation of a planet: just as important are the relative abundance of the different elements and the way in which the electrons are grouped in their atoms, both of which determine the compounds that they form. The properties of the compounds—densities and melting points—differ from those of the pure elements and govern the distribution of the elements.

**Differentiation Slows the Engine Down** One very important consequence of chemical zoning is the concentration of the heavy elements uranium and thorium in the Earth's crust as oxides and silicates. In the early stages, when these radioactive elements were evenly distributed throughout the Earth, they were probably highly effective in raising the temperature to the melting point of iron. Following the catastrophic formation of the core, however, when the interior was hot and mobile, the radioactive "fuel" began to concentrate in the outermost layers, where the heat that was generated could be conducted through much shorter distances to the surface and so be lost more easily. In this way the differentiation of a planet may act as a governor to slow down the operations of the heat engine.

**Formation of Continents, Oceans, and Atmosphere**   We have only the most general notion of how the first continents may have formed; the story goes like this. Lava flowing from the partially molten interior of the Earth spread over the surface and solidified to form a thin crust. This primeval crust melted and solidified repeatedly, and the lighter compounds were successively separated from the heavier ones and distributed at the top. **Weathering** by rainwater and other components of the atmosphere broke up and altered the rocks; **erosion** led to the formation of sediments—the residue of broken-down rock particles. As the sediments accumulated, they were in turn penetrated by hot gases and solutions from below, heated up and altered—or "cooked," in a sense—into new rocks, or were entirely resorbed and recycled. The end product of such processes as these was the primitive nucleus of a continent, which continued to grow by a continuation of the process. As a rough check on this theory, if we take the amount of lava ejected each year by modern volcanoes, and multiply that amount by a few billion years, the volume of rock produced corresponds fairly well with the volume of the continents.

If we start with the premise that the Earth accreted from cold planetesimals, then the oceans came from the interior as a product of the processes of heating up and differentiation. Originally the water was locked up, chemically bound as oxygen and hydrogen in such minerals as potassium-aluminum mica, $(KAl_3Si_3O_{10}(OH)_2)$. As the Earth warmed up and partial melting occurred, the water was released and carried to the surface along with lava. As the lava reached the surface, much of the water escaped in the form of hot vapor clouds. Even at today's rate of volcanic activity, the lavas that reached the surface in the past would have contained enough water vapor to fill the oceans in the course of geological time, though much of that water may have been recycled by the melting of oceanic sediments containing hydrated (water-bearing) minerals.

How and when the atmosphere began to develop is a more difficult matter to determine. We can, however, give one self-consistent explanation that agrees with the known facts. There is little doubt that the earliest atmosphere was entirely different from the primarily nitrogen-oxygen one we live in now. The cold planetesimals, which aggregated to form the Earth, could have carried no atmosphere because they were too small to hold any gases by gravitation. Thus **outgassing** as part of the process of differentiation—that is, the release of gases from the interior due to internal heat and chemical reactions—is generally thought to be the primary source. From the chemical composition of lava and from studies of gases released by modern volcanoes, we can surmise that volcanic gases consisted mainly of water vapor, hydrogen, hydrogen chlorides, carbon monoxide, carbon dioxide, and nitrogen. But the light hydrogen molecules could not have been held by the Earth's gravity, and the hydrogen would have escaped into space as it does steadily today. With energy supplied by sunlight, some water vapor in the upper atmosphere may have broken down to hydrogen and oxygen by a process called **photolysis** ("breaking" by "light"). Any oxygen so formed would not have remained free (uncombined) because it is quite reactive and would have quickly combined with gases like methane and carbon monoxide to form water and carbon dioxide. It would also have combined with crustal materials—with metals like iron in olivines and pyroxenes to form iron oxides like **hematite** ($Fe_2O_3$).

The production of significant amounts of free oxygen and its persistance in the atmosphere probably came only after life evolved at least to the complexity of green algae, which, like all higher forms of green plants, convert sunlight to organic matter by **photosynthesis,** using carbon dioxide and water to make organic matter and oxygen. Not until the production of oxygen exceeded its loss by chemical combination with other gases and metals could this by-product of photosynthesis begin accumulating in the atmosphere.

The big question about the atmosphere concerns life itself. How did life develop in the "poisonous," primitive atmosphere and evolve the green plants that "purified" the air with free oxygen and cleared the way for the higher forms of life to develop? Some answers are given in Chapter 14.

Much carbon dioxide was removed from the atmosphere by chemical combinations with calcium, hydrogen, and oxygen, which formed limestone, coal, and petroleum, the great bulk lying buried in the crust. These are the great reserves of fossil fuels that we have relied on for energy to power our industrial societies. The few hundredths of one percent of the atmosphere that remains as $CO_2$ is of course highly important to us in that it is an essential raw material that plants use for photosynthesis.

**The Account of the Earth's Evolution—Fact or Fiction?** The preceding story is, of course, speculation based on physical plausibility and extrapolation from experiment. *In all fairness, however, it should be mentioned that we have presented but one of several possible courses for our planet's evolution from among the several hypotheses currently extant. There are alternative views for the nature of each stage, but about the general course of events there is fair agreement.* We are like children making a house out of a variety of blocks: the constraint is that each block has to fit with the ones under it and

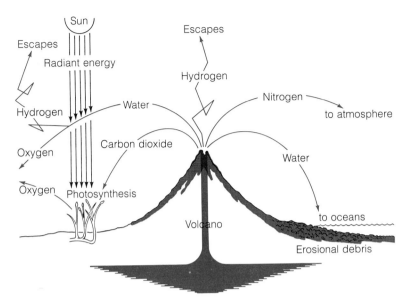

**Figure 1-15**
Volcanism has contributed enormous amounts of water, carbon dioxide, and other
gases to the atmosphere. Sunlight broke water molecules into its components,
hydrogen and oxygen. Photosynthesis by plants removed carbon dioxide and
added oxygen to the primitive atmosphere. Volcanism was an important factor in
the growth of continents.

over it and that the whole structure has to stand up. But there
may be more than one way to build the house.

Even if the details are obscure, we believe that the general sequence of events just described may be taken as a tentative solution
(by no means a unanimous one) to the problem of the early history
of the Earth, because it explains several facts of paramount importance. The Earth has an iron core and a silicate mantle. In order
for this separation to take place the interior would have had to
become hot enough for iron to melt. The Earth has a crust that is
chemically distinct from the mantle, and it contains such high
concentrations of certain elements (for example, uranium and
thorium) that they must have been extracted from the whole of
its body. This implies a mobility that only extensive melting could
provide. Finally, there is a source of thermal energy to which we
can attribute the melting—radioactivity.

In the remainder of this book we will be concerned mainly
with what happened to the Earth following the stage of primary
differentiation. The mantle solidified, a primitive crust and continents developed, oceans and atmospheres were produced by
outgassing, and the processes that we know today were set in
motion. Differentiation slowed down the internal heat engine of
Earth, leaving enough heat in the interior for the planet to continue
to evolve to this very day.

# THE EARTH MACHINE SET IN MOTION, MUCH AS WE KNOW IT TODAY

**Plate Tectonics — A Unifying Theory** Geology books written in the 1970's have a distinct advantage over their predecessors. A hundred years ago, another idea held sway — that of a contracting Earth. For the first time since then, a single, all-encompassing concept is available that interprets many of the major geological features of the Earth. Such topics as the classification and distribution of rocks, the history of sedimentary rock sequences, the positions and characteristics of volcanoes, earthquake belts, mountain systems, deep-sea trenches, and ocean basins were formerly described in more or less unrelated fashion. We have the advantage of being able to treat these and other topics in the context of a unifying theory in which the large-scale pattern, if not the underlying mechanism, is recognized. To be sure, some geologists do not now accept certain parts of the theory, and some reject it entirely. Revolutions rarely proceed by unanimous consent!

The outermost shells of our concentrically zoned Earth are illustrated in Figure 1-16, which is a representation of what we think we know of the Earth today. Why we propose this picture will of course be justified in later chapters, rather than here. The **lithosphere** is depicted as the strong, solid, outermost shell riding on the weak, partially molten **asthenosphere**. The continents are raftlike inclusions embedded in the lithosphere. A thin crust tops the lithosphere.

The central idea of plate tectonics is remarkably simple. The lithosphere is broken into about ten large rigid plates, each plate moving as a distinct unit. The major plates and the directions in which they move are sketched in Figure 1-17. Many large-scale geological features are associated with the boundaries between the plates.

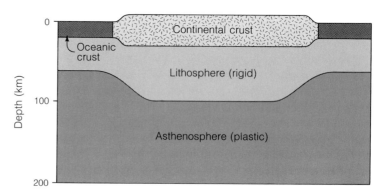

**Figure 1-16**
The outermost shell of the Earth is the strong, solid lithosphere, riding on the weak, partially molten asthenosphere. The lithosphere is topped by a thin crust under the oceans and a thicker continental crust. [After "Plate Tectonics" by J. F. Dewey. Copyright © 1972 by Scientific American, Inc. All rights reserved.]

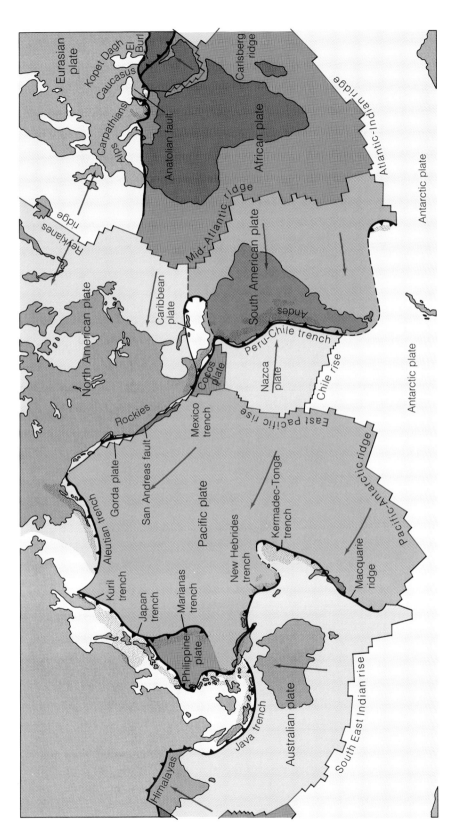

**Figure 1-17**
Earth's lithosphere is broken into large, rigid plates, each moving as a distinct unit. The relative motions of the plates, assuming the African plate to be stationary, are shown by the arrows. Plate boundaries are outlined by earthquake belts. Plates separate along the axes of mid-ocean ridges, slide past each other along transform faults, and collide at subduction zones. [After "Plate Tectonics" by J. F. Dewey. Copyright © 1972 by Scientific American, Inc. All rights reserved.]

—⏜— Subduction zone

— Transform

— Ridge axis

----- Uncertain plate boundary

→ Direction of plate motion

Areas of deep-focus earthquakes

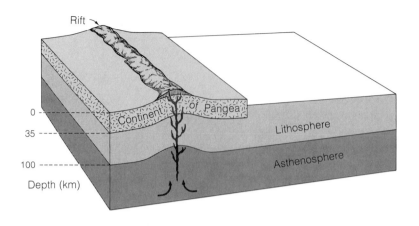

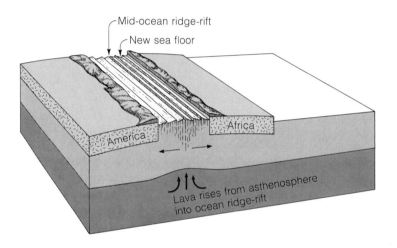

**Figure 1-18**
Stages in the development of a divergent junction and sea-floor spreading.
(Top) The lithosphere breaks, and a rift develops under a continent. Molten
basalt from the asthenosphere spills out. (Bottom) The rift continues to
open, separating the two parts of the continent—in this example, America
and Africa. A new ocean basin, the Atlantic, is created between the
separating land masses. The active rift is marked by a mid-ocean ridge,
with earthquakes and volcanism as characteristic features. [After "The
Breakup of Pangaea" by R. S. Dietz and J. C. Holden. Copyright © 1970
by Scientific American, Inc. All rights reserved.]

Plates spread apart along **divergent junctions** typified by the
cracklike valley at the crest of the mid-Atlantic ridge. This particular feature is the contact between the American plate on the one
hand, and the Eurasian and African plates on the other. The
divergent junction is characterized by earthquake activity and
volcanism. The void between the receding plates is filled by melted,
mobile material that rises from below the lithosphere. The material
solidifies in the crack, and the plates grow as they separate. Since
new sea floor is created, this part of the process, illustrated in
Figure 1-20, has been dubbed **sea-floor spreading**.

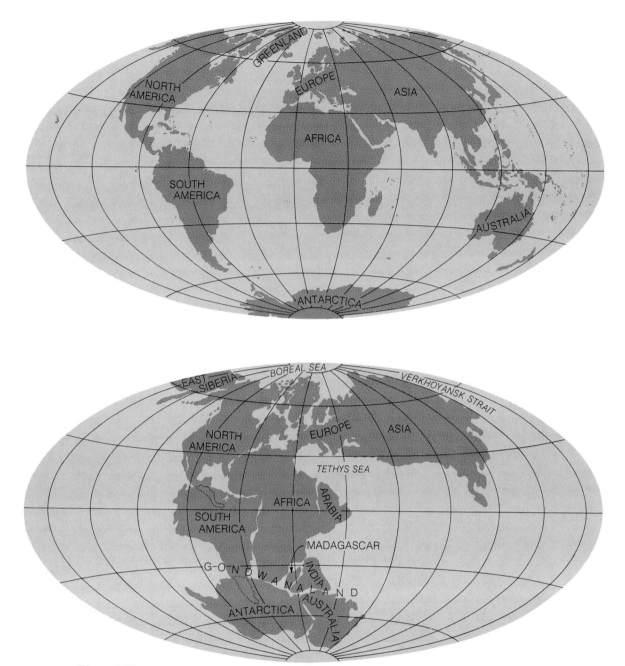

**Figure 1-19**
Single supercontinent of Pangaea, presumed to have existed 200 million years ago, may have appeared as shown below. Present-day map is shown at the top. [After "Continental Drift" by J. Tuzo Wilson. Copyright © 1963 by Scientific American, Inc. All rights reserved.]

There seems to be little doubt that the Atlantic Ocean did not exist 150 million years ago. Instead, the large, single hypothetical land mass reconstructed in Figure 1-19, named **Pangaea,** contained Eurasia, Africa, and the Americas. Pangaea was split by a crack

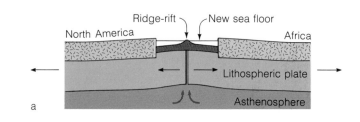

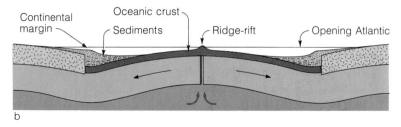

**Figure 1-20**
As a rift widens and a new sea floor is created (a), the trailing edges of the receding continents receive erosional debris from inland, deposited as thick wedges of sediment (b). [After "Geosynclines, Mountains, and Continent-Building" by R. S. Dietz. Copyright © 1972 by Scientific American, Inc. All rights reserved.]

Pangaea is not the only possible reconstruction of former continental positions that can be made on the basis of plate-tectonic theory, but it has the sentimental value of resurrecting an idea and a name first proposed almost fifty years ago by *Alfred Wegener* on the basis of evidence then available to him.

that subsequently opened to form the Atlantic Ocean. The trailing edges of the receding continents became a new repository for erosional debris brought to the newly formed sea from inland sources. The sedimentary deposits were laid down in wedge-shaped bodies along submerged continental margins. Figure 1-20 shows how thick deposits of sediments along continental margins can be explained by this mechanism.

If plates separate in one place, they must converge somewhere else, and they do. Plates grind together head-on along **convergent junctions**. It is no surprise that crumpled mountain ranges, deep-sea trenches, shallow and deep earthquakes, and volcanoes are all associated with convergence. Examples are the contact between the Nazca plate and the American plate (Fig. 1-17), where one finds the Andes Mountains, the Chilean deep-sea trench, and where some of the world's great earthquakes have been recorded. Another example is the Kuril-Kamchatka-Aleutian arc system, along which the Pacific plate runs into the Eurasian plate. We will see that convergent zones take on different manifestations, depending on spreading rates and whether the leading edges are continental or oceanic.

It has been said that one picture is worth a thousand words. This is certainly true of Figure 1-21, which shows the profusion of geological activity associated with but one plate-collision scenario, where the abutting edges are ocean floor and continent. The heavy, oceanic lithosphere descends into the mantle, below the lighter continental lithosphere. Downbuckling is marked by an offshore trench. Great earthquakes occur adjacent to the inclined

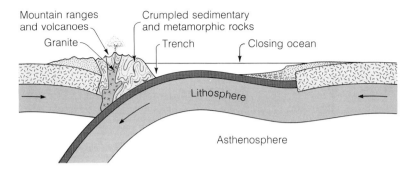

**Figure 1-21**
When plates collide, one plate usually buckles downward. The overriding plate is crumpled and uplifted. Deep-sea trenches (sites of the greatest of ocean depths), high mountain chains, volcanoes, and the greatest earthquakes are associated with such regions of plate convergence. [After "Geosynclines, Mountains, and Continent-Building" by R. S. Dietz. Copyright © 1972 by Scientific American, Inc. All rights reserved.]

contact between the two plates. The edge of the overriding plate is crumpled and uplifted to form a mountain chain that runs parallel to the trench. Deep-sea sediments may be scraped off the descending slab and incorporated into the adjacent mountains. What a complicated mess! As the mountains are raised, erosion by wind, ice, and water wears them down. The debris eventually ends up in the oceans to repeat the cycle of rock formation and destruction over and over again.

Such regions of convergence, where lithosphere is consumed, have been named **subduction zones**. If the divergent zones are sources of new lithosphere, then the subduction zones are sinks where material is consumed in equal amount. Otherwise the Earth would change in size, but neither geological evidence nor physical theory support this conclusion.

Rocks caught up in a subduction zone will be squeezed and heated. A new assemblage of minerals will appear as a result of this "pressure-cooking" treatment, depending on the pressure, temperature, and deformation to which the rocks are subjected by the subduction process. **Metamorphism**, the cumulative effect of all of these processes, leads to the formation of new mineral grains and therefore new rocks. The sequence of sedimentary and metamorphic rocks in and next to convergent zones may provide the key to unraveling the history of a collision between plates.

As the oceanic plate descends into the hot mantle, parts of it may begin to melt. The rock melt, or **magma**, thus formed would float upward, some of it reaching the surface as lava erupting from volcanic vents. In a manner not fully understood, the formation of magma in the subduction zone may be a key element in the creation of granitic rocks, the major rock from which continents are formed.

Plates can separate and collide, as we have just seen. They can also slip past one another along a **transform fault** (Fig. 1-22).

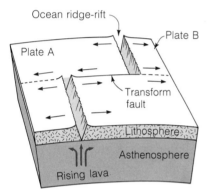

**Figure 1-22**
A transform fault is a plate boundary
along which two plates slide past each
other. In the example shown, Plates A
and B are separating. The ridge crest,
which marks the zone of spreading, is
offset by the transform fault. The arrows
indicate the opposite plate motions on
the two sides of the transform fault
between the ridges.

**Figure 1-23**
View northwest along the San Andreas fault in the Carrizo Plains of California.
This fault is an example of a transform fault. It forms a portion of the sliding
boundary between the Pacific and American plates. [Photo by R. E. Wallace,
U.S. Geological Survey.]

The famed San Andreas fault of California is an example of such a
boundary. There the Pacific plate slides past the American plate
in a northwesterly direction. This contact is characterized by contrasting geology on both sides and by large, shallow-focus earthquakes of the kind that destroyed San Francisco in 1906.

What we have just described is the general pattern of the
internal heat engine's work output as we can see it today. The
process probably has been going on since the large-scale differentiation of the Earth ended some 4 billion years ago. A description,
however, is not an explanation. We have unraveled the kinematics,
or motions, of the process, but the dynamics, or forces, that are
responsible, have yet to be understood (Fig. 1-24): so far we can
do no more than make vague reference to convection serving as
an internal heat engine. Furthermore, there is a wealth of major
geologic features that lie within the continents and whose relations
to plate tectonics, if any, have yet to be established. Nevertheless,
a grand scheme of things has been recognized, and the course of
geology has undoubtedly changed.

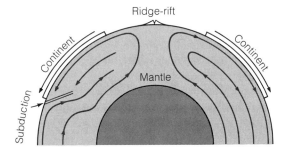

**Figure 1-24**
A simple model showing how convection currents in
the deep interior might be the driving force of sea-floor
spreading and continental drift. Hot matter rises under
the ocean ridge and flows apart, carrying the plates
along. [After "The Origin of Ocean Ridges" by
E. Orowan. Copyright © 1969 by Scientific American,
Inc. All rights reserved.]

A parallel development of the 1960's was the increasing sophis-
tication of our awareness of the external heat engine of the Earth.
That engine, powered by the Sun's energy, drives the circulation
of the atmosphere and oceans, provides the rain, wind, ice, and
running water that erodes the mountains of the Earth. As soon as
the atmosphere formed by outgassing, the external heat engine
started to grind out the products of erosion and chemical decomposition; rivers began to run to the ocean, delivering sediments
to beaches and deltas; ocean currents carried the finest of the
sediments farther from land, where in quiet waters they accumulated on the margins of continents and in the deep sea. The external
heat engine is of immediate consequence to us, for aside from its
importance in most of the geological processes we see operating
at the surface of the Earth, we are living in the midst of it and
depend completely on its workings for our food and our energy.

The many ways in which the internal and external heat engines
are linked provide exciting new territory to explore for geologists
and oceanographers. As we find out more about each engine, we
find out more about their interactions. Their interactions, too,
can affect us drastically—as when an earthquake, which happens to
occur during Peru's rainy season, triggers a massive mudflow that
kills thousands of people (1970), or when a volcano like Mount
Agung in Bali erupts (1963) and sends so much volcanic ash into
the atmosphere that the quantity of atmospheric dust over the
whole world is significantly increased.

In what follows we will deal with continents and ocean basins
and interpret the information in the different kinds of rocks in
the conceptual framework provided by plate tectonics. Though
that concept does not explain everything, and may not even relate
to some kinds of phenomena, it gives us by all odds the best framework on which to lay the story of the Earth.

# SUMMARY

1. The Sun and its family of planets formed when a primeval cloud of gas and dust condensed about 5 billion years ago.

2. Earth probably grew by accretion of small chunks of matter to reach its present size. At its birth it was probably a homogeneous and relatively cool body.

3. Because of radioactivity, Earth began heating up and within about a billion years or less the temperatures probably reached the melting point of iron. Drops of molten iron sank to the center, and lighter matter floated up to form the outer layers that became the continents. Outgassing gave rise to the oceans and a primitive atmosphere. In this way the Earth was transformed to a differentiated planet with chemically distinct zones — an iron core, a magnesium-iron-silicate mantle, and a crust enriched in oxygen, silicon, aluminum, calcium, sodium, and radioactive elements.

4. The lithosphere, or outermost shell of the Earth, is broken into about ten large, rigid plates, which jostle each other because of their individual motions. The boundaries of these plates are zones of intense activity; associated with them are many of the large-scale geological features produced by such processes as mountain-building, volcanism, creation and destruction of the sea floor, and earthquake activity.

# EXERCISES

1. What factors have made Earth a particularly congenial place for life to develop?

2. Summarize the important facts about the solar system which must be explained by any theory of its evolution.

3. If you were an astronaut exploring another planet, what evidence would you look for to decide whether the planet was differentiated?

4. How does the Earth's crust differ in elemental composition from that of the whole Earth? In general terms, what are the reasons for this difference?

5. What are the sources of Earth's internal heat? Speculate about what life would be like if Earth had no internal heat sources.

6. Describe the central idea of plate tectonics and the large-scale geological features associated with plate boundaries.

7. Speculate on what life would be like today if the ancient continent of Pangaea remained a single land mass instead of breaking up into Eurasia, Africa, and the Americas.

Brandt, J. C., and S. P. Moran, *New Horizons in Astronomy.* San Francisco: W. H. Freeman and Company, 1972. (Chapters 7, 8, 15.)

Gass, I. G., P. J. Smith, and R. C. L. Wilson (editors), *Understanding the Earth.* Cambridge: M.I.T. Press, 1971.

Jastrow, Robert, and M. H. Thompson, *Astronomy: Fundamentals and Frontiers.* New York: John Wiley & Sons, 1972. (Part Three, The Planets.)

Laporte, L. F., *The Earth and Human Affairs.* San Francisco: Canfield Press, 1972.

Leveson, David, *A Sense of the Earth.* New York: Doubleday and Company, Inc., 1972.

Takeuchi, S., S. Uyeda, and H. Kanamori, *Debate About the Earth.* San Francisco: Freeman, Cooper & Company, 1970.

Wilson, J. Tuzo, (compiler), *Continents Adrift: Readings from Scientific American.* San Francisco: W. H. Freeman and Company, 1970.

Unconformable contact of upper Old Red Sandstone beds on vertical Silurian rocks at Siccar Point, Berwickshire, Scotland. It was from a study of this locality that James Hutton, in 1788, first realized the meaning of unconformity. The geologist is seated at the contact. [Photo courtesy of Frederick Schwab.]

# 2

# The Geological Time Scale and the Rock Record

*Concepts of time are central to geology. Most geologic processes that shape the surface of the Earth and give structure to its interior operate over long times, up to millions and billions of years. The igneous, metamorphic, and sedimentary rocks exposed at the surface are the visible records of the geologic processes of the past. From the time and space relations that those rocks manifest, geologists have built the geologic time scale, which is used for placing the geologic events of Earth history in sequence according to relative age. Spontaneous decay of radioactive atoms in rocks gives absolute ages that date the geologic periods and the origin of the Earth.*

## HOW LONG IS A LONG TIME?

The key difference between geologists (and astronomers) and most other scientists is in their attitude toward time. Though many chemical reactions and physical processes that we measure in the laboratory operate over time periods—or what we will call **time scales**—of seconds or fractions of seconds, geological processes that we can observe directly take place over a much broader time range, from relatively short times counted in seconds to much longer ones, measured in days or years. The time scale for earthquake tremors may range from seconds to minutes, and the times for the seismic waves generated by that earthquake to travel through the Earth and along its surface are reckoned in minutes and hours. Erosion and transportation of an immense amount of boulders, sand, silt, and clay by a major river in flood operates over a period of a few days. Sand bars move in and out from a beach in days or weeks. These are processes we can see or

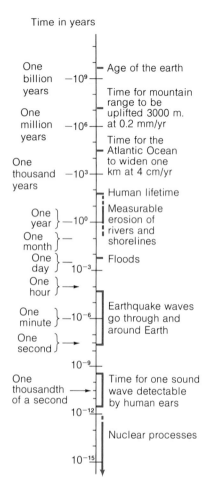

Time in years

One billion years —10⁹ — Age of the earth

Time for mountain range to be uplifted 3000 m. at 0.2 mm/yr

One million years —10⁶ —

Time for the Atlantic Ocean to widen one km at 4 cm/yr

One thousand years —10³ —

Human lifetime

One year —10⁰ — Measurable erosion of rivers and shorelines

One month —

One day 10⁻³ — Floods

One hour —

One minute —10⁻⁶ — Earthquake waves go through and around Earth

One second —

10⁻⁹ —

One thousandth of a second — Time for one sound wave detectable by human ears

10⁻¹² —

Nuclear processes

10⁻¹⁵ —

**Figure 2-1**
Orders of magnitude of times for some common processes and events.

In many geological books and journals, long time units have come into general use as abbreviations: *m.y.* for millions of years, *b.y.* for billions of years. A billion years is commonly called an *eon*.

feel actively happening. Others cannot be observed directly; for example, the Atlantic Ocean is spreading apart at the Mid-Atlantic Ridge on a much longer time scale—at a rate of a few centimeters per year. The gullying of overgrazed or carelessly farmed land by accelerated erosion takes place over a time scale of a few years. But when time scales are 50 years or more in duration, our memories start to fail us and we rely on historical records in order to measure how much a marshy tideland has filled in, how much a hillside has eroded, or how much a river has gradually changed its course.

Despite an occasional earthquake or volcanic eruption, the Earth seems to provide us with a reasonably stable foundation upon which to build a civilization. This holds true on a time scale important to organized society—namely, hundreds or thousands of years. That stability is absent on a geological time scale of millions to hundreds of millions of years, during which continents, oceans, and mountain chains have moved horizontally and vertically through large distances. Although the evidence for long-term instability is all around us, only in recent years have scientists begun to recognize a world-wide pattern in these movements. Working out that pattern requires a means of timing the movements relative to each other and to a known time scale.

In Chapter 1 we discussed the time scales of millions of years, that elapsed during the formation of the Sun and the planets—times measured by radioactive clocks and inferred by reasoning from the rates of the physical and chemical processes involved. We concluded that the Earth is about 4.7 billion years old. What happened during all of the time between then and now? How long does it take geologic processes to make mountains and destroy them? What is the lifetime of a river? Time scales ranging from a few tens of years to a few billions of years are characteristic of most of the processes discussed in this book.

Why do we care about time scales? One of the most important reasons, one that comes from the heart of historical geology, the story of the evolution of the Earth as we now see it, is to be able to reconstruct just what happened when. When were the Appalachians or the Rocky Mountains formed? When did the Atlantic Ocean start spreading apart? Equally important in the value of time scales, however, is the help we get in trying to figure out *how* something happened. This comes from an old rule of physical science: if two things formed within different time scales, they are likely to have been made by different processes. Most of the time we use this rule of thumb in an unconscious way, but it is sometimes worthwhile to spell it out explicitly, as a guide to thinking about a specific problem.

**How We Estimate Very Slow Rates of Earth Processes**  A few simple, rough calculations give surprisingly good estimates. Rocks older than about 200 million years have not been found on the deep sea floor, even with the help from the many holes drilled there in the past few years by the JOIDES (Joint Oceanographic Institutions Deep Earth Sampling) project sponsored by the National Science Foundation. We may therefore take the 200 million years as some sort of an upper limit to the age of an ocean basin. If 10,000 kilometers is used as a representative width of an ocean—that is, the distance the plates have spread to form the basin—the spreading rate comes out to be 10,000 kilometers/200 million years = 5 centimeters/year.

The well-known San Andreas fault in California is a transform fault along which the North Pacific plate slides past the North American plate. At some precisely located places along the fault, measurements have been made by survey for almost a century: during this period, slip along the fault, due both to earthquakes and to steady creep, has amounted to about 4 to 6 centimeters/year. On a longer time scale, the rate of movement can be determined by matching up distinctive geological formations that have been split by the fault, the separate parts moving away from each other. The average movement over the past several million years can be inferred in this way to be about 1 centimeter/year in northern and central California. At that rate, about 25 million years ago the block containing the present San Francisco coast would have been at the latitude of the present Los Angeles! Later we will describe more precise methods for obtaining rates of plate movements in different parts of the world, including dating of magnetic anomalies and dating of cores obtained in the JOIDES project.

Vertical movements can be evaluated by dating marine deposits that are now above sea level. Mountains made up of rocks that contain marine fossils have been uplifted 3000 meters in 15 million years, giving a rate of 0.2 millimeters/year. About 40,000 years ago, during the last ice age, Fennoscandia, the area including Norway, Sweden, and Finland, was covered by 2 to 3 kilometers of ice. The tremendous weight of the ice sheet loaded and depressed that part of the lithosphere. When a warming trend set in and the ice melted away, the lithosphere there rebounded some 500 meters, as evidenced by the existence of ancient beaches at that elevation above sea level. Thus postglacial uplift of Fennoscandia took place at a rate of about 1 centimeter/year. This particular evidence gives us important data for determining the viscosity of the mantle.

Erosional processes are continuously at work wearing down the land surface; without mountain building and other vertical rejuvenation, such as the type just described, all land would eventually be reduced to sea level. Erosion rates can be estimated by adding up all of the disintegrated and dissolved products of erosion being carried away by rivers and wind. The rate for the North American continent has been estimated to be about 0.03 millimeter/year (about 1 foot per 10,000 years), whereas the great plateau of Africa is being lowered by about 1 foot in only a few

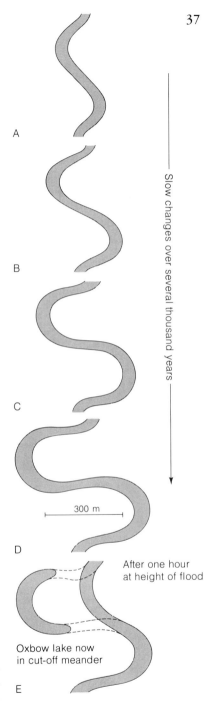

A

B

C

D

E

Slow changes over several thousand years

300 m

After one hour at height of flood

Oxbow lake now in cut-off meander

**Figure 2-2**
A meander loop of a river slowly accentuates its bends over hundreds or thousands of years (*a* to *d*), but in one hour's time, at the height of a flood, may break through its banks (*d* to *e*) to cut off a meander and isolate an oxbow lake.

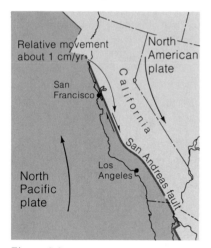

**Figure 2-3**
The North American plate has been
moving south relative to the North
Pacific plate along the San Andreas
transform fault at a rate of about
1 cm/yr.

**Figure 2-4**
Helicopter view of a raised beach northeast of Thule, Greenland. This beach
has risen above sea level as a result of the crust rising after the melting of
glaciers whose great weight had depressed the crust. [Photo by R. B. Colton,
U.S. Geological Survey.]

thousand years. Thus it takes hundreds of millions ($10^8$) of years
to open an ocean basin, about 20 million ($2 \times 10^7$) years to raise
a mountain, and $10^8$ years to cut it down to sea level.

Even these are relatively short periods on the geological time
scale of the history of the Earth. The Earth has experienced many
cycles of mountain-building and erosion in the past 4 billion years.
In the next sections we explore how we can date all of the events
that are parts of those cycles.

## THE ROCK RECORD

The only record we have of things that happened on Earth in the
geological past is the rocks that were preserved from erosional
destruction.

**What Is a Rock?**   A rock is many things. It is a collection of
the particular chemical elements that make it up. Those elements
are not found randomly mixed up in a rock but are distributed
among an assemblage of **minerals,** which are chemical compounds,
usually in crystalline form. The minerals make up the rock just
as bricks make up a brick wall, in a great variety of arrangements.
In coarse-grained rocks the minerals are large enough to be seen
with the naked eye. In some rocks the minerals can be seen to

have **crystal faces,** smooth planes bounded by sharp edges; in others, such as a typical sandstone, the minerals are in the form of fragments without faces. In fine-grained rocks, the individual mineral grains are so small that they can be seen only with a powerful magnifying glass, the hand lens that the field geologist carries. Some are so small that a powerful microscope is needed to make them out.

Of the thousands of different kinds of rocks that can be classified on the basis of the kinds of minerals in them, the **mineral composition,** or on the basis of the size and arrangement of mineral crystals, the **texture,** geologists recognize three major classes on the basis of origin: **igneous, sedimentary,** and **metamorphic.** Igneous rocks form by cooling and solidification of a hot molten liquid, or **magma.** This origin is reflected in their texture, which is that of a mosaic of crystals, arranged in much the same fashion as a flagstone walk. They are further divided into two major types, **intrusive** and **extrusive.** Intrusive igneous rocks are those that were emplaced within the rocks that surround them, deep in the Earth; they originated as hot magma that pushed its way into cracks and other openings, separating and sometimes melting the surrounding rocks. Extrusive rocks form from **lava** that flowed out of volcanoes or from volcanic **ash** that was exploded high into the air. The origin of intrusive rocks is demonstrated by mapping the connection between an intrusive and the extrusive lava flows that erupted at the surface from the same magma. The intrusives are coarse grained, or **phaneritic** (from the Greek, *phanero-,* meaning visible), and the extrusives fine grained, or **aphanitic** (from the Greek *aphan-,* meaning invisible).

The second major class, the sedimentary rocks, are made up of the hardened, compacted, and cemented sands, silts, and clays that we call sandstones, siltstones, and shales. Other sediments are limestones, made up of calcium carbonate ($CaCO_3$), much of it in the form of fossils. One of the most obvious characteristics of sedimentary rocks is their **layering, stratification,** or **bedding,**

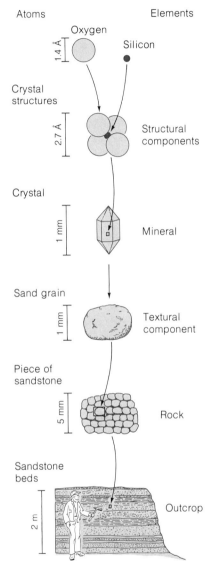

**Figure 2-5**
How atoms are put together to form rocks. Atoms make small structural components that combine to form minerals, which in turn combine to form rocks.

**Figure 2-6**
The texture of an igneous rock, as seen through a microscope, is a mosaic of intergrown crystals. [From *Petrography* by Williams, Turner, and Gilbert. W. H. Freeman and Company. Copyright © 1954.]

**Figure 2-7**
Bedding in sedimentary rocks. In this outcrop the beds are all parallel and tilted by structural deformation that took place long after deposition of the layers. [Photo by R. Siever.]

derived from the settling of particles from water or air to form layers of sediment (hence the name, from the Latin **sedimentum**, meaning settled). Sediments form at the surface of the Earth as a consequence of the **erosion** and fragmentation of pre-existing rocks, **transportation** of the erosion products, and **deposition** in a variety of **sedimentary environments**, such as the sea, a river bottom, or a desert sand dune.

The third major class, the **metamorphic** rocks (from the Greek, meaning to change form or to transform), includes rocks of the other two classes that have been altered by the action of heat and pressure after formation. Metamorphic rocks, too, may be coarse or fine grained, though the large majority are coarse. A major textural attribute is **foliation**, a platy or leafy structure of slates, schists, and gneisses, whose minerals are aligned along straight or wavy planes. Sedimentary and igneous rocks may become meta-

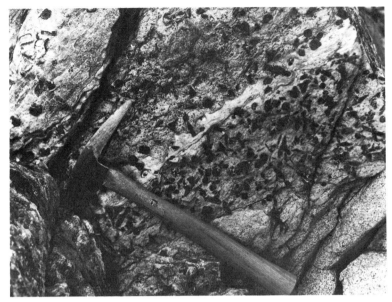

Figure 2-8
Close-up view of a schist. This metamorphic rock, composed largely of quartz
and white (muscovite) mica, contains large crystals of two silicate minerals,
garnet and staurolite, and abundant specks of black (biotite) mica near the
hammer point and butt. The wavy planes running from upper right to lower
left are the foliation that is characteristic of schists. [Photo by W. R. Hansen,
U.S. Geological Survey.]

morphosed by being buried deeply in the earth and being subjected
to heat and pressure over wide volumes, a process called **regional
metamorphism**, or by being invaded by magmas that bake the
rocks near their contacts, a process called **contact metamorphism**.

The kinds of minerals and the textures of the rocks are clues
to the origins of each rock group. In igneous rocks, the minerals
are a guide to the chemical composition of the molten magma
from which the rock crystallized and the temperature of the
magma. The texture is a guide to the speed with which it cooled.
In sedimentary rocks the mineral composition reflects both the
nature of the pre-existing rocks that were fragmented and the
chemical nature of the sedimentary environment in which some
of the minerals were precipitated, such as salt from an evaporating
body of water. The texture of sediments is mainly a guide to the
kinds and intensity of the currents that laid them down. The minerals of metamorphic rocks are used mainly as thermometers and
pressure gauges to indicate the temperatures and pressures to
which the rock was subjected.

Though we usually refer to a rock's age as the time of its formation, a rock's history only begins with its birth. It may subsequently
be deformed mechanically, become chemically and minerally
changed, and moved laterally or vertically in response to external
forces. Man can collect rocks on the surface of the continents,

The deepest wells drilled are those for exploration for oil and gas. Recently, some holes were drilled to depths of more than 9 km (29,500 ft) in Oklahoma and Texas, and it seems likely that we have sufficient knowledge about how to tackle problems of deep drilling to go at least another few km.

Color of rocks, though sometimes spectacular and frequently useful in telling one rock or formation apart from another, is rarely of much use in deducing the details of a rock's origin. A student was once on a field trip with N. L. Bowen, a great authority on igneous rocks, and asked him to identify a particular rock specimen. He answered with a twinkle in his eye, "That's a *black* rock."

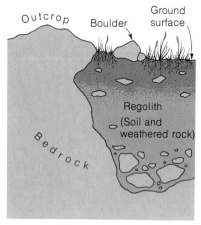

**Figure 2-9**
Cross section through surface of the ground, showing relations among outcrop, boulders, and soil.

dredge them up from the deep sea, or drill them to a depth of about 9 kilometers, where his ability to drill runs out, at least with the technology currently in use. Wherever these rocks are found now, they may have originated somewhere else in the upper few hundred kilometers of the Earth's outer shell, on the surface, on the sea floor, in the lithosphere or asthenosphere, or even deeper. Finding fossils of marine organisms in sediments at the top of a mountain showed early geologists that rocks have sometimes moved far from their place or origin. Geologists have been learning for the past century, and recently more than ever, how to unlock the information that tells the how, where, and when of the origin of rocks, even from just a sample. In the course of this book we will show how to classify rocks in order to discuss the conditions of their formation, using mineral content, texture, and other features as guiding properties. An important part of the knowledge of a rock's minerals and textures is learned from microscopic examinations of transparent slices of rocks only 0.03 mm thick and determination of the chemical composition of the whole rock.

**Rocks in the Field**  Rock samples are the raw material for analysis in the laboratory, but another whole class of knowledge of rocks is learned from **field relations**. Before the sample ever gets put into the geologist's bag, the acute field observer notes the geometric characters of the rock masses in an **outcrop**, the parts of the bedrock exposed at the surface. The first way to do this is to group the rocks into **formations**, which contain materials that for the most part have the same physical appearance and properties, a combination called **lithology**. Formations are a convenience for mapping. Sometimes they are all of one rock type or lithology, such as a granite or a limestone. Others may be interlayered thin beds of different lithologies, such as sandstone or shale. However they vary, each one comprises a distinctive group of rocks that can be recognized and thus mapped as a unit. From this stems the definition of the formation as the smallest mappable lithologic unit.

The geometry of a formation vis-a-vis adjacent formations is an important clue to understanding a rock in the context of the geological history of a region. Sedimentary formations show a series of layers called a **lithologic** or **stratigraphic sequence** that can be interpreted as a succession of depositional and erosional events. Igneous intrusions injected as a mobile magma may show **discordant** or **cross-cutting** contacts with surrounding rocks, typified by thin **dikes** that cut across original structures in the surrounding or **country rock;** or they may show **concordant** contacts, as in the case of **sills,** which may follow the bedding of the sediments into which they were intruded.

Many rocks show evidence of physical deformation. Sedimentary layers, once horizontal, are in some places **folded**, that is, bent into a wavy structure, and **faulted**, that is, broken and displaced

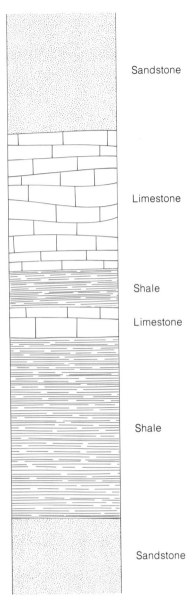

Sandstone

Limestone

Shale

Limestone

Shale

Sandstone

**Figure 2-10**
A stratigraphic sequence of sedimentary
layers of several lithologies. Such
sequences may be exposed in one
outcrop or may be composites inferred
from different parts of the sequence
exposed in a number of outcrops.

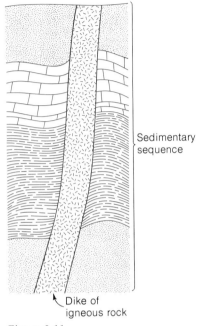

Sedimentary
sequence

Dike of
igneous rock

**Figure 2-11**
Discordant contact. A dike of igneous
rock has intruded a sedimentary
sequence by cutting across the bedding.

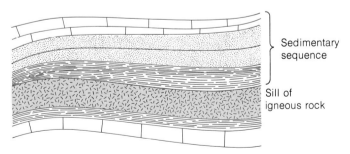

Sedimentary
sequence

Sill of
igneous rock

**Figure 2-12**
Concordant contact. A sill of igneous rock has intruded a sedimentary
sequence along bedding planes.

44

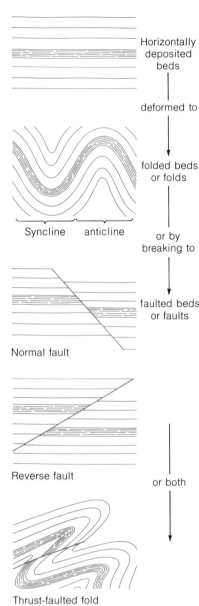

Horizontally deposited beds

deformed to

folded beds or folds

Syncline    anticline

or by breaking to

Normal fault

faulted beds or faults

Reverse fault

or both

Thrust-faulted fold

**Figure 2-13**
How originally horizontal beds can be deformed into folds and faults. Though easiest to see in sediments, all kinds of rocks may be folded and faulted.

along fractures, as in Figure 2-13. The same structural features, sometimes less easily recognized, are found in igneous and metamorphic rocks. Some of these structures are the result of compressional forces, such as those that produce folds and faults and thus shorten the crust parallel to the direction of force. Tensional forces may produce faults and result in elongation of the crust parallel to the direction of force.

Folding may range from slightly undulating open folds, in which the beds depart only slightly from the horizontal, to severely compressed strata whose beds are vertical or even overturned. Similarly, faults may range from simple small breaks, in which the rocks have moved only slightly with respect to each other, to great complex faults, such as the San Andreas, hundreds of kilometers long.

**How to Use a Rock Section to Tell a Story: The Grand Canyon**
There are a great variety of places where the bedrock, which everywhere underlies the surface, is exposed, not obscured by soil or loose boulders. Such outcrops range in size from small projections of weathered rock on a hillside and ledges in the beds of small streams to high cliffs that make the walls of canyons in mountainous terrain. Geologists also know how to use man's engineering works—formerly, railroad cuts, but for the past half century, road cuts—as ideal places to study well-exposed sections of rock. But it might take a crew of 1000 men with bulldozers, dynamite blasting, and heavy earth-moving machinery more than a lifetime to even begin to excavate a highway cut that would touch

**Figure 2-14**
The Colorado River at 224-mile rapids. The boat is of the Carnegie Institute of Washington Geological Expedition of 1937. [Photo by Edwin D. McKee, U.S. Geological Survey.]

**Figure 2-15**
The Grand Canyon of the Colorado: a view south from the bordering Kaibab Plateau. Though the topography is rugged, the horizontal stratification of the rocks in the canyon is clearly displayed. [Photo by Edwin D. McKee. U.S. Geological Survey.]

the dimensions of a place like the Grand Canyon of the Colorado River. And it wouldn't have anything of the beauty and grandeur of that fantastic gorge in the Earth, more than a mile (1.6 kilometers) deep in places, 4 to 18 miles wide (6.5 to 29 kilometers), and 280 miles (450 kilometers) long. The main part alone is 56 miles (90 kilometers) long. It is still an adventure to go down the Canyon of the Colorado River in a small boat, repeating the first trip, made in 1869. That first perilous journey, 3 months long, was made by Major *John W. Powell,* a government geologist, and his small party of nine men in four small rowboats! Powell went on to help found the U.S. Geological Survey and later became a director of the agency.

The first thing we notice about the rocks of the Grand Canyon is that there is a great thickness of horizontal sedimentary rocks whose bedding planes can be traced along the walls of the canyon for great distances. The stratification is the simple basis for the **stratigraphic time scale,** a clock that we can use to measure time and events. It is a simple matter to observe that such sediments as sands, clays, and gravels are laid down at the surface of the Earth in more or less horizontal layers. We can see layers of sand being deposited on beaches or sandbars, and muds and silts laid down after river floods. In the same way the mixtures of calcareous debris made up of the calcium carbonate shells of clams, corals, and other invertebrate animals are deposited in the beaches and shallow lagoons of the Bahama Islands, Bermuda, and the Pacific coral reef islands.

A glimpse of Powell's experience is given by an excerpt from his report on this survey: ". . . Now the danger is over; now the toil has ceased; now the gloom has disappeared. . . The river rolls by us in silent majesty: the quiet of the camp is sweet; our joy is almost ecstasy. . ."

**Figure 2-16**
Freshly deposited river sand and silt, showing original horizontality of sedimentary deposits. View is of a cut spaded through two layers of a sand bar in the Vermillion River, Illinois; the upper horizontal layer is made up of many inclined cross-beds and shows a ripplemarked surface. [Photo by Paul E. Potter.]

Once we make this generalization (which has held up for the three hundred years since it was first enunciated) we can go on to the obvious next thought: since it is absurd to think that a sedimentary layer can be formed by lifting up a bed previously deposited and sliding in a new one underneath, it must be that in any series of layers new ones are always added on top. Naturally we have to add the proviso that the whole series has not been deformed and completely overturned at some later time.

Simple as these generalizations are, they are a beautiful example of a well-known dictum: the truly great discoveries are the ones that are perfectly obvious *after* someone has pointed them out to us. It was *Nicolaus Steno,* a Danish court Physician living in post-Renaissance Italy who in 1669 formulated the *Principle of Original Horizontality* and the *Principle of Superposition,* which we have given above in modified form. He also stated the *Principle of Original Continuity,* which states that a sedimentary layer forms at the time of its deposition a continuous sheet that ends either by thinning to disappearance, by gradually changing to a bed of different composition, or by abutting against a wall or barrier, such as a shoreline that confines the depositional area. From the law of continuity we intuitively grasp the idea that the face of a bed, as we see it in the excavation for a highway cut, or in the walls of the Grand Canyon, is the broken or eroded edge of a once-continuous sheet.

From these principles we get the rudiments of our stratigraphic clock; that is, we establish a total length of time necessary for all of the rocks to be laid down and a time interval within the whole span for each layer. So if we had some idea of how long each bed took to be laid down, and *if* all of the time span were accounted for by the time to lay down the sum of all of the beds, our clock would be constructed. But that last *if* is a barrier, and the reason we suspect that a certain amount of time is not represented by rock is from the observation of river floods and other kinds of sedimentation. The silts laid down on flood plains of rivers, such as the historic ones of the Nile River in ancient Egypt, do not accumulate steadily and uniformly. We have seen that the time scale of flood deposits is about days long, but we have to remember that there is also a time scale for the times between floods, a time interval that may be a few years to several decades. In other words, a hiatus in which there was no sedimentation may be 2 to 3 orders of magnitude (powers of ten) greater than the time for sedimentation of a layer of flood silt.

A special kind of sediment, called *varved clay,* forms in lakes that are frozen over in the winter. The varves are layers of the clay that form couplets; one part of each couplet is a relatively thick, coarse-grained, light-gray silty clay that grades upward into a thin, fine-grained dark-gray clay. The light layers form in the summer and the dark ones in the winter, each pair marks one year. By counting the annual banding in lakes formed after the last ice age in Northern Europe, Baron G. DeGeer of Sweden was able to determine that about 8,700 years had passed since the glacial ice had retreated from all of Southern Europe. The accuracy of this method was confirmed by radioactive age determinations made much later, and only slight revision of the date had to be made.

Figure 2-17
Varved clay from Ogden, Utah. [Photo by R. M. Leggett, U.S. Geological Survey.]

**Figure 2-18**
Modern and ancient cephalopods. The white shell is that of a modern
chambered nautilus; the dark shell is that of a closely related form that
lived about 200 million years ago. Such similarity of form was the basis
for early recognition of the significance of fossils as evidence of ancient
life. [Photo by R. Siever.]

**Figure 2-19**
Plant fossils from Pennsylvanian formations in Illinois (*Alethopteris Serlii*)
[Photo by E. S. Barghoorn.]

**How Old Is a Fossil?** We have another more powerful tool that can be used to establish the time sequence of the fossil remains of ancient organisms found in the sediments. Most limestones and a good many shales and sandstones contain shelly materials, many of which can easily be identified by comparison with similar kinds of shells found on beaches today. Many others can be found that look vaguely like some animals that live today but are obviously different. And others are obviously some sort of organism's shell but not like anything that we see alive today. Not all fossils are shells of invertebrates like clams, oysters, cowry shells, and periwinkles. If you dig in the hillsides of sandy rock in the badlands of South Dakota (and some other places too) you may uncover a vertebrate bone of an early piglike, cud-chewing mammal. Fish skeletons and shark teeth are found in other rocks. Plant fossils are abundant in some rocks, particularly the rocks in and above coal beds, where one can recognize fernlike and other leaves, plus twigs, branches, and even whole tree trunks.

What could be so natural as to jump to the conclusion that these fossils represent life of former times and that from them we could deduce the flow of evolution from the most primitive organisms to such a complicated one as man. The jump took some time to make, though. One of the people who made it first in modern times (some Greeks had known it long before) was Leonardo da Vinci. He was followed later by Nicolaus Steno, who in the seventeenth century compared teeth from modern sharks with the so-called "tongue-stones" of Malta in the Mediterranean and concluded that they must have come from the same kind of shark. Plenty of people objected to Steno's conclusion, but the similarities between the forms of modern animals, especially the hard parts, such as teeth, bones, and shells, soon piled up in such numbers that the evidence was overwhelming and could not be reasonably dismissed either as "some accident of form" or as an obscure expression of God's wisdom in creating the Earth.

A recent controversy—one that still has some life in it—contains some of the same ideas. At the center of the controversy were certain peculiar-looking objects in meteorites, which some argued could be remnants of life elsewhere in the solar system. The answer, however, seems to be that they are inorganic structures, many of which have strange and irregular shapes, made during crystallization of the minerals of the meteorites. Another discovery of recent decades was that of single-celled organisms that were among the first to have evolved on Earth, about 2.5 to 3.0 billion years ago. They were identified as primitive plant fossils by a Harvard botanist turned geologist, who was able to identify in those early forms not only similarities in general form but structures of undoubted physiological significance, such as cell walls (see Fig. 14-6).

Figure 2-21
Faunal succession in a sequence of
lithologic units of Pennsylvanian age
in the coal-bearing strata of Illinois. The
assemblages of fossils are characteristic
of each rock type.

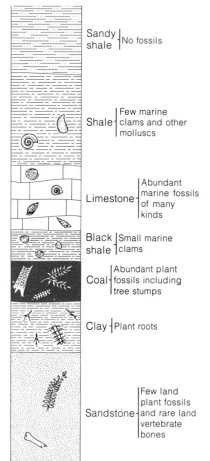

Sandy shale | No fossils

Shale | Few marine clams and other molluscs

Limestone | Abundant marine fossils of many kinds

Black shale | Small marine clams

Coal | Abundant plant fossils including tree stumps

Clay | Plant roots

Sandstone | Few land plant fossils and rare land vertebrate bones

But what has all this to do with making a time scale? The beginning of an answer can be found in the rocks of the Grand Canyon. For there we can find fossils in many of the exposed rocks, particularly in the limestones. Each limestone has a number of fossils of different species, but the assortment, or assemblage, of fossils in one limestone may differ sufficiently from those in another as to be readily distinguishable. The vertical arrangement of fossils, the **faunal succession**, corresponds to the lithologic succession and has the same order. This is the rule in fossiliferous formations of all ages since the beginning of the Cambrian Period, when shelled animals evolved. Thus fossil assemblages can be used as "fingerprints" of formations; each assemblage has slightly distinguishable characteristics, even though individual species may be present in several different formations.

It was this attribute of fossiliferous sediments that William Smith, an engineer and surveyor who had worked in coal mines and canal building, began to see so clearly when he set about collecting fossils in southeastern England in 1793. Smith knew nothing of the idea of organic evolution that was to come from Charles Darwin some decades later. He did note, however, that different fossils came from different formations, and he was able to tell one formation from another by the differences in the fossils. As Smith extended his mapping all over southern England in the early part of the nineteenth century, he was able to draw up a stratigraphic succession of rocks that appeared in different places at different levels — a composite showing how the complete section would have looked if all of the rocks from those different places and levels were brought together in one place.

Let us see how this idea works if we forget about names of fossils and formations but simply call them by different letters and numbers. At one place, **I**, a geologist sees the following horizontal succession with its characteristic fossil assemblage.

I

| Rock | Fossil assemblage |
|------|-------------------|
| A | a |
| B | b |
| C | c |

In another place, miles away, he sees another horizontal succession.

II

| Rock | Fossil assemblage |
|------|-------------------|
| B | b |
| C | c |
| D | d |

By simple use of the principle of superposition we infer that the composite sequence is:

| | |
|---|---|
| A | a |
| B | b |
| C | c |
| D | d |

But now, at place **III**, he discovers another sequence:

| | |
|---|---|
| A | a |
| C | c |
| F | f |

Here *B* is missing for some reason, but where does *F* belong—above or below *D*? We cannot tell from this information alone. Thus this is a problem of making an ordered sequence with respect to time from different fragments or partial records. To resolve the problem, the **stratigrapher** will have to hunt for an outcrop that shows both *F* and *D*.

**Figure 2-22**
Sequence of events involved in the making of an unconformity.

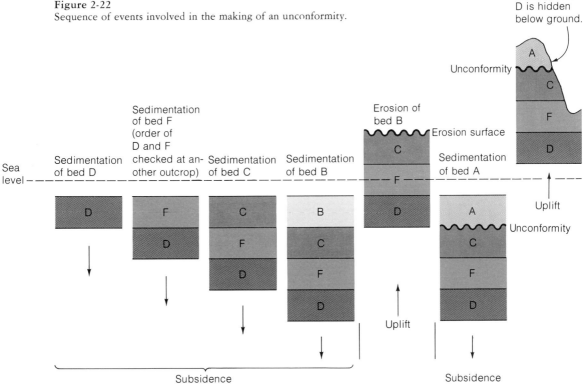

**Figure 2-23**
At the top of the inner gorge, the Cambrian Tapeats Sandstone and Bright
Angel Shale overlie the Precambrian Vishnu Schist and the Grand Canyon
Series, which here make up the walls of the inner gorge. The closeup (below)
shows the major unconformity between the jointed, foliated Vishnu Schist
and the overlying, horizontally bedded Tapeats Sandstone. [Photos by Edwin D.
McKee, U.S. Geological Survey.]

The missing *B* at **III** may be explained by erosion. Place **III**
apparently may once have had *B* deposited after *C,* but it may
have been eroded away before *A* was laid down. Or *B* may never
have been laid down in the first place. For example, there are places
in the world today, typically in shallow marine waters, where there
may be neither deposition nor erosion, but a constant balance
between the two, so that no material accumulates.

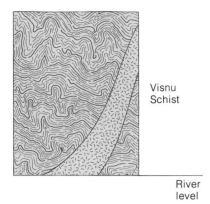

Visnu
Schist

River
level

**Figure 2-24**
The bottom of the Grand Canyon stratigraphic sequence: the Vishnu Schist.

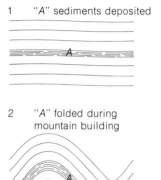

1    "*A*" sediments deposited

2    "*A*" folded during
      mountain building

3    Surface of "*A*" is eroded

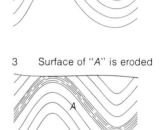

4    "*B*" sediments deposited
      on erosion surface

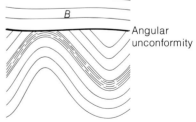

Angular
unconformity

**Figure 2-25**
The order of geologic events involved in the making of an angular unconformity.

**The Grand Canyon Sequence Interpreted** Let us go back again to the Grand Canyon and through the sequence from the base up. At the bottom is a complex group of igneous and metamorphic rocks, called the Vishnu Schist. This formation is severely deformed by folds and faults, and metamorphism by heat and pressure has obscured the original nature of the rocks. Nowhere are these rocks of the canyon horizontal. The Vishnu has no fossils, and geologists long ago learned that schists and other rocks look alike if they were formed by the same processes at different times. In other words, you cannot tell how old a rock is by looking at its minerals and their texture. But the Vishnu is at least the oldest rock we can see in the Grand Canyon inner gorge: it's at the bottom.

Above the Vishnu, and separated from it by a sharp line of discontinuity, an **angular unconformity**, is another series of rocks. An angular unconformity is a surface of erosion that separates two sets of beds whose bedding planes are not parallel. It thus signifies that the originally horizontal lower set was tilted, folded, or faulted, then eroded to a more or less even surface before the upper set was deposited horizontally upon it. The rocks above the Vishnu are interlayered sandstones, shales, and limestones. Many of them are characteristically reddish, which in combination with the varied hues of other formations in the Canyon creates a striking display of color—the extraordinary visual experience that impressed Major Powell and has awed the millions who have followed him into what is today Grand Canyon National Park.

The Grand Canyon Series, as the upper rocks of the inner gorge are called, contains no fossils, and all that we can say about it from this inspection is that it is younger than the Vishnu, older than the rocks above, and is tilted from its originally horizontal position. Rocks like this are perfectly ordinary sediments in every way except for two characteristics. They contain none of the fossils so typical of younger rocks, and they are typically associated with deformed and metamorphosed rocks like the Vishnu. As early

Figure 2-27
The Cambrian Tapeats Sandstones, unconformably overlying the contorted, structurally deformed Precambrian rocks in the Eastern Grand Canyon. This is one of the major angular unconformities of the region. [Photo Edwin D. McKee, U.S. Geological Survey.]

Figure 2-26
The Grand Canyon sequence. Above the Vishnu Schist, and separated from it by an angular unconformity, lies the Grand Canyon Series of moderately deformed sediments.

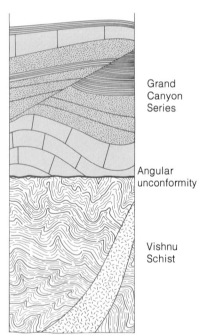

as the eighteenth century, rocks of this kind were set apart from younger fossiliferous strata, and three great classes were distinguished: Primary, Secondary, and Tertiary rocks. The main difference at that time was thought to be the complexly folded and faulted nature of the Primary as contrasted with the more-or-less horizontal, overlying Secondary. The importance of fossils was to come later.

Another clearly discernible unconformity separates the Grand Canyon Series from the overlying, pebbly, brown Tapeats Sandstone. The Tapeats, about 70 meters (225 feet) thick, contains no fossils, but it can be dated by reference to the overlying formations that it blends into without any break, thus forming a **conformable succession.** Farther up the canyon wall, the Tapeats gradually gives way to a formation consisting dominantly of shale, a hardened equivalent of a muddy sediment, called the Bright Angel Shale. The Bright Angel Shale contains a few fossils, most of which are **trilobites,** extinct arthropods related to modern crayfish. The differences among the various trilobites of different ages can be used by paleontologists to date these rocks. By matching fossil trilobite species in different stratigraphic sequences in different parts of the world, a composite succession was worked out. One of the consequences was the discovery that the part of the Bright Angel Shale that lies just above the nonfossiliferous Tapeats Sandstone in the western part of the Canyon is older than the part of the Shale that occupies the same stratigraphic position in the eastern part of the Canyon. This means that the sea in which the

Bright Angel was deposited flooded the land in the east at a later date. This is evidence of a **transgression,** meaning that as the Bright Angel Shale was deposited the sea was gradually moving landward, from west to east. Once again, it is simple geometric evidence that leads to this conclusion: as the sea advanced slowly from west to east, it continuously laid down sand along beaches and mud in deeper water. The reverse, the withdrawal of a sea and the inverse distribution of sediments in relation to shorelines, is called a **regression.**

The Bright Angel Shale is about 140 meters (450 feet) thick and grades upward into the overlying Muav Limestone, which is about the same thickness and contains trilobites of the same general age as those in the Bright Angel Shale. Studies of the trilobites in the Bright Angel Shale and the Muav Limestone reveal something about the age relations of the two formations—or the transition between them. The transition is lower in the succession and older in the west than it is in the east. This means that the Muav is also a part of the transgressive sequence.

Even from a great distance, most of the formations of the canyon wall can be distinguished easily. For example, the Tapeats Sandstone forms an obvious rim at the top edge of the inner gorge, called loosely the "Granite" gorge. In contrast, most of the Bright Angel Shale, just above that rim, is hidden beneath rubble, which slopes gently upward from the top of the Tapeats to the base of a

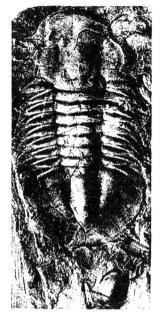

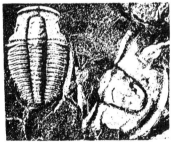

**Figure 2-28**
Cambrian trilobites from the Bright Angel Shale, Grand Canyon. [From *Cambrian History of the Grand Canyon Region* by Edwin D. McKee and Charles E. Resser, Carnegie Institution of Washington, 1945.]

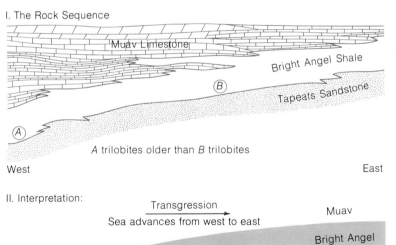

**Figure 2-29**
I. The stratigraphic relations of the Bright Angel Shale in the Grand Canyon region. The fact that trilobites at *A* are older than those at *B* leads to the interpretation shown in II.

**Figure 2-30**
The Grand Canyon sequence. Above
the Vishnu Schist and the Grand
Canyon Series are horizontally bedded
sediments: the Tapeats Sandstone,
the Bright Angel Shale, and the Muav
Limestone.

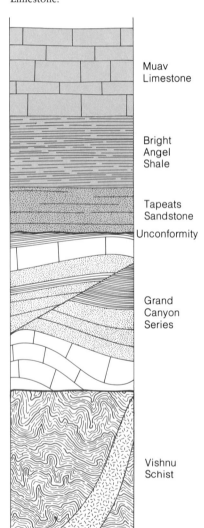

Muav
Limestone

Bright
Angel
Shale

Tapeats
Sandstone

Unconformity

Grand
Canyon
Series

Vishnu
Schist

**Figure 2-31**
Formations exposed in Diamond Creek, Grand Canyon. The Cambrian
Bright Angel Shale, which forms the rubbly slopes at the bottom, is overlain
by the Cliff-forming Muav, Temple Butte, and Redwall limestones. Above the
main cliff is the thinner Supai Formation, which crops out with gentler slopes
than the cliff formations below. [Photo by Edwin D. McKee, U.S. Geological
Survey.]

higher topographic bench, or step, which is the base of the Muav
Limestone.

The next formation up, the Temple Butte Limestone, is one
that could easily be missed; it is thin, and in places along the
Canyon wall it is missing entirely; in these places, the Redwall
Limestone, which elsewhere overlies the Temple Butte, rests
directly on the Muav. The Temple Butte forms the base of a steep
cliff, but the important thing about it is that it contains fossil skeletons of primitive fishes, which we know from the general succession
of fossil animals lived a long time after the trilobites of the Muav.
Fossils of a great many marine animals that lived between the times
of deposition of the Muav and the Temple Butte are known from
other formations in various parts of the world—but they are all
missing here. Here, then, is evidence of a great gap in the record,
for there is an unconformity between the Muav and the Temple
Butte. If any sediment had been laid down during the time represented by this unconformity, it was later eroded without leaving
a trace.

Another time gap is represented by an unconformity between
the Temple Butte and the overlying Redwall Limestone; still
another is represented by an unconformity between the Redwall
and the Supai Formation. The Redwall's age we know from its
sparse content of fossil marine animals. The Supai contains no
marine fossils, but it does contain fossils of land plants. Thus it

can be dated with reference to the worldwide succession of plant fossils, a succession that has been worked out in much the same way as the succession of animal fossils. The fossil plant remains are similar to those of the famous bituminous and anthracite coal beds in the United States and the Ruhr Valley in Europe. Of even greater interest are the fossil footprints of primitive reptiles found in the Supai. Thus it turns out that we can tell something of time not only from marine sediments but from those deposited on land — the **terrestrial**, or **continental**, **deposits**.

Continuing up the canyon walls, we find that there is also an unconformity at the top of the Supai. Above the Supai is a sandy, red shale, the Hermit Shale, which is succeeded by the Coconino Sandstone. Not only does the Coconino contain more vertebrate animal tracks, but it has a distinctive form of bedding, which is not uniform and horizontal like that of many sediments, but is composed of many sets of interfering wedges of bedded material inclined at various angles up to 35° from the horizontal. This form of bedding, called **cross-bedding**, is a characteristic structure of sand dunes on land — at least that is what geologists have thought for a long time. It is now known, however, that this form of bedding is also characteristic of dunes formed by currents in rivers and under the sea. The Coconino Sandstone may have been formed by wind, as many geologists believe, but the recent observations that such dune structures and bedding are also characteristic of some large river bars and submarine dunes have given reason to question that belief.

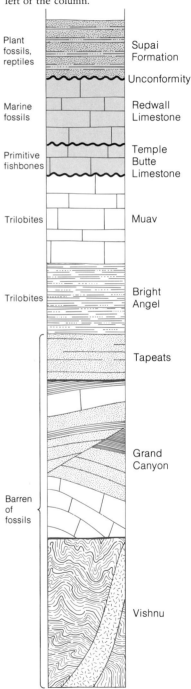

Plant fossils, reptiles — Supai Formation

Unconformity

Marine fossils — Redwall Limestone

Primitive fishbones — Temple Butte Limestone

Trilobites — Muav

Trilobites — Bright Angel

Tapeats

Grand Canyon

Barren of fossils

Vishnu

Figure 2-33
Cross-bedding in the Coconico Sandstone. The cross-bedding of different layers interferes to produce such a complex pattern that the true horizontal bedding planes are not easily distinguished. [Photo by Edwin D. McKee, U.S. Geological Survey.]

Why the Redwall Limestone got its name is obvious from any look at it in the Canyon. But the "red" in "Redwall" is in at least one way undeserved, for when a piece of the Redwall is cracked the fresh rock turns out to be gray. A close look at the cliff shows that the limestone is stained, as if it had been rubbed with a reddish pastel chalk. The stain is washed down by rain from the overlying sandy red shale of the Supai Formation.

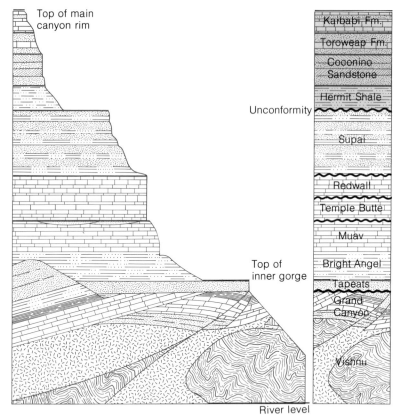

**Figure 2-34**
The Grand Canyon sequence. The top of the canyon wall is made up of the Hermit Shale; Coconino Sandstone; Toroweap Formation, composed of limestone and shale; and Kaibab formation, composed of limestone and sandstone. An approximate topographic profile is shown at left.

The Coconino appears to be conformable with the underlying Hermit; that is, there is no evidence that the beds at the top of the Hermit have been eroded, nor do the few fossils available indicate any missing time interval. Similarly, the Coconino is conformable with the overlying Toroweap, a formation of limestone and red, sandy shale, which is succeeded by a massive formation of limestone and sandstone that forms the top of the cliffs at the upper rim of the Canyon—the Kaibab Formation.

If we were to inspect highlands above the canyon rim in the general region of the Grand Canyon, we would find formations younger than the Kaibab; from their fragmentary successions we could build a composite that would include about 1830 to 2120 meters (6000 to 7000 feet) of red, brown, yellow, and gray sandstones, conglomerates, and shales that contain the famous petrified forest of tree trunks and, in some places, dinosaur remains.

**Figure 2-35**
South rim of the Grand Canyon, showing formations from the Supai below
to the Kaibab at the rim. The lowest slopes are those formed by the Supai,
followed upward by the more gentle, vegetated slopes of the Hermit Shale.
Above the Hermit is the lighter-colored, cliff-forming Coconino Sandstone,
overlain by the Toroweap Formation, which forms gentle, vegetated slopes.
Above them are the light cliffs of the rim formed by the Kaibab Formation.
[Photo by Edwin D. McKee. U.S. Geological Survey.]

The rocks of the Grand Canyon have many stories to tell: of
the advance and retreat of the seas over the continent at this place,
of the appearance and disappearance of different kinds of organisms, and of the different kinds of marine and terrestrial environments in which this remarkable variety of sediments was deposited.
But one of the most important is the tale of time — the time that
is represented by the rocks of the Canyon and the time that is
recorded by the unconformities between so many of the formations. From the radioactive time scale, which is based upon the
decay of radioactive elements in minerals, we know that the oldest
formation, the Vishnu Schist, is about 1400 to 1500 million years
old, the base of the Bright Angel Shale about 550 million years
old, and the top of the Kaibab about 225 million years old. An
enormous amount of time is represented by these rocks and
unconformities.

## ROCKS AS RECORDS OF EARTH MOVEMENTS

Angular unconformities not only date erosion intervals, but
they also are a record of ancient earth movements. Beds below
such unconformities were folded, tilted, faulted, and uplifted
before erosion produced the more-or-less even, unconformable

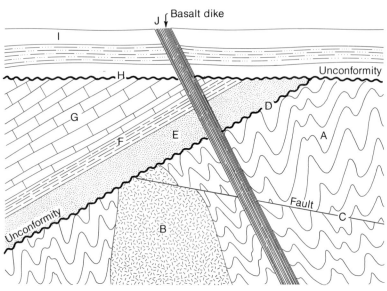

**Figure 2-36**
Relative time dating by field relations. The sequence of events shown must
have been: (1) sediments *A* deposited, (2) sediments *A* deformed, (3) granite *B*
intruded into *A*, (4) faulting along *C*, (5) erosion produced unconformity at *D*,
(6) sediments *E, F,* and *G* deposited, (7) tilting and erosion to produce
unconformity at *H*, (8) sediment *I* deposited, (9) basalt dike intruded at *J*.

surfaces that we observe, Erosion was, in turn, followed by additional earth movement, for only crustal subsidence could account
for a change from erosion to further sedimentation. Unconformities, therefore, are records of periods of mountain-building,
even though the roots of the mountains are all we can see today.
**Disconformities**—time gaps between two units whose bedding
planes are parallel—are less dramatic, but still imply the same
general sequence of uplift, erosion, and subsidence.

There are other ways of telling a time sequence too. Though
igneous rocks are not layered, as sediments are, they too have
characteristics that may be used to place them in time. The concordant and discordant field relations between them and adjacent
sedimentary, metamorphic, and other igneous rocks can be used
in just the same way that Steno's laws of original horizontality
and superposition can be used to figure out relative ages of sediments. Similarly, faults can be fitted into time sequences as well.

Shown in Figure 2-36 is the record of a series of events that can
be interpreted in only one way. Of course, once such a time-event
jigsaw puzzle has been put together, as in this diagram, it looks
easy. But the job of sorting out the pieces and seeing how they
fit can sorely tax the imagination of a field geologist. *To sum up,
the ways of ordering geological events with respect to a relative time*

*scale are based on interpretations of sedimentary successions, igneous field relations, such as cross-cutting, and tectonic deformation, such as folding and faulting and angular unconformity.* (The word **tectonic** refers to the deformation of the rocks of the Earth's crust and the structure of the crust in general.) By piecing together the information gleaned from the study of such field relations, geologists during the nineteenth century worked out the entire stratigraphic time scale.

**Hutton and Uniformitarianism** Although the basis of the reasoning used here in interpreting the Grand Canyon sequence may seem obviously correct, it was not until the close of the eighteenth century that geologists, who were then struggling to find ways to explain the field relationships of rock formations, broke free of the static concept that the Earth was created just as we see it today, with all of its river valleys, mountains, and plains placed where they are by God. The new way of looking at the Earth was one that recognized constant change takes place as geological forces modify the surface and the interior.

A well-known Scottish gentleman farmer, *James Hutton,* led the way with a book carrying the bold title: *Theory of the Earth with Proof and Illustrations,* first presented to the Royal Society of Edinburgh in 1785. Hutton's greatness lies in his recognition of the cyclical nature of geological changes, and of the way in which ordinary processes, operating over long time intervals, can effect great changes. He reasoned from observation that rocks slowly decay and disintegrate under the action of water and air. This process—**weathering**—produces debris in the form of gravel, sand, and silt. Water and air also act to transport the debris, most of which ends up near or below sea level. The deposits are compacted, cemented, and ultimately become sedimentary rocks. At a later time, according to Hutton, a **plutonic** episode may occur in which subterranean heat and thermal expansion produce an intrusion of igneous rock accompanied by upheaval of the sediments and by mountain-building, called **orogeny**. Marine sediments emerge as land, bringing deposition to a halt, and then erosion of the newly emerged highlands initiates the cycle all over again.

Hutton observed and learned from the modern counterparts of each stage of his cycle: mountains being eroded, rivers carrying debris to the sea, ocean waves pounding rocks, sands and muds settling to the bottom and then being buried on the sea floor. Nature behaves in a uniform fashion throughout time, so that by studying the present one can infer the behavior of past processes. Hutton, followed by *Charles Lyell (Principles of Geology,* 1830) used and publicized this **Principle of Uniformitarianism**. Actually what remains "uniform" are the physical and chemical laws that govern geological activity. Volcanism may have been

Pluto was the Greek god of the lower world. Hutton's appeal to subterranean heat earned his theory the sobriquet *Plutonism.*

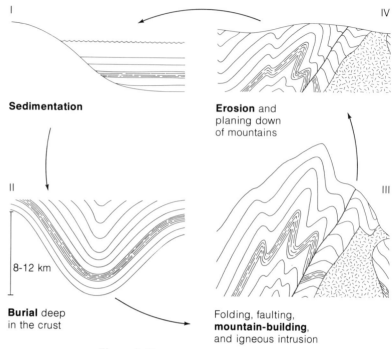

**Figure 2-37**
The geologic cycle as deduced by Hutton.

more frequent in the past than it is now. Nevertheless, ancient volcanoes should have released gases and deposited ash layers and lava flows, just as modern ones do when they erupt. Uniformitarianism, as we understand it today, does not hold that the rates of geological processes or their precise nature had to be the same.

Many of the concepts used in modern field interpretation go back to discoveries made by late eighteenth-century and nineteenth-century geologists, among them: *William Smith*, an engineer and father of stratigraphy; *George Cuvier*, the great zoologist who pioneered in stratigraphy and paleontology in France; and *Jean Guettard* and *Nicolas Desmarest*, the geologists who explored the Auvergne region of France and showed that basalts were ancient lava flows. But it was Hutton who recognized that igneous bodies must be younger than the rocks they intrude. Hutton also pointed out that debris fragments in a sedimentary or igneous formation must have been derived from a parent rock older than the formation itself, and he was the first to grasp the idea that a cycle of upheaval, erosion, subsidence, and sedimentation would show as an *unconformity* in the stratigraphic record.

With these principles established, nineteenth-century geologists opened a new era. The history contained in geologic formations could at last be deciphered, and man could travel backward in time to view ancient landscapes. It became possible to reconstruct the interrelations between mountains, oceans, climates, animals, and plants long since gone, as if the geologist were a magician who could conjure the past. By this time, geography and geology had their historical counterparts, *paleogeography* and *paleogeology*. Geology entered a period of discovery and glamour. It is not surprising that one of the major scientific organizations, the American Association for the Advancement of Science (A.A.A.S.), was formed in 1847 by the enlargement of the American Society of Geologists and that fifteen of the first twenty presidents of the A.A.A.S. were distinguished geologists.

## EVOLUTION AND THE TIME SCALE

In 1859 Charles Darwin's *On the Origin of Species by Means of Natural Selection* was published, and the theory of organic evolution was launched. Along with it was launched one of the great controversies in the history of science: the ideas of evolution were denounced by many as monstrous and antireligious, if not at least downright silly. But geologists and biologists under the leadership of Lyell and Thomas Huxley (grandfather of Julian Huxley, an eminent biologist, and Aldous Huxley, the author of *Brave New World*) moved in a few years to acceptance of the theory. The concept of evolution had enormous impact, for its theoretical framework gave support to the idea that the time-related changes in fossil species could be used to set up a stratigraphic time scale.

Armed with a refined science of paleontology, based on Darwinian evolutionary theory, and with expanded opportunities for travel during the tremendous period of world exploration that accompanied European economic expansion and imperialism, geologists mapped the surface of the earth and fitted together what we now call the **Phanerozoic** (known) time scale. The names of the time periods are taken either from the geographic locality in which the formations were best displayed or first studied or from some characteristic of the formations. For example, the Jurassic is named from the Jura Mountains of France and Switzerland, and the Carboniferous is named from the coal-bearing sedimentary rocks of Europe and North America.

Each time period of the stratigraphic time scale is represented by its appropriate system of rocks, and we differentiate time units, the **periods**, from time-rock units (the rocks that represent time), the **systems**. Each major unit is divided: the period into **epochs**, and the system into **series**. Epochs and series have geographic names, except for the older names of many of the epochs, which are simply called Upper, Middle, or Lower. Thus the Upper Jurassic series comprises the rocks of the Upper Jurassic epoch of time.

An illuminating sidelight to the enunciation of Darwin's new theory is that the discovery was actually made by both Darwin and Alfred Russell Wallace, a young, unknown naturalist working in the East Indies (Indonesia). Darwin had in 1857 explained his theory in a now-famous letter to the great American botanist Asa Gray at Harvard; Darwin's theory had been germinating in preliminary drafts for almost 20 years. In 1858 Wallace sent Darwin a manuscript that he had written in the midst of a bout with intermittent fever in the Molucca Islands. Darwin's reaction was close to panic, for Wallace's manuscript outlined his theory in essential details. On the advice of two friends, Charles Lyell, then the most famous geologist in England, and Joseph Hooker, a leading English botanist, Darwin and Wallace presented their views in a joint essay before the Linnaean Society of London on July 1, 1858. Independent simultaneous discovery of an idea whose time has come, and the rush to publish it, is not a new development in science!

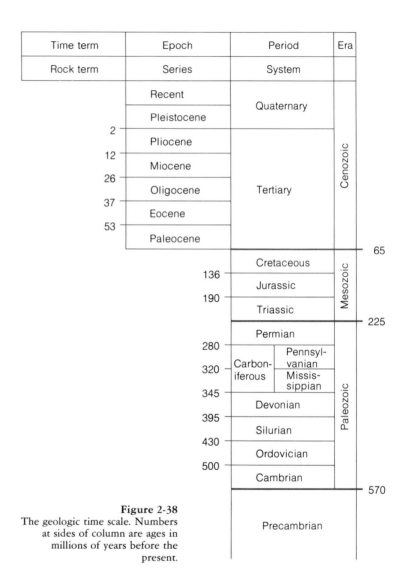

| Time term | Epoch | Period | Era |
|---|---|---|---|
| Rock term | Series | System | |

(The geologic time scale figure shows, with ages in millions of years:)

| | Epoch/Series | Period/System | Era |
|---|---|---|---|
| | Recent | Quaternary | Cenozoic |
| | Pleistocene | Quaternary | Cenozoic |
| 2 | Pliocene | Tertiary | Cenozoic |
| 12 | Miocene | Tertiary | Cenozoic |
| 26 | Oligocene | Tertiary | Cenozoic |
| 37 | Eocene | Tertiary | Cenozoic |
| 53 | Paleocene | Tertiary | Cenozoic |
| 65 | Cretaceous | Cretaceous | Mesozoic |
| 136 | Jurassic | Jurassic | Mesozoic |
| 190 | Triassic | Triassic | Mesozoic |
| 225 | Permian | Permian | Paleozoic |
| 280 | Pennsylvanian | Carboniferous | Paleozoic |
| 320 | Mississippian | Carboniferous | Paleozoic |
| 345 | Devonian | Devonian | Paleozoic |
| 395 | Silurian | Silurian | Paleozoic |
| 430 | Ordovician | Ordovician | Paleozoic |
| 500 | Cambrian | Cambrian | Paleozoic |
| 570 | Precambrian | Precambrian | |

**Figure 2-38**
The geologic time scale. Numbers
at sides of column are ages in
millions of years before the
present.

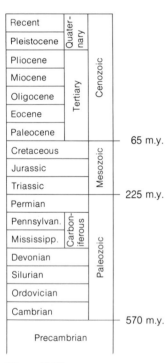

**Figure 2-39**
Abbreviated geologic time
scale.

## ABSOLUTE TIME AND THE GEOLOGIC TIME SCALE

The question of just how many years are represented by the rocks
in the stratigraphic time scale has been around for at least 2500
years. As far as we know, Xenophanes of Colophon (570 – 470
B.C.) was the first of the early philosophers to focus upon this problem. He recognized the significance of fossils as remnants of former
life and correctly inferred that sedimentary rocks originated as
sediments deposited on the sea bottom. Moreover, he concluded
that such rocks and fossils must be of great age. Xenophanes was
not particularly loved by some of the erudite establishment Greek
philosophers, most of whom had deduced that there was no beginning and no end to the Earth. Around 450 B.C., Herodotus, the

great Greek historian, travelled through the lower Nile River valley. His observations led him to reason that the Nile delta must have been made from a series of floods. It then quickly followed, by his reasoning, that if a single flood were to lay down only a thin layer of sediment, it must have taken many thousands of years to build up the Nile delta.

Aristotle and other Greek and Roman philosopher-naturalists strengthened and expanded this scholarly approach of observation combined with deduction and left a legacy to early Christian scholars, such as St. Augustine, who in all essentials continued the tradition. But the thread of that kind of thinking was temporarily lost in late medieval outgrowths of the revolution of Christian scholasticism and theological idealism, which found a solution to the problem of the antiquity of the Earth and all things on it in one book: *Genesis.* The literal interpretation of this book of The Bible gained increasing devotion by Churchmen, not in the early Middle Ages, when the teachings of the Greeks were still largely accepted in the secular world and the New Testament was still new, but in the later Middle Ages and the beginnings of modern times in a reaction to the scientific explorations of the Renaissance. Literalism, to many, was as explicit as the pronouncement by Archbishop Ussher of Ireland in 1664 that the Earth was created at 9:00 A.M., October 26, 4004 B.C. (presumably Greenwich mean time!). Other dates could be and were calculated for Noah's flood and various other Biblical events.

When modern geology started gathering momentum, through the work of James Hutton and others, no attempts were made to guess at the age of the Earth; instead, there was an obvious return to the old Greek way of looking at things. The evidence demanded that a long time was necessary for Earth processes to have had any effect in carving mountains and accumulating sediment. By looking at the time scale of processes and using the idea of uniformitarianism, geologists came to recognize that rocks are very old, and the Earth much older.

At the same time, physicists were busy making various sorts of calculations, based on the ideas of Galileo and those of Newton, who in 1687 established the basis for gravitational theory in his *Principia.* After Laplace proposed his theory of the origin of the solar system and Cavendish determined the gravitational constant (see Chapter 3), there was a sound foundation for calculating the relation between time and planetary orbits. The times necessary for the formation of the Sun and planets, and for planetary motions, seemed much too long compared with the shortness of time demanded by the Bible. Before the nineteenth century, however, physicists did not challenge religious orthodoxy: Sir Isaac Newton was a devoutly religious man—one whom we would call today a "fundamentalist." Even so, in the middle of the eighteenth century, Comte de Buffon of France analyzed rates of melting and cooling rates of iron balls and put what he learned together with his guess

It was the Greek historian Herodotus who invented the name *delta* for the large body of sand, silt, and mud that accumulates where a river empties into a lake or sea. The shape of the lower Nile floodplain valley, below Thebes, looked to him like the Greek letter delta Δ, roughly triangular in shape.

"And God said, Let the waters under the heaven be gathered together in one place, and let the dry land appear: and it was so. And God called the dry land Earth; and the gathering together of the waters called he Seas: and God saw that it was good. And God said, Let the Earth bring forth grass, the herb yielding seed, and the fruit tree yielding fruit after his kind, whose seed was in itself, after his kind, upon the Earth: and it was so. And the Earth brought forth grass, and herb yielding seed after his kind, and the tree yielding fruit, whose seed was in itself, after his kind, and God saw that it was good. And the evening and the morning were the third day." How long a day!

Helmholtz's argument that the Sun underwent gravitational collapse held unexpected implications. For example, if the Sun began to collapse some 20 million years ago, it would at that time have to have been much larger than it is today—so much larger that its radius would have extended to the Earth's present orbit around the Sun. Geologists were not prepared to accept this.

One who would not give way to Kelvin was a fellow Scotsman, Sir Archibald Geikie, who occupied the professorship of geology at Edinburgh while Kelvin was the professor of physics at Glasgow. Maybe it took an Edinburghian not to be overwhelmed by the immense forcefulness and power of the Glasgwegian Kelvin.

that the interior of the Earth had to be like iron (it was so dense) and calculated how long it would take the molten earth to cool. His result, 75,000 years, made no one happy; for the fundamentalists it was much too long, and for many geologists it was much too short.

The plot thickened in 1854, when Herman von Helmholtz, one of the founders of the science of thermodynamics, seized on the problem of the Sun's luminosity. Not long before, Immanuel Kant had calculated that if the Sun's light came from ordinary combustion, it would have burned up in only 1000 years. Helmholtz realized that the way out of this bind was to infer that the Sun's light came from the heating required by the gravitational contraction of the immense mass of the Sun. His notion was that particles would literally tend to fall into the center and that the potential energy released in that fall would be converted to heat. Helmholtz came up with estimates of 20 to 40 million years.

Helmholtz's argument was taken up by another great physicist of the nineteenth century, William Thomson, better known as Lord Kelvin. In his first paper of a long career, and in many following articles, he expanded and revised Helmholtz's estimates, put them together with Buffon's cooling-rate figures and other more ingenious arguments, and pronounced that the most probable value was something like 20 to 40 million years, probably closer to 20 million.

Many geologists found themselves unable to adjust their notions of the length of geologic time to accord with Kelvin's estimate, and continued to argue vehemently that not everything was yet known about physics and that it was foolish to contradict the obvious evidence of the rocks in such composite sequences as those in the Grand Canyon, for they clearly must have taken many times 25 million years to accumulate. Others, such as *Clarence King,* one of the founders of the U.S. Geological Survey, who had done much of the geological exploration of the West, made his own calculations, compared them with those of Kelvin, and concluded that Kelvin was about right. Charles Walcott, one of the most famous paleontologist-stratigraphers of the day, added up the whole known stratigraphic sequence and concluded that it could reasonably well account for as much as 75 million years, one of the figures given by Kelvin at various times. As late as the turn of the century, Lord Kelvin gave an important address to the American Association for the Advancement of Science in which he reiterated his faith in his calculations and in the physical constraints they imposed on the age of the Earth. A long answer followed from one of the geological unbelievers, *T. C. Chamberlin,* the head of the Department of Geology at what was then the brand new University of Chicago. Chamberlin, who held many agnostic opinions, added his speculation that there might yet be discovered new sources of energy (unknown to him then) within the particles of matter that would eliminate burning and gravitational contraction as the sole possible causes of the Sun's luminosity.

Curiously enough, the debate of 1899–1900 came four years after a discovery that was to be of world-shaking importance (Both figuratively *and* literally, as it turned out!). In 1895, Henri Becquerel, a French physicist, discovered radioactivity in uranium salts, almost at the same time that the German physicist Wilhelm Röntgen discovered X-rays. Soon after, Marie Sklodowska-Curie made the crucial discovery and isolation of radium, a radioactive element. Thus in the space of a few months, the stage was set for the use of the clocks that are built into the nuclei of radio-active atoms — the clocks that were eventually to resolve the bitter argument about the age of the Earth. Yet neither Kelvin nor Chamberlin, to say nothing of their audiences, seemed to be aware of the extraordinary implications of these discoveries, though the journals of physics and chemistry and even the Sunday newspaper supplements were popping with the news.

It was not until 1905 that *Ernest Rutherford,* a famous English physicist who had been studying radioactive processes, suggested that radioactive minerals could be used to date rocks. He dated a uranium mineral in his laboratory the next year. In the same year, B. B. Boltwood of Yale discovered "ionium," which turned out to be an isotope of thorium — the first isotope to be isolated. Both Rutherford and Boltwood published ages of dated minerals, but it was not until F. Soddy, in 1913, clarified the nature of isotopes that the methods could be refined and made more accurate, for most of the early ages turned out to be too high. When the full series of decay products of radioactive disintegration was firmly established, it became clear that the Earth had to be not millions, but *billions,* of years old.

A typically long-delayed award of distinction came to Boltwood, who was credited with discovering the first isotope: a mineral was named after him, appropriately enough a uranium-bearing mineral. In this way science incorporates into its vocabulary the names of its great ones. Not all, of course, have minerals named after them. Kelvin's name was given to the absolute temperature scale (we refer to "degrees Kelvin"); Helmholtz's to a number of things, from electrical coils to equations and types of energy (Helmholtz free energy). Geologists have been immortalized not only by mineral names but by fossil names, like *Oryctocara geikiei,* a trilobite, whose specific name was given in honor of Geikie.

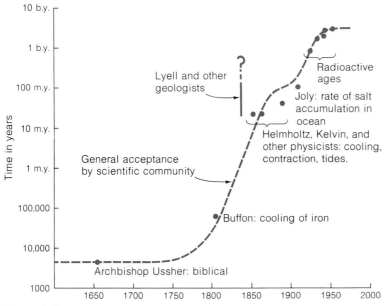

**Figure 2-40**
How Man's estimate of the age of the Earth has changed in the last few centuries.

In a few decades all of the apparent contradictions were cleared up. In the 1920's and 1930's, astronomers and physicists recognized that the immense energies liberated by nuclear processes must be responsible for the luminosity of the sun and that the heat given off by radioactive decay was the explanation for the heat trapped in the Earth's interior.

## THE CLOCKS IN ROCKS: RADIOACTIVE ATOMS

What the pioneers of nuclear physics discovered at the turn of the century was that atoms of certain elements, the radioactive ones, spontaneously disintegrate to form atoms of a different element, liberating energy in the process. The important reason why radioactive decay offers a dependable means of keeping time is that the average rate of disintegration *is fixed and does not vary* with any of the typical changes in chemical or physical conditions that affect most chemical or physical processes. This means that once a quantity of a radioactive element is made somewhere in the universe, it starts to act like the balance wheel of a clock, steadily "firing" off one atom after another at a definite rate.

To tell time, we cannot simply look at a balance wheel or listen to ticking. We need some kind of face with hands and numbers on it or the numbered counting wheels of modern digital clocks. The numbers that we use to read the radioactive clock are supplied in the form of the new atoms, the **daughter** elements, that are created from the old disintegrated ones, the **parent** elements. If we can identify and count the number of daughter element atoms, and if we know the rate of decay, we can work back to the time when there were no daughters but only parents. The idea is simple, but its practical implementation has required a major effort on the part of those geologists who combine their knowledge of nuclear physics with that of geology—the specialists in **geochronology**, which is sometimes called a subdivision of **nuclear geology** or **isotope geology**.

**What Happens to a Radioactive Atom?** All atoms have a dense nucleus that contains practically all of the mass of the atom. Surrounding the nucleus is a cloud of orbiting **electrons**. The nucleus contains two kinds of particles, each with mass 1: a **proton** has a positive electrical charge of +1, and a **neutron** is electrically neutral. The number of protons in the nucleus, which is balanced by the same number of electrons in the cloud outside, each of which has a negative charge of −1, is unique for any element, and is called the **atomic number** (usually symbolized by $Z$). The sum of the masses of protons and neutrons give the **atomic weight** of the atom. All atoms of the same element—for example, carbon—have the same atomic number; the atomic number of carbon is 6. Different **isotopes** have different numbers of neutrons. Carbon

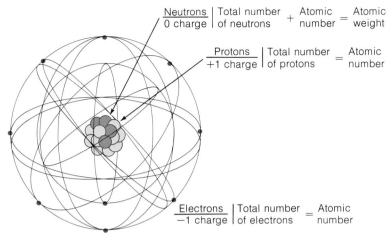

| Neutrons / 0 charge | Total number of neutrons | + | Atomic number | = | Atomic weight |

$$\frac{\text{Protons}}{\text{+1 charge}} \;\Big|\; \frac{\text{Total number}}{\text{of protons}} = \frac{\text{Atomic}}{\text{number}}$$

$$\frac{\text{Electrons}}{\text{−1 charge}} \;\Big|\; \frac{\text{Total number}}{\text{of electrons}} = \frac{\text{Atomic}}{\text{number}}$$

**Figure 2-41**
The structure of an atom. The nucleus of protons and neutrons is surrounded by a cloud of orbiting electrons.

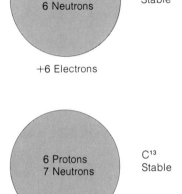

6 Protons
6 Neutrons

$C^{12}$
Stable

+6 Electrons

isotopes exist with 6, 7, and 8 neutrons, giving atomic masses of 12, 13, and 14. Of these isotopes, $C^{12}$ and $C^{13}$ are stable; that is, they do not change or spontaneously disintegrate. But $C^{14}$ is radioactive and spontaneously decays by emitting $\beta$-**particles** (electrons shot from the nucleus when a neutron is split into a proton and an electron). This changes its atomic number to 7, which is characteristic of the element nitrogen, but does not change its mass (because an electron has practically no mass). So $N^{14}$ is produced from $C^{14}$. In the shorthand notation of equations:

$$C^{14} - \beta \rightarrow N^{14}$$

Atomic weights of naturally occurring elements, which chemists spent much of their time measuring accurately in the nineteenth century and early twentieth century, turn out to be odd numbers. Thus the atomic weight of carbon given in tables is 12.01115 [relative to oxygen atomic weight as 15.999 (essentially 16)]. The reason for this is that the natural elements are mixtures of the various isotopes. The very small amount of $C^{13}$, and even smaller amounts of $C^{14}$, in most samples of carbon is enough to affect the average weight.

Another element that decays spontaneously is Rubidium$^{87}$ ($Rb^{87}$), which emits a $\beta$-particle in the same way as $C^{14}$ to produce Strontium$^{87}$ ($Sr^{87}$):

$$Rb^{87} - \beta \rightarrow Sr^{87}$$

An important difference between the decay of $C^{14}$ and $Rb^{87}$ is the *rate* at which the atoms decay. That rate is given in terms of the number of atoms, $n$, that decay in a period of time (usually

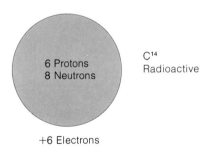

6 Protons
7 Neutrons

$C^{13}$
Stable

+6 Electrons

6 Protons
8 Neutrons

$C^{14}$
Radioactive

+6 Electrons

**Figure 2-42**
Carbon isotopes. All have the same number of protons, and thus atomic number 6, but different numbers of neutrons.

per second or per year, depending on which is more convenient) relative to the total number of atoms of that element ($N$) in any given amount of it. The proportion $n/N$, which is constant no matter what $N$ is, is called the decay or **disintegration constant,** and is usually symbolized by the Greek letter lambda, $\lambda$. Another measure of the rate is the time necessary for the original number

---

**Box 2-1   Dating Rocks by Means of Radioactive Minerals**

The general way in which we solve for the age of a rock is as follows:

Let $N^0$ = the original number of radioactive atoms in a sample of the parent mineral.

Let $N$ = the number of radioactive atoms remaining in the sample after the lapse of a certain amount of time.

Let $\lambda$ = the decay constant, the fraction that disintegrates per unit of time.

At the end of one time unit, the number left is:

$$N^0 - \lambda N^0 = N_1 \qquad (1)$$

or

$$N^0(1-\lambda) = N_1. \qquad (2)$$

At the end of a second unit of time,

$$N_1 - \lambda N_1 = N_2. \qquad (3)$$

Substituting from (2) into (3) gives

$$N^0(1-\lambda) - \lambda[N^0(1-\lambda)] = N_2. \qquad (4)$$

Factoring out $N^0(1-\lambda)$ gives

$$N^0(1-\lambda)(1-\lambda) = N_2 \qquad (5)$$

or

$$N^0(1-\lambda)^2 = N_2. \qquad (6)$$

At the end of the third unit of time,

$$N^0 - \lambda N_2 = N_3. \qquad (7)$$

Substituting from (6) gives

$$N^0(1-\lambda)^2 - \lambda[N^0(1-\lambda)^2] = N_3. \qquad (8)$$

Factoring out $N^0(1-\lambda)^2$ reduces the equation to

$$N^0(1-\lambda)^2(1-\lambda) = N_3 \qquad (9)$$

or

$$N^0(1-\lambda)^3 = N_3. \qquad (10)$$

Consequently, it can be seen from (2), (6), and (10) that where $y$ is the number of time units,

$$N^0(1-\lambda)^y = N_y. \qquad (11)$$

Transposing gives

$$\frac{N_y}{N^0} = (1-\lambda)^y. \qquad (12)$$

In logarithmic form this becomes

$$\log \frac{N_y}{N^0} = y[\log (1-\lambda)]. \qquad (13)$$

Since $[\log (1 - \lambda]$ is a constant (because $\lambda$ is), it follows that $\log N/N^0$ is directly proportional to the amount of time that the radioactive element has been disintegrating. Hence if we plot $\log N/N^0$ against time we should obtain a straight line with a slope of $[\log (1-\lambda)]$. If we choose the unit of time for $\lambda$ to be the half-life of the element, $T$, then $\lambda = 0.5$ by definition. So, $\log (1-\lambda) = \log (1-0.5) = \log 0.5 = -0.3$. Substituting in equation (13) we get

$$\log \frac{N}{N^0} = -0.3y,$$

where $y$ is the number of half-lives that have elapsed since the mineral was formed. The age of the rock is determined by measuring $N$ and $N^0$ and solving for $y$; then $y$ multiplied by $T$, the half-life, gives the age in years. (The solution of the radioactive decay problem is shortened to one equation by using the calculus, which was invented to solve such problems. It is, $dN/dt = -\lambda N$, or in integrated form, called the age equation, $N = N_0 e^{-\lambda t}$.)

of atoms to be reduced by one-half, called the **half-life**, usually represented by $T$. The half-life is simply related to $\lambda$: it is always

$$T = 0.693/\lambda$$

We can now compare the rate of decay of carbon[14], which has a half-life of 5570 years, with that of $Rb^{87}$, which has a half-life of 47 billion years. It is the half-life that dictates that $C^{14}$ can be used for timekeeping for only the last 30,000 years or so of Earth history, about equal to a little more than 5 half-lives (at which point all except the last 1/32 of the original amount is gone). In contrast, 3 billion years, about the age of some very old rocks dated on earth, is only about 1/16 of one half-life of $Rb^{87}$, which makes it easily the choice for dating many old rocks.

The two examples we have given so far have been relatively simple decay processes. The major decay scheme that was first developed and is still used extensively is the decay of the element uranium, which has two radioactive isotopes that decay to lead and helium:

$$U^{235} \longrightarrow Pb^{207} + 7He^4$$

$$U^{238} \longrightarrow Pb^{206} + 8He^4$$

In addition, another element, thorium, can also decay to lead:

$$Th^{232} \longrightarrow Pb^{208} + 6He^4$$

The half-lives of these decays amount to hundreds or thousands of millions of years, which makes them eminently suitable for dating the oldest objects in our solar system.

One other radioactive element is of great importance in dating rocks, potassium, $K^{40}$. It decays by a scheme that has two paths. In one path, the $K^{40}$ decays to a calcium isotope, $Ca^{40}$. About 89 percent of the $K^{40}$ atoms in any group of atoms will follow this route. The remaining 11 percent of the $K^{40}$ atoms decay to form the inert gas argon, $A^{40}$. The latter decay path is the one that is used, for the daughter, $A^{40}$, can be easily distinguished from ordinary argon formed in other ways, whereas $Ca^{40}$ cannot be distinguished from ordinary calcium.

**Reading the Clocks**   Once isotopes were discovered and instruments were invented that could make the chemical analysis for the ratios of parents to daughters, the business of dating rocks could begin. It was uranium-thorium decay that received first attention and which is, in a sense, the easiest to use, for it requires only an ordinary chemical analysis for uranium and lead. Though the analysis does not take into account the presence of $Pb^{204}$, it is good enough for getting approximate dates of many uranium minerals that occur in rocks that are inferred to have had little or no original lead. It also does not distinguish between the lead that originates from the different uranium isotopes and from thorium.

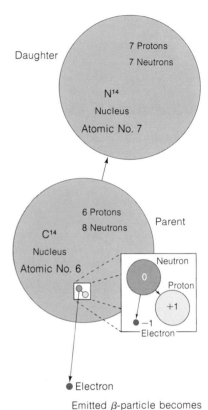

**Figure 2-43**
A radioactive carbon atom, $C^{14}$, spontaneously decays to a nitrogen atom, $N^{14}$, by emitting a $\beta$-particle that becomes an orbiting electron. This is the same process by which radioactive rubidium, $Rb^{87}$, decays to strontium, $Sr^{87}$.

The next important development came in the 1920's and 1930's, when the **mass spectrometer** was invented. That machine was designed to produce a beam of electrically charged atoms from the sample to be studied. The beam is deflected by electrical and magnetic fields in such a way that the atoms are deflected by an amount proportional to their masses. Thus isotopes of elements can be separated. The precision and sensitivity of these instruments has been steadily improving, so that today even minute amounts of individual isotopes can be analyzed. During World War II, the techniques of mass spectrometry and related instruments were developed during the Manhattan Project's drive to develop the atom bomb. A by-product of that work was a series of studies on geochronology, done right after the war by Harold Urey and many of his students. It was at that time that the potassium-argon method was first developed as an accurate dating tool by one of Urey's students, G. J. Wasserburg.

Not all radioactive decays are analyzed by mass spectrometry. Most $C^{14}$ ages are determined from the carbon in dead plant material. During growth, plants steadily incorporate a small amount of $C^{14}$ along with the other carbon isotopes contained in the carbon dioxide molecules in the atmosphere. When plants die, photosynthesis stops, and no new $C^{14}$ is taken up. The relative amount of $C^{14}$ at this point is approximately the same as the ratio in the atmosphere, but steadily decreases with age as the $C^{14}$ radioactivity decays. The measurement of parent and daughter of $C^{14}$, which has a relatively short half-life, is done indirectly, by counting the $\beta$-particles being emitted by the $C^{14}$ still present undecayed in the sample. The result is a count of what is called **$C^{14}$ activity**, from which one can calculate the age, since the number of $\beta$-particles produced is proportional to the number of $C^{14}$ atoms still present, and that count can be compared with the higher activity of a contemporary sample of $C^{14}$.

**When Were the Clocks Started?**   Once the minerals in a rock are formed, any radioactive elements in those minerals keep ticking away all of the time. What we actually measure is the time elapsed since the radioactive parent elements become part of a rock from which their daughter elements could not escape. For example, when uranium becomes incorporated into rock-forming minerals that congeal from a molten state, there is no lead from previous decay. Once the decay process commences, however, daughter elements are trapped and quantities of lead eventually produced. The amounts of parent and daughter present in a sample of rock are a measure of the time interval between now and the time that the rock crystallized. Thus the methods based on the decay of uranium, rubidium, and potassium all give the date of crystallization of the rocks in which those minerals are found—and by geologic inference any other rocks that bear a definite age relation to the rock analyzed. Thus when we date the crystallization of a granite, we know also from geology that the absolute age of the surrounding sedimentary rocks, into which the granite was intruded when it crystallized, can be no younger than the granite. If the geologic ages of those sedimentary rocks are known by fossils and those strata in turn overlie, and thus are younger than, other radioactively dated rocks we can bracket the absolute age of the sedimentary rocks, even though they may contain no datable minerals.

There are several things that can cause dates based directly on radiometric methods to be incorrect. First, one must be able to show that there has been no removal of a daughter element. For example, if groundwater solutions had dissolved some of the lead produced by uranium decay, the age would be underestimated. Second, one must recognize that a geological event, such as heating or partial melting of the rock in a later metamorphic episode, may have reset the clock to zero by allowing earlier-formed daughter elements to escape. The interpretation of potassium-argon ages is complicated by the fact that the mass-spectrometric analysis is for argon, a gas that can diffuse out of the solid mineral and so give a falsely young age for the rock. Since the argon diffusion rate depends strongly on the temperature, what is actually being dated may be interpreted as the time when the rock cooled sufficiently to slow argon leakage to a stop. That time has been shown to be appreciably later than the original formation of the rock. In this way, the difficulties in interpreting radioactive dating can be turned to our advantage and used to learn more about the complex history of the rocks, such as cooling times or later metamorphic episodes.

## AGES OF METEORITES AND THE MOON

One of the triumphs of radioactive dating only gradually emerged as more and more workers dated meteorites by various methods. Many problems in dating techniques came to light as a result of this effort, but it also became surprisingly apparent that all meteorites are of the same age, somewhere in the vicinity of 4.5 billion years old. This is true regardless of their composition—whether they are of the stony kind, made up of rock-forming minerals like those of the Earth's crust, or of the iron variety, made up mostly of alloys of iron and nickel. That there are no meteorites of any other age, regardless of when they fell to Earth, suggested strongly that they originated in other bodies of the solar system that formed at the same time that the Earth did. If this is true, then the meteorites give the age of the Earth too. Though the story is now corroborated by $Rb^{87}$–$Sr^{87}$ and $K^{40}$–$A^{40}$ methods, the pioneer work was done by Claire Patterson of Caltech, who used the various methods based on the different paths by which uranium and thorium decay to lead.

In the past few years, Patterson has applied his ability to make precise measurements of extremely small amounts of lead isotopes to studies of the worldwide distribution of lead in the atmosphere. He has shown that much of atmospheric lead comes from the tetraethyl lead added to automobile gasoline as an antiknock agent. In at least one way, then, the radioactive dating of rocks has indirectly provided what could be vital information on a possible environmental hazard—an example of the unanticipated dividends that the pursuit of basic science has repeatedly shown.

**Figure 2-44**
A nickel–iron meteorite sawn and polished to show the pattern of intergrown
metal crystals. The rough external appearance is typical of the pitting and
erosion that a meteorite is subjected to as it falls through the atmosphere.
[Photo by R. Siever.]

**The Moon and Sixpence of Science**  The heading used here is
the title of a lecture given by one geochronologist in 1971 in discussing the role of radioactivity in studies of the Moon. The title
was not intended to call attention to the novel by Maugham but to
the importance of lunar science in relation to the smallness of the
research funds apportioned to the study of the Moon as compared
with the costs of getting there. The importance of lunar science
is shown in many ways, but what we have learned about the ages
of Moon rocks is one of the most crucial, for some of the very
first rocks dated turned out to be much more than 4 billion years
old, which is thoroughly consistent with the ages of the meteorites
and is proof that some of the original primordial rocks of the moon
were still at the surface, relatively undisturbed. Because such rocks
on Earth would long ago have been weathered, eroded, and transformed many times over into other rocks, we have no such relics.

The great significance of Moon-rock dates is what they tell us
about the time span of such dynamic activities as mountain-building and volcanism, for it is these kinds of events that are chronicled
by radioactive dating. An interesting contrast between the Moon
and the Earth is that the ages of Moon rocks range from 3.1 to
4.5 billion years old, but none are younger, whereas the rocks on
Earth range from zero to 3.7 billion years old, but none are older.
As far as we know, no rocks less than about 3 billion years old were

ever made on the Moon, and rocks older than 3.7 billion years on Earth have probably all been destroyed (but we may yet find some a little older than those found so far). As of now it seems that the Moon's internal heat engine shut down about 1.5 billion years after it was formed, whereas Earth's engine is still going strong.

## HOW TO SYNCHRONIZE CLOCKS: THE STRATIGRAPHIC AND ABSOLUTE TIME SCALES

Geologists working with fossiliferous sediments have a fine watch by which to measure time. It is accurate enough to enable geologists to distinguish the relative ages of formations that are only a few feet or meters thick and which may represent periods of time far less than a million years in duration. A million years, we have to remember, is only about 1/5000 of the history of the Earth. The entire world rock record of fossiliferous sediments has been mapped and subdivided into the scheme of eras, epochs, and ages that is shown here. A geologist studying fossiliferous rocks in the field needs only an average knowledge of paleontology in order to make a fairly accurate estimate of the epoch in which the rocks he is studying belong. But he may have to be more expert to make a precise estimate of their ages or of the time equivalence of zones, finer subdivisions of the time scale. Using the stratigraphic time scale is like reading a watch that enables you to tell one time from another but gives you no idea of just how long a minute really is.

It was not surprising, then, that the discovery of radioactive age dating was immediately seen and welcomed by some geologists as the way to make a combination clock that would function as an absolute as well as a relative timekeeper. Only ten years after the discovery of radioactivity, the first rocks were dated; eight years after that, *Arthur Holmes,* a young English geologist who had not yet received his doctor's degree, published the first edition of what was to become a classic work, *The Age of the Earth.* Holmes plotted radioactive age dates opposite the stratigraphic time scale by carefully figuring out as closely as he could the age relations of the sediments, which were dated by fossils, and of the cross-cutting igneous rocks, which were dated by radioactivity. His estimates were remarkable, for most of them have held up in a general way. His first estimate of the beginning of the Cambrian placed it at about 600 million years before the present. His last estimate, published in 1959, a few years before his death, was the same as his first—and that in the face of thousands of more recent and more accurately determined dates (see Fig. 2.39).

Authorities still differ on precise details, but the general outlines are undisputed. The Phanerozoic time scale of rocks containing fossils of higher organisms represents about 600 million years, and is divided into somewhat unequal divisions, the eras: the Paleozoic, about 350 to 400 million years long; the Mesozoic, about 150 million years long; and the Cenozoic, about 70 million years long. The epochs are similarly unequal in length. The inequalities are the result of accidents of choice. The stratigraphers of the nineteenth century divided parts of the geologic column according to what seemed convenient or appropriate to the area they were studying. If the Chinese and Indians had done most of the stratigraphic work, the column would have been far different, though the boundary between the Phanerozoic and Precambrian would still have been the same, for the evolution of shelled, higher forms of life did not take place at different times in different places; once those forms evolved, they spread rapidly (geologically speaking).

**Telling Precambrian Time**  Because there are no fossils to rely on, the Precambrian has always been somewhat of a mystery to stratigraphers. Although they have been able to unravel complicated sequences of sedimentary and igneous and metamorphic rocks in small areas where it was possible to correlate from one outcrop to the next (just as we could get along without fossils in correlating from one wall of the Grand Canyon to the other), sheer guesswork was required in attempting to correlate between one part of a continent and another. With the advent of radioactive dating, however, the situation has changed—so much so that well-established notions about the way the Precambrian fit together have been thoroughly upset.

There are two major differences between radioactive dating of the Precambrian and the stratigraphic dating of the Phanerozoic. First, the Precambrian "events" that can be dated are mainly episodes of igneous intrusion, or metamorphism, or mountain-building, whereas the Phanerozoic is dated by the ages of the sediments. Because of these differences in the two kinds of rock timekeepers, the Precambrian will give a much more discontinuous record, for the occurrence of intrusion, metamorphism, and mountain-building is spasmodic compared to the almost continuous record of sedimentation. Second, the resolution, or accuracy, of radioactive dating in the Precambrian is much lower than that of stratigraphic dating. As we have noted before, stratigraphers can divide their columns into units that may be as little as a million years long, and they can usually estimate relative time to the nearest 10 million years. In contrast, the imprecision involved in sampling and analyzing Precambrian rocks for radioactive dating is such that even with the delicate and sensitive instruments now being used, old dates have to be given with a large range of probable error, expressed as plus or minus as much as 100 million years. Such a range of possible uncertainty in age—200 million years from

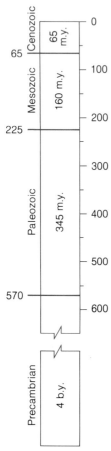

Figure 2-45
Unequal duration
of the major eras
of the geological
time scale.

**Figure 2-46**
Precambrian-age map of Canada, showing provinces drawn on the basis of
radioactive age determinations. [After J. A. Lowdon and others, Geological
Survey of Canada, 1963.]

plus to minus — is equivalent to all the time of the Cambrian, the
Ordovician, the Silurian, plus part of the Devonian periods of the
Paleozoic. It is therefore not surprising that Precambrian stratig-
raphy is still lacking in detail.

Figure 2-46 shows the kind of geologic map that has been
worked out for Precambrian rocks in Canada. The provinces are
drawn on the basis of radioactive ages. The oldest, the Superior
province, is one in which most of the rocks have ages clustering
about 2500 million years, corresponding to an orogeny called the
Kenoran. The rocks of the Churchill province, which partly surrounds the Superior, are about 1750 million years old, corresponding to the Hudsonian orogeny; and those of the Grenville province,
are about 950 to 1000 million years old, corresponding to the
Grenville orogeny. The same general kind of map can be made
for other Precambrian areas of the world, such as those in Scan-
dinavia or Africa. That such maps can be made on the basis of
radioactive dating alone is the triumph of the method.

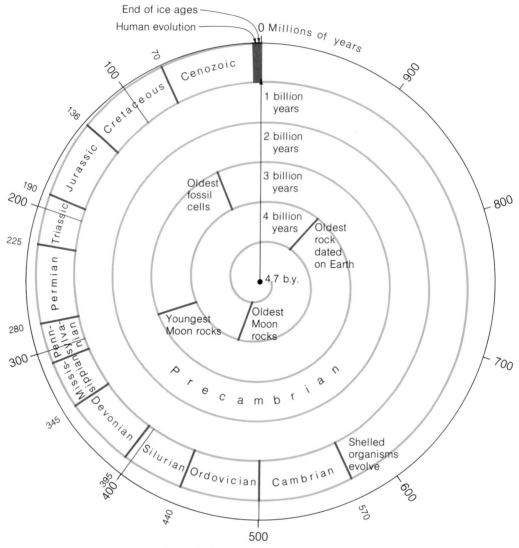

**Figure 2-47**
The geologic clock in time before present.

## AN OVERVIEW OF THE GEOLOGICAL CLOCK

We can now put together a clock for the whole history of the Earth. The clock is shown in the form of a spiral (Fig. 2-47); each revolution of the clock's hand corresponds to 1 billion years; each subdivision, the "hours," corresponds to 100 million years; and the "minutes" are 10 million years long. Geologists work backwards from the present as zero to the beginning at 4.7 billion years ago. From a look at this clock, we can see how short a period of the total of Earth history is taken up by the Phanerozoic eras, and what a minuscule amount of time has elapsed since man evolved.

History, as recorded by man's artifacts, his buildings, and his writings and drawings, is only the last instant of all of this time. One of the minor miracles of how people think is demonstrated by the way they can quickly shift mental gears and consider different time scales—from "Did the earth form 4.5 or 4.7 billion years ago?" to "Will this class hour never end?"

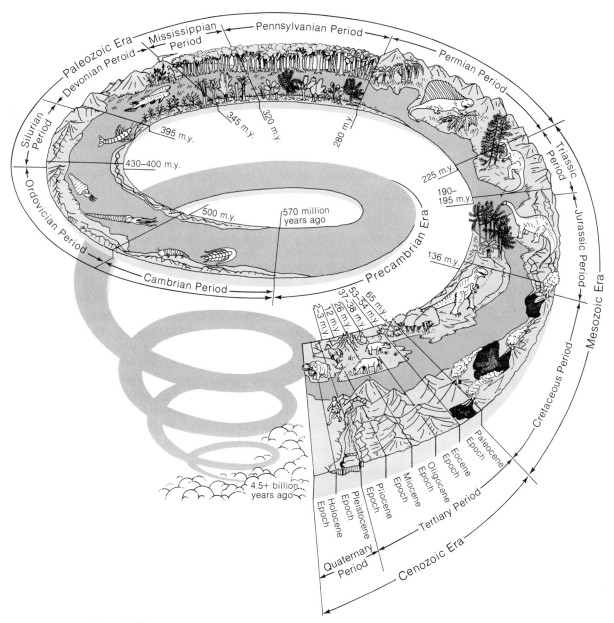

**Figure 2-48**
Relative and absolute time and the history of the Earth. [After *Geologic Time*, U.S. Geological Survey publication.]

1. Geological time scales tend to be long; most are measured in thousands to millions of years. These slow rates of processes are measured by geological change.

2. Rocks are made of minerals, which are chemical compounds, usually in crystalline form. Igneous rocks form by cooling of a magma. Sedimentary rocks show stratification and form by settling of erosional debris and chemical precipitates. Metamorphic rocks form by heat and pressure, which transforms pre-existing rocks into new ones with different mineral compositions.

3. The field relations of rocks, such as stratigraphic sequences, concordant and discordant contacts with igneous rocks, and structural deformation, can be used to deduce a history of sedimentation, uplift, deformation, igneous activity, and erosion.

4. A stratigraphic time scale can be devised from sedimentary rocks by matching the evolutionary changes in ancient life as revealed by fossils found in the strata with the stratigraphic sequence. The rocks of the Grand Canyon illustrate how field relations and fossils are used to infer a complex history of geological episodes.

5. Absolute time measurements can be made by using spontaneous decay of radioactive elements. Such dates of minerals containing radioactive substances provide an absolute age framework for the stratigraphic time scale and allow the dating of unfossiliferous Precambrian rocks. Radioactive age dating of meteorites gives the time of Earth's origin.


## EXERCISES

1. What are the time scales and how might you measure the duration of the following man-made "geologic" phenomena: (1) digging a hole to plant a shrub; (2) excavating a highway road cut; (3) building a large dam on a major river; (4) building a pyramid in ancient Egypt. What does this tell you about the relationship of time scale to the geometric scale, or size, of the phenomenon?

2. Make up a diagram similar to Figure 2-36 to show the following sequence of events: (1) extrusion of thick lavas; (2) erosion; (3) deposition of sediments; (4) faulting; (5) erosion; (6) deposition of sediments; and (7) intrusion of igneous dikes.

3. From the geologic and absolute time scales and the data given on the Grand Canyon sequence, estimate the proportion of time since the Grand Canyon Series was laid down that is represented by sedimentary rocks and the proportion that is represented by unconformities.

4. Abraham Werner, an important figure in late eighteenth-century geology, believed that almost *all* kinds of rocks were deposited on the sea floor. What lines of evidence from the Grand Canyon or elsewhere could you use to show that this is not so?

5. Radioactive decay of which elements would be most appropriate for dating formations of the following ages? (1) early Precambrian; (2) Mesozoic; (3) Miocene; (4) late Pleistocene.

6. What different kinds of geological event are dated by flakes of mica in (1) a granite, (2) a schist, and (3) a sandstone?

7. For the past few hundred years, Man's estimates of the age of the Earth have grown longer and longer. Might the present estimate of 4.7 billion years change too? If so, what geologic theories that pertain to geologic time might have to be re-evaluated?

Berry, W. B. N., *Growth of a Prehistoric Time Scale*. San Francisco: W. H. Freeman and Company, 1968.

Deevey, E. S., Jr., "Radiocarbon Dating." *Scientific American,* February, 1952. (Offprint No. 811.)

Dott, R. H., Jr., and R. L. Batten, *Evolution of the Earth*. New York: McGraw-Hill Book Company, 1971.

Eicher, D. L., *Geologic Time*. Engelwood Cliffs, New Jersey: Prentice-Hall, Inc., 1968.

Eiseley, L. C., "Charles Lyell." *Scientific American*, August, 1959. (Offprint No. 846.)

Hurley, P. M., *How Old is the Earth?* New York: Doubleday and Company, Inc. (Anchor Books), 1959.

Kummel, B., *History of the Earth*. San Francisco: W. H. Freeman and Company, 1970.

Simpson, G. G., *The Life of the Past*. New Haven: Yale University Press, 1953.

A strain seismograph in a tunnel under the ground. The long pipe-like part of the system is a quartz tube, anchored in piers at each end. An electronic motion detector, placed at a gap in the tube, measures and records changes in the distance between the two piers. These variations are responses to tides in the solid Earth and to the passage of seismic waves. This system is so sensitive that it could detect a change of 1 mm in the distance between New York and California.

# How We Find Out About the Earth: Observation, Experiment, and Deduction

*Not only does the geologist study rocks in their natural setting, he collects them and subjects them to physical and chemical examinations. He also models the Earth in his laboratory, analyzes seismic waves, observes the Earth's magnetic field, and conducts other experiments to obtain information about the deep interior he can never reach himself.*

## EXPLORE A PLANET

A few years ago a group of scientists was called together and charged with designing a program for the exploration of the Moon. The historic nature of the occasion did not escape these men as they planned the sequence of observations and experiments that was later to become the scientific blueprint for Project Apollo. In detailing a program for man's first survey of another planet, however, the advisors were caught in a dilemma. On the one hand, they could rely upon their experience with diverse methods that have been developed for studying the Earth and give highest priority to those techniques that reveal the most information. On the other hand, the Moon could have evolved differently, so that the use of some procedures proved to be reliable on Earth might give misleading results on the Moon. Although the dilemma produced much heated debate, there was no opposition to assigning highest priority to the return of lunar rocks for analysis in laboratories, for

**Figure 3-1**
Historic photograph of the Apollo 11 rock box being opened in a vacuum chamber. These are the first rocks returned from the Moon. [From National Aeronautics and Space Administration.]

**Figure 3-2**
Comparison of a photograph from an orbiting space craft with a geological map of the same region. Geologists are learning to use the new tool of space photography not only to expedite mapping but also to discover features unrecognizable from the surface. [Map after U.S. Geological Survey; photo from National Aeronautics and Space Administration.]

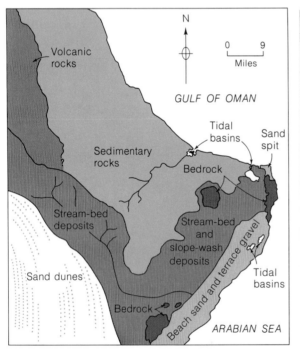

it is from rocks that much of the story of the Earth has been gleaned. Photography was placed high on the list in order that the lunar surface and the pattern of its rocks and its topography could be mapped with economy; the idea of using hundreds of geologists to survey the moon was out of the question. A key feature of the recommended program was the emplacement of a lunar observatory containing a seismograph to monitor moonquakes, a magnetometer to measure the magnetic field, and a heat-flow probe to measure the escape of heat from the interior. The decision to go with methods that paid off on Earth proved to be a wise one. Let us examine how these and other methods help us reconstruct and interpret the history of the Earth and how it works — which is of course the ultimate goal of geology.

In the previous chapter we saw that geological science is founded on field observations — mapping and explaining the distribution of the diverse rocks that man can sample on or near the Earth's surface. Our purpose in this chapter is to show something of the diversity of additional methods used by the geologist in the laboratory, rather than to develop any one branch of the subject in depth.

## THE ROLE OF EXPERIMENTS IN GEOLOGY

A physicist who recently won a Nobel Prize once commented in jest that geology is a special case of physics: since all matter is composed of atoms, if we knew all the laws governing their interaction we could explain minerals, rocks, evolving planets, even the origin and evolution of life. Although this is true in a sense, we would wait a long time before enough was known to explain large-scale natural phenomena by working upward from atoms. We simply do not know enough physics and chemistry to predict the outcome of most complex geological processes from the initial state, nor could we perform the trillions of calculations needed even with the best of computers if we did know the laws. Nevertheless, we can often observe the results and gain a deeper understanding of what happened by invoking the principles of physics and chemistry.

Geologists rely heavily on direct observation of natural systems. Some types of geological activity, such as volcanism and sedimentation, can be studied as they occur. Igneous intrusion is an example of a process that can only be viewed in the field after the fact — millions of years following solidification of the molten rock. But even field observations and physical and chemical reasoning, as important as they are, are not enough when dealing with a system as complicated as a rock. It is essential to perform controlled laboratory experiments that duplicate the conditions under which geological materials form to provide the basic data on how rocks and minerals behave chemically and physically.

Much of what a geologist does is like detective work. To interpret past events, climates, and processes, the geologist constantly searches for clues and evidence in the rocks. This searching is not unlike going into a laboratory at night—after the chemists have gone home and the experiments are done—to look at the sink, inspect the sewage, note the bottles of chemicals on the shelves, and then reconstruct the experiments that were done that day.

**Experimental Petrology*** Take an igneous rock, for example. At one time it was molten and mobile. We can melt igneous rocks in the laboratory and determine how melting-point temperatures depend on chemical composition and pressure. We can measure the **viscosity**, or fluidity, of the molten material and discover, for example, that granites melt at a lower temperature and are more viscous than basalts. Simple laboratory experiments show that lava, cooling rapidly at the temperatures of the Earth's surface, turns to a glass, lacking individual grains, or to a rock made up of many minute crystals mixed with glass. Slow cooling leads to the formation of large crystals, as can be seen in intrusions that were thermally shielded by the surrounding rock.

Natural rocks are often too complicated even for controlled laboratory experiments. Petrologists therefore resort to making their own rocks. They design simple experiments and use synthetic systems made up of fewer chemical components than are present in natural systems. A typical procedure is to pack small amounts of a few elements or compounds into a platinum capsule that is then squeezed and heated. The contents react in a specific way depending on the composition, pressure, and temperature. The capsule is then cooled, or "quenched," and the contents are examined to see which minerals formed under the physical conditions that prevailed before quenching. The experiment is repeated for many different pressures and temperatures.

*N. L. Bowen,* one of the originators of the experimental approach to petrology, began using this procedure in the early part of this century at the Geophysical Laboratory in Washington, D.C. His work was followed by that of others who performed more complicated experiments, all of which paid off handsomely. The total effort must be reckoned as a successful assault on one of the most important problems of geology, for we now have a pretty good idea of how the major chemical elements distribute themselves among the different minerals during the crystallization of a magma and how different minerals form from the same chemical elements at different temperatures.

Bowen wondered why magmas having the same composition can crystallize into different types of rocks. He postulated that the rate of cooling and the fate of the early crystallizing minerals— that is, whether they remain in the liquid or settle out—are important factors that influence the end product. His experiments led him to introduce the **reaction series** as a fundamental concept in determining why igneous rocks end up with specific groupings of minerals. Simply stated, $A \rightarrow B \rightarrow C$ may represent three minerals that form at different stages in the crystallization of a cooling magma. Mineral $A$ is a high-temperature mineral in that it is the first one to crystallize (solidify as a crystal) as the magma,

*****Petrology**: a specialty concerned with the composition and origin of rocks.

Box 3-1   How Basalt from the Volcano Kilauea on
Hawaii Melts at a Depth of 20 km (12 miles) in the Earth.
A Laboratory Experiment Performed by *H. S. Yoder* and
*C. E. Tilley* in 1956.

Listed below are the minerals present at each stage of melting;
for example, olivine and magnetite are the last minerals present
before melting is complete.

| | |
|---|---|
| Melting begins: | 750–800°C |
| Hornblende, plagioclase, sphene, magnetite + melt | |
| Hornblende, sphene, magnetite + melt | 815°C |
| Hornblende, magnetite + melt | 875°C |
| Pyroxene, hornblende, magnetite + melt | 925°C |
| Pyroxene, olivine, magnetite + melt | 975°C |
| Olivine, magnetite + melt | 1090°C |
| Melting complete | 1125°C |

or melt, cools. But *A* can react with the remaining melt on further
cooling and transform to mineral *B,* in which case the rock ends
up composed mostly of *B.* If *A* is removed before the *B* reaction
sets in (for example, if it is heavy and settles to the bottom of the
intrusion), then the composition of the remaining melt is changed
by the removal of the elements that go with *A,* and the reaction proceeds to stage *C*; or beyond to further stages.

Facing the city of New York on the west bank of the Hudson
River is the Palisades, a massive cliff some 50 miles long and in
places more than 1000 feet from top to bottom. The Palisades is
a formation of basalt that was intruded almost horizontally into
nonmarine sedimentary rocks as a melt. The variation in mineral
composition from top to bottom of this formation makes it a classic
example of how laboratory experiments serve to explain field
observations. After it was emplaced at a temperature near 1200°C,
which is the melting temperature of a rock of this composition,
the parts of the intrusion within a few feet of its upper and lower
contact cooled rapidly to become a fine-grained rock, preserving
or "quenching in" the chemical composition of the original magma.
The molten interior cooled more slowly, so that large crystals
could form. We can again use the letters *A, B,* and *C* to represent
the actual minerals of this field example. The first mineral to crystallize was *A* (in this example, olivine), which was heavy and sank
through the melt to the bottom of the intrusion, where it can be
found today in the olivine-rich layer just above the chilled, or
fine-grained basalt zone. The reaction proceeded to stage *B*
(pyroxene) in some parts of the intrusion, and the pyroxene accumulated in the lower third, probably by gravity-settling. After

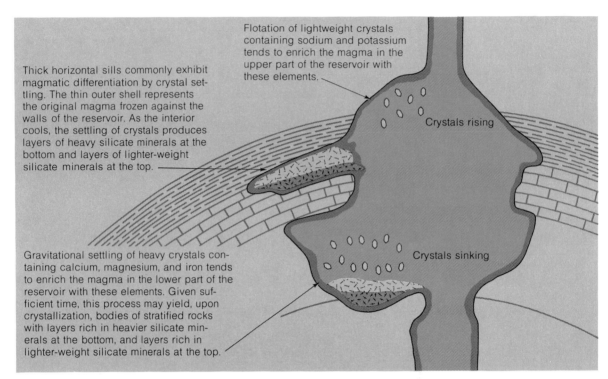

Thick horizontal sills commonly exhibit magmatic differentiation by crystal settling. The thin outer shell represents the original magma frozen against the walls of the reservoir. As the interior cools, the settling of crystals produces layers of heavy silicate minerals at the bottom and layers of lighter-weight silicate minerals at the top. ——

Flotation of lightweight crystals containing sodium and potassium tends to enrich the magma in the upper part of the reservoir with these elements.

Crystals rising

Crystals sinking

Gravitational settling of heavy crystals containing calcium, magnesium, and iron tends to enrich the magma in the lower part of the reservoir with these elements. Given sufficient time, this process may yield, upon crystallization, bodies of stratified rocks with layers rich in heavier silicate minerals at the bottom, and layers rich in lighter-weight silicate minerals at the top.

**Figure 3-3**
Crystal settling and differentiation. Laboratory experiments tell us which minerals crystallize as magma cools, and which sink or float up, accounting for the final distribution of minerals in the solidified mass. [After U.S. Geological Survey.]

the olivine crystals settled out, changing the composition of the remaining magma, the reaction proceeded to stage *C* (plagioclase feldspar) in the cooling remainder of the melt. This explains the enrichment of plagioclase feldspar found in the upper third of the intrusion. A simplified schematic illustration of the process is depicted in the "sill" of Figure 3-3. The importance of laboratory experiments is manifest in our ability to reconstruct an event that took place 200 m.y. ago in Triassic time, and to account for such major features as the distribution of grain sizes and minerals in the formation as we find it today (Fig. 3-4). A more detailed explanation of this kind of crystallization is given in Chapter 16.

**An Experiment with a Bucket of Sea Water** Evaporites are salt deposits, most of which form from evaporation of the most obviously salty water on Earth — sea water. Although they represent a small percentage of the total volume of sedimentary rocks, they are nevertheless important economically as sources of chemicals and because they often block the flow of oil, trapping it in pools. Evaporites also reveal much about ancient environments and geography. Evaporites form in isolated arms of the sea or more rarely in desert lakes as a result of extensive or total evaporation of the water. The geologist is interested in the relation between the temperature at which evaporation takes place and the sequence of crystallization of the various salt minerals, so that he can explain the succession of minerals found in natural salt deposits. The process can readily be reproduced experimentally in the laboratory. The first person to do so was an Italian chemist, *Usiglio*, back in 1849. Usiglio simply took a bucket of water from the Mediterranean Sea and let it evaporate into the atmosphere. Although he succeeded in crystallizing several minerals found in natural salt deposits, his results were erratic, since he could not maintain a constant temperature and he was using sea water, an extremely complex solution containing many components. This is another example of why simplified laboratory experiments are needed to unravel complex natural systems.

*Van't Hoff*, a Dutch chemist working in the early part of this century (known by most as the scientist who discovered the laws of osmosis), understood this and worked with simple two-component systems. He studied the conditions of crystallization of one salt at a time, making the system more complicated in sequential stages so that each step could be understood. Van't Hoff's experiments constitute one of the first successful applications of physical-chemical methods to geological problems. We now know that as sea water evaporates, **calcite** or **aragonite** (calcium carbonate, $CaCO_3$) is the first solid to precipitate. If evaporation continues, **gypsum** or **anhydrite** (calcium sulphate, $CaSO_4 \cdot 2H_2O$ or $CaSO_4$) may form, depending on the temperature and salinity of the body of water. When 90 percent of the water has been evaporated, **halite**, or rock salt (sodium chloride, $NaCl$), separates out. Finally, potassium and magnesium salts precipitate when only 1.5 percent of the original volume is left, but this rarely occurs in nature, which is unfortunate as these are economically important minerals.* As a result of such experiments, we can interpret the conditions of ancient environments from the succession of minerals found in natural salt deposits (Fig. 3-5).

**Scale Models** The dimensions of some geological processes are so large that experiments designed to model them must be scaled down before they can be done in the laboratory. For some, only the sizes need be scaled down; for others, sizes, masses, and time factors must be changed in a consistent way to preserve the validity of the experiment.

An example of an experiment for which only size needs to be scaled down is one designed to study the movement of unconsolidated sediments by flowing water. In nature this movement

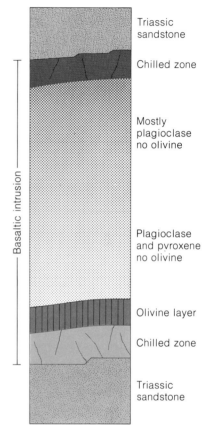

Triassic
sandstone

Chilled zone

Mostly
plagioclase
no olivine

Plagioclase
and pyroxene
no olivine

Olivine layer

Chilled zone

Triassic
sandstone

Basaltic intrusion

**Figure 3-4**
Diagrammatic section of the Palisades, on the west bank of the Hudson River. The formation is a basalt that intruded sediments as a melt some 200 million years ago. The interpretation of the vertical variation in texture and mineral composition in this formation is a classic example of the way in which laboratory experiments help explain field observations. [After F. Walker, 1940.]

---

*Although magnesium is abundant in rocks, it is more easily extracted from sea water by electrolysis of magnesium salts. It is one of the most important resources that the United States obtains from the sea.

**Figure 3-5**
Core of halite, including dark bands of anhydrite, recovered by deep drilling
in the Western Mediterranean Sea. On the basis of this evidence, scientists of
the Deep Sea Drilling Project postulate that the Mediterranean Sea dried up
5 to 10 million years ago. [From National Science Foundation.]

occurs on river bottoms and on the sea floor, where bottom currents are active. Nature can be duplicated realistically in a **flume**,
whose dimensions are small compared to a river or ocean. A flume
is an elongated tank equipped with adjustable pumps to control
the velocity of moving water. Real sediments are placed at the
bottom of the tank. For silts and fine sands, flume experiments
show that flow velocities of less than 10 centimeters per second
(1/4 mile/hour) are ineffective in eroding and transporting sediments. At velocities in the range 10 to 50 centimeters per second,
the sediments begin to move, and ripples develop spontaneously
in the previously flat bed of unconsolidated particles. Once formed,
the ripples migrate in the direction of flow as a result of erosion
on the upstream side and deposition on the downstream side.
At the higher velocities, similar but larger features develop, such
as stream dunes or sand waves. Above about 50 centimeters
per second all of these features disappear and the bed becomes
flat again, because the velocities keep the entire bed in motion
along the bottom. Flume experiments can be made more useful
by adding distinctively colored grains to the system while ripples
or stream dunes are forming. At certain velocities of flow, the
colored grains show another feature, the formation of layers,
which become prominently marked by the "tracers" of colored
grains. Depending on the velocity, the layers may be either
flat lying or inclined. Cross-bedding, a variety of inclined bedding,
turns out to be an excellent measure of the direction in which
ripples or dunes migrate in a depositional environment. As a
result of such experimental studies of particle movement in
flumes, when we find this kind of bedding in rocks, we can estimate the velocities and directions of currents in rivers and oceans
long since vanished. Their directions can be mapped in the field
to give a reconstruction of ancient current systems.

**Figure 3-6**
A ripple made by water flowing over coarse silt in a laboratory flume. The ripple, migrating from right to left, is about 10 cm high. Cross-bedding is clearly evident. Experiments like this help to reconstruct ancient conditions from ripple marks and cross-bedding in sedimentary rocks. [Photo by J. Southard, MIT.]

Experiments in laboratory modelling can be designed for even larger-scale processes. For example, we think that convection currents in the Earth's mantle may supply the driving force for plate movements, perhaps by exerting frictional drag on the bottom of the plates. Unfortunately, the mathematics for modelling convective flow is formidable, and progress in understanding what happens will be slow. Can experiments shed some light on this important problem? Here dimensions of thousands of kilometers must be reduced to tens of centimeters, and times measured in millions of years must be scaled down to hours so that the experiment can be done conveniently. To preserve the validity of the experiment, the viscosity of real Earth materials and the velocity of plate movement must be scaled in a manner physically consistent with the reduction in length and time dimensions. The American geophysicist *David Griggs* performed one of the first dimensionally valid experiments of this kind in 1939. In simulating the response of a plastic crust to a subcrustal convection current, Griggs reduced the time factor by about $10^{12}$*, so that one million years in nature corresponded to one minute in the laboratory. Length was reduced by $10^6$, density by 0.6, strength of material by $10^6$. Nature's rock was replaced by materials like glycerine or thick oil and sand mixtures in order to lower the viscosity by the required $10^{18}$ so that convection could be observed. Temperature differences of thousands of degrees would have been needed to drive the laboratory convection cell by heat alone, and this forced Griggs to substitute rotating drums instead to simulate the flow. The results of Griggs'

*$10^n$ = 1 followed by $n$ zeroes.

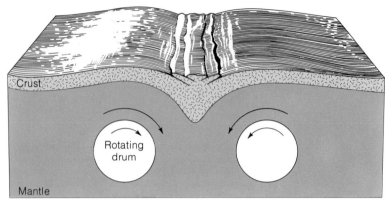

**Figure 3-7**
Diagram of a laboratory experiment performed by D. Griggs in 1939 to model
the effect of convection currents in the Earth's interior. In this laboratory
simulation, rotating drums set up convection-like currents that drag the crust
downward in a manner suggestive of the subduction process.

experiment, shown in Figure 3-7, are remarkably close to modern
notions of plate movements driven by convection currents. To
this day his experiment is cited to buttress the case for mantle
convection as a key factor in the deformation of the lithosphere.

**Computer Simulation** With the availability of large computers,
it is increasingly becoming more convenient to make numerical
models rather than to construct physical analogs in the laboratory. Such problems as convective flow in the mantle, evolution
of planetary orbits, the thermal history of a planet (allowing for
melting and differentiation) are now being tackled by solving
numerically the equations that govern the process. For most of
these problems the computer can be programmed to simulate a
more complex and therefore more realistic situation than can be
simulated by a scaled laboratory model. The appropriate equations
have been known for decades, but the solutions were intractable
until the arrival of the computer age.

For example, a long fracture in the Earth's crust—the San Andreas Fault—is to California what the sword was to Damocles.
An earthquake can occur on this fault at any time; in fact, if the
historical records serve as any guide, a destructive tremor may be
overdue in California. Fault breaks are now being modeled numerically to see if there is a pattern to the occurrence of earthquakes.
The computer is fed the ground displacements or strain increases
in California, which are obtained by repeated surveying. The
strength of rock, as well as the frictional properties of rock sliding
against rock obtained in laboratory studies, is placed in the computer memory. When the computer detects a place on the fault
where a critical stress has been reached (sufficient to produce a
fracture), it signals an earthquake. This simulated earthquake
releases strains that have been stored up in some places but also

concentrates strains elsewhere on the fault. In the meantime the crust continues to be deformed, the regional strain builds up, and more earthquakes occur—induced not only by the regional strain accumulated but also affected by the previous earthquake history. A realistic pattern of earthquakes is obtained in which the time sequence of events as well as the magnitudes of the "numerical" earthquakes appear amazingly realistic. It may well be that in the future, a computer, simulating the earthquake process and updating its memory automatically with data from sensors on the fault, will enable geologists in California to predict earthquakes, and thereby reduce the danger that hangs over those who live there.

**Floating Laboratories**  Plate tectonics, the conceptual framework of this book, could not have been substantiated without oceanographic expeditions. Some 80 percent of the Earth's surface is covered by water, so that the 20 percent available for direct observation is in a real sense atypical. Oceanographic ships now map the sea floor even more efficiently than land surveys. An example of the variety of work carried out on an oceanographic expedition is shown in Table 3-1. All of these methods help us determine the structure and origin of the suboceanic crust and mantle. With the development of deep submersibles of the

Table 3-1
Summary of operations on a cruise of research vessel *Conrad* December 1, 1965–October 16, 1966.

*Length of Cruise:* 320 Days

*Areas covered:* Western North Atlantic, Caribbean Sea, South Pacific, North Pacific, Caribbean (2nd time), North Atlantic (2nd time)

*Ports:* New York, Bermuda, Jamaica, Panama, Mexico, Tahiti, Australia, Tokyo, Yokosuka, Adak, British Columbia, Panama, New York

*Miles Steamed:* 53,914

*Underway Measurements* (Entire cruise; Continuous Measurements):
  Precision Depth Recordings (Topographic mapping of sea floor)
  Magnetic Total Field Intensity (Magnetization of oceanic crust)
  Seismic Reflection Profiling (Thickness of sediments)
  Relative Gravitational Attraction (Subsurface structure)
  Surface Temperature, Air Temperature

*Station Measurements or Collections:*

| | |
|---|---:|
| Seismic Refraction Profiles (Determination of Crustal Thickness) | 27 |
| Cores (Samples of deep-sea sediments) | 290 |
| Large-Volume Water Samples for Particulate Matter | 189 |
| Hydrograd (Continuous Water Temperature to Bottom) | 135 |
| Thermograd (Temperature Gradient in Sediments) | 135 |
| Bottom Camera Stations (Photographs of the sea floor) | 138 |
| Nephelometer Stations (Water Transparency) | 225 |
| Rock Dredges (Collect rocks from ocean ridges) | 6 |

*Biological Collections:*

| | |
|---|---:|
| Surface Plankton Tows (Recovery of organisms in nets) | 252 |
| Oblique and Vertical Tows (0–300 m) | 52 |
| Opening-and-Closing Plankton Tows (5 Depth Ranges to 2000 m) | 128 |
| Nanoplankton (Electron-Microscope Studies) | 455 |

**Figure 3-8**
View of the deep submersible vessel Alvin, designed to conduct geological and biological exploration of the sea floor. [Courtesy of Woods Hole Oceanographic Institution.]

type shown in Figure 3-8, geologists have begun to travel as aquanauts to the sea floor where they can photograph the bottom, collect samples not blindly, but knowing where they come from, and install automatic observatories. Some examples of deep-sea experiments are depicted schematically in Figure 3-9.

## PROBING THE DEEP INTERIOR

**Journey to the Center of the Earth** So Jules Verne titled his great adventure story of the last century. Although most of Verne's science-fiction demonstrated his foreknowledge of many things, this particular book was off the mark in its description of caves, oceans, serpents, forests of mushrooms, herds of mastodon deep in the Earth. Of course, we can never directly examine the deep interior of a planet, and unanswered questions will thus always remain. In fact, the surface of the Moon or of Mars is an easier target for direct examination by man than the Earth's interior — at least beyond more than a few miles. It is surprising, however, how much we can confidently say about the interior of the Earth from experiments, observations, and physical reasoning.

Our knowledge about the composition and physical state of the interior comes primarily from astronomical observations, observations made on the surface of the Earth, and laboratory experiments. Some examples are discussed in the next few pages, and are developed in greater detail in later chapters.

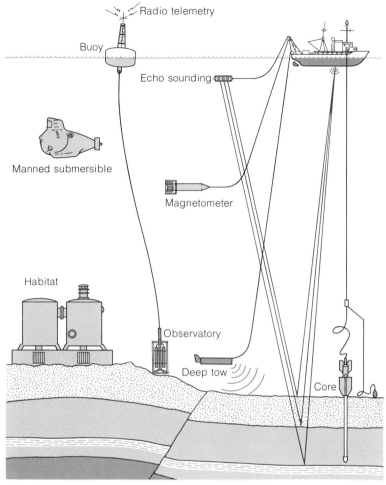

**Figure 3-9**
Schematic representation of the diverse deep-sea observations made during
cruises of oceanographic research vessels.

**The Relative Abundance of the Elements**    We know about the
chemical composition of the universe from the spectroscopic
examination of the light of the Sun and stars, from the analysis
of meteorites, and the composition of the rocks found near the
surface of the Earth. Table 3-2 shows our current knowledge of
elemental abundances—the results of many measurements. The
data support the notion that the relative abundance of elements in
the universe is uniform except for local variations, such as the
dissipation into space of hydrogen and helium from the terrestrial
planets or the internal differentiation of a planet. It has been postulated that a large part of the Earth's interior consists of iron. But
why iron, and not some other heavy metal? We have no direct
way of determining that it is iron. We do know, however, from

Table 3-2
Abundance of the principal elements in the Universe expressed as the
numbers of atoms relative to a base of 10,000 atoms of silicon.

| Element | Atomic number | Atomic weight | Abundance in the universe |
|---|---|---|---|
| Hydrogen | 1 | 1 | 400,000,000 |
| Helium | 2 | 4 | 31,000,000 |
| Oxygen | 8 | 16 | 215,000 |
| Neon | 10 | 20 | 86,000 |
| Nitrogen | 7 | 14 | 66,000 |
| Carbon | 6 | 12 | 35,000 |
| Silicon | 14 | 28 | 10,000 |
| Magnesium | 12 | 24 | 9,100 |
| Iron | 26 | 56 | 6,000 |
| Sulphur | 16 | 32 | 3,750 |
| Argon | 18 | 40 | 1,500 |
| Aluminum | 13 | 27 | 950 |
| Calcium | 20 | 40 | 490 |
| Sodium | 11 | 23 | 440 |
| Nickel | 28 | 59 | 270 |
| Phosphorus | 15 | 31 | 100 |
| Chlorine | 17 | 35 | 90 |
| Chromium | 24 | 52 | 78 |
| Manganese | 25 | 55 | 69 |
| Potassium | 19 | 39 | 32 |
| Titanium | 22 | 48 | 24 |
| Cobalt | 27 | 59 | 18 |
| Fluorine | 9 | 19 | 16 |

Source: After H. Suess and H. Urey, 1956.

physical calculations made on the basis of the mean density of
the Earth and from seismological data that the innermost part of
the Earth's interior has a high density. The choice of iron seems
appropriate because among all the candidates with high atomic
weight, it is by far the most plentiful. Our confidence in this choice
is reinforced when we note that 58 percent of all meteorites found
consist of iron.

**The Mass and Mean Density of the Earth**  The mass ($m$) of
the Earth can be obtained by measuring the gravitational acceleration ($g$) at the surface and solving for $m$ in the equation for the
famous law of gravitational attraction.

$$g = G \frac{m}{R^2},$$

where $G$ is the universal gravitational constant and $R$ is the
Earth's radius. The value of $g$ can be obtained quite accurately from
the period of a swinging pendulum or the acceleration of a falling
weight. Dividing $m$ by the volume of the Earth gives the mean

## Box 3-2   The Universal Constant $G$

The measurement of Earth's density presupposes a knowledge of the universal constant $G$. *Henry Cavendish* made the first determination of $G$ in 1798 in London. It was the first measurement of a "fundamental constant," and perhaps the most important experiment of eighteenth-century science. Cavendish mounted two solid 2-inch lead spheres on the ends of a lightweight rod, which is suspended in the horizontal plane by means of a fiber. Two 12-inch diameter lead spheres are brought near the suspended spheres. The gravitational attraction between the masses causes the rod to rotate by an amount that can be measured and used to determine $G$. The difficulty in conducting this experiment stems from the extreme weakness of gravitational interaction, and as a result is known only to about 1 part in 100. Since $G$ is important in determining the density distribution in planets and stars, experimenters in England, Italy, and the United States are working toward a more precise value.

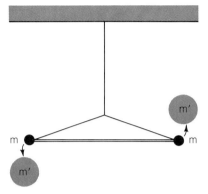

Schematic of Cavendish's 1798 experiment to determine the universal gravitational constant $G$. Two small, solid spheres are mounted at the ends of a light, stiff rod, suspended by a fiber. If two large masses $m'$ are brought near the spheres, the gravitational attraction between the masses $m$ and $m'$ produces a rotation that twists the suspension fiber. The value of $G$ can be determined from the amount of rotation.

density, which turns out to be 5.52 grams per cubic centimeter. Rocks at the Earth's surface have an average density of 2.7 grams per cubic centimeter, much less than the planetary mean. Since the overall density is what it is, a large part of the interior must have a density greater than 6 grams per cubic centimeter in order to compensate for the lightness of the outer rocks. Here is direct evidence that the Earth's density distribution is not uniform — that it is a vertically differentiated planet. Knowledge of average density alone, however, is a weak constraint on possible Earth models — one that admits too many possibilities to enable us to speculate about physical state and composition of the interior. Fortunately there are other ways, as we will see, to delimit the possible density distributions in the Earth.

**Deep Seismic Sounding**   This is what Russian scientists call the most powerful method available to us for exploring the deep interior. Conceptually the method is very straightforward. An underground explosion or an earthquake excites seismic waves akin to sound waves, which travel into the interior and emerge at the surface at distant places. Seismographs, which are basically very sensitive motion detectors attached to the ground, pick up the seismic waves, even though the ground motion may be as small as 1/10,000,000 of a centimeter.

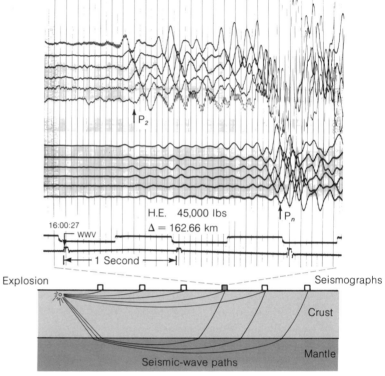

**Figure 3-10**
Seismic waves recorded at a distance of 163 km from an explosion of 45,000 lb
of TNT. The top six traces and the bottom six traces are from the same seismographs, but the amplification of ground motion is 1,000,000 for the top traces
and 150,000 for the bottom traces. The waves $P_2$ traveled through the Earth's
crust. The waves $Pn$ traveled mostly through the mantle below. The six seismographs, spaced 100 m apart, are placed in line with the explosion, as shown
at the bottom.

There are almost a thousand seismograph stations spread over
the surface of the Earth, each with a precise clock that can record
the time of arrival of the seismic waves to 1/10 second. Special
arrays of seismographs can be laid down for specific experiments.
Some seismographs feed their data directly to computers for analysis, and others make photographic records of the ground motion, as
shown in Figure 3-10. Knowing the time and place of the explosion
or earthquake, the time of travel of the waves through the interior
can be calculated and used to measure the velocity of the waves
in the interior. Some rays go directly to the Earth's core and bounce
back, in radar-like fashion. These reflections can be timed to give
a precise depth to the Earth's core. Seismic waves have also shown
that the core is liquid. We will see how deep seismic sounding can
be used to map the thickness of the Earth's crust, to find zones of
partial melting, and to explore the deeper mantle and core.

**High Pressure and Shock Experiments**   Even if we knew in detail how density and seismic velocities change with depth in the Earth, we would still want to identify the materials and describe their physical conditions. To do this we also need information on the densities of different materials and the velocities with which seismic waves travel through them under the high pressures and temperatures that exist in the interior of a planet. The pressure at the center of the Earth is nearly 4 million times greater than atmospheric pressure,* and temperatures range in several thousands of degrees. Using a hydraulic press, geophysicists can squeeze rocks in the laboratory to pressures of about 100 kilobars, heat them to temperatures of about 1000°C, and at the same time measure many of their properties. This procedure, however, only duplicates the conditions at depths of about 300 kilometers—far short of our goal, the very center of the planet. Fortunately a new technique is available for compressing rocks to the necessary pressures; it happens to be the very same method used to compress uranium when triggering an atomic explosion. An ordinary chemical explosive, such as dynamite, is wrapped around the rock. When the dynamite is detonated, the shock wave squeezes the rock, raising the pressure and temperatures to the high values needed to duplicate conditions at great depths. The rock is destroyed in the process, but in the few millionths of a second before it falls apart, data needed to calculate the density, pressure, and shock velocity (which is simply related to the seismic wave velocities) are obtained electronically from sensors on the rock. Other types of experiments are used to determine such things as strength, thermal, electrical, and elastic properties at elevated pressures and temperatures.

**Other Methods**   Many other methods are available, though some are just in the initial phase of development. Electrical currents can be used to measure the variation of electric conductivity with depth. This in turn depends on temperature and composition. The solid Earth responds to the gravitational pull of the Moon just as the oceans do. This response depends on the mean rigidity of the Earth, which can be estimated from the amount the surface moves due to the Moon's gravity. Both of these methods may be particularly important on the Moon, where seismological experiments are difficult to carry out because (1) there are no big moonquakes to excite deeply penetrating seismic waves, (2) large chemical explosions are awkward to carry out because the explosives weigh too much, and (3) nuclear explosions are forbidden by international treaty.

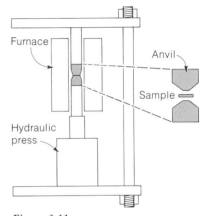

Figure 3-11
The Bridgman "squeezer," a device for subjecting minerals to pressures of a few tens of thousands of atmospheres and temperatures of several hundred degrees. The low pressure of the hydraulic press is amplified by concentrating the total force on the small area of the anvil. A furnace surrounding the anvil supplies heat. This apparatus simulates environments deep in the Earth's crust.

---

*Pressure is measured in atmospheres (atm), bars, or pounds per square inch (p.s.i.). 1 atm = 1.01 bars = 14.7 p.s.i. Geologists tend to use bars, kilobars ($10^3$ bars) and megabars ($10^6$ bars).

Most of these experiments have been performed on Earth. We know the density of the Earth to within a few percent. Using nuclear explosions detonated for quite different purposes, the travel times of seismic waves have been measured to 1/10 sec. Light from many stars and the Sun has been analyzed spectroscopically, and the abundance of the elements is reasonably well known. Hundreds of elements and compounds, minerals and rocks have been squeezed and shocked to determine their high-pressure and high-temperature properties. The next step is to find models of the Earth that fit all these data, which we will do later in Part III of this book.

## MONITORING THE DYNAMICS OF THE EARTH

Before the birth of modern geology, curious and worried men speculated about the forces that give rise to earthquakes, volcanic eruptions, and mountains. In the fourteenth century it was thought that "earthly vapours" collected in underground caverns, and when the pressure became great enough, earthquakes occurred. This supposedly happened at astrologically unfavorable times, when Mars, Jupiter, and Saturn were in certain positions. In the eighteenth century, underground vapors were still voguish, but the source to which they were attributed was a more reasonable choice. Subterranean fires were thought to exist, perhaps caused by underground combustion of coal or shale. It was also believed that when water contained in underground fissures happened to make contact with rocks heated by these fires, steam and vapor formed and triggered an earthquake. Volcanoes were explained as another manifestation of subterranean fires. Perhaps Jules Verne drew his inspiration from these theories when he wrote of storms raging over heated oceans deep beneath the surface.

Earthquakes, volcanoes, mountains—manifestations of dynamic processes—are objects of intensive study by modern geologists who have the advantage of modern instruments, computers, and a knowledge of twentieth-century physics and chemistry.

**Seismicity, or Where Earthquakes Occur**  We have mentioned that seismic waves generated by earthquakes are important in probing the deep interior. Here we refer to the use of earthquakes as indicators of stresses in the Earth and the pattern of deformation, or **tectonic** movements, that stresses produce.

The magnitudes of earthquakes can be measured and their place of origin located by means of the seismograms recorded at nearly a thousand observatories throughout the world. The place of origin is located when the earthquake's **epicenter** (latitude and longitude), depth beneath the surface, and time of occurrence are specified. If we know in advance the **travel times** of seismic waves through the Earth, which we do know with good precision, it takes four independent observations of seismic-wave arrival times to solve for these four unknown parameters. For most earthquakes, tens or hundreds of arrival times are fed into a computer, which takes only seconds to locate the site of origin with a precision of several kilometers.

Although precision of epicenter location was not as good in the past as it is today, the correlation of **seismicity,** the frequency of earthquakes, with the configuration of the surface (or **geomorphic** evidence) has been recognized for a long time. Fifty years ago *Montessus de Ballore* laid down the rule: Where there is high relief, expect active faults and high seismicity. With the modern world-wide array of seismographs, not only can we locate epicenters with pinpoint precision, we can track down many more

**Figure 3-12**
An aerial view taken just after the installation of an array of 25 seismometers along ditches that form the radiating arms of a star, each arm 3.5 km long. LASA (Large Aperture Seismic Array) is a large system that comprises twenty-one such arrays (525 seismometers) installed in Montana to conduct research on distinguishing nuclear explosions from earthquakes, the purpose being to establish a basis for an underground nuclear test-ban treaty. LASA is also used to explore the Earth's deep interior using seismic waves. *Ecological note*: the ditches shown in the photograph were dug after a snow storm, were filled in, and are now fully overgrown. [ From MIT Lincoln Laboratory.]

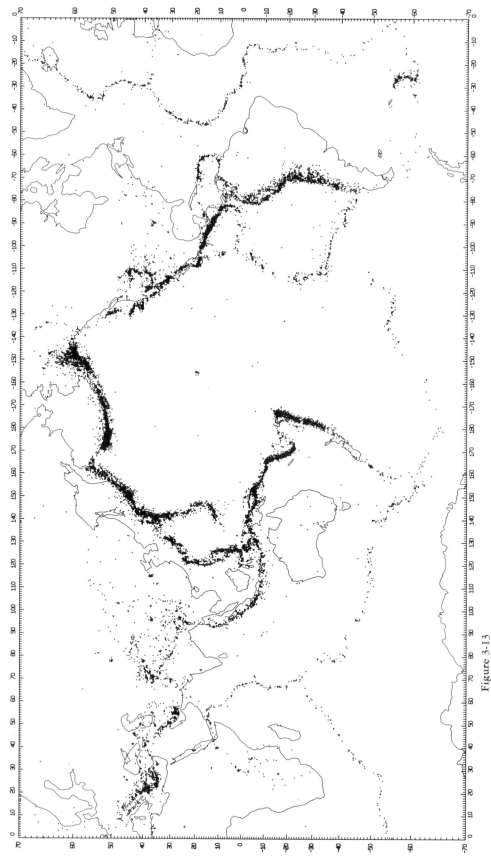

Figure 3-13

Seismicity of the Earth 1962–1967. A plot of shallow earthquakes recorded in this period shows that most such earthquakes originate in narrow belts coincident with plate boundaries. [After M. Barazangi and J. Dorman.]

earthquakes because of the increased instrumental sensitivity, the larger number of observations, and the efficient processing of large quantities of data by computers. This enables us to produce seismicity maps of the quality shown in Figure 3-13. One sees at a glance that most earthquakes are confined to narrow belts, which seem to demark the boundaries of rigid plates within which only minor seismicity occurs.

The coincidence between the world's active seismic belts and the boundaries between plates is easily explained by postulating that the plates move with respect to each other. The differential motion at plate boundaries is impeded by friction, strain builds up, and eventually the strain is released by sudden slipping, causing an earthquake. Seismologists have found that the earthquake mechanisms are different at the different kinds of plate boundaries, or junctions, where plates converge, diverge, or simply slip past each other. The differences are revealed by studying the appearance of seismic waves from these different sources, as we will see in Chapter 19.

**Volcanoes—Windows to the Mantle**  Merely plotting the locations of volcanoes tells us something important—that most volcanic activity tends to occur near plate junctions of the divergent or convergent type. Where plates move apart, mantle material flows in, and the lightest, most mobile fraction eventually reaches the surface as lava. Samples of this material should tell us about the parent rock, but not without some difficulty, since some of the components have been left behind and the composition of these must be inferred from the fraction of materials that squirt out. Actually most lavas extruded at divergence zones are released through linear fissures rather than volcanoes.

Volcanism associated with convergence zones is related to the process of plate resorption, and the kind of lava erupted is a clue to help unravel this very complicated business. For example, one hypothesis holds that convergence-zone volcanism is related to melting at the contact between the descending plate and the overriding plate. If this is so, chemical composition of lava should vary in a consistent way with distance from the trench as the depth to the inclined zone increases. Such correlations have now been found, as Figure 3-14 shows. Sampling and chemical analysis of lavas in the context of plate tectonics is a promising field for aspiring young geologists to get into. It requires a combination of field work and laboratory analysis.

Volcanoes can be instrumented to study the process of eruption itself. On Hawaii, arrays of seismographs and tiltmeters (which record ground tilts) monitor volcanic episodes. Deep earthquakes indicate when magma enters the conduit system some 55 kilometers below the volcano. Upward swelling of the volcano shows on the tiltmeters when the magma slowly wells up from the depths and accumulates a few kilometers below the surface.

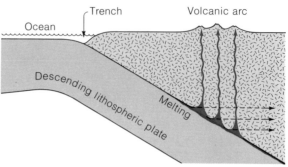

**Figure 3-14**
The chemical composition of lava is an important tool for exploring the interior.
One hypothesis suggests that where plates collide, volcanoes derive their lava
from remelted rocks at the contact between the descending plate and the over-
riding plate. If so, the chemical composition of lava should change with increasing
distance from the trench, because the depth to the zone of melting increases.
Such a change has been found in the ratio of potassium to silicon. [After W. R.
Dickinson and T. Hatherton.]

Numerous shallow earthquakes signal an imminent break-out of
the lava, as the eruptive fissure spreads open toward the surface.
After extrusion, the surface subsides again and the fissure closes.
Studies of this kind are not only scientifically important, they
may one day be used as a warning system for destructive eruptions.

**The Earth's Magnetism and the Dynamics of the Core** That
the Earth behaves like a huge magnet has been known and used
for navigation since the sixteenth century. The Earth's magnetic
field behaves as if a small but powerful permanent bar magnet
exists near the center of the Earth, inclined slightly from the geographic axis. Continuous magnetic records show that the magnetic
field drifts slowly relative to the surface. For example, the compass
needle in London has wandered from 15° east of true north to
25° west of north and back 5° west of north in two centuries.
Unfortunately, the permanent-magnet theory of the Earth's field
has a fatal defect—magnetic materials lose their permanent magnetism at high temperatures, so that permanent magnetism cannot
exist in the mantle or core of the Earth where the temperature is
high. Another way to make a magnetic field is by electric currents.
About a billion amperes of electric current is required to produce
the Earth's magnetic field, but where inside the Earth is there an
electric current generator or dynamo with this capacity? Since a
dynamo requires a moving electrical conductor, the obvious place
to look is in the Earth's molten iron core. Iron is a good conductor,

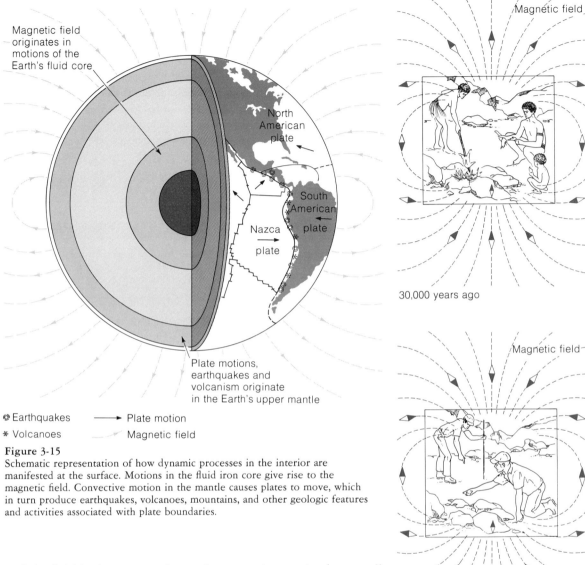

Magnetic field
originates in
motions of the
Earth's fluid core

North
American
plate

South
American
plate

Nazca
plate

Plate motions,
earthquakes and
volcanism originate
in the Earth's upper mantle

⊕ Earthquakes     ⟶ Plate motion

✳ Volcanoes     ⟋ Magnetic field

**Figure 3-15**
Schematic representation of how dynamic processes in the interior are
manifested at the surface. Motions in the fluid iron core give rise to the
magnetic field. Convective motion in the mantle causes plates to move, which
in turn produce earthquakes, volcanoes, mountains, and other geologic features
and activities associated with plate boundaries.

and the fluid in the core can be set in convective motion by a small
amount of radioactively generated heat. Thus large-scale motions
in the fluid core seem to offer the best possible basis for explaining
the Earth's magnetism, and we now have a means of studying these
motions—namely, the record of slow changes in the field.

**Fossil Magnetism** What was the Earth's magnetic field like a
million years ago, or a hundred million years ago? Surprisingly
enough, Earth scientists have found a way to answer this question.
We have mentioned that permanent magnetism is lost when magnetizable material is heated. Another feature of magnetizable
material is that in cooling from a heated, unmagnetized state it
becomes magnetized in the direction of the surrounding magnetic
field. Think, for a moment, of an ancient volcano erupting lava 100

Magnetic field

30,000 years ago

Magnetic field

Today

**Figure 3-16**
Earth's magnetic field 30,000 years
ago was reversed from that of today's.
We know this from the discovery
of reversely magnetized rocks found
in the fireplace of an ancient campsite.
The rocks, cooling down after the
last fire, became magnetized in the
direction of the ancient magnetic
field, leaving a permanent record of
it, just as a fossil leaves a record of
ancient life.

million years ago. When its lava solidified and cooled, the iron-containing minerals became magnetized in the direction of the Earth's field existing at that time, leaving us with a permanent record of the ancient magnetic field, just as a fossil leaves a record of ancient life. Equipped with this new means of exploring the Earth's magnetic field backward in time, geophysicists have roamed the surface of the Earth, collecting old rocks on every continent in order to reconstruct the history of changes in the Earth's magnetic field (Figure 3-16). This great effort has led to many discoveries, which we will take up later, such as pole wandering, reversals of the Earth's magnetic field, and the record of magnetic reversals preserved in the sea floor.

## SUMMARY

1. Almost every branch of geology has its field, laboratory and theoretical aspects. This chapter gives something of the flavor of the diverse methods which modern geologists use to explore and understand their planet.

## EXERCISES

1. Why are experiments so important in helping to interpret geological phenomena?

2. Geologists can study land areas by direct observation. Some 80 per cent of the surface is covered by water and hidden from direct view. How would you explore these important regions?

3. How would you model or otherwise investigate the following:
   a) formation of craters by meteorite impact;
   b) ability of wind to transport unconsolidated surface sediments;
   c) rate of erosion of mountains;
   d) rate of sediment deposition on sea floor;
   e) melting point of rocks?

4. Rocks near the surface of the Moon have densities that are close to the average density of the whole Moon. Iron is more than twice as dense. The Moon has no planetary magnetic field like Earth's. Do these facts imply anything about the existence of a liquid iron core in the Moon?

## BIBLIOGRAPHY

American Chemical Society, *Chemistry and the Environment: The Solid Earth, The Oceans, The Atmosphere.* Washington, D.C.: American Chemical Society, 1967.
Menard, H. W., *Anatomy of an Expedition.* New York: McGraw-Hill Book Company, 1969.
Phillips, O. M., *The Heart of the Earth.* San Francisco: Freeman, Cooper & Company, 1968.
Runcorn, S. K. (editor), *International Dictionary of Geophysics.* New York: Pergamon Press, 1967.

# 4

# Rocks and Minerals

*Rocks and the minerals that make them up are the tangible record of geologic processes. The varied minerals of the Earth are understood in terms of their architecture—the way in which their atoms are arranged to make crystal structures. The kinds of atoms and their type of chemical bonding determine not only the crystal structures but the chemical and physical properties of minerals, all of which are used for their identification. Rocks are divided into the three major groups, igneous, metamorphic, and sedimentary, on the basis of origin. They are further subdivided within each group according to mineral composition and texture, which provide the data that allow us to interpret details of their origin.*

Picking up pretty stones and showing or wearing them must go far back into man's prehistory. The earliest records of man's practical use of stones, though, are of arrows and spears equipped with points made of **flint** (a sedimentary rock) or **obsidian** (volcanic glass), both of which are hard materials that break with sharp edges. From the practical use of individual stones as tools, weapons, and decorations, man moved to the wholesale mining or quarrying of rocks and minerals for building, for making clay for pottery, and then for the ores that contain metals. Today our mining is done so expertly and intensively that geologists have come to concern themselves with the exhaustion of the world's valuable mineral resources; minerals of economic significance and their reserves are covered in Chapter 24.

108

**Figure 4-1**
A pre-Columbian artifact carved from volcanic rock: conch-blower Costa Rica, approximately A.D. 1000. [Photo by R. Siever.]

**Figure 4-2**
One of the largest copper mines in the world is an open-cut mine in Bingham Canyon, Utah, near Salt Lake City. Since operations started in 1905, more than 10 million tons of copper have been produced here. The vertical distance between top and bottom levels of the mine is 720 m. [Courtesy Kennecott Copper Corporation.]

Because of the many uses of rocks and minerals, we have a practical curiosity about where they are found and how they were made: we want to be able to find more. Yet there are other reasons, too, for rocks, as we have seen, are the only records of how the Earth evolved, and they are an important guide to how the Earth works today. For this reason, **mineralogy** and **petrology**—studies of the nature and origin of minerals and rocks—are important sub-fields of geology. Finally, there is the intrinsic interest in the extraordinary range of the mineral kingdom, with its immense variety of color, form, and texture. Minerals and rocks, after all, give us the marble and alabaster of sculpture, the jade of Eastern carvings, and the pigments used by Rembrandt.

**What Information Do We Want from Rocks?**  If the nature of rocks can give us a clue to many of the things we want to know about the Earth, how do we go about it? We need an interpretive key, just as ancient historians needed the Rosetta stone to crack the "code" of Egyptian hieroglyphics before they could read that part of man's history. First of all we want to find out just what the minerals are made up of and how the rock is put together from its constituent minerals. From its composition we should be able to say something about where the parent material came from and what it was like. What was the magma like? Or, what were the

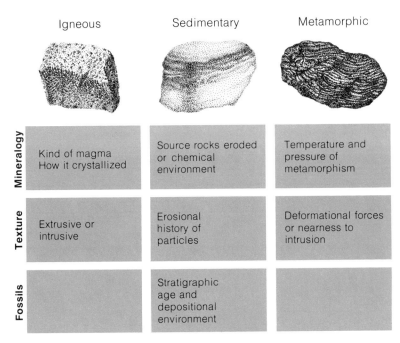

| | Igneous | Sedimentary | Metamorphic |
|---|---|---|---|
| **Mineralogy** | Kind of magma How it crystallized | Source rocks eroded or chemical environment | Temperature and pressure of metamorphism |
| **Texture** | Extrusive or intrusive | Erosional history of particles | Deformational forces or nearness to intrusion |
| **Fossils** | | Stratigraphic age and depositional environment | |

Figure 4-3
The mineralogy and texture of a rock are the keys to inferring its origin.

source rocks of a sediment? Or, what were the pre-existing rocks that were heated and compressed to make a metamorphic rock? From the composition and the texture of the rock we should also be able to tell something of the pressures and temperatures at which the rock was formed by comparing them with the artificial rocks and minerals made by **experimental petrologists** in the laboratory.

From the radioactive elements in a rock, or from the fossils in it, we can also tell its age. But the age of a rock bears little relation to its composition. Rocks of the same kind form at different times in the Earth, and age is no clue to their mode of formation.

**Rocks Are Made Up of Minerals** We have already touched on some of the most obvious characteristics of rocks in Chapter 2; most of a rock's properties, those given in Figure 4-4 are easily seen with the naked eye, though the details are better revealed with a low-power magnifying glass (magnification of 5× to 10×) — the hand lens that field geologists usually have hanging around their necks.

From the characteristics shown, particularly the physical and chemical properties, we can distinguish several thousand minerals, each defined by its unique set of properties. Thus all grains or crystals of quartz have just about the same qualities, regardless of the kind of rock in which they are found. Some minerals, particularly those that have a more complex mixture of atoms, vary slightly in their properties, depending on their precise composition. A mineral like garnet, for example, has a number of varieties, each with its own range of composition, such as the proportions of iron and other elements, and hence, properties.

Rocks are not as uniquely defined by their properties as minerals are. Because of the immense number of ways in which the thousands of minerals can be combined, the geologist is faced with a bewildering array of rock types. The only way for us to make order out of this array is to classify like with like and to sort out by general type. The major division of rocks into igneous, sedimentary, and metamorphic is just such an aid. Within each major division there are many groups and types. Using characteristic properties, we can divide the rock kingdom into several hundred general types, each with its own more-or-less distinctive earmarks.

Despite all of these numbers, it is remarkable how much can be done even if only a small number of the most common minerals and rocks are known. In most parts of the world a field geologist can make an accurate geologic map by knowing only a few dozen major minerals and even fewer common rock types. This simplification is possible because most of the thousands of known minerals are either rare or unusual on the one hand or subspecies or varieties on the other. Thus the geologist who can recognize garnet will do well, even though a mineral sophisticate who can distinguish the

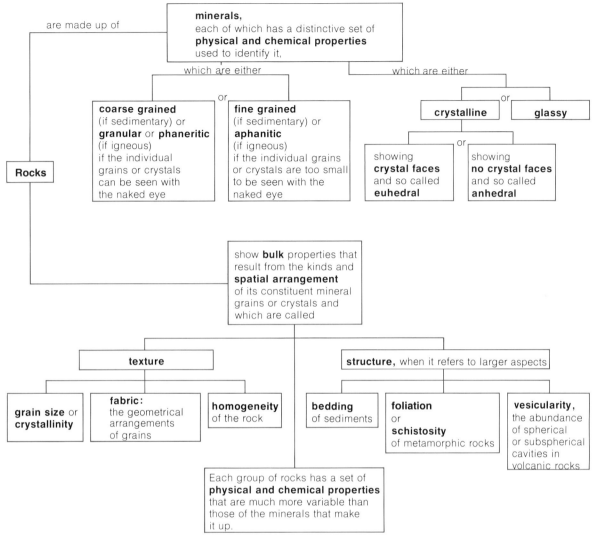

**Figure 4-4**
Flow chart of rock properties.

many varieties of garnet by their slightly different chemical compositions might do better. Naturally, the more we can distinguish, the more the information gleaned, and the greater the power of our theories of explanation. That is why professional petrologists have to know a great deal about mineralogy.

The next section tells how we go about identifying minerals by means of their properties, which also guide us to an understanding of how matter is put together to make crystals and how the conditions of their formation may be deduced from their geometry and composition.

## CRYSTALS: FACES AND SYMMETRY

The regularity of crystal faces is a major aspect of the external form of minerals, and for many years they were studied and identified mainly by analyzing their symmetry. In addition to his earlier contributions—enunciating the laws of stratigraphy and recognizing fossils—Steno wrote in 1669 that quartz crystals, wherever found, always showed the same angle between similar crystal faces. By the late eighteenth century, his **constancy of interfacial angles** became accepted as a generality applicable to all minerals. By 1801, *René Haüy* was able to summarize the laws of **crystal symmetry**, the regularities of crystal faces. Shortly afterwards, in 1809, crystallography was revolutionized by the invention of a new instrument, the **reflecting goniometer**. The inventor was *Samuel Wollaston,* an English physician who had abandoned medicine for research in chemistry, mineralogy, and many other sciences. Before that, the angles between crystal faces could be measured only on large crystals, against whose faces a plane could be laid; the new Wollaston instrument made it possible to measure angles between minute faces that one could see only with a powerful lens. The early part of the nineteenth century was a time of intense exploration of the external forms of minerals in relation to their chemical composition, a development that paralleled the great geological exploration of the Earth and the growth of the geological time scale.

The major work of the great crystallographer Haüy was accomplished in the midst of the great upheaval of the French Revolution. Though it may seem strange, some of the greatest scientific advances have been made during times of great social and political turmoil. Although Haüy survived the revolution, one of his contemporaries (born in the same year, 1743), did not. Antoine Lavoisier—the man who has been called the greatest of all chemists, the man who discovered oxygen and formulated the metric system—was guillotined in 1794.

**Figure 4-5**
Quartz exhibits a variety of crystalline forms ranging from groups of crystals with well-developed crystal faces, such as those on the left, to masses of crystals intergrown in such a way that no crystal faces are developed, such as in the broken pebble of vein quartz on the right. [Photo by R. Siever.]

Figure 4-6
Halite crystals, showing cubic form. [Courtesy of Mineralogical Museum, Harvard University.]

**How to Measure a Crystal** In the large, a crystal is a piece of matter whose boundaries are naturally formed plane surfaces. The geometry of a crystal may be relatively simple, as in the cubes of common salt, the mineral **halite**, or they may be beautifully complex, as in snowflakes, the crystals of ice. Crystals are usually formed when a liquid solidifies or when a solution becomes **supersaturated** — that is, too enriched in dissolved material to hold it any more — and the dissolved substances precipitate, or "drop," out of solution. When some liquids, such as molten silicates, congeal very quickly, the solid that forms is not crystalline but **glassy**, a condition in which crystals with plane faces do not form, but only masses with curved, irregular surfaces. This absence of crystallinity is typical of many lavas.

The most useful measurements that can be made on crystals are those of the angles between faces. From these measurements, an idealized drawing can be made of the geometry of all of the faces in relation to each other; such a drawing eliminates all of the imperfections of the natural faces, such as scratches or breaks. Haüy and other mineralogists discovered as a result of such work that, in addition to constancy of interfacial angles, each kind of crystal exhibits other regularities: definite symmetrical relationships exist among faces, and there are certain simple mathematical

A *solution* is made by dissolving a relatively small amount of a substance, called the *solute,* in a much larger amount of *solvent.* Salt, NaCl, is the solute and water the solvent of a salty solution. Water, of course, isn't the only solvent. Alcohol is an excellent solvent for many organic substances. Tincture of iodine is a solution of iodine in alcohol. A solution is *saturated* when it holds all of a solute that it can; this amount is called the *solubility.* The amount of solute at saturation depends on temperature: most warmer solutions will hold more than colder ones. If a warm, saturated solution is cooled, it will then contain more than it can at the lower temperature, and will be *supersaturated.*

**Figure 4-7**
Photomicrographs of snowflake crystals. These were described as among the choicest specimens of crystal architecture in a 25-year search by Wilson Bentley, pioneer photographer of snow crystals. [Photographs by Wilson Bentley, courtesy of Duncan Blanchard, State University of New York at Albany.]

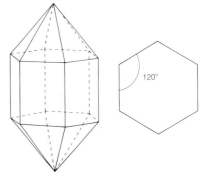

**Figure 4-8**
A fragment of obsidian, a volcanic glass. The curved, sharply terminated fracture surfaces are typical of what is called *conchoidal fracture,* which characterizes glasses and a number of other materials, including many minerals. [Photo by R. Siever.]

**Figure 4-9**
Drawing of quartz crystal. A section at right angles to the long axis, showing a regular hexagon with faces at 120° angles.

relations between the angles of all faces. The marvel of finding such simplicity in the midst of apparent complexity is what led Haüy and others to infer that there must be an underlying order to the arrangement of the atoms in crystals—an idea that could only be verified a century later when X-rays were discovered and beamed through crystals.

Another element of order appeared after hundreds of different kinds of crystals were measured. All crystals can be put into seven general classes of symmetry, as shown in Table 4-1. A further grouping of thirty-two **crystal classes** can be distributed among the seven systems. Mathematical analysis has shown that there are only these thirty-two different ways of arranging atoms about a point that will allow the building of a crystal that obeys symmetry rules and fills space in three dimensions. The combined effort of measuring angles between crystal faces and analyzing the symmetry of crystals has led to a simple and orderly classification that consists of a limited number of major crystal systems and classes and which includes all crystals.

## Box 4-1   Crystal Symmetry

One can see how a simple crystal is analyzed by taking a cube and noting, first, the obvious relation that all faces are always at 90° to each other and, second, that one can draw three mutually perpendicular imaginary planes such that each is a *plane of symmetry*; that is, each face on one side of the plane is mirrored on the other side. Or, one can hold opposite faces of the cube on an axis between thumb and forefinger and spin it around to find that there is a *fourfold axis of symmetry*; that is, in one complete rotation of 360° a face will be

repeated four times. Another axis of rotation between opposite corners is a *three-fold* axis of symmetry. The essence of symmetry is this: a simple geometrical operation can be performed that will repeat a face in another position. To put it another way, if one performs an operation such as a rotation, a new face will occupy the same position that was occupied by another face before rotation—and one cannot tell that the operation has been performed from the final appearance alone.

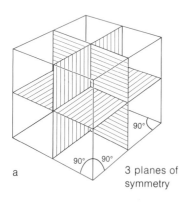

a   3 planes of symmetry

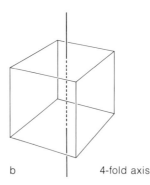

b   4-fold axis

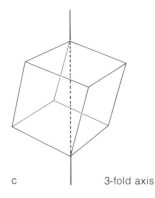

c   3-fold axis

Table 4-1
Crystal systems.

| Crystal system | Minimum symmetry | Reference axes |
| --- | --- | --- |
| Isometric | 4 3-fold axes | 3 mutually perpendicular; all of same length |
| Trigonal | 1 3-fold axis | 4 axes, 3 horizontal with 120° intersections, 1 vertical. Horizontal axes all of same length; vertical axis any length |
| Hexagonal | 1 6-fold axis | Same as trigonal |
| Tetragonal | 1 4-fold axis | 3 mutually perpendicular; 2 of same length, 3rd of any length |
| Orthorhombic | 3 2-fold axes or 3 symmetry planes | 3 mutually perpendicular; each of any length |
| Monoclinic | 1 2-fold axis and/ or 1 symmetry plane | 2 axes at oblique angle, 3rd axis perpendicular to plane of the other 2; each axis of any length |
| Triclinic | 1-fold axis or center of symmetry | 3 axes at oblique angles; each of any length |

In some classifications, hexagonal and trigonal are considered to be divisions of one crystal system, the hexagonal.

In 1858, at the end of the half century during which the modern laws of crystallography were worked out, an English geologist, *Henry Sorby,* published a memoir, *On the Microscopical Structure of Crystals.* Sorby's memoir was the fruit of almost ten years of work spent studying thin, transparent slices of rock with the polarizing microscope, which had been invented about thirty years earlier but hardly used. For geologists, it was the beginning of a new subject, **petrography,** the study of rocks with the polarizing microscope, known more familiarly to most geologists as the **petrographic microscope.**

The inspiration of Sorby was to use the knowledge that even the darkest, most opaque-looking rocks might be made transparent if slices were thin enough, and that the way crystals affect polarized light passing through them could provide a means of identifying minerals and of studying the mineral compositions and textures of a wide variety of rocks. For the first time he could identify small grains or crystals that could be seen only with a high-power microscope.

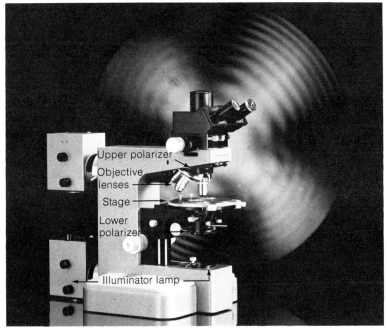

**Figure 4-10**
A modern petrographic microscope. Light produced by a lamp in the base shines vertically into the lower polarizer below the specimen stage, passes through the thin section on the stage, and then through the upper polarizer, situated above the objective lenses. Shown in the background is one kind of optical pattern that minerals produce under the polarizing microscope and which serves to identify them uniquely. [Courtesy of E. Leitz, Inc.]

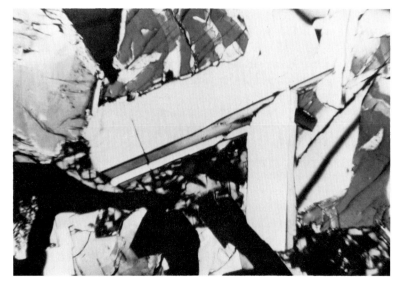

Thin sections are made by cutting off a slice of rock about 1 mm thick with a saw containing embedded diamonds. One side is then ground smooth with silicon carbide grinding powders and cemented to a glass slide. Next, the open side is ground down until the desired thinness is reached, about 0.03 mm, the last gentle stages of polishing being monitored with a microscope. Some minerals are opaque no matter how thin they are ground. These are the metallic minerals, sometimes called the *opaque minerals*. An example is pyrite, $FeS_2$.

**Figure 4-11**
Thin section of a coarse-grained basalt under crossed polarizers. The white to light-gray crystals are plagioclase feldspar; they are surrounded by pyroxene crystals. This basalt is from the Moon, but its mineralogy and texture are typical of those on Earth. [Photo by W. von Engelhardt and D. Stöffler, Tübingen University; courtesy Carl Zeiss, Inc.]

**Figure 4-12**
The angle of refraction of any substance depends on its index of refraction, which is constant for any substance, and on the angle of incidence.

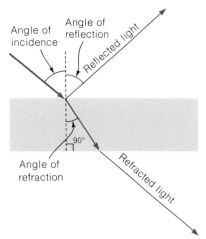

**How Light Is Transmitted Through Crystals** When light travels from air into a transparent material such as water or a crystal, it gets bent, or **refracted**, in proportion to the density of the substance. We have all seen examples of this phenomenon, such as the apparent bending of the part of a pencil or stick that is immersed in water. The degree of refraction is a property of the substance and is called the **index of refraction**. In an **isotropic** substance, the index of refraction is the same in all directions. All fluids are isotropic, but the only solids that have this property are glasses and isometric crystals, both of which are made up of atoms arranged in the same pattern in all crystallographic directions. The other solids, the great majority, are **anisotropic**; that is, they refract light by different amounts in different directions, a property that is determined by the different arrangement of atoms in different crystallographic directions.

Ordinary light, traveling through air or some other isotropic substance, consists of waves that vibrate in all directions perpendicular to the light ray. Anisotropic minerals have the property of polarizing, or splitting, ordinary light into two polarized light rays that vibrate at right angles to one another. Certain materials, such as the substance used to make polaroid sunglasses and the various synthetics used to make polarizing filters, transmit light that vibrates in only one direction. If you look at an object with polaroid sunglasses held horizontally and then interpose another pair held vertically, all light will be absorbed, and you will see

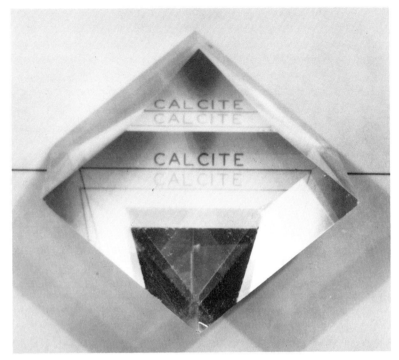
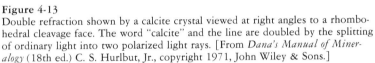

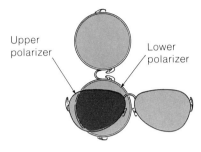

**Figure 4-13**
Double refraction shown by a calcite crystal viewed at right angles to a rhombo-hedral cleavage face. The word "calcite" and the line are doubled by the splitting of ordinary light into two polarized light rays. [From *Dana's Manual of Miner-alogy* (18th ed.) C. S. Hurlbut, Jr., copyright 1971, John Wiley & Sons.]

blackness. The polarizing microscope is equipped with two polar-izing filters at right angles to one another. One is situated between the microscope light and the thin section; the other one is situated between the thin section and the eye, and is removable. When the removable polarizer is inserted into the microscope tube, the polarizers are said to be "crossed." Under crossed polarizers the microscopist can instantly detect whether a crystal is isotropic, because it turns black.

Anistropic minerals transmit polarized light in a more complex way, such that they turn dark only at certain positions, every 90°, called **extinction** positions. It is in these positions that the direc-tions in which the anisotropic crystal polarizes light coincide with the directions of the two polarizers in the microscope. Extinction occurs at fixed angles, called **extinction angles**, to crystal faces or lines of cleavage that may be shown by the mineral. From extinc-tion angles and the values of the indices of refraction in the dif-ferent directions, optical crystallographers have constructed tables for the identification of minerals. Using these detailed tables as he studies thin sections of rock under the microscope, the geologist can tell just what kinds of minerals are put together in a particular pattern to make up the rock. The optical properties shown by

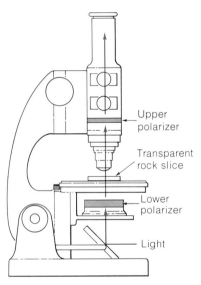

**Figure 4-14**
Two polarizing sunglass lenses crossed at right angles block the passage of light. All of the light that passes through the first lens is filtered out by the second. A polarizing microscope has polarizers above and below the rock slice to analyze the polarizing properties of minerals.

crystals formed the basis for speculating on the atomic structure
of crystals even before X-rays were discovered. Recently, these
properties have come to be of interest to geophysicists, who are
trying to use them to measure how thermal energy is transmitted
through materials under extremely high pressures in the Earth.

## THE ATOMIC STRUCTURE OF MINERALS

Though many nineteenth-century mineralogists, following the
lead of earlier scientists, had speculated about how the atoms
that make up all matter might be internally arranged in crystals,
there was no way to confirm any relation between external form
and internal structure. By the beginning of the twentieth century,
many mineralogists were becoming increasingly convinced that
crystal form, chemical composition, and such physical properties
as color and hardness, might be explained in terms of some hypothetical atomic pattern, but they had no proof. That proof was
provided in 1912 by the German physicist Max von Laue and two
of his students, when they irradiated a crystal of copper sulfate
with X-rays and produced an **X-ray diffraction pattern** on a film
placed behind the crystal. It had already been realized that X-rays
are a variety of electromagnetic radiation, like ordinary light, except
that they have very short wavelengths, which give extreme penetrating power. Von Laue reasoned that if ordinary light could
be **diffracted** — that is, apparently deflected around corners to
give a pattern of fringes when it was directed through very tiny
openings such as pinholes or slits — so could X-rays. It was known
that the smaller the wavelength of the visible light — that is, the
closer the wavelengths are to the blue and violet end of the spectrum — the smaller the pinhole had to be to produce diffraction
patterns. Von Laue guessed that if the mineralogists were right
in their speculations about crystals being orderly arrangements
of atoms, the smallest "pinholes" he could use to make the very
short wavelengths of X-rays diffract would be the spaces between
the atoms of a crystal, which must be of the order of one ten-
millionth ($10^{-7}$) of a millimeter. Not only that, but the pattern of
the X-rays could reveal the way the "pinholes" — and therefore the
atoms — are arranged in space. It took only one year after that
remarkable and crucial experiment for an English father-and-son
team, William H. Bragg and William L. Bragg, to publish the first
crystal-structure analysis of a mineral, halite. The Nobel Prize
in physics was awarded to the Braggs for this work in 1915, when
the younger, William L., was only twenty-five.

In the next dozen years a great many minerals were structurally
mapped. This paved the way for the Norwegian mineralogist and
petrologist *V. M. Goldschmidt* to outline, in 1926, all of the general
principles governing the ways that the atoms of different elements
were put together to form crystals. Finally, some explanation could

Forty years after von Laue and the
Braggs opened up X-ray crystallography as a science, two young men
working in the laboratories at Cambridge University in England, under
the direction of the younger Bragg,
who was then in his sixties, brilliantly
applied their knowlege of X-rays and
molecular architecture to deduce the
crystal structure of the genetic substance DNA (desoxyribonucleic acid).
The detective story of the search for
that structure by J. D. Watson and
F. M. Crick, later described by Watson
in *The Double Helix,* involved a race
among some of the leading figures in
X-ray diffraction studies.

be given for the long-known facts of the external appearance of crystals and their properties. The explanation lies in the structure of atoms, their systematic changes with increasing atomic number and atomic weight, their atomic size, and their ability to form various kinds of chemical bonds with other atoms.

**The Electronic Structure of the Elements**   Let us start by again noting the structure of an atom as given in Chapter 2, the nucleus of protons and neutrons and the surrounding cloud of electrons orbiting the relatively small nucleus. An atom may gain or lose one or more electrons, thus forming an **ion**, charged positively or negatively depending on the relative numbers of protons in the nucleus and electrons surrounding the nucleus. The number of protons remains the same, even though electrons may be lost or gained. A positively charged ion, a **cation**, has lost electrons. A negatively charged ion, an **anion**, has gained electrons.

The hydrogen atom is the smallest and simplest of all atoms, having just one proton in the nucleus and just one orbiting electron (atomic number 1). Electron orbits correspond to "shells" that surround the nucleus, for electrons do not follow a simple, single path as the planets do around the Sun but can be considered to occupy, at various instants, points in a sphere, or spheres, around the nucleus. As atomic number increases, so does the number of electrons and so does the number of shells. Table 4-2 lists the shells and the maximum number of electrons each can hold. The single hydrogen electron occupies the *K*-shell, the innermost of the seven shells. The element of next higher atomic number is helium, which has two electrons, both in the *K*-shell (atomic number 2). The *K*-shell can be occupied by only two electrons. The element of next higher atomic number, lithium, has three electrons (atomic number 3), one of which occupies the next shell outward, the *L*-shell. This shell can hold eight electrons, The first two shells, *K* and *L*, fill up in order as the atomic number of the elements increases from hydrogen to neon. After that, from sodium (atomic number 11) to argon (atomic number 18), the next eight electrons start to fill up the *M*-shell. Then the regular order is interrupted and the *N*-shell starts to fill up before the *M*-shell is completed. Then electrons are added to both *M* and *N* shells, until *M* is filled up, whereupon with increasing atomic number, the *O*-shell starts filling up and *N* continues being completed. This complication of shell filling helps to explain why the chemistry of the heavier elements is somewhat more varied and complex than that of the light elements, those of the first two or three shells.

The elements can be divided into groups on the basis of their electronic structure (Table 4-3). One small group includes the elements whose outermost shell is populated by eight electrons. They are the inert, or noble, gases, neon, argon, krypton, xenon, and radon, together with helium, which has its *K*-shell filled with two electrons, all that is possible. Because this is a stable con-

The nucleus, containing almost the entire mass of an atom, has a diameter between $10^{-13}$ and $10^{-12}$ cm, thus having a volume only one-thousandth to one ten-thousandth of the entire atomic volume. For the convenience of using small integral numbers, the *Ångstrom* (= $10^{-8}$ cm) is used as a unit of measurement for atomic diameters.

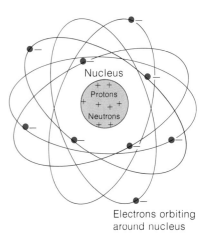

Electrons orbiting
around nucleus

**Figure 4-15**
Structure of an atom. The mass is concentrated in the nucleus. Each electron carries one minus charge.

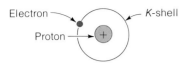

Hydrogen atom
Net charge (1+) + (1−) = 0
Atomic number 1

Hydrogen ion, a cation (*K*-electron lost)
Net charge (1+)
Atomic number 1

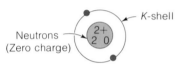

Helium atom
Net charge (2+) + (2−0) + 2(−) = 0
Atomic number 2

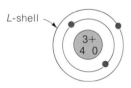

Lithium atom
Net charge (3+) + (4−0) + (3−) = 0
Atomic number 3

Lithium ion, a cation (*L*-electron lost)
Net charge (3+) + (4−0) + (2−) = (1+)
Atomic number 3

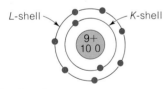

Fluorine atom
Net charge (9+) + (10 − 0) + (9−) = 0
Atomic number 9

Fluorine ion, an anion
(*L*-electron gained)
Net charge (9+) + (10−0) + (10−) = 1−
Atomic number 9

**Figure 4-16**
As atomic number increases, electrons
fill up the innermost, or *K*-shell,
first, and then start to fill the *L*-shell.
Ions are formed when atoms lose
or gain electrons.

**Table 4-2**
Electron shells of atoms.

| Shell | Maximum number of electrons possible in shell | | Example |
|-------|:-----------------------------------------:|---|---------|
| *K* | 2 | | Hydrogen *K*-1 |
| *L* | 8 | | Carbon *K*-2, *L*-4 |
| *M* | 18 | | Silicon *K*-2, *L*-8, *M*-4 |
| *N* | 32 | | Iron *K*-2, *L*-8, *M*-14, *N*-2 |
| *O* | 50 | | Silver *K*-2, *L*-8, *M*-18 *N*-18, *O*-1 |
| *P* | 72 | | Gold *K*-2, *L*-8, *M*-18 *N*-32, *O*-18, *P*-1 |
| *Q* | 98 | | Uranium *K*-2, *L*-8, *M*-18, *N*-32, *O*-21, *P*-9 *Q*-2 |

Table 4-3
Types of electronic structure of the elements (other than rare earths or actinides).

| Type of element | Common elements | Characteristics |
| --- | --- | --- |
| Noble gases | helium, argon | Outer shells of elements have stable, filled configurations and no tendency to form chemical compounds |
| Alkali metals<br>Alkaline Earth metals | sodium, potassium<br>calcium, magnesium | Outer shells have one or two valence electrons and strong tendency to lose them and form cations |
| Halogen group<br>Oxygen group | fluorine, chlorine<br>oxygen, sulfur | Outer shells need one or two electrons to assume stable configurations and have strong tendency to gain electrons and form anions |
| Boron group<br>Carbon group<br>Nitrogen group | boron, aluminum<br>carbon, silicon<br>nitrogen, phosphorus | Outer shells tend to assume stable configurations by sharing electrons with other atoms and have weak tendency to gain or lose electrons |
| Transition elements | Iron, copper, zinc | Valence electrons tend to be gained or lost from an inner shell rather than the outer shell |

figuration, these elements have no tendency either to gain or to lose electrons and so do not form any important crystalline compounds.

A second group of elements is made up of those whose outermost shells may gain or lose electrons. This group includes most of the common abundant elements, such as sodium, potassium, magnesium, and calcium. These elements have a tendency either to gain or to lose electrons in order to assume the stable configuration of the inert gases. When a chlorine atom gains an electron, it becomes the chloride anion with 2 $K$-electrons, 8 $L$-electrons, and 8 $M$-electrons, thus achieving the same structure as that of atomic argon. Similarly, the element potassium has a strong tendency to lose an electron and become a cation with exactly the same structure as the chloride ion. The electrons that are gained and lost, called **valence electrons**, determine the chemical behavior of the elements.

A third group of elements, commonly called the **transition elements**, consists of those whose valence electrons are in the next to the outermost shells. In many elements of higher atomic number, the shells are not filled up uniformly. These elements, which may, for example, have their outermost electrons in the $P$-shell, nevertheless gain and lose electrons from the $O$-shell to form ions. They may form several different kinds of ions; iron for example, forms two cations, $Fe^{2+}$ and $Fe^{3+}$.

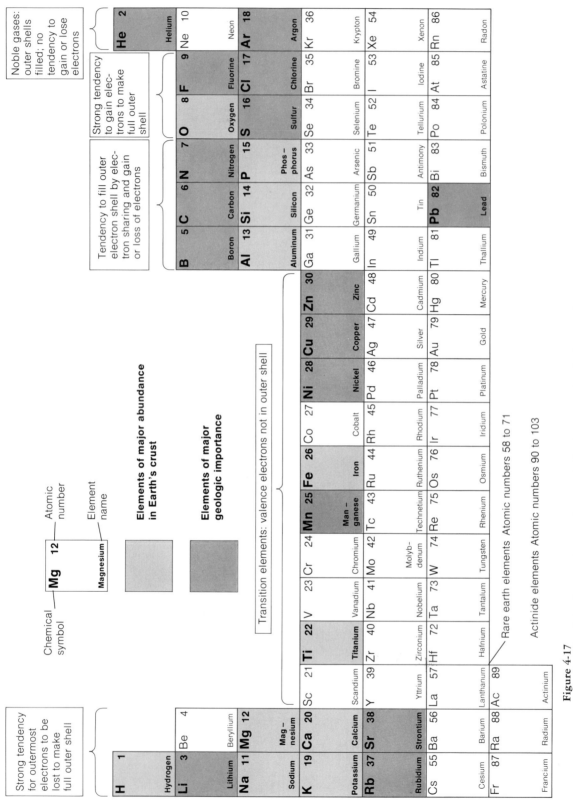

**Figure 4-17**
Periodic table of the elements (rare earth and actinide elements omitted), showing electronic structure and geologic importance.

The fourth group, those whose valence electrons are in the third from the outermost shell, includes the rare earth and actinide elements, none of which is of great abundance in the Earth or of much importance to our story.

The importance of electronic structure, shells of valence electrons, and stability of the outer shell is that they all determine the nature of **chemical bonding** of one element to another in crystals. There are a number of bond types, and they are not mutually exclusive. A particular bond may be largely of one type but have some characteristics of another. The simplest form of chemical bond, in some ways, is the ionic bond.

**Ionic Bonds** Bonds of this type are formed by electrostatic attraction between ions of opposite charge, which is of exactly the same nature as the attraction that makes hair "stand up" when some synthetic fabrics are drawn over it. Thus ionic bonds form most strongly between elements like sodium and chlorine, which tend to gain or lose electrons and become cations or anions. In fact the simplest of all ionically bonded crystalline substances is sodium chloride, whose crystal structure was the first to be worked out. From a look at Figure 4-20 one can see the symmetry of the ionic arrangement, that each ion of one kind is surrounded by six of the other, and that there is no "molecule" of NaCl as such. The six neighbors give rise to a new measure, the **coordination number**. Both sodium and chloride ions are in six-coordination. Or we can say that NaCl is **octahedrally** coordinated, from the regular octahedron described by lines connecting the six coordinated ions.

One can also see from Figure 4-20 that, although there is no molecule of NaCl, there is what is called a **unit cell**—the geometric form that is outlined by the minimum number of ions necessary to show the basic building block that is repeated a great number of times to form a visible crystal. The unit cells of NaCl and the other chlorides are cubic because of the equidimensional spacing of ions. X-ray diffraction studies show that there is a systematic variation in the sizes of the other chloride unit cells, that they increase in size with the atomic number of the successively heavier metals, lithium, sodium, potassium, rubidium, and cesium. From this kind of information it is possible to deduce a structure of packed spherical ions of different sizes. When the exact size of an ion, the **ionic radius**, is calculated from such data, tables of ionic radii can be made that show certain trends (Table 4-4). The ionic radii of elements in the same chemical group, such as lithium, sodium, rubidium, and cesium, increase with atomic number, in relation to the addition of electrons and electron shells. This change is not in exact proportion to the increase in atomic number because, as we have seen, electrons are added irregularly to shells. Other regularities are related to the charge on the ion.

Electrostatic attraction is described by *Coulomb's Law*, which states that the attraction (or repulsion) between two charged particles is directly proportional to their product and inversely proportional to the square of the distance between them, or:

$$E = k(q_1 \times q_2)/d^2$$

where $E$ is the attractive force, the $q$'s the number of charges on the ions or electrons, $d$ the distance, and $k$ a proportionality constant associated with the medium in which the attraction takes place. The value of $k$ is 1 in a vacuum.

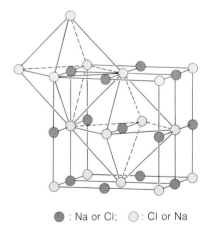

● : Na or Cl;  ○ : Cl or Na

**Figure 4-18**
Cubic structure of sodium chloride (ions not drawn to true scale). [From R. C. Evans, *Crystal* Chemistry (2nd ed.), Cambridge Univ. Press, 1966.]

Table 4-4
Ionic radii of alkali metal group of elements.

| Element | Ion symbol | Atomic number $(Z) = $ no. of electrons | Ionic radius (Ångstrom units)* | |
|---------|-----------|------------------------------------------|-------------------------------|---|
| Lithium | Li⁺ | 3 | 0.60 | |
| Sodium | Na⁺ | 11 | 0.95 | |
| Potassium | K⁺ | 19 | 1.33 | |
| Rubidium | Rb⁺ | 37 | 1.48 | |
| Cesium | Cs⁺ | 55 | 1.69 | |

*One Angstrom unit $10^{-8}$ cm. See marginal note, p. 121.

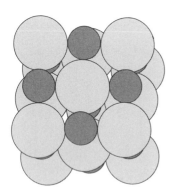

**Figure 4-19**
Cubic structure of sodium
chloride, showing the ions in
their correct relative sizes.
[From R. C. Evans, Crystal
Chemistry (2nd ed.), Cambridge
Univ. Press, 1966.]

Cations are small, most of them less than $10^{-8}$ centimeters in
radius. But most anions are large, particularly the most common
anion, oxygen. From this it is apparent that most of the space of
a crystal is occupied by the anions and that the cations tend to fit
into interstices between them. Figure 4-9 shows the NaCl structure with ions of approximately the correct relative size. A convenient way to measure the relative sizes of anions and cations is

by **radius ratio**, the ratio of cation radius to anion radius. For some structures the radius ratio shows that the cations are too large to fit into interstices and that the anions are held slightly apart. If a slightly smaller cation were put into the space, the anions would be held less far apart and the closer packed structure would be more stable because the electrostatic attraction holding it all together would be stronger. The structure would be most stable if the cation were just small enough to fit exactly into the spaces between anions that touch each other. If a cation is smaller than the space, it will have no effect on the anions, for they cannot get any closer than contact. Detailed analyses of this sort allow crystallographers to predict what kinds of structures are possible with different elements and to explain why minerals make the kinds of crystals they do.

**Covalent Bonds** Compounds that make a stable electronic configuration by **sharing** electrons rather than gaining or losing them are held together with **covalent bonds**. The formation of such bonds depends on the number and distribution of shared electrons in the outer shells, and so the kinds of compounds and crystal structures formed are determined by more complex factors than the simple geometric ones of dominantly ionically bonded crystals. Elements that do not readily gain or lose electrons to form ions—for example, carbon—form bonds of this kind. The simplest covalent structure is that of diamond, in which every carbon atom (not an ion) is surrounded by four others (coordination number 4) arranged in a regular tetrahedron. When coordinated in this way, each carbon atom shares one electron with each of its four neighbors and thus achieves a stable octet of electrons in its outer shell.

**The van der Waals Bond** Weak bonds that have far less strength than ionic or covalent bonds exist between all ions and atoms in solids. They are named after the man who inferred their existence from weak attractive forces shown by atoms and molecules in gases. This form of bond is a weak electrical attraction that is related to the asymmetry of certain atoms and ions. Van der Waals bonds play a role of moderate importance in some silicate minerals, where their presence is not masked by the more powerful ionic and covalent links.

**The Structure of Some Common Minerals** Though the variety of known structure types in the universe is great, geologists seldom encounter more than a relatively small number, largely because most rocks are made up of silicate minerals, which combine the two most abundant elements in the Earth's crust, oxygen (O) and silicon (Si). Silicates can be divided into a few major structure types. The basis for all silicate structures is the radius ratio of silicon to oxygen, 0.30, which allows each silicon to be four-coordinated to four surrounding oxygens in a regular tetrahedron.

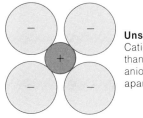

**Unstable**
Cation larger than opening: anions held apart

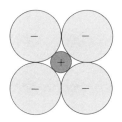

**Most stable**
Cation exactly fits opening: anions in contact

**Stable**
Cation smaller than opening: anions in contact

**Figure 4-20**
Radius ratio of cations to anions determines the stability of ionic bonds.

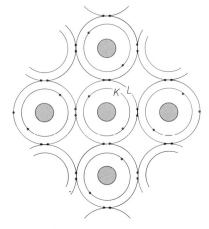

**Figure 4-21**
A carbon atom in diamond may share one outer electron with each of four neighbors and so achieve the stable number of 8 electrons in the *L*-shell even though each individual atom would have only 4 electrons by itself.

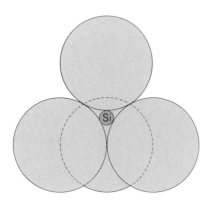

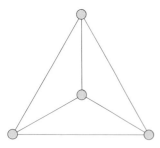

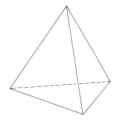

**Figure 4-22**
Silica ($SiO_4$) tetrahedron in three
different representations.

Silicate structures are made up of these tetrahedrons arranged in different ways, with such cations as sodium ($Na^+$), potassium ($K^+$), calcium ($Ca^{++}$), magnesium ($Mg^{++}$), ferrous iron ($Fe^{++}$) and ferric iron ($Fe^{+3}$) in the interstices between. The bond between Si and O is about half ionic in character and half covalent, the Si sharing one of its outer electrons with each oxygen ion. Adjacent silicons may share oxygen ions, allowing networks of several kinds to be built up of tetrahedra.

Many silicate minerals also contain aluminum, the third most abundant element in the Earth's crust after oxygen and silicon. The radius ratio of aluminum to oxygen is 0.36, a value that is on the boundary between stable four- and six-coordination. As a result, it is found in both coordinations. When its coordination number is 4, it can take the place of a silicon ion in a tetrahedral structure. When its coordination number is 6, it is octahedrally coordinated, just like the ions in NaCl, and may join adjacent tetrahedra by a dominantly ionic bond. Such cations as $Na^+$, $K^+$, $Ca^{++}$, $Mg^{++}$, and $Fe^{++}$ may have 6-, 8-, or 12-fold coordination in silicate structures.

Cations that have similar coordination numbers and have similar ionic radii tend to substitute for each other and make mixed compounds that we call **solid solutions**, which are analogous in every way to common liquid solutions. For example, $Mg^{++}$ and $Fe^{++}$ have radius ratios of 0.66 and 0.74, respectively, and natural olivines are solid solutions of variable amounts of Fe and Mg silicates. The pure Mg olivine is $Mg_2SiO_4$, forsterite; the pure Fe olivine is $Fe_2SiO_4$, fayalite. The composition of the natural solid-solution mineral is represented by the formula $(Mg, Fe)_2SiO_4$.

Silicates are classified and named according to the way the tetrahedra are linked, as shown in Table 4-5. **Isolated tetrahedra** are linked by mutual bonding to a cation between. **Rings of tetrahedra** are formed by bonding of two oxygens of each tetrahedron to adjacent tetrahedra in closed rings. **Single chains** form by the same linkage. Two single chains are combined in some minerals to form **double chains**, in which the chains are linked by cations. **Sheets** are structures in which each tetrahedron shares three of its oxygens with adjacent tetrahedra to build planar lattices. In three-dimensional **frameworks**, the tetrahedra are linked by sharing oxygens with other tetrahedra. Aluminum in 4-coordination may substitute for silicon in chains, sheets, and frameworks.

A different sort of building block is the carbonate ion. In this group of ions the carbon atom is surrounded by three oxygen atoms in a planar triangle. Groups of carbonate ions are arranged in sheets in a manner somewhat like that of the sheet silicates. The mineral calcite is made up of carbonate sheets and intervening planes of calcium ions. The mineral dolomite is made up of the same carbonate sheets separated by alternating sheets of calcium and magnesium ions.

Glasses lack regularity of the structure that is typical of the crystalline state. In silica glass, for example, the atoms are arranged

Table 4-5
Major silicate structures.

| Geometry of linkage of $SiO_4$ tetrahedra | Si/O ratio | | Example mineral | Formula |
|---|---|---|---|---|
| *Isolated tetrahedra:* linked by bonds sharing oxygens only through cation | 1:4 | | Olivine | $(Mg,Fe)SiO_4$ |
| *Rings of tetrahedra:* joined by shared oxygens in 3-, 4- or 6-membered rings | 1:3 | | Beryl | $BeAl_2(Si_6O_{18})$ |
| *Single chains:* each tetrahedron linked to two others by shared oxygens. Chains bonded by cations | 1:3 | | Pyroxene | $(Mg,Fe)SiO_3$ |
| *Double chains:* 2 chains joined by shared oxygens as well as cations | 4:11 | | Amphibole | $(Ca_2Mg_5)Si_8O_{22}(OH)_2$ |
| *Sheets:* each tetrahedron linked to 3 others by shared oxygens. Sheets bonded by cations or aluminia sheets | 2:5 | | Kaolinite | $Al_2Si_2O_5(OH)_4$ |
| *Frameworks:* each tetrahedron shares all its oxygens with other $SiO_4$ tetrahedra (in quartz) or $AlO_4$ tetrahedra | 3:8 | | Feldspar (Albite) | $NaAlSi_3O_8$ |
| | 1:2 | | Quartz | $SiO_2$ |

in silica tetrahedrons, but the tetrahedrons are not linked up in a regular repetitive manner but are more or less randomly arranged, as they would be in a liquid.

This brief tour through the world of crystal structures is intended to give some idea of the power of the principles of atomic physics, crystallography, and mineralogy, for it is from these disciplines that we have gained the extraordinary details that enable scientists

Silica tetrahedron
$(SiO_4)^{-4}$
$Si^{++++}$ and $O^{--}$

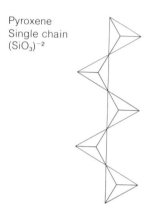

Pyroxene
Single chain
$(SiO_3)^{-2}$

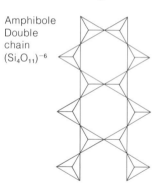

Amphibole
Double
chain
$(Si_4O_{11})^{-6}$

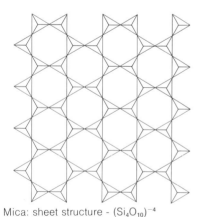

Mica: sheet structure - $(Si_4O_{10})^{-4}$

**Figure 4-23**
Silicate structures. Single and
double chains and sheets.

to explain how crystals are put together. Now that you have some idea of what is known about structure, it is possible to show how X-ray diffraction was used to learn it all.

## X-RAYS: THE DIAGNOSTIC TOOL OF MINERALOGY

Soon after Laue discovered X-ray diffraction, W. L. Bragg found out that the patterns of diffraction corresponded to one that would be formed by reflection of the X-ray beam from planes within the crystal, the reflection angles following certain mathematical requirements that we now call Bragg's Law. Figure 4-24 shows how the planes can be visualized as made up of sheets of atoms in a regular array. As one can see from looking at any crystal structure diagrams, there are many possible planes that can be drawn. Each one reflects in accordance with Bragg's Law and so contributes its share to an X-ray pattern. The many lines of reflection that are found on X-ray patterns come from all of the planes. The X-ray crystallographer knows the wavelength of the X-rays used and measures the angle of the crystal to the beam to calculate the spacing of the planes. The patterns of all of the spacings can be harmonized to make a probable crystal structure. Today, sixty years after the Bragg's did their first work on NaCl, crystal structures are analyzed by automatic computers controlling X-ray diffraction equipment. The results are fed into another computer that spins out a complete structure analysis, possibly even with crystal drawings.

X-rays are being used as a rapid means of identifying minerals by their structures. That they can be so used is because no two minerals have precisely the same X-ray pattern, even though they

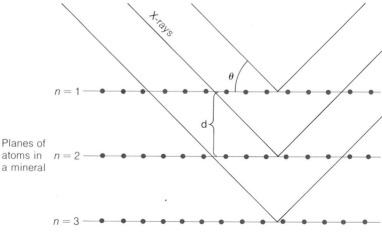

Planes of
atoms in
a mineral

**Figure 4-24**
X-rays are reflected from the planes of atoms in a crystalline mineral when the angle of incidence, $\theta$, is in a certain proportion to the spacing between planes, $d$.

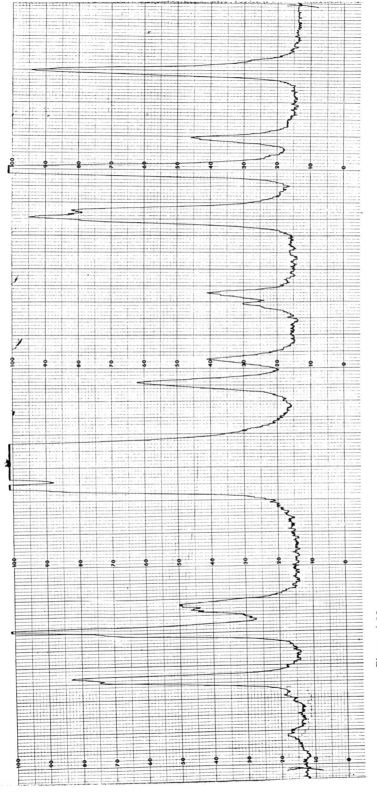

Figure 4-25
An X-ray diffraction pattern of a feldspar produced by an X-ray spectrometer. The horizontal axis represents the angle of reflection of the X-ray beam from various planes of atoms in the crystal as the crystal is rotated in the X-ray beam. The vertical axis is a measure of the intensity of the reflection.

The usual statement of Bragg's Law is $n\lambda = 2d \sin \theta$, where $\lambda$ is the wavelength of the X-rays used, $d$ the distance between planes of atoms (sometimes called "$d$-spacing"), $\theta$ the angle between the X-ray beam and the crystal planes, and $n$ the "order" of a reflection. The order of a reflection that measures the distance between two successive planes is 1, that of the distance between every other plane, 2, and so on. A crystal of the clay mineral kaolinite is built up of planes 7 Å (1 Å, angstrom unit $= 10^{-8}$ cm.) apart. Its first-order reflection is 7 Å, second order at 14 Å, third at 21 Å, and so on. Normally we work only with first-order reflections.

may have the same structure type, such as the micas, muscovite and biotite, which have the same sheet structure, but whose precise patterns differ. The spacings and the intensity of the reflections depend on the kinds of atoms as well as their arrangement, so minerals are distinctive. In practice, the three most intense reflections on an X-ray pattern are compared to a list, sometimes in a card file, of minerals arranged by the spacing of their reflections. Today the X-ray machine is used for simple mineral identification almost as routinely as the somewhat crude physical and chemical tests were used in the last century. X-ray machines, however, are not easily carried into the field, even though some semiportable units have been invented, so it is still necessary to learn the simple diagnostic features of the common minerals (Appendix IV).

## PHYSICAL AND CHEMICAL PROPERTIES OF MINERALS

A few years ago in Quincy, Massachusetts, a scratched message was discovered on the window of a house that John Hancock had owned in the eighteenth century. Speculation was that Hancock had expressed his feelings of love toward a woman by scratching their initials on the glass with his diamond ring. Whoever it was perhaps knew that a diamond was the hardest mineral known and could easily scratch glass. What made diamond so hard was then unknown. People also knew that mica, the sheety mineral commonly used for oven and lantern windows at that time, was especially easy to cleave into paper-thin sheets as transparent as window glass, but they did not know why. They were aware of many distinctive and useful properties of minerals. The earliest and still the simplest ways of identifying many common minerals are by testing for hardness, cleavage, and other physical properties.

**Physical Properties, Bond Types, and Structures** We now know that the way atoms and ions are bound has a major effect on their physical properties, as shown in Table 4-6. Few bonds are exclusively of one type; instead, most are hybrids. For this reason, the correlation with properties can be only a weak generalization. Particular properties of a specific mineral can be explained on the basis of the kinds of atoms and their structures, together with the bond types.

A good example is a mineral that has been used for ages as a cosmetic, talc, which has given us talcum powder. Talc is made up of silicate layers that are weakly bonded to each other by van der Waals forces, and so slip and break apart easily. It is that easy slippage of one layer over the other that gives talc its soapy feel and that makes it grindable to a fine, smooth, cosmetic powder. A contrary example is given by beryl, whose clear, deep-green variety is emerald. Beryl is also made up of sheets, but the bonds between them are strong, predominantly ionic; as a result, the mineral is hard and breaks with difficulty in any direction.

Table 4-6
Bond types and physical properties.

| Property | Ionic | Covalent | van der Waals |
|---|---|---|---|
| Structural types | Bond strength of any ion is uniform in all directions with high coordination numbers | Interatomic or ionic bonds are strong only in a few set directions, hence coordination numbers low | Bond strength of any ion is uniform in all directions with high coordination numbers |
| Hardness | Strong bonds give high hardness | Stronger bonds, in general, giving higher hardness than ionic bonds | Weak bonds give low hardness |
| Cleavage | Ease of cleavage, poor to good, depends on distance between planes of atoms and individual bond strengths | | Excellent cleavage |
| Melting points | Moderate to high; ions are present in the melt | Very high; molecules are present in melt | Low |

Source: After R. C. Evans, An Introduction to Crystal Chemistry, 2nd ed., Cambridge University Press, Oxford, 1966.

The various physical properties and the ways in which bonding and structure influence them are illustrated further in the next section.

**Hardness** Just as a diamond scratches glass, so will a quartz crystal scratch a feldspar crystal, because one is harder than the other. Mohs, an Austrian mineralogist, devised a scale of hardness in 1822 that spanned the spectrum from the softest to the hardest mineral known (Table 4-7). Even with all of the elegant instruments, called sclerometers, that are now available to determine hardness, the **Mohs Scale of Hardness** remains the best practical way to identify an unknown mineral. A combination of common objects and a few of the minerals on the hardness scale are all one needs to bracket an unknown mineral between two points on the scale. The hardness of any mineral depends on the strength of the bonds between ions or atoms; the stronger the bonds, the harder the mineral. Because bond strengths may be different along the various crystallographic axes, hardness too may vary slightly in direction.

Mohs arranged the scale so that the hardness difference between any two adjacent minerals was about the same, except for the difference between corundum and diamond, which is much greater. Mohs could not find any mineral with an intermediate hardness, nor has anyone else.

**Cleavage** This term is used for breakage along definite planar surfaces, typified by that of mica. The number of planes of cleavage varies among minerals. The cleavage planes of a mineral have some of the same characteristics as its crystal faces, though they are not to be confused with them. They always occupy definite and constant positions with respect to the symmetry or crystallographic axes of the crystal and are parallel to a possible crystal face.

Table 4-7
Mohs scale of hardness.

| Scale number | Mineral | Common objects |
|:---:|:---|:---|
| 1 | Talc | |
| 2 | Gypsum | |
| | | Fingernail |
| | | Copper wire or coin |
| 3 | Calcite | |
| 4 | Fluorite | |
| 5 | Apatite | |
| | | Pocket knife blade |
| | | Window glass |
| 6 | Orthoclase (feldspar) | |
| | | Steel file |
| 7 | Quartz | |
| 8 | Topaz | |
| 9 | Corundum | |
| 10 | Diamond | |

Cleavage is the expression of differing bond strengths along the various planes of a crystal. The bonds across some of those planes may be very weak, like the van der Waals bonds that are responsible for the easy breakage across the cleavage planes of mica. Calcite has two excellent cleavages parallel to the faces of a rhombohedron, the result of weak bonds. Quartz is strongly bonded in all directions and has no cleavage. Cleavage can be scaled on the basis of ease of cleaving with a chisel, from the most perfect cleavages of the micas to the poor cleavage of beryl (emerald). The most expert cleaving in the world is done in Amsterdam, where the center of the diamond "cutting" industry is; it is the excellent cleavage of diamond along many planes that makes possible the shaping of the hardest mineral known, for there is nothing harder with which to cut it!

**Fracture** The way in which a mineral breaks other than along cleavage planes also serves to group them and help in identification. Fracture may be **conchoidal**, showing smooth, curved surfaces like those of a thick piece of broken glass (see Fig. 4-8); **fibrous** or **splintery**, **hackly**, and **uneven** or **irregular**, all of which are self-explanatory, though imprecise terms. Geometric properties of such variability and irregularity as fracture still defy the attempts of scientists to devise simple, quantitative measures to make the description objective and uniform. The type and ir-

**Figure 4-26**
Calcite crystals. (Top) A cleavage rhombohedron whose surfaces were made by cleaving, or breaking, the crystalline material along zones of weak bonding strength in the crystal. (Bottom) Faces formed by growth of the crystal from a solution. [From *Dana's Manual of Mineralogy* (16th ed.) by C. S. Hurlbut, Jr., copyright 1952, John Wiley & Sons.]

regularity of fracture bears a complex relation to the breaking of bonds in directions that cut across crystallographic planes.

**Streak**   Streak is the name given to the color of the fine powder that is produced when a mineral is scraped, or rubbed, across a tile of unglazed porcelain, called a **streak plate**. The color is diagnostic for many minerals; for example, hematite, an iron oxide, always gives a reddish-brown streak regardless of the color of the particular mineral aggregate being streaked, which may be black, red, or brown.

**Luster**   The way in which the surface of a mineral reflects light gives it a characteristic luster. Mineral lusters are described by more or less self-explanatory terms (Table 4-8). The quality of light reflected from mineral surfaces is controlled by the index of refraction and the absorption of light by the elements of the mineral.

Table 4-8
Mineral luster.

| |
|---|
| *metallic*:  strong reflections produced by opaque substances. |
| *vitreous*:  bright, as in glass. |
| *resinous*:  characteristic of resins, such as amber. |
| *greasy*:  surface has the appearance of being coated with an oily substance. |
| *pearly*:  the whitish iridescence of materials like pearl. |
| *silky*:  the sheen of fibrous materials like silk. |
| *adamantine*:  the brilliant luster of diamond and similar minerals. |

Those properties, in turn, are related not only to the kinds of atoms but to their bonding. For example, covalently bonded minerals tend toward adamantine luster, whereas ionically bonded minerals are more vitreous. Other lusters are related to irregularities of the surface.

**Color**   Many minerals show a characteristic color on freshly broken surfaces; others show characteristic colors on weathered or altered surfaces. We use not only the colors reflected from mineral surfaces but also the colors transmitted through minerals in microscopic thin sections. Color seen in thin sections is a good, quick guide to the identification of certain minerals. Some minerals—for example, precious opals—show a stunning play of colors on reflecting surfaces. Others will change color slightly with a change in the angle of the light shining on the surface. Minerals show a variety of other unusual optical properties, most of which are of chief interest to the gem trade, such as the **cat's eye** of chrysoberyl. Color may be a property of the pure substance or it may be the result of impurities. The color of pure substances is dependent on the presence of certain ions, such as iron or chromium, which are strong absorbers of certain colors of light. Impurities, many too small to be seen except by the most powerful microscope, such as small dispersed flakes of hematite in a quartz crystal, impart a general color to an otherwise colorless mineral.

**Specific Gravity and Density**   Though the obvious difference in weight of a piece of hematite iron ore and a piece of sulfur of the same size is easily felt by hefting the pieces in the hand, a great many common rock-forming minerals have about the same range of **density** (= mass/volume). Consequently, it became necessary to devise methods that would make it easy to measure that property of minerals accurately. The standard measure of density is **specific gravity** (= weight of the mineral in air/weight of an equal volume of water at 4°C). This is usually measured by taking the ratio of the weight of the mineral in air to the difference between the weight of the mineral in air and the mineral in water, which is the equivalent of the definition. Because density is not only useful in identifying minerals but is also a property that depends strongly on the internal structure and composition of the substance, it has been measured accurately for most minerals. Density is dependent on the atomic weight of the constituents and the tightness of packing of the atoms in the crystal structure.

Density is also affected by pressure, and pressure-induced increases in density affect the way rocks and minerals transmit light, heat, and elastic waves, such as those associated with earthquakes (see Chapter 19). Some chemical substances form more than one kind of crystal; the different kinds of crystals are called **polymorphs**. Calcium carbonate, for example, forms both a trigonal polymorph, calcite (density 2.71), and an orthorhombic one, aragonite (density 2.93). The denser aragonite is known from experimentation to form and remain stable at high pressures, whereas calcite forms and is stable at low pressures. The very high pressure polymorphs of silica ($SiO_2$), called coesite and stishovite, have been identified as products of shock-wave experiments (see Chapter 3, p. 000) and as natural minerals from impact craters of meteorites or nuclear explosions.

**Chemical properties** The distinctive chemical compositions of minerals are the basis for the major classification of the mineral kingdom. The chemical criterion used to divide the mineral kingdom into classes is the anion of the mineral. For example, halite, NaCl, is classed as a **chloride**, as is its close relative, sylvite, KCl. In this way, all minerals have been grouped into eight classes, as shown in Table 4-9.

Most of what we know about the chemical composition of minerals has been gained through the use of ordinary wet chemical methods, by which materials are dissolved, separated into their constituent elements, and their weight or volume then measured. Various other methods have been added over the past few decades as chemical instruments became more capable of measuring small quantities of elements in new ways. Most recently the **electron probe**, a device that beams electrons at a sample mineral and analyzes how the elements in the sample generate X-rays, has been used to get good chemical analyses of very small crystals.

Mineralogists have also relied for many years on quick, simple chemical tests that can be made in the field. Perhaps the best known is the "acid test," by which a dilute hydrochloric acid (HCl) solution

The names "anion" for negative ion and "cation" for positive ion date back to early experiments with electrochemical cells, the ancestors of modern wet-cell automobile batteries. Because the *cat*hode, the negatively charged electrode, attracts positive ions, those ions became called *cat*ions. Similarly, the *an*ode, the positively charged electrode, attracts negative ions, hence *an*ion. Since then, hundreds of mnemonic devices (clues to memorizing) have been thought up to make it easy to remember the pair. One example: AP is not only Associated Press but anode = positive.

Table 4-9
Chemical classes of minerals*

| Class | Defining anions | Example |
|---|---|---|
| Native elements | None: no charged ions | *Copper*, Cu |
| Sulfides and similar compounds | Sulfide: S$^{--}$ and similar anions | *Pyrite*, FeS$_2$ |
| Oxides and Hydroxides | O$^{--}$ OH$^{--}$ | *Hematite*, Fe$_2$O$_3$ *Brucite*, Mg(OH) |
| Halides | Cl$^-$, F$^-$, Br$^-$, I$^-$ | *Halite*, NaCl |
| Carbonates and similar compounds | CO$_3^{--}$ | *Calcite*, CaCO$_3$ |
| Sulfates and similar compounds | SO$_4^{--}$ and similar anions | *Barite*, BaSO$_4$ |
| Phosphates and similar compounds | PO$_4^{-3}$ and similar anions | *Apatite*, Ca$_5$F(PO$_4$)$_3$ |
| Silicates (see Table 4-5 for details of silicates) | SiO$_3$ | *Pyroxene*, MgSiO$_3$ |

*This classification, derived originally by Berzelius in the nineteenth century and used extensively by Dana, is a simplified form of the scheme used by Berry and Mason in *Elements of Mineralogy*, W. H. Freeman and Company, 1968.

is dropped on a mineral to see if it fizzes. If it does, it is likely to be calcite, a carbonate mineral. A full series of tests can be made with the **blowpipe**, a narrow tube through which one blows air into the flame of a gas or alcohol burner to make the flame hotter. Using this simple instrument, minerals can be heated in various ways to perform diagnostic tests for the presence of specific elements.

There are impurities in all natural minerals, enough of them so that it is frequently hard to decide when an element is an integral part of the mineral or is merely an extraneous contaminant. Elements that make up much less than 0.1 percent of the mineral are reported in analyses by the word "trace," and many elements are called **trace elements**. But most trace elements we rarely hear about, for they are neither abundant in the earth (or the solar system) nor have they come into any important industrial use. Typical of these are lanthanum (La), one of the rare earth elements, and scandium (Sc), an element that has affinities with aluminum and boron. Trace elements are analyzed by means of an **emission spectrograph**—an instrument that vaporizes the mineral with an extremely hot electric arc and then analyzes the light emitted by the burning substance. A newer analytical method is **neutron activation**: an atomic reactor is used to bombard the mineral with the fast-moving neutrons that are formed from the radioactive decay process of uranium-235 in the reactor. The bombardment changes some of the originally nonradioactive elements in the mineral into new, radioactive elements. The abundance of the new elements can then be analyzed by counting radioactivity; one different atom among a billion others can be detected in this way. Thus we can obtain an accurate measure of the abundance of the originally nonradioactive elements that were transmuted by bombardment. There are enough research reactors in universities all over the country that can be used by visitors to make this method practical for all those who learn how to do it.

Another part of chemical analysis is the determination of the relative proportions of different isotopes, the elements with the same atomic number (protons) but different atomic weights. In Chapter 2, we have discussed the use of radioactive isotopes in dating rocks, but there are other isotopes of some elements that are stable—that is, nonradioactive. Three of the most important are the heavy stable isotope of carbon, $C^{13}$, the heavy stable isotope of oxygen, $O^{18}$, and the heavy stable isotope of sulfur, $S^{34}$. The study of these isotopes in minerals, in the atmosphere, and in natural waters has contributed greatly to geology. As was discussed in Chapter 2, isotopes are analyzed by use of a mass spectrometer.

An important example of the geological value of stable isotopes is the use of the ratio of $O^{18}$ to $O^{16}$ to determine ancient temperatures. The $O^{18}/O^{16}$ ratio of the oxygen atoms in calcium carbonate is dependent on the temperature of the water from which it was precipitated. Thus a well-preserved fossil oyster shell can be analyzed to find the temperature of the sea in which the animal lived.

## ROCKS AS MINERAL AGGREGATES

The architecture of a rock is not unlike that of a building—a design of building blocks put together in certain ways. The building blocks of a rock, of course, are minerals. Without too much difficulty, we might be able to think of all imaginable designs—all conceivable ways of putting minerals together—but the great majority of arrangements would bear little relation to any we find in nature, just as doing the same for all of the products in a building-supply warehouse would probably not bear much relation to a building like the Parthenon. The problem, then, is to find out the designs of nature and to find out why there are certain ways in which minerals are put together to form rocks and why there are many, many other ways in which they are not. Much of this book is devoted to questions such as these.

The initial subdivision of rocks is made on the basis of origin, and the three major categories are igneous, metamorphic, and sedimentary. Origin is determined by a combination of rock characters, such as bedding, the layering that indicates a sedimentary origin; foliation, the preferred orientation of crystals that indicates a metamorphic origin; and a variety of other textures and structures. Although this distinction is based partly on mineral composition, it is better to discuss such mineralogical and textural attributes within each major group and not try to make an all-inclusive definition of rock classification. For example, a pure quartz sandstone (sedimentary) and a vein of quartz (igneous) may both be 100 percent quartz, yet there are no igneous and metamorphic rocks composed of alternating bands of halite and gypsum, and there are no sedimentary rocks made up largely of pyroxene. Once the igneous, metamorphic, and sedimentary rocks have been described, it will be easier to see how the three groups can be distinguished from each other.

**The Igneous Rocks**   One major systematic grouping of igneous rocks is made on the basis of a combined chemical and mineralogical classification, the chemistry of the rocks being revealed largely by their dominant minerals and serving as a clue to the composition of the magmas from which they solidified. The names formerly used come from one of the first criteria used when the igneous rocks were first being studied in the last century — that is, the amount of silica, $SiO_2$, in the chemical analysis. During that pre-modern period of chemistry, the silica was though to be derived from silicic acid, and so the more silica in the rock, the more "acidic" the magma. **Granite**, rich in silica, is the most abundant acidic rock. The rocks lower in silica were called basic. **Gabbro**, poor in silica, is the basic counterpart of granite. Though we now know that silica content is not a measure of acidity as that word is used in chemistry, the terms persist even though what we really mean is "more or less **silicic**." The amount of silica is not necessarily related to the amount of quartz, for much of the silica may be combined in other silicate minerals. In the classification by silica content, the coarse-grained igneous rocks range in sequence from granite on the more silicic side, through **granodiorite** and **diorite**, to **gabbro** on the less silicic side.

Chemistry and mineralogical composition form the basis of the modern system of classifying the major groups, which turns out to be much the same as the classification by silica content. The two terms most commonly used come from a broad division into **light** and **dark** minerals, called, respectively, **felsic** and **mafic**. These terms were used because the dominant minerals of the light group are quartz and feldspars, both rich in silica (hence "felsic," from *fel(s)*, feldspar, plus *ic*) and those of the dark group are pyroxenes, amphiboles, and olivines, all of which are rich in magnesium and iron (hence "mafic," from *ma*gnesium and *f* (ferrous), for iron, plus *ic*). The varieties of feldspar are most important in the classification of the igneous rocks, both because they are abundant and because the proportions of different kinds of feldspar vary systematically from felsic to mafic rocks. Granite is rich in potassium feldspar (mainly the mineral orthoclase), whereas the more mafic rocks are dominated by sodium and calcium feldspars, the plagioclases. The dark rocks are dominated by biotite mica and amphibole at the felsic end and pyroxene and olivine at the mafic end. Pyroxene and olivine are the major minerals of the **ultramafic** rocks, which are even lower in silica than the basalts and gabbros; **peridotite** is dominated by olivine and pyroxene and **dunite** by olivine.

The other major basis for classification of the igneous rocks is, as already described in Chapter 2, a textural one, whereby we differentiate between the coarse-grained granular rocks, or **phanerites**, and the fine-grained **aphanites**. A textural property that is in some ways a crystallographic property, is the degree of **glassiness** or, more commonly, by the opposite quality, the **crystal-**

linity. Both grain size and crystallinity are products of the speed with which a magma cools. The more rapid the cooling, the finer the grain size and the poorer the crystallinity. The slowest-cooling magmas congeal in the last stages of intrusion, when magmas become less viscous and contain more dissolved gases. These give rise to **pegmatites**, very coarse-grained rocks that may include huge crystals several meters across. The fastest-cooling magmatic materials of course, are those that are thrown high into the sky, where they instantly freeze to glass.

The granular or coarse-grained igneous rocks are subdivided on the basis of the abundance of their characteristic minerals, which are easily identifiable in coarse crystalline form in the field. The fine-grained igneous rocks, the volcanics, are more difficult to subdivide in the field because their minerals and glasses are not so easily identified. Their textures, however, are a good basis for grouping. Some volcanic rocks (covered in more detail in Chapter 17) may show the texture of **pyroclastic** rocks. These result from volcanic explosions that expel magma that quickly cools to glass. The finest fragments, which make **volcanic ash** and **dust**, are mixed with particles a few millimeters or centimeters in diameter, called **lapilli**, and larger pieces, called **blocks and bombs**. The resultant texture is one of an agglomeration of pieces of glass, crystals that had started to form before the explosion, and fragments of previously cooled lava. The glass may be in the form of fragments of **pumice**, a frothy mass of glass with a great number of bubbles formed by gas escaping from the melt. The sharp, spiky glass fragments are called **shards**. The bubble holes in pumice and other extrusive rocks are **vesicles**, which if they become filled with other minerals at a later time are called **amygdules**. A very vesicular rock is called **scoria**. The solidified rocks that harden from **ash falls** and **ash flows** and all of the other varieties of ejected material are lumped under the term **tuff**.

The other major textural class of volcanics includes the lavas that flow from volcanoes or large fissures in the crust. The lavas have been given Hawaiian names: **pahoehoe** for the smoother, ropy kind, and **aa** for the sharp, spiky, more jagged kind. The difference in appearance and structure are related to the ease of flowage, which depends on the lava's viscosity, which in turn is determined by the magma's composition.

The lavas and pyroclastics are named for their mineral and chemical compositions in the same way as the coarse-grained series, but fewer subdivision names are in common use. **Basalts**, the most abundant of volcanic types, are chemically the equivalents of gabbros, and **andesites** correspond to diabases and other rocks intermediate between granite and gabbro (Fig. 4-27). On the felsic side, rhyolite is the aphanitic counterpart of granite, and **dacite** the aphanitic counterpart of granodiorite. Because of the glass in the volcanics, mineral compositions are not exactly the same as in corresponding granular rocks.

142

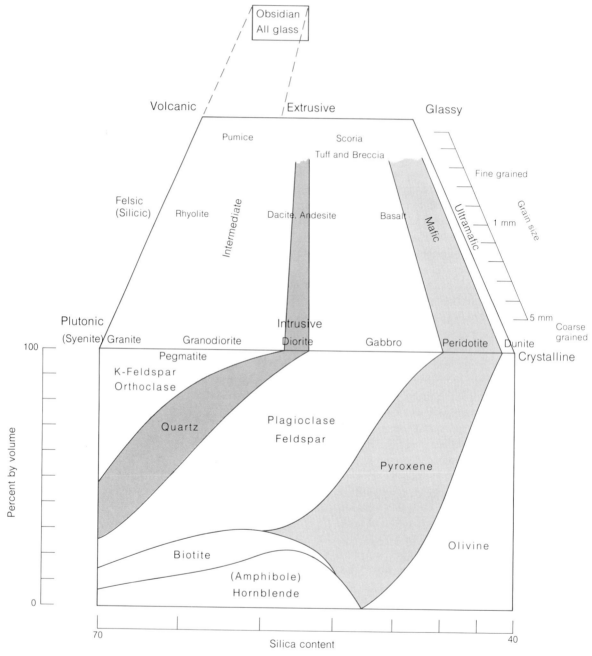

**Figure 4-27**
Igneous rock classification.

Mineral composition loses meaning in the entirely glassy rock **obsidian**, which is the ultimate of the fine-grained equivalents of granites and granodiorites. The glassy, sharp, broken edges of this highly silicic glass make it perfect for Indian arrowheads. The pumices, frothy glasses, are commonly silicic also.

Many igneous rocks are made up of a mixture of both large and small crystals. If some crystals are distinctively larger than the surrounding mass, or **matrix**, they are called **phenocrysts**. A rock with many phenocrysts is called a porphyry. The matrix of a **porphyry** may consist of coarse or fine crystals or, in volcanics, glass.

Putting together mineralogy and texture gives us a scheme of classification like the one shown in Figure 4-27. The mineralogical parameter measures the proportions and kinds of felsic and mafic minerals; and the textural parameter, grain size. As is true of most classifications, the rocks do not all fall neatly into pigeonholes, for composition and texture vary continuously, and we draw arbitrary dividing lines between them to preserve as nearly as possible the traditional meanings of names that have accumulated over two centuries of geological usage.

**The Sedimentary Rocks**   Mineralogy and texture are also useful in subdividing the sedimentary rocks. They are used in combination to set apart two main groups, the **detrital** and the **chemical**. The detrital sediments are those that carry the earmarks of the mechanical transportation and deposition of the debris of erosion, **detritus**, by currents. The minerals are fragments of rocks or minerals broken and eroded from pre-existing rocks, and so are called **clastic** (from the Greek *klastos*, to break). The rocks of ancient mountains worn down by erosion can be reconstructed from the minerals of detrital rocks. Quartz, feldspar, and the clay minerals make up the bulk of that contribution. The fragments tend to wear and abrade during transportation and so become **rounded**. During sedimentation, currents **sort** the minerals by size and weight with variable efficiency; the stronger the current, the larger the particle size carried. **Size** and **sorting** of clastic sedimentary particles are characteristic of the nature of the currents that carried them. They also form the basis for subdividing the detrital sediments into (1) coarse-grained, the **gravels** and their hardened, or **lithified**, equivalents, the **conglomerates**; (2) medium-grained, the **sands** and **sandstones**; and (3) fine-grained, **clays** and **muds** and their lithified equivalents, the **shales**. Coarse sedimentary rocks composed of sharp, angular pieces of rocks and minerals are **breccias**, which contrast with the rounded pebbles and cobbles of conglomerates.

The chemical sediments are precipitated from solution, mostly in the ocean, and so their minerals reflect the composition of the parent solution. The most abundant chemical rocks are **limestone** and **dolomite**, made up largely of calcium and magnesium carbonates, the minerals calcite and dolomite. Limestones may be made up in large part of calcareous fossils—shells formed by **biochemical** precipitation of calcium carbonate that animals extract from sea water. Other chemical sediments are also characterized by their chemical composition in relation to origin. The **evaporites** are composed largely of gypsum and halite, some including a complex group of other salts crystallized from evaporated sea water. The chemical rocks show a texture of crystal intergrowths resembling that of intrusive igneous rocks, on whose surfaces crystals can be seen to fit together like a carefully constructed flagstone walk.

Mineralogy is an important criterion for distinguishing varieties of detrital rocks, particularly the sandstones. **Quartzose sandstones** or **quartz arenites** (arenite is a frequently used word for sand-sized sedimentary rocks) are mainly composed of quartz grains. This composition results from the erosional disappearance of feldspars and mafic minerals, leaving quartz, which is the most stable and resistant, as the sole residual mineral. **Arkose** is a sandstone that contains much feldspar in addition to quartz. **Graywackes** are poorly sorted dark sandstones that contain much feldspar and sand-sized rock fragments of metamorphic or volcanic rocks. Because the angular grains of graywackes are set in an abundant clay matrix, the rocks are sometimes called "dirty" sandstones, a somewhat imprecise term.

Shale, sandstone, and limestone, the three most abundant sedimentary rock types, account for more than 95 percent of the total sedimentary part of the crust: shale accounts for about 70 percent, sandstone about 20 percent, and limestone about 10 percent of the total of the three.

The classification of sediments is based on mineralogy, texture, and chemical composition. The main criterion for subdivision of the detrital rocks is grain size, given as the diameter of the grain in millimeters. The boundaries are at 2 mm for gravel/sand, 1/16 mm for sand/silt, and 1/256 mm for silt/clay. The detritals are further grouped by their mineral content, mainly the relative amounts of quartz, feldspar, micas, and clay minerals. The chemical sediments are grouped by chemical composition, which is reflected in their mineralogy, into the **carbonates**, limestone and dolomite; **evaporites**, including chlorides (halite) and sulfates (gypsum); **cherts**, or **siliceous rocks**, containing much silica either as quartz or other varieties of $SiO_2$; **iron formations**, containing much iron in silicate, carbonate, or iron oxide minerals; **organics**, the organic carbon sediments including coal, gas, and oil; and **phosphates**, rocks containing phosphate as a variety of the mineral apatite or in clay minerals.

Many limestones and dolomites—those made up of pieces of shells and grains of carbonate minerals moved by currents—look much like detrital rocks, and they are classed according to grain size in much the same way as the sands and muds.

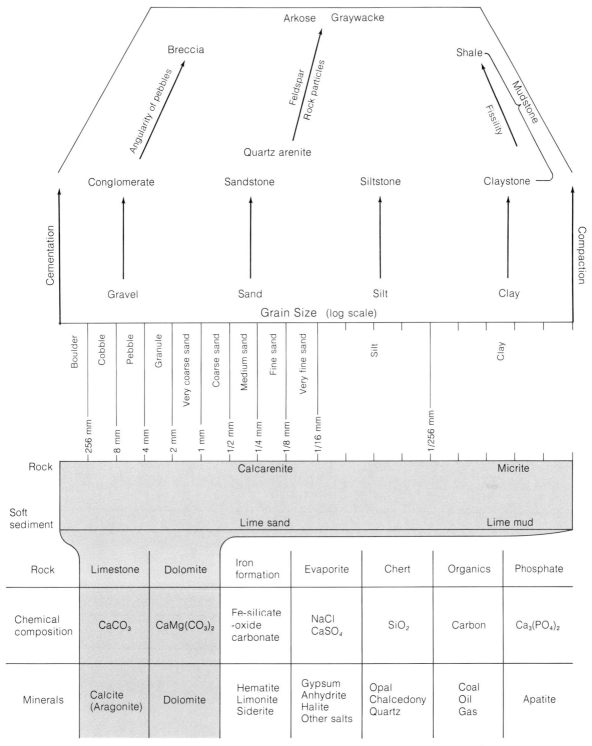

**Figure 4-28**
Sedimentary rock classification.

**The Metamorphic Rocks** Just as igneous rocks are divided
into intrusive and extrusive and sediments into detrital and chemical, so are the metamorphic rocks divided into two broad genetic
classes. They are either the result of **regional metamorphism**
on the one hand or **contact metamorphism** on the other. Regional
metamorphic rocks are produced by heat and pressure that transforms deeply buried rocks of all kinds, igneous, sedimentary and
metamorphic. Contact metamorphics are made by the alteration
of rocks near an igneous intrusion, largely from the heat but also
from pressure. The characteristic textures produced are the clues
to these two modes of origin. Most regional metamorphics show
**foliation**—a platy, wavy, or leafy structure imparted to the rock
by the parallel alignment of minerals, particularly the sheety ones
like mica. Some contact metamorphics are also foliated, but most
tend to be granular, such as the **hornfelses**, which are very fine-
grained silicate rocks of varied composition.

Foliation type and grain size are used, in combination, as the
basis for subdividing the metamorphics into **schist, slate, gneiss,**
and **granulite.** The schists are characterized by partings along
well-defined planes of medium-grained platy micaceous minerals.
The slates have more perfect planar partings and are finer grained,
so that individual minerals cannot be easily seen. The gneisses

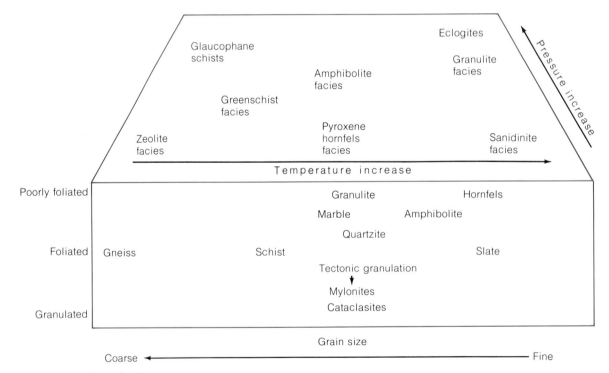

**Figure 4-29**
Metamorphic rock classification. The gradation in grain size, shown at front of diagram, is not necessarily correlated with pressure and temperature, shown on top of diagram.

are coarse grained and show much broader and less distinct foliation, and they do not split or cleave in the way schists and slates do. The planes of foliation of slates and schists are called **slaty cleavage**. Granulites are like their textural equivalents the granular igneous rocks, in having a mosaic of interlocking, more-or-less equidimensional crystals. They show only faint foliation, if any.

Within these textural groups, mineral assemblages form the basis for further dividing the rocks into smaller groups, or **facies**. The metamorphic facies have a genetic basis, for the minerals are determined by the temperatures and pressures required to form them. For example, **albite-epidote-amphibole schists** are the product of moderate temperature and pressure in regional metamorphism, whereas a **pyroxene hornfels** is the result of high temperature and moderate pressure in contact metamorphism. In normal field usage, metamorphic rocks are prefixed by the name of an abundant or prominent mineral constituent—for example, **mica schist**. Others are named for a mineral constituent that is greatly predominant, such as **amphibolite**. **Marbles**, metamorphosed limestones, are largely made of calcite. **Quartzites**, metamorphosed quartz arenites, are mainly quartz.

Some metamorphic rocks have characteristic textures produced by the crushing and mechanical deformation of grains as the rocks are folded and faulted. The broken, pulverized grain texture is called **cataclastic**, and fine-grained rocks produced by this kind of frictional action are called **mylonites**.

The diagram of metamorphic rocks shows the relationship between the different textural groups and the facies within them. Most facies can be found in a variety of textural types, except for the highest pressure and temperature groups, in which slates and phyllites are not preserved as such.

**The Information in Rocks** The characteristics of igneous, metamorphic, and sedimentary rocks can be used as a guide to identification (Fig. 4-30). Sedimentary rocks are identified as such mainly by their stratification, but also by their mineralogy and texture, as are igneous and metamorphic rocks. The identification of a metamorphic rock depends on recognition of foliation and other textures. In the same way, the earmarks of igneous rocks are their mineralogy, textures, and structures. Within the major groups, identifications are made on the same dual basis of mineralogy and texture.

From this brief summary of rock types, it should be apparent that mineralogy and texture are the major languages by which the information in rocks can be understood. The rounded quartz grains of a sandstone tell us about a history of erosion and transportation of fragments of a quartz-bearing igneous or metamorphic rock. The feldspar crystals in a granite inform us of the chemical composition of the magma and perhaps of the rate at which it cooled. The mica flakes of a schist are bits of information on the

**Figure 4-30** Rock identification flow chart.

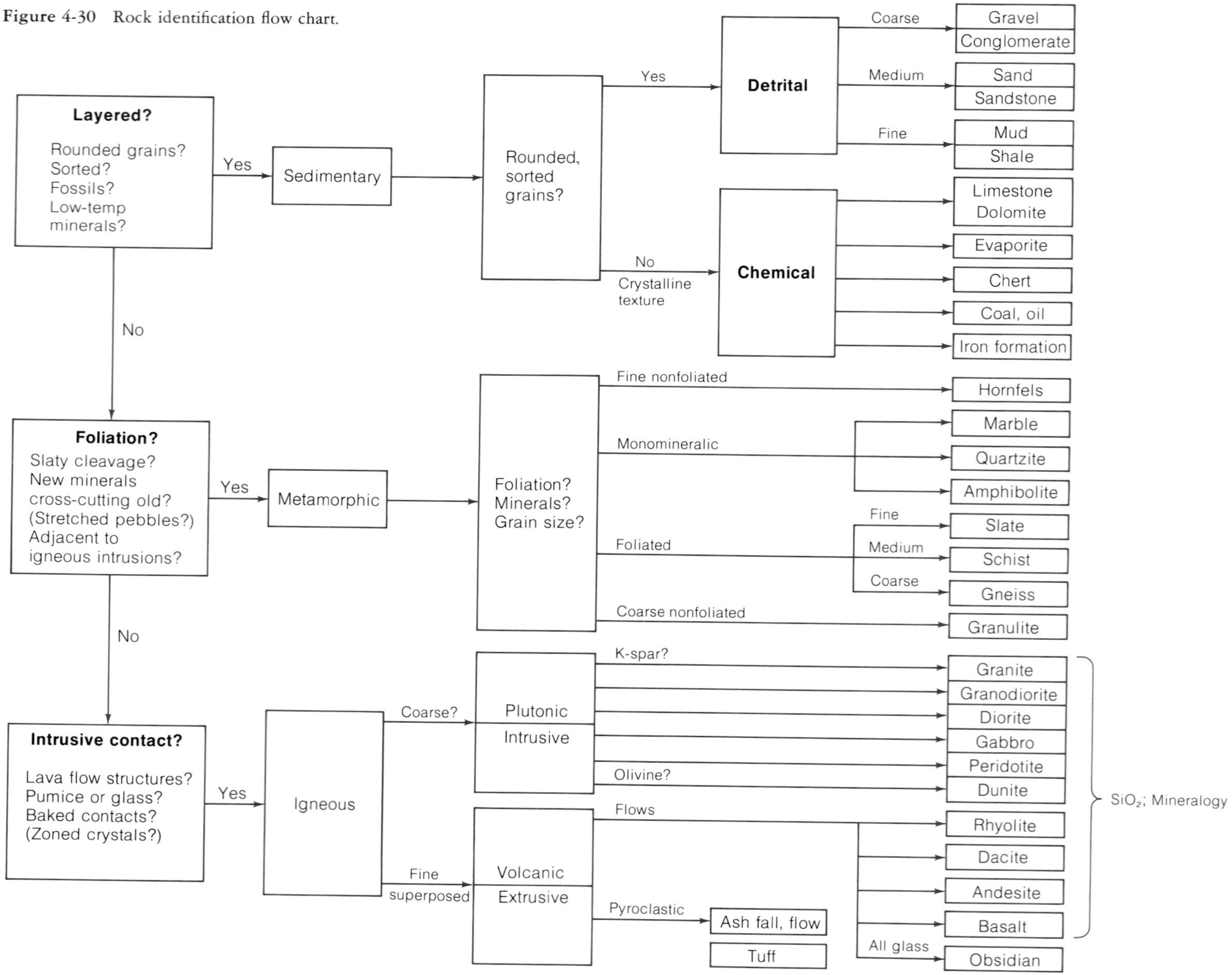

temperatures and pressures at which a shale was baked to make the schist. To get this information we need to know what the processes are by which they are made, where they operate in the Earth, and their relation to the larger architectural patterns of the Earth. It is these subjects that we move on to in the remainder of this book.

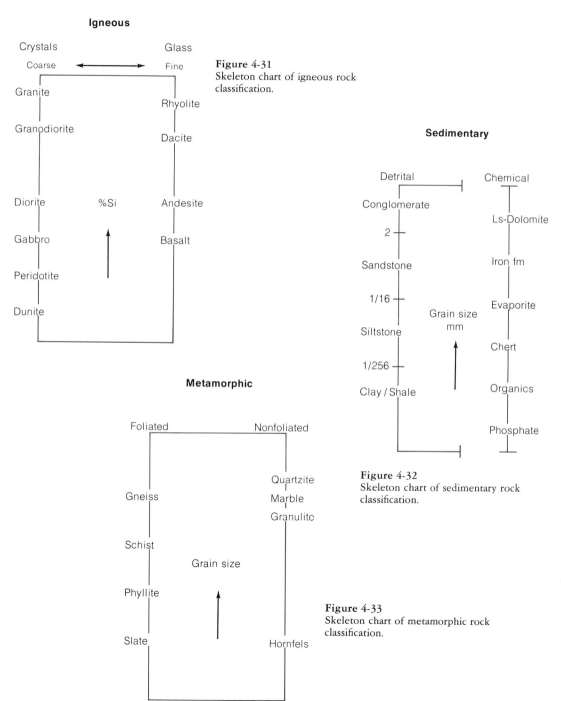

**Figure 4-31**
Skeleton chart of igneous rock classification.

**Figure 4-32**
Skeleton chart of sedimentary rock classification.

**Figure 4-33**
Skeleton chart of metamorphic rock classification.

1. Rocks, and the minerals that make them up, are the records of the geological events of the past and contain information that is used to infer their origin.

2. Crystals exhibit external regularities of shape, constancy of interfacial angles, and symmetry relations that allow them to be divided into thirty-two crystal classes distributed among seven general symmetry groups.

3. The optical properties of minerals—the distinctive ways in which they transmit light—can be analyzed by means of a polarizing microscope to identify them.

4. The atomic structure of minerals has become known largely as a result of X-ray diffraction studies, in which X-rays are reflected by planes of atoms in the crystal.

5. The electronic structure of the elements explains the different kinds of chemical bonding of atoms in minerals. Bonding may be ionic or covalent, depending on whether valence electrons are gained, lost, or shared. A weaker type of bonding is van der Waals, which does not depend on valence electrons being gained, lost, or shared.

6. The radius of an ion in relation to those of the ions to which it is bonded determines the coordination number and so the nature of the structures into which different ions can be organized.

7. Silicate mineral structures are made up of basic building blocks of silica tetrahedrons bonded to each other and to other ions to form isolated tetrahedrons, rings, chains, sheets, and three-dimensional frameworks.

8. Physical properties of minerals—hardness, cleavage, fracture, streak, luster, color, and density—are determined by the kinds of elements making up the minerals, their bond types, and crystal structures.

9. Chemical properties of minerals are determined by the kinds of elements present and their chemical bonding; minerals are classified on the basis of their anions.

10. Igneous rocks are divided on the basis of origin into intrusives and extrusives. The former are usually coarse grained; the latter, fine-grained or glassy. They are also classified as felsic or mafic, depending on the kinds and relative amounts of light and dark minerals. Volcanic rocks are further subdivided by texture.

11. Sedimentary rocks fall into two main groups, the detrital, those formed from particles eroded from pre-existing rocks, and the chemical, those formed by chemical precipitation of minerals from salt or fresh water. The detrital sediments are subdivided on the basis of grain size and mineralogy. The chemical sediments are subdivided primarily on the basis of chemical composition.

12. Metamorphic rocks are the products of either regional metamorphism, in which large volumes of rock are transformed by regional increases in pressure and temperature, or contact metamorphism, in which rocks close to igneous intrusions are transformed primarily by heat. Metamorphic rocks are classified by the nature of foliation or cleavage and mineral assemblages or facies.

1. Name several properties of minerals that can also be used to describe a rock made up of those minerals. How would you go about averaging the values of each mineral to get the value for the rock?

2. Noteworthy for their absence from the mineral kingdom are crystals with 5-fold, or pentagonal, symmetry. Describe and draw some symmetries possible in a "pentagonal system." How would you describe the reference crystal axes?

3. Draw the electronic structures of magnesium ion ($Mg^{++}$) and oxygen ion ($O^{--}$). The atomic number of magnesium is 12 and that of oxygen 8. How would these ions differ from sodium ion ($Na^+$) and fluoride ion ($F^-$), which have atomic numbers 11 and 9, respectively?

4. What property of what common mineral is the basis for sandpaper? For what kind of jobs might you prefer to use emery paper (emery is a granular variety of the mineral corundum ($Al_2O_3$).

5. Choose three minerals (other than diamond) from Appendix IV that you think might make good gemstones, and point out which physical properties might make them suitable for that purpose.

6. Name the properties of each of the following rocks, and describe the features that you would look for in order to identify them in outcrop: (1) a diabase dike, (2) a rhyolite flow, (3) a mica schist, (4) a conglomerate, (5) a hornfels, and (6) a marine shale.

7. In the text it was stated that no igneous rocks are composed of beds of halite and gypsum and that no sedimentary rocks are made up largely of pyroxene. Give an example of other combinations of minerals, or groups of minerals, together with their outcrop appearance, that would be incompatible with (1) igneous, (2) sedimentary, (3) metamorphic origins. Can you think of an example of a "forbidden" association of minerals and outcrop appearance — that is, a combination that could not be found in any natural rock?

## BIBLIOGRAPHY

Berry, L. G., and B. Mason, *Elements of Mineralogy*. San Francisco: W. H. Freeman and Company, 1968.

Bragg, Sir Lawrence, "X-Ray Crystallography." *Scientific American*, July, 1968. (Offprint No. 325.)

Ernst, W. G., *Earth Materials*. Englewood Cliffs, New Jersey: Prentice-Hall, Inc., 1969.

Hurlbut, C. S., Jr., *Minerals and Man*. New York: Random House, 1969.

Mott, Sir Nevill, "The Solid State." *Scientific American*, September, 1967.

Pauling, L., *College Chemistry*. San Francisco: W. H. Freeman and Company, 1964.

Simpson, B., *Rocks and Minerals*. New York: Pergamon Press, 1966.

Smith, C., "Materials." *Scientific American*, September, 1967.

Williams, H., F. J. Turner, and C. M. Gilbert, *Petrography*. San Francisco: W. H. Freeman and Company, 1954.

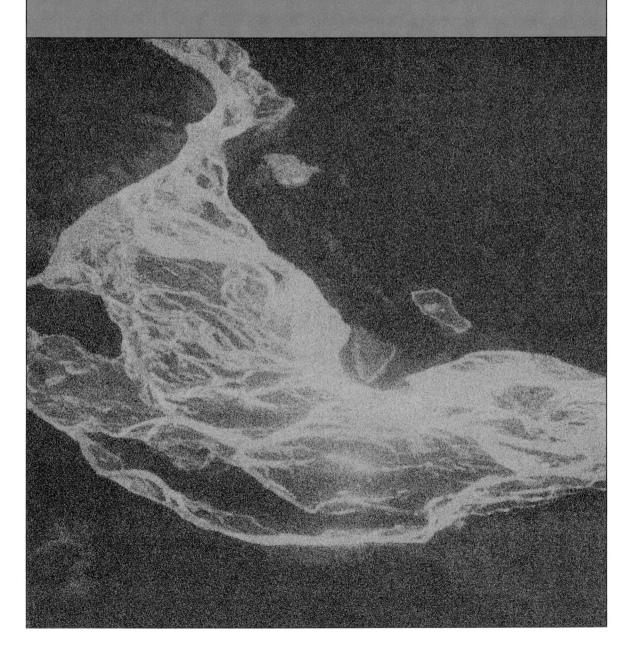

# Part II

# THE SKIN OF THE EARTH:
# SURFACE PROCESSES

The surface of the Earth is what we explore all of our lives. It is the interface between the solid planet and the gaseous envelope of air around it. Though we may go down into mines and fly through the air, it is the surface upon which we live and upon which we depend for our existence. The interface between the solid Earth and the vast bodies of water that cover much of it is being explored too; our curiosity about the sea floor has produced a crescendo of activity over the past two decades, but in satisfying our curiosity about this interface we must rely on instruments to do our seeing and feeling.

The dominant element of the dynamics of the surface is the Sun, which irradiates it. Solar energy drives the atmosphere in a complex pattern of winds to give us our climates and weather, and it drives the ocean's circulation in a pattern that is coupled to the atmosphere. The water and gases of the oceans and atmosphere chemically react with the solid surface and physically transport material from one place to another.

The processes that operate on the surface, then, are results of interactions between phenomena caused by the external heat engine and surface manifestations of the internal heat engine: mountains, volcanoes, and rocks brought up from the interior.

# 5

## The Face of the Earth: Landscape and Sea Floor

*The shape of the land surface of the Earth can be illustrated by three distinctive examples, the Grand Canyon, Yosemite Valley, and the Ozark Mountains. The forms and slopes of the valleys are clues to the agents of erosion that gave them their shapes—the water and ice that act in opposition to the tectonic forces that elevate mountains. The evolution of landscape is a balance between uplift and erosion over the course of time. The sea floor is shaped by volcanism and the construction of ridges, rises, and trenches created by the forces of plate tectonics and by the accumulation of sediments.*

## THE FORM OF THE SURFACE: HIGH AND LOW

What does the surface of the land look like from an orbiting satellite? Most prominent are the mountains and the canyons, the surfaces that depart most dramatically from the more average, undulating contours of the plains or lowlands that characterize much of the Earth's surface. Lakes and rivers fleck the continents, but it is the world-encircling ocean that makes Earth the blue planet. If that ocean were siphoned off, the sea floor would show some of its own kinds of mountains, plains, and valleys. Equally prominent mountains are hidden by the South Polar ice cap on the continent of Antarctica, and by the Greenland ice cap near the opposite pole. How do we describe this varied surface of Earth?

The heights and depths that give shape to the Earth's surface are called its **topography**. The standard level, or **datum plane**, to which we compare all heights or depths is usually **mean sea level**, the average between high and low tides. On land, vertical

The topography of the sea floor is also called bathymetry, from the Greek "bathos," for depth, and is measured in depth below sea level. Most oceanographers use the words topography and bathymetry interchangeably.

distances above mean sea level are called **elevations,** or **altitudes;** at sea, vertical distances below mean sea level are called **depths.** A **topographic map** shows the distribution of elevations in the area covered: in its two dimensions, the map gives a representation of the three-dimensional view we would get from looking down from the air. On most such maps this is done by means of

**Figure 5-1**
The varied aspects of continental topography are shown by this satellite photo of the Pacific Coast of South America at Antofagasta, northern Chile, looking across Chile toward Argentina. In the foreground, an irregular coast is dominated by mountainous terrain. Inland are plains and plateaus, and in the distance are the Andes Mountains. In the middle background are large basinal depressions filled with salt flats. [From National Aeronautics and Space Administration.]

contours—lines that connect points of equal elevation. Less commonly used in topographic maps for professional use, but often used in atlases and road maps, are **shadings**. **Hachures**, short lines that indicate slope, give a much more realistic but less precise idea of the land surface. Both contours and hachures bring out another measure of topography, its roughness, by illustrating the

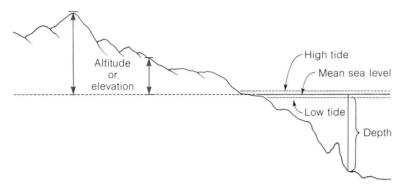

**Figure 5-2**
Altitude, or elevation, and depth below water are measured with respect to mean sea level, the average between high and low tides.

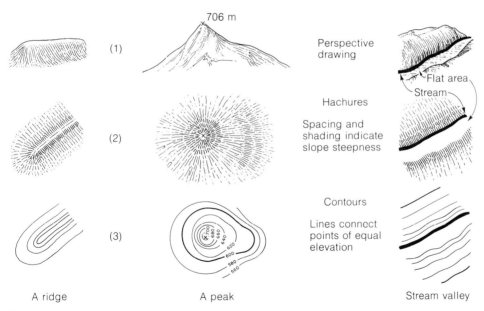

**Figure 5-3**
Three topographic features, a ridge, a peak, and a stream valley, as represented (1) by perspective drawings as they might look from the ground, (2) by hachures, short, straight lines that indicate the degree and direction of slopes as they might look from the air, and (3) by contours, lines that connect all points of equal elevation and thus most accurately represent the configuration of the surface. The more closely spaced the contour lines, the steeper the slope. Normally, every fifth line is heavier in weight and is numbered.

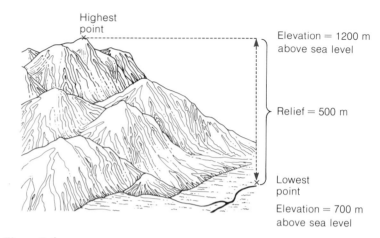

Highest point

Elevation = 1200 m above sea level

Relief = 500 m

Lowest point

Elevation = 700 m above sea level

**Figure 5-4**
Relief is the difference between the highest and lowest elevations in a region.

**relief**—the vertical distance between the highest and lowest points in the map area. One estimates relief on a contour map by subtracting the elevation of the lowest contour, down in a river bottom, from the highest, on a hill or mountain top. Climbing a mountain is the best way to experience what relief means. It is one matter to climb to a 3600-meter (12,000-foot) peak from a high plateau of 3000 meters (10,000 feet) but quite another to climb a 2400-meter (8000-foot) peak starting from a few hundred meters above sea level.

## THREE SAMPLES OF TOPOGRAPHY

There is much more to landscape than the grossest features seen from the great distance of a satellite. The infinite variety of form and detail that we look upon as scenery holds much of the information we need to deduce how it was made. Again we "visit" the Grand Canyon, this time to explore the natural architecture of its surface.

**The Grand Canyon Landscape**  The canyon is deep, broad, rocky, and ruggedly mountainous. The Colorado River can rarely be seen from the rim, though one can make out the edge of the river gorge in which it has been cutting its way downward for the past 15 to 20 million years. Steep cliffs compete with the more gentle rubbly slopes for our attention. What we see are the effects of **erosion**, the breaking up and removal of rock by the physical and chemical processes that nature has at its disposal. Although the river has removed enormous quantities of rock, we can easily trace such formations as the cliff-forming Kaibab Limestone from

**Figure 5-5**
Grand Canyon near mouth of Tuckup Canyon, showing the rugged topography
of the canyon and the steep rim of the inner gorge, here cut in the Cambrian
formations. [Photo by Edwin D. McKee, U.S. Geological Survey.]

one rim to the other. Once a continuous sheet, as it still is under
the plateau back from the Canyon, it has been breached, broken
up, and taken away.

Merely by sampling its water, one can demonstrate the power
of the Colorado River as an agent of erosion. Colorado River
water, as those engineers in charge of purifying it for drinking
well know, is turbid with suspended fine clay. Sand in the river
channel is constantly in movement. A tremendous volume of sand,
silt, and clay is being transported all the time. One measure of
this volume is the amount of this sediment that is dropped by the
river as it enters Lake Mead, created by the building of Hoover
Dam. Since the completion of the dam in 1936, many millions of
tons of sediment have accumulated on the lake bottom. A close
look at this sediment shows sand grains of the kind found in the
Coconino and other sandstone formations, mica flakes that could
have come from the Vishnu Schist, and small pebbles of almost
every kind of rock found in the Canyon. It is clear that the river
is a prime agent of removing rock material from the Canyon.

Now focus on the topography of the canyon. In Chapter 2, in

One unhappy consequence of the
erosional power of rivers is the silting
up of reservoirs. Sediment that would
normally be carried downstream is
deposited in the artificial lake formed
by a dam. Some reservoirs are filling
with sediment at so great a rate that
the useful life of the dam for flood
control and irrigation purposes is
shortened to a few decades, hardly
enough to have made such expensive
projects worthwhile.

**Figure 5-6**
The varied nature of the slopes of the Grand Canyon are expressions of the
resistance of the different kinds of rock to erosion; the steeper cliffs are formed
in sandstones and limestones and the gentler, rubbley slopes formed on shale.
[Photo by Edwin D. McKee, U.S. Geological Survey.]

summarizing the stratigraphic sequence, we used the nature of
slopes and cliffs to pick out formations. Steep cliffs are made of
hard rocks that are **resistant** to erosion. In the Grand Canyon these
rocks are limestones, and to a lesser extent sandstones. The gradual slopes, usually covered with a jumbled mass of broken rock
that we call **talus**, are underlain by shales. This differential resis-

tance to erosion shown by the various rock types is one of the major controls on topography. Cliffs are formed as soft, less resistant, weak rocks erode and undercut overlying resistant rocks, which then break off in larger blocks.

The valley of the Grand Canyon is much broader than the River itself. In the river gorge it is several times wider at the top than at the bottom, and a cut through it would show a simple **V-shaped** profile. Above the inner gorge, in the main Canyon valley, the profile is much more complex, but can be generalized to a much broader V-shaped valley. As the river cuts downward, the valley gradually widens. The metamorphic rocks of the inner gorge are much more resistant to erosion than the weaker Paleozoic rocks above and so support narrower canyon walls. Above, formations like the Bright Angel erode rapidly and broaden the valley.

The Grand Canyon teaches us about the nature of erosion of a river valley in terrain where most of the rocks are horizontal sediments. Another part of the story emerges when we take up another example, the Yosemite Valley and the Sierra Nevada.

Figure 5-7
The inner gorge of the Colorado River in the eastern Grand Canyon shows the characteristic V-shape of a river valley actively eroding its channel. The rocks here are Precambrian. [Photo by Edwin D. McKee, U.S. Geological Survey.]

**Figure 5-8**
The eastern slope of the Sierra Nevada, a great fault escarpment nearly 2 miles (3.2 kilometers) high. The view is from the Owens Valley, Inyo County, California. [Photo by W. C. Mendenhall, U.S. Geological Survey.]

**Figure 5-9**
View down Yosemite Valley from Glacier Point Trail above Union Point. At right is El Capitan. The U-shape of the valley is typical of glacially eroded terrain. [Photo by F. E. Matthes, U.S. Geological Survey.]

**The Sierra Nevada and Yosemite** The greatest relief of the Sierra Nevada can be seen along the eastern slope, where the front of the range abruptly drops as much as 2700 to 3000 meters (9000 to 10,000 feet) to the floor of the Great Basin of eastern California and western Nevada. **Basins** are lowlands between mountain ranges, sometimes called **intermontane basins**.

The steep Sierra slope coincides with a major fault along which the mountains on the western block have been raised relative to the basin on the eastern block. The association of fault and topography immediately suggests the kind of causal relation between deformational forces and mountain-building that Hutton perceived in the Scottish highlands. It is the tectonic forces operating within the Earth that elevate mountains.

Erosion creates landforms in the granite-walled valley of the Yosemite that are different from those in the Grand Canyon. The profile of Yosemite is that of a **U-shaped valley** with steep walls and flat floor. Another difference is that the wide part of Yosemite valley comes to an end near Half Dome on the east, whereas the Colorado Gorge maintains a fairly narrow and uniform width for hundreds of miles. The Merced River, which flows along the floor of Yosemite Valley, is a relatively modest stream that seems disproportionately small in comparison with the width of the valley. In contrast, the Colorado is a vigorous, major river that seems quite capable of having cut its own gorge.

The tributaries of the Colorado, like those of almost all rivers, enter the main stream at the same elevation, a not surprising generality first enunciated by John Playfair, a contemporary of Hutton's who did much to popularize his ideas. Put another way, tributaries do not ordinarily enter their rivers by waterfalls. But in Yosemite some do. Bridal Veil Falls plunges from what is called a **hanging valley** — a tributary (Bridal Veil Creek) that joins the main stream (the Merced River) at an extreme difference in elevation.

What does this catalogue of differences mean? It seems strange that the erosive action of rivers would exhibit such differences in Yosemite. Scientists do not like to look upon such differences merely as "exceptions that prove the rule." It is more likely that a different agent of erosion was responsible for Yosemite. That agent is identified by comparing the topography of Yosemite with similar topography in Alaska and the Alps, where solid rock is sculptured by rivers of ice — **glaciers**. Yosemite, it turns out, was excavated in part by a **valley glacier** that flowed downhill on the western slope of the Sierras, heading up near Half Dome. The mechanisms by which slowly creeping ice masses erode rock — by scouring, scraping, and breaking it off in blocks — result in U-shaped valleys. Because the erosive action of glaciers is related to the ice thickness, the major glacier will cut down deeper than the thinner tributary glaciers, leaving the hanging valleys when the ice melts after a change in climate. The Merced River did not make the valley by itself. It just occupies it now.

**Figure 5-10**
The hanging valley of Bridal Veil Creek, which empties into the main Yosemite
Valley by Bridal Veil Falls. The valley of the creek has not been strongly
affected by glacial erosion and retains much of its original V-shape. [Photo by
F. E. Matthes, U.S. Geological Survey.]

What we learn from the Yosemite adds more to our deductions
about how landscape forms. Tectonics makes the mountains that
erosion wears away. The forms that erosion produces can be related
to the agent of erosion, those produced by ice differing from those
produced by water. In contrasting ice with water, we imply another determining factor: the temperature. Climate, after all, was
what made the glacial ice that was responsible for the magnificent
sculpture of Yosemite.

**The Ozarks**   Among other things, the Grand Canyon and Yosemite have in common their youth, for both were developed relatively
recently (Yosemite much more recently than the Grand Canyon).
Age is always such an important factor in geology that we ought
to look at old mountains. One example in the middle of the continent, nowhere near the Rockies in the West or the Appalachians
in the East, is the Ozark Mountains (sometimes called a plateau).

The Ozarks, situated in Missouri and Arkansas, are partly granite, like the Sierra Nevada, but elsewhere are composed of relatively flat-lying sedimentary rocks of many of the same lithologies as those in the Grand Canyon. The Ozarks, not in the same league with the Rockies in height or relief, nevertheless include some fairly hilly topography.

The slopes of the Ozarks are gentler and more rounded than the Grand Canyon or Yosemite. The valleys are broad but V-shaped, and there is no sign of glacial activity. There are few steep cliffs, and the streams meander along fairly wide valley bottoms that are called **floodplains**, from the frequency with which the rivers flood over their banks and occupy the entire width. The dominant aspect is one of less vigorous erosion; the streams carry much less detritus than either the Colorado or mountain streams that carry immense loads of rock debris down the steep eastern slopes of the Sierra Nevada that stand so high above the Great Basin.

The granite of the Ozarks is Precambrian, about a billion years older than the granite of the Sierra Nevada. The sedimentary rocks surrounding the granite are early Paleozoic. The lowest

Figure 5-11
View overlooking Taum Sauk Mountain in the Ozark Mountains of Missouri. The gently rounded hillsides are typical of old mountains that have long been eroded. [Photo courtesy of Missouri Tourism Commission.]

**Figure 5-12**
The wide flood plain of the Tennessee River in the Appalachian Mountains. In older, deeply eroded landscapes such as this, the rivers widen their valleys, and floor plains become extensive. [Photo by A. Keith, U.S. Geological Survey.]

sedimentary rock formation lying just above the granite on the flanks of the central area, a Cambrian sandstone, is an arkose, a feldspar-rich sandstone that was formed from the disintegration of the granite. The coarseness and abundance of the feldspar grains is a sign that the Ozarks were still rapidly eroding granite hills as the first Paleozoic seas lapped up around them. The Ozarks were later lowered further and covered by those seas, for there are sedimentary rocks found on them. Later the mountains were exhumed as the soft sediments were stripped away in another, later episode of erosion, at the end of the Paleozoic.

Wherever we find older mountains, we see low elevations, low relief, and gentler slopes and broader river valleys. They are also all tectonically "quiet." They are rarely the sites of earthquakes or volcanoes, nor do they show any evidence of recent rock deformation or metamorphism. The conclusion: once the tectonic machine that created the mountains in a region has run down, erosion operating over a long time will wear down mountains to low hills.

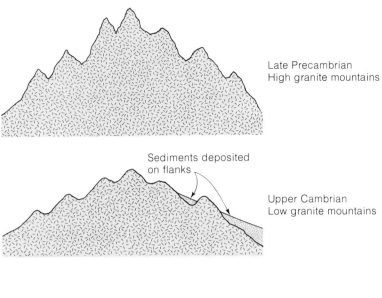

Late Precambrian
High granite mountains

Sediments deposited
on flanks

Upper Cambrian
Low granite mountains

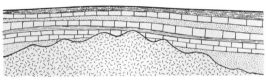

Mid-Paleozoic
Granite hills buried
by sediment

Present
Granite hills exhumed
as erosion stripped
away most sediment

**Figure 5-13**
Evolution of the Ozark mountains. (1) In the Precambrian, granite was intruded
during a mountain-building episode and then exposed by erosion. (2) By Upper
Cambrian time, the mountains were extensively eroded and sediments were
being deposited on their flanks. (3) During the Paleozoic, the granite was buried
by a series of sediments. (4) From the Paleozoic to the present, erosion stripped
away the sediments, exhuming the ancient topography.

## THE OVERALL PATTERN

We might put it that "tectonics proposes and erosion disposes."
In this chapter and those that follow, we will see how the mechanisms display an overall pattern that we can call a **negative-feedback** process. This kind of process is one in which the results of
a first or primary action or operation induce a proportionate secondary action that reduces the first action. A familiar example is
what happens when a thirsty person drinks a glass of water. The
more thirsty the person, the faster the gulping at first. Just as the
thirst induces the drinking, so the drinking acts to decrease the

168

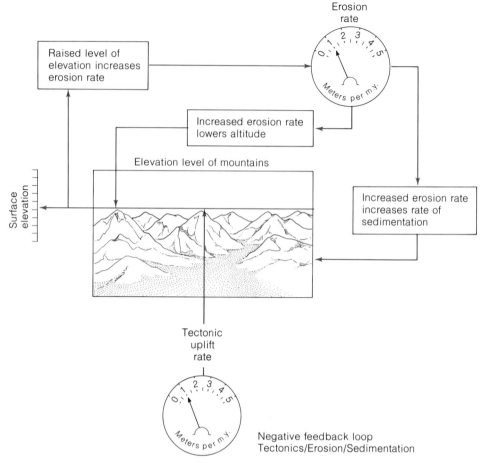

**Figure 5-14**
The negative-feedback loop that relates uplift, erosion, surface elevation, and sedimentation. Tectonic uplift contributes to increased erosion rates, which in turn lowers the surface elevation and increases the sedimentation rate. The elevation is thus a balance between tectonic uplift and erosion rate.

thirst. As the thirst decreases, the rate of drinking decreases; when thirst is satisfied, drinking stops.

The general pattern is one in which strong tectonic action elevates mountains, provoking intense erosion. The higher the mountains, the faster the rate at which erosion wears them down, but tectonics prevails and altitudes increase. As tectonic movement slows down—not because of any important effect of erosion but because its own machine starts running down—the mountains keep going up at a slower pace and for a time erosion keeps up with uplift and the mountains do not change in elevation. Then as uplift slows, erosion becomes dominant and the elevations begin to lower. As the lowering proceeds, the erosion slows too, the whole process eventually tapering off. The relief is constantly lowered by wearing away of the mountain tops and filling in of valleys and low spots by sedimentation of the erosional debris. Sedimentation, the consequence of erosion, acts to depress relief.

**The World Distribution of Elevations** The first impression from the satellite photographs was that most of the Earth's surface is flat. To check this impression we can draw from accurate maps of Earth's topography a **hypsometric diagram**, a representation of the relative proportion of the surface lying at each elevation, by plotting a graph of elevation against the total area lying at that elevation. From these diagrams we can see that the maximum height of mountains above sea level, about 8900 meters (29,000 feet), is not much different from the maximum depth of the deepest ocean trenches below sea level, about 10,800 meters (35,000 feet). In comparison to the wide areas of both land and sea floor that are relatively close to sea level, such extremes are few. The great bulk of the continental land surfaces lie within a few hundred meters of sea level, and much of the ocean floor lies at intermediate depths. Put this together with the relation of erosion to uplift, and there is a strong suggestion that on the continents the tendency is to reduce the land surface to a little above sea level. On the sea bottom the tendency is to maintain some intermediate depth below sea level; this tendency must represent some balance of forces acting on the sea floors other than erosion coupled to uplift. Another inference that comes from this diagram is that the recent

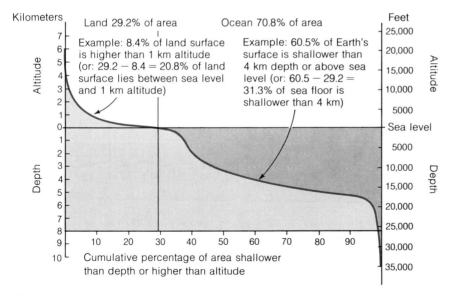

**Figure 5-15**
Hypsometric diagram of the surface of the Earth. The curve shows the relative amounts of land and subsurface areas lying at various elevations with respect to sea level by a plot of cumulative percent of area shallower than a given depth or higher than a given altitude.

tectonic activity that has caused great changes in elevation on land or in the sea, such as mountain ranges or trenches, is distributed in relatively narrow, restricted belts, a small proportion of the total surface. Those belts are the edges of lithospheric plates of the present and recent past. The face of the Earth is scarred by plate margins. The older plate boundaries of the geologic past are now healed, and largely obliterated by later erosion and sedimentation, but here and there, in places like the Appalachian Mountains, the old scars show through.

## LANDFORMS: THE PHYSIOGNOMY OF THE EARTH

The shapes that give variety to Earth's landscape have long been subjects of interest to artists and scientists, both of whom may look with analytical eyes. In his poem "New Hampshire," Robert Frost says:

> *The Vermont mountains stretch extending straight;*
> *New Hampshire mountains curl up in a coil.*

Even though he did not use technical terms, Frost was being a **geomorphologist**, a student of the morphology (form) of the Earth. Poets, however, are free to use words in ways that scientists are not. A poet's aim is not necessarily unambiguous description, nor is a poet required to use language with precision. Those who understand poetry know this. Scientists, however, must be able to communicate with each other with as much precision as possible, hence the language of science is one of description and analysis, and much of the vocabulary is of the scientist's invention. Most of the names of landforms, however, come from the common language. We categorize landforms according to their elevation relative to that of the surrounding country. A hill is a **positive elevation**, and a valley is a **negative elevation**. Broad regions may be referred to as highlands or lowlands.

**Positive Landforms: From Mountains to Molehills**  As it turns out, even some of the most ordinary names of landforms are not easily defined with precision. We have used the word **mountain** so many times in this book, yet we can define it no more precisely than to say that a mountain is a large mass of rock that projects well above its surroundings. A mountain can be a single peak, like the Devil's Tower, which is the preserved neck of a former volcano (see Chapter 17), or a combination of peaks, as in most high mountain ranges, where it is hard to pick out distinct separate mountains but only peaks of various heights (see Fig. 22-49). How do we distinguish mountains from **hills**, other than by size? We do not. The gamut of size is continuous, from a 5-meter (16-foot) knoll good for playing "king-of-the-hill" to Mt. Everest, elevation

9524 meters (29,028 feet); precise limits are only arbitrary. We learn to accept that the Black "Hills" are higher than some Appalachian "Mountains." In general, positive landforms more than a few hundred meters in elevation are called mountains.

A broad, flat area of appreciable elevation above its surroundings, at least on one side, is a **plateau**. Smaller plateaus are sometimes called **table lands**, and in the western part of the United States, some are called **mesas. Butte**, a French word for hill, is used, as is "mesa," for smaller elevations with flat or rounded tops surrounded by steep slopes on all sides. Few plateaus have elevations much higher than 3000 meters (10,000 feet). Something in the nature of erosion produces peaked mountains when they get higher than that. Many plateaus are relatively flat because they are floored with undeformed sediments or layers of lava flows (see Chapter 17 for a discussion of plateau basalts).

Though all positive elevations are reflections of upward tectonic or magmatic movements, their forms are dictated mainly by the erosional process, the kind of rock, and its structure. As in the Grand Canyon, shales almost always form gentle slopes, much of the bedrock being completely covered either by a rock rubble or a soil, the difference depending upon climate and vegetation. Limestones are resistant to erosion in dry climates and tend to form steep cliffs. But in the wet tropics, limestone erodes quickly, mainly by dissolving in water (see Chapter 6), and even small outcrops may be hard to find, much less steep cliffs. From these comparisons it is apparent that an important variable of landscape is the nature of the hillslopes. Are they steep or gradual, rock

Figure 5-16
Mesas in Monument Valley, Navajo County, Arizona. [Photo by I. J. Witkind, U.S. Geological Survey.]

**Figure 5-17**
An aerial view of Crown Butte. An igneous rock resistant to erosion, a sill, caps
the more easily eroded sandstone that makes up the main body of the butte.
[Photo by J. R. Balsley, U.S. Geological Survey.]

covered or mantled with soil, convex or concave? A little further
on in this chapter we will discuss how slopes can be used as important criteria for understanding landscape evolution. The relation
of slopes to rock structure we discuss here.

**Structural Control of Topography**  The folds and faults produced by rock deformation in the course of mountain building
leave their mark on the configuration of the Earth's surface. These
topographic expressions of deformation are often a guide to the
structures that control them. Even the shapes of features of somewhat smaller scale, such as hills, valleys, and stream courses, are
controlled by a complex interaction of erosion and structural
elements.

Mesas, buttes, and the complex pattern of erosion of the Grand
Canyon are forms that develop in horizontal, undeformed sedimentary rocks or lava beds with alternately resistant and easily
eroded layers. The resistant layers support the elevated flat-topped
hills and make the cliffs; the softer layers below may erode to form
gentler slopes or undercut the capping strata. In horizontal beds
of uniform resistance, the topography is much more homogeneous,
all the hills and valleys having about the same slope.

In a tilted and eroded series of alternating resistant and soft
beds, asymmetrical ridges, or **cuestas**, tend to form. One side of
a cuesta has a long, gentle slope determined by the angle of inclination of the resistant bed; the other side is a steep escarpment
formed at the erosional edge of the resistant bed, where it is undercut by erosion of a weaker bed beneath. Much more steeply inclined or vertical resistant beds erode to form **hogbacks**, narrow
steep ridges of about the same shape as cuestas but more accen-

tuated. Escarpments are also produced by near-vertical faults along which one side has been raised relative to the other. Faults can displace topography in such a way that the sides of hills appear sheared off as by a knife, river valleys show sudden detours, and ridges are offset (see Fig. 22-59).

**Figure 5-18**
Cuestas formed by gently inclined beds of hard sandstone, New Mexico. [Photo by W. T. Lee, U.S. Geological Survey.]

**Figure 5-19**
Aerial view of the eastern edge of the Front Range of the Colorado Rockies, showing the prominent hogback of the steeply dipping Dakota Sandstone in the center and the lower hogback of the Niobrara Limestone to the right. [Photo by T. S. Lovering, U.S. Geological Survey.]

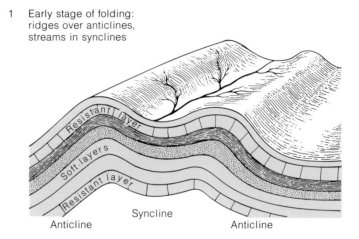

1 Early stage of folding:
ridges over anticlines,
streams in synclines

Resistant layer

Soft layers

Resistant layer

Anticline    Syncline    Anticline

2 Later stages of erosion:
ridges may overlie synclinal axes
if capped by resistant beds

Resistant

Soft

**Figure 5-20**
Two stages in the development of ridges and valleys in folded
mountains. The original positions of ridges may be determined by
the shape of the structure, as in (1). At later stages, however, as in
(2), anticlines may be breached and eroded more rapidly through nonresistant rocks, and ridges may be held up by caps of resistant rocks.

In the early stages of folding, the upfolds form ridges, or **anti-**
**clines**, and the downfolds form valleys, or **synclines**. Streams flow
down the flanks of the anticlines and follow the course of the
synclinal troughs. As continuing erosion bites deeper into the
structure, the presence of resistant and soft beds becomes a controlling factor. For example, if the core of an anticline is soft, it may
be eroded away to form an anticlinal valley. If further erosion
reaches a hard layer, an anticlinal mountain may once again form.
In a region that has long been eroded, a pattern of linear anticlines
and synclines produces a series of ridges and valleys, such as those
of the Appalachian Mountains in Pennsylvania and adjacent states.
The positions of the ridges versus valleys are determined by the
erosional resistance of the rocks rather than the original anticlinal
highs or synclinal lows formed by structural deformation.

**Figure 5-21**
Valley and ridge topography formed on a folded terrane of sedimentary rock. The deformation is so recent (Pliocene) that erosion has not yet modified significantly the original structural forms of anticlines (ridges) and synclines (valleys). Circular dark areas in foreground are salt domes formed when deeper salt formations were squeezed upward through overlying sediments as cylindrical masses. The scene is in Zagros Mountains, Iran, looking west over the Persian Gulf. [From National Aeronautics and Space Administration.]

**Elevations Above the Sea Floor** Mountains on the sea floor, even though they may be as large and high in relief as those on land, can only be made out from a shipboard instrument, the **echo-sounder**. This instrument, sometimes called a **fathometer,** * or **depth-recorder**, works by sending sound waves from the ship to

---

*From "fathom," the traditional nautical measure of depth, equal to 6 feet.

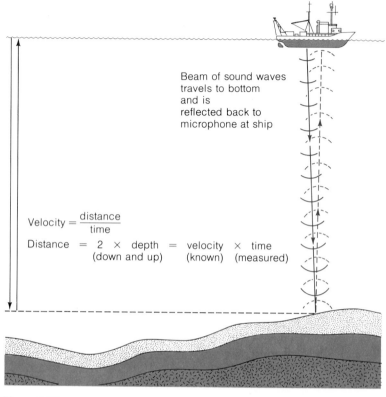

Beam of sound waves
travels to bottom
and is
reflected back to
microphone at ship

$$\text{Velocity} = \frac{\text{distance}}{\text{time}}$$

Distance = 2 × depth = velocity × time
(down and up) (known) (measured)

**Figure 5-22**
Echo sounders sense underwater topography by beaming sound waves to the bottom and measuring the time required for the beam to be reflected back to the ship.

the bottom. The sound waves reflected from the bottom are picked up by sensitive detectors on the ship. From the measured delay in time and the known speed of sound in water, oceanographers can determine the depth. As the ship cruises along a straight course, a two-dimensional profile of the topography is traced out automatically; from many such profiles a bathymetric map is constructed. A limitation of the accuracy of such maps is the assumption that the places between widely spaced profiles have something like the same shapes and forms that we are used to seeing in three-dimensions on land. This limitation is partly overcome by a series of very closely spaced profiles in a regular grid. One view of ocean-bottom topography drawn on this basis is given in Figure 5-25, a section of a map prepared by geologists at the Woods Hole Oceanographic Institution. In the last few years submarine geologists have used new instruments, such as echo-sounders towed along the bottom, to get a better idea of the topography.

High mountain chains rise from the ocean floors, the most famous being the mid-Atlantic **ridge**. A broader, lower feature that

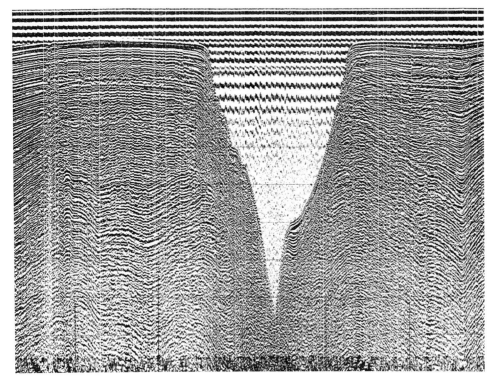

**Figure 5-23**
An echo-sounding profile of the Congo submarine canyon off the west coast of Africa (Republic of Zaire). The bottom of the canyon at this point is about 3000 meters below the sea floor of the continental shelf; at the top, the canyon is more than 10 kilometers wide. The heavy black horizontal lines in the water at the top of the recording are artifacts of the process. The wavy lines below the sea-floor surface are sound reflections from bedding planes in continental shelf sediments, somewhat deformed because of mild tectonic disturbance. [From K. O. Emery, Woods Hole Oceanographic Institution.]

arches upward from the sea floor in the southeastern Pacific Ocean is the East Pacific **rise**. The terms "ridge" and "rise" are used in preference to "mountain" for these large topographic units. Some mountain tops of ridges protrude above the ocean surface, forming islands like Iceland and the Azores. **Seamounts**, isolated mountains, most of volcanic origin, dot the ocean floors. Many seamounts have eroded, flat tops; such seamounts are called **guyots** (named after the first professor of geology at Princeton University). Great areas of the ocean floors are covered with small hills several tens to hundreds of meters high. These **abyssal hills** vary in shape, size, and origin. Some consist of volcanic rock thinly mantled with sediment; others are mostly sedimentary. Their modes of origin remain mysteries.

The contrast between submarine and land topography is largely an expression of the relative inefficiency of erosion under the sea.

**Figure 5-24**

An artist's representation of the floor of the North Atlantic Ocean based on bathymetric studies of Bruce C. Heezen and Marie Tharp of the Lamont-Doherty Geological Observatory. Depths shown in feet below sea level. [Painted by Heinrich C. Berann; copyright National Geographic Society.]

Water currents do run in places on the bottoms of the oceans, but they pale in significance when compared to the eroding power of a mountain stream on land. So even though there are talus slopes of broken rock and erosional valleys, the dominant aspect is the constructional one: ridges pushed up by rising convection currents from the mantle, seamounts formed by volcanoes, and abyssal hills produced by a combination of lava and sediment.

The largest constructional topographic features of the ocean floor are the shallowest, the aprons of the continents. A **continental shelf** is the gently sloping marginal area that extends from most continental regions from sea level—the shoreline—to a depth of about 550 meters (1800 feet). It is largely built of sediment from the continents. At the edge of the shelf there is an abrupt pitch off into greater depths. The slope at the edge of the shelf is called the **continental slope**, and though it is usually shown on maps or diagrams as being very steep, it pitches at an angle of only a few degrees. At the foot of the slope is the **continental rise**, another gently sloping apron of sediment that merges at its outer edges with the floor of the deep ocean.

Exaggeration is a favorite device of explanation, and we use it constantly in our topographic and geologic profiles and the imaginary cuts through the Earth that we call **cross sections**. We use **vertical exaggeration** to accentuate differences in elevation by

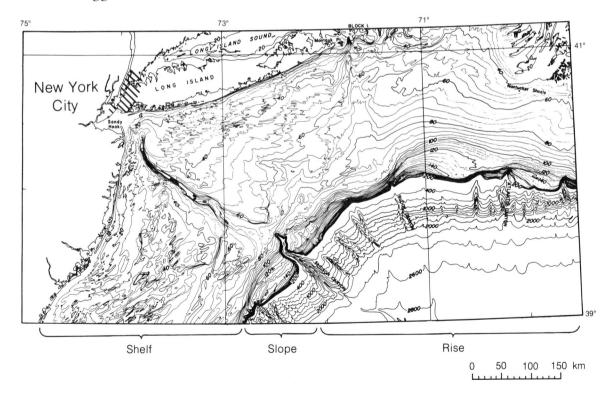

**Figure 5-25**
Map of continental shelf, slope, rise, and Hudson Canyon off New York City. [From K. O. Emery, Woods Hole Oceanographic Institution.]

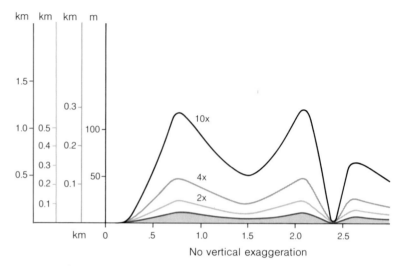

**Figure 5-26**
Vertical exaggeration is a device for revealing details of topography. Each profile
is matched with the appropriate vertical scale. [After D. E. Easterbrook,
*Principles of Geomorphology*, McGraw-Hill Book Company, 1969.]

stretching out the scale of vertical representation compared to the
horizontal. A true (unexaggerated) drawing of topography shows
little obvious detail and contrast. Just as an artist may distort and
exaggerate to give a more meaningful though less representational
view of a subject, so does the geologist, but in a more formal and
uniform way. And just as artistic distortion can be excessive, so
too can vertical exaggeration, hence it is used with restraint.

**Negative Elevations: The Low Spots**   What is the opposite of
a positive elevation? Is it a lowland plain or a valley? Those two
different shapes, broad versus narrow in area, correspond to two
ways that negative elevations form. A **plain** is a relatively flat low
area that is not being pushed up by tectonic activity. We can generalize further that **lowlands** are broad flat areas of two kinds. They
are either **stable**, neither moving up nor down, or they are unstable,
truly the opposite of mountains. Unstable lowlands are moving
down, a tectonic process we call **subsidence**. **Valleys, canyons,
arroyos, gulches,** and **gullies** all have the same general kind of
geometry, and all are topographic lows that form in mountains and
plains, highlands and lowlands. As we learned from the Grand
Canyon — and as can be seen clearly by watching a stream in flood
stage eroding its banks — valleys are excavated by streams; most
of these excavations are made by water, some, as in Yosemite,
by ice. These low spots are smaller than the large land areas of
plains and lowlands. A closer look at lowlands shows that much
of their large area is made up of smaller, broad, flat river valleys.
Thus valleys and plains are distinguished on the bases of scale as
well as process.

Valleys can be steep-sided or bounded by broad, low slopes. They can be shallow, like the little gullies in a field, or as deep as the Grand Canyon. A special kind of valley—the tectonic valley—is long, narrow, relatively flat-floored, and bounded on both rims by faults. Some of these valleys are occupied by major rivers, the upper Rhine being one of the most famous. The Great Valley of California, the route of the San Joaquin and Sacramento rivers, is also a tectonic valley. A distinctive variant of the tectonic valley is the **rift valley**, formed by incipient or active spreading apart of lithospheric plates. The great African rift valleys are occupied by large lakes, such as Lake Rudolph and Lake Tanganyika. The River Jordan and the Dead Sea occupy another rift valley. Beneath the oceans, great rift valleys form the axes of mid-ocean ridges. In fact, it was the discovery of the mid-ocean rift system in the late 1950's that sparked much of the exploration of sea-floor ridges that led to the concepts of sea-floor spreading and plate tectonics. It was topography that gave the first clue.

A walk around the Mammoth Cave district in Kentucky would convince you of the existence of an entirely different kind of negative topography, for in walking across a meadow you might come upon a deep, round hole, a **sinkhole**. Some of such holes are filled

A great valley on Mars is about ten times as long and three times as deep as the Grand Canyon. It runs from west to east along the Martian Equator, covering nearly 4000 km (2500 mi); it is strong geomorphic evidence that erosion by water has taken place on the planet according to the interpretations of some. Others think it is of more complex origin.

**Figure 5-27**
Large sinkhole, about 130 by 100 by 45 m (425 × 350 × 150 ft) that formed in 1972 in central Alabama as a result of collapse of cavernous limestone below the surface. This may be the largest sinkhole to have formed in recent years. [From U.S. Geological Survey.]

with water; others quickly "drink up" all the water from a rainfall and remain open. Streams disappear into the ground at some sinkholes, and great gushing springs well up from others. This kind of land surface, called **Karst topography**, is produced in areas of plentiful rainfall where the bedrock is limestone. As in Kentucky, it is associated with caves in the limestone formations. The conclusion is inescapable that the major process of erosion in karst regions is solution of limestones by water. It is a chemical topography.

Low areas, whether they are stable or subsiding, rarely persist for long times, because they are the natural dumping grounds for sediment. Many intermontane basins of the Rocky Mountains would be far deeper than they now appear if all of the gravels, sands, and clays eroded from the surrounding mountains were removed. The Gulf Coast country of Mississippi, Louisiana, and eastern Texas has been subsiding steadily for many millions of years, but since sedimentation there keeps pace with subsidence, the land stays above sea level.

The great flat areas of the ocean floors are the **abyssal plains** (Fig. 5-24). Everywhere on those plains, the surface is made of

**Figure 5-28**
Continuous seismic profile, showing a continuous cross-sectional view of the geological structure of the deep ocean floor flanking the mid-Atlantic ridge. The profile was obtained in June of 1967 by means of a seismic recording system carried on board the Teledyne Exploration Company's vessel *Stranger*. The profile shows a part of the western abyssal plain at the flank of the ridge. In this region the sediments are comparatively flat and uniform, estimated to be from 1000 to 1500 m (3300 to 5000 ft) thick. The depth from surface to sediments is about 5300 m (17,500 ft). The irregular basaltic basement protrudes through the sediments at left and dominates the topography to the right, closer to the ridge. [Courtesy Teledyne Exploration Company.]

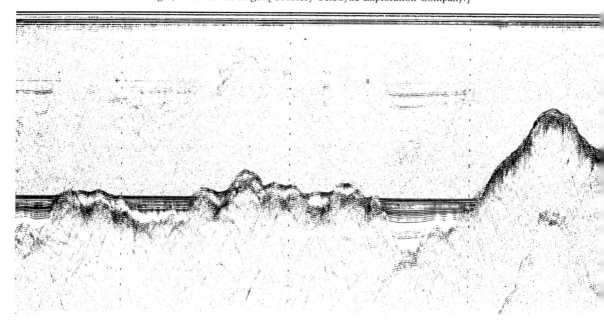

muds and sandy muds. **Continuous seismic profiling**, an operation very similar to echo-sounding but in which stronger sound impulses are used to penetrate soft sediment and reflect from harder rocks below, has revealed that the sediments bury a varied and rough topography.

The most pronounced and deepest of the negative topographic forms on the deep-sea floor are the **trenches** and the slightly wider **troughs** — narrow, long depressions that geophysicists have discovered to be the sites where oceanic plates dive into the mantle. Subduction is subsidence with a vengeance, but differs in mechanics and in topographic expression from any of the modes of continental subsidence.

Deep erosional valleys, **submarine canyons**, are abundant on many continental shelves (Fig. 5-23). Since they were first mapped in detail in the 1930's by pioneers in marine geology, these canyons have been a source of incessant controversy. Many look much like canyons on land, and are among the most intensively explored features of the sea. Thousands of depth recordings have been made of these canyons; scuba divers have swum in and around the shallowest ones, and they were explored in some of the first descents of the deep-diving oceanographic submarines, like Jacques Yves Cousteau's *Calypso* and the *Trieste*. There is no question that the canyons are erosional, and it now seems likely that a combination of current types is responsible. **Turbidity currents** — dense suspensions of mud and sand that move down submarine slopes, are most likely one of the chief agents (see Chapter 12).

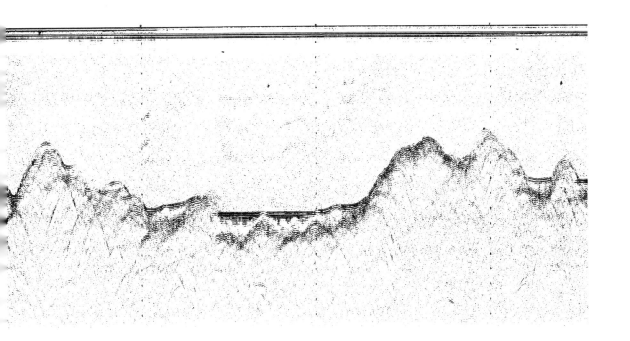

**The Face of North America** The landform map of North America shows the major relief patterns of the large topographic units and how clearly they relate to tectonics and sedimentation (see Fig. 22-51). The major mountain chains are in the East and in the West; the older Appalachians are much lower and less rugged than the Rockies. Between are the prairies and plains; their gentle southward slope is well indicated by the drainage pattern. All rivers run to the Gulf of Mexico. A low **coastal plain** lies along much of the central and southern Atlantic coasts and the Gulf coast. North of the Great Lakes are the low-relief plains and lakes of Ontario and Quebec. West of the Rocky Mountains is the Basin and Range province, a region of many smaller chains of mountains alternating with elongate basins. This is the area that some see as a potential huge rift valley that may someday separate the continent into two parts. West of the Basin and Range come the Sierra Nevada, the Cascades, and, to the north, the British Columbian Rockies. The western edge of the continent is marked by the Coast Ranges on the far side of the Great Valley of California, the valley of the Willamette River in Oregon, and its northward extensions. Alaska and the Yukon of Canada are extensions of the Rockies, or, as the whole western mountain belt of the Americas is called, the **Cordillera** (from the use of the Spanish word for "rope" to describe long narrow mountain ranges).

The history of this whole continent is dimly shown by its geomorphology, each event being seen through the veil of later episodes of mountain-building, erosion, and sedimentation. If geomorphology is to be of historical value, however, it has to be based on a theory of landscape evolution. Is there some orderly sequence by which mountains are worn down to plains?

## EVOLUTION OF LANDSCAPE

What is the evolution of the topography of a mountain range or a low plain? If it weren't always there, then it must have existed in some preceding state or stage, and necessarily will give way to a succeeding one. Perhaps the best way to answer the question is to study some young mountains; their characteristics can then be compared with those of older mountains that have been exposed to erosion for a long time. The youngest individual mountains are recently erupted volcanoes, but the youngest large mountainous region is the Himalayas, pushed up by the collision of the Indian subcontinental plate with the Eurasian plate. We can date the major uplift as Pleistocene because that is the age of the youngest sedimentary layers uplifted, but uplift is undoubtedly still going on. The mountains are the highest in the world, the relief is great, and the slopes and peaks are steep and rugged. Though the height of a volcano like Mt. Pelée may not be so great, it is similar in ruggedness and relief, and its slopes are as steep as the

**Figure 5-29**
The Himalayan range, photographed from *Apollo* 7 in October, 1968. The view is toward the west.
Nepal is at the left of the range, and China is at the right. Mount Everest is near the S-shaped lake
near the middle of the picture. The river at the left is the Ganges. The creation of the Himalayas
began 45 million years ago when India collided with the underside of Asia. [From National
Aeronautics and Space Administration.]

material can hold up (see "angle of repose" of volcanoes in Chapter 17, p. 572).

When we study old mountains, like the Appalachians, we find that their heights and relief are much lower. Although there may be many cliffs that are tricky to climb over, most slopes are more gradual, and the mountain tops have a softer, rounded appearance.

Here, then, is the germ of an idea for a time ordering of the sequence of stages in wearing down mountains. *William Morris Davis,* a Harvard geologist, brought this kind of study of topography to its fullest expression at the end of the nineteenth century, when he studied mountains and plains all over the world. He characterized the **geomorphic cycle** of topography as progressing from the high, rugged mountains of **youth** to the rounded forms of **maturity** to the worn down plains of **old age**. Perfect plains, the theoretical end-products, are never realized; **peneplains,** relatively flat erosion surfaces that approach being perfect plains

**Figure 5-30**
View of the Great Smoky Mountains National Park in the Appalachian Mountains of Tennessee. The generally gentle slopes are those of a tectonically quiet mountain belt in an advanced stage of erosion. [Photo by W. B. Hamilton, U.S. Geological Survey.]

One way to indicate the difference between a batch process and a continuous-flow process is to contrast two ways of baking bread. In the home (or in an old-fashioned bakery), a batch of dough is prepared, formed into loaves, and then baked. In a modern, computerized bread "factory," the flour and other ingredients are continuously being mixed in complicated vessels which allow the finished dough to flow out at one end, while the raw ingredients enter at the front. The dough is on a continuous assembly line while being formed into loaves and baked.

do result from very long periods of erosion. Flat surfaces of unconformities, such as the one at the top of the Precambrian in the Grand Canyon, must represent something like a peneplain. Here and there, standing uneroded above such a plain, would be erosional remnants, perhaps of more resistant rocks. Davis believed Mt. Monadnock in Southern New Hampshire to be such a hill and named all such **monadnocks** after it.

The assumptions underlying this concept do not stand up quite so well as the geomorphic sequence itself, which has proved useful in deciphering geomorphic histories. The process is looked upon as one that a chemical engineer might describe as a "batch process," in which there is a start and stop to the operation, rather than a "continuous-flow" process, in which all parts of the operation are going on simultaneously. The mountains were thought to have been elevated over a very short time period, almost instantaneously from a geological point of view (a batch), and then to have stayed fixed tectonically as erosion slowly wore them down through the stages of the cycle. Today our knowledge of the long, drawn-out, intermittent, and uneven nature of mountain-building makes that assumption untenable. It is more of a continuous process by which mountains keep on moving up while they erode, the evolution being determined by shifts in the balance between the two. Other assumptions that flawed the theory were that the sequence was characteristic of all rocks in all climates, and it is now generally agreed that progression from youth to old age can produce a variety of other forms, particularly in arid climates.

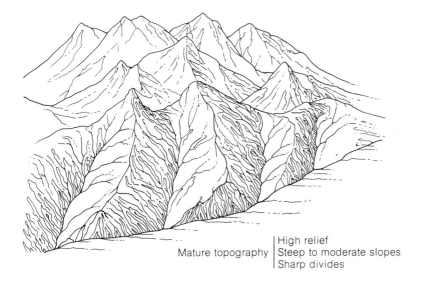

Mature topography | High relief
| Steep to moderate slopes
| Sharp divides

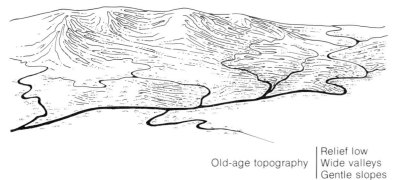

Old-age topography | Relief low
| Wide valleys
| Gentle slopes

**Figure 5-31**
Two types of topography thought by W. M. Davis to be mature and old-age
stages of an evolving landscape as it is worn down by rivers.

**Figure 5-32**
A monadnock, an elevated erosional remnant left after erosion has worn down
most of the region to a low-relief plain. [Photo by J. R. Balsley, U.S.
Geological Survey.]

Current views of geomorphic evolution as a continuous process emphasize the **steady state**, wherein a constant balance of forces results in a topography unchanging with time, though rocks are steadily being uplifted and eroded. Such a concept tends to deny much possibility of reconstructing a geomorphic history of the region, for it says that what a region is like is the product of currently active operations, not what went before. What it will be next has little to do with what it is now. Still other models incorporate all of the operating variables: tectonics, rock type, extent of rock weathering, and drainage in relation to climate and vegetation, together with the possibility of some randomness in the course of events. This leads to a low degree of predictability and some sense of disillusion for those who had hopes for a thoroughly orderly and predictable pattern.

Isn't it strange that the part of the Earth that we see and walk over all the time is so difficult to make a simple order of? Perhaps the very wealth of information and the complexities of all of the variables that we know have an effect on landscape are what make it so resistant to analysis. Perhaps the subject waits for those who would do as Einstein did: in describing his thought patterns about relativity, he wrote that "I had to divest myself of my intuition — then I could think clearly."

Our knowledge of how the landscape beneath the ocean evolved is in a rudimentary state of hypothesis making, partly because of the dearth of information and partly because of our inability either to observe directly erosive forces at work or to reason from historical records of change. The lack of information is not surprising; the oceans are immense, and the overwhelming majority of its area has never been crossed by an oceanographic research ship. The information itself, two-dimensional profiles, is not as useful as the three-dimensional view we see of the land.

The information we do have suggests that sedimentation, tectonics and volcanism play a major role in shaping the topography of most of the ocean basin areas, with only slight modification by weak erosion. Where erosion does work effectively is in the relatively shallow waters of the continental shelves and continental slope. There the submarine canyons have been cut deeply into the continental-shelf sediments by currents. The evolution of deep-sea floor topography starts from an early or youthful stage of hilly or mountainous volcanic topography on active plate-divergence ridges. As the topography rides away from the ridge on its plate, the slow but constant rain of sediment from the sea water above gradually buries the rough topography with a blanket that fills in the lowest spots and drapes and smooths the rough surface. If the topography rides far enough and the sedimentation is great enough, a gently hilly, rounded topography is formed, the abyssal hills.

As the plates move apart, there is some tendency for them to sink slightly relative to sea level. Large areas may become the lowest broad areas of the deep sea and become filled with sediment, to become the featureless abyssal plains. The currents that fill these low spots are turbidity currents, the mud-laden streams of water flowing on the bottom. Guyots, volcanoes whose flat tops were eroded at sea level, also gradually drop as the plates separate, so that their tops get deeper, all the while receiving their share of sediment from above.

On the continental shelves there is more erosion, but sedimentation is also more rapid, for the shelves are the dumping ground for much of the detritus brought by rivers from the land. The net result is construction rather than destruction: the shelves build upward and outward to a smooth apron, broken here and there by submarine canyons.

Most of the clues to the analysis of landscape and the topography of the sea bottom will come from an understanding of the mechanisms of erosion and transportation of detritus, and that is what we turn to in the next chapters.

## SUMMARY

1. Topography on land and under the sea are the areal distributions of the heights and depths of the Earth's surface, usually represented by contour maps.

2. The Grand Canyon is a landscape developed by river erosion of a series of flat-lying sediments of varying resistance to erosion. The valley is V-shaped.

3. Yosemite Valley is a landscape produced by glacial ice scouring a U-shaped valley in a tectonically elevated block of granite, the Sierra Nevada mountains.

4. The Ozark Mountains are a landscape of gentler forms and low relief on a very old granite uplift that has been eroded for long geologic times and partly overlapped by sediment.

5. The overall pattern of landscape development is one in which tectonic uplift, which tends to produce high altitudes, is countered by the erosional wearing down by water and ice.

6. Landforms, such as hills, plateaus, and valleys, are controlled by the erosional process acting over geologic times on a variety of resistant and nonresistant rock types that are arranged in different structural patterns.

7. Underwater topography, as recorded by echo-sounding, includes many constructional features. One group includes the ridges and rises associated with zones of plate divergence. Another group is volcanic, the seamounts and guyots. Sedimentation forms the shelves, slopes, and rises bordering the continental platforms.

8. Karst topography is formed by chemical solution of limestones in areas of moderate to high rainfall.

9. The landform map of North America reveals the diverse effects of tectonic and erosional activity, modified by sedimentation.

10. Evolution of landscape, formerly perceived as simple erosional stages of youth, maturity, and old age following a single uplift, are understandable as a series of states of balance between structural uplift and erosional wearing away.

## EXERCISES

1. In some areas the relief of the natural topography is less than that of man's architecture or construction. What kinds of topographic areas are they, and what is the magnitude of relief?

2. If you were to see the topographies of the Sierra Nevada and the Grand Canyon 20 million years from now, assuming no great change in climate and no tectonic activity, would you expect greater or lesser differences between them than is shown now? Describe what each might look like at that time in the future.

3. Describe the changes in topography that might take place if the Ozark Mountain region were to be uplifted by 5000 meters (1) very rapidly, and (2) very slowly.

4. In flying across the country, what topographic features would you look for to tell (1) whether you were flying over structurally deformed or flat-lying sedimentary rocks, and (2) whether you were flying over beds or rock bodies of uniform resistance to erosion or varying resistance to erosion.

5. How does a mountain peak on land differ in origin from a seamount rising from the sea floor?

6. Describe three different geologic processes by which deep valleys can form on land. Which one(s) of these may also be responsible for deep valleys on the sea floor?

## BIBLIOGRAPHY

Bloom, A. L., *The Surface of the Earth.* Englewood Cliffs, New Jersey: Prentice-Hall, Inc., 1969

Davis, W. M., *Geographical Essays.* Boston: Ginn and Company, 1909. (Reprinted 1954, Dover Publications, Inc., New York.)

Heezen, B. C., "The Origin of Submarine Canyons." *Scientific American,* August, 1956. (Offprint No. 807.)

Hunt, C. B., *Natural Regions of the United States and Canada.* San Francisco: W. H. Freeman and Company, 1973.

Shepard, F. P., *Submarine Geology* (3rd ed.). New York: Harper and Row, 1973 (Chapters 6, 9–11, 13).

Shepard, F. P., and R. F. Dill, *Submarine Canyons and Other Sea Valleys.* Chicago: Rand-McNally, 1966.

# 6

# Weathering: The Decomposition of Rocks

*The primary mechanism of erosion is weathering—chemical decay, under Earth surface conditions, of minerals that formed, for the most part, at high pressures and temperatures in the interior of the Earth. Such minerals as feldspar partially dissolve after reacting with air and water, leaving a solid residue of clay. Other minerals, such as calcite and some mafic minerals, dissolve completely. The intensity of weathering is dependent on climate, tectonics, original rock composition, and time. Given enough time, factors other than climate are relatively unimportant. The weathering process produces the clays of the world, all soils, and the dissolved substances that are carried by rivers to the ocean. The salt water of the ocean and the chemical sediments on the sea floors are themselves end products of chemical weathering of the surface of the continents.*

Both the lowering of mountains by erosion and the origin of the raw materials of sediments can be explained in terms of weathering, which is a twofold process. It is both fragmentation, or **mechanical weathering**, and decay, or **chemical weathering**, operating together, each helping and reinforcing the other. The smaller the pieces, the greater the surface area available for chemical attack and the faster the pieces decay; the faster the decay, the more the pieces become weakened and susceptible to breakage. In this chapter we focus on the chemical aspect, for it is in some ways the more fundamental driving force of the whole process; the effect of mechanical weathering, always important, is the main subject of Chapter 8.

**Figure 6-1**
A weathered rock. The surface is corroded and pitted and divided by enlarged cracks and joints, in some of which plants grow, thereby helping further to weather the rock. [Photo by R. Siever.]

## HOW LONG DOES A ROCK LAST?

The first person to make a tool of iron must have been the first to notice that it would rust, and perhaps wondered why. An old nail in soil will usually be so badly rusted that it can be snapped like a match, yet nails in wood in some of our colonial houses are still strong and covered with only a thin brownish film. A list of the kinds of places where nails rust quickly would easily show that exposure to air and moisture are important factors. Rusting is a chemical process in which oxygen and water convert metallic iron to its **oxidized** form, ferric iron.

In contrast, some materials seem unchanged by exposure to air and water. Anyone who has hiked a trail to a lonely spot and discovered an old beer bottle knows that glass survives long exposure to the elements. Old bottles that have lain buried in moist soil do get a sometimes beautifully colored coating added to the glass, but the main part of the glass itself seems immune from chemical attack.

We can see the same range of response to wind, rain, snow, sun, and cold in old monuments and cemeteries. Some gravestones seem to be made for the ages, but others have lost almost all of their inscriptions. In a temperate climate limestone goes fast, but granite lasts. In the aridity of desert lands, everything lasts. Today we can see well-preserved details on stone monuments left by early civilizations in the Middle East; that these details still remain is directly attributable to the dry climate. Alabaster (gypsum) sculpture could never have lasted outdoors in Wisconsin as it has in North Africa.

Though we usually think of organic materials as being quick to decay or putrefy, these kinds of substances have as wide a range of responses to erosion as do inorganic minerals. Some tough resistant tissues of organisms, such as pollen grains, can survive longer than many inorganic minerals. Man, the animal that learned to synthesize thousands of new chemicals, contributes an undue share to the mass of decay-resistant organic materials by producing huge quantities of nondegradable compounds, among them being the older detergents, DDT, and the ever-present plastic products that are found scattered all over the land surface and floating in the middle of the ocean. We rely on the rapid decay of natural organic matter, which is primarily the work of bacteria, to return nutrients to soil and to produce fodder for farm animals from the fermentation of plant materials in silos. Some materials, however,

**Figure 6-2**
Two gravestones of the early nineteenth century in a cemetery in Wellfleet, Massachusetts. The lighter stone is limestone that has been ravaged by chemical weathering; the darker stone is slate, which remains practically untouched; even the delicate etching at the top of the stone is preserved. [Photo by R. Siever.]

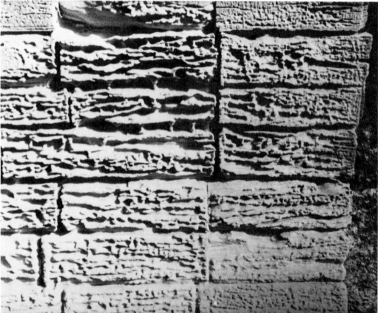

**Figure 6-3**
Weathered limestone blocks of a Roman aqueduct about 2000 years old, showing the pitted, etched surface caused by chemical solution. (a) One of the arches. (b) A closeup of a number of blocks. Pont du Gard, Provence, France. [Photo by R. Siever.]

just don't decay as rapidly as we would like. For example, in the past decade we have become concerned about just how long it will take natural processes to eliminate the millions of gallons of oil that get into the oceans as a result of spills of various kinds and leaks at offshore oil-drilling sites. All of that oil originated as organic matter that accumulated in sediments on the Earth's surface and was later preserved and altered by burial (see Chapter 24). The process of oil formation made the material much more resistant to decay.

Why do some inorganic materials weather so quickly and others so slowly? From everyday observations we can contrast the differences among rocks of different composition, those in different climates, and those in different surroundings, from soil to bare mountainside. Mere age of a rock is of no consequence, for we can find fresh-looking Precambrian rocks and badly weathered Pleistocene ones. How long they have been exposed to erosion is a more important matter, as anyone can testify by comparing a newly blasted road cut with one 20 or 30 years old. *Soil itself is both a factor in weathering and a result of it. The production of soil is evidently a* **positive-feedback process,** *one in which the product of the process works, by its presence, to increase the output of the process. Once soil starts to form, rock weathers more rapidly and more soil is formed.*

At the heart of the whole process are the ways in which individual kinds of minerals react with water and air. We turn now to some important examples of the major rock-forming minerals, the first being the most abundant mineral in the Earth's crust, feldspar.

## HOW TO DISSOLVE FELDSPAR

Feldspar is a key mineral in a great many igneous, metamorphic, and sedimentary rocks. Understanding its weathering behavior should contribute much to our grasp of the weathering process in general. The place to start is in the field, making observations that give the clue to what the natural process is; then come the laboratory experiments, which enable us to quantify and identify the intermediate and final products of decomposition under controlled conditions.

**Observing Decomposition in the Field** Much as people in temperate regions are used to thinking of granite as the most permanent of rocks, those in the humid tropics know that many granite boulders in soil can easily be kicked into a heap of mineral grains. Close examination of the particles, and comparison with a fresh piece, shows that the crystals of feldspar are all punky and chalky with clay, many of them so soft that they can be gouged with a fingernail, a condition conspicuously in contrast with that of the clear unaltered quartz crystals. The rock falls apart because the original interlocking crystal network of quartz and feldspar no longer holds together when the feldspar weathers to a loosely adhering clay. The white to cream-colored clay is the mineral kaolinite (sometimes just called kaolin), used in pure form as raw material for pottery and china.

The word "kaolin" is a French modification of the Chinese "kao-ling," named for the place where the Chinese mined it for centuries before the finished product was exported as "china" to Europe. If Europeans had explored the state of Georgia before they visited China, they would have found one of the large kaolin deposits in the western hemisphere, and the name might have been different.

**Figure 6-4**
Disintegrating boulder of granite partially embedded in a glacial soil, shown
partly excavated by digging. The outer rim shows the original size of the boulder.
[Photo by F. E. Matthes, U.S. Geological Survey.]

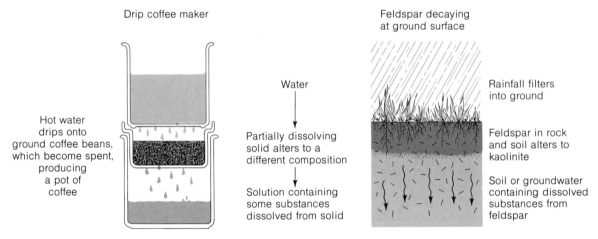

**Figure 6-5**
The process by which feldspar decays is analogous to the way in which coffee is made. In both processes
water dissolves some of the substance of the solid, leaving behind an altered material and producing
a solution containing the substances drawn from the original solid.

Decay is not nearly so rapid in temperate climates, but any outcrop of a feldspar-bearing rock will show the beginnings of the
same weathering to kaolin on some grains. Only in severely arid
climates typical of some deserts does feldspar stay relatively untouched. From this we infer that water is essential to the chemical
reaction by which feldspar alters to kaolin. The kind of reaction
can be illustrated by a useful analogy to a familiar chemical reaction
of the same kind: coffee making. Fresh ground coffee beans plus
hot water make a solution—the coffee—that is extracted from the
solid, leaving behind spent coffee grounds as a residue. How much

can be extracted from the ground coffee depends on how much water is used, how hot it is, and how long it percolates. In the same way, the feldspar-kaolinite conversion depends on the amount of rainfall, the temperature, and the length of time water stays in contact with the feldspar-containing rock, such as granite. These are the controlling factors, whether the rock is a boulder in soil or bedrock beneath a soil.

We can write the beginnings of an equation for the weathering of the common feldspar of granite, orthoclase, which is made up of potassium (K), aluminum (Al), silicon (Si), and oxygen (O):

feldspar + water $\longrightarrow$ kaolinite

| $KAlSi_3O_8$ | $H_2O$ | $Al_2Si_2O_5(OH)_4$ |
|---|---|---|
| 1K | | |
| 1Al | | 2Al |
| 3Si | | 2Si |
| 8O | 1O | 6O |
| | 2H | 4H |

But chemical equations are like those of algebra: they have to add up to the same total on both sides. This requirement is imposed by the conservation of matter: the total number and kinds of atoms of the starting materials, the **reactants**, have to be the same as those of the **products** of the reaction. In what we have written, potassium appears on one side, as a component of the feldspar, but not on the other side. To make an equation of what we have written, we must add potassium to the right side. Since no potassium minerals are found in the weathered products, we presume the potassium was carried away by a water solution (like the coffee extracted from the grounds). Now notice the silicon in relation to aluminum. On the left side there is one Al atom and 3 Si atoms, but on the right there are 2 Al atoms and 2 Si atoms. If we multiply the left side by 2 (written as $2KAlSi_3O_8$, which means multiplying the entire formula by two, equivalent to writing $K_2Al_2Si_6O_{16}$), we get the same number of Al atoms on both sides. We are still left without equal numbers, however, for there are 4 Si atoms not included in the kaolinite, and the O and H atoms do not balance.

2 feldspar + 1 $H_2O$ $\longrightarrow$ 1 kaolinite + 2 dissolved potassium

| $2\ KAlSi_3O_8$ | | $1\ Al_2Si_2O_5(OH)_4 + 2K^+$ |
|---|---|---|
| 2K | | 2K |
| 2Al | | 2Al |
| 6Si | | 2Si |
| 16O | 1O | 9O |
| | 2H | 4H |

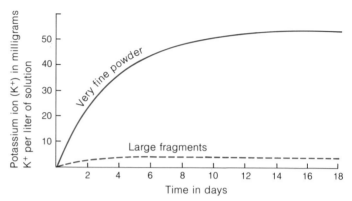

**Figure 6-6**
The rate at which potassium feldspar partially dissolves when put into water. The amount of feldspar that dissolves is measured by the amount of potassium ion liberated to the solution. The feldspar dissolves at a fast rate in the first time intervals and then slows, finally until no further change can be detected. Grinding the feldspar to a fine powder makes it dissolve faster and allows more of it to dissolve.

To make the balance for Si at the same time as we do it for Al means adding 4 Si atoms in some form to the right as well as balancing the oxygen and hydrogen. Familiarity with these kinds of equations enables the geochemist to guess intelligently what the exact form of the equation might be, but he cannot be sure until he does the experiment.

**Dissolving in the Laboratory** When trying to dissolve any of a great many materials, the thing to use is an acid. Strong mixtures of acids are used as cleaning solutions for laboratory glassware to make use of their dissolving power. The experiment of choice, then, is to immerse feldspar in a readily available acid like hydrochloric acid, HCl. The feldspar is ground to a fine powder first, to make the dissolution faster by exposing more surface area to the solution—an effect we take advantage of when we use ground coffee instead of whole coffee beans. As the feldspar dissolves, the experimenter takes small samples from the solution and analyzes them for the expected dissolved potassium ion and, he guesses, dissolved silicon in some form. At first the feldspar dissolves rapidly, as measured by the steady increase in dissolved $K^+$ in the solution, but then the rate of dissolution starts to level off as the feldspar reacts more and more slowly and finally undergoes no further change. If the solution continues not to change, it is presumed to be in **equilibrium** with feldspar powder. The analysis of the solution at this point gives the exact balance of the equation, which turns out to be:

feldspar + hydrochloric + water $\rightarrow$ kaolinite + dissolved + dissolved + dissolved
                 acid                                   silica    potassium    chloride

$2KAlSi_3O_8$     $2(H^+ + Cl^-)$     $H_2O$    $Al_2Si_2O_5(OH)_4$   $4SiO_2$      $2K^+$       $2Cl^-$

We write the acid as ($H^+$ + $Cl^-$) to emphasize that strong acids in solution are completely ionized—that is, present as individual ions, not molecules. The dissolved silica is written as $SiO_2$.

*Three major points about this equation are important. (1) The K and Si produced by dissolving the feldspar appear as dissolved material. (2) Water is used up in the reaction. It is absorbed, in the form of one of its constituent ions, hydroxyl ($OH^-$), into the kaolinite structure. (3) Hydrogen ion ($H^+$) is used up in the reaction, and since acidity is measured by the amount of $H^+$, the solution becomes less acidic as the reaction proceeds.*

This simple experiment includes the three main chemical effects of chemical weathering on silicates: it leaches, or dissolves away, cations and silica; it adds water, or **hydrates** the minerals; and it makes the solutions less acidic—more basic. If we were to dissolve feldspar in distilled water, the reaction would still proceed, though very much more slowly, and produce a very dilute basic solution, like that of a weak caustic soda or lye solution. It is because $H^+$ is used up that makes this reaction look to a chemist like one in which an acid mixes with a base to produce a salt, as when HCl mixes with potassium hydroxide (KOH), a strong base, to form potassium chloride (KCl) and water. In our experiment, the feldspar acted as the base.

But nature does not pour HCl on an outcrop of a granite. If we think again—now with the benefit of laboratory experience—about how feldspar dissolves in nature, we are led to the most common acid on the Earth's surface, carbonic acid ($H_2CO_3$). This weak acid is formed by the solution in rainwater of a small amount of carbon dioxide ($CO_2$) gas from the atmosphere.

**Carbon Dioxide and Carbonic Acid**  We are familiar with everyday solutions of $CO_2$ in water: soft drinks made by pumping pressurized $CO_2$ gas into the liquid. When a carbonated beverage is left to stand until bubbling has stopped, the dissolved $CO_2$ has decreased to a stable value, one we usually characterize as "flat" to the taste. The remaining dissolved $CO_2$ is in equilibrium with the small amount of $CO_2$ gas in the atmosphere. We measure the amount of any gas by the pressure it exerts, using as a standard the pressure exerted by the Earth's atmosphere at sea level, called one **atmosphere** of pressure. The pressure of $CO_2$ in the atmosphere is about 0.0003 (or 0.03 percent) of an atmosphere, which may seem small, but actually makes it one of the four most abundant gases, just behind argon (0.9 percent), oxygen (21 percent), and nitrogen (78 percent).

Carbon dioxide gas dissolves in water to form a new compound, carbonic acid, by a **reversible reaction**, which is written with a double arrow $\rightleftarrows$, to indicate that the reaction goes both ways:

carbon dioxide gas + water $\rightleftarrows$ carbonic acid in solution

$$CO_2 \qquad\qquad H_2O \qquad\qquad H_2CO_3$$

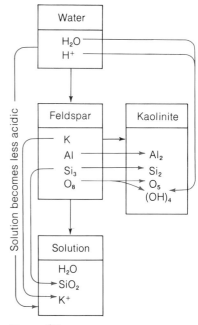

Figure 6-7
The flow of material in the alteration of feldspar to kaolinite by the dissolving action of water. Some water and hydrogen ions ($H^+$) are absorbed into the kaolinite structure while all of the potassium ($K^+$) and some of the silica ($SiO_2$) end up in solution.

The pressure exerted by the Earth's atmosphere at sea level, one atmosphere, is equal to the pressure exerted by a 760 mm high column of mercury. For the much higher pressures characteristic of the interior of the Earth, we use the term *bar*. One bar is equal to 0.987 atmospheres. A *kilobar* is 1000 bars.

The reverse reaction by which $CO_2$ is formed from carbonic acid, is the one in which bubbles of $CO_2$ form from a pressurized solution of a carbonated beverage.

Once formed, carbonic acid spontaneously ionizes by another reversible reaction, in the same way that HCl or KCl does, to form a hydrogen ion, $H^+$, and a **bicarbonate** ion, $HCO_3^-$:

$$\text{carbonic} \underset{}{\overset{}{\rightleftharpoons}} \text{hydrogen} + \text{bicarbonate}$$
$$\text{acid} \qquad \text{ion} \qquad \text{ion}$$

$$H_2CO_3 \qquad H^+ \qquad HCO_3^-$$

This reaction is different from the ionization of the HCl and KCl because far from all of the carbonic acid molecules are ionized; only one molecule out of a thousand is ionized at equilibrium. But even such a weak acid can still dissolve great quantities of rocks over a long time. An even smaller amount of $HCO_3^-$, only one ion out of 100,000, further ionizes to form another hydrogen ion and a **carbonate ion**, $CO_3^{--}$

$$\text{bicarbonate} \underset{}{\overset{}{\rightleftharpoons}} \text{hydrogen} + \text{carbonate}$$
$$\text{ion} \qquad \text{ion} \qquad \text{ion}$$

$$HCO_3^- \qquad H^+ \qquad CO_3^{--}$$

The $H^+$ provided when $CO_2$ dissolves in rainwater makes it slightly acid, not enough to be noticed by plants and animals but slowly corrosive for feldspars. In areas where the air is greatly polluted with sulfurous gases from industries, rainwaters may be much more acid, for the sulfurous gases dissolve in rain in the same way as $CO_2$ to form sulfuric acid, a far stronger acid than carbonic acid. Although rainwater contaminated with sulfuric acid is still much too weak to sting our skins, it does damage our fabrics, paints, and metals. It also weathers our stone monuments and outdoor sculptures at such a rapid rate that public officials are beginning to take notice of their deterioration.

**How Does Weathering Proceed in Nature?** Now that we have a workable model for the major chemical reaction by which rainwater weathers feldspar, we have to return to field observations to see how the process works on outcrops and in the soil. Just as a nail rusts more quickly and deeply in the soil than it does lying on a rooftop, we find that feldspars on bare rock surfaces seem much better preserved than those in pebbles buried in wet soils, which may be crumbly and coated with soft rinds of kaolinite. The chemical equation gives us a clue to this difference. First, in the absence of water the reaction will not proceed; during all of the dry periods, the bare rock is untouched. In moist soil the feldspar is constantly bathed by a corrosive acid-water solution. Second, there is more acid in the soil than in rainwater, and the stronger the acid the more extensive the decay. Some of the acids

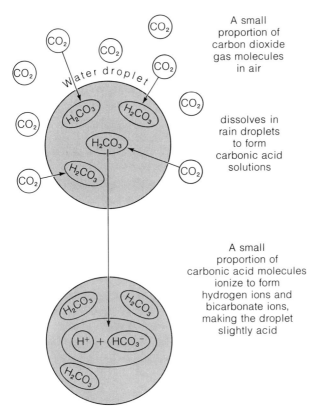

A small
proportion of
carbon dioxide
gas molecules
in air

dissolves in
rain droplets
to form
carbonic acid
solutions

A small
proportion of
carbonic acid molecules
ionize to form
hydrogen ions and
bicarbonate ions,
making the droplet
slightly acid

**Figure 6-8**
The formation of slightly acid rain by the dissolution of
carbon dioxide gas molecules in water droplets in the
atmosphere. Once dissolved, the carbon dioxide reacts with
water to form carbonic acid, which then ionizes to form
hydrogen and bicarbonate ions.

in soil are organic compounds produced by bacterial decay of
plant and animal remains. Some bacteria seem able to grow even
on bare rock particles and eat away at them. The acid content of
soil water gets an extra boost from plant roots and bacterial decay, which produce $CO_2$ by **respiration** processes similar to the
one upon which we depend. They oxidize organic matter and give
off $CO_2$; we breathe oxygen and exhale $CO_2$. The amount of $CO_2$
in soil can be as much as ten times that in rainwater, and makes
soil water an efficient dissolver of feldspars.

Weathering does not proceed at the same rate in all climates.
One reason is the contrast between hot and cold: the speed of
chemical reactions increases as the temperature goes up, and,
even more importantly, plants and bacteria grow and multiply
much faster in warmer climates. The other reason is the contrast
between wet and dry: water is needed for the weathering reaction,
and vegetation grows more lushly in humid climates. It is not

**Figure 6-9**
Cleopatra's Needle, an obelisk of granite (left) in Egypt and (right) after being removed to Central Park, New York City. After several decades of exposure to the elements in New York, the obelisk had weathered more extensively than it had after thirty-five centuries of exposure in Egypt. [Photos courtesy of the Metropolitan Museum of Art, New York.]

surprising, then, that weathering is most intense in tropical climates, which are both wet and warm.

We will see in Chapter 8 how mechanical erosion is related to the speed of weathering and to the topography. Just as the fresh feldspar of the laboratory experiment reacted rapidly at first and then slowed down, a fresh piece of feldspar exposed at the surface of an outcrop by mechanical breakage will chemically alter faster than one in an unbroken rock. Because the weathering products in mountains are carried away quickly by rains and melting snows, we cannot see the effects as easily as in areas of lower relief where much of the clays and other products accumulate on gentle slopes and in valleys to form soil.

**Making the Clays of the Earth** We have given the details of how one kind of feldspar weathers to make one kind of clay. Some other silicate minerals, such as mica and volcanic glasses, weather in the same general manner, some changing to kaolinite and others to different clay minerals. Feldspar can alter to different kinds of clay minerals under different weathering conditions. **Montmorillonite** is an abundant clay formed by weathering in many warm and semiarid climates. It is also the main clay product of the weathering of volcanic ash. Yet another abundant clay mineral typically develops from the weathering of sediments in temperate regions. Ralph Grim, one of the pioneers of modern clay mineralogy in the 1930's and 1940's, worked at the Illinois Geological Survey and named that mineral **illite** for Illinois. Under the extreme weathering conditions of the tropics, kaolinite can partially dissolve to give silica in solution, leaving a solid residue of the mineral **gibbsite**, aluminum hydroxide $(Al(OH)_3)$. Gibbsite makes up the deposits of **bauxite**, the earthy ore that is the major source of economically extractable aluminum in the world.

All of the clay minerals have the basic structures of sheet silicates; sheets of silica tetrahedra alternate with sheets of alumina octahedra. In and between the sheets are cations, mainly $K^+$, $Na^+$, $Ca^{++}$, and $Mg^{++}$, which give each clay type its distinctive characters. Montmorillonite, for example, has a structure and cation composition that gives it the ability to soak up large quantities of water, which spreads the layers apart and makes them easily cleavable — to the point of slipperiness. If you ever drive on a montmorillonite road and get caught in a rain, watch out; you may slide helplessly off the road into a ditch as your tires lose all traction on the gooey mass of watery clay. Kaolinite absorbs less water, but is just right for making pottery and ceramics.

Because feldspar and other silicates that weather to form clays compose the large bulk of the igneous and metamorphic rocks of the crust there is an enormous amount of clay produced by the weathering of those rocks. Just how large is that figure? The total amount of sediment now existing over the whole Earth is about 3 times $10^{25}$ grams. Since about one-third of all sediments (most of it in shales) is made up of clay minerals, there must be about 1 times $10^{25}$ grams of clay in the Earth! Obviously the importance of chemical weathering as a source of sedimentary materials cannot be overemphasized. Other kinds of weathering, however, do not result in the formation of clay, but in the complete disappearance of minerals.

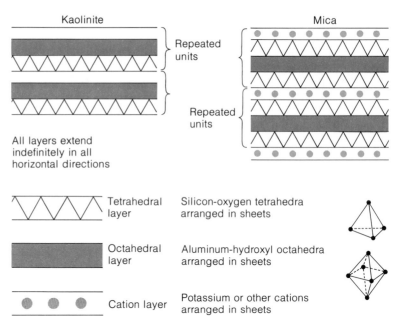

**Figure 6-10**
Two basic types of clay mineral structure. The basic building blocks are sheets of silica tetrahedra and alumina octahedra. Kaolinite is composed of a simple two-layer unit of one tetrahedral and one octahedral sheet that is repeated many times to build up a crystal. Clay minerals, such as illite and montmorillonite, are closely related to the mica structure, which is composed of a "sandwich" of two tetrahedral units on both sides of an octahedral sheet, each "sandwich" separated from the next by a layer of cations, such as potassium in the mineral muscovite or the clay mineral illite.

## DISSOLVING WITHOUT A TRACE

Whereas feldspar and other aluminosilicate minerals form clay by incomplete dissolution, minerals of other chemical compositions may dissolve completely, just like salt in water. Quantitatively, the most important of these minerals are the carbonate minerals, calcite ($CaCO_3$) and dolomite ($CaMg(CO_3)_2$), which together make up the limestones of the world.

**Limestone, the Soluble Sediment** The fast disappearance of limestone is an obvious fact to the farmer or gardener who spreads finely ground limestone to reduce the acidity of his soil. The application has to be repeated every year or two because the limestone dissolves. Caves are characteristic of many limestone formations in humid climates. Their patterns of extensive passageways and the multitude of pits, holes, and irregular flutings on cave walls and ceilings are witness to the extensive dissolving power of groundwater. Caves and sinkholes are the indicators of abundant water acting on limestones.

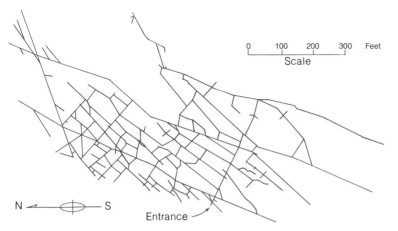

Figure 6-11
Map of Mark Twain Cave, Hannibal, Missouri. The many passageways in this complex cave dissolved along joint patterns in Paleozoic limestone. [From J. Harlen Bretz, "Vadose and Phreatic Features of Limestone Canyons," J. Geol vol. 50, 1942.]

Possibly the most celebrated description in American literature of the maze of passageways in a limestone cave is in Mark Twain's *Tom Sawyer*. Tom and Becky Thatcher got lost in what is now known as Mark Twain cave, near Hannibal, Missouri, the scene of the novel.

The dissolution of limestone gets us back to the $CO_2$ system. The carbonate ion ($CO_3^{--}$) is $CO_2$ in another form, made by carbonic acid ($H_2CO_3$) ionizing first to bicarbonate ion ($HCO_3^-$) and then to $CO_3^{--}$. Calcite dissolves in pure water by the separation of the calcium ion ($Ca^{++}$) from $CO_3^{--}$:

$$\text{calcite} \rightarrow \text{calcium ion} + \text{carbonate ion}$$
$$CaCO_3 \qquad Ca^{++} \qquad CO_3^{--}$$

In the solution the $CO_3^{--}$ tends to combine with $H^+$ ions from the water to form $HCO_3^-$. This has the effect of making the water less acidic, which is why it is used for agricultural purposes. At the same time, if $H^+$ keeps on being supplied by $CO_2$ dissolving in the water, more and more $CO_3^{--}$ is used up. As the $CO_3^{--}$ converts to $HCO_3^-$, the solution remains undersaturated with solid calcite, which then keeps dissolving. The overall reaction is described in shorthand by:

$$\begin{array}{cccc} \text{calcite} + \text{carbonic} & \rightarrow & \text{calcium ion} + & \text{bicarbonate} \\ \text{acid} & & & \text{ion} \end{array}$$
$$CaCO_3 \quad H_2CO_3 \qquad Ca^{++} \qquad 2HCO_3^-$$

Compare this reaction with that for feldspar weathering. In both reactions $H^+$ is used up as $CO_2$ reacts with water, and $HCO_3^-$ is formed as a product. In this way $CO_2$ is extracted from the atmosphere and put into water. The more the weathering, the more $CO_2$ is taken out of the atmosphere. Both reactions supply dissolved cations.

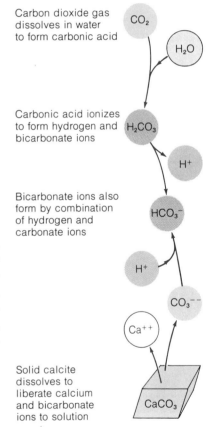

Carbon dioxide gas dissolves in water to form carbonic acid

Carbonic acid ionizes to form hydrogen and bicarbonate ions

Bicarbonate ions also form by combination of hydrogen and carbonate ions

Solid calcite dissolves to liberate calcium and bicarbonate ions to solution

Figure 6-12
The general scheme by which calcite or other carbonate minerals dissolve in water containing dissolved carbon dioxide. The reactions are all reversible, and the system should be visualized with all reactions operating at the same time to produce the general net tendency shown by the arrows.

In a given quantity of water, limestone dissolves faster and in greater amounts than silicates. For both reasons, the chemical weathering of limestone accounts for more of the total chemical erosion of the land surface than any other rock, even though much larger areas are covered by silicate rocks. Some mafic silicate minerals dissolve completely, like limestone, but not as rapidly. Although they by no means rival limestone in quantity eroded from the land surface, they do account for a major fraction of the total amounts of certain elements weathered from rocks — in particular, iron and magnesium.

**Mafic Minerals** The beating of ocean waves against rocky cliffs is one of nature's ways of hurling solvent at a rock. On the rocky coast of New England the chemical effects of the attack are seen when the geologist hunts for pyroxene crystals in a granodiorite. All that can be found are little cavities in the shape of pyroxene crystals. Olivine behaves the same way. Like limestone, both minerals wholly dissolve. A magnesium pyroxene, $MgSiO_3$, dissolves to give $Mg^{++}$ and dissolved silica. An iron pyroxene $(FeSiO_3)$ dissolves to give $Fe^{++}$ and dissolved silica. The iron in pyroxene is **ferrous** $(Fe^{++})$, in contrast to the iron in hematite, $(Fe_2O_3)$, which is **ferric**. When iron pyroxene weathers, not only does the silicate structure dissolve, but the ferrous iron is oxidized by oxygen to ferric iron. Ferric iron in the presence of oxygen is very insoluble in most natural surface waters, and it precipitates to form a poorly crystallized mineral that looks like rust, limonite $[FeO(OH)]$, a ferric oxide that contains water (in the form of hydroxyl). The overall reaction, which includes oxidation and hydration of the iron and the dissolution of the silica, is:

Ferric iron is more *oxidized* than ferrous iron—that is, it has fewer electrons. Oxidation can be defined as a loss of electrons in a chemical reaction, regardless of whether the element combines with oxygen itself. Reduction is the opposite of oxidation; it is a gain of electrons. In comparison with ferric iron, ferrous iron is more reduced.

Limonite is actually a mixture of two minerals of the same chemical composition but different crystal structures, goethite and lepidocrocite. Goethite is far more abundant. Magnetic varieties of these iron oxide minerals give us our magnetic recording tapes.

| iron pyroxene | + | oxygen from atmosphere | + | water | $\rightarrow$ | limonite (hydrated iron oxide) | + | dissolved silica |
|---|---|---|---|---|---|---|---|---|
| $4FeSiO_3$ | | $O_2$ | | $2H_2O$ | | $4FeO(OH)$ | | $4SiO_2$ |

For every four iron atoms that are weathered after being brought to the surface of the earth by mountain building or volcanism, two atoms of oxygen are extracted from the atmosphere. In the long run, millions of years, if it were not for the rapid replenishment of oxygen by the photosynthesis of plants, the weathering of mafic rocks would have eliminated all of the oxygen from the air. Liberated by the weathering of mafic minerals such as olivines, pyroxenes, amphiboles, and biotite mica, the iron precipitates as fine-grained, poorly crystalline limonite coatings and encrustations that color soils and stain rocks shades of red and brown. They also have served as the raw materials of artists' pigments — the ochres, siennas, and umbers.

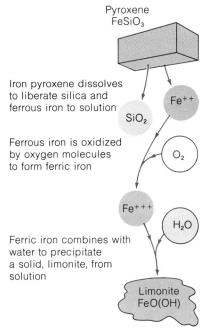

Pyroxene
$FeSiO_3$

Iron pyroxene dissolves
to liberate silica and
ferrous iron to solution

$SiO_2$ $Fe^{++}$

Ferrous iron is oxidized
by oxygen molecules
to form ferric iron

$O_2$

$Fe^{+++}$ $H_2O$

Ferric iron combines with
water to precipitate
a solid, limonite, from
solution

Limonite
$FeO(OH)$

**Figure 6-13**
The general scheme by which an iron-rich silicate mineral, such as pyroxene, weathers. In the presence of oxygen and water, as at the Earth's surface, these reactions proceed sequentially and irreversibly. Limonite formed in this way is typically a poorly crystallized yellow-brown mass of the mineral geothite, sometimes mixed with another mineral, lepidocrocite, of the same composition but different crystal structure.

One more dissolution process is important because so little of it happens. Quartz, $SiO_2$, dissolves completely without any residue, as does calcite, but its solubility is so extremely low and it dissolves so slowly that it is for practical purposes insoluble in water. As a result, most of the quartz in rocks remains chemically unaltered and survives as a solid residue of weathering, even though it may be mechanically modified by breakage or rounding. Quartz is thus a stable mineral under the conditions that prevail at the earth's surface. But exactly what do we mean by stability of minerals?

As an experiment, some ground quartz was kept in water for three years, and at the end of that time only one part per million—one thousandth of a gram in a liter of water—had dissolved.

## CONCEPTS OF STABILITY

Stability can be demonstrated rather easily by trying to balance this book on one of its corners. You may succeed temporarily, but the slightest touch will topple it. It is **unstable** in that position because the slightest change in its state—its position in this example—will cause a larger change to another state: the book will fall flat. When it is lying flat, lift it by one edge, and as soon as you let it go it will immediately return to its flat, or **stable** position. The book can be placed in an intermediate state by standing it upright; in this position it is **metastable,** because it will return to its upright position if pushed very slightly, but will fall to its stable state if given a bigger push.

The chemical stability of minerals can be described in much the same way by reference to their tendency to go to other states. Other states of minerals are the products of reactions, simple or complex. A simple change is that shown by the different crystalline structures of calcium carbonate. Aragonite, a tightly packed structure, is stable at high pressures, but if it remains for very long at the low pressures of the Earth's surface, it will tend to change to the more open structure calcite, which is stable at low pressures. Iron pyroxene is unstable at surface conditions partly because it tends to dissolve in water. But that is not a good enough criterion, for salt dissolves even more easily in water and it is stable. The missing criterion is that after dissolving, the stable material will reprecipitate or crystallize as the same mineral, whereas the unstable one will form something else. The dissolved salt will form new salt crystals after evaporation, but the dissolved pyroxene will form a mixture of crystals of hydrated iron oxide and silica. We can rank the stability of such minerals as salt and quartz under weathering conditions on the basis of their solubility: the less soluble, the more stable. Quartz wins hands down.

Given these criteria, we can use a combination of our knowledge of weathering in the field and chemical experimentation to deduce a ranking of minerals in the order of stability under weathering conditions, including alteration and solubility behavior, as in Table 6-1. Included in the table are the products of weathering, clays and oxides, which are among the most stable.

Table 6-1
Stability of some common minerals under weathering conditions at the Earth's surface.

| Most stable | Fe-oxides |
|---|---|
| | Al-oxides |
| | Quartz |
| | Clay minerals |
| | Muscovite |
| | K-feldspar (orthoclase) |
| | Biotite |
| | Na-feldspar (albite) |
| | Amphibole |
| | Pyroxene |
| | Ca-feldspar (anorthite) |
| Least stable | Olivine |

We can use these relative stabilities in a general way to show how drastically different surface conditions on Earth are from the environments in the Earth where the minerals were crystallized and remained stable. Muscovite, a mica, is formed at moderate temperatures and pressures, and hence has a small tendency to alter to clay minerals. Pyroxene is formed at higher temperatures and pressures, and so has a greater tendency to weather. Both the order of crystallization and the order of stability under weathering conditions are related to the stabilities of chemical bonds and crystal structures under different temperatures and pressure, subjects covered in the chapters on igneous rocks (16) and metamorphism (18). The order of stability of the minerals found in igneous rocks is roughly the reverse of the order of crystallization from a granitic melt as it cools; the first to crystallize, olivine, is also the first to weather.

It is the instability of silicates that leads to the formation of clays, which are a major constituent of soils. So far we have kept our focus on individual minerals and how they weather. Soils are a mixture of those minerals; in a sense a soil is like a rock — an assemblage of mineral particles that gives us information about its formation. Understanding soils is of the utmost practical interest, for the world's agriculture depends on maintaining their fertility.

## SOIL: THE RESIDUE OF WEATHERING

**The Soil Profile**   A good place to start with soils is to think about them as forming a thin layer that rests on bedrock, analogous to a coating of rust on iron. The way to see its structure, its vertical profile, is to cut a trench through the soil down to bedrock. The top section, or **A-horizon**, is usually the darkest, and contains most of the organic matter, or **humus** — tiny particles of decayed leaves, twigs, and animal remains. It is a zone of intense biological activity, populated by everything from worms and insects to gophers, plant roots, and abundant microorganisms, all living in an ecological microcosm. In a thick soil that has developed over a long time, the minerals of the top layer are mostly clays and residual insoluble minerals like quartz. The soluble minerals are absent. Lower in the soil profile, in an intermediate layer called the **B-horizon**, there is relatively little organic matter, and soluble minerals and iron oxides are present. The lowest part, the **C-horizon**, is slightly altered bedrock, broken and decayed, mixed with clay. At its base it merges imperceptibly into solid bedrock.

Geologists and soil scientists have put together many observations that show that the soils are thicker where they are older and where they receive warmth and moisture. These are expected conclusions from what we know about the weathering of minerals. From them we can construct a model for soil formation.

**Soil Formation in Relation to Climate**   We can compare a soil to a vertically oriented pipe containing fixed solid chemicals that react with water. Water runs through soils just as it would the

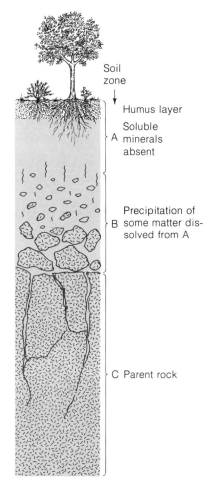

Soil
zone

Humus layer

Soluble
A minerals
absent

Precipitation of
B some matter dis-
solved from A

C Parent rock

**Figure 6-14**
Typical soil profile developed in temperate regions on granite bedrock. The thickness of the soil profile depends on the age of the soil and the climate. Transitions from one zone to another are normally indistinct.

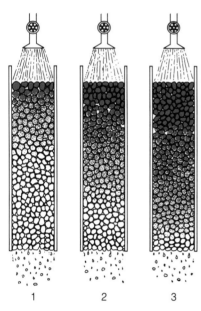

Figure 6-15
The extent of mineral alteration in a
soil as illustrated by water flowing
through a pipe packed with chemicals
that react with the water. Columns 1, 2,
and 3 represent successive stages in the
downward progression of alteration
zones under a steady flow of water.
Alternatively, they can be thought to
represent the results, at the same time
stages, of three different rates of flow of
water, 1 the lowest; 3, the highest.

pipe, for soils are permeable, and rain water runs through them
before infiltrating the pore spaces of the bedrock below. As rain
water passes over the solids in the pipe, it reacts chemically with
them. The more fresh water that runs through the pipe, the more
the solids will react with the fresh undersaturated solvent. The
longer the time the water runs through, the more the solids will
be altered. The higher the temperature, the faster the reactions.
All these effects combine to make the part of the solid closest to
the fresh-water inflow react the most because it is there that the
water is most undersaturated. The water is less efficient a solvent
as it passes on through the rest of the solids, for it is starting to
approach saturation. As a result, the alteration proceeds along a
"front" that moves slowly along the pipe as time goes by.

This is a simplified view of what is essentially the same kind of
process in soils. It explains why the tropical soils are thick and
devoid of most unstable minerals and why arid soils tend to be thin
and rich in unstable minerals. The complexities of clay composition also result in differences in soil mineralogy under different
climates. Just as temperature speeds up mineral alteration, so does
it stimulate the growth of vegetation and bacteria, which speed up
the rate of soil formation. In the extraordinarily dry, frigid environment of some bare ground areas of Antarctica, there are
practically no microorganisms, much less larger organisms, and
weathering is at a minimum. The whole process can be diagrammed
as a complex positive-feedback process involving the response of
mineral decay to the fundamental variables of time, temperature,
and rainfall and the secondarily induced variables of biological
activity (Fig. 6-16). If all this is true, then we should find a fair
correspondence between soil type and thickness and climatic
regions — and we do.

**The Major Soil Groups**   At least three major kinds of soil can
be classified on the basis of the minerals present in their A- and
B-horizons. The expected A- and B- zones of a soil in a high-rainfall
area would be **leached;** that is, the soluble or quickly altered
materials would be carried away by the water. The resulting soil
would be rich in the insolubles, quartz, clay minerals, and iron
oxide alteration products. Calcium carbonate would be absent.
From their abundance of aluminum and iron, the soils are called
**pedalfers** ( *ped* from the Greek word for soil; *al* and *fer* from the
English and Latin names of aluminum and iron). Much of the eastern half of the United States and most of Canada are covered with
fertile soils of this kind.

A second major group of soils are those that contain soluble
minerals, indicating formation in dry, warm climates, such as those
of the western United States. Since calcium carbonate is an important constituent of these soils, they are called **pedocals.** Some of
the calcium carbonate may be dissolved by the occasional rainfall,

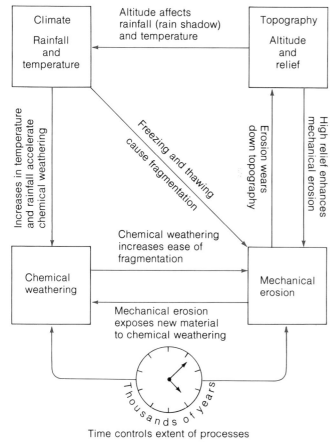

**Figure 6-16**
Interrelations among the major factors controlling weathering and erosion.

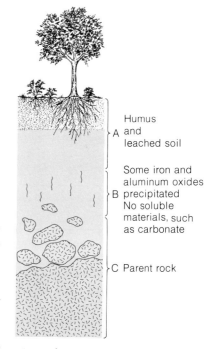

**Figure 6-17**
Pedalfer soil profile developed in a region of high rainfall on granite. The only mineral materials in the upper parts of the soil profile are iron and aluminum oxides and silicates such as quartz and clay minerals, all of which are very insoluble. Calcium carbonate is absent.

but in a warm, dry climate, much of the soil water evaporates, leaving precipitated pellets and nodules of calcium carbonate, typically in the B-horizon of the soil. These nodules may form even where there is no limestone; in such soils the calcium carbonate may come from a combination of the calcium leached from silicates and bicarbonate formed by weathering reactions of carbonic acid. The pedocals contain less clay and more unaltered silicates than the pedalfers, again a result of the low rainfall. These soils are normally less fertile, as a result of the combination of mineralogy, texture of the material, and climate, which does not permit a rich population of organisms in the soil.

A third soil type is the **laterite**, a deep red soil of the tropics in which all silicates are completely altered, leaving mostly aluminum and iron oxides, with not only calcium carbonate leached, but silica as well. So much water has passed through these soils

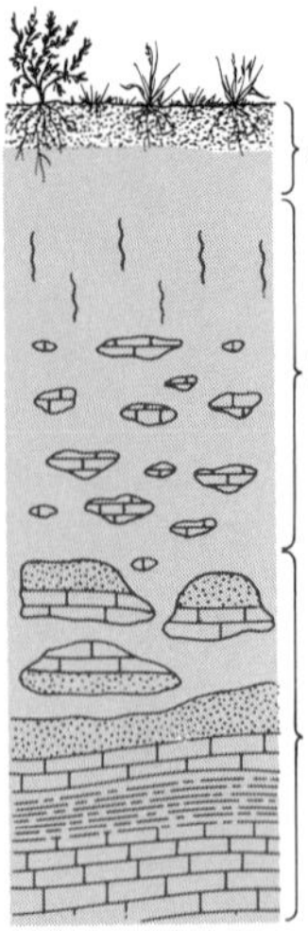

Figure 6-18
Pedocal soil developed in a region of low rainfall on sedimentary bedrock. The A-zone is leached; the B-zone is enriched in calcium carbonate precipitated by evaporating soil waters.

A — Humus and leached soil

B — Calcium carbonate pellets and nodules precipitated

C — Parent rocks: sandstone, shale, and limestone

Thin or absent humus

Thick irregular masses of limonite and other insoluble oxides

Iron-rich clays and aluminum hydroxides

Thin leached zone

Mafic igneous rock

Figure 6-19
Laterite soil profile developed on a mafic igneous rock in a tropical region. In the upper zone, only the most insoluble precipitated iron and similar oxides remain, plus occasional quartz. All soluble materials, even including relatively insoluble silica, are leached; thus the whole soil profile may be considered to be an A-zone directly overlying a C-zone.

that all except the most insoluble compounds are gone. If there is little iron in the parent bedrock, then bauxite, a special variety of this kind of soil, is formed. These tropical soils, though they may support lush vegetation in places, are not very fertile soils for crop plants, because clearing and tilling cause the organic matter, the humus, to be quickly oxidized under the warm climate, so that it does not accumulate to give a rich black soil. For this reason, many lateritic soils can be used for only a few years before they become barren and have to be abandoned.

## WEATHERING: SOURCE OF SEDIMENT

Weathering is not only the major process by which rocks are destroyed, but it is also the supplier of the raw material of sediment. By weakening rocks so that they are more easily fragmented, chemical weathering provides the rock and mineral particles of sedimentary detritus. The detritus bears the marks of the kind and extent of weathering in its composition. If chemical weathering is extensive, virtually the only products will be clays, iron oxides, and quartz grains. Some will remain as soil on the weathering surface; the rest will be transported by rivers and the wind to sedimentary environments where muds and quartz sands will be deposited. But such extensive weathering is an extreme and most of the Earth's surface is not so intensely affected by the coupling of rapid chemical attack in hot, wet climates and the long time periods of crustal stability required for the process to reach an advanced stage. The intermediate stages of weathering are more common; therefore, it is more important for us to see what the sedimentary products of such weathering are.

1    High granitic mountains: feldspar and quartz rapidly fragmented by mechanical erosion

Transportation rounds fragments and sorts by size, producing arkose, a sandy sediment rich in feldspar

Granite               Arkose

Feldspar           Quartz

2    Low hills and plains: most feldspar weathered to clay before mechanical erosion; quartz remains intact

Transportation rounds and sorts quartz, producing sands composed almost entirely of quartz

Granite            Quartz Arenite

**Figure 6-20**
Two different kinds of topography on the same granitic bedrock weather to produce sandy sediments of differing feldspar content.

**Incomplete Weathering: Quartz and Feldspar**    If we compare the survival rates of the two most abundant rock-forming minerals, feldspar and quartz, we can get an index of weathering processes. The total amount of quartz stays relatively constant because it is so insoluble and hard, and therefore resistant to erosion. Feldspar, however, is gradually changed to clay; the longer and more intense the weathering, the less of it there will be. Conversely, if chemical weathering is slight and mechanical erosion is the major process, feldspar grains will be contributed to sediments in virtually the same proportions as they are found in the parent, or source, rock. Thus the ratio of feldspar to quartz should be a general indicator as to which of the two weathering processes is most active.

For example, the rapid mechanical erosion of granitic mountains results in the accumulation of unweathered sand and gravel river deposits that have about the same mineral composition as the original granite. Arkoses, feldspar-rich sandstones that form in this way, can be distinguished from their source rocks only by the nature of the sorting and other sedimentary structures, such as bedding. As the highlands are worn down, perhaps a change in climate will take place; chemical weathering will then play a dominant role, and the feldspar will be increasingly altered to clay. The sand transported and deposited by the rivers becomes dominated by quartz, with much associated clay. The rocks that are formed from such deposits are quartz-rich sandstones, or quartz arenites. In contrast with an arkose, which provides detailed information about the mineral composition of the source rock being eroded, the quartz arenite tells us mainly that the source rock contained quartz, even though an occasional grain of feldspar or some other silicate may give a bit more information on the source rocks. It is in this way that geologists working with sedimentary rocks can do detective work on the topography, climate, and rock composition of the past.

Implicit in the interpretation of climate and topography from weathering products is the control that parent rock composition has on soil composition and sedimentary detritus. It is again a matter of time and intensity. Under intense, long-continued weathering, the soils and detritus will reflect mainly climate and topography; the soil developed on a granite may look much like one on a shale. The shorter the time and the less intensive the weathering, the more the compositional differences of the source rocks show up in the soil and detritus. A young soil will still show some remains of unweathered minerals from which the parent rock composition can be inferred. But we gain one kind of information at the expense of another. The more we are able to deduce about the parent rock, the less likely it is that we will be able to get much information on climate, because weathering will probably have been too slight to leave a strong impression. As in our automobile junkyards, the older and rustier and more dismantled the cars are, the harder it is to tell the make, model, and year. The younger they are, the easier it is to estimate how rapidly they have rusted.

## CHEMICAL WEATHERING AS THE REACTION BETWEEN ATMOSPHERE, OCEAN, AND CRUST

If chemical weathering is important as the maker of detrital sediments, it is equally important as the source of the dissolved materials that filter into the ground, that go to the oceans in rivers, and that make sea water the complex salty solution it is. The weathering processes that produce those solutions are chemical reactions in which atmospheric gases (carbon dioxide and oxygen) combine with water to attack exposed crustal materials of the earth. Thus by describing the overall effects of weathering, we can see how the atmosphere, the waters of the continents, the oceans, and the surface of the crust are all linked in a giant chemical system.

All the gases of the atmosphere and all the water on the Earth's surface originated from within the interior of the Earth and escaped via volcanoes, hot springs, and other zones of connection between interior and surface. Probably most of this gas came out early in the Earth's history and has been added to slowly throughout geologic time. The water has accumulated primarily in the oceans, but much is stored as groundwater, and in hydrated minerals of sedimentary origin. The great bulk of the carbon dioxide has not accumulated in the atmosphere, but has been transformed by dissolving in water to form bicarbonate ion and then combining with calcium to make the limestones of the crust. Another fraction of the carbon dioxide was transformed to organic carbon by photosynthesis, buried in the form of organic matter in sediments, and ultimately altered to coal and oil. Oxygen accumulated in the atmosphere as the by-product of the photosynthesis that produced the organic carbon. Rocks elevated to high mountains by tectonic forces and exposed to the atmosphere by erosion have also come from the interior.

All of these components, the rocks, the water, and the gases, were once together at various places deep in the Earth at high temperatures and pressures. The movements of solid lithosphere and magmas combine to allow the gases and water to separate from the rock-making materials from the interior and leak out into the atmosphere as the rocks are pushed to the surface. In the erosional processes that take place at the surface of the Earth, all those components come back into contact at much lower temperatures and pressures. A new set of low-temperature reactions— the weathering process— takes place between the rocks and gases. Water and carbon dioxide combine to form carbonic acid, which weathers silicate minerals. Oxygen oxidizes iron and other metals. Because the chemical stabilities of minerals that can be formed by these components are so different at low temperatures and pressures, an entirely new set of minerals, dominated by clays and carbonates, is formed instead of minerals like olivines and pyroxenes. Thus the earth is like a giant chemical reaction chamber made up of two subchambers, a hot one in the interior and a cold one at the exterior. The conveyor belt between the chambers is volcanism and tectonics. Weathering is the reaction in the cold chamber.

The ocean basins are the receptacles for all of the dissolved products of weathering. These materials do not simply accumulate there, for in the sea they come together to precipitate sedimentary minerals. The most abundant is limestone, formed in the sea by the combination of calcium and bicarbonate ions released by weathering. In this way the calcium is precipitated—mostly in

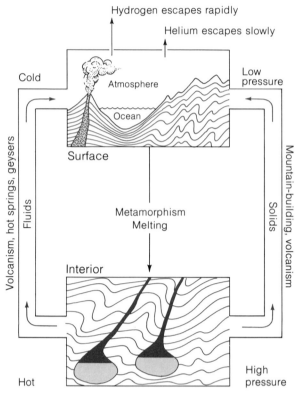

**Figure 6-21**
The chemical reaction system of the Earth is composed of two coupled subsystems, one in the interior, where igneous and metamorphic rocks are formed, and the other at the surface, where atmosphere, ocean, and crust react. Nothing escapes from this system except the lightest gases, hydrogen and helium, which have atoms small enough to escape the Earth's gravity field. [After R. Siever, "Sedimentological Consequences of a Steady-State Ocean Atmosphere", *Sedimentology,* vol. 11, copyright International Association Sedimentology.]

the form of shells of organisms — at the same rate that it is brought in by rivers, and the ocean stays in a steady state, unchanging in the amount of calcium, even though ions are constantly coming in and going out. So it is with all of the other elements. Because of these steady-state reactions, the oceans have probably remained at about the same composition throughout much of the history of the Earth. The ocean and its sediments are the products of the last stage in the reactions between the atmosphere and the crust.

1. Chemical and mechanical weathering complement and reinforce each other in erosion.

2. Chemical reactions of weathering, such as oxidation of iron or decomposition of organic material, vary widely with climate and material.

3. Potassium feldspar weathers by partially dissolving to form solid kaolinite. Evidence for this reaction comes from field observation, its exact nature from laboratory experiment.

4. Feldspar weathers by using up water and hydrogen ion to form kaolinite, leaving potassium and silica in solution.

5. Carbon dioxide dissolves in water to form carbonic acid, an effective chemical weathering agent.

6. Chemical weathering is most active in warm humid regions and least active in cold or arid regions.

7. Clay minerals — kaolinite, montmorillonite, illite, and others — are the products of weathering of various materials under different climatic regimes.

8. Limestone is completely dissolved during chemical decay and is the rock type most rapidly attacked during weathering in humid climates.

9. Mafic minerals may dissolve completely. Iron contained in them is oxidized by the oxygen of the atmosphere and then precipitated in soils and sediments as limonite.

10. Relative chemical stabilities of various minerals at the Earth's surface are ranked on the basis of the kind and degree of solubility.

11. Soils, the residues of weathering over long periods of time, form distinctive types in response to climatic regimes: pedalfers, pedocals, and laterites.

12. The quartz-feldspar ratio of a clastic sediment is a useful guide to the rate and kind of weathering in the eroding source area.

13. Chemical weathering is a reaction between the solid crust of the Earth and the atmosphere and oceans, all of which originated within the interior of the Earth.

14. The ocean keeps its chemical composition in a steady state as it precipitates chemical sediments while new dissolved substances are brought in by rivers.

# EXERCISES

1. Rank the following rocks in order of their rapidity of weathering in a humid climate: granite, limestone, quartz sandstone, basalt. Give reasons for your choice.

2. Would you expect a basalt to weather more quickly than a granite in a semiarid climate? Would you expect a basalt to weather more quickly than a gabbro, its coarse-grained equivalent, in a semiarid climate? Why?

3. If you were a mountain climber and had to drive pitons (metal spikes used to hold ropes of climbers) into bare rocks at high altitudes, would you find it easier in desert mountains or mountains that receive high rainfall?

4. Describe how you might expect a calcium-iron garnet ($Ca_3Fe_2(SiO_4)_3$) to weather chemically and what the weathering products might be.

5. Venus has much more carbon dioxide in its atmosphere than Earth, but little water. Assuming that igneous rocks are much the same there as on Earth, would you think weathering on Venus would be more or less intense than on Earth?

6. If you were a conservation engineer and had to design a plan for making soil form as quickly as possible on a fresh, broken rock rubble left by construction or mining operations, what would you do (without importing any soil from elsewhere)?

## BIBLIOGRAPHY

Goldich, S. S., "A Study in Rock Weathering." *Journal of Geology,* vol. 46, pp. 17–58, 1938.

Keller, W. D., *The Principles of Chemical Weathering.* Columbia, Missouri: Lucas Brothers, 1955.

Kellogg, C. E., "Soil." *Scientific American,* July, 1950. (Offprint No. 821.)

Krauskopf, K. B., *Introduction to Geochemistry.* New York: McGraw-Hill Book Company, 1967. (Chapters 4 and 7.)

Loughnan, F. C., *Chemical Weathering of the Silicate Minerals.* New York: American Elsevier Publishing Company, 1969.

McNeil, M., "Lateritic Soils." *Scientific American,* November, 1964. (Offprint No. 870.)

# 7

## The Natural Water Cycle and Groundwater

*Water dissolves minerals during weathering, then carries the dissolved material away—into the ground or into rivers, most of which ultimately empty into the ocean. The movement of the Earth's waters from one place to another and the dissolved loads carried by them are parts of a continuous overall pattern: the hydrologic cycle. Groundwater accumulates by infiltration of water into soils and bedrock, and reappears at the surface in springs and stream beds. Groundwater levels, and thus water supplies, are a balance between the rate of infiltration and the rate of loss by springs, streams, and pumping from wells. The evolution of surface waters and the ocean are related to the escape of gases from the interior.*

### WATER, WATER, EVERYWHERE . . .

Since the first men rocketed to the moon, the Earth has had a new name: "the Blue Planet." Although most of us had heard or read the dry statistic that about two-thirds of the Earth's surface is covered with water, it had never been brought home as spectacularly as it was at the time of the first space flight. Given the same view of Earth, a non-Earthman space scientist might have speculated on what kind of life could possibly exist in the midst of all of the clouds of water vapor in the sky and the immense areas of the blue seas on the surface. Geologists, of course, have long wondered what other planets could possibly be like without this liquid that sustains life and takes part in almost every process on the surface of Earth. Thus it is not surprising that the study of water, the science of **hydrology,** is an important part of geology. The study of water is also a central part of oceanography and meteorology.

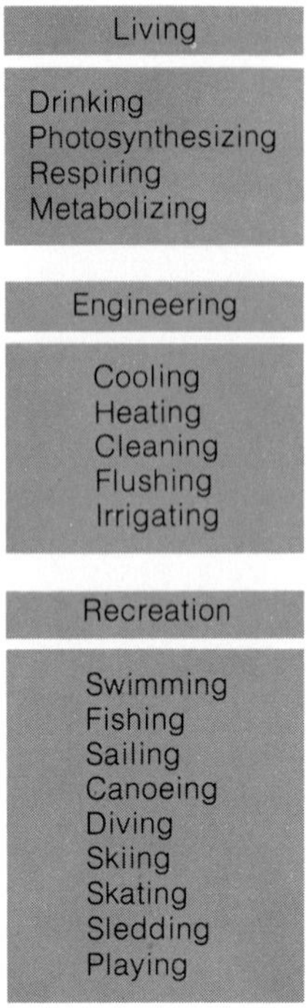

| Living |
| :---: |
| Drinking |
| Photosynthesizing |
| Respiring |
| Metabolizing |

| Engineering |
| :---: |
| Cooling |
| Heating |
| Cleaning |
| Flushing |
| Irrigating |

| Recreation |
| :---: |
| Swimming |
| Fishing |
| Sailing |
| Canoeing |
| Diving |
| Skiing |
| Skating |
| Sledding |
| Playing |

**Figure 7-1**
Some of the value
of water.

The radius of the hydrogen atom is so small compared to the large oxygen atom that the hydrogens appear as small bumps on the surface of the oxygen. Partly as a result of this, the electrical charge is unevenly distributed and the molecule acts as a little magnet. The magnetic behavior of water is an important reason why it dissolves many solids so easily: the magnetism attracts ions of the material being dissolved, in a sense by "plucking" them out of the solid and keeping them in solution.

In this chapter we will concentrate on how water moves from one place to another on and in the crust and, in so doing, acts as the great transporting agent of dissolved material weathered from rocks. Ice and snow, different forms of the water molecule, play their role too in hydrology, and so we start with the basic building block, the chemical compound $H_2O$.

**Two Atoms of Hydrogen and One of Oxygen**   Two centuries ago *John Dalton,* one of the fathers of modern chemistry, puzzled over the results of experiments that showed that the characteristic elements of a chemical compound were always combined in the same relative proportions. The ratio of the weights of hydrogen gas and oxygen gas that would combine to form water, each gas being used up completely, was always the same. It was his recognition of this **law of constant proportions** that laid the basis for the atomic theory of matter, which holds that everything is made up of small building blocks, or atoms, and that each chemical element consists of atoms of a distinctive kind and weight. The atomic theory led to the formula for water, $H_2O$, which is shorthand for the chemical compound that always consists of two atoms of hydrogen combined with one atom of oxygen.

Some of the properties of $H_2O$, though they may be familiar enough to us, are actually unusual in comparison with those of many similar liquids. One characteristic whose effects are far-reaching in our lives is the response of water to heat or cold—in particular, its **thermal expansion,** the way it expands or contracts when the temperature is changed. As most substances do, liquid water contracts when cooled (though the amount is small). But between 4°C (about 39°F) and its freezing temperature, 0°C (32°F), it abruptly changes its course and expands. When the liquid freezes, it again behaves contrary to most other materials and expands further. The expansion that accompanies freezing exerts a powerful force: it can crack automobile engine blocks if sufficient anti-freeze has not been added. Water seeps into cracks in rocks and expands as it freezes, thus acting to wedge apart the cracks, eventually splitting them. In this way, one important property of water is responsible for much mechanical fragmentation of exposed rock (see Chapter 8).

Water is the compound that is so well known to us that we use it to exemplify the three states of matter, solid (ice and snow), liquid (water), and gas (water vapor). We explored the solid state in Chapter 4. Matter in this state has a definite volume and is rigid; it holds its own shape, however formed. This is so because the atoms are held together strongly by chemical bonds arranged in a three-dimensional network, making the structure of a crystalline or semicrystalline solid. Ice is a crystalline mineral whose individual crystal forms are displayed in the enormous variety of snowflakes. A glacier is, in a sense, a "rock" body made up of an enormous quantity of interlocking crystals of the mineral ice.

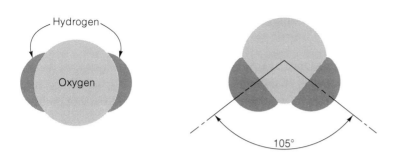

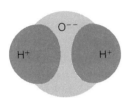

**Figure 7-2**
Three views of the water molecule, in which two hydrogen atoms are bonded to one oxygen atom. Each hydrogen loses an electron, becoming a positively charged ion. The oxygen atom gains the two electrons, becoming a negatively charged ion.

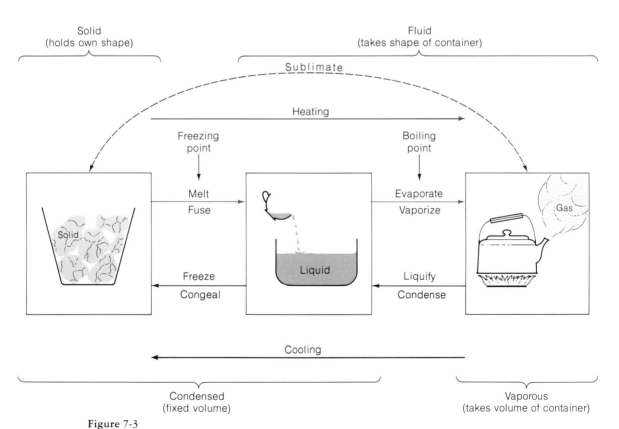

**Figure 7-3**
The three states of matter, as exemplified by water: ice (solid), water (liquid), and water vapor or steam (gas).

Raising the temperature of ice above 0°C causes it to turn into water, liquid $H_2O$. Though water has a definite volume, even though it may expand or contract with change of temperature, it is fluid; it does not hold its own shape. The molecules are loosely bound to each other and easily slip over and around each other, something like couples on a packed dance floor. It is as a liquid that water is such a good solvent and, when in motion as strongly flowing currents, able to carry mud, silt, sand, and boulders. It is at the temperatures at which $H_2O$ is liquid that life on earth exists, for water is the carrier of dissolved food that is vital for every form of life, from humans to bacteria.

Raising the temperature of water to 100°C (212°F) causes it to boil and form water vapor. But in contrast to the melting of ice, which takes place at only one temperature, water can slowly evaporate to form vapor at *any* temperature below boiling. Even ice can "evaporate" to vapor—a process we call **sublimation**. Pure water vapor condenses to liquid water at or below the boiling temperature. Our everyday experience is the clue to the behavior of water vapor in air, where it is not pure but is mixed with other gases, nitrogen and oxygen. The **relative humidity** is the amount of water vapor that is in the air relative to the total amount it could hold if it were saturated—a concept similar to that of the saturated liquid solution. Because the degree of saturation changes with temperature, air saturated with water vapor at one temperature will condense some of its vapor to water droplets when it is cooled to a lower temperature, as anyone who has mowed grass on a dewy morning knows intuitively.

A gas is fluid and has no fixed volume; it expands and contracts to fill containers. Of course, the gaseous envelope that surrounds the Earth has no container; its volume and distribution are determined by the pull of gravity. Gases in general are not efficient solvents for solid substances, as are liquids, so that they cannot carry the dissolved products of weathering, as do rivers. They are not the most efficient transporter of solids, either, for the currents —the winds of our atmosphere—have to be strong to carry any particles coarser than clay or fine silt any great distance above the ground. Vapor is the state in which water is transported through the Earth's atmosphere—from the oceans to the continents, where it is condensed and precipitated as rain or snow.

Water vapor is an invisible gas. The white clouds of the sky or the billows of steam that we associate with vapor are actually composed of tiny droplets of liquid water formed by the cooling of the vapor below the boiling point. A good way to see this is to note the spout of a boiling teakettle. At the spout, the stream of vapor is colorless, but within a short distance the white clouds appear as the vapor mixes with the cool air outside and condenses to droplets.

When an occasional molecule of $H_2O$ in the upper atmosphere splits up into separate hydrogen and oxygen atoms under the effect of strong radiation from the Sun, some of the hydrogen atoms may escape permanently from Earth. Thus there is actually a tiny loss of water to outer space as a result of the steady loss of hydrogen from the atmosphere. Hydrogen is the lightest element, and some of its little atoms can move fast enough to escape from Earth's gravity—just like a rocket.

## WATER TRANSPORT: THE HYDROLOGIC CYCLE

Not only is this planet very watery, but gravity ensures that it will stay that way. Although much water is moved around from one place to another, the total is conserved. The **hydrologic cycle** is a simplified description of the ways in which waters move from one place to another and how much is transported. Knowledge of the cycle informs the judgement of water conservationists who tell us that there are limits to how much natural fresh water we can ever get to satisfy the voracious "thirst" of our industrial society. The limits are determined by the total amount of rainfall that reaches the continents.

The external heat engine of Earth, powered by the sun, is what drives the hydrologic cycle at the surface. It does so mainly by evaporating water from the surface of the warm oceans of the tropics and transporting it by winds, themselves driven by the differences between the temperatures of the hot and cold parts of the globe. The water in the atmosphere condenses to clouds and eventually falls as rain or snow. Much of the rain soaks into the ground by **infiltration**—hence **groundwater.** What does not soak in collects as **runoff**, which finds it way into streams and rivers and runs back into the oceans. Some of the water in the ground may return directly to the atmosphere by evaporating through the soil surface. Another part may be absorbed by plant roots, carried up to leaves, and returned to the atmosphere by **transpiration** —the release of water vapor by plant foliage. A major fraction of the groundwater stays in the ground, most of the shallow subterranean waters slowly moving through near-surface formations and ultimately exiting into stream beds, springs, or the oceans.

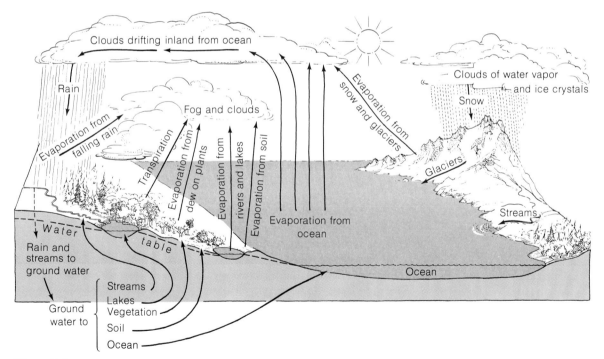

**Figure 7-4**
The hydrologic cycle. [After *Plant Science* by Janick, Schery, Woods, and Ruttan. W. H. Freeman and Company. Copyright © 1969.]

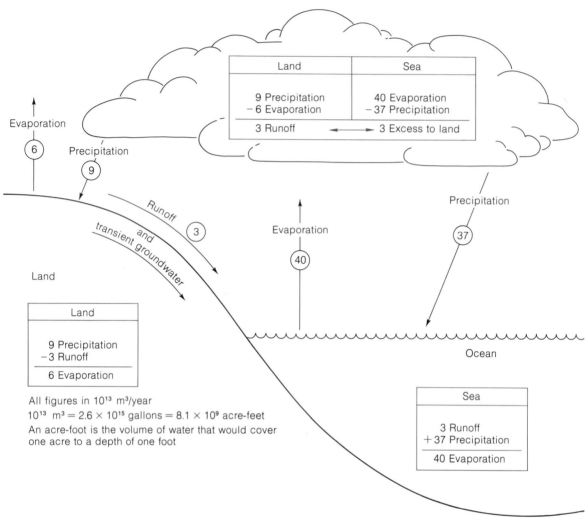

**Figure 7-5**
The mass balance, or water-flow budget, of land, seas, and atmosphere.

Now it is clear why it is the total rainfall that determines the absolute limit of fresh water available on the continents: there is no other source. The only way man could increase the total supply would be to make artificially fresh water by modern chemical methods of desalting sea water—a different way of accomplishing what nature does by evaporation and precipitation. How the total flow from one place to another adds up—what we call a **mass balance**—is shown in Figure 7-5. From this elementary "water-flow budget" for land and sea, we can see that more water evaporates from the oceans than falls on it as rain. This is exactly balanced by the return of water via runoff from the continents, which itself exactly balances the excess of precipitation over evaporation on land.

The United States and many other developed countries of the world face the prospect of carefully planning the use of these large, but not infinite, water resources for the ever-growing demands of industry, urban population concentrations, and agriculture. A more detailed breakdown of the water budget for the United States (Fig. 7-6) gives the "road map" for the decisions that will have to be made. Though only about 7 percent of the total rainfall is used for agriculture, industry, and cities and towns, another 71 percent is unavailable because of loss by evaporation, transpiration, and infiltration (though much of this may go to usable groundwater supplies). It is the 22 percent of untapped stream water that has one of the biggest potentials. That is why the pollution of rivers poses a much greater threat than the loss of their esthetic and recreational values. We may need all the good water we can get as the demands of our population increase. As of 1970, the United States was using 1.4 billion cubic meters of water per day ($5.1 \times 10^{11}$ cubic meters per year) from surface and groundwater sources. Total potential water supplies are only about three times this present use. Unfortunately, abundant river water supplies are not always near the areas of high demand, and piping water for long distances is expensive. Thus, in areas with limited supplies of river water, it will become imperative to maintain—or restore—a flow of unpolluted, usable water.

**Surface Runoff** The direct relation between rain and stream runoff is obvious—especially to anyone who has ever experienced a flash flood following a torrential rainstorm. The yearly amount of water flowing in streams in any region is directly related to the annual rainfall. Most of the streams in arid or semiarid regions are dry except right after the rare rains. The difference between wet and dry is particularly conspicuous when going over a mountain range into a "rain shadow"—an area of low rainfall on the leeward slope of a mountain range. Moisture-laden air rises as it crosses high mountains, cooling and precipitating much of its water vapor on the windward slope. By the time the air reaches the opposite leeward slope, it has lost much of its moisture. As it drops and warms, it becomes lower in relative humidity, and little rain falls. The Cascade Mountains of Oregon offer a striking example of the contrast. The western, windward slopes have high rainfall, the abundant water draining into the Willamette River valley with its lush natural and agricultural vegetation. But just a few tens of miles to the east, over the crest of the range, the countryside is dry and relatively barren. The Sierra Nevada, an even broader rain-shadowing mountain range, keeps the Great Basin to the east a semiarid region swept by dry winds. In this way, we arrive at a perhaps unexpected relation between climate and tectonics—the forces that are responsible for the mountains. The Cascades, a range of extinct volcanoes that extruded enormous volumes of basalt, are the surface expression of ancient tectonic activity; they

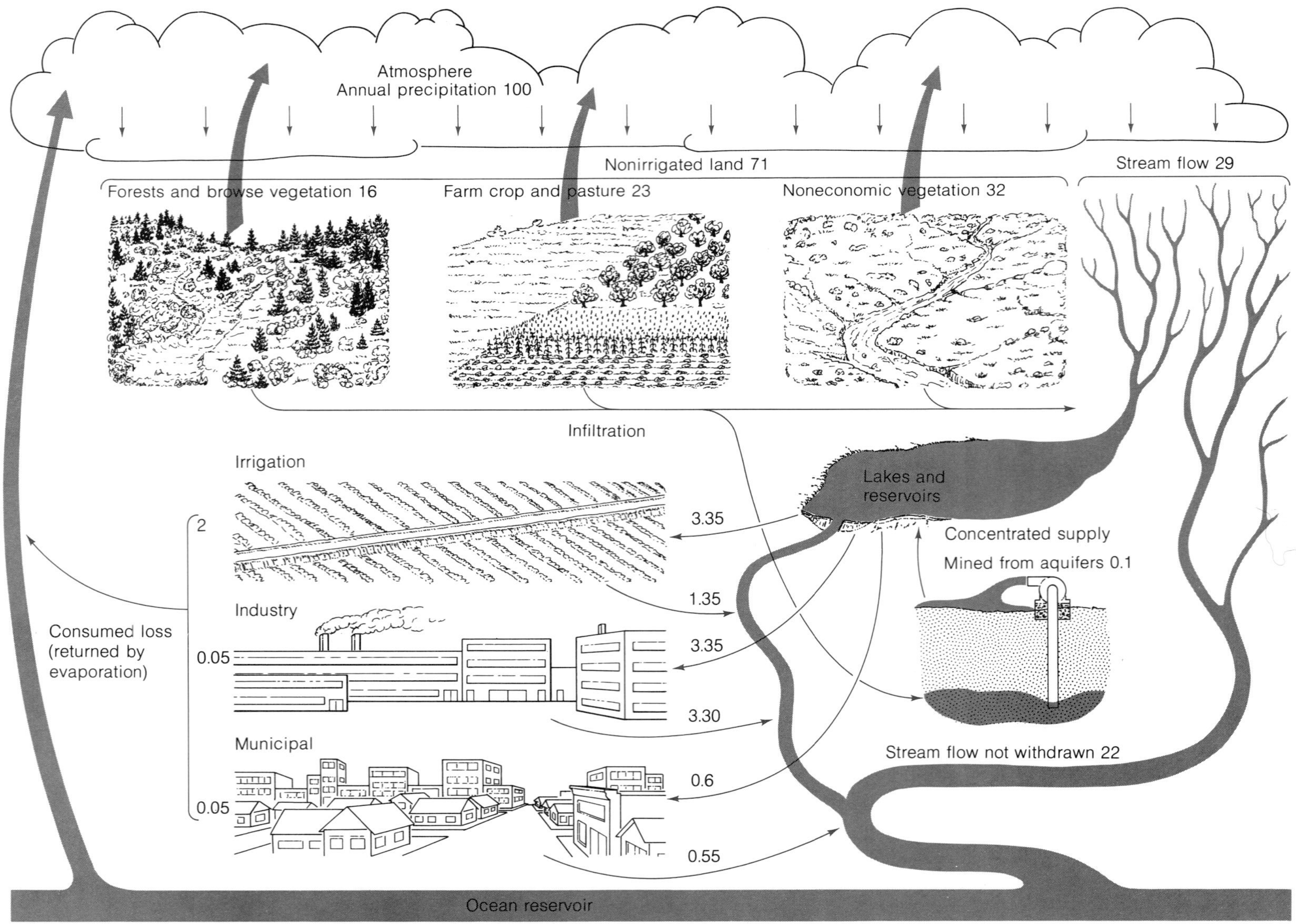

Atmosphere
Annual precipitation 100

Stream flow 29

Nonirrigated land 71

Forests and browse vegetation 16

Farm crop and pasture 23

Noneconomic vegetation 32

Infiltration

Lakes and reservoirs

Concentrated supply
Mined from aquifers 0.1

Stream flow not withdrawn 22

Irrigation 2

Industry 0.05

Municipal 0.05

3.35

1.35

3.35

3.30

0.6

0.55

Consumed loss
(returned by evaporation)

Ocean reservoir

Figure 7-6 *(facing page)*
The hydrologic cycle for the United States shows the fraction of annual precipitation used in a highly developed nation. Twenty-nine percent of the rainfall arrives at the oceans (bottom) via stream flow; 71 percent falls on various types of non-irrigated land, returning directly to the atmosphere (top) by transpiration and evaporation. Water withdrawn for irrigation, industry, and municipal use is shown at left to constitute only 7.3 percent of the total.

may represent a zone along which the Pacific plate was subducted beneath the western edge of the North American plate, throwing up a belt of basaltic and andesitic volcanoes above the descending oceanic plate. The Sierra Nevada is made up of granitic rocks that were thrust up as a large mass during a major orogeny of the Cenozoic (see Chapter 22).

Most of the surface runoff is dominated by the large stream networks of major river systems. Of the enormous number of streams on the continents, only about seventy major systems carry about one-half of the entire runoff from the land areas of the world. Even more dominant are a few major rivers, such as the Mississippi and the Amazon.

Rivers ebb and flow in discharge from season to season, mainly as a response to seasonal precipitation patterns. In some tropical regions it rains heavily only for a few months of the year. In some maritime climates, like parts of the Pacific Northwest of this continent and most of the British Isles, the rain is abundant and more constant through the year. A different pattern of stream flow is shown by rivers draining snow or ice-covered country, where

In his 1967 study of the Amazon River, a member of a U.S. Geological Survey team stressed how huge the Amazon River was by using a "Mississippi" as a unit of measurement. Halfway upstream, the Amazon was only a few "Mississippis" in discharge but by the time it reached the Atlantic Ocean it was seven "Mississippi's."

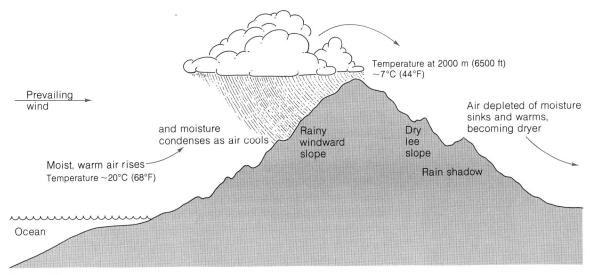

Figure 7-7
A mountain range may produce a rain shadow on its lee slope by forcing warm, moist air to rise with a prevailing wind from the ocean. The rising air cools and causes precipitation on the windward slope and dryness on the lee slope.

Table 7-1
Water flows of some great rivers (in m³/sec).

| | |
|---|---|
| Amazon, South America | 113,300 |
| La Plata, South America | 79,300 |
| Congo, Africa | 39,600 |
| Yangtze, Asia | 21,800 |
| Brahmaputra, Asia | 19,800 |
| Ganges, Asia | 18,700 |
| Mississippi, North America | 17,500 |

spring melt waters can swell the streams to flood stage. Here the response of runoff to precipitation lags over the winter months, the precipitation that fell as snow being stored temporarily in snow and ice fields. The runoff jumps when the store is released by warming in the spring. Similarly, lakes and large areas of swamp and marshland, sometimes called "wetlands," act as storage depots for runoffs. These reservoirs by their large volume smooth out the big variations in discharge of rivers and release a steady outflow downstream. This is also the basis for some flood control dams. It is also the reason why some geologists have worked to stop the artificial draining of wetlands by real estate developers, which if unchecked could wreck much of the natural flood regulation capacity in some areas.

But all of these storage areas for precipitation and runoff are minor compared to the water stored in the ground. Why do rivers continue to run day after day after several weeks of no rain—and no melting snow to feed them? How do springs continue to flow day after day to feed the rivers and lakes? Groundwater flow is the answer.

## WATER IN THE GROUND

How do we know about the water in the ground? We can literally see the rain disappear into the ground in many places. Watch rain falling on a beach sometime. Some soils act like porous sand in sponging up rainwater; others have dense, tough clay layers, or "hardpans," that impede the infiltration of water and are poor absorbers of water. Poorest of all are the rocky surfaces of mountains, which have little or no soil. There, little water runs into the ground; what does, goes through joints and cracks.

Bubbling springs show us where some of the water that soaks into the ground comes out; good places to look are on the lower slopes or the base of a hill or valley wall. If you drill a well, and have picked a good place to do it—for example, where porous sandstone beds are near the surface in a temperate climate—water will fill the bottom of the well. If you are lucky, you might even get water flowing up and out of the hole: in that event, you have

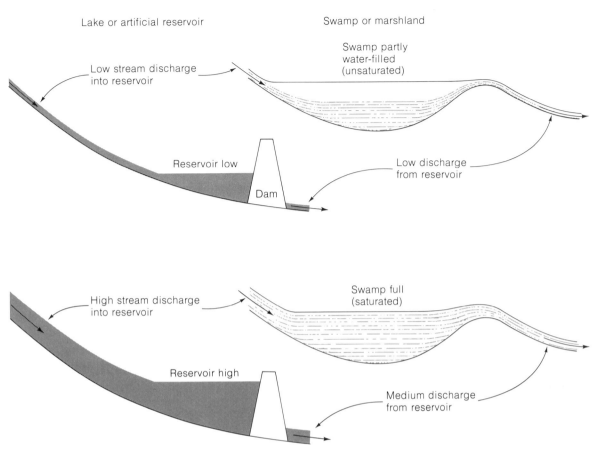

**Lake or artificial reservoir**

**Swamp or marshland**

Swamp partly
water-filled
(unsaturated)

Low stream discharge
into reservoir

Reservoir low

Low discharge
from reservoir

Dam

High stream discharge
into reservoir

Swamp full
(saturated)

Reservoir high

Medium discharge
from reservoir

**Figure 7-8**
A swamp or marshland acts much like a natural lake or an artificial reservoir behind a dam in storing water during times of rapid runoff and slowly releasing it during periods of little runoff.

drilled what is called an **artesian** well. A few more facts and you will have enough to get the general picture. After long periods of dryness, wells run dry, the ground becomes dried out well below the surface, and smaller rivers shrink and may dry up entirely. There has to be an underground connection between the water that sinks in from rain on the one hand and rivers, springs, and wells on the other.

The underground connection has to be through soil and rock. There are no open spaces under the ground with rivers running like they do on the surface, except in cavernous limestone networks and some open lava tubes in volcanic terranes. But if you fill a short length of pipe with clean sand, put a screen at the bottom, and pour water in at the top, in a short time you will see the water dripping out at the bottom. The water travels by twisting and turning through the pore spaces between the sand grains. The smaller the pore spaces and the more tortuous the journey, the slower the water goes through the sand. This property of

allowing fluids to pass through a solid we call **permeability.**
Sands are permeable; hard, compacted muds and shales are not.
The permeability depends largely on the amount of pore space
between the grains or crystals of the rock, the **porosity.** Rocks
vary widely in porosity. Most igneous and metamorphic rocks are
not very porous, but many sedimentary rocks, particularly sandstones, are. Because in some rocks the pores are all connected to
give an easy path and in others the pores are disconnected and do
not allow any exit, all porous rocks are not equally permeable.
Clay particles obstruct paths of flow between many pore spaces
in some sandstones and so may significantly decrease their permeability. This is an important factor in obtaining a good flow of oil:
many such sandstone beds are saturated with oil, but the permeability is too low to allow economic pumping.

Thus the ground acts like a sponge, soaking up rain in some
places and leaking it out at other places. When we add the geological information about the structure of formations and their depth
below the surface to data on the depth of water in wells dug at
various places, we get the picture in Figure 7-9. The sandy or other
kinds of beds that produce waters are called **aquifers.**

**The Groundwater Table** If we connect the levels of water
in the wells shown in Figure 7-9, we get a surface that we call the
**groundwater table,** or just **water table.** That surface is the
boundary between two zones of rock or soil: below the water
table, the pores are completely filled with water, and the zone is
called the **saturated,** or **phreatic,** zone; above the water table is

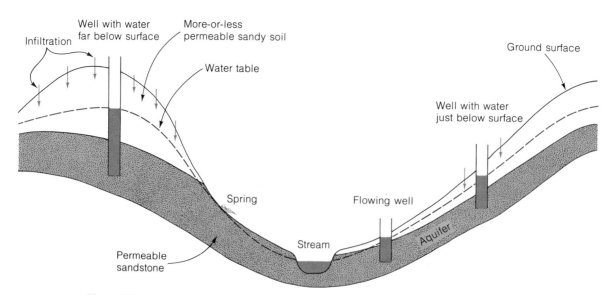

**Figure 7-9**
The groundwater table lies at varying depths beneath the surface, as shown by the water level in
wells. The water table intersects the ground surface at springs and river beds.

an **unsaturated,** or **vadose,** zone. The unsaturated zone is also called the **zone of aeration** because the pore spaces are partly filled with water and partly filled with air. Some of the water in this upper, unsaturated zone is at times in transit down to the water table, but much is held in smaller pore spaces in the same way that sand in a sandbox or beach stays moist long after it has been wetted. Evaporation is slowed almost to nothing for these little drops buried in pores because of the effects of surface tension. Surface tension is shown by a water surface that will support a needle or razor blade gently placed on it. This effect, whereby a skin-like surface layer of the fluid, a few molecules thick, is in a state of tension that gives it supporting strength, reduces evaporation strongly in small droplets.

Knowing the depth and shape of the water table is one of the important aims of the groundwater hydrologist and the driller of water wells. How to predict it from a combination of general understanding and detailed information on a specific area is becoming increasingly important in North America, particularly in the broad, thirsty areas of the Great Plains but also in many other parts of the West and Southwest. Even in regions of relatively abundant rainfall, it is no longer easy to put down a well almost anywhere and get water. Dowsers claim that they can somehow divine where water is underground with a forked stick or other inanimate object rather than with geological knowledge. But scientists remain firm in their conviction that there has been neither proven demonstration of success other than chance nor proposal of any verifiable mechanism that could give an objective explanation of how a forked stick could be attracted by groundwater.

Geological observation teaches us that the water table generally follows with a more subdued form the contour of the surface topography. The "outcrops" of the water table are springs and the beds of rivers; it is in these places where the water drains out of the land. That drainage gives the clue to the fact that most groundwaters near the surface are in motion. The depth of the water table is another example of the workings of a steady-state process: it is a balance between the rate of infiltration, usually called **recharge,** and the rate of discharge at rivers and springs — or pumped water wells. The water table stays at the same place when rain falls frequently enough to stay in balance with the river, spring, and well outflow. Any imbalance, typically by seasonal fluctuation of rainfall, raises or lowers the water table. A long-term imbalance results if there is either a decrease in recharge, such as from a prolonged drought, or from increase in the discharge. That is why water tables are slowly dropping in many regions where discharge has steadily climbed from constantly increasing pumping of water wells. In such areas the imbalance has depleted the reservoir. For an aquifer is just as much a reservoir as any surface lake or pond that can be filled up or depleted.

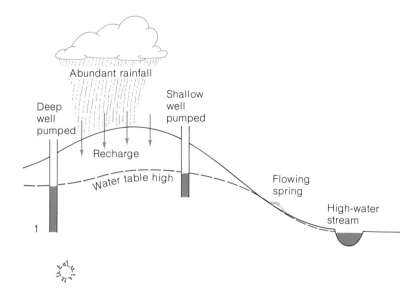

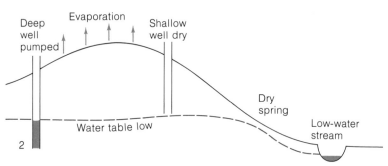

**Figure 7-10**
The depth of the water table fluctuates in response to the balance between what is added by precipitation and what is lost by evaporation plus discharge from wells, springs, and streams.

The flow of groundwater and the position of the water table may be complicated by alternating permeable aquifers and relatively impermeable beds, called **aquicludes,** that hinder or prevent water movement. When aquicludes lie both over and under an aquifer, they produce a **confined** water reservoir. The water pressure in such an aquifer depends on the difference in height between it and the recharge area. If the difference is great enough, water will flow spontaneously out of a well drilled into it; such wells are called artesian wells. If enough artesian wells flow long enough to deplete the reservoir and lower the water table in the recharge area, the pressure will drop. Eventually the wells will stop flowing and have to be pumped. An aquiclude underlying a discontinuous aquifer may support a **perched water table** above

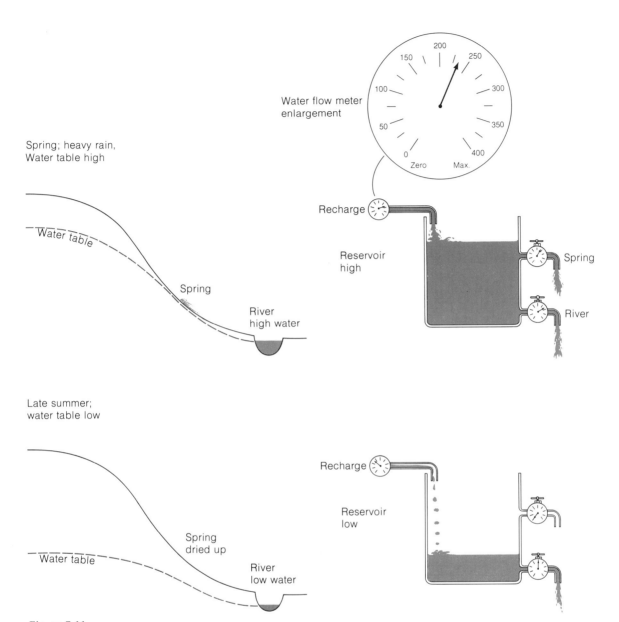

Water flow meter
enlargement

Spring; heavy rain,
Water table high

Water table

Spring

River
high water

Recharge

Reservoir
high

Spring

River

Late summer;
water table low

Water table

Spring
dried up

River
low water

Recharge

Reservoir
low

**Figure 7-11**
The changing height of the water table is a reflection of the volume of water stored in the ground, and can be analyzed as a reservoir with inputs and outputs. The flow meters indicate the relative changes typical from spring to late summer.

the main water table, as in Figure 7-13. Some perched water tables may extend over large areas of hundreds of square kilometers or miles.

Yet another complication is faced by the people who live at the edge of the ocean. A few summers ago, one of us had a visit from a neighbor on the seashore who wanted to know what he could do about a salty taste in the water from his well. He didn't realize

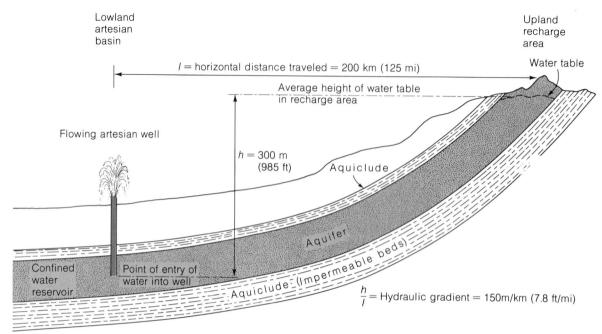

**Figure 7-12**
A confined water reservoir is created where water enters an aquifer situated between two confining aquicludes. The artesian well flows in response to the pressure difference between the height of the water table and the bottom of the well, equivalent to the pressure of a water column 300 m high in this example.

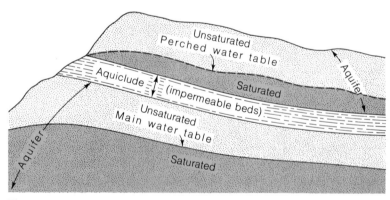

**Figure 7-13**
A perched water table is separated from the main water table by an aquiclude composed of shale or other impermeable beds.

that it was the extensive pumping of his and all the other wells that was causing an incursion of salt water. There is a sloping underground boundary between the fresh water under land adjacent to the sea—or under an island—and the salt water in the sea, as shown in Figure 7-14. If withdrawals by pumping are not excessive,

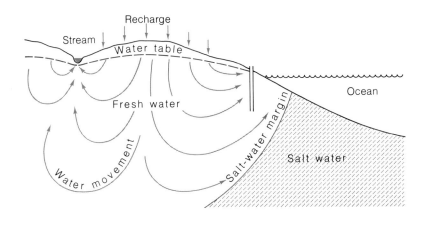

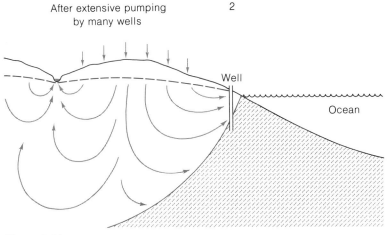

**Figure 7-14**
The boundary between fresh groundwater and salt water along shorelines is determined by the balance between recharge and discharge to the ocean or wells. Normally, as in (1), the pressure of fresh water keeps the salt-water margin slightly offshore. Extensive pumping, as in (2), lowers the pressure on the fresh water, allowing the salt-water margin to move inland.

then the recharge of the fresh water on the land is enough to keep the water table elevated above sea level and maintain a reservoir of fresh water bulged into the sea water at depth. But if withdrawals exceed recharge, the water table is lowered and the reservoir depleted; the pressure of fresh water on the bulge decreases, and sea water moves in. Those closest to the shore are the first affected. Cape Cod, Long Island, Bermuda, and many other near-shore communities and islands have this problem to contend with. There is no solution to the inexorable incursion of seawater except to reduce the pumping or artificially increase the recharge, as will be discussed later in this chapter.

Studies on the rate of groundwater movement have been confirmed and made more precise by the application of radioactive dating techniques to groundwater, using radioactive carbon, $C^{14}$, to date the dissolved carbon dioxide in the waters. On entering the ground the water had in its $CO_2$ the minute amount of $C^{14}$ that is characteristic of rainwater, but as it slowly traveled through the ground the $C^{14}$ decayed radioactively. The velocity is obtained by measuring the distance from the well in which the water was sampled to the recharge area and dividing this by the age as determined by $C^{14}$.

**How Fast Does Groundwater Move?** From studies of response of wells and springs to rainfall recharge, we have learned that the rate of movement of groundwater is fairly slow. Here is another fact of nature that is lucky for us. It is that slowness that keeps the underground reservoirs full, for fast discharge would mean that water wells would run dry after just a few days of no rain. What makes some groundwaters travel rapidly and others slowly? The details of the answer were worked out in the middle of the nineteenth century by *Henry Darcy*, the town engineer of Dijon, France. From observations of the heights of water in wells, the distance traveled, and how permeable the rocks through which the water flowed came what we now call Darcy's Law:

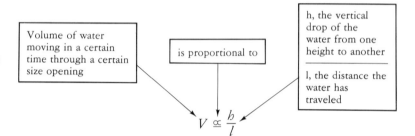

Darcy reasoned that the permeability of a rock was what more or less slowed down the flow for a given drop of height ($h$) in a certain distance ($l$) and so made this into an equation by multiplying the right-hand side by a proportionality factor, called $K$. Darcy identified $K$ as a measure of the permeability of the rock, or in other words, how easily it transmitted water.

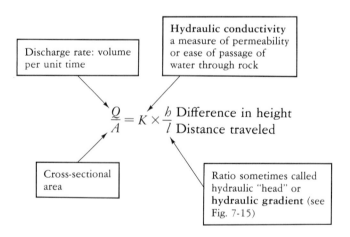

From this equation, we can either determine the velocity of flow or, knowing the velocity, the hydraulic conductivity. Velocities

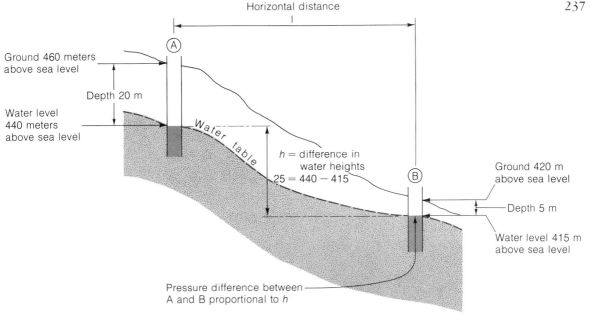

Figure 7-15
Darcy's Law.

have been found by using this equation and dating groundwater by radioactivity. Velocities can also be determined by experimental observations in which dyes are introduced into groundwater at a well used as an artificial recharge point and picked up at a nearby pumping well. Movement in deeper formations is rarely faster than 100 centimeters/day or slower than 0.5 centimeters/day. In most aquifers, groundwater moves at rates of a few centimeters/day. In very permeable gravel beds near the surface, groundwater may travel as much as 15 centimeters/day.

**How Far Does Water Travel Underground?** The length of time that groundwater takes from recharge to discharge depends not only on its velocity but on how long its path is. Routes of travel vary widely, and in a typical flow, such as shown in Fig. 7-16, some routes may be many times longer than others.

So far we have been concentrating on flow over relatively small areas, such as a hillside or valley. Flow is also demonstrable over regions the size of counties or groups of counties hundreds of square kilometers in area. Over such broad areas, flow is dominated by the geological position of permeable formations, the aquifers. Typically the pattern is like that shown in Figure 7-12, the essential condition being a continuous aquifer with a recharge area in an upland area so that the water can run "downhill" underground. Many areas in the great plains and parts of the central midwest

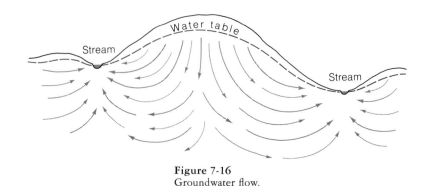

**Figure 7-16**
Groundwater flow.

are underlain by sandstones of various ages that transport waters for hundreds of kilometers. Recall from Darcy's Law that the discharge and velocity are proportional to the change in height over the distance. The vertical height is rarely more than a few hundred meters (in most areas it is much less), but the horizontal distances are measured in hundreds of kilometers, hence the rates of water movement are slow and the recharge takes a long time. What are the consequences for us? If extensive water-well pumping removes the water faster than the slow recharge can fill it, the reservoir will gradually become smaller.

**Mining Ground Water**  A baby born when the first wells were drilled into the Ogallala, a Tertiary formation of sands (and gravels) in western Texas and New Mexico, would be about ninety years old now. He or she would have lived to see the water pressure of those wells steadily diminish and the water table drop drastically, as much as 30 meters (100 feet). The Ogallala aquifer has been the water support of a population that has gradually grown to almost half a million people in this century, providing ample water for irrigation for the agriculture that is the economic base of the region. Intensive research has been going on for the past few years by the United States Geological Survey on how to improve the recharge of this aquifer, which is very slow to recharge naturally because of evaporation and some impermeable soils. It is a hard task, for water has been pumped so extensively that at present rates of withdrawal and recharge the water table would take several thousand years to recover its original position. Other aquifers in the high plains tell the same story in varying degrees.

Because of the extreme slowness of recharge and slow rates of water movement the groundwater in such extensive areas is in every sense being mined, just like coal, so that the amount left in the ground steadily diminishes as production continues. For all practical purposes, this water is an exhaustible resource that, once gone, cannot be replenished. Yet we are not in a crisis situation nationally, though certain communities, in areas where deple-

Table 7-2
Distribution of world's estimated water supply.

| Mode of storage | Locations | Water volume (m³) | Percentage of total water |
|---|---|---|---|
| **Surface** | | | |
| | Fresh-water lakes | $1.3 \times 10^{14}$ | 0.006 |
| | Saline lakes and inland seas | $1.0 \times 10^{14}$ | 0.008 |
| | Average in stream channels | $1.3 \times 10^{12}$ | 0.0001 |
| **Subsurface** | | | |
| | Vadose water (includes soil moisture) | $6.7 \times 10^{13}$ | 0.005 |
| | Groundwater within depth of half a mile | $4.2 \times 10^{15}$ | 0.31 |
| | Groundwater at great depths | $4.2 \times 10^{15}$ | 0.31 |
| **Other** | | | |
| | Icecaps and glaciers | $2.9 \times 10^{16}$ | 2.15 |
| | Atmosphere (at sea level) | $1.3 \times 10^{13}$ | 0.001 |
| | World ocean | $1.3 \times 10^{18}$ | 97.2 |
| Totals | | $1.4 \times 10^{18}$ | 100 |

*Source:* Data from U.S.G.S.

tion is in an advanced stage, are in for tough times. The amount of usable fresh water in the ground is staggering (Table 7-2). Recent estimates are that it amounts to much more than 90 percent of all of the fresh water on the earth. Rivers, lakes, reservoirs, and soil moisture account for the remainder. Glacial ice is the equivalent of seven times the groundwater supply.

The problem is to find untapped supplies and plan for the recharging of aquifers that can be replenished, particularly those

In 1972 and 1973, after unusually wet seasons on Long Island, the water table was so high that water seeped into the basements of many homes unwisely built in low, poorly drained areas in times when the water table was low.

that can be recharged in a short time. Long Island, in New York, offers one example of success at groundwater recharge. Five million people there depend on wells for almost half their water supply. By the early 1940's, extensive pumping had depleted the reservoir to the extent that salt water was encroaching on the aquifer, and a state-regulated program of artificial recharge was started to reverse the deterioration of the water supply. By drilling an extensive system of recharge wells so that used water could be put back into the aquifer, and by constructing large basins to help accelerate infiltration from surface waters, including storm and industrial waste drainage, the aquifer was built back up to usable levels.

The Long Island story is one of success up to now, and is an encouraging example of a stitch in time. But the situation remains precarious, for population and water demands grow daily. One of the unlooked-for results of rapid growth in urban and suburban areas such as Long Island is that natural recharge of the aquifer through seepage into soil becomes decreased as buildings increase in number and more and more paved streets, sidewalks, driveways, and parking lots appear. Our expanding urban areas not only deplete the underground reservoirs by pumping—they interfere with its natural refilling.

## THE QUALITY OF WATER

When Portia speaks of mercy in the Shakespeare that so many have learned, she uses a common metaphor:

> *"The quality of mercy is not strain'd.*
> *It droppeth as the gentle rain from heaven."*
> MERCHANT OF VENICE, ACT 4, SCENE 1

Rain water is our usual standard for purity, and in parts of the Earth where the air is still clean, so is the water. A great many, if not most, water wells produce fine water of excellent quality, just as good as rainwater to drink. Yet centuries before anyone thought of chemical contamination of our air and water, people who drank some well waters made wry faces at the taste and tended to prefer wine or beer. Some groundwaters taste of "iron"; others may have a disagreeable odor. Every few months a newspaper somewhere will feature a story about water that froths out of the faucet like detergent. What makes the difference?

**Rainwater to Groundwater** One place where we can sometimes find groundwater about as pure as rainwater is under a bare dune made of pure quartz sand. Analyses of such waters show that they have about the same amount of carbon dioxide dissolved in the water as the rain—and little else. Compare this geological environment with one where the groundwater is drinkable but is known to be a **hard water**—one that contains some dissolved calcium and usually some magnesium, which makes it difficult to make a lather or soap suds. Such water may also have a slight taste given by other ions in solution. One place to find such groundwater is below the water table in a soil-covered terrain underlain by limestone and shale.

These kinds of comparisons suggest the same conclusions that we drew from observations of rock weathering (Chapter 6): rainwater infiltrating the ground chemically alters rock and soil and in so doing becomes a groundwater and contains dissolved ions and other components contributed by the plant and animal life on and in the soil. The water under the quartz sand dune is like rainwater because quartz is so insoluble that for practical purposes it is inert, and neither soil nor vegetation contribute dissolved substances to the water. In contrast to quartz sand, surface materials of soil, limestone, and shale react with rain to produce dissolved materials that contribute to the taste and hardness of the groundwater.

**How Good Is the Water?**   The amounts of dissolved material in **potable** groundwaters—that is, waters usable for drinking—are very small, usually measured in parts per million, ppm, a unit made up in the same way as the more familiar percent, which is just parts per hundred. For example, 10 ppm is the same as 0.0001 percent, but it is much easier to use parts per million because we can almost always avoid decimals and all those zeroes. One hundred and fifty ppm might be a typical figure for the total dissolved material in a good water, for no natural water is as pure as a carefully distilled water. The upper limit for potable water for human consumption is usually 500 to 1,000 ppm; for watering livestock, it is usually 2000, though both humans and animals can get used to more in some circumstances, A water can measure in the upper range of permissible dissolved ions and yet still taste good if its dissolved materials consist largely of a mixture of calcium and bicarbonate ions. Yet a water containing only several hundred ppm of sodium and chloride ions will taste slightly salty. The worst offenders to taste are the dissolved organic compounds, many introduced by human activities. Some dissolved components have no taste but beneficial qualities. For example, the movement for fluoridation of water supplies was sparked by the discovery that in some regions small amounts of naturally occurring fluoride in groundwater are responsible for great resistance to tooth decay. Conversely, however, some toxic elements have no taste, but in more than minute quantities may be dangerous to health. Concentrations of such poisonous elements as lead and arsenic must be watched carefully. No matter how good the water may taste, it is unsafe to allow these elements to exceed a tenth to a twentieth of a part per million.

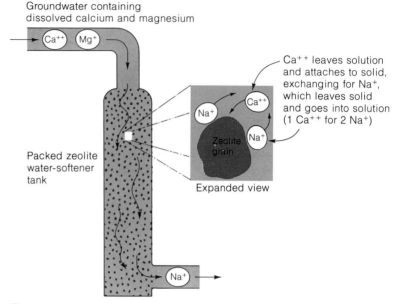

Groundwater containing
dissolved calcium and magnesium

$Ca^{++}$ leaves solution
and attaches to solid,
exchanging for $Na^+$,
which leaves solid
and goes into solution
(1 $Ca^{++}$ for 2 $Na^+$)

Packed zeolite
water-softener
tank

Expanded view

**Figure 7-17**
A zeolite water softener operates by exchanging sodium ($Na^+$), or other ions, in
the solid zeolite for the ions that make the water hard, calcium ($Ca^{++}$) and
magnesium ($Mg^{++}$). The exchanging ions always balance total electrical charge;
thus 1 $Ca^{++}$ will exchange for 2 $Na^+$.

A few groundwaters are naturally **softened** as they pass through
and react with formations containing zeolite minerals, which remove calcium and other ions. Zeolites, hydrous silicates, have a
strong chemical tendency to absorb certain ions from water in
exchange for other ions bound in the solid. If, for example, a water
carrying calcium ion travels through a zeolite formation that ex-
changes sodium ion for the calcium, the water comes out with
sodium instead of calcium, and the zeolite becomes richer in calcium and poorer in sodium. This process has been put to home and
municipal use by companies that attach tanks of zeolite to the
water inlet as a water softener. The commercial zeolite eventually
gets used up—saturated—and has to be changed.

Groundwaters are always clear of solid materials as they come
from the aquifer. The intricate passageways of the aquifer act
as a fine filter and remove small particles of clay or any other
solids. They even strain out bacteria. The dirtiest sewage will be
clear of suspended solid matter once it has gone through a thick
bed of sand. The dissolved matter is another story. It is not
strained out.

**Contamination of Water Supplies** In the past few decades
the tremendous increase in the use of septic tanks for home sewage
disposal has contributed a new source of dissolved material to
groundwater. The septic tank waters seep into the soil and, where

water-supply aquifers are shallow, will contaminate groundwater with phosphate and foaming agents from detergents and a variety of other substances that are undesirable or potentially harmful to health, such as nitrates. Still another source of groundwater contamination is nitrate fertilizer, which is spread over soil. The nitrate is very soluble, and although some is used by plants—the purpose of the spreading—much dissolved nitrate escapes unused into deeper parts of the soil and into groundwater. In combination, these sources (sewage and fertilizer) have increased nitrate levels in some aquifers to the worry point, for nitrate is toxic to humans even in amounts as small as 10 to 15 ppm.

It must be obvious that the faster the recharge of the aquifer, the more quickly it might become contaminated from these sources. But the fast recharge also means that the aquifer can recover in a short time, once the sources of contamination are closed off. Contamination of slow-recharge reservoirs is more serious, for by the time the results of many years of contamination start showing up, it will be too late for quick recovery. Because of the definite links between surface waters and groundwaters, we cannot afford to view the surface and the subsurface as separate entities. Integrated planning for groundwater reservoirs in relation to pollution of and from surface sources will become increasingly necessary in order to maintain the quality of our water supplies.

**Deeply Buried Waters** Below the groundwater table, rocks are almost everywhere saturated with water, especially below depths of a few hundred meters. No matter how deeply oil wells are drilled, geologists always find water in permeable formations. In most places the water below the upper few hundreds of meters becomes increasingly concentrated in dissolved materials; for this reason they are frequently called **brines.** They may vary widely in chemical composition, just as surface waters do, but they are much older than the near-surface waters and presumably have had a much more complex history. These waters move very slowly, probably at speeds of less than a centimeter per year. Many are rich in ordinary salt, NaCl, particularly where a water has been steeped in buried salt beds. For example, in Southern Illinois there is a series of (Paleozoic) sedimentary formations arranged in a spoon-shaped basin a few kilometers thick; the deeper one drills in this area, the more salty and concentrated the waters become. Because these deep waters are in some ways similar in composition to sea water, they used to be considered as remnants of original sea water trapped in the pores of sediments at the time of deposition. But detailed studies of the chemistry of the waters of the Illinois basin have shown that all of these waters have become salty by reacting with various rocks over a long time interval. The deeper and the older the waters, and the more time they have had to react with rocks, the greater is the content of dissolved solids.

**How Deep Does Water Go?** A deep well may penetrate an unconformity under a thick sequence of sedimentary formations and enter what is called "basement"—by definition, a highly complex mixture of igneous and metamorphic rocks that everywhere underlies the sediments of the continents. Basement may range in depth from a few meters to ten kilometers or more. In the Grand Canyon, the basement is Precambrian, including the Grand Canyon Series and the Vishnu Schist. Somewhere else the basement may be Paleozoic, as it is along parts of the Eastern Seaboard of the United States, where flat-lying Mesozoic and Cenozoic sediments are separated by a widespread unconformity from deformed and metamorphosed rocks of Devonian age and older. Because igneous and metamorphic rocks have low porosities and permeabilities, and so contain minor amounts of water, the basement is relatively dry. Some sedimentary sequences, notably those along the Gulf coasts of eastern Texas, Louisiana, and Mississippi, are so thick that the depth of basement is known only from seismological evidence, and it lies well beyond our present deep-drilling capabilities. Here the sedimentary formations show a decrease in permeability with depth, caused partly by compaction due to the immense weight of overlying rock. Another cause of the decrease in permeability is the strong tendency for pore spaces to be filled up by such minerals as quartz or calcite, which are precipitated under the chemical conditions of deep burial. Though the deepest wells drilled so far—more than 9 kilometers (29,500 feet)—still penetrate permeable rocks, we can nevertheless assume that at much greater depths these sedimentary rocks will be as dense and dry as basement rocks.

The conclusion is inescapable: the accessible waters of the Earth are limited to the surface and the near-surface part of the crust. Most of it is surface water—in oceans, rivers, and lakes; the remainder is stored in rocks, to depths of not much more than 10 kilometers. The overwhelming majority of the world's *fresh* waters are in the ground rather than on the surface. Although many deeply buried waters are salty (brines), most near-surface groundwater is fresh. Rivers and lakes account for only a small fraction of the total fresh water.

**Thermal Waters** In such places as Marienbad, Germany, Aix-les-Bains, France, Karlsbad, Czechoslovakia, Hot Springs, Arkansas, the restorative properties of mineral baths in hot spring pools have been known (or assumed) from the earliest days of civilization. They also are a part of our story, for the activity of these mineral waters, or **hydrothermal** waters (literally, "hot water" in Latin), is responsible for extensive deposits of travertine and other minerals that encrust rocks around the pools. Those that do not reach the surface are responsible for the formation at depth of many of the richest metallic ore deposits of the world (see Chapter 24).

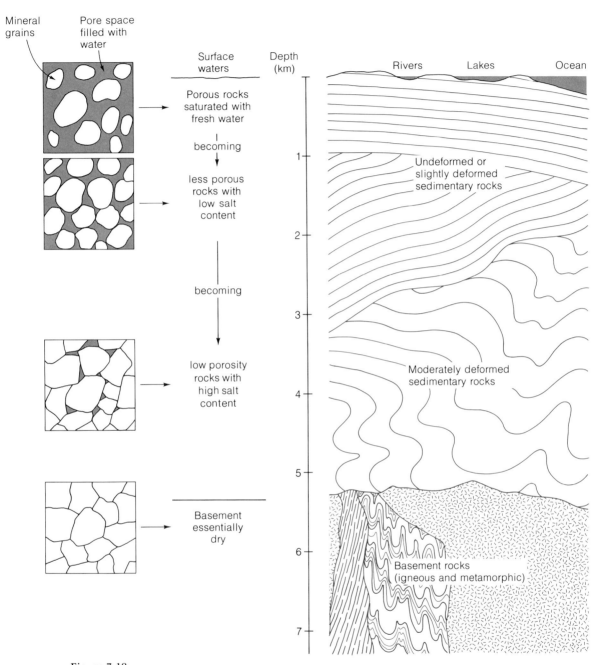

**Figure 7-18**
A typical section through continental crust, showing the distribution of water. Most is at the surface or in sedimentary rocks buried at shallow depths. Depth to basement varies greatly, but is rarely more than 10 km. Porosity and water content generally decrease with increasing depth and greater structural deformation.

Thermal waters are found in areas of current or recent igneous activity—deep-seated, or intrusive, activity and surface, or extrusive, activity. The water itself is mostly **meteoric** (meaning derived from rainwater, and having nothing to do with meteors except for the derivation of the name). It is mixed with water originally dissolved in the magma itself, called **magmatic** water. In areas of igneous activity, surface waters moving downward encounter hot masses of rock, become heated, and mix with the magmatic waters at depth. Return circulation takes place along fracture zones or other channels that intersect the surface at continuously fed hot springs or geysers, some of which spout intermittently like the spectacular Old Faithful in Yellowstone Park (see Fig. 17-39).

Hot water is a much more efficient dissolver of rocks than cool rainwater, and so these hydrothermal solutions are more concentrated in many ions than ordinary groundwaters. As the hot solutions rise to the surface, they cool. If these hot solutions are saturated at the higher temperature, they will become supersaturated as they cool and will deposit the excess as mineral precipitates. The extensive whitish to grayish, muddy flats around

**Figure 7-19**
Circulation of water over a magma that produces geysers or hot springs. Cold rainwater soaks into the soil and filters down through permeable rocks. As it approaches the magma, it heats, therefore becoming less dense, and thus sets up a circulation system that returns it to the surface.

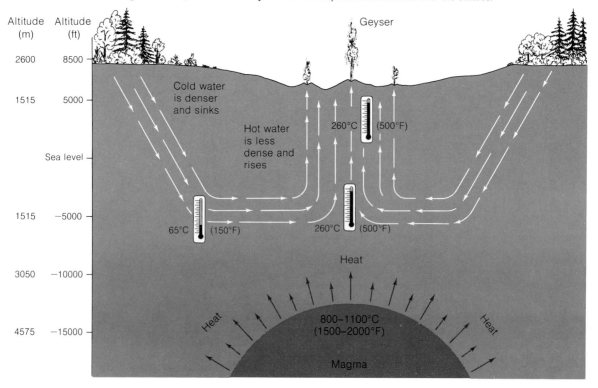

some of the hot springs in Yellowstone Park are made up of fine particles of silica precipitated from the solutions that had originally, when hotter, dissolved it from silicate rocks at depth, near the hot igneous rock. **Travertine** is a calcium carbonate deposit commonly precipitated at cooling hot springs from solutions that passed through limestone formations at depth. Other hot waters may contain large amounts of sulfur. Our ancestors were right when they referred to such smelly, sulfurous waters as escapes from the "hellish" interior of Earth.

In the search for new and clean sources of energy, scientists have turned their attention to hot water areas as natural sites for the development of steam turbines and other devices to convert the heat into electricity. Though this is not a generally applicable solution to the problem of our rapidly climbing energy needs, the potential for efficient harnessing this source of energy is good in areas that have hot springs or geysers. California, Iceland, Italy, and New Zealand are already exploiting their hot spring areas for power in a practical way (also see Chapter 24 and Figs. 15-1 and 17-46).

## WHERE DID ALL THE WATER COME FROM?

Sooner or later after geologists began to question seriously how the Earth evolved as a planet, someone had to tackle the origin of all the water on Earth, for if it were not always here it had to come from somewhere. In the late 1940's a remarkably versatile geologist who was equally at home doing field mapping, calculating chemical equations, and studying the physics of how water currents carried sediment, found himself attracted to this problem. *William W. Rubey,* working for the U.S. Geological Survey on the origin of marine phosphate rocks, started wondering whether the saltiness of the ocean had ever changed much in the past. As frequently happens to people of insatiable curiosity, that question led to others. Why was the ocean salty in the first place, and where did all the water and salt come from?

Rubey compared the average chemical composition of igneous, metamorphic, and sedimentary rocks and reconfirmed what geologists had long known—that sedimentary rocks are much richer in water and carbon dioxide than the others. He followed by adding up the amounts of all of the water, carbon dioxide, chlorine, and a few other substances, such as sulfur, now present in sedimentary rocks, atmosphere, and sea waters (Table 7-3). He then compared those totals with estimates of the amounts of these substances that could have been derived by the weathering of igneous rocks. This attempt at a mass balance was reasonable; after all, we know that the process of chemical weathering of igneous rocks produces sediments and the dissolved ions in ground and surface waters that eventually end up in the oceans.

Table 7-3

Where some volatile substances are found at or near the Earth's surface.

| Compound or element | Where found |
|---|---|
| Water ($H_2O$) | As liquid in oceans; as vapor in atmosphere; chemically bound in hydrous minerals, such as clays, in sedimentary rocks. |
| Carbon dioxide ($CO_2$) | Dissolved in oceans; as gas in atmosphere; as carbonate minerals, such as calcite($CaCO_3$), in limestones and other sedimentary rocks; in reduced form as organic carbon in oil, gas, coal, and other organic matter. |
| Chlorine (Cl) | As chloride ion ($Cl^-$) dissolved in oceans; as component of saline minerals, such as halite ($NaCl$), in evaporite deposits. |
| Sulfur (S) | As sulfate ion ($SO_4^{--}$) dissolved in oceans; as calcium sulfate ($CaSO_4$) in the form of gypsum and anhydrite in evaporite deposits; as sulfide ion ($S^-$) in iron sulfide ($FeS_2$) in the form of pyrite in sedimentary rocks. |

Rubey found an enormous discrepancy. There was far too much water, carbon dioxide, and chloride in the atmosphere, oceans, and sedimentary rocks to be accounted for by the weathering of igneous rocks, which contain little of these substances. The only reasonable way to make the budget balance was to revive an idea first proposed in the nineteenth century—that all of the gases released by volcanoes, hot springs, and geysers, lumped together as **volcanic emanations**, could make up the difference. A check on the composition of gases from volcanoes quickly showed that in fact their ratios of water to carbon dioxide and other elements was very crudely in proportion to the ratios calculated for the whole crust. Because these materials all tended to form gases (water vapor, carbon dioxide and other carbon gases, hydrogen chloride, chlorine, hydrogen sulfide, sulfur dioxide, and others) and were present in such abundance on or near the surface or buried in sedimentary rocks, Rubey called them **excess volatiles**. Thus he concluded that almost all of the water vapor and the other gases were released from deep in the earth by volcanism. Since chemists use the term **outgassing** to refer to the exhalation of gases from a solid or liquid as a result of heating or some other cause, it was natural to refer to release of these gases as **outgassing of the Earth**.

Outgassing does, indeed, seem to be the process responsible for the formation of the oceans and the atmosphere. But the question remains, when did it happen: did it happen all at once or over a long period of time? Rubey looked first for evidence in the geo-logic record that might indicate whether any change in the salinity of sea water had ever taken place. He reasoned that marine organisms are known to have little tolerance for such changes. From fossil evidence dating back to the Cambrian, he argued that marine life has never been drastically affected by variations in the salt content of the oceans. This led him to conclude that neither the oceans nor the atmosphere have changed much in composition over geologic time. Other lines of evidence led him to conclude that the volcanic gases from the interior accumulated slowly and steadily throughout geologic time, having always been released in about the same relative compositions as they are today.

## THE CONSTANCY OF SEA WATER COMPOSITION

Since Rubey's paper opened up this whole line of thought, there have been many new ideas on it. In 1959 the world's first international oceanographic congress was held in the U.N. Building in New York. For a week the delegates' lounges and bars were alive with talk about the outpouring of new information on the oceans instead of the usual diplomatic activities of ambassadors. At one session a brilliant Swedish physical chemist, the late *Lars Gunnar Sillén*, brought his deep knowledge of the quantitative data on inorganic chemical reactions to bear in his first essay on how the chemical composition of the oceans could be explained. Sillén started out with pure water mixed with dissolved carbon dioxide and oxygen in equilibrium with the amounts present in the atmosphere. He then calculated how that water could chemically react to equilibrium with all of the clay minerals, carbonates, and other minerals characteristic of the sediments flooring the oceans. He further calculated the same kinds of reaction between water and all of the unweathered igneous minerals that are transported to the oceans as detritus eroded from continental rocks. He concluded that the final product would be a solution much like sea water. This analysis made even more probable Rubey's estimate that sea water was always of the same composition, and emphasized even more strongly that sea water was the expectable ordinary consequence of weathering and sedimentation.

At about the same time, the work of other geochemists showed that the chemical elements dissolved in sea water were being precipitated onto the bottom as sediments at about the same rate as they were being supplied to the ocean in solution by the world's rivers. Thus we now see the oceans as the product of a steady-state process rather than the result of some early, single, cataclysmic event. Our view of Earth history has been sharply altered by the gradual recognition that steady-state processes are the rule rather than the exception.

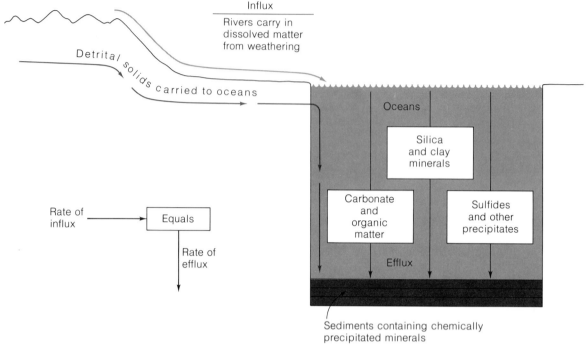

Dissolved matter
leached from rocks

Influx

Rivers carry in
dissolved matter
from weathering

Detrital solids carried to oceans

Oceans

Silica
and clay
minerals

Rate of
influx

Equals

Carbonate
and
organic
matter

Sulfides
and other
precipitates

Rate of
efflux

Efflux

Sediments containing chemically
precipitated minerals

**Figure 7-20**
A model of the steady-state ocean. The rate of influx of dissolved substances by rivers matches the
rate of efflux by chemical precipitation of sedimentary minerals. Detrital solids formed by erosion
accompanies the dissolved material and is deposited as sands and muds in various mixtures with the
chemical precipitates.

**How Long an Atom Stays in the Ocean** The steady state of
the oceans leads to another concept that relates the inflow or outflow of an element to its total amount in the sea. A good example
might be a crowded cocktail party, where many more people have
been invited than fit comfortably in the rooms. As people come
in, the rooms quickly fill up, and then as more arrive, the crush
gets so much that people start to leave. At its most active, the
party is in a steady-state, arrivals and departures balanced, and the
rooms saturated. Now, even though some come early and stay
late, and others arrive, see the crowd, and leave, there is an average
length of time between each person's arrival and departure that
we can call **residence time**. The length of stay at the party is determined by the number of people that can be jammed into the rooms,
which we will call the capacity, and the rate of arrivals, which we

will call the rate of influx (this, of course, is equal to the rate of departures, or efflux).

$$\text{Residence time} = \frac{\text{capacity}}{\text{rate of influx}}$$

A simple numerical example: if a room will hold thirty people and a new person arrives every two minutes ($1/2$ per minute), you can verify that the average residence time has to be one hour.

Applying all this to the oceans allows us to calculate residence times for the dissolved elements in sea water. You can visualize this as the average length of time that elapses between the entry of an atom into the sea and its removal by sedimentation. Calcium is carried in abundance by rivers, and even though it is readily taken up by many kinds of organisms that build shells composed of calcium carbonate, it has a long residence time, 8 million years. Sodium has one of the longest residence times, 260 million years. Iron, on the other hand, stays in the ocean only 140 years. Knowing the residence time of elements is important not only for working out the general scheme of things but for predicting the behavior of toxic or radio-active elements in the ocean. It is important today, for example, to determine the residence time of crude oil spilled on the sea by tankers before it is decomposed or sedimented out on the sea bottom or its beaches. Likewise, as we will discuss in Chapter 14, it is crucial that we know the residence time or carbon dioxide if we are to be able to predict the effects of excess $CO_2$ in the atmosphere and oceans.

Yet another way of looking at the ocean has been proposed by some geochemists, who have presented a kind of cookbook recipe for making sea water that gives much insight into the details of the process. They calculated how much and what kinds of sediment would be deposited if you took average river water, evaporated it until it contained as much chloride salt as the ocean, precipitated excess calcium carbonate, silica, clays, and other sedimentary minerals—all to make the final solution like sea water. They compared existing sedimentary rock accumulations with their results and found good agreement. This whole field of study is most active, and we are now beginning to see how plate tectonics and geochemical connections between the surface and the interior of the Earth may have a bearing on the compositions of sea water and sediments.

**When All the Water Came Out**  Rubey thought that the outgassing of the interior took place gradually, but information that has accumulated during the years since the late 1940's has led some scientists to believe that the great bulk of the Earth's atmosphere and oceans may have been outgassed early in its history and have been partially recycled ever since. Part of this evidence is in the isotopic composition of the gases from modern volcanoes, which indicates that they are of surface or crustal origin and are recycled only through the outer 25 to 30 kilometers of the Earth rather than outgassed from the deep interior. Other lines of reasoning that point to early outgassing come from current ideas on the timing of the astronomical and geological events in the early history of the Earth. For instance, the Moon and the Earth were much closer during the early part of their joint history, and the Earth would have been stressed by the strong pull. Another factor is the early timing of differentiation, for the melting of iron and the formation of an iron core would have caused a profound disturbance that would also have promoted outgassing (see Chapter 1). Finally, the theory of plate tectonics holds that continuous mixing and recycling of material goes on between the crust and the mantle, and provides a rationale for the distribution and kinds of volcanic activity that contribute gases to the atmosphere and oceans. Thus we have evidence that forces promoting extensive outgassing would have been strongest in the earliest stage of Earth history, and we have a mechanism — plate tectonics — that would allow for continuous recycling of some of the gases through the crust and mantle and also account for the gaseous volcanic emanations of today.

Putting all of these lines of evidence together, we refer again to what was called in Chapter 1 the "big burp." It is likely that there was a huge outgassing of the excess volatiles at the time when much of the earth became molten as a result of the accumulation of radioactively produced heat. At the same time, the body of the earth would have been being distorted by large earth tides caused by the nearby moon. That too would have promoted the escape of gases from the interior. Since that early time of outgassing, the excess volatiles have been cycling through the crust and mantle. The temporarily dense atmosphere, the newly formed oceans, and the large volume of primitive sediments that would have formed by the condensation of water clouds and reaction of all of the gases with igneous rocks would long since have been deformed or subducted by plate and other tectonic movements.

Thus our concept of the history of the oceans and atmospheres includes a stage of rapid evolution early in the Earth's history, during which the steady-state "machine" was set up, followed by a second stage, almost all of geologic time, during which the "machine" has been steadily operating at more or less the same rate.

Knowing the importance of recycling of excess volatiles we can integrate the movements of water in the hydrologic cycle with the routes of the dissolved ions and gases, such as carbon dioxide, that it carries (Fig. 7-23). The water is recycled at the surface of Earth by evaporation from land and sea, circulation as vapor and clouds in the atmosphere, and precipitation on land

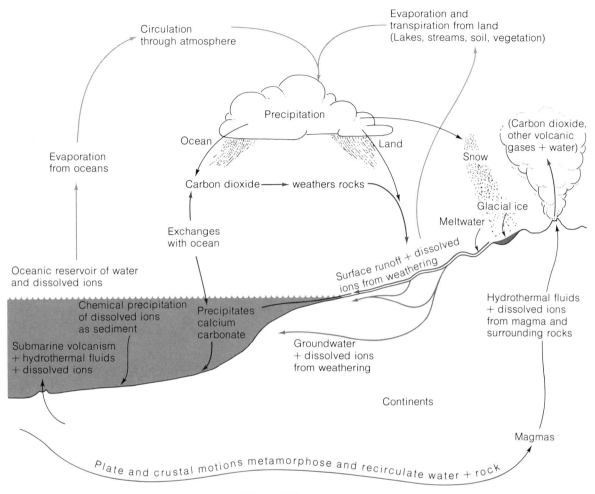

**Figure 7-21**
Overall hydrologic cycle.

and sea. On land the precipitation of snow may lead to glaciers, whose meltwaters join ordinary rainwater runoff in bringing land precipitation back to the sea. Surface runoff is connected to shallow, subsurface groundwater flow, much of which eventually returns to the surface in the beds of streams and lakes. Some surface water is incorporated into hydrous minerals and reaches the interior, where it is recycled along with solids by the tectonic, magmatic, and metamorphic processes of exchange that take place between the crust and upper mantle of the Earth. In this way the surface hydrologic cycle, powered by the external heat engine, the Sun, is linked to the Earth's internal heat engine, powered by radioactivity. Water from the interior eventually returns to the surface in volcanic emanations, which may escape at the surface or in the deep sea. Volcanism also helps recirculate carbon dioxide, one of the most abundant of the volcanic gases next to water vapor. Carbon dioxide dissolved in rain and surface waters is important in the weathering of rocks, for it releases ions from decomposed rock. Surface water and groundwater constantly carry the dissolved substances to the oceans, where they steadily recombine as the chemically precipitated fraction of sediments. Calcium carbonate forms in this way, and effectively buries some of the carbon dioxide from the atmosphere. The carbon dioxide of the oceans and atmosphere are in equilibrium with each other by exchange across the surface of the ocean. All of the operations are inter-related and combine to keep the ocean constant in composition as water evaporates from the surface and as precipitates settle to the bottom.

Thus we can visualize the operation of this great chemical processing plant as a coupling of two subunits powered by the two heat engines of Earth. In the surface subunit, water is the pipeline carrying the products of weathering to the sea, where they are removed by sedimentation. The water is purified by evaporation and recycled to continue its weathering work. In the subunit operating within the Earth, the water is recycled with solid rock as part of interior movements of crust and mantle, ultimately to reach the surface again as volcanic emanations. In the surface subunit—that is, in what geologists call the hydrologic cycle—a molecule of water may be recycled in a few years, but in the interior subunit recycling may take millions of years.

## SUMMARY

1. The properties of water are unique indeed; existing as gas, liquid, and solid on the Earth's surface, it participates in all geological processes.

2. The hydrologic cycle is a description of the pathways by which water evaporates from land and sea surface and returns by precipitation as rain, ice, and snow.

3. Surface runoff is directly related to the amount of precipitation, and thus reflects the climate.

4. Groundwater forms by infiltration of precipitation through soil into permeable aquifers.

5. The groundwater table closely follows the contour of the land surface and changes in level in response to precipitation.

6. Groundwater moves at slow to moderate rates through aquifers, the rates depending on the permeability and the hydraulic gradient. Water may move for long distances underground.

7. Groundwater supplies may be depleted by excessive pumping and insufficiently rapid recharge. The water supply in some places can be replenished by increasing recharge.

8. The quality of water is dependent on the kinds of dissolved materials picked up by the water on its passage through soils and aquifers. Hard waters form when abundant calcium dissolves. Potable waters contain only small amounts of dissolved material.

9. Groundwaters may be contaminated by improper sewage disposal and by seepage of chemical nutrients applied to soils.

10. Groundwaters become more concentrated in dissolved solids at greater depths. Waters fill pore spaces of rocks down to depths where porosity decreases to the vanishing point.

11. The occurrence of thermal waters is related to the presence of igneous intrusions. Thermal waters precipitate a variety of deposits around hot springs and geysers.

12. The waters of the surface evolved from the escape of volcanic gases from the Earth's interior.

13. Sea water has stayed constant in composition throughout most of geologic time by a balance of inflow of dissolved material from rivers and precipitation of sediments. Residence time is the average length of time an atom stays in the ocean before precipitating.

14. Much of the gases and water at the surface escaped from the interior early, setting in motion the hydrologic cycle much as we know it today.

## EXERCISES

1. If a thin layer of oil were to form at many places on the ocean and reduce evaporation of water from the sea surface to half of its present value, what changes might you expect in the hydrologic cycle?

2. If there were a vast increase in the number of desalination plants making fresh water out of sea water all over the world, so that a significant volume of fresh water were produced, would it affect the workings of the hydrologic cycle, and if so, how?

3. Assume that in the future, motions along the San Andreas fault, a plate boundary, were to change to divergence, separating the San Joaquin Valley from the coastal mountains to the west and changing it to a low coastal plain bordering a new arm of the Pacific. What changes in the hydrologic budget of the valley might follow?

4. Several hundred meters under the city of Chicago are permeable aquifers of sandstone, which crop out in Southern Wisconsin, 100 to 200 miles north of the city. One-hundred years ago, Chicago got much of its water from flowing wells drilled into these aquifers, but water is now difficult to pump from the wells, and the city gets its supply from Lake Michigan. What do you think might have been the history of Chicago's water supply?

5. The government of the island of Bermuda, far out in the Atlantic Ocean, does not allow private water wells, and most residents use the rainwater that collects on their roofs. Why do you think the government has this rule?

6. If, as recently happened in one State, it were discovered that radioactive waste from a nuclear processing plant had seeped into groundwater in the immediate vicinity of the plant, what kind of information would you want in order to predict the length of time before that waste might move into a groundwater aquifer many kilometers away?

7. Why do most water-well drillers not recommend drilling too deeply, many hundreds of meters, to get pure water supplies?

Bretz, J. H., "Vadose and Phreatic Features of Limestone Caverns." *Journal of Geology,* v. 50, pp. 675–811, 1942.

Davis, S. N., and R. J. M. DeWiest, *Hydrogeology.* New York: John Wiley & Sons, 1966.

Garrels, R. M., and F. T. Mackenzie, *Evolution of Sedimentary Rocks.* New York: W. W. Norton Co., 1971. (Chapter 4.)

Peixit, J. P., and M. A. Kettani, "The Control of the Water Cycle." *Scientific American,* April, 1973. (Offprint No. 907.)

Revelle, R., "Water." *Scientific American,* September, 1963. (Offprint No. 878.)

Sayre, A. N., "Ground Water." *Scientific American,* November, 1950. (Offprint No. 818.)

# 8

## The Beginning of Erosion and Transportation: Breakup of Rocks and Downhill Movement

*The breakdown of outcropping rocks to fragments ranging from boulders to clay is the product of chemical weathering combined with the forces of physical disintegration. Once fragmented, the particles start downhill in many kinds of gravitational movement, both as individual particles and as large masses.*

Not a year passes without the appearance of headlines telling the news of a flood, a mass of mud oozing down a hillside, or a catastrophic landslide. In 1972, for example, Rapid City, situated at the foot of the Black Hills of South Dakota, was hit by a flash flood that shocked the country because of the suddenness with which it caused such vast destruction and tragic loss of life. Everywhere in the devastated city one could see rocks, mud, and silt brought in by the floodwaters. Disastrous as this particular event was, it should not have been unexpected. The flood, predictable enough once the rain began to fall so rapidly and continuously from a most unusual storm of long duration, was only the final transporting agent of the debris that invaded Rapid City. Much of that debris began its downhill journey a long time before, having been produced at a normal rate by chemical and mechanical weathering in the high parts of the central core of the Black Hills. The prevailing erosional process was suddenly augmented by the rapid downhill movement of water-soaked soil and rock debris into the flooding tributary streams of the Cheyenne River, which rushed

**Figure 8-1**
Jackson Boulevard in Rapid City, South Dakota, on June 10, 1972, after torrential rains caused a sudden flash flood that deluged the city. [Photo by Perry H. Rahn.]

the masses of detritus onto the plains. Rock fragments and clay that had taken tens to hundreds of years to produce and build up on the slopes were caught up in fast downhill movements that were provoked by a storm that might occur only once in a hundred years.

This chapter is concerned with the beginnings of active erosion and transportation that follow the slow, sometimes imperceptible, decay of chemical weathering. One part of the story is the fragmentation of rock masses into boulders, sand, and finer material and how breakup is related to topography and climate. The other part is the downslope transport by slow creep and rapid slides and flows that makes the debris of erosion available to the major long-distance transport agents: rivers, wind, and glaciers, which will be covered in following chapters.

## FRAGMENTATION

Fragmentation, the breakup of rock masses into boulders, pebbles, sand, and silt, is a physical process that is inextricably linked to the chemical processes of rock weathering, which causes feldspars to decay to clays and causes other minerals to dissolve. Chemical decay is not always apparent when one looks at a mass of broken rock at the foot of a cliff. Pieces may look freshly broken and unaltered. But however minutely, weathering has played some role in making the rock break. Chemical weathering helps fragmentation by an interaction in which the physical breakup of the rock opens channels for water and air to penetrate the rock farther and promote chemical decay. The rate at which chemical reactions take place increases when more surface area of the solid is exposed to the gas or liquid reacting with it. For example, a single log will react with air to burn slowly, but the same log split into kindling will ignite quickly and burn rapidly. Similarly, because the ratio of surface area to volume of rocks and minerals increases as the average particle size decreases, chemical weathering becomes more efficient as rocks are broken into small sand and silt particles. Chemical and physical processes are both related to the action of organisms, from bacteria to tree roots, which play their own roles in the destruction of rock.

In studying such complexly intertwined processes the geologist looks for special places where one or another effect is absent so that the others can be studied alone. Up to a few years ago, some geologists went to study mechanical fragmentation in severely cold polar regions, where there was little active plant or bacterial life and little liquid water, which it was thought would tend to minimize chemical weathering. Even in such hostile regions, however, there was enough chemical decay to be significant. The study of the moon gave a new and better opportunity to study fragmentation in the absence of ordinary chemical weathering as we understand it on Earth and so to compare and contrast mechanical erosion with and without chemical action.

**Rock Fragmentation on the Moon**  Even before the landing on the Moon, the shapes and slopes of surface features suggested that they were made up of piles of broken rock, but with the close-up photographs of the first Surveyor it became certain that the surface is composed of loose pieces of rock, from large boulders to fine dust (see Figs. 23-8, 23-18, 23-20). Now that we have Moon rocks to examine, we can be sure that there has been no chemical weathering of minerals in the basalts and other igneous rocks. These materials are as fresh as might be expected, having never been exposed to an atmosphere containing oxygen, carbon dioxide, and water. Fragmentation must result completely from physical mechanisms, such as the breakup of large parts of the surface rocks by the crashing impact of meteorites, as is discussed in Chapter 23. The boulders appear to have been broken by the impact of meteorites and thrown up and showered all around the craters that have long been familiar features of the Moon's surface (see Fig. 23-10). But what about the dust? The accumulation of fine dust particles distributed so widely over the Moon's surface is only partly the product of large impacts; much of it comes from the continual spray of small meteorites ranging in size down to the finest micrometeorites. Because the rocks and dust have been dated to be billions of years old, the fragments that geologists have been able to examine have not changed in size or shape throughout most of the Moon's history.

To a minor extent there has been some rounding of the sharp edges by a sort of "sandblasting" process—the constant bombardment by micrometeorites. The effect is not unlike the smoothing of sharp edges of broken glass on a sandy dune on Earth. Rounding takes millions of years on the Moon; a piece of glass on an Earth dune can be rounded in one year.

Other than fragmentation by these processes, Moon rocks show nothing of the jointing, cracking, spalling, and slabbing of rocks that we are so used to on Earth. Our "laboratory" investigation of another planet has ended by showing us that fragmentation on Earth has no component that does not depend in some part on chemical reaction with our atmosphere. What, then, are the Earth's particles like, and how do they form?

**Kinds of Fragmentation**   A look at an outcrop of a well-bedded sandstone or a schist with pronounced cleavage will show how important the original character of the rock is in determining the course of breakage. The sandstone breaks into slabs or plates along bedding planes, and the schist breaks into smaller, sometimes splintery pieces whose flat sides parallel the cleavage. The cleavage, schistosity, bedding, or other structural planes in rocks become zones of weakness, and cracks form along them.

**Figure 8-2**
Weathering of different rock types at different rates emphasizes the bedding planes in this series of alternating sandstones and shales. The series has been structurally deformed, so that bedding planes are now approximately vertical. Erosion is rapid on this beach, a wave-cut terrace exposed at low tide. Zumaja, Spain. [Photo by R. Siever.]

Figure 8-3
Joint patterns in igneous rocks near Saint Rafael, France. [Photo by R. Siever.]

On a larger scale, unbedded rocks — or what we call **massive** rocks, like granite — can be seen to have broken up along regularly spaced cracks, called **joints**, at intervals of a meter to a few meters apart. Joints are produced by structural deformation in the course of mountain building, and to some degree are characteristic of almost every rock type. In some igneous rocks the joints take the form of closely spaced, more-or-less plane surfaces, which in some exposures are curved. The resulting layered structure is called **sheeting**.

All these zones of weakness open as cracks when the rocks, once deeply buried in the hearts of mountain ranges, are gradually

**Figure 8-4**
A disrupted glacial boulder in Wallace Canyon, Sequoia National Park, California. This boulder, originally one piece 2.5 m wide and 1.5 m high when it was deposited by a glacier, has been split into a series of parallel slabs by the freezing action of water that penetrated originally tight joints. [Photo by F. E. Matthes, U.S. Geological Survey.]

brought to the surface and the weight of tons of overlying rock is removed. The process is probably not unlike the way in which poorly glued joints in a wooden chair separate as the clamps that held them together during glueing are released. A sidelight—and confirmation—is given by the sometimes severe jointing that spontaneously occurs in deep mines as tunnels are opened up and the pressure on the rock in those passageways is released.

Once the joints, bedding planes, or other cracks are there, the infiltration of water, bacteria, and plant roots start to do their work in enlarging and wedging the openings apart (see Fig. 6-1). Plant roots play some role in this, probably mostly by promoting chemical weathering but also by exerting physical pressure as they grow. One of the most efficient physical mechanisms is freezing and thawing of ice. As we noted in Chapter 7, water expands as it freezes, and the expansive force exerted during freezing is strong enough to crack cast iron engine blocks—and rock too. The process has the same effect as a wedge being pounded into the crack, enlarging it and finally splitting it open. During warmer periods, chemical weathering by water and organisms helps the process along by weakening the minerals holding the rock together.

If freezing is so important, maybe heating in the hot sun and cooling at night might be effective too—or so everyone thought for many years until someone tried it in the laboratory. Pieces of granite were alternately heated and cooled to temperatures

Long before power drills and saws were invented, freezing was the traditional method stonemasons used to quarry building blocks. They drilled small holes at critical points, such as at incipient cracks, filled them with water, and waited for freezing to do its work.

far more extreme than those they would be subjected to on any desert. The number of rapid alterations of hot to cold was equivalent to hundreds of years of days and nights, yet such rough treatment resulted in no visible fragmentation of the rock. Other experiments of the same kind have been tried with little success. In one experiment, water was sprayed on basalt to simulate dew at each cooling. The result was still negative, even though some chemical solution was demonstrated. Most geologists who have observed fragmentation in the desert still tend to believe that the stresses induced by thermal expansion and contraction in rocks may be significant. Certainly the laboratory experiments so far are a poor copy of what happens in nature. We will have to wait for better experiments or observations to get a firm answer.

Perhaps thermal expansion combines with frost action and chemical weathering to produce two commonly observed kinds of mechanical disintegration, **exfoliation** and **spheroidal weathering**. Exfoliation is the peeling off of large, curved sheets or slabs of rock from the weathering surface of an outcrop. Exfoliating outcrops have the appearance of onion layers. Spheroidal weathering has the same general appearance but on a much smaller scale, in which rounded boulders spall off layers or "shells" from the surface. Large-scale exfoliation may be partly the consequence of relief of pressure. As each slab breaks off, it releases weight

**Figure 8-5**
Exfoliation on a rocky slope, Yosemite National Park, California. The partings produced by weathering are nearly planar, and strongly resemble inclined joints. [Photo by F. E. Matthes, U.S. Geological Survey.]

**Figure 8-6**
Spheroidal weathering of a boulder of gabbro, Mesa Grande, San Diego County, California. [Photo by W. T. Schaller, U.S. Geological Survey.]

from the underlying mass, which then may expand enough, however slightly, to break the rock. In spheroidal weathering the pressure released by unloading is so minute that it cannot be important. Chemical weathering is thought to be of some importance in these processes, producing mechanical stresses by uneven volume changes in the surface as minerals decay. This leads to the breakage of curved plates from the rock mass.

Though people are rarely around to see such things happen, rocks can break up into smaller pieces when a boulder-size chunk falls from a rock outcrop down a cliff or talus slope. But most of fragmentation is the result of the same processes that cause chunks of rock to be split from outcrops in the first place, and the processes continue until the big pieces become small pieces. Impacts of one pebble against another in the course of being carried by a rapidly flowing stream or being tossed about by waves on a shore is another cause of breakage. Less obvious than breakage as a result of impacts are the abrasion and wearing away of particles by streams and wind-blown sand. As we will see in Chapter 11, glaciers are effective in grinding transported rocks into a fine powder by dragging them across the surfaces of buried rock masses.

**Sizes and Shapes of Fragments** The various processes that cause fragmentation produce a bewildering array of sizes and shapes. Careful study of the shapes of fragments that can be traced to their parent outcrop, even those kilometers distant, have shown

**Figure 8-7**
Erosion produces rock debris of all sizes, from large cobbles to sand and clay.
These accumulations of gravel, sand, and clay have been transported only a short
distance from the terrain where they were eroded. [Photo by F. Press.]

that their shapes are largely inherited from the patterns of joints,
bedding, cleavage, and other structures rather than being produced
by the transporting medium. An experienced geologist can guess
just from the shapes of pebbles on a beach what kinds of rock they
came from. This inheritance extends to many sand grains, whose
shapes, modified only slightly by erosion and transportation, are
derived from the original crystals or grains in the eroded source
rocks.

The sizes of fragments are a good clue to the intensity of mechanical erosion. A good general rule of thumb is that the higher

**Figure 8-8**
A microscopic view of a thin section of the Recluse Formation (Precambrian) in Canada. The grains are angular: they retain the original shapes they had when fragmented from their source rocks. Grains are poorly sorted by size, one of the characteristics of many graywacke sandstones transported by turbity currents. (Magnification: × 44.) [From *Sand and Sandstone* by F. J. Pettijohn, P. E. Potter, and R. Siever, copyright 1972, Springer-Verlag New York, Inc.]

or steeper the topography, the larger the fragments. Regardless of the rates of mechanical and chemical weathering, that govern the break-up of rocks in a particular place, the fragments will be largest close to their source. Once fragments have been gradually worked downhill from their source, they enter new domains of weathering, and they undergo further fragmentation. Moreover, once boulders and pebbles get into streams they quickly break and abrade, and the downstream decrease in size of pebbles becomes dramatic. Much of this size decrease is the result of the inability of most streams to carry large boulders very far, except in rare floods, as we will see in Chapter 9.

Sand grains, however, are not produced by mere wear — that is by the gradual abrasion of pebbles. Nor is any one sand grain the only remnant of what was once a pebble: this would require the disappearance, or wearing away, of too much material. A simple calculation shows why: if each pebble in a loaded freight car were replaced by a sand grain, the volume would be reduced to a cupful. If gradual abrasion has such results in nature, the volume of sand in the world should be much less than the volume of pebbles. But just the opposite is true: there are enormously greater volumes of sand than there are of pebbles. The similarity in size, shape, and mineral composition of sand grains to the crystals that make granites and other coarse-grained igneous and metamorphic rocks points to the origin of most sand grains. It is the mechanical fragmentation of rocks, not their gradual wearing away, that produces

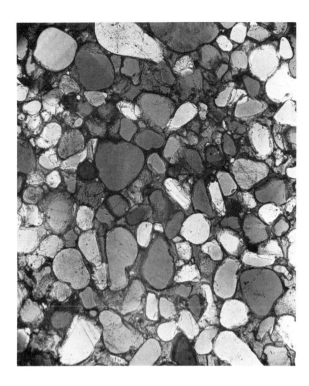

**Figure 8-9**
A microscopic view of a thin section of St. Peter sandstone (Ordovician) of Illinois. All of the grains are well rounded and well sorted by the action of transporting currents. Almost all of the grains are quartz. (Magnification: $\times$ 44.) [From *Sand and Sandstone* by F. J. Pettijohn, P. E. Potter, and R. Siever, copyright 1972, Springer-Verlag New York, Inc.]

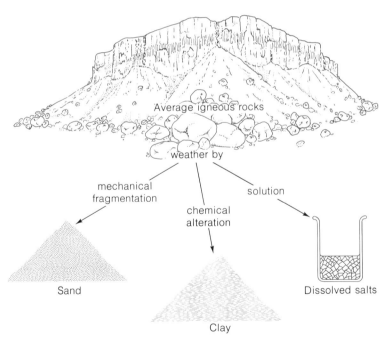

**Figure 8-10**
The chemical and physical processes of weathering transforms the average igneous rock into sand and clay particles and dissolved salts. The total volume of erosional debris produced is somewhat greater than the original, depending on the degree of chemical weathering, which adds carbon dioxide, water, and oxygen to the original composition.

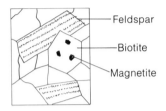

1  Fresh

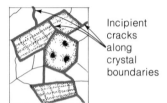

2  Feldspar,
biotite, and
magnetite
start to
decay

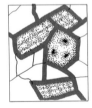

3  Feldspar,
biotite, and
magnetite
decay
extensively

**Figure 8-11**
Microscopic views of stages in the disintegration of an igneous rock, such as
granite. As chemical decay proceeds,
grain boundaries weaken, and the rock
begins disintegrating into fragments.

most sand. As the crystals of feldspars, micas, and other weather-
able minerals get eaten away at the edges, the interlocked crystals
get loosened, and what was once whole rock starts to crumble
into individual grains or goups of grains. In some humid regions
of the tropics, where granites weather rapidly, a small granite boulder can easily be kicked apart into a heap of grains. Where weathering is slower, rocks get smaller bit by bit as grains chip off one by
one.

Silt particles, smaller than sand grains, are crystals or broken
fragments of crystals produced in the same way as sand grains.
Clay minerals, on the other hand, are the chemical products of
feldspar weathering, as was discussed in Chapter 6. Clay-size particles are by definition those less than 1/256 millimeter in diameter
(about 0.004 inch). Not all particles of clay size, however, are produced by chemical weathering; there are also some clay-size particles of quartz, feldspar, and other minerals that are the result of
abrasion, disintegration, and physical fragmentation of larger
particles. The action of glaciers produces much material of these
sizes, sometimes called **rock flour** (see Chapter 11).

Thus the different sizes and shapes of eroded particles—ranging
from huge boulders to microscopic clay particles—are attributable
to certain characteristics of their source rocks. The sizes and shapes
of larger boulders, cobbles, and pebbles are largely inherited from
patterns of jointing, bedding, and foliation, along which breakage
occurs. Sand grains are largely the constituent crystals of disintegrated coarse-grained rocks; silts are broken and abraded finer
crystals. Clays are the chemical breakdown products of unstable
minerals.

## CLIMATE AND TOPOGRAPHY

If freezing, thermal expansion, and water are causes of mechanical
weathering, then some overall relation between climate and fragmentation is implied. The story leads on, for climate also influences
topography—and is affected by it as well. It is then logical to
suppose that topography has some control on mechanical weathering. Here is another interesting example of the way in which
the results of one process have an effect on other processes (see
Fig. 6-16).

**Climatic Effects**  As we saw in Chapters 5 and 6, rainfall and
temperature strongly affect chemical weathering and topography.
Indirectly they affect mechanical weathering too, for rainfall is
the source of river water, which helps break up and abrade rock
and mineral particles. In addition to the obvious cooling effect
of high altitudes, topography also has various effects on climate
(see Chapter 10). For example, mountains cause rain shadows—
dry areas on the lee slopes—because warm, moist air is cooled as

it moves up the windward slopes, and the moisture is lost there as precipitation before the air reaches the lee slopes. Chemical weathering is hastened by abundant rainfall and warm temperatures, and to the extent that decay promotes mechanical weathering, these climatic factors indirectly affect mechanical weathering. In more rigorous climates, frost action and extreme daily variations in temperature are direct causes of rock disintegration. Some of the most important ways in which climate works is by its control on vegetation, which is the direct agent of much chemical alteration of rocks. Lichens—leafless, primitive plants—encrust rocks in many areas where other plants do not grow, even at high altitudes, and they contribute in their small way to chemical weathering. A dry climate promotes physical weathering relative to chemical weathering by slowing the decay produced by the solvent action of water and, equally important, by preventing the growth of much vegetation. Polar climates also inhibit the growth of plants and bacteria; in addition, chemical decay is minimized because the extreme cold prevents much moisture from remaining in the air.

**Topography and Tectonics**  If climate influences topography, and vice versa, it is also true that topography strongly controls weathering. High altitude and relief enhance mechanical weathering, partly through the promotion of freezing and thawing, but also because the fragmented debris on mountains quickly moves downhill, exposing new fresh rock to the attack of the weather.

A climb up a typical high mountain will show the evidence of high rates of mechanical erosion. As one goes higher the soil

Figure 8-12
A rocky mountain slope above timberline. [Photo by F. Press.]

thins. Above timber line, it becomes spotty, and much of the land-scape is bare rock. Slopes of blocky talus alternate with steep cliffs and craggy points. The rocks lose the discolored, sometimes lichen-covered, rounded look that chemical weathering produces at lower elevations; instead, the angular blocks have the appearance of freshly quarried stone and quickly wear out the boots of the climber. Talus slopes here may be unstable: sometimes a small disturbance, one careless step, can set off a roar of cascading blocks that can crush almost anything in its path. In the cold, hostile en-vironment at the top, the wind drives snow and ice across bare rock.

Field geologists have long been well acquainted with the high rates of mechanical erosion at high elevations; from the evidence of these high rates, they have concluded that if mechanical weath-ering is at a maximum in high, rugged mountains, then chemical weathering must be at a minimum in such places. Ronald Gibbs, one of a large group of geologists studying the Amazon River basin, decided to check this by simultaneously tracing the origins of the solid detritus from mechanical weathering and the dissolved products of chemical weathering in the river water. He traced the detritus and chemically dissolved material back to the headwaters of the Amazon in the eastern high Andes Mountains. Gibbs's find-ings were the cause of much surprise and some dismay among those who had not questioned "accepted wisdom." He found that the high Andes were not only the scene of accelerated mechanical weathering, but that they were the site of most of the chemical decay that took place as well—in spite of the cool, dry climate of those high altitudes. The extensive chemical decay was just not apparent in the high outcrops because the products were so quickly stripped away by mechanical erosion. The flatter lowlands of the Amazon jungle, where rotten granite cobbles and thick soils show the dominance of chemical weathering, are eroded much more slowly, both chemically and mechanically, than the uplands. In lowlands, where detritus is not removed so rapidly, the mechanical effects are masked by the chemical.

If topography is so important in weathering, a search for even more fundamental controls on weathering leads us from topo-graphic effects to the deep-seated tectonic movements that deter-mine the heights of mountains and depths of lowland basins. Ultimately it is those tectonic forces, which push up mountains, that influence the nature of erosion, which in turn affect the kind of debris produced. Thus we can infer historical tectonic events from the evidence of erosion provided by the rock and mineral fragments of ancient sediments. This chain of reasoning has already been touched on in Chapter 6. In the following section we explore more fully how tectonics can be deduced from sediments.

**Ratio of Mechanical to Chemical Weathering**   We can express the course of erosion in terms of a ratio of mechanical to chemical weathering. For example, in arid and very cold climates the ratio

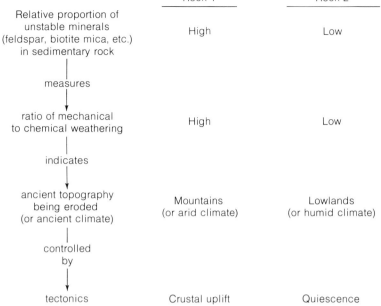

|  | Rock 1 | Rock 2 |
|---|---|---|
| Relative proportion of unstable minerals (feldspar, biotite mica, etc.) in sedimentary rock | High | Low |
| ↓ measures | | |
| ratio of mechanical to chemical weathering | High | Low |
| ↓ indicates | | |
| ancient topography being eroded (or ancient climate) | Mountains (or arid climate) | Lowlands (or humid climate) |
| ↓ controlled by | | |
| tectonics | Crustal uplift | Quiescence |

**Figure 8-13**
The relative proportion of stable and unstable detrital minerals in sediments can be used to infer the erosional origin of the detritus and thus the topography and ultimately the tectonics of the source area.

is high: mechanical breakup dominates. In high mountainous terrain the ratio is also high, for even though chemical weathering is active, mechanical fragmentation strongly predominates. In low plains, the ratio is low, for even though chemical decay is somewhat slower than in high mountains, mechanical erosion is so minor that chemical weathering is predominant.

We can gauge the ratio of past weathering processes by looking at the mineralogy and size of the detrital particles that make up the sedimentary rocks produced by erosion of lands that have long since disappeared. The more unstable the minerals and the larger the size, the higher the ratio must have been. So a coarse sandstone with abundant feldspar, biotite, and other minerals that are easily chemically weathered speak to us of rapid erosion in mountains. Rocks containing large angular cobbles and pebbles are even more eloquent. Fine-grained quartz sandstones, siltstones, and clay-rich shales suggest low-lying source areas where chemical alteration was dominant.

But how do we sort out the effects of climate from those of topography? A conglomeratic sandstone might have come from a low-lying arid region instead of a high mountain, and shales and siltstones might have come from wet, warm hills. The problem is one that is not yet completely settled, but most of the weight of evidence seems to support the notion that topography is more important than climate in determining the ratio of mechanical to chemical weathering. Moreover, it has become clearly recognized that topography much more strongly and directly influences climate than the other way around. Finally, because topography strongly controls the rate at which debris is removed from eroding terrain, it thus promotes both chemical and mechanical weathering. No matter how wet the climate and how great the discharge of rainwater through rivers, mechanical erosion will not have much effect on a low-lying plain.

The logic of tectonic control of erosion, and thus sedimentation, is now apparent. Tectonic movements create mountainous topography and associated climates. Mountains are the places where the ratio of mechanical to chemical weathering is high, producing detritus rich in coarse, unstable rock and mineral fragments. The ancient sedimentary rocks made up of such debris are the key to the mountains of the past.

## MASS MOVEMENTS

Downhill movement of the debris of erosion is not an obscure process to people who suddenly find themselves the victims of a disastrous mudflow or landslide. Almost every year the world's television news cameras are rushed to another spot on the globe that few people had ever heard of to record the tragedy of death and destruction. Earth movements are sudden, the accompanying volumes of sliding debris often enormous, and events seem unpredictable. Yet when geologists hear about some of these disastrous mass movements, many of them react strongly to the lack of foresight and poor engineering planning that are often responsible for them.

A few years ago a Japanese geologist's observations of certain conditions on a hillside convinced him that a landslide would take place within a short period of time and bury a nearby village. Because he warned the people of the village, predicting the imminence of the landslide, he was instrumental in saving thousands of lives. The landslide came soon after and bore out his prediction. In gratitude, the villagers erected a shrine in his honor.

In Quebec the stage was set for a mudflow when excessively steep slopes were cut in soft formations of silt and clay in the course of widening a highway. In 1955, after a period of heavy rain, the steepened cliffs became saturated with water and suddenly gave way in a flowing mass of debris that carried away buildings, roads, and people. The deaths of three people can be traced to poor highway engineering. In 1963 more than 200 million cubic meters of rock debris roared down steep slopes into the waters of a deep reservoir impounded by a dam built in the Italian Alps. The enormous volume of debris plunged into the reservoir and created a giant spillover of water at the dam. Downstream 2600 people lost their lives in the catastrophic torrent. The dam, 265 meters (870 feet) high, was built in a steepsided valley. The engineers who designed it ignored three things: the weakness of the cracked and deformed layers of limestone and shale that walled the reservoir, an obvious geologic history of landslides, and a premonition of danger

**Figure 8-14**
A large mudflow that ran like a river down mountain slopes. Hinsdale County,
Colorado. [Photo by W. Cross, U.S. Geological Survey.]

signalled by a small rockslide three years before. Though the land-slide was not preventable and would have caused some damage to roads and buildings, and perhaps a few injuries, the reservoir, whose waters caused the flood and multiplied the damage enor-mously, might have been placed in a geologically safer place. Fortu-nately, mass movements like this are infrequent, and most of them do not affect many people; many are known from the geological forms they leave behind rather than from first-hand observations.

Though downhill movement of debris is obviously induced by gravity, there are other reasons why some movements are fast on some slopes and slow on others; and why some slopes are stable and why others move if disturbed. First in importance is oversteep-ening of slopes, illustrated best by loose sand. Sandboxes have made nearly everyone familiar since childhood with the fact that a pile of sand will assume a characteristic slope; the angle is the same whether the pile is only a few centimeters high or three meters high. If some sand is scooped from the base of the pile very slowly and carefully, the angle of slope will steepen a little and hold. But if someone jumps on the ground near it, the slope cas-cades down and again assumes the original angle, which is called the **angle of repose**. That angle varies with the size and shape of the material; larger, flatter, and more angular pieces of loose mate-rial will remain stable on steeper slopes.

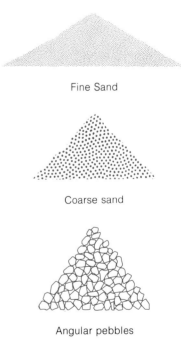

Fine Sand

Coarse sand

Angular pebbles

**Figure 8-15**
Angles of repose of particle mounds increase as particle sizes increase and shapes become more angular.

Next in importance is the effect of a lubricant, usually water. An extreme way to observe this effect is to try to build a sand pile with dripping wet sand: it just flows until it becomes as flat as a pancake. Another way is to try to walk down a clayey slope in the rain without sliding. Water filling the pores of permeable material allows the grains to slide past each other with little friction. The effects of slope and lubricant, of course, become more complex in other materials, depending on the nature of the mixture, the irregularities of slopes, the structurally stabilizing role of vegetation roots, and the amounts of water required to saturate the materials. Nevertheless, these effects are at the root of all downslope movements of boulders, sand, and soil, and they work in different ways to produce the variety of what we call **mass movements**: landslides, mudflows, rock slides, slumps, and creep.

**Landslides**　The terms **landslide** and **rockslide** refer to the rapid downhill movement of masses of rock, weathered erosional debris, soil, and anything else that may be in the way, such as houses,

**Figure 8-16**
Air view of the landslide in Madison Canyon that slid off the mountainside in the left background after an earthquake in August, 1959. The slide dammed the Madison River, forming the lake in the foreground. [Photo by J. R. Stacy, U.S. Geological Survey.]

fences, and roads. These names are used to distinguish slides on the basis of the main material forming them. Rockslides are movements of blocks of bedrock along joints or cracks, usually along the water-lubricated bedding planes of tilted sediments. The preconditions for landslides are slopes that are too steep for the frictional forces to hold the material in place against the force of gravity. Typically, small landslides are formed in places where slopes are first steepened by erosion, either natural or artificial (such as excavations), and then soaked by rain until saturated; once saturated, the slope is in a critical stage of instability, ready to give. Once the stage of extreme instability is reached, a slide is the inevitable result. Vibrations of some kind are responsible for triggering many slides. The more delicately poised something is, the smaller the jar needed to knock it over. Vibrations of strong earthquakes will trigger slides in material that might otherwise have stayed fixed for a long time. Most of the damage that resulted from the Alaskan earthquake of 1964 was caused by the slides it produced.

Melting snow and heavy rains made water abundant in the valley of the Gros Ventre River in the spring of 1925. That was the year of the Gros Ventre slide, a classic among the landslides of the twentieth century—perhaps because it was witnessed by so many people and perhaps because several geological studies were made soon afterward. The Gros Ventre River flows through the Jackson Hole area of Wyoming, which is noted for its natural beauty and is situated near Yellowstone Park. Though the valley is sparsely settled, it is visited by many. One eyewitness, a rancher who lived there, barely escaped from the path of the slide on horseback. Warned by the roar of the slide, he looked up to see a huge mass

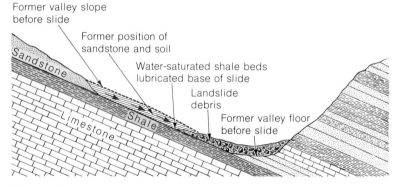

**Figure 8-17**
Cross section of the Gros Ventre slide, Wyoming. Arrows show the direction of sliding of a mass of soil-covered sandstone down a water-lubricated surface of shale underlying the sandstone. The conditions for this slide were set when the Gros Ventre River cut its valley floor through the sandstone formation, leaving unsupported the sandstone beds higher on the valley slopes. [Modified from W. C. Alden, 1928, Trans. Am. Inst. Mng. Met. Eng., v. 76, pp. 345–361.]

**Figure 8-18**
The scar of the Gros Ventre slide. This photo, taken in 1966, more than forty
years after the slide, shows the still-unforested slopes of the scar left by the slide.
[Photo by R. Siever.]

of valleyside hurtling toward his ranch. From the gate to his property he watched the slide race past him at about 50 miles an hour
and bury everything he owned.

It has been estimated that about 37 million cubic meters of
rock and soil slid down one side of the valley, surged more than
30 meters (100 feet) up onto the opposite side, and then fell back
to the valley floor, damming the Gros Ventre River. Most of the
slide was a confused mass of blocks of sandstone, shale, and soil,
but one large block, covered with soil and a forest of pine, slid
down as a unit. Two years later, the large lake that formed behind
the slide overflowed it, draining the lake and causing a torrential
flood in the valley below.

The forces that caused the Gros Ventre slide were all natural.
The stratigraphy and structure of the valley (Fig. 8-18) made it
almost inevitable. On the side of the valley where the slide occurred, a permeable, erosion-resistant sandstone formation dipping
about 20° toward the river rested on a soft shale. The stage was
set for the event to happen, sooner or later, after the channel of
the Gros Ventre River cut through most of the sandstone and
left it with virtually no support where it intersected the river
channel. Only friction along the bedding planes between shale

and sandstone kept it from sliding. The removal of this support by channel erosion was the equivalent of oversteepening. The heavy rains and melting snow saturated not only the sandstone but the surface of the underlying shale. Once the force of gravity overcame friction, almost all of the sandstone slid toward the bottom of the valley along the lubricated surface of the wet shale. This general pattern is characteristic of many other slides, except that some slides take place along joints or other zones of structural weakness instead of bedding planes.

In **unconsolidated** material — that is, in materials not held together by a cement or by a strong interlocking crystal structure — landslides start after a significant part of the whole mass is saturated with water and lubricated. Again, some shock, however small, can trigger the downslope movement of a whole unstable hillside of such material. Some such movements take the form of a **slump**, in which the mass travels as a unit, in most places slipping along a surface that is concave upward. Rockslides can also move as slumps. Many landslides, however, rather than slumping as a single unit or as several, will move as an incoherent jumbled mass flowing almost with the characteristics of a fluid. Ronald Shreve, a geomorphologist at the University of California, Los Angeles, has spent much time studying landslides. He thinks that some of them, once they start, race along on an undersurface of trapped, compressed air that provides an almost frictionless speedway.

Figure 8-19
Aerial view of slump in a hillslope underlain by weak shale. The slump is about 140 m (450 ft) wide. Stanley County, South Dakota. [Photo by D. R. Crandell, U.S. Geological Survey.]

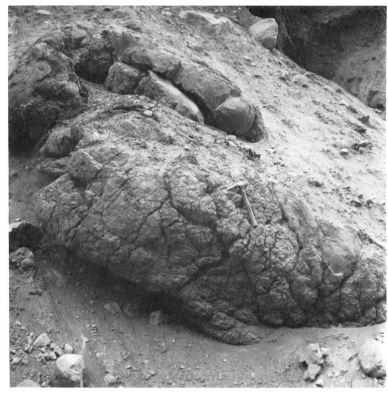

**Figure 8-20**
A mud-flow deposit on the slopes of Mt. Rainier, Washington. The mudflow consists of a mixture of volcanic ash and rock fragments, including a large volcanic bomb (above the hammer). [Photo by D. R. Crandell, U.S. Geological Survey.]

**Mudflows** are particularly fluid forms of landslides that move as viscous tongues of mixed mud, soil, rock, and water. Most common in hilly and semiarid regions, mudflows are generated by infrequent, sometimes prolonged, rainstorms. The flows start down upper valley slopes and merge on valley floors. Where mudflows exit from confined valleys into broader, lower valley slopes and flats, they may splay out to cover large areas with a mixture of wet debris. Mudflows can carry large boulders, trees, and even whole houses. **Earthflows** are something like mudflows, but are less fluid and thus tend to be confined to slopes of weathered shale or clay, where they move short distances downslope. **Debris-avalanches** are another kind of flow associated with humid mountainous regions. Water-saturated soil eroded down to bedrock may move rapidly, carrying everything in its path with it. In 1962 a debris avalanche in the Peruvian Andes traveled almost 15 kilometers (9 miles) in about 7 minutes, engulfing most parts of eight towns and killing 3500 people! Mudflows and debris-avalanches are easily triggered on slopes of volcanic cinder cones as their uncon-

**Figure 8-21**
An aerial view of the Peruvian mountain region affected by the giant debris
avalanche of May 31, 1970. The mass of ice, water, mud, and rock debris, shown
to the left of center, started with the sliding of a mass of glacial ice and rock,
about 1 km wide and more than 1.5 km long, that swept downslope at about 15
km/hr. It buried two towns and killed more than 20,000 people. [Servicio
Aerofotografico Nacional; courtesy U.S. Geological Survey.]

solidated accumulations of ash and other erupted material become
saturated during rains. Eruptions trigger some flows by associated
earthquakes and tilting movements, cloudbursts, and sudden falls
of volcanic debris. More than fifty-five such flows, many of them
associated with eruptions, have occurred on the slopes of Mt.
Rainier, Washington, in the past 10,000 years, according to U.S.
Geological Survey geologists.

All of these rapid mass movements of mud, soil, debris, and
rock are expected, though infrequent, natural occurrences — the
consequences of varying combinations of processes that steepen
slopes and lubricate near-surface materials with water. They cannot be eliminated, but humans can avoid provoking them by unwise
engineering and can learn better ways of predicting them. We
may also have to learn not to build houses in particularly dangerous
places where natural movements cannot be controlled.

**Figure 8-22**
Talus produced by a rockslide at the base of a rocky valley wall in Yosemite National Park, California. The trees in the foreground were defoliated by the blast of air that accompanied the slide. [Photo by F. E. Matthes, U.S. Geological Survey.]

**Rock Falls, Soil Creep, and Solifluction** Talus slopes are built up by the accumulation of many individual falls of talus blocks—rocks broken from the outcrop by physical and chemical forces, Though each rock falls suddenly, the accumulation of talus is slow and steady, building up blocky slopes everywhere along the base of a cliff. Some talus slopes have been accumulating for so long that they almost completely cover the bedrock. The talus slopes themselves are eroded at their lower slopes by being further broken down in size and being carried away by mudflows and other mass movements or streams.

**Soil creep** is a downhill movement of soil that takes place so slowly that its rate of movement is difficult to measure over short time periods, much less notice, from day to day. Yet such slow movements are the reason why trees, telephone poles, fences, and other object apparently fixed in the soil tend to lean or be displaced slightly downslope. By comparing careful measurements of the precise positions of marked stakes fixed in bedrock with those of others in the soil, geologists have learned that most creep ranges from a few millimeters per year to a few centimeters per year, depending on the kind of soil, steepness of slope, rainfall, and vegetation. The mechanics of creep are not fully understood, but it is fairly certain that it is the result of a combination of minute movements caused by freezing and thawing, wetting and drying, thermal expansion, activities of burrowing organisms, and root movements of growing and decaying plants.

**Figure 8-23**
A rock glacier in the Copper River region of Alaska. The glacier is supplied with rock from the talus-covered slopes above. [Photo by F. H. Moffit, U.S. Geological Survey.]

Depending on the degree of saturation, soil may move slowly or rapidly. In cold regions, where saturated ground becomes frozen, the movement of soil, called **solifluction**, takes place slowly, induced by freezing and thawing. Because ground in frigid areas may be frozen to great depths, the lower zones do not thaw. But when the surface zones of the soil thaws, the water has no place to go. The water-saturated upper layers then ooze downhill, carrying broken rocks and other debris. A related phenomenon is the movement of **rock glaciers**, which are like lobed rivers of rock that have moved down slopes and valleys in cold, mountainous regions. The pores between the rocks are partly filled with ice, and the motion is in many ways like that of true glaciers.

Though unspectacular, all of the motions of creep and related slow downslope travel of rock and soil account for a large part of the initial transportation of fragmented and weathered material from the primary sites of erosion to lower slopes, where they are further eroded and transported by streams.

**Slope Wash** The prelude to organized river-channel transportation of detritus is **slope wash**, the generalized surface runoff of rainwater on the surface of the ground. Included in this category are all of the little rills and threads of water that gradually come together, ultimately to form rivulets, then streams, then river. If you have the patience and don't mind the rain, you can observe how water running down a hillside moves small particles of silt and sand, gradually undermines clayey soil from beneath pebbles, and makes them tilt and move. A rain heavy enough will produce a **sheet wash** on even slopes—a continuous sheet of flowing water. Sheet wash does not normally persist very far down a slope before it gets organized into tiny channels or rills, but a good place to see it is on sloping concrete embankments, such as those near bridges or culverts. The impact of raindrops and hailstones is sufficient to erode clay and silt and help lift it into the moving slope wash. Slope wash operates most effectively in such places as freshly graded highway shoulders and around new houses. The best way to avoid excessive erosion by slope wash is to plant grass and other vegetation. Leaves and stems shelter the surface from raindrops and retard the flow of slope wash; roots hold the soil together to resist erosion.

Though it is hard to get reliable figures on the quantity of material moved by slope wash and mass movements in general, almost all of the enormous load of erosional debris that eventually gets into the rivers of the world must have first moved downhill by various combinations of those processes. The removal of earth materials, at different rates and by different processes, creates the forms of landscape. Fragmentation, weathering, and mass movements sculpture the land and feed detritus to the major water-transportation networks of the Earth—rivers.

1. Fragmentation of rock into boulders, pebbles, sand, and silt is a mechanical process promoted by chemical weathering.

2. Studies of rock fragmentation on the moon show that little of the normal mechanical erosion that we observe on Earth takes place there.

3. Zones of weakness open up as cracks and joints when rocks formed in the interior are exposed at the surface. Water seeps into the openings and causes weathering: when water freezes, it wedges the rocks apart.

4. Sizes and shapes of fragmented rock particles, from boulders to sand grains, are largely inherited from such characteristics of source rocks as jointing, bedding, and crystalline texture.

5. Tectonics creates mountains, and topography strongly affects erosion. Mechanical weathering dominates in areas of high topographic relief, but is relatively less important in low-lying regions. The ratio of mechanical to chemical weathering is a function of both climate and topography, and ultimately of tectonic activity.

6. Mass movements, including rock falls, landslides, slump, and creep are slow to rapid downhill movements induced by gravity when a slope is oversteepened by natural erosion or construction.

7. Mudflows, debris-avalanches, and earthflows are water-saturated, nearly fluid flows.

8. Slope wash is the generalized unchanneled flow of rainwater on slopes — the prelude to organized rills, streams, and rivers.

## EXERCISES

1. Which would you expect to fragment more rapidly into small pieces, a granite or a schist? What effects might climate have on the difference? How might the shapes of the pieces differ?

2. Of the various mechanisms of rock breakage described, which would you expect to be more important in the rainy tropics as opposed to cold polar regions?

3. What are the erosional processes that produce these three different sizes of material: blocky boulders up to several meters across, sand grains about 1 millimeter across, and clay particles a few thousandths of a millimeter in diameter? How far from the eroding outcrop might pieces of each size get in a given time period?

4. Which kinds of mass movements tend to follow periods of heavy rains and why?

5. What geologic conditions would you want to check on before you built or bought a house on the side of a steep, soil-covered hill?

6. What role do earthquakes play in landslides and related phenomena? Would you advise people living in areas where there are few, if any, earthquakes, to show more, less, or the same awareness of these mass movement hazards than those in areas where earthquakes are frequent?

Crandell, D. R., and H. H. Waldron, "A Recent Volcanic Mudflow of Exceptional Dimensions from Mt. Rainier, Washington." *American Journal of Science,* 254, pp. 349–362, 1956.

Eckel, E. B. (editor), *Landslides in Engineering Practice.* Washington, D.C.: Highway Research Board Special Report 29, National Research Council, 1958.

Legget, R. F., *Geology and Engineering.* New York: McGraw-Hill Book Co., 1962.

Sharpe, C. F. S., *Landslides and Related Phenomena.* New York: Columbia University Press, 1938.

Wahrhaftig, C., and A. Cox, "Rock Glaciers in the Alaska Range." *Geological Society of America Bulletin,* v. 70, pp. 383–436, 1959.

Young, A., *Slopes.* Edinburgh: Oliver & Boyd, 1972.

# 9

# Rivers—Currents, Channels, and Networks

*Most of the debris of erosion is carried downhill by running water in the streams of the Earth. Turbulent currents excavate river valleys by eroding channels and moving particles downstream. The ability of a river to transport particles in different amounts and sizes is dependent on current velocity, the total amount of water carried, and the downhill slope of the channel. Currents form ripples, dunes, and bars in the channel sediment. A river can be analyzed as a complex system of drainage patterns that form in response to the particular conditions of climate, bedrock, and particles produced by weathering.*

Rivers grip our imaginations. They bring out cliches like "mighty waters" and "raging torrents" and, at more serene moments, others like "majestic flow" and "babbling brook." The variety of cliches, most of which center on the idea of constant flowing, is matched only by the variety of forms themselves. Rivers may be straight, but they may bend and loop around too. A river may flow in a broad low swale that barely has the form of a valley, but that same river in another place may rush between the nearly vertical walls of a narrow gorge. A high mountain stream in the Rockies will be a hazard to cross because of the jumble of rocks through which the water rushes, but at times of drought you can walk almost entirely across the fine sand and mud on the bed of the Missouri River at Omaha. Rivers are clear, and rivers are turbid. Some are fast, and others slow.

The factors responsible for these differences in how water moves, what it carries, and its effects on the landscape make up

Figure 9-1
A bend in the North Platte River, Converse County, Wyoming. In the course of its meandering, the river has eroded its banks (left background) and deposited point bars on the insides of the bends (center). [Photo by N. H. Darton, U. S. Geological Survey.]

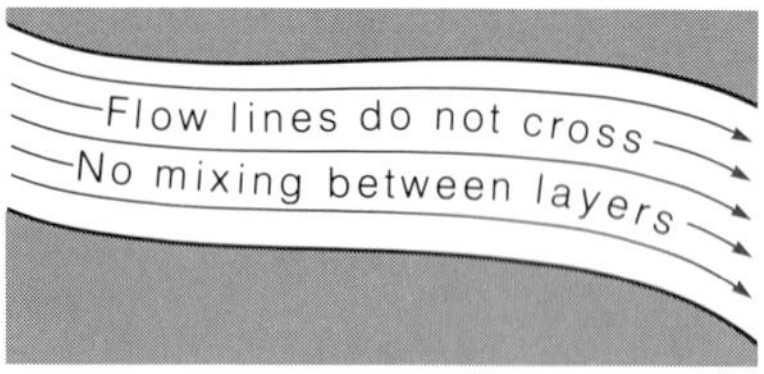

Figure 9-2
Laminar flow of a fluid between two solid channel walls. Flow lines in this type of flow do not cross, indicating that there is no mixing of fluid between layers.

the subject of this chapter. We start with some notions about how water moves in currents and how the movement enables it to carry things of all sizes and shapes, from mud, sand, and pebbles to whole houses ripped from their foundations by floods.

## HOW CURRENTS MOVE PARTICLES

If you slowly pour cold syrup over melting butter on pancakes, you can observe a most basic form of fluid flow: a pattern of **streamlines**—lines of flow—is formed by thin strands of melted butter flowing along with the syrup. As long as the syrup is cold, and therefore thick and sluggish, the streamlines will show the two fluids moving steadily without mixing. This is the essence of what is called **laminar flow**, in which the particles of the fluid move in parallel layers with no mixing of material across the boundaries between layers. Some, however, like their syrup hot, and that changes the picture. The property of fluids that makes them move slowly or rapidly when poured is viscosity; the viscosity of such fluids as automobile lubricating oils is usually determined by measuring the time that a certain quantity of fluid takes to flow through a small opening, like the bottom of a funnel. Viscosity is a function of temperature, so that the viscosity of most fluids is lower—that is, they flow faster—at higher temperatures. Thus hot syrup, which is less viscous than cold syrup, may not move in obvious, slow laminar flow but in a faster, more confused pattern, one of **turbulent flow**. In a quiet room, tobacco smoke, another fluid, will at first rise from a cigarette in a straight, vertical plume of laminar flow and then break up into a turbulent, eddying, complex flow in which the streamlines cross, interfere, and eventually become dissipated by completely mixing with air. The essence of turbulence is in the mixing of the flow in complex patterns. The general behavior is the same whether the fluid is a gas, like smoke in air, or a liquid, like syrup or water.

The most important factor that determines whether the flow of water is laminar or turbulent is the velocity. Only very slow flows are laminar; flow in practically all of the streams we see is turbulent. Secondary factors are the roughness of the stream bottom and the depth of the stream. The shallower the stream and the smoother the bottom, the more likely the flow will be laminar. Laminar flow in nature can be seen in thin sheets on nearly flat slopes, and occasionally in very cold water in streams running over smooth bottoms. Turbulent flow varies in the intensity of eddying and swirling movements, again mainly a matter of stream velocity.

Strange behavior is shown by a fast-flowing, turbulent stream where it suddenly steepens and starts to flow even faster than before. The rush of water smooths to what looks like straight, uneddied movement, even though it is still basically turbulent; its surface smoothness is deceptive, for the flow is actually so fast that it is commonly called **shooting flow**, as opposed to the more **tranquil**, or **streaming flow**, at lower velocities. Shooting flow can be seen typically in rocky rapids or where water flows down steep narrow chutes.

A startling development in petroleum technology came some years ago when chemists were able to produce modern engine oils whose temperature-viscosity behavior is the opposite of normal, so that the oils would be more viscous in the summer and less viscous in the winter—just what automotive engineers had hoped for.

**Figure 9-3**
Turbulent flow of a fluid between two solid channel walls. Flow lines in this type of flow cross in a complex pattern; unlike laminar flow, no definite layers of fluid can be discerned, which indicates extensive mixing of the entire fluid mass.

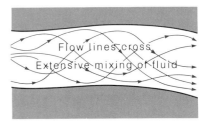

**Figure 9-4**
Streaming (tranquil) and shooting (rapid) flow are determined by slope angle. Beyond a certain steepness of slope, slower, wavy-surfaced streaming flows abruptly change to faster, smooth-surfaced shooting flows. Where the slope again becomes gentle, the flow returns to streaming, the transition being marked by a turbulent roller, or eddy.

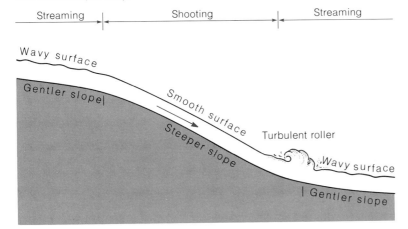

**Figure 9-5**
Sand in movement in a laboratory flume. Finely powdered aluminum injected into the flow reveals the movement of the fluid above the sand—particularly, the turbulent eddies in the lee of small sand ripples. Flow is from right to left. [Photo by A. V. Jopling, from *Sand and Sandstone* by F. J. Pettijohn, P. E. Potter, and R. Siever. Copyright 1972 by Springer-Verlag New York, Inc.]

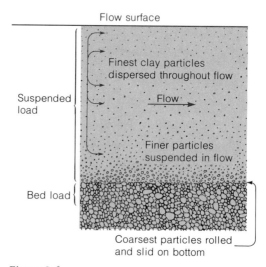

**Figure 9-6**
A current flowing over a bed of sand, silt, and clay particles transports particles in two ways: (1) as bed load, which moves by sliding and rolling along the bottom, and (2) as suspended load, which moves by being temporarily or permanently suspended in the flow itself. The finest clay particles remain permanently suspended in all but extremely sluggish flows.

**Movement of Particles** These three forms of flow—laminar, turbulent, and shooting—differ enough in velocity and in other characteristics that they may be likened to fine, medium, and coarse sandpaper in their ability to erode stream bottoms. Laminar flows are gentle; they erode the bottom slightly if at all and carry only the smallest of particles—those of clay size. Turbulent flows, depending on their speed, can move particles from clay size up to pebbles and cobbles. The fastest streams, including all shooting flows, wear away solid rock and carry along coarse particles at high speeds. All of the debris of erosion that rivers deposit as sediment is called **alluvium**. The valley floors of rivers are covered with such sediment, and many ancient sedimentary rocks originated in this way.

Turbulent currents move particles downstream mainly by lifting them up into the flow or by rolling or sliding them along the bottom. Smaller particles, clay and silt, are easily pulled up into the flow by upward-moving eddies. Once in the stream, the particles are carried along in suspension; all particles so carried by a stream at any one time are its **suspension load**. The forward force of the moving current acts more directly on the larger grains at the bottom, pushing, rolling, and sliding them along. Particles moved in this way constitute the **bed load** of a stream. Intuitively we grasp the idea that the stronger the current, the larger the particles it can carry in suspension or as bed load. The measure of this ability to carry particles of different size we call **competence**. The greater a stream's competence, the larger the particle size it can carry; the straightforward way to measure a stream's competence is by the largest particle sizes it can carry.

Just as dump trucks of various sizes can all carry sand, but some more than others, so do streams carry similar loads, but in different amounts depending on their size. The load of particles of any size that a stream can transport is called its **capacity**. Although the capacity of a stream is related to its size, more important than the width and depth of a stream is the amount of water that flows in it—the **discharge**, or the volume of water that flows past a given point in a specified unit of time. Discharge is commonly measured in cubic meters per second or cubic feet per second. A typical small stream may vary in discharge from 10 to 10,000 cubic feet per second (0.28 to 280 cubic meters/second). The Mississippi River discharge may be as low as about 50,000 cubic feet/second (1400 cubic meters/second), but in flood may amount to more than 2,000,000 cubic feet/second (57,000 cubic meters/second). The Mississippi, though it can ordinarily carry only clay, silt, and sand, carries enormous quantities of those small particles; in contrast, a fast mountain stream may move boulders, but only a few of them.

A stream's competence for suspended load is a balance between the uplifting forces that turbulence exerts on particles and the

Size of particles carried

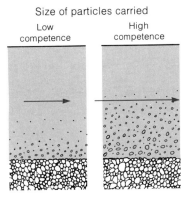

Number of particles carried

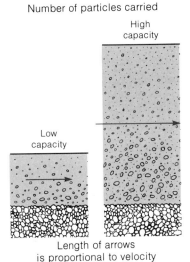

Length of arrows is proportional to velocity

**Figure 9-7**
The competence of a current to carry particles is a measure of the largest sizes it can carry. The capacity of a current is a measure of the number of particles it can carry. Although both depend on velocity to some extent, competence depends much more directly on velocity and capacity depends largely upon discharge, which is the volume of water flowing past a point in a given time.

Discharge is measured in practice at a gaging station, where the height, or *stage*, of a river is measured together with the velocity of the flow, from which the discharge can easily be calculated. There are now more than 7000 gaging stations in the United States.

force of gravity, which pulls the grains back down to the stream bed. Particles of silt and clay are so easily lifted and drop back so slowly that they tend to remain in suspension in all but the most still waters. The **settling velocity**—the speed with which particles fall through the fluid to the bed—is much greater for sand grains than for silt and clay particles. The typical movement of sand grains is an intermittent jumping, called **saltation**, in which they get sucked up by eddies into the flow, travel for a while, and then fall back to the bottom. The bigger or denser the grain, the shorter its travel time and the longer it rests on the bottom before being lifted again by a particularly strong eddy. The lighter the grain, the more frequently it is picked up and moved along above the stream bed.

The measurement of the exact relations between current velocity and competence has been a major concern of geologists and hydraulic engineers who study transport of sediments by streams. Their objective is to use these measurements to obtain some idea of the nature of the sediment that is produced by a particular current so that they can work backwards to infer the nature of ancient currents from the sizes of grains in sedimentary rocks. Engineers, for example, frequently want to predict what kind of sediment and how much of it will be brought in by a river to a lake behind a dam. Figure 9-9 shows the results of experimental measurements. From such a graph one can quickly read off the size of a particle that can be carried at a given velocity, or vice versa. This graph is applicable only to flows of one meter depth. The band indicates the lack of precision of the data from experi-

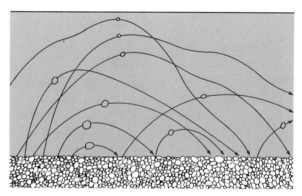

**Figure 9-8**
Saltation is an intermittent "jumping" motion of grains. Turbulent eddies pull grains up into the flow, where they travel with the flow for a distance and then fall back to the bed. In general, the smaller the particle, the higher it will jump and the farther it will travel. Turbulence keeps some very small particles suspended throughout a long travel path before they fall back to the bed.

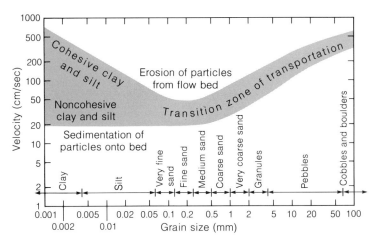

**Figure 9-9**
Experimental data on the relation between grain size of particles and velocity of
a flow 1 m deep over a flat, granular bed. For a specific grain size, the lower line
is the velocity at which all particles of that size will fall to the bed; the upper line
is the velocity at which all particles of that size will continue to be picked up
from the bed. The transportation zone is a transition zone in which particles
already suspended will remain in the flow and in which there is no tendency for
either erosion or sedimentation to take place. Cohesive behavior of most fine-
grained natural materials makes them more resistant to erosion than noncohesive
clays and silts. [After F. Hjulstrom, as modified by A. Sundborg, 1956, "The
River Klarälven," *Geografisk Annaler.*]

ments in which it is impossible to control all factors completely
and rigidly. The unexpected upturn that the curve shows for
clay-size particles does not mean that they are harder to carry but
that they are harder to lift from a smooth clay bottom because the
particles cling together with strong cohesive forces.

**Bedding and Sedimentary Structures** The movement of
sand grains in currents creates not only the familiar horizontal
bedding planes but also ripples and dunes on the stream bed.
**Ripples** in sand are the low narrow ridges that are separated by
wider **troughs**, both of which can be seen on the surfaces of wind-
swept sand dunes, on underwater sandbars in shallow streams,
and under the waves at beaches. They come in an assortment of
shapes, sizes, and patterns that are characteristic of the currents
that form them. Our main source of information is data from
experiments done in laboratories where the streams of nature
are simulated in flumes—long tilted channels in which the flow
of sand and water can be controlled. Flumes allow careful measurement of velocity and other variables plus direct observation,
through transparent side walls, of the ways in which grains move
and accumulate on the bed to form ripples and stream dunes.
Ripples, dunes, and a variety of other forms produced in the
process of sedimentation are called **sedimentary structures.**

The Skin of the Earth:
Surface Processes

Figure 9-10
Ripples in sand form in response to current action of wind or water. Their size and spacing increase as the current velocity increases. [Photo by P. E. Potter.]

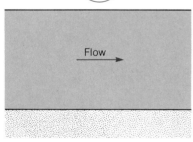

Figure 9-11
Flat bed in a flume experiment of steady flows at very low velocity over a sand bed. At this flow stage, sediment transport takes place by sliding and rolling motions of single grains. [After D. A. Simons and E. V. Richardson, 1961, "Forms of Bed Roughness in Alluvial Channels," *Am. Soc. Civil Eng. Proc.* v. 87, Hy3, pp. 87–105.]

One of the best ways to obtain good data from a dynamic experiment is to hold all variables constant except one and to change that one variable systematically and measure the change in the result. In 1961, members of a U.S. Geological Survey group reported the results of their experiments on the mechanics of river flow with flumes. Using sensitive equipment and interpreting their results in terms of modern theory, they repeated a simple, classic experiment that was done in 1914 by one of the great geologists of that time, *G. K. Gilbert.* Imagine that you are redoing this experiment. Beginning with a flume containing a preformed, flat bed of sand grains, all of the same size, you start the water flowing as a very weak current by opening a faucet just a crack. At this low velocity of flow, the current is too weak to move any grains, and the bed remains flat and undisturbed. Upon turning the water up a bit more to make the current a little stronger, you begin to see a few grains here and there roll or slide short distances, and a rare one saltate briefly. The bed stays flat. After another increase in flow, however, you notice that many grains are saltating, and after a little while at this flow velocity small ripples a few centimeters high start to form all over the bed. Soon you notice that the ripples themselves are moving like waves almost imperceptibly downstream whereas the particles themselves move much more rapidly.

The ripples have a gentle slope upstream and a steep slope downstream: they are **asymmetrical ripples**. At the transparent sides of the flume, you can see the ripples in cross section and observe the inclined bedding that we call **cross-bedding**. Here, of course, the cross-bedding is on a very small scale compared to that seen in such rocks as the Coconino Sandstone of the Grand Canyon (Chapter 2). The angle of the cross-bedding is the angle of the downstream, or lee, slope of the ripple.

The next higher stages of velocity make the ripples move faster and grow larger, until they get many centimeters high, large enough to be called **dunes**. Dunes have the same general form and structure as ripples but are larger; they form under water as commonly as those that people more frequently observe forming under air. As the dunes grow larger in the flume, smaller ripples form and climb up their backs and disappear over the lee slope. Increasing the velocity still further changes the character of flow from streaming to shooting flow, and as it does the bed responds quickly: ripples and dunes disappear, and the bed becomes flat again. But this time all the grains in it appear to be in rapid motion. Increasing the flow still more, to a velocity seldom met in rivers, produces a more irregular bed with some dune-like forms that move upstream and, finally, a complete washout of the whole bed of sand, which becomes suspended in the flow.

Figure 9-12
Rippled bed in a flume experiment. Major transport is by bed load and saltation. Ripples migrate downstream and have cross-bedded structure. [After D. A. Simons and E. V. Richardson, 1961, "Forms of Bed Roughness in Alluvial Channels," *Am. Soc. Civil Eng. Proc.* v. 87, Hy3, pp. 87–105.]

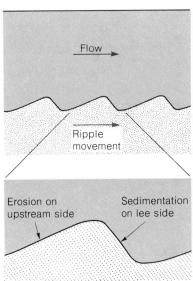

Enlarged view of ripple cross section

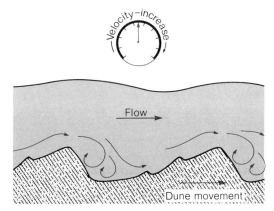

Figure 9-13
Rippled dune transport at moderate velocity in a flume experiment. Dunes have the same cross-bedded structure as ripples. Because ripples migrate downstream faster, they tend to climb over the backs of the dunes. Strong reverse eddies form in the lee of dunes. [After L. A. Simons and E. V. Richardson, 1961, "Forms of Bed Roughness in Alluvial Channels," *Am. Soc. Civil Eng. Proc.* v. 87, Hy3, pp. 87–105.]

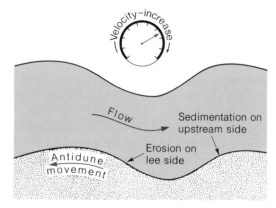

**Figure 9-14**
Antidune transport at high velocity in a flume experiment. Antidunes migrate upstream while there is extensive sand-grain movement downstream by suspension and bed load, almost the entire bed surface being in motion. [After D. A. Simons and E. V. Richardson, 1961, "Forms of Bed Roughness in Alluvial Channels," *Am. Soc. Civil Eng. Proc.* v. 87, Hy3, pp. 87–105.]

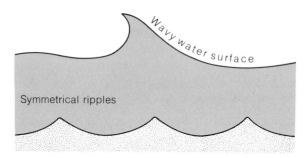

**Figure 9-15**
Oscillation ripples are symmetrical forms produced by waves. They commonly have much sharper crests than current ripples.

Sedimentologists—geologists who specialize in the study of sediments (there is hardly any subject that has not become a specialty by now)—have worked hard to infer from the ripples and cross-bedding preserved in rocks the kind and intensity of current that formed them. They have relied on flume studies and a knowledge of how fluids behave, coupled with observations of how modern sediments, such as river sand bars and other sand accumulations are constructed. Such studies have shown us how to recognize ripples formed by wave action for they are symmetrical and have much sharper crests than current ripples.

Formed by the back-and-forth movement of waves, they are called **oscillation ripples**. Different cross-bedding forms are diagnostic of other environments. In river bars, for example, the current scoops out long, spoon-shaped troughs that get filled up by cross-beds, as in Figure 9-16, and geologists can recognize this pattern in ancient rocks.

Currents of wind act in the same way as those of water, except that wind is not confined to channels. And because air is so much less dense than water, its competence and capacity are on the average lower than those of water flows. Winds also produce a variety of kinds of sedimentary structures, which are covered in Chapter 10.

## EROSION BY CURRENTS

Though it is obvious from observations of sediment transport that currents can erode a bed of soft sediment by picking up particles and carrying them downstream, one cannot so easily observe the erosion of solid rock by a stream, because it happens much more slowly. Here the analogy to the action of sandpaper (or sandblasting) is more apt, for much of the erosive power of a stream comes from the abrasion of the bottom by the sand and gravel it carries. Common on some river bottoms are pot-holes—rounded depressions and deep holes carved into solid

Figure 9-16
Trough cross-bedding is produced on river bars as troughs are scoured and then filled with sand, building up a series of interfering structures as earlier ones are partly eroded. The open end of the trough always faces downcurrent, but the direction of cross-bedding inclination may be variable, depending on the orientation of the exposure with respect to the troughs.

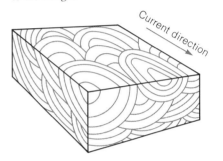

Figure 9-17
Potholes in granite ledges exposed at low water in James River, Virginia. The pothole in the right foreground shows gravel at the bottom, which is the scouring agent that erodes the holes in solid rock. [Photo by C. K. Wentworth, U. S. Geological Survey.]

Niagara Falls has been carefully
studied to learn the rate of its
headward erosion. Historical records
show that the main (horseshoe)
part of the falls has been moving
upstream at a rate of more than
1 m/yr.

rock by sand grains and pebbles swirling in fast eddies. We rarely
see the bedrock floor of a river except in rocky streams in mountains or where a stream crosses rapids, and even there most of it
is covered by boulders of all sizes. The floors of most rivers that
flow in broad valleys in plains or lowlands are covered with the
sand, gravel, and mud that they carry. These rivers usually scour
their bedrock bottoms only at high flood stages. The bedrock
channels and walls of a river are eroded physically by a variety
of other mechanisms. Just as in weathering on the outcrop,
chemical decay along joints and cracks helps the breakup of
bedrock, but here it is aided by the sledgehammer action of
boulders slamming against solid rock. Chemical action is helped
by a variety of organisms, among them the aquatic algae that
attach themselves to rock surfaces (and make them slippery to
cross). Once large blocks have been loosened by decay along
joints, they are dislodged by strong upward eddies that literally
pull them up and out by what amounts to violent, sudden suction.
Channels at the bottoms of waterfalls are eroded at an enormously
rapid rate by the plunging action of the water and transported
rocks, which fall with tremendous impact. Waterfalls are eroded
and move upstream as the cliff that forms the falls is undercut
and the upper beds collapse. This situation occurs where a nearly
horizontal, erosion-resistant bed overlies more easily erodible
formations.

Erosion of soft unconsolidated material by a river is much more
obvious than that of bedrock channels and falls. Scouring and
slumping of soft river banks is commonly seen at high water.
Gullies in soils erode headward at a rapid pace. On a small scale,
one can observe this kind of erosion happening during a strong
rainstorm. Evidence of intense erosion of river banks, channels,
and surrounding areas is widespread after a major flood.

Strong as this direct evidence is to us, it was not the only thing
that convinced Hutton almost two hundred years ago that valleys —
some deep and narrow, others low and broad — were eroded by
the streams that flow in them. For Hutton, a key question was:
where does all the sediment carried by a stream come from?
The match between the rocks and debris carried by a stream and
the bedrock of the valley could be interpreted reasonably only
as an indication of erosion. A clinching point was the stratigraphic
matching of sedimentary beds on opposite valley walls. Because
those beds were once continuous sheets, the discontinuity represented by the valley had to come later by erosion. The many
lines of evidence, all supporting each other, lead to only one
conclusion: valleys are cut into rock by streams. This does not
always mean that the stream now in the valley was the one that
cut it. As we saw in Chapter 5, mountain glaciers — rivers of ice —
cut U-shaped valleys, and these are usually occupied by streams
that inherited the valley after the ice melted.

**Figure 9-18**
The Badlands of South Dakota, a region underlain by easily eroded horizontal sandstones and shales, which have been heavily gullied by slope wash, rills, and small streams. [Photo by N. H. Darton, U. S. Geological Survey.]

## THE RIVER AS A SYSTEM

Rivers have been personified as living things by various writers, who, like most of us, are affected by their constant motion and change. Rivers are dynamic systems that are some kind of balance between input and output. Overall, the inputs are all the water that reaches a river as surface runoff and groundwater plus the erosional debris from the entire area drained by it. The outputs are water and sediment, which are both ultimately carried to the ocean. In detail, any unchanging stretch of river has assumed a shape and a size in response to the input, so that output can match input. Changes in a river are the result of imbalance. Floods are obvious examples of a river channel's inability to carry off a suddenly increased input of water and sediment; the river responds by overflowing and "inventing" new channels or flow paths to carry away the water. When input decreases to normal, the river shrinks to its normal size.

As we noted earlier, the amount of water flowing past a given point per unit of time is the discharge, which is by definition equal to the cross-sectional area of the river multiplied by the velocity:

$$
\begin{array}{ccc}
\text{Discharge} & = & \text{Cross-sectional} \\
\text{volume} & & \text{area of stream} \\
(\text{ft}^3 \text{ or } \text{m}^3/\text{sec}) & & (\text{usually width times} \\
& & \text{depth in ft}^2 \text{ or } \text{m}^2)
\end{array}
\times
\begin{array}{c}
\text{Velocity} \\
(\text{ft or m/sec})
\end{array}
$$

**Figure 9-19**
Floodwaters of the Feather River at Nicolaus, Sutter County, California, after a break in the river's levee during a flood in December, 1955. The river channel is to the left (note broken bridge), and the break in the levee is in the left foreground. [Photo by W. Hofmann, U. S. Geological Survey.]

Because rivers are generally wide in relation to their depth, and are fairly easily describable as simple rectangles in cross section, the width-depth relationship is more useful than cross-sectional area. Now we ask: what happens to width and depth as discharge increases—for example, as a response to increased rainfall runoff? Just looking at this relationship, one can see that the water will have to run faster, or the width or depth will have to increase, or some other combination of factors will have to change. We could think about what *might* happen for a long time, but the way to find out is to measure what actually does happen. Figure 9-20 shows some typical graphs made at a measuring station; the graphs show that there is a systematic increase in all three factors, width, depth, and velocity, as discharge increases. The proportion of increase varies from one stream to another, depending on the debris load of the river and the ease with which it erodes its banks and channel bed, for erosion is the way a channel can widen and deepen.

The amount of sediment that the stream carries—its transport capacity—is also related to the discharge. The higher the discharge, the larger the load of sediment, as is shown in Figure 9-21. There is also a relationship between discharge and the size of the material eroded or deposited, for the flow picks up or deposits material in accordance with the graph in Figure 9-9.

Discharge in most rivers increases downstream as more and more water is collected from tributaries. Inevitably, then, the width, depth, and velocity must change too. The progressive

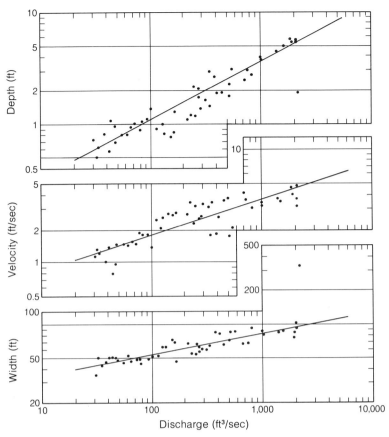

**Figure 9-20**
Changes of width, depth, and velocity with discharge at a specific river station: Seneca Creek at Dawsonville, Maryland. The drainage area of this stream is about 260 km². Discharge is given in cubic feet per second. [After *Fluvial Processes in Geomorphology* by L. B. Leopold, M. G. Wolman, and J. P. Miller. W. H. Freeman and Company. Copyright © 1964.]

downstream change in these factors has been observed to follow a pattern similar to that shown by an increase in discharge at one place: width, depth, and velocity all increase systematically, width far more than depth. The downstream changes in river channels at high and low water are shown in Figure 9-22. Velocity does not increase downstream as much as one might expect from the increase in discharge, because the decrease in the downhill slope of a channel is accompanied by a decrease in velocity. After all, water does run downhill, and Galileo showed the world 350 years ago that the steeper the slope, the faster the motion. The **longitudinal profile** of a river—that is, a plot of the relative elevation of its stream bed from headwaters to mouth—is concave upward, which is in accord with the steady downstream decrease in slope. Although the decrease in slope tends to slow the current,

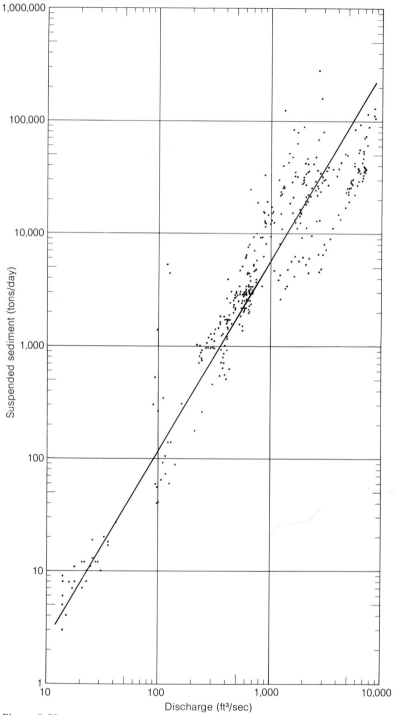

**Figure 9-21**
Increase of suspended sediment carried by a river in relation to increased discharge at a specific river station: Rio Grande near Bernalillo, New Mexico. Discharge is given in cubic feet per second. [After *Fluvial Processes in Geomorphology* by L. B. Leopold, M. G. Wolman, and J. P. Miller. W. H. Freeman and Company. Copyright © 1964.]

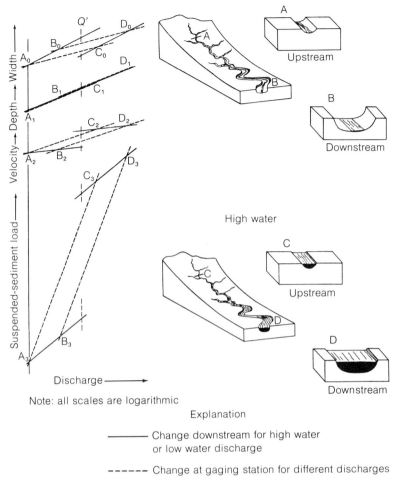

Note: all scales are logarithmic

Explanation

———— Change downstream for high water
or low water discharge

- - - - - Change at gaging station for different discharges

**Figure 9-22**
Changes in width, depth, velocity, and suspended sediment load in relation to
discharge at upstream and downstream stations. As discharge increases, at either
the upstream or the downstream station, the channel dimensions increase at
different rates. [After *Fluvial Processes in Geomorphology* by L. B. Leopold, M. G.
Wolman, and J. P. Miller. W. H. Freeman and Company. Copyright © 1964.]

its influence may be overcome by an increase in discharge that
makes the stream flow faster. In many rivers the discharge does
not increase greatly downstream, and the lower downstream slope
reduces the velocity. Because the slope is itself a product of the
river's capability of downward erosion and the opposing capability
of depositing sediment in its channel, the longitudinal profile
is a response the river makes to the balance between input and
output of water and sediments.

The concave upward longitudinal profile is remarkable for
its generality; it evolves in small rills and large rivers, in natural
streams and artificial water courses. Something so general must

**Table 9-1**
Normal changes in flow properties in the downstream direction and effects on river flow.

| Property of flow | Normal change in downstream direction | Effect on river downstream |
|---|---|---|
| Discharge | Increases | Velocity increases |
| Width | Increases | Ratio changes so that |
| Depth | Increases | friction is decreased |
| Velocity | Increases or decreases depending on changes in discharge and slope | Increases capacity to erode and transport |
| Frictional resistance to flow | Increases or decreases depending on perimeter and load | An increase causes decrease in velocity; a decrease causes increase in velocity |
| Sediment load (suspension + bed load) | Increases | Increases friction |
| Size of sediment particles (gravel, sand, silt, clay) | Decreases | Decreases flow resistance |
| Slope of river bed | Decreases | Decreases velocity |

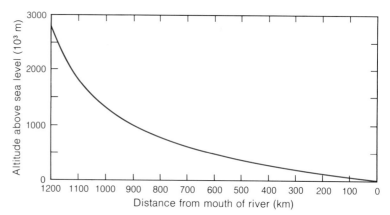

**Figure 9-23**
Longitudinal profile of the Platte and South Platte rivers from the mouth of the Platte at the Missouri River in Nebraska to the headwaters of the South Platte in central Colorado. This profile shows the typical concave-upward slope of rivers, much steeper at the headwaters than near the mouth. [Generalized from data of H. Gannett, in *Profiles of Rivers in the U.S.,* U.S. Geological Survey Water Supply Paper 44, 1901.]

be bound up with some basic dynamics common to all streams. The downstream increase in discharge gives one clue. As it increases, we have seen, so in general do the width and depth of the stream. But as the width and depth increase, so does the ratio of volume to cross-sectional perimeter. The effect of this increase is the same as that produced by increasing the diameter of a pipe: it allows the volume of flow to increase. The cross-sectional area, which partly governs the discharge, increases in proportion to the square of the radius ($\pi r^2$), whereas the circumference increases in proportion to the first power of the radius ($2\pi r$). The key here

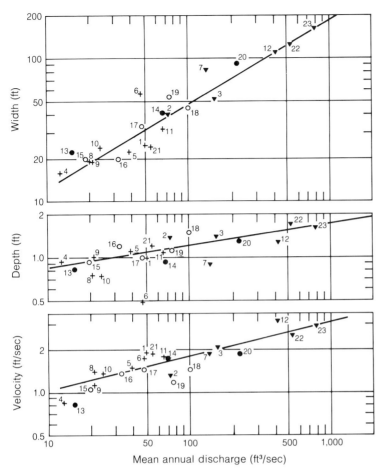

**Figure 9-24**
Downstream in stream width, depth, velocity, and discharge on the Powder
River. Stations are numbered in ascending order going downstream. Though
there is much scatter in the points, the trend toward increasing discharge down-
stream is strong. [After *Fluvial Processes in Geomorphology* by L. B. Leopold, M. G.
Wolman, and J. P. Miller. W. H. Freeman and Company. Copyright © 1964.]

is that the circumference of a pipe or the perimeter of a stream
is what controls the friction between the flowing water and the
walls of the pipe or the banks and floor of a stream. As the volume
of water increases, proportionally less of it is in contact with the
sides, and so friction, which retards the flow, is relatively decreased.
The less the friction, the more rapid the flow velocity. By these
same kinds of analyses we can list all of the factors that change
in the downstream direction that influence a river's longitudinal
profile (Table 9-1). Putting all of these interrelated factors together
in a single formula that would express quantitatively how a concave
profile develops is a formidable task — one that has been only
partially done. What is clear in a general way is that as they move

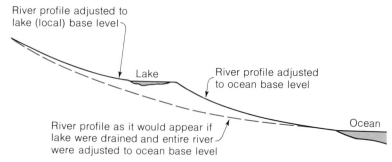

**Figure 9-25**
Regional and local base level as illustrated by a river flowing into a lake and from the lake into the ocean. In each river segment, the longitudinal profile adjusts to the lowest level it can reach. The lake may be a natural one or an artificial reservoir behind a dam. If the lake were drained and the river allowed to erode downward, the riverbed would adjust eventually to a single concave-upward profile.

downstream, rivers are able to transport more material of finer size at a higher discharge but at a lower slope. The concave upward profile seems to be another one of the steady-state equilibrium patterns that characterize so many Earth processes.

The longitudinal profile is also responsive to **base level** — the level or elevation at which the mouth of a stream enters a large standing body of water, a lake or the ocean, and so disappears as a river. Streams cannot cut below base level, for base level is "the bottom of the hill" — the lower limit of the longitudinal profile. A river may have a number of local base levels; for example, lakes or waterfalls along the watercourse form limiting levels that affect the flow of stretches immediately upstream. Changes of base level will cause a stream to change its characteristics, so that the water, sediment, and channel geometry come to a new balance. The most common adjustment to a raising of base level is predictable from the variables involved. The slope will be lowered, causing a reduction of velocity, which in turn decreases the stream's sediment-transporting ability. The stream then responds by dropping some of the sediment on the bed, which causes a change in the concavity of the profile, making it somewhat shallower than before. Lowering the base level increases the slope, the velocity, and the capacity to transport and erode the stream bed, causing a steepening of the concavity. The precise course of the change is predictable only when all of the flow-sediment relationships are known. For example, a stream greatly choked with sediment might be practically unaffected by a slight steepening of profile caused by a lowered base level, because the steepening would not be enough to make a significant increase in the load-carrying capacity in relation to the sediment supply.

The ways in which flow, sediment transport, and channel geometry adjust to changes in base level, so that a stream always

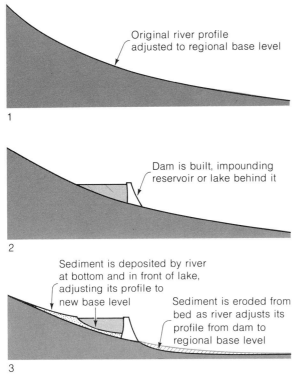

**Figure 9-26**
Change of base level and its consequences on a river profile.
The original profile, adjusted to be in equilibrium with the
regional base level, is altered when a dam is built. The dam
impounds a lake behind it, which raises the local base level.
The river responds upstream by adjusting its bed to a new
profile as sediment is deposited in and in front of the lake,
and it responds downstream by eroding its bed to a new
profile with respect to the original regional base level.

seems to stay in balance, leads to the concept of the **graded
stream**. An old idea, the concept was neatly expressed by the
late J. Hoover Mackin, a leading American geomorphologist.
He described a graded stream as "one in which, over a period of
years, slope is delicately adjusted to provide, with available
discharge and with prevailing channel characteristics, just the
velocity required for the transportation of the load supplied from
the drainage basin. The graded stream is a system in equilibrium;
its diagnostic characteristic is that any change in any of the controlling factors will cause a displacement of the equilibrium in
a direction that will tend to absorb the effect of the change."
How rivers do this is by forming a variety of depositional and
erosional patterns and by altering the shape of the channel.

**River Channels and Their Deposits**  If you were to test a river
severely to see how it would adjust to a change, you might pick

**Figure 9-27**
Alluvial fans in the Mohave desert, California. Each cone-shaped fan has been
deposited where the slope changes abruptly at the base of the mountains, and the
fans have grown together at their feet, completely obscuring the original mountain slopes. [Photo by J. R. Balsley, U. S. Geological Survey.]

out a place where a stream comes out of a mountainous area and
suddenly enters a relatively flat valley or plain. The change from
the high slope of the mountain stream to the much lower slope of
the plain is abrupt, and so is the response — an immediate dumping
of sediment in a cone or fan-shaped accumulation called an **alluvial
fan**. The fan itself normally shows the characteristic concave upward profile. On the steep upper slopes, alluvial fans are typically
dominated by coarse materials, boulders and cobbles, gravel and
sand; on their lower, gentler slopes, by finer sands, silts, and muds.
Fans from many adjacent streams along a mountain will merge to
form a long wedge of sediment whose external appearance may
mask the outlines of the individual fans that make it up. As these
compound fans grow upward and outward from a mountain front
over thousands to millions of years, they build up extensive sedimentary accumulations many cubic kilometers in volume.

**Channel Patterns** Streams running down alluvial fans will
frequently break up into small multiple channels that branch into
distributary networks, and this too is a consequence of the sedi-
ment-flow relationship. Because the sediment load is large in
relation to discharge, slope, and channel depth, the shallow
channels tend to become filled, and the streams then break through
their walls and form new channels. Many of the subsidiary channels
will rejoin farther downstream, giving the stream as a whole
a familiar pattern from which it gets the appropriate name **braided
stream**. Braided streams are not restricted to alluvial fans by any

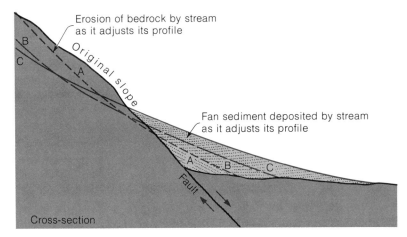

**Figure 9-28**
Formation of an alluvial fan as a stream adjusts its profile to a fault scarp. The original fault slope is eroded above the fan as the fan is built upward and outward from the base of the scarp in successive positions *A, B,* and *C.*

**Figure 9-29**
Braided stream choked with erosional debris, near the edge of a melting glacier. Muddy River, near junction with McKinley River, Alaska. [Photo by B. Washburn.]

**Figure 9-30**
Meanders of the Mississippi River, looking north from Alexandria, Louisiana, to Jackson, Mississippi.
The tributaries in the foreground meander as well, but with a much smaller distance between loops.
Sand point bars of the Mississippi show up as white areas on the insides of bends. [From National
Aeronautics and Space Administration.]

means; they are also typical of rivers running in plains and low
valleys — anywhere, in fact, where discharge is low in relation to
sediment load and the stream cannot transport enough to maintain
a single, deep channel. The braided pattern of multiple channels
develops within the confines of a single larger channel, and there
may be several scales of braiding, or braids within braids.

The word **meander** comes from the name of a river in Turkey
fabled in ancient times for its twisting, winding course of looping
bends. The word in geology refers to the more-or-less regular
bends that may form in any river. Meanders are normal for rivers
flowing on low slopes in plains or lowlands where the sediment

**Figure 9-31**
Incised meanders of the San Juan River, San Juan County, Utah. [Photo by E. C. La Rue, U. S. Geological Survey.]

is fine sand, silt, or mud, but in almost all kinds of terrain they are more common than long straight stretches of river. Nor are meanders restricted to rivers that carry sediment. They are common in meltwater streams on glaciers that carry no sediment. The Gulf Stream of the Atlantic Ocean meanders, and there appears to be a kind of meander pattern to jet streams in the stratosphere. Some meandering rivers have eroded deeply into bedrock. Meanders are common in streams flowing in limestone caverns, where they erode mainly by dissolving the limestone.

The three kinds of channel pattern—braided, meandering, and straight—each blend from one to the other without sharp division. Few long stretches of any river can be counted as truly straight; there are always some curves and bends. We call the measure of how straight or curved a river is its **sinuosity**, the ratio of its actual channel length to the straight-line distance down valley. A sinuosity of 1 corresponds to the perfect straight stretch. Meandering streams have sinuosities of 4 or more.

Rivers in straight channels do not flow precisely straight. The **thalweg**—the line of maximum depth of a river (usually close to the line of strongest current)—winds from one side to the other of straight stretches. Where the thalweg comes close to one bank, sand bars or mud banks tend to accumulate on the opposite bank, and a series of such bars may be found alternating from one side to the other along the channel. Straight streams also tend to have undulating beds in which shallows, called **riffles**, regularly alternate with deep spots, **pools**, usually at a repeating distance of five to seven times the channel width. Pools and riffles tend to form where there are many different sizes of sediment in the stream; the coarser sizes collect on riffles, where they may

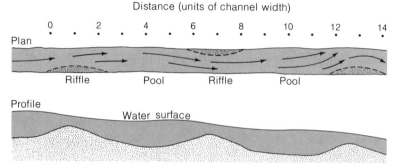

**Figure 9-32**
Straight reach of a river has a more or less uneven bed that consists of alternating deeps and shallows, known to trout fishermen as riffles and pools. The humps in the stream bed that give rise to the riffles tend to be situated alternately on each side of the stream at intervals roughly equal to five to seven times the local stream width. As a consequence the stream at low flow seems to follow a course that wanders from one side of the channel to the other, in a manner having an obvious similarity to meandering. [After "River Meanders" by Leopold and Langbein. Copyright © 1966 by Scientific American, Inc. All rights reserved.]

become gravel bars, and the finer sizes floor the pools. Riffles and pools, bars and mudbanks along straight rivers, and meanders all seem to show a wave-like regularity in their spacing, and seem to be different forms of the same general behavior pattern. These forms exhibit an even stronger similarity to wave-like behavior, for they all migrate downstream in much the same way that waves travel down a whipped rope. The patterns are stable. A river does not tend to meander one week and straighten the next. All of these facts support the idea proposed by Luna B. Leopold and W. B. Langbein, two of our most acute analysts of streams, that a river tends to take the course of least resistance—to minimize the work it must do in running downhill with a certain sediment load along a certain slope. The characteristic channel patterns are appropriate "solutions" that rivers find for different sets of prevailing conditions.

Meanders, because they are so easily seen, have been studied the most. They move both downstream and to the side, and we have the results of numerous river surveys to show just how they do. The sideways movement is accomplished by erosion along the outside of the bend, where the current is strongest, and by deposition of a curved bar on the inside of the bend, called a **point bar**, where the current is weakest. Some meanders on the Mississippi shift as much as 20 meters per year. As the meanders move sideways and downstream, so do the point bars, building up an accumulation of sand and silt that covers the part of the valley floor over which the channel has migrated. Characteristic patterns of cross-bedding in these deposits have been recognized to be the same as those in ancient sandstones, in which the down-current direction of cross-bedding has proved to be a major clue to the direction of flow of river systems that date back hundreds and millions of years.

As meanders move, they may progress unevenly, so that the relatively straight parts between bends get closer and closer to one another. When they get close enough, the river may cut across to the next loop, shortening its course and leaving behind the abandoned bend as an **oxbow lake** (Fig. 2-2). These lakes soon fill up with mud and silt, which support the growth of reeds and other vegetation. The cut-off may take place anytime, but it is most likely to happen during a flood stage, when the stress on the banks is high. The shortening is only temporary in comparison with the many thousands or millions of years of a river's lifetime, for new meanders are constantly evolving and enlarging, lengthening the course. Over a long time period on a given slope, the average length of the channel tends to remain constant. This is another way of stating that the river tends to maintain the same concave longitudinal profile.

**The River Floodplain** As meanders migrate back and forth across a river valley, laying down deposits of gravel, sand, and silt in the channels as they go, they create a wide belt of almost flat plain (Fig. 9-34), called a **floodplain**. The name is appropriate, for it is this part of the stream valley that is covered with water when a river overflows its banks at flood stages. The areal extent of a floodplain may be large in proportion to that of the river channel itself. For example, the floodplain of the Mississippi River stretches from the junction of the Ohio and Mississippi rivers at Cairo, Illinois, south to New Orleans at the delta, embracing about 80,000 square kilometers (about 30,000 square

In *Life on the Mississippi*, Mark Twain humorously expressed his version of a common misunderstanding of scientific extrapolation: "In the space of one hundred and seventy-six years the Lower Mississippi has shortened itself two hundred and forty-two miles. That is an average of a trifle over one mile and a third per year. Therefore, any calm person, who is not blind or idiotic, can see that in the Old Oolitic Silurian Period, just a million years ago next November, the Lowest Mississippi River was upward of one million three hundred thousand miles long, and stuck out over the Gulf of Mexico like a fishing rod. And by the same token any person can see that seven hundred and forty-two years from now the Lower Mississippi will be only a mile and three-quarters long, and Cairo and New Orleans will have joined their streets together and be plodding comfortably along under a single mayor and a mutual board of aldermen. There is something fascinating about science. One gets such wholesale returns of conjecture out of such a trifling investment of fact."

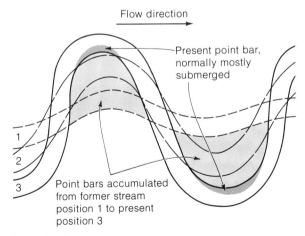

**Figure 9-33**
Lateral movement of river meanders gradually enlarges point bars: 1, 2, and 3 are three stages in the meander movement.

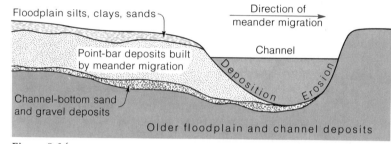

**Figure 9-34**
Formation of river valley deposits by a meandering channel. As the meanders swing from one side of the valley to the other, they leave a train of point bar and channel deposits from former channel positions, overlain by finer-grained silts, clays, and occasional sands deposited by floods. Point bars are usually cross-bedded, and floodplain deposits are horizontally bedded and ripplemarked.

miles). Floodplains of the Nile, the Tigris, and the Euphrates, were important in history and prehistory for their rich agriculture on fertile silt and clay deposited at times of flooding. Floodplain deposits are of two kinds, coarse and fine. The coarse material, usually gravel and coarse sand, is deposited as channel beds or bars, which tend to be smeared out by lateral migration of the river. The fine material, silt and clay, may be deposited partly by settling action as flood waters spread over the floodplain, but much of it may be from the lateral migration of muddy banks and bars of the river channel. Some of the fine material comes from the fill of oxbow lakes. Some of the coarse material is channel sand and gravel that flows through breaches in the banks during floods, forming fans or linear accumulations that extend away from the channel, called **splay** deposits.

During floods, when a river overflows its banks, the flow of sediment-laden water rapidly decreases as it spreads out over the floodplain. Along the strip bordering the river banks, where the decrease is particularly rapid, much coarse sediment is deposited; lesser amounts of finer sediment are distributed more widely over the floodplain. In this way, successive floods build up ridges on both sides of a river channel, to form **natural levees** that confine the river within its banks between flood stages. Some levees have been built up many meters above the surrounding plain, so that the plain may be lower than the river surface. Rivers that carry mostly fine-grained sediment in suspension do not build natural levees, or at most deposit very low ones.

Rivers regularly overflow onto their floodplains, as any river-bottom native can tell you. Rivers reach the **bankfull** stage, that level in which the water completely fills the channel, every year or two on the average. They break their banks and flood every 2 or 3 years, again on the average. Most of these frequent floods are small. The larger ones come much less frequently. We

1 Before flood

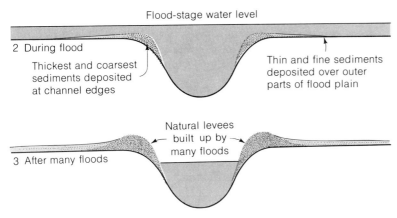

Flood-stage water level

2 During flood

Thickest and coarsest
sediments deposited
at channel edges

Thin and fine sediments
deposited over outer
parts of flood plain

Natural levees
built up by
many floods

3 After many floods

**Figure 9-35**
The formation of natural levees by river floods. As a river in flood stage overflows its banks, it rapidly decreases in velocity away from the channel and so
drops most of its sediment, the coarser fraction near the channel and the finer
fraction as a thinner layer of silt and clay over most of the floodplain. Successive
floods build up the natural levees to ridges many meters high.

speak of a "ten-year flood" as a flood of a certain discharge that
has a probability of recurring every ten years. A fifty-year flood
might be twice as big as the ten-year flood, but will recur on the
average only once every fifty years. These are probability statements, not predictions, and the people who live in river towns
have learned not to relax after one large flood in the false security
that the next one will not recur for another so many years. The
U.S. Geological Survey has published maps of the potential for
recurrence of floods of given discharges in the United States.

A ten-year flood is a moderate event, normally only one and
a half to two times the low flood that recurs on the average every
year, the "mean annual flood." The fifty-year flood is more likely
to be a catastrophic event, and may be as much as three times the
mean annual flood. The floodplain is overflowed by a depth of
water proportional to the magnitude of the flood. The fifty-year
flood, for example, may cover the floodplain with a depth of water
equal to about eight-tenths the bankfull depth, whereas the ten-
year flood will be only about half the bankfull depth. The depths
vary widely with the width of the floodplain and the details of
its topography.

Does a river do most of its work of erosion and transport during
everyday stages, during small or moderate floods, or during

the major events that come only once in a generation? It appears that the great bulk of the sediment in the few rivers studied is carried by floods that recur at least every five years — events of moderate intensity that happen often. Though the most infrequent events are very intense, they do not recur often enough to add much to the total. The amount transported by rivers in their everyday stages is too small to contribute significantly.

## DRAINAGE NETWORKS

About fifty miles west of Denver, U.S. Highway 40 crosses Berthoud Pass at an altitude of 3450 meters (11,314 feet) above sea level. At the crest of this pass is the **continental divide**, the line west of which all water flows eventually to the Pacific Ocean and east of which to the Atlantic. The continental divide is a continuous line that runs north and south the length of North America, dividing the continent into two enormous areas that contribute water to the major river systems on either side. In northern Wisconsin, about 50 miles southeast of Duluth, on Lake Superior, near the town of Cable, U.S. Highway 63 crosses a fairly low east-west rise at an elevation only a few hundred meters above sea level. This rise is another continental divide, separating all of the water that flows to the north, to the Great Lakes and the St. Lawrence River, from the water that flows to the south via the Mississippi and its tributaries. These **divides** and all of the smaller ones, down to the ridges that separate small ravines, separate **drainage basins**, the areas that funnel all their water into the streams draining them.

A good example of the hierarchy of drainage basins and divides can be found at Springfield, Illinois, in the drainage basin of the Sangamon River (Fig. 9-36). To the south is the divide between it and the smaller drainage basin of Macoupin Creek. The rivers of both drainage basins empty into the Illinois River, and so Springfield is also in the larger Illinois River drainage basin, separated by a divide some sixty miles east of Springfield from the drainage basin of the Wabash River, which drains Eastern Illinois and Western Indiana to the Ohio River near Shawneetown, Illinois. And so it goes. The Ohio and Illinois both drain to the Mississippi, which is the master river of the whole drainage basin east of the north-south continental divide and south of the east-west continental divide.

Divides are not immune from change. By virtue of discharge, slope, or other factors, the stream on one side of a divide may be able to erode and transport much more rapidly than the stream on the opposite side. It may, as a result, break down the divide between the two at some place along it and take part or all of the drainage of the slower stream. Victorian geologists of the last century used a

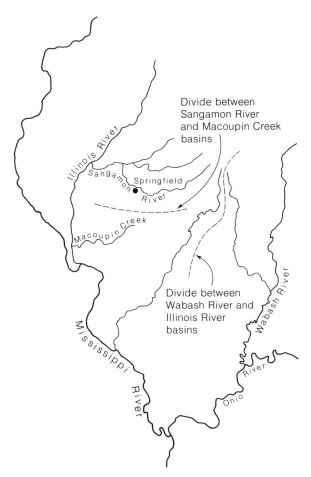

**Figure 9-36**
River drainage basins of central and southern Illinois,
showing divides between the rivers discussed in the text and
the hierarchical nature of drainage networks.

moralistic word for this kind of "capture" and called it **stream
piracy**. The breaching of divides and capture of streams is common
in small rills, becomes less common as streams grow larger, and is
rare in large rivers. Piracy explains such oddities as narrow valleys
or gorges that have no active streams running in them: the part of
the stream that cut the valley was drained away, its course having
been changed by the pirate stream.

The streams in all drainage basins follow certain rules. They all
are connected by a one-way network by which smaller tributaries
drain into larger ones with a definite pattern. The number of
streams and their distance apart follows a fairly orderly distribution. Most tributaries of about the same size are about the same
length, and the intervals between the mouths of tributaries are

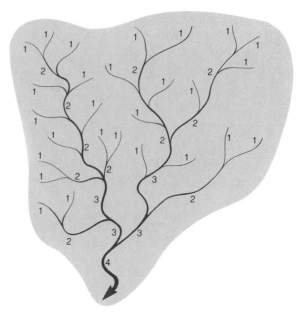

**Figure 9-37**
Order numbers of streams are designated by their position in
the hierarchical network pattern. Streams of order 1 have no
tributaries; those of order 2 have tributaries of order 1;
those of order 3 have tributaries of order 2; and so on. The
number of streams of a given order in a drainage basin
decreases strongly as order number increases. Thus, in the
illustration above, there are 27 streams of order 1, 7 of order
2, 2 of order 3, and 1 of order 4.

fairly uniform. The larger the drainage area, the longer the stream,
the ratio between the two being constant for similar terrains.
Robert Horton, an American hydraulic engineer, used as a basic
measure of the hierarchy of streams their **order**—that is, their
position in the tributary network. A stream of order 1 has no tributaries; a stream of order 2 has tributaries of order 1; a stream
of order 3 has tributaries of order 2; and so on. Thus the order
is defined by the order of tributaries: we count up as we move
downstream. As the order of streams increases, the following
changes are systematic (see Fig. 9-38): the length of main streams
increases, the number of main streams decreases, and the drainage
area increases. These general characteristics of drainage networks
seem to be exhibited by many kinds of systems. For example, even
the blood circulatory systems of mammals seem to have some of
the same characteristics. Here, as in the meander story, there are
some underlying laws that govern a wide range of phenomena that
at first sight seem unrelated.

Though all drainage networks branch in the same way, the shape
of their patterns varies greatly from one kind of terrain to another,
mainly as a response to the rock type or structural pattern of
folds and faults. The most common is **dendritic** drainage, named

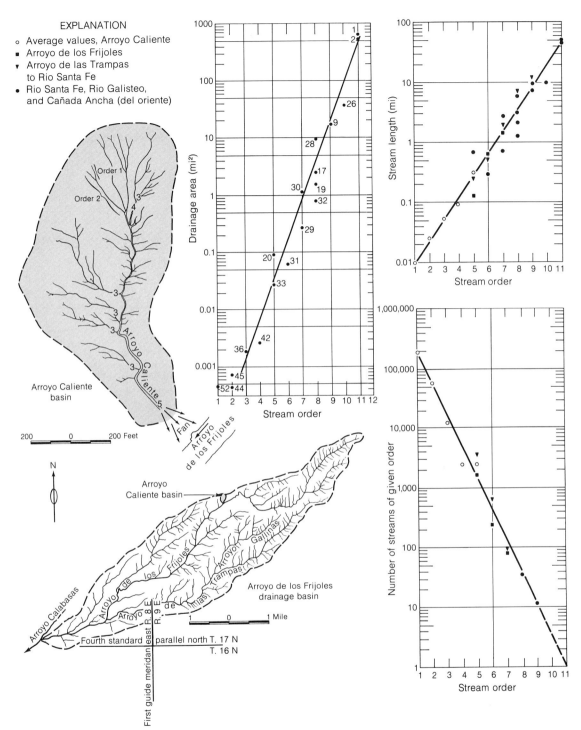

**Figure 9-38**
Relation of stream characteristics to stream order in drainage network of arroyos near Santa Fe, New Mexico. (Left) Detailed map of small tributaries. (Upper right) Relation of stream length to stream order. (Lower right) Relation of number of streams to stream order. (Center) Relation of drainage area to stream order. [After *Fluvial Processes in Geomorphology* by L. B. Leopold, M. G. Wolman, and J. P. Miller. W. H. Freeman and Company. Copyright © 1964.]

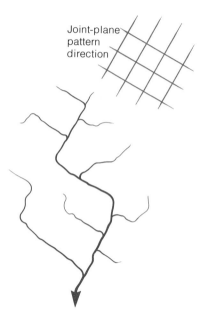

Joint-plane
pattern
direction

**Figure 9-39**
Typical dendritic drainage pattern,
characterized by a branching pattern
similar to that of the limbs or roots
of trees.

**Figure 9-40**
Typical rectangular drainage pattern
developed on a strongly jointed rocky
terrain. Drainage tends to follow the
joint pattern.

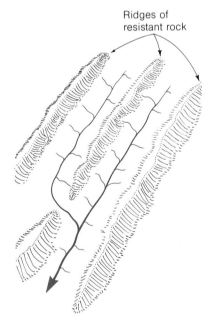

Ridges of
resistant rock

**Figure 9-41**
Typical trellis drainage developed in
valley and ridge terrain, in which rocks
of varying resistance to erosion are
folded into anticlines and synclines.

for the characteristic pattern shown by most branching deciduous
trees. This fairly random pattern is typical of terrain floored by
uniform rock types such as horizontal sedimentary rocks, or
massive igneous or metamorphic rocks. Differential weathering
of fractures or joint systems in bedrock localizes stream flow,
producing a more ordered and geometrical **rectangular** drainage
pattern. A special variety of rectangular drainage, the **trellis**
pattern, is formed where bands of rock resistant to weathering
alternate with bands that erode more rapidly, typically in a region
where rocks have been deformed into a series of parallel folds.
Drainage from a central high point, such as a volcano or a domal
uplift is **radial**. From this series of examples one can see that
drainage patterns reflect mainly the underlying structure of the
rocks and their varying resistance to erosion, though the density
of drainage depends more on climate. Areas of high rainfall tend
to have many more streams than arid regions. One rock type,
limestone, may virtually eliminate surface drainage in areas of
high rainfall, because it dissolves to form caves and underground
drainage. The northern part of the humid, rainy Yucatan Peninsula
of Mexico has no streams because all of the water sinks into porous,
cavernous limestone.

Geological events may strongly control the course of a river.
Streams flowing in a dendritic pattern on flat-lying sediments
may erode through the sediments to expose underlying, strongly

folded and faulted rocks of varying resistance to erosion. Because these **superposed** streams, preordained in their courses by their former bed of sediments, cannot break out easily to form a wholly new drainage pattern appropriate to the new bedrock, they adjust by cutting relatively narrow valleys or gorges through ridges of resistant rock, and elsewhere meandering through broad valleys in softer rock.

Drainage networks can be used for geological detective work. Near the mouth of one major river, pollution by a toxic chemical was discovered. The problem: to find the polluting source in a large drainage basin. By a statistical analysis of the branching of the different order streams, geochemists were able with a minimum of time-consuming sampling and analysis to eliminate quickly large sections of the basin and concentrate on the stream segment from which the pollution came.

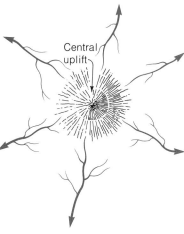

**Figure 9-42**
Typical radial drainage pattern developed on a large single peak, such as a large dormant volcano or a domal uplift.

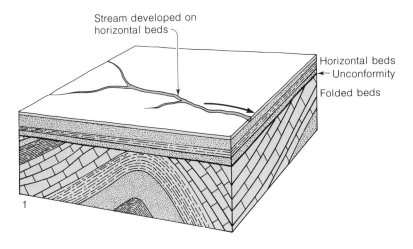

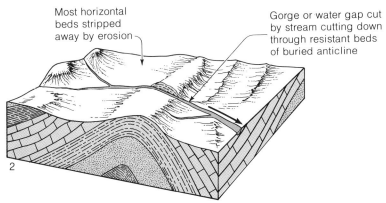

**Figure 9-43**
The development of a superposed stream by erosion of horizontal beds unconformably overlying folded beds of varying resistance to erosion. As the downcutting stream encounters the buried anticline, it erodes a narrow gorge or water gap in the resistant beds of the anticline.

## THE END OF THE LINE: DELTAS

Sooner or later, the end comes to all rivers. But why? If a stream runs downhill on land, why doesn't it continue downslope under the waters of a lake or ocean until it hits the lowest spot? That rivers do stop when they empty into large standing bodies of water is so well known as to appear intuitively obvious. The answer comes from watching the behavior of a river as it enters the sea and mixes turbulently with the surrounding water. The mixing gradually transfers the powerful forward momentum of the flowing river to the surrounding water and dissipates the current. It does so in much the same way that a billiard ball shot into a large cluster of balls on a small pool table will lose its forward motion and transfer it to more or less random movements of all of the balls as they break apart and rebound from the cushions and from each other, gradually coming to a stop as friction overcomes their motion.

The current decays more or less gradually, depending on its volume and velocity. The discharge of the Amazon River, the world's largest, was recently measured at 4 billion gallons per minute (15 million cubic meters/second) by a team of U.S. and Brazilian hydrologists. The flow from the river maintains its integrity many kilometers out to sea. In contrast, small streams entering a turbulent, wave-swept coast may mix so rapidly that the current disappears almost immediately.

Had the Ancient Mariner—crying out, "water, water, everywhere, nor any drop to drink,"—been becalmed off Brazil, he could have drunk fresh water from the sea.

**Delta Sedimentation** The dissipation of the river currents sets in motion a train of events that accounts for some major sedimentary deposits: deltas. The dying current loses its competence for transporting particles and so deposits the series of sedimentary layers that make up the delta. The simplest deltas are formed by modest streams entering fresh-water lakes. Because the density of the entering river water is the same as that of the surrounding lake water, the current mixes in all directions in a cone-like pattern and rapidly slows to a halt. The coarsest material is dropped first, followed by medium and fine materials farther out. In the usual situation, where the lake bed slopes away from the floor of the river channel, the coarse material builds up a depositional platform like that shown in Figure 9-45. **Foreset** beds, inclined downcurrent from the delta front in every way analogous to cross-bedding, are covered by thin, horizontal **topset** beds and preceded on the lake bottom by thin, horizontal **bottomset** beds. This sequence has been reproduced, on a small scale, in laboratory flumes.

Deltas formed in the sea are generally the same in form but tend to be stretched out more in the horizontal dimension. The reason for this is that the fresh water of the river, (density about 1 gram/cubic centimeter) enters sea water (density 1.02 grams/cubic

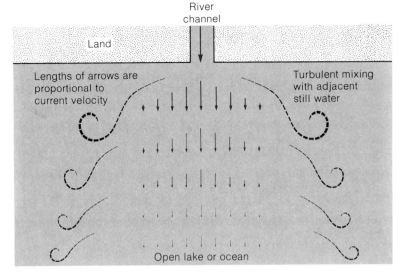

**Figure 9-44**
Decrease in the velocity of flow as a river enters an open body of water, such as a lake or an ocean. The current persists for the longest distance along the main line of flow and drastically decreases to either side. The current mixes with still water around the edges of the flow, losing momentum as it does so. [Data in part from C. C. Bates, "Rational Theory of Delta Formation," *Bull. A.A.P.G.*, v. 37, pp. 2119–2162, 1953.]

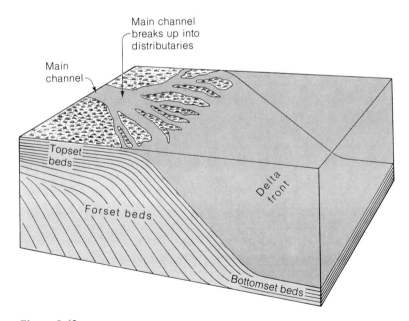

**Figure 9-45**
A typical fresh-water delta, characterized by well-defined topset, foreset, and bottomset beds. The slope of the delta front is fairly steep, up to 25°. The main river channel may or may not break up into distributaries, depending on the stream discharge and the sediment load: the greater the load in relation to discharge, the greater the tendency to form distributaries.

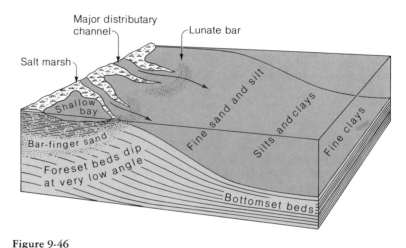

**Figure 9-46**
A typical marine delta, in which the foreset beds are fine-grained and deposited at a very low angle, normally only 4 to 5° or less. Lunate bars—crescent-shaped sand accumulations—form at the distributary mouths, where the current velocity suddenly decreases. As the delta builds forward, the bar-finger sand is built up by the advance of the lunate bar and the distributary channel. Between channels, shallow bays fill with fine-grained sediment and become salt marshes. This general structure is found on the Mississippi River delta.

centimeter). The lighter fresh water tends to "float" on the sea water and consequently mixes with it only in two dimensions— horizontally but not vertically as in a fresh-water lake. A lower mixing rate means that the current dissipates more slowly, so that coarse, medium, and fine material spreads out along a much longer path. The slope of foreset beds on large marine deltas may be only a few degrees.

Major rivers, such as the Mississippi and the Nile, form large deltas thousands of square kilometers in area with complex structures. The Mississippi has been studied intensively for generations by hydraulic engineers responsible for the operations of the waterway and more recently by geologists trying to work out the relations between river, sediment, and sea that make it like it is. The Mississippi delta, like most others, reverses the normal dendritic drainage pattern in which smaller tributaries join to form bigger ones. As it approaches the sea and lowers its slope, the river branches to form **distributaries** that funnel off the water in various paths to the sea, many small ones and three large ones near the mouth. At the mouth of main distributaries the coarsest sediment, sand, is dropped at crescentic bars, and the silt and mud are carried farther out. The sediments at these distributaries grow seaward as long sand fingers, **bar finger sands**, whereas the finer muds and silts get washed by shore currents and waves into the bays between, gradually filling them up with mud and swamp deposits. In this way the whole delta is built out into the sea.

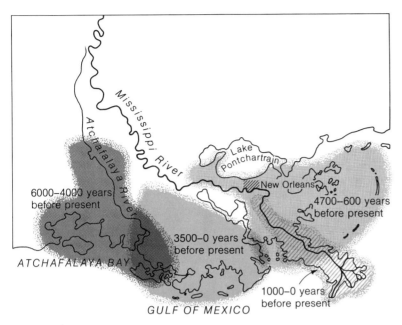

**Figure 9-47**
The Mississippi delta complexes of the past 6000 years are a series of major
distributary centers that shifted with time as the river built its delta first in one
direction and then in another. The oldest complex that has been mapped underlies
the present area of Atchafalaya Bay and the mouth of the Atchafalaya River. It
was succeeded by a series of deltas deposited to the east and west of the modern
birdfoot delta. For more than a decade before 1955, the Mississippi River was
diverting an increasing amount of its discharge to the Atchafalaya, an indication
that the river was again about to shift its major site of delta building. At that
time, the Army Corps of Engineers began construction to prevent major diversion
into the Atchafalaya. [Map compiled from various sources, including C. R. Kolb,
and J. R. Van Lopik, 1966, *in* M. L. Shirley (ed.) *Deltas in their Geological Framework,* Houston Geol. Soc.; D. E. Frazier, 1967, *in Recent Deltaic Deposits of the
Mississippi River: Their Development and Chronology,* Gulf Coast Assn. Geol. Soc.
Trans., v. 17, pp. 287–315.]

The delta of the Mississippi has been growing for many millions
of years. It started out around Cairo, Illinois, in Cretaceous times
and has advanced about 1600 kilometers since then. The pattern
of its growth for the past few thousand years has been mapped
by $C^{14}$ dating of wood found by drilling into the different kinds
of deposits at various depths (Fig. 9-47). Deltas grow in one direction for a while and then shift to another, seeking a shorter path
to the sea. The individual subdeltas interfinger and partially pile
up on top of each other to form a rather complex mass of sediment.
Because of this complexity, recognition in the rock record of
major ancient deltas formed hundreds of millions of years ago is
no mean task, but the geologist has one good clue: the close
juxtaposition of coarse sandy beds containing land plant or animal
fossils with finer beds of the same age containing marine fossils.
Parts of the Catskill Mountains of New York State have been

deduced to be just such an ancient delta, formed in Devonian times. Coarse, land-deposited sandstone and conglomerate in the east, near the Hudson River, interfinger with finer marine sandstones and shales farther west. The river forming that delta came from a belt of high mountains roughly paralleling the Taconic and Green Mountains and ancestral to them.

**Rivers Without Deltas** Not all rivers, even those that carry large loads of sediment, deposit deltas. Some enter the sea at places where a combination of waves, shore currents, and tides work together to move the sediment along the shore as rapidly as it is dropped by the river. In this way, the river's detritus is spread out along the coast, sometimes for hundreds of kilometers. A delta is formed only when the sediment load is great enough and the waves and ocean currents weak enough to permit it to form. Some rivers form small deltas of coarse material, the finer material — the bulk of the load — being swept away by waves and currents.

However rivers end their journey, they drop two kinds of load, the solid particles of rock detritus and the dissolved salts coming from weathering. The rivers of the Earth constitute the major transport system for the products of weathering, most of which end up in the sea. Water, however, is not the only agent that erodes, transports, and deposits. In the next chapter, we discuss another, the wind.

## SUMMARY

1. Water flows in laminar and turbulent currents. Most natural streams flow turbulently and erode, transport, and deposit solid materials. Depending on turbulence, velocity, and slope, streams have competence to carry particles of a certain size and a certain capacity of total sediment load. Particles move by suspension, saltation, and rolling.

2. The sandy sediment deposited by stream currents typically shows crossbedding and ripples. Ripples, dunes, and flat surfaces form in an orderly sequence on sand beds as the velocity is increased.

3. Currents effectively erode unconsolidated material and hard rock by undercutting, scouring, sandblasting, and plucking.

4. The river is a dynamic system characterized by discharge, velocity, and slope. Changes in one are accompanied by mutual changes in the others as the river adjusts its longitudinal profile.

5. Longitudinal profiles tend to be graded in adjustment to base level.

6. River deposits are alluvial fans formed in response to a sudden change in slope. Channels form meanders and pools and riffles with regular patterns of point bars and other channel deposits.

7. Most of the buildup on floodplains happens during the moderately frequent, moderately high floods that come every few years.

8. Drainage networks show hierarchical branching patterns that follow rules of distribution. The pattern of drainage may be dendritic, rectangular, trellis, or radial, depending on topography, rock type, and structure.

9. Deltas are the major sites of deposition of sediment by rivers. They show a structure of topset, foreset, and bottomset beds that differs from lake delta to ocean delta. Not all rivers have deltas; the sediment of these rivers is dispersed by strong shoreline currents or tides.

## EXERCISES

1. Predict the probable sequence of events that would follow the building of a dam on a river that has an ample supply of sediment and that is graded in equilibrium.

2. What could you infer about the changing current conditions that produced a series of sand beds in which the lowest beds are cross-bedded in a series of large interfering troughs, the middle beds finely rippled, and the highest beds horizontal with no ripples or any other structures?

3. Would you think that the ratio of first- to second-order streams in a trellis drainage system would be much greater than, about the same, or much less than the ratio of first- to second-order streams in a radial drainage system? Why?

4. In the middle part of its length, a major stream runs through a belt of cavernous limestone with Karst topography. How would its discharge-velocity relations be affected, and how might the drainage network pattern change?

5. You are the city engineer of a city of 100,000 population, built partly on the floodplain of a major river. Two years ago the city suffered the most disastrous flood in its 200-year history, and the city council now says that you need not worry about contingency planning for another major flood for the next few years. What would be your response to the council?

6. One popular engineering project used to control rivers is channelization—the artificial straightening and widening of the river channel. What natural changes might you expect to take place over a number of years after the river is left free to adjust its course naturally after channelization?

7. Both alluvial fans and deltas have drainage patterns in which a channel breaks up into a number of distributaries. Is there a common cause for this behavior? If so, what is it? If not, what are the causes in each case?

326    BIBLIOGRAPHY

Leopold, L. B., and W. B. Langbein, "River Meanders." *Scientific American,* June, 1966. (Offprint No. 869.)

Leopold, L. B., M. G. Wolman, and J. P. Miller, *Fluvial Processes in Geomorphology.* San Francisco: W. H. Freeman and Company, 1964.

Matthes, G. H., "Paradoxes of the Mississippi." *Scientific American,* April, 1951. (Offprint No. 836.)

Morisawa, M., *Streams — Their Dyamics and Morphology.* New York: McGraw-Hill Book Company, 1968.

Morgan, J. P. (editor), *Deltaic Sedimentation, Modern and Ancient.* Tulsa, Oklahoma: Society of Economic Paleontologists and Mineralogists Special Paper 15, 1970.

Sundborg, Ake, "The River Klarälven. A study of Fluvial Processes." *Geografiska Annaler,* Uppsala University, vol. 38, pp. 125–316. 1956.

# 10

## Wind, Dust, and Deserts

*Wind, though less powerful than water currents, can erode sand and silt effectively, particularly in arid regions. The deserts of the world, sites of intensive wind action, display special varieties of topography and erosional and depositional processes that are intenser than those of humid regions. Sand dunes, depositional landforms created by wind, are heaps of sand that accumulate in different shapes and move in response to the abundance of the sand supply, the strength and direction of the wind, and the nature of the bedrock surface.*

All of us have been caught, at one time or another, in a high wind so strong that it could have blown us over if we hadn't leaned into it or held on to something solid. A wind strong enough to move a body weighing more than 50 kilograms (110 pounds) is easily capable of blowing sand grains into the air, as anyone who has ever been in a sandstorm can attest. Wind is a turbulent stream of air, and its ability to erode, transport, and deposit sediment is much like that of water, for the same general laws of fluid motion that govern liquids govern gases as well. There are differences, of course, and they are traceable to two properties of wind: its low density and the fact that its flow is unconfined, not restricted to channels. The low density of air limits its competence to move larger particles, and the fact that its flow need not be confined enables it to spread over wide areas and high into the atmosphere. In contrast to rivers, whose discharge is dependent upon rainfall, it is the lack of rain that allows wind to work most effectively. In this chapter, we will discuss arid environments, the deserts of the Earth, because so many of the geologic processes of the desert are related to the work of the wind. Winds are also important because they are coupled to the waters of the ocean as they blow waves into being and influence the general circulation of the oceans. A common term used to describe the activity and deposits of the wind is **eolian**, from Aeolus, Greek god of the winds.

## HOW MUCH CAN THE WIND CARRY?

Winds are highly variable in direction and power (see Table 10-1). Though the average wind on a breezy day might be about 15 kilometers (9 miles) per hour, gusts up to twice that speed occur intermittently and may spawn momentary blowing up of dust or sand clouds. Most people in temperate climates are used to winds that come mainly from one direction, the prevailing westerlies. Those in the tropics are familiar with the equatorial easterlies. Yet within these belts the winds will be variable in direction and power, depending on the movement of air masses and storms. Many of us live with wind fluctuations during the day, such as the sea breeze that blows during warm summer days and dies down in the evening, or the daytime valley breeze and nighttime mountain breeze in high-relief terrain. More constant are the steady, strong winds that blow for days without letup, like the dry Chinook winds of the eastern slopes of the Rocky Mountains and the Mistral that howls down the Rhone Valley of France in the winter.

Table 10-1
Terminology of wind speeds, according to 1939 international agreement and U.S. Weather Bureau, with the effects of winds on the sea.

| Beaufort number* | Wind speed (km/hr)† | International description | U.S. Weather Bureau description | Effect of wind on the sea |
|---|---|---|---|---|
| 0 | <1 | Calm | Light wind | Mirror surface to small wavelets |
| 1 | 1–5 | Light air | | |
| 2 | 6–11 | Light breeze | | |
| 3 | 12–19 | Gentle breeze | Gentle–moderate | Large wavelets to small waves |
| 4 | 20–28 | Moderate breeze | | |
| 5 | 29–38 | Fresh breeze | Fresh wind | Moderate waves, many whitecaps |
| 6 | 39–49 | Strong breeze | Strong wind | Large waves, many whitecaps, foamy sea |
| 7 | 50–61 | Moderate gale | | |
| 8 | 62–74 | Fresh gale | Gale | High waves, foam streaks, and spray |
| 9 | 75–88 | Strong gale | | |
| 10 | 89–102 | Whole gale | Whole gale | Very high waves, rolling sea, reduced visibility |
| 11 | 103–117 | Storm | | |
| 12–17 | >117 | Hurricane | Hurricane | Sea white with spray and foam, low visibility |

*Source:* Based in part on N. Bowditch, 1958, "American Practical Navigator," U.S. Navy Hydrographic Office Publ. 9.
  *The Beaufort scale was devised near the beginning of the nineteenth century as a way of measuring winds at sea by noting types and heights of waves, spray, and visibility.
  †Today, wind speed is usually measured in knots (nautical miles per hour; 1 nautical mile = 1.15 statute miles or 1.85 kilometers).

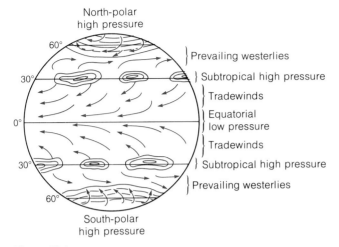

North-polar
high pressure

Prevailing westerlies

Subtropical high pressure

Tradewinds

Equatorial
low pressure

Tradewinds

Subtropical high pressure

Prevailing westerlies

South-polar
high pressure

**Figure 10-1**
The global pattern of belts of surface winds and barometric-
pressure cells. The subtropical high-pressure cells move somewhat
and change size from summer to winter. Arrows show generalized
wind directions.

The distribution and intensity of winds in combination with
climate has much to do with the location of wind erosion and
windblown deposits on Earth. The prevalence of sea breezes,
for example, is responsible for the sand dunes blown back from
sandy beaches along coasts. Most of the geologic work of the wind
is done by the moderately infrequent strong winds of long duration, just as the major part of a river's geological work is done by
floods. Everyday mild winds and the rare tornado, whose winds may
exceed 160 kilometers (100 miles) per hour but which covers only
a narrow strip on the Earth's surface for a short time, are responsible for relatively little geologic change. Hurricanes and typhoons
are important agents because of their frequency in certain maritime
regions, but, because of the rain they bring, they do their work
by causing floods and stirring up waves rather than by blowing
sand or dust: their rain washes dust particles out of the air and
wets down the ground surface, which prevents pickup of further
particles. This points to dryness of the wind as a crucial pre-
requisite for erosion and transport.

**Sand Transport** The wind exerts the same kind of force on
particles on the land surface as a river current exerts on its bed,
the turbulence and forward motion combining to lift particles
up into the windstream at least temporarily. **Saltation**, the bounding and jumping movement of grains that we explored in the
last chapter, operates in the air in the same way that it does under
water, but in the air it is much more effective: partly as a result
of the lower frictional and retarding force of the air, saltating
grains will frequently rise to heights of 50 centimeters over a sand

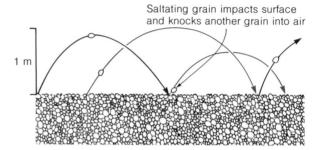

**Figure 10-2**
Saltation transport of sand grains in wind. As a saltating
grain falls to the ground, it may strike another grain lying
on the surface with sufficient force to throw the struck
grain into a saltation trajectory.

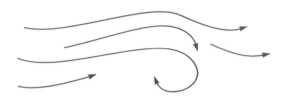

Air clear above saltation layer

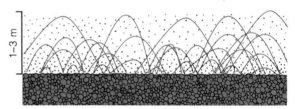

**Figure 10-3**
Saltation layers formed during many windstorms will be
so dense that the ground cannot be seen, yet the top of
the layer may be sharp and the air above it clear.

bed and up to 2 meters over a pebbly surface. In a strong wind,
there can be so many saltating grains that there will be a cloudy
layer near the ground dense with blowing particles and capable of
sandblasting any object in its path. The saltating grains falling
back to the ground hit the surface with all the force of the wind
combined with gravity, hardly cushioned at all by the air. This
strong impact induces saltation of some of the grains on the surface as they are struck. Its much wider effect is the pushing
forward of struck grains, not enough to throw them up into the
air but enough to cause a general forward creep of sand particles
along the surface as the rain of saltating grains falls on it. A sand
grain striking the surface can move another grain up to six times
its own diameter. Because saltation blows smaller grains more

quickly and surface creep moves larger particles more slowly, the two will sometimes separate: the fine sand blows away, leaving behind a pavement of coarser sand and gravel. The fine sand, up to 0.1 millimeter in diameter, accumulates in dunes and sheets downwind.

The amount of sand that can be moved by winds of various strengths is shown in Figure 10-4, taken from a famous book on sand and sand dunes by *R. A. Bagnold,* a British officer in the desert forces in Egypt and Libya. Before and during World War II, Bagnold pioneered the study of the forms and movements of desert sand. As can be seen from the graph, one half ton of sand per day can be moved over a meter-wide strip of ground by a strong wind of 48 kilometers (30 miles) per hour. As wind increases to gale force, about 80 kilometers (50 miles) per hour, the rate of sand movement increases more rapidly. No wonder that a whole house can be buried in a long sandstorm driven by strong winds.

**Dust Transport** Turbulent winds can easily sweep small dust particles high into the atmosphere, once they are lifted off the ground surface. But it is not so easy for the wind to pick up small particles of clay or dust from the ground, mainly because, in a thin layer less than one millimeter thick right next to the ground surface, the air is practically still. Sand grains protrude through this layer into the windstream above, and thus are caused to roll or saltate; smaller particles, however, are directly affected only at high wind speeds. But once a saltating sand grain or other particle hits the ground and kicks up a bit of dust, the wind catches it. Smaller dust particles, many of which are only a few thousandths of a millimeter in diameter, settle so slowly that the slightest air currents keep them suspended. Violent volcanic explosions can inject pyroclastic dust high into the stratosphere, and such dust may take years to settle.

The capacity of air to hold dust is enormous. In large dust storms, one cubic kilometer of air may carry up to 1000 tons; if such a storm covers hundreds of square kilometers, it may carry more than 100 million tons of dust. In the worst dust storms of the 1930s in the "dust bowl," the western plains of the United States, whole farmhouses were buried in the dust. The entombment of the Roman city of Pompeii was largely caused by a torrential fall of volcanic dust from Mt. Vesuvius.

**Detritus Carried by the Wind** Sand grains carried by the wind may be of almost any mineral, but most are quartz, which reflects the dominance of that mineral in most sands and sandstones. The grains are typically frosted in appearance, like the ground-glass focusing screens of many cameras. It has been shown by experiment that much of this frosting results from the long continued action of dew. This moisture, even the tiny amount that is found in

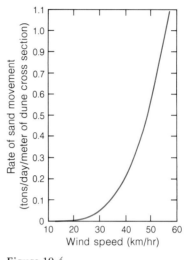

**Figure 10-4**
The amount of sand moved across each meter of width of a dune cross section in relation to wind speed. High-speed winds blowing for several days can move enormous quantities of sand and change dune positions markedly. [After R. A. Bagnold, *The Physics of Blown Sand and Desert Dunes,* London, Methuen, 1941.]

332

**Figure 10-5**
Dust storm in Owens Lake, Inyo County, California. This intermontane valley
floor becomes very dry, and dried-up lake sediments get caught up by storm
winds. [Photo by T. P. Thayer, U. S. Geological Survey.]

**Figure 10-6**
Aeolianite formations on Bermuda, formed by the erosion of lithified Pleistocene sand dunes of
calcium carbonate. The dipping beds are large cross-beds. [Photo by R. Siever.]

arid climates, is enough to dissolve away little pits and holes on the grain, creating a matte surface. This frosting, which is limited to the grains lying on the surface of the ground, is quickly smoothed and polished when the grains are blown into a river and transported by water. This experimental and observational evidence, which includes studies showing that wind alone could not produce the frosted surface, contradicts a long held inference that the matte appearance was the result of sandblasting, which is known to produce frosting on larger glass objects. "Common sense," here based only on one kind of observation, doesn't always give the right answer in science.

Give the wind enough sand-sized rock or mineral fragments of any variety and it will blow them to form dunes. In Bermuda, for example, winds from the ocean have produced dunes made up of the calcium carbonate fragments of shells and coral. The White Sands National Monument area in New Mexico is largely covered by dunes of gypsum grains eroded from evaporite bedrock in a dry climate. In some places, clay and silt particles will clump together to form sand-size grains that are blown into dunes.

Dust, the fine-grained particles carried by the wind, is of every conceivable origin. Microscopic rock and mineral fragments of all kinds are found, but most common are the silicate minerals, as might be expected from their abundance as rock-forming minerals. Among these, two types are most important: clays from soils and volcanic dust. Pollen, hair, bacteria, and a great variety of fragments of plants and animals are also components of dust. Fragments of charcoal can be seen in dust blown from areas with forest fires. Since the industrial revolution, civilization has been producing a rapidly multiplying dust component consisting of particles of ash from burning coal and many solid chemical compounds from manufacturing processes. The burning in automobile motors of the tetraethyl lead that is added to gasoline causes the accumulation of large quantities of tiny lead particles in the atmosphere. Lead from that source may be a hazard to health in areas densely congested with automobiles.

## WIND EROSION

Winds need chemical and mechanical weathering coupled to dryness to assist them in eroding and transporting materials. Wet materials are cohesive, the water binding the particles together enough to resist the wind's tendency to pull them apart. By themselves, winds can do little to erode most solid rock exposed at the surface; but once given some fragmentation of mineral particles, the wind can act.

**Deflation** As dry, loose particles of dust and silt are lifted and blown away, the surface of the ground is gradually eroded and

**Figure 10-7**
Desert pavement in Death Valley, California. Weathering of boulders and
cobbles at the surface has produced a new mantle of blocks, slabs, and flakes,
forming a smooth pavement in which the stones are closely spaced but barely
or not at all shingled. [Photo by J. R. Stacy, U. S. Geological Survey.]

lowered, or **deflated.** This occurs on dry plains and deserts, on
temporarily dried-up flood plains of rivers, on tidal flats, or on
lake beds. Deflation is characteristic, too, of cold dry periods on
the plains of mud, silt, and sand formed by melting glaciers.
Where vegetation is firmly established, the effect of wind is
small, both because the roots bind soil together and because stems
and leaves break up the wind and shelter the ground. But where
the vegetation cover is breached naturally by drought, or artificially by construction, deflation may attack and scoop out shallow

depressions or hollows. Once started, the hollows grow as wind stress on the surface becomes more efficient, usually helped by rapid water erosion during storms.

Deflation can attack rock surfaces, detaching small fragments, especially from sandstones that weather to form loose grains, but formation of depressions in rock surfaces is slow from this cause alone, except in some desert areas underlain by soft, easily disaggregated sandstones. Deflation, by removing the fines from a heterogeneous mixture of gravel, sand, and silt, can produce a remnant surface of gravel too large for transport. Over thousands of years, the gravel accumulates as a layer of **desert pavement**, which protects the soil or beds below from further erosion. Deflation may be only part of the process that forms such pavements, for there is some evidence that pebbles just below the surface work their way up to the top in some manner, leaving behind a thin substrate of sand and silt.

**Sandblasting**  Armed with blown sand, a wind can effectively wear away and shape solid rock surfaces by the constant impact of grains. Such sandblasting accounts for some erosion and rounding of the parts of rock outcrops close to the ground, the frosting of smooth rock surfaces (and glass bottles), and the shaping of **ventifacts**, or wind-facetted pebbles. Such pebbles, which show several curved or almost planar faces that meet at sharp ridges, are almost invariably composed of a fine-grained

Figure 10-8
Ventifacts—pebbles and cobbles pitted and faceted by wind-blown sand—from Sweetwater County, Wyoming. [Photo by M. R. Campbell, U. S. Geological Survey.]

rock, such as chert or quartzite. The pebbles form by the erosion of the windward side, followed by their being moved during a storm to lie on that sandblasted face, allowing another windward face to be sandblasted. As with frosted surfaces, ventifacts buried in deposits give us clues to strong wind action at the time of their formation. The main finds of ventifacts in most parts of the world have been in relatively recent glacial deposits; they are also common in deserts. Ventifacts are rare in older deposits, possibly because the combination of strong winds, sand for sandblasting, and pebbles to be shaped is a rare association that has been limited to regions affected by the few glaciations of the earth's history and to rarely preserved pebbly desert deposits.

## DESERTS: MUCH WIND AND RARE WATER

The hot, dry deserts of the world are among the most hostile environments to humans, yet, with their strange forms of animal and plant life, and their barren forms of bare rock and sand dune, they exert a strong fascination for many people. If you get out into the desert, away from automobiles and trailbikes, you may find one of the stillest, quietest places on earth. You may also find yourself in a sudden windstorm, with sand stinging and penetrating everywhere. The desert is where the wind is best able to do its work of eroding and depositing. It does so in partnership with river action that, however infrequent, still does the major part of the work.

**Figure 10-9**
A Libyan desert landscape. In the foreground, a rocky pavement; in the background, sand dunes are gradually covering a rocky ridge. [Photo by G. H. Goudarzi, U. S. Geological Survey.]

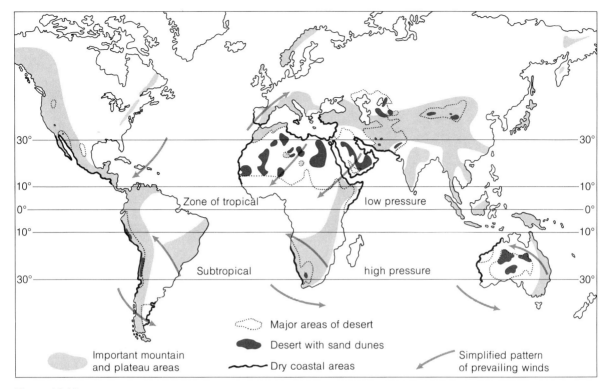

**Figure 10-10**
Major desert areas of the world in relation to prevailing wind directions and major topographic elevations. Sand dune areas are only a small proportion of the total desert areas. [After K. W. Glennie, *Desert Sedimentary Environments,* Elsevier, 1970.]

**Where Deserts Are** Maps that show average rainfall throughout the world show the great tropical deserts, such as the Sahara and Kalahari of Africa and the Great Australian desert, all lying at about the Tropics of Cancer and Capricorn (latitudes 23°30′N and 23°30′S, respectively). These deserts are under virtually stationary areas of high atmospheric pressure. The stable high-pressure areas are characterized by continuously subsiding air that is warmed up as it sinks, producing low humidity. Tropical deserts normally receive less than 25 millimeters (1 inch) of rainfall each year, in some places less than 5 millimeters (0.2 inch). They are all on the westward sides of continents.

Another belt of deserts and semiarid lands are those of the middle latitudes — between 30° and 50° in each hemisphere — such as the deserts of central Asia and the Great Basin and Mohave deserts of the western United States. Here, dryness is a result either of distance from the great reservoir of moisture, the ocean, or of being in the rain shadow of a mountain range. The mountain ranges of the western coastal states of the United States and Canada cause moist air coming east from the Pacific to rise and cool, precipitating rain on their western slopes. When the air descends again on the eastern side, it warms and dries, with less total moisture than before. Thus, rainfall is limited in the large arid and semiarid regions of the western interior intermontane basins. So the deserts, in a sense, are a matter of tectonics, for the rain-shadow deserts of middle latitudes are there by virtue of mountains produced by forces within the Earth, and tropical deserts are there because continental drift has placed the continents on which they occur astride the Tropics of Capricorn and Cancer. We may now see another way in which the external and internal heat machines of the Earth are linked. Not only are the sands of the great Saharan–Saudi Arabian desert a function of global-wind circulation patterns, they are products of a geologic history of formation of ancient sandstones, tectonic stability, and drifting motions of the northern part of the African continental plate.

All told, arid regions amount to one-fifth of all the land area of the Earth. Semiarid steppe and plains regions account for an additional one-seventh. Because the dynamics of the Earth that produce deserts have been operating through known geologic time, we must infer that deserts have always been present, though we cannot be sure that they always covered areas as extensive as those covered today. At certain times in the past, perhaps the Earth was even more extensively covered by deserts.

The land area of the world is about 145,000,000 km². Desert and near desert areas are about 27,500,000 km² (about 19 percent) and semiarid regions about 21,200,000 km² (about 14.6 percent).

**Erosion and Deposition in the Desert**   As different as deserts are, there are no geologic processes that are unique to them. Weathering, downslope movements, and transportation operate in basically the same way in deserts as they do elsewhere on Earth, but with a different balance. Soils are thin or absent because of the lack of abundant vegetation and slowness of chemical decay of rocks, both of which are traceable to the lack of moisture. What little fine soil is formed is likely to be blown away by the strong winds. Sand, gravel, rock rubble of many different sizes, and bare rock are characteristic of much desert surface.

The coarser, more heterogeneously sized fragments produced by desert weathering form steeper slopes than do the products of weathering in more humid, soil- and vegetation-covered regions. Desert slopes look, in some ways, more like mountainous terrain above the timber line. Everywhere, the landscape is without the softening effects of vegetation and chemical weathering, which are characterized by rounded slopes and rounded rock edges. Instead, one sees steep cliffs and valley walls with masses of angular talus blocks at their bases. Yet chemical weathering is not completely absent: Many iron silicate minerals, such as pyroxene, will weather slowly in desert climates, producing the oxidized rusty-iron color of some desert detritus. Clays and soils do form, but slowly.

**Figure 10-11**
Desert mountain slopes, partly covered with alluvium, looking down into Death Valley, California. In the distance is the Panamint Range. [Photo by H. Drewes, U. S. Geological Survey.]

Even the driest desert gets occasional rain. Rain falling on most desert areas may infiltrate permeable rock and surface debris to become groundwater. Some eventually reaches the groundwater table far below—as much as hundreds of meters below, in extreme cases—but most stays in the unsaturated zone, some of it very slowly evaporating. An oasis in the desert generally owes its existence to the groundwater table coming close enough to the surface for roots of palms and other plants to reach it. Some oases are fed by intermittent streams; being dependent on rainfall and runoff, they are seasonal.

Infiltration can be relatively slow; it may be virtually absent in areas with much impermeable rock. The rare heavy rain storm produces so much water in so short a time that it cannot all soak in; thus, the desert intermittent stream is born. The runoff is rapid, being unhindered by vegetation, and may cause flash floods along valley floors that have been dry for years. Because the loose debris on the surface is not held by vegetation, the erosive power of such streams is great, and the streams become choked with sediment, sometimes so much so that they look more like mudflows than rivers. Because such a high proportion of stream flows in the desert are floods or nearly so, they are efficient eroders of valleys in bedrock. Valleys in the deserts have the same range of shapes, depending on the rock type, as those found elsewhere, but far more of them have steep valley walls caused by the rapid erosion. A normally dry valley called a **dry wash** (in the Middle East, a **wadi**), is floored with a flat alluvial fill of coarse debris left from the

**Figure 10-12**
A desert dry wash. Such valleys are river channels only briefly during infrequent desert rainfalls. [Photo by F. Press.]

last water flow; the ordinary features of levees and floodplains are absent. Alluvial fans (see Chapter 9) play a significant role in runoff in mountainous deserts because of the high sediment load of the streams coupled with the fast infiltration and loss of water in the normally permeable fan itself.

The drainage systems of deserts are widely spaced because of the relatively infrequent rainfall, but their patterns may be similar to those of other terrains—with one difference: Because of water losses by evaporation and infiltration, many of the streams die out, never reaching across the desert to join larger master streams or the ocean. Because many deserts are associated with mountains, this discontinuous or **interior drainage** is compounded by dead-ended intermontane basins called **bolsons**. Such basins — whose drainage exits are blocked by faults in combination with alluvial fans or landslides — may be the sites of temporary lakes, **playa lakes,** formed by storm runoff. Playa lakes may last days, weeks, or longer before complete evaporation leaves a **playa,** a flat lake bed of clay, sometimes encrusted with salts precipitated during evaporation. Playa lakes that are highly alkaline may precipitate sodium bicarbonate or sodium carbonate. Others contain a variety of unusual precipitated salt deposits, such as borax, a sodium borate, $(Na_2B_4O_7 \cdot 10H_2O)$, which is quarried in California from ancient playa-lake deposits. Desert streams tend to contain a higher concentration of dissolved salts than those in humid regions because they redissolve salts deposited by evaporation from earlier runoff. The playas may start out fairly fresh and then become saturated with salts, the particular variety depending on the kinds of rocks from which they were weathered. Because of the unusual nature and great variety of minerals they contain, playa lakes are natural hunting grounds for geochemists interested in the ways in which salts are formed.

Deserts do not necessarily respect shorelines. Along the coast of the Persian Gulf, for example, on the shores of Saudi Arabia, the desert marches right up to the water's edge. There, the interplay of the tides, winds from the water, and sediment produces large flat areas just above the high-water mark that are crusted with a layer of sandy detritus cemented with salt, gypsum, and calcium carbonate. This kind of flat, called by the Arabic word **sabkha,** is formed by the evaporation of salt water that has seeped upward by a wicklike action from a shallow sea-water table below, augmented by the drying of sea spray blown inland.

Erosion and sedimentation patterns are responsible for **pediments,** one of the characteristic landforms of the desert. These are broad, gently sloping areas of bedrock, spread as aprons around the bases of mountains. A cross section of a typical pediment and its mountain would show a fairly steep mountain slope abruptly changing to the gentle slope of the pediment. A pediment is an erosional surface covered with scattered detritus, which is normally more plentiful downslope, the lower surface of the pediment being covered with compound alluvial-fan deposits merging with the sedimentary fill of the lower valley or basin. The pediment is formed by running water that builds up an apron of alluvium below as it continues to cut back an erosional platform above. As such, it is an evolutionary product of long-continued headward erosion of rock types (such as sandstones, shales, and volcanic tuffs) that are easily eroded in the desert environment. The precise way in which pediments evolve has been a subject of argument for most of this century, and there is as yet no widely accepted explanation.

**Figure 10-13**
Panorama of a playa, Soda Lake, California. Surrounded by mountains and filled with debris from them, the playa is occupied by a periodically dessicated playa lake. Strong winds blow sand from the surface of the playa sediment to form sand dunes. [Photos by D. G. Thompson, U.S. Geological Survey.]

**Figure 10-14**
A variety of sabkha flat (salt flat) formed in depressions in the desert interior. The white is a crust of salt formed by evaporation. Similar sabkhas form at desert sea margins where sea water laps onto tidal flats. [From G. H. Goudarzi, U. S. Geological Survey; photo by Harry F. Thomas.]

For generations, Hollywood and TV westerns made in southern California and Nevada have shown us the kind of arid-land topography that is formed in regions of flat-lying sedimentary rocks: Tablelike uplands capped by flat erosion-resistant beds and bounded by steep-sided—sometimes vertical—erosional cliffs rise up prominently above pediments and desert flats. A **mesa** is a large upland of this kind, ranging in area from a few to many square kilometers; a **butte** is a small version of a mesa, being only a few acres in area. The capping bed of a mesa tends to maintain the upland level. If it is breached, however, erosion cuts down rapidly and creates the cliffs. As the cliffs retreat, the upland area shrinks, finally leaving only a few small, isolated mesas above the surrounding lowland.

Large, small, or in between, the valleys and uplands of the dry regions of the world, like those of humid regions, are forms produced by the eroding action of rivers. Even where water is scarce, it is running water—however infrequent—that does most of the basic work of erosion. The wind helps, but it rarely controls. What the wind does control is the transport and deposition of sand, shaping all the mounds, ridges, and swirling forms of dunes.

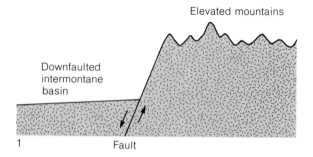

Elevated mountains

Downfaulted
intermontane
basin

1       Fault

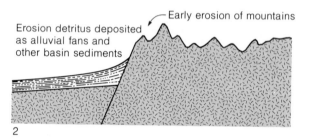

Early erosion of mountains

Erosion detritus deposited
as alluvial fans and
other basin sediments

2

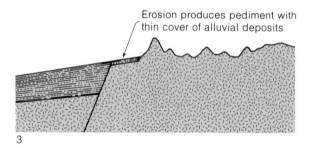

Erosion produces pediment with
thin cover of alluvial deposits

3

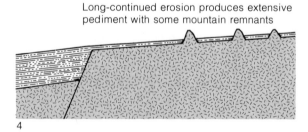

Long-continued erosion produces extensive
pediment with some mountain remnants

4

**Figure 10-15**
Stages in the evolution of a typical pediment, an erosional
form produced in arid climates. The surface of the pediment
is covered by thin deposits of alluvial sands and gravels,
which merge downslope with the depositional surface of the
intermontane basin sediments. The mountain slopes retreat
steadily but keep about the same slope angle, in contrast
to the rounded slopes in humid climates.

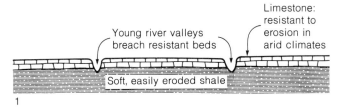

1

Continued erosion: broadened valleys isolate extensive mesa

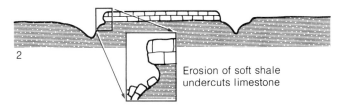

2

Erosion of soft shale undercuts limestone

Long continued erosion: small mesa remains as erosional remnant

3

**Figure 10-16**
The evolution of a mesa is controlled by an erosion-resistant bed overlying weak, easily erodable rocks. The border of the mesa retreats as the lower beds are eroded, undercutting the resistant beds, which then break off as large talus blocks.

## LANDFORMS OF THE WIND: DUNES

Erosion sculptures the rocks of the desert into crags, angles, cracks, and blocks, but sand dunes provide the soft forms of undulating slopes and rounded lows, sometimes framed by sinuous, sharp crestlines. The overwhelming impression one gets standing in the middle of a dune field is one of an incomprehensible mass of forms with no order, all merging together. Yet, as with everything else, there are patterns, but sometimes they can be seen clearly only from the air. The patterns are not accidental: they depend upon the kind and amount of sand available to the wind, upon the underlying rocks, and, most of all, upon the direction, duration, and strength of the wind. The same factors determine where dunes form.

**Where Do Dunes Form?** Though we may think of deserts as being unending expanses of sand dunes, only a small fraction of most desert land is covered by sand. Only a little more than one-tenth of the Sahara is sand-covered, and sand dunes are far

**Figure 10-17**
Desert sand "mountains" in the Saudi Arabian desert may reach heights of as much as 250 m.
[From Arabian American Oil Co.]

less common in many of the arid lands of the southwestern United
States. Most people are more familiar with the dunes that form
behind beaches, such as the dunes of Provincetown on Cape Cod,
the Indiana dunes on the south shore of Lake Michigan, and the
coastal dunes of Oregon. Dunes may also form along the banks
of some large rivers.

These places all have in common a ready supply of loose sand.
Beach sand deposited by tidal action is the source of coastal
dunes; in river valleys, it comes from the sandy channels and
floodplain deposits of rivers at low water. In the desert, sand is
produced by the weathering of sandy bedrock formations. Linked
with the supply of sand is wind power. Shorelines are likely places
for dunes to form because strong sea or lake winds tend to blow
sand inshore away from the beaches. As we have noted, strong
winds, sometimes of long duration, are common in deserts.

**Dynamics of Dunes**  Given adequate sand and wind, how does a
dune start? Any obstacle to the wind can be enough to start the formation of a dune, be it a large rock, a clump of vegetation, or even
a slightly humped-up pile of pebbles. Wind streams, just like water

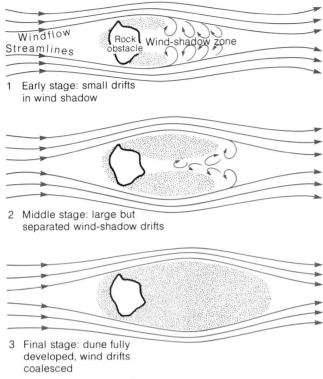

1 Early stage: small drifts
in wind shadow

2 Middle stage: large but
separated wind-shadow drifts

3 Final stage: dune fully
developed, wind drifts
coalesced

**Figure 10-18**
Formation of sand drift in the lee of an obstacle, such as a rock.
The obstacle, by separating the flow streamlines, creates a wind
shadow in which the eddies are weaker than the main windflow,
thus allowing saltating grains to settle and build up a drift. The
drift is a streamlined body built in response to the nonstreamlined
obstacle. [After R. A. Bagnold *The Physics of Blown Sand and
Desert Dunes*, Methuen, 1941.]

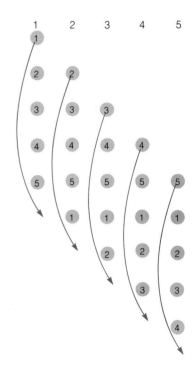

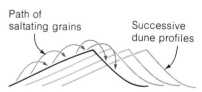

**Figure 10-19**
The movement of a column of people
by sequential movements of individuals
from rear to front (above) is roughly
analogous to the movement of a ripple
or dune by individual grain movements,
though all the sand grains do not move
sequentially in as simple a pattern.

streamlines, separate around obstacles and continue slightly deflected, creating a wind-shadow zone in which the wind velocity is
much less than in the deflected streams. Blowing sand grains are
deflected with the wind stream. Many will, as they saltate, come to
rest by chance in the wind shadow. There they stay, unable to
continue their travels, because of the lower wind velocity. Thus, the
pile-up of sand in the wind shadow, a sand **drift**, continues to the
point downstream at which the parted wind stream merges to
form a single flow again, as in Figure 10-18. In a strong wind, you
can perform this experiment yourself to see exactly how it happens.

If an obstacle is big enough to get a sanddrift started, the drift
soon becomes an obstacle itself. This is another example of positive
feedback, in which the result of an action continues to enhance
that action. If there is enough sand, and if the wind continues to
blow in the same direction for a sufficient time, the drift grows into
a larger mound of sand — a dune. Dunes may also form by growth

from sand ripples, much as underwater dunes form in flumes and streams. Desert and coastal dunes are normally covered by ripples moving over them, just like the ripples that form on the surface of large underwater dunes.

As a dune grows, the whole mound starts moving downwind: The sand grains constantly saltate up the low-angle windward slope to the top, then fall over into the wind shadow on the lee slope. As grains accumulate on the lee slope, they spontaneously cascade down in little groups to keep a more or less constant angle of repose, which is the hallmark of this **slip face** of the dune. Erosion on the windward slope and deposition on the slip face cause the forward movement of all dunes, whether under wind or water. The slip face, deposited at the angle of repose, creates the crossbedding of dunes. The crossbedding is normally irregular in windblown dunes, because even slight changes in wind direction will change the direction of dune movement, first making the slip face grow one way, then another.

As dunes accumulate, interfere with each other, and then become buried in a sedimentary sequence, the shapes of the dunes as they appeared under the wind are lost but the crossbedded structure remains. The irregularity of the crossbeds of ancient dunes may give us a clue to whether they were formed by wind or by water. Very thick crossbedded sandstone units were once very high dunes; if those units are many tens of meters thick, the dunes that formed them were most likely windblown, because underwater dunes are rarely that high. A statistical average of the directions of maximum inclination—the dips—of the crossbeds will give us the average direction of the ancient winds, or **paleowinds**. From this and other kinds of information, we can start to reconstruct ancient climates—**paleoclimates**—that can be harmonized with former positions of continental plates relative to the poles.

The height of dunes is determined by a switchover from positive to negative feedback. Positive feedback makes the little ones grow bigger by more rapid pileup of saltating grains on the windward slope than on the slip face, but only to a point. When the dune reaches a certain height, the windstream lines become "squeezed" upward and together by the mound of sand. As more air has to rush through a smaller space, the wind velocity increases, ultimately to the point at which sand grains blow off the top to the advancing slip face as fast as they are brought up, and a steady state is reached. Any tendency to further height would be countered by the negative feedback of increased windspeed. Exactly what that steady-state height is, of course, depends on the wind speed and the size of the sand grains, assuming that the wind blows long enough to build up to the steady-state height. Dune heights of 30 meters (100 feet) are not uncommon, and some huge dunes in the Sahara may reach 100 meters (330 feet).

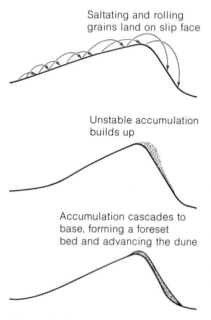

Saltating and rolling grains land on slip face

Unstable accumulation builds up

Accumulation cascades to base, forming a foreset bed and advancing the dune

**Figure 10-20**
The formation of the slip face of a dune. The slip face accumulates an unstable slope as a result of the deposition of saltating, rolling, and sliding grains. Intermittently, this accumulation becomes so unstable that it spontaneously slips—that is, cascades—to the base, and a new, lower-angle, stable slope is formed.

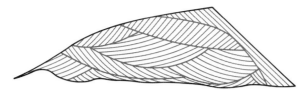

**Figure 10-21**
Cross section of a sand dune, showing interfering sets of crossbeds, a response to variable wind directions. The appearance can be similar to the interfering trough cross-bedding of Figure 9-16.

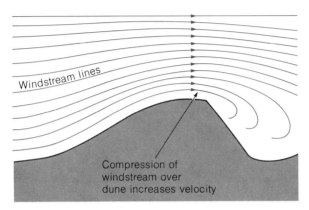

Windstream lines

Compression of windstream over dune increases velocity

**Figure 10-22**
Limitation of the height of a sand dune by a compressed windstream. As the dune grows higher, the windstream (shown by streamlines) becomes more compressed and so travels at a greater velocity, which makes it more competent to transport sand grains. Eventually, a height is reached at which the wind speed is so great that all of the sand is transported and the dune stops growing vertically.

**Dune Types** In an attempt to make some order out of immense diversity, dunes have been lumped into several major types, though there are gradations among them and many irregular shapes that are hard to fit into the scheme. The solitary crescent-shaped dune, the **barchan**, which moves over a flat surface of pebbles or bedrock, is a product of limited sand supply. Barchans cluster as groups, and sometimes two or three of them coalesce into a compound dune, or they may form long lines. Characteristic of deserts, they may form elsewhere, given a flat surface and a limited sand supply. The slip face and windward slope are easily seen in this simplest of all dune types.

Where sand is abundant, vegetation is absent, and sand dominates the landscape, dunes form in long, wavy ridges that lie transverse to the prevailing wind—hence, **transverse dunes**. These dunes have the same easily distinguishable slip faces as barchans,

**Figure 10-23**
Barchan sand dunes in Sherman County, Oregon. Winds blow predominantly
from left to right; the "horns" of the barchans point downwind. [Photo by
G. K. Gilbert, U. S. Geological Survey.]

**Figure 10-24**
Transverse dunes of the Saudi Arabian desert, aligned in a general direction at right angles to the
prevailing wind direction. [From Arabian American Oil Co.]

and they may grade into them at the edges of the dune field, where the sand supply is more limited. Typical sandy-beach dunes are transverse dunes; in some places, a series of them may form back from the shore. The sand is supplied by the beach. In more humid regions, there is a balance between the rate of sand movement and the growth of vegetation, which binds the sand and stabilizes it, preventing further movement. If the vegetation of such stabilized dunes is overwhelmed by sand at some point along the shore, the dune migrates inland, forming a **blowout**, a dune shaped like a parabola. Blowouts may be natural, but these days, when beach-front vegetation is sometimes bulldozed away by home or road builders, blowouts are often the result of a crude upsetting of the sometimes fragile balance between vegetation and sand. Those who have tried it can speak with feeling of the difficulty of planting vegetation and getting it to grow and stabilize the dune once a blowout is on the move. Over many years, vegetation gets a chance to stabilize the dunes as they move inland, farther from their source of sand on the beach. The hilly wooded or shrub-covered topography of shorelands in many places is one of overgrown, older, stabilized dunes.

**Figure 10-25**
Dunes in the Arabian desert, showing the slip face of a large dune advancing on a sand-rippled surface. The grooved appearance of the slip face is formed by cascades of sand grains assuming the stable angle of repose. [From Arabian American Oil Co.]

**Figure 10-26**
Longitudinal dunes parallel to the prevailing wind direction in the Saudi
Arabian desert. The upper photo is a view of the dunes as seen from the great
height of a satellite (Gemini IV); the lower is a low-altitude airplane photograph.
[From Arabian American Oil Co.]

Wind can blow sand into long, straight ridges more or less
parallel to the prevailing wind—**longitudinal dunes** or, as
Bagnold called them, **seif dunes**. These dunes may reach heights
of 100 meters and may extend many kilometers. Moderate sand
supply, a rough pavement, and winds of several directions (but

all of them in the same general quadrant of the compass) combine to form the long ridges. Desert geologists are not sure why seifs form rather than barchans or transverse dunes. Bagnold thinks they form by the gradual stretching out of one side of a barchan to link up with one downwind and slightly off to one side from it, en echelon, in a system that includes subsidiary winds at angles to the main wind-transport direction. Though we know much about the shape and orientation of the dunes, our quantitative records of wind direction and velocity in relation to sand movement are too sparse to allow us to pin down the explanation with any certainty.

## DUST FALLS

No one had to teach the geology of dust deposits to western dust-bowl farmers in the 1930s. Their houses and croplands were buried by dust. The extent of those dust falls was dramatic, but geologists had years earlier deduced such falls from thick deposits of unstratified fine-grained silt and clay called **loess**. Loess is composed of angular particles of quartz, feldspar, mica, and mafic minerals, mixed with abundant clay minerals. In North America, the windblown origin of Pleistocene loess was originally demonstrated by its pattern of distribution in the upper Mississippi Valley as a blanket of more or less uniform thickness on hills and valleys alike, all in or near formerly glaciated areas. Even more conclusive was the much greater thickness of loess on the eastern sides of major river floodplains than on the western sides and the rapid fall-off in thickness and grain size farther east of the valley. The facts were consistent with deposition from prevailing westerly winds blowing over stream valleys carrying abundant silt and clay from the outwash of melting glaciers.

Loess is a distinctive sediment. It tends to form vertical cracks and walls when pieces break off during erosion, and it is frequently filled with thin tubes and nodules of calcite, most of them oriented vertically. The vertical cracking may be caused by a combination of vertical root penetration and uniform downward percolation of groundwater, but the details still await a good explanation. Loess is sometimes picked up and redeposited in river deposits, which, if not too well stratified, may easily be confused with the original windblown material. The correlation of the distribution of thin loess over parts of Illinois, Iowa, and adjacent states and the crop-producing capacity of those areas is not accidental. Soils formed on loess are some of the most fertile. They are also easily eroded into gullies by small streams and deflated by the wind when poorly cultivated. Ultimately, after the tragedy of the dust bowl, the fertility of the land was increased—no consolation to the thousands who had been driven from their farms, like the Joads of John Steinbeck's *Grapes of Wrath.*

**Figure 10-27**
Pleistocene loess, showing prismatic jointing, in Colorado. [Photo by H. E. Malde, U. S. Geological Survey.]

Loess is found in deserts or downwind of deserts, as in China, where loess was deposited to thicknesses of 30–100 meters over wide areas by winds blowing over the arid lands of central Asia. Measurements of airborne dust far out to sea have been made by oceanographic research vessels and the mineral composition compared with the sediment at the sea bottom. Thus, dust has been discovered to be an important constituent of deep-sea sediments, perhaps accounting for as much as twenty percent. Much of the dust is traceable to volcanic activity and, as we noted before, volcanic dust falls are important eolian deposits. Volcanic

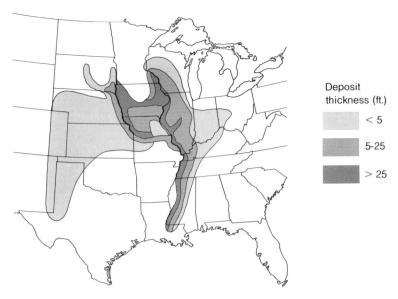

**Figure 10-28**
Loess and other eolian deposits in the central United States. The loess thins
leeward from its sources along the major river valleys that discharged outwash in
Pleistocene time. Westward, toward the Rocky Mountains, the eolian deposits on
the Great Plains are sandy. [From *Geology of Soils* by C. B. Hunt. W. H. Freeman
and Company. Copyright © 1972.]

Deposit
thickness (ft.)

< 5

5-25

> 25

dust may be the most important source of particulate matter in the
atmosphere. Since particulate matter increases the power of Earth's
atmosphere to reflect sunlight, it can affect our worldwide climate.
There has been some speculation that modern civilization's pollution of the atmosphere by tiny particles of lead, soot, and thousands
of other materials might raise the level of dust in the atmosphere
enough that the Earth might cool by a small amount. People then
envision the catastrophe of a new ice age. A sober look at the data,
however, has convinced most scientists who have looked into the
question that natural volcanic dust is far more important, and that
the amount of it in the atmosphere has probably been fluctuating a
good deal for much of geologic time. If doom befalls us in the form
of a new ice age, at least it won't be the fault of modern civilization.

Last but not least, dust in the atmosphere provides us with some
marvelous red sunsets.

# SUMMARY

1. Winds are important agents of erosion and transportation, particularly in deserts.

2. The distribution of the wind belts of the Earth and the relation of winds to mountain ranges and shorelines are closely related to the locations of such windblown deposits as sand dunes.

3. Wind transports dust by suspension and sand grains by saltation. Air can hold enormous quantities of dust extending high into the atmosphere, but sand is transported in smaller quantities near the ground. Wind-deposited sand grains are frosted by the dissolving action of dew.

4. Wind erosion deflates arid regions and may produce desert pavements. Sandblasting facets and polishes rocks into ventifacts.

5. Deserts are located in a few tropical regions of stable high pressure and in the rain shadows of mountain ranges.

6. Erosion and deposition operate in much the same way in deserts as they do elsewhere but with different intensity, as a result of dryness, lack of vegetation to bind soil, and rate of river flows. Much of desert erosion is caused by rivers.

7. Drainage of deserts is widely spaced and normally interior, and may include playas and bolsons. Shoreline deserts form sabkha flats, in which carbonate and evaporite deposits may form.

8. Pediments are characteristic erosional forms of arid lands.

9. Dunes, the constructional products of wind deposition, form by wind action in deserts, behind beaches, and along some sandy river valleys. All dunes are crossbedded, and they fall into a few general types: barchans, transverse dunes, and longitudinal or seif dunes.

10. Loess is a deposit of windblown dust in deserts and in regions that were glaciated during the Pleistocene.

# EXERCISES

1. Do you think a study of the shorelines of continents would show a tendency for coastal sand dune belts to be more abundant and wider on the eastern than on the western coasts or vice versa. Why?

2. In terms of visibility and sandblast, which vehicle would be more affected by a sandstorm—a low slung sports car or a high freight truck? Do you think the same would hold true in a dust storm?

3. You are asked to estimate requirements for hauling away windblown sand from a two-lane highway over which a dune is advancing in a region where the wind blows an average of 40 km/hr all day. What tonnage trucks would you provide, and how many trips would you make each week?

4. Describe a set of conditions that might again produce widespread dust storms similar to those of the 1930's in the western plains.

5. Describe the possible changes in the Sahara desert if the African continent were to move toward the north pole approximately 2000 km.

Bagnold, R. A., *The Physics of Blown Sand and Desert Dunes.* London: Methuen Publishing Company, 1941.

Bryan, K., *Erosion and Sedimentation in the Papago Country, Arizona.* Washington: U. S. Geological Survey Bulletin 730, 1932.

Glennie, K. W., *Desert Sedimentary Environments.* New York: Elsevier Publishing Company, 1970.

Hadley, R. F., "Pediments and Pediment-Forming Processes. *J. Geol. Education,* v. 15, pp. 83–89, 1967.

Starr, V. P., "The General Circulation of the Atmosphere." *Scientific American,* December, 1956. (Offprint No. 841.)

Folds in the Malaspina glacier, Alaska. Seward glacier appears in the right rear. The scale of folding is in miles; in places, as much as 10 miles. [Photo by Austin Post, U.S. Geological Survey.]

# 11

# The Flow of Ice—Glaciers

*Low temperatures and precipitation of snow contribute to the formation of glaciers and snow fields in cold polar regions and in high mountains. As snow accumulates, it becomes compacted, gradually changing from snowflakes to granules to solid, massive ice. As glacial ice moves down valleys, it sculptures the topography by eroding rock and transporting the debris to the terminus, where melting takes place. Glaciers of continental size, like those in Greenland and Antarctica, produce a variety of erosional and sedimentary landforms. The glacial landforms of the recent past are evidence of the Pleistocene glacial epoch, during which huge areas of North America and Eurasia were covered by ice. Advances and retreats of the ice fronts caused large fluctuations in sea level, alternately flooding and exposing shallow ocean margins of the continents. The causes of past ice ages and any predictions of future ones are equally uncertain.*

Ice is a rock, a mass of crystalline grains of the mineral ice. That idea should not seem surprising; after all, it is a solid substance that occurs naturally on the Earth. It is hard like most rocks, but its composition makes it much less dense. It shares with igneous rocks an origin as a frozen fluid. Like sediments, it is deposited in layers at the surface of the Earth, and can accumulate to great thicknesses. Like metamorphic rocks, it is transformed by recrystallization under pressure. Masses of ice may creep, flow, or slide downhill, and just like other masses, they may be folded and faulted. A large mass of ice that is on land and shows evidence of being in motion or of once having moved is called a glacier. The motion of glaciers is the clue to the effective work they do in eroding the surface of the Earth into distinctive sculptural forms and in transporting rock

**Figure 11-1**
Thrust faulting in glacial ice, showing its rock-like characteristics. The slanting lines are a series of thrust faults along which the ice on the right was pushed to the left and upward. Jointing and fracturing are also prominent in this photograph of Susitna glacier, Alaska. [Photo by B. Washburn.]

debris and depositing it in various ways and forms. The movement of glaciers is now invested with a new and practical interest for Man: early warning of global climatic changes may be indicated by advances or retreats of glaciers. In this chapter we explore how and where ice accumulates, the mechanics of its movement, and the ways in which it helps shape the surface of the Earth.

## THE FORMATION OF GLACIERS

High altitudes and high latitudes have something in common, for they are both cold. High altitudes are cold because the lowest 10 kilometers of the atmosphere steadily cools with distance from the Earth, and high latitudes are cold because the angle of inclination of the Sun's rays increases toward the poles. In parts of the Earth that are high enough, or north or south enough, not all of the snow that falls in a year melts. It accumulates in **snow fields** and becomes transformed into ice. In the warm regions of the Earth near the equator, glaciers form only above about 4500 meters (about 15,000 feet). The altitude necessary for glacier formation decreases toward the polar regions, where ice is stable the year round at sea level. Because moisture-containing winds tend to drop most of their snow on the windward side of a coastal mountain range, the lee side is likely to be cold but desert-like, and lacking in glaciers. For example, glaciers in the high Andes Mountains are limited to the western slopes; the dry eastern slopes are barren of ice. Inland mountains far from sources of moisture are not likely to have glaciers unless, as in Antarctica, it is so cold all year round that there is no melting and all snow is preserved. Thus, even arid climates, if cold enough, will promote the formation of glaciers. A few places along the coast of Antarctica, however, are so devoid of moisture that no snow falls, and the ground remains dry and free of ice. These dry valleys of Antarctica, in which no indigenous life can grow, have been investigated by scientists as a possible model of the inhospitable conditions on Mars.

Most of us are familiar with ice-making from the common experience of putting water into an ice-cube tray and then freezing it in a refrigerator. Those of us who live in northern regions know that rivers and lakes freeze over in winter. Glaciers, however, do not just freeze; they grow by a different process—the gradual transformation of snow into glacial ice. A fresh snowfall is a fluffy mass of loosely packed snowflakes, which are small, delicate ice crystals, grown in the atmosphere. As the snow ages on the ground for weeks or months, the crystals shrink and become more compact,

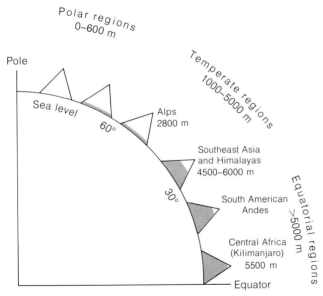

**Figure 11-2**
The height of the snow line, the altitude above which snow does not completely melt in summer, varies with latitude from at or near sea level in polar regions to heights of more than 6000 meters at the equator. The exact height of the snow line varies with local and regional climate.

Many of the terms used in glacial geology come from words used by the Swiss and French people of the Alps. The word "glacier" itself comes from the old French of the Alpine province of Savoy. *Firn* is a German-Swiss word used for "last year's snow." The word *névé* means the same as "firn," but comes from a French-Swiss dialect word derived from the Latin word for "snow."

**Figure 11-3**
Stages in the transformation of snow crystals to granular ice as snow changes to nevé and glacial ice. Accompanying the change of individual crystals is an increase in density by elimination of air. [After H. Bader and others, "Der Schnee und seine Metamorphose," *Beiträge zur Geologie der Schweiz,* 1939.]

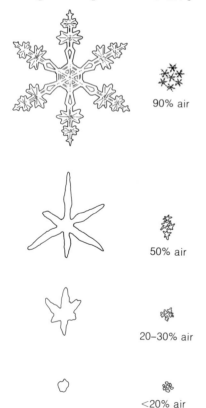

90% air

50% air

20–30% air

<20% air
as bubbles

and the whole mass becomes squeezed together into a more dense form, **granular** snow. As new snow falls and buries the older snow, the layers of granular snow further compact to **firn**, a much denser kind of snow, usually a year or more old, with little pore space. Further burial and slow cementation—a process by which crystals become bound together in a mosaic of intergrown ice crystals—finally produce solid glacial ice. One can think of snow as a sediment that becomes transformed by burial into the metamorphic rock ice by a process of **recrystallization**, the growth of new crystals at the expense of old ones. The whole process may take as little as a few years but more likely ten or twenty years or longer. The snow is usually many meters deep by the time the lower layers are converted to ice.

The formation of a glacier is complete when ice has accumulated to a thickness (and thus weight) sufficient to make the ice move slowly under pressure, in much the same way that rock deforms plastically deep in the Earth. Once that point is reached, the ice flows downhill, either as a tongue of ice filling a valley or a thick ice cap that flows out in all directions from the highest central area where most snow accumulates. The trip down leads to the eventual melting of ice. In medium and low latitudes, ice melts and evaporates as it flows to lower elevations. In polar climates the ice breaks off in huge chunks at the ocean's edge, and the chunks float out to sea as icebergs, which finally melt.

**Figure 11-4**
Icebergs calving from the Miles glacier, Alaska. As the ice meets the sea, it breaks into large irregular masses, many as big as a small island, that become icebergs. [Photo by B. Washburn.]

**Figure 11-5**
Melting of ice at the terminus of a glacier, showing the dark accumulation of debris at the wasting ice and the formation of meltwater streams (lower left). [Photo by B. Washburn.]

**The Budget of a Glacier** During winter, the typical glacier grows slightly as snow falls everywhere on the ice surface. In summer, the glacier shrinks, mainly as the snow on the surface of the lowermost reaches melts and evaporates to uncover solid ice, while the upper reaches stay snow-covered. The annual growth budget of a glacier is the amount of solid water added by snow, the **accumulation**, minus the amount lost, called **ablation**, which in temperate climates takes place primarily by melting and in polar climates mainly by evaporation and sublimation—the transformation from solid to vapor without melting—as well as by breaking off of icebergs. The difference between accumulation and ablation is a measure of either growth or shrinkage of the glacier. When accumulation minus ablation, the net budget, is zero over a long time period, the glacier is in a steady-state, accumulating new snow in its upper reaches and constantly moving down slope to the lower reaches, where ablation takes place. Glaciers' budgets fluctuate from year to year, and many show long-term trends of growth or shrinkage in response to climatic variation over periods of many decades. Geologic evidence suggests that many glaciers have tended to stay very roughly in a steady-state over the past several thousand years.

Because snow and ice vary so much in internal air space, accumulations are measured by the height of a column of liquid water equivalent to the total amount of precipitation falling over each square centimeter per year. In the cold, dry interior regions of Antarctica, the accumulation may be only 5 centimeters; in the wet maritime climate of Iceland, however, a glacier may accumulate more than 300 centimeters each year.

In most temperate climates, ablation occurs primarily by melting under the Sun's rays and less by evaporation. Warm air may blow over the lower reaches of the glacier and speed melting further. The air becomes chilled in the process. If the air is humid, it may precipitate rain over the lower glacial slopes, causing even more ablation. The meltwaters of glaciers in some areas have assumed increasing importance as a possible source of fresh water for irrigation and other purposes. The city of Tacoma, Washington, gets hydroelectric power from a meltwater-fed river that originates in the Nisqually Glacier. In the cold of the Antarctic, where melting is minor, ice ablates more by the breaking off of icebergs and evaporation.

## HOW ICE MOVES

Once the ice on a slope builds to a great enough thickness, it begins to flow downhill. The effect of thickness on flow can be demonstrated with a viscous fluid like honey. A thin layer of honey on a slightly tilted piece of bread will flow very slowly, but if more honey is poured on to increase the thickness, it flows much more rapidly. The speed of the flow can also be increased by tilting the bread more. The greater the thickness of glacial ice, and the greater its slope, the faster the movement of the glacier. The difference between the way ice flows and the way a brick slides down an inclined board is that the brick slides only along its base, whereas the ice moves throughout its bulk by internal sliding or flowing movements, as well as along its base.

This internal flow throughout the ice accounts for much of its motion. Under the stress of weight, individual crystals of ice slip tiny amounts over distances of about a ten-millionth $(10^{-7})$ of a millimeter in short time intervals, but the sum total of all those small movements in the enormous number of ice crystals over longer time periods amounts to much larger movements of the whole mass. It could be visualized as a large bulk made up of many randomly arranged packs of playing cards, each pack held together by rubber bands, the whole thing moving as the sum of many small slips between cards in the individual packs. This kind of movement is similar to the movement shown by metals, which slowly creep when subjected to a stress—or to the movement of hard rocks at high temperatures and pressures when buried deep

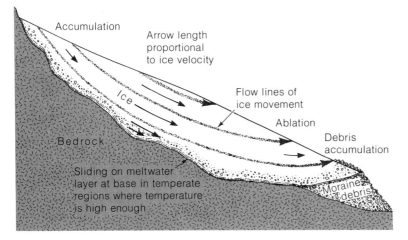

**Figure 11-6**
How ice flows in a typical valley glacier of temperate regions. The rate of
movement decreases toward the base. In temperate climates, where temperatures
at the base of the glacier are sufficiently high that the ice pressure will cause
melting, the entire thickness of the glacier will move by sliding along the liquid
layer next to the ground.

in the Earth. This movement is not the whole story. Other processes are at work also. Ice crystals tend to melt and recrystallize
a microscopic amount farther downslope, and other crystal
distortions result in movement.

The sliding of a glacier along its base accounts for an important
part of the total movement; the ice at the base of some glaciers
is near the melting temperature, and much of the movement
takes place there. Some of the sliding is caused by the melting
and refreezing of the ice at the base. The melting results from some
combination of the pressure of the overlying ice and the flow of
heat upward from the interior of the Earth. If enough melting
takes place, a layer of liquid water forms at the bottom; such a
layer was found when the bottom of the Antarctic ice sheet was
penetrated by a drill at a depth of more than 2 kilometers below
the surface.

In the deeper layers of ice, the solid material that might otherwise tend to be pulled apart by the forces of movement is held
together by the compressive force of the overlying mass of ice.
But the upper layers of a glacier have little pressure on them, and
so the surface ice behaves as a rigid, brittle solid, becoming
cracked as it is dragged along by the movement of the plastic
ice below. These cracks are the **crevasses** that break up the surface
ice into many small blocks near bedrock walls and at places where
the ice surface steepens sharply in the direction of flow. The movement of the surface at these parts of a glacier is a "flow" resulting
from slipping movements between these irregular blocks.

**Figure 11-7**
Photomicrograph of typical crystalline mosaic of glacier ice. Sample is from the Antarctic Ice Sheet, depth 193 m, Byrd Station, Antarctica. Each grid measures 1.0 cm. Each small area of uniform white, gray, or black is a single crystal as seen in polarized light. The tiny circular and tubular spots are bubbles of air. [From A. J. Gow, U.S. Army Cold Regions Research and Engineering Laboratory.]

Well over a century ago, the first measurements were made of the speed of Alpine glaciers by placing stakes in the ice and noting the changes in their position over a few years. The stakes showed clearly that the center of the tongue of ice moved much faster than the edges, where friction of the ice against rock walls hindered the flow. The most rapid movement was about 75 meters (250

**Figure 11-8**
Crevasses in a valley glacier are deep cracks in the ice. They typically form along the sides of the glacier or, as in right foreground, across the glacier at a bedrock hump in the valley floor. Upper basin of Muldrow glacier, Alaska. [Photo by B. Washburn.]

feet) in one year. Since then, a wide range of ice speeds have been measured, ranging from a few centimeters to a meter per day. It has also been shown that the base of a glacier moves more slowly than the upper parts, as shown in Figure 11-9. This was done by sinking a straight tube vertically into the ice and measuring its bending as the upper parts moved faster than the lower.

Sudden movements of glaciers, called **surges**, sometimes occur after long periods of little movement. Surges last more than two or three years and may travel at rates of more than 6 kilometers (3.7 miles) in one year. The movement comes largely from a sudden redistribution of ice from upper to lower parts of the glacier with little change in distribution rate or in the total size of the ice mass. Debate goes on about whether surges are caused by periodic melting of the base of the glacier, allowing it to slide rapidly for

One of the pioneers in the measurement of glacier flow was *Louis Agassiz,* a Swiss zoologist and geologist. As a young professor, not yet thirty, he built a hut on a glacier and, with the help of his students, mapped its movements. Twelve years later, he emigrated to the United States and became instrumental in founding glacial geology here.

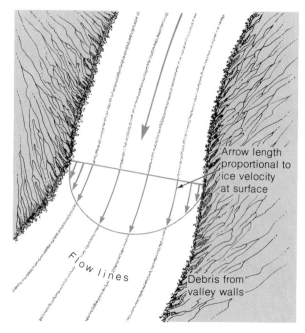

**Figure 11-9**
A valley glacier moves most rapidly at its center and with greatly reduced velocity along the valley walls.

**Figure 11-10**
Valley glaciers showing high snowfields and cirques in the background, tributary glaciers flowing down steep valleys, and their junction with the master glacier in the foreground. Harvard glacier, Alaska. [Photo by B. Washburn.]

a while until freezing slows it, or whether they are caused by intermittent releases of ice that gradually piles up in middle parts of the glacier while the lowest parts are melting. A few geologists think that the large number of surges that have been observed in recent years herald an era of glacial advance, a new ice age, but most glacial specialists think that these movements are peculiar to a particular variety of glacier that we have come to know better in recent years and whose movements have no global climatic significance.

## SHAPES AND SIZES OF GLACIERS

On the basis of size and shape, we can easily divide glaciers into two general types: the **valley glacier** and the much larger **continental glacier**.

**Valley Glaciers**   The glaciers best known to skiers are valley glaciers, rivers of ice flowing downhill in narrow ribbons. They start as snow fields in the high slopes in amphitheatre-like hollows, **cirques**. As the ice forms from firn in the cirque, it flows down the course of the valley, in some places passing over uneven slopes where the smooth surface of the ice may break up into crevasses. As the glacier moves down its valley, it may be joined by tributary valley glaciers. Unlike the waters of tributary rivers, however, the ice of tributary glaciers does not mix with that of the main glacier but runs along in separate streams, side by side. Another difference between rivers and glaciers is that although the ice surface is at the same level at the junction of glacier tributaries, the floors of the valleys at that place may be at very different depths below the ice surface, which is why there are **hanging valleys** where former glacial tributaries entered a main glacial valley, as in Yosemite (Fig. 5-10).

Where valley glaciers terminate on land, streams of meltwater flowing on and under the glacier coalesce downstream to form a single river. Sometimes melting at the ice margin is uneven, and larger blocks of ice may be left isolated as the main part of the glacier melts back in warm weather. Glaciers ending at the water's edge form cliffs of ice, many of them standing from 30 to 60 meters or even more above the lake or sea level. Because ice is less dense than water, it floats; the buoyant force exerted upon it, combined with the melting action of sea water and tidal movements, breaks off icebergs from the edge of the ice. The icebergs then drift with ocean currents (a hazard to ships), many ending up in warmer waters where they gradually melt.

The position of the terminus of a glacier, which is a response to the balance between accumulation and ablation, can be a sensitive indicator of climate. In the 1930's, for example, many glaciers were observed to shrink fairly steadily by retreat of the

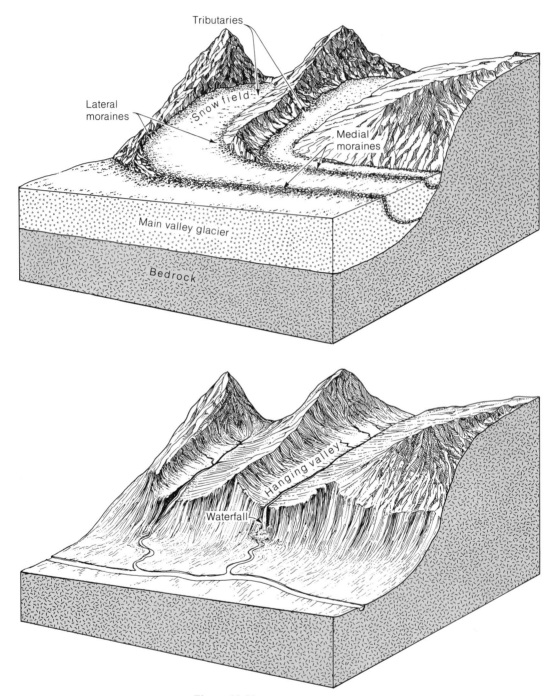

**Figure 11-11**
The evolution of hanging valleys and their waterfalls by tributary valley glaciers.
(Top) During glaciation, tributary ice enters a major glacier at different levels.
(Bottom) The region after deglaciation.

ice margin up the valley. This was just a part of a general warming trend that had gone on since the middle of the nineteenth century. Since 1940 there has been evidence in glacial advances of a slight cooling trend—a trend confirmed by world weather records.

**Continental Glaciers** Enormous, thick sheets of glacial ice cover nearly all of Greenland and the continent of Antarctica. These glaciers are not confined to valleys, but cover all the land surface. The 2,800,000 cubic kilometers (672,000 cubic miles) of Greenland ice covers an area of 1,726,400 square kilometers (666,560 square miles), about 80 percent of the total area of the land. In cross section, the ice has the shape of an extremely wide lens, convex on both the smooth upper surface and the rough lower boundary with the ground. At the center of the ice sheet, in the middle of the island, the ice is more than 3200 meters (10,500 feet) thick.

It is in the central area that the snow accumulates in a huge field of firn that gradually transforms to glacial ice. The ice surface is thus built up at the center and slopes to the sea on all sides. The glacier moves down and out in all directions. At the edge of the sea, where much of the coast is rimmed by high mountain ranges, the ice sheet breaks up into narrow tongues resembling valley glaciers that wind through the mountains to the sea. This glacier is the major source of icebergs in the North Atlantic.

As large as the Greenland glacier is, it is dwarfed by the Antarctic glacier, which covers more than 90 percent of the entire continent. The area covered by ice is about 12,500,000 square kilometers (4,830,000 square miles), and ice thicknesses exceed 3000 meters (10,000 feet) in much of the central area. In some ice-buried valleys, the rock floor of the ice is more than 2500 meters (8200 feet) below sea level. The ground beneath the ice is mountainous over much of the continent; in many places along the margins of the continent, mountains rise up through the surrounding ice. A conspicuous body of ice in the Antarctic, the Ross Ice Shelf, is a thick layer of ice about the size of Texas. The ice shelf floats on the water of the Ross Sea, but is attached to the main part of the glacier on land. In Antarctica, as in Greenland, the ice domes in the center and flows out to the margins, where it ends in much the same way as the Greenland glacier.

## GLACIAL LANDFORMS

As both valley and continental glaciers move over their bedrock floors, they erode large quantities of material and transport it downstream. There it is deposited as the ice melts. An accumulation of rock, sand, and clay on the ice or left on the ground after the ice melts is a **moraine**. Glacial erosion and deposition produce

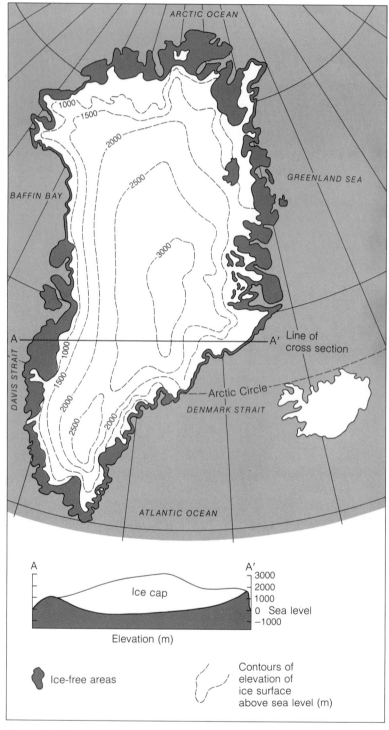

**Figure 11-12**
(Top) Map of Greenland, showing the extent of the glacial ice cap and the elevation of the ice surface. (Bottom) A generalized cross section of south-central Greenland, showing the central depression of the rock floor and the lens-like shape of the ice cap. The ice moves down and out from the thickest section. [Information from J. Haller, *Geology of the East Greenland Caledonides,* Figure 2. Copyright 1971 by John Wiley & Sons.]

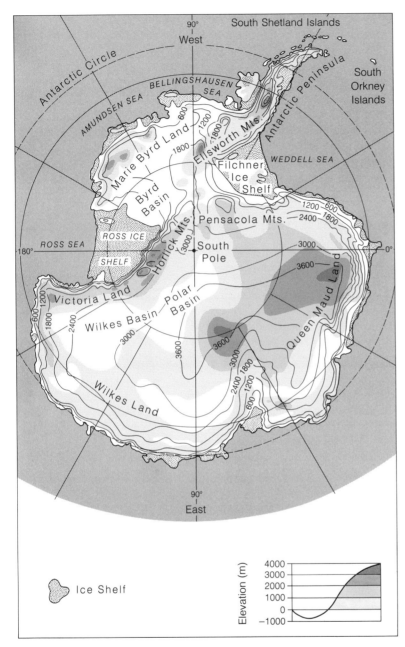

**Figure 11-13**
Contour map of Antarctica, showing the relief of the continental ice and the
land beneath it. The relief of the ice is indicated by the black contour lines, and
the height of the land is indicated by various shades of brown. The continental
land that lies below sea level is shown in white. The black stippled areas are ice
shelves. The black circle around the south pole is at 80° latitude; the next circle
out is at 70° latitude. [From "The Antarctic," by A. P. Crary. Copyright © 1962
by Scientific American, Inc. All rights reserved.]

**Figure 11-14**
Moraines of the Northern Iliamna glacier, Alaska. In the middle background is the present ice margin, and in the foreground a series of semicircular terminal moraines, called recessional moraines, deposited as the ice melted back from its farthest extent. [Photo by B. Washburn.]

a variety of distinctive topographic landforms made up of sedimentary debris, which are left behind as clues when the ice melts. From the study of those forms and materials and their comparison with those of modern glaciers came the suggestion that great polar ice caps may have extended far south into Europe and North America during a former epoch.

**Abrasion and Sediment Transport by Ice**  To anyone who has been in an alpine valley near the end of a glacier, the evidence is plain. As the ice melts, rock material of all sizes is dropped, from the finest clay to huge boulders. The ice itself can be seen to contain all of these materials, with by far the greatest part of the load being near the base and along the valley walls. As the ice melts back from the bare rock pavements over which it moved, it reveals parallel scratches and grooves aligned in the direction of ice movement. In many valleys of the Alps, granitic rocks crop out high in the mountains, and sedimentary rocks make up the lower stretches of the valley. Yet in the lower stretches, many of the boulders at the melting ice edge are granitic blocks, some of them as big as a house. All signs point to the conclusion that the

**Figure 11-15**
A huge erratic boulder carried to Cape Cod during the Pleistocene glaciation. The boulder is a metamorphic rock type typical of areas in southern New Hampshire. Doan Rock, Eastham, Massachusetts. [Photo by R. Siever.]

ice is the scraper that plucks, grinds, and tears rock particles from the walls and floor of the valley and drags them down slope.

Flowing ice does just as much work as running water does, only much better. The capacity of ice to erode is great because it engulfs jointed, cracked blocks in its solid mass and breaks them away from bedrock. As it carries these rocks along, it drags them against the rock pavement below, grinding it into smaller grains, much of it into a fine pulverized **rock flour**. The scratches, or **striations**, are the relic of that action. Striations are clues to the direction the ice moved at that point. By mapping striations over wide areas, we can reconstruct the flow patterns of glaciers. Small hills of bedrock are smoothed on the up-current side and plucked to a rough shape on the down-current side. They too give us an indication of the direction of ice movement.

As a transporter of debris, ice is most effective because once the material is picked up by the ice, it does not settle out like the load carried by a river. Thus ice can carry huge blocks that no other transporting agent can budge. The carrying capacity of ice is tremendous: some ice is loaded with so much rock material that it looks more like a sediment cemented together with a little ice. Just how far the material will be carried depends on the length of the glacier.

Ice erodes and transports material from the sides of its valley with a special efficiency. Not only does it scrape the sides below the ice level, but the ice acts as a conveyor belt for any debris that happens to fall or slide from higher up on the valley walls onto

Photographs of the surface of Mars suggest the possibility that glaciers may once have been active near its South Pole. Deep grooves may have been gouged by glaciers, and step-like terraces of rocky debris may be moraines.

**Figure 11-16**
Glacial polish, striations, and grooves formed on a fine-grained granite surface. Sequoia National Park, California. The direction of ice movement was diagonally toward the right and away from the camera. Since the glaciation, the rock has been disrupted into angular blocks by repeated frost action. [Photo by F. E. Matthes, U.S. Geological Survey.]

**Figure 11-17**
Forks of Susitna glacier, Alaska, showing how lateral moraines (dark bands), formed at the valley walls of the tributaries, coalesce to become medial moraines in the merged stream of ice. Irregular distortion and folding of bands reflects differential deformation of ice as it flows. [Photo by B. Washburn.]

**Figure 11-18**
Alpine sculpture of mountainous terrain by glaciers on the approaches to Mt. McKinley, looking up Ruth glacier. This region was completely glaciated in the Pleistocene; since that time the shrinkage of the ice has revealed the sharp, knifelike divides between adjacent valleys and the generally rugged topography. [Photo by B. Washburn.]

the surface of the glacier. **Lateral moraines**, strips of dirt and rock that flow with the ice along the sides of valley glaciers, are remarkable tracers of the flow of individual tributaries. The lateral moraines of joining ice streams merge and form a single **medial moraine** in the middle of larger flows. The ice arrives at the end of the glacier carrying its load, some of it as clearly marked segregations of lateral and medial moraines, much of it in the basal layers, and the rest dispersed throughout the bulk of the ice.

**Sculpture of Topography**   Much of the kind of rugged mountainous terrain that has been celebrated for its scenic beauty is the product of glacial erosion. The Sierra Nevada, the Cascades, the Rockies, and the Alps all have high valleys that were filled with glaciers just a little more than ten thousand years ago, as evidenced by the signs we have learned to recognize: striations, large blocks carried from high in the mountains, moraines, and cirques. In the mountains of northern California, Oregon, and Washington, there are now only small relic glaciers left in the few places that receive enough snow and stay cold enough to preserve ice all year round. The rest of the terrain is exposed, and we can see the effects of Pleistocene glaciation. Farther north, in Canada and Alaska, glaciers become more numerous, but some of the ice-sculptured topography is still exposed.

As glaciers ate away at the mountains, the topography acquired a distinctive appearance. Cirques were formed at the heads of glaciers as the ice plucked and undercut the highest peaks and ridges, and adjacent cirques gradually met to form knife-sharp divides. The result is a jagged, serrated, linear crest. The valleys were excavated to the characteristic U-shape as the glacier scraped and rounded off irregularities. As the ice retreated, hanging valleys of glacial tributaries were exposed. Lakes formed in depressions in cirques. Most spectacular are the **fiords**, arms of the sea that occupy U-shaped valleys that were cut below sea level by valley glaciers descending from coastal mountains.

**Depositional Landforms**  If glacial erosion is accountable for rugged alpine topography, deposition of glacial debris is responsible for some of the most tried and true country scenes that decorate calendars: rolling hills, small lakes, and fertile fields. Plains in the Dakotas, rolling hills in Iowa and Wisconsin, Finger Lakes in New York, rocky fields in New England are all depositional products of a large continental glacier of a past so recent that its depositional landforms have not been obliterated by erosion. The many different elements that compose such varied landscapes, both in North America and in Europe, led Louis Agassiz and others to the hypothesis, now fully accepted as theory, that these parts of the Earth had recently gone through a great ice age.

In many of their characteristics, glacial deposits differ greatly from river, wind, or ocean deposits. In regions formerly covered by ice, the general topography tends to lack organized drainage networks, as though rivers had not had a chance to shape divides and valleys into a systematic, branching tributary pattern. Instead there are hollows with no outlets, some of which are filled with lakes. In some places there are isolated hills, in others long and sinuously winding ridges, none of which seem to have much relation to the underlying bedrock.

The bedrock is rarely seen in some of these regions. It is deeply buried, sometimes 50 meters (160 feet) or more. The ground over the bedrock is mostly a heterogeneous mixture of sand, clay, pebbles, and boulders rather than ordinary soil. Many of the pebbles and boulders differ entirely in composition from the bedrock. Some rare fossils in this material turn out to be the bones of extinct, large, arctic-dwelling elephants, the woolly mammoths. But mostly there are no signs of life at all.

**Figure 11-19**
Generalized geographic map of North America showing the maximum extent of glaciation in Pleistocene time. Arrows indicate direction of ice movement. [After U.S. Geological Survey.]

There was no way in which the geologists of Agassiz's time could relate all of these characteristics to erosion, transportation, and deposition by water or wind. It was this and the striking similarity of some landforms to the moraines and other products of modern valley glaciers that led to Agassiz's conclusion: glacial ice was responsible, but the areal extent of the various landforms is so great that the ice that produced them must have been a continental glacier like those of Greenland and Antarctica. Once the idea of a former continental glaciation gained acceptance, the many geologists who lived and worked in the northern provinces of Europe and North America began to realize that many of the landforms around them fitted the new concept of continental glaciation.

Glacial moraines proved to be important pieces of the puzzle. The hilly ridges composed of heterogeneous rock fragments, sand, and clay were recognized as **terminal moraines**, similar to those of valley glaciers, marking the farthest advance of the ice.

Long before there was a glacial theory, the heterogeneous materials that make up moraines were thought to have "drifted in" somehow from other areas. The name applied then is still in use: **drift** is the name for *all* material of glacial origin found anywhere on land or at sea. The drift of moraines, heterogeneous and unstratified, was inferred to be the direct deposit of melting ice and was given the name **till**, a Scottish word for much of the rocky soil of Scotland. Mapping an ice front then becomes a matter of finding the limit of till, usually marked by a hilly ridge. Because a terminal moraine consists of till built up at the melting front of the ice, its height is a rough measure of how long the ice front remained in a certain position, its rate of ablation exactly balancing the rate of advance.

Terminal moraines may be deposited in the relatively shallow waters of continental shelves wherever ice pushes out from land at a level below that of the sea, as does the Antarctic Ice Sheet today. Cape Cod and Long Island are two of the best known of these moraines. Built up many meters from the sea floor by debris dropped from melting ice, such moraines are left as islands or capes when the ice melts. Because the areas behind some of these moraines are built up less by glacial deposition than the terminal moraine, they become flooded by the sea. Long Island Sound and Cape Cod Bay are two such areas.

In front of terminal moraines, no till is found. Instead, drift of another kind may be found — stratified deposits made up of the same kinds of materials as the till of the moraine, but well sorted by size into distinct layers of gravel, sand, or finer materials. The gravels are rounded, and the sands are cross-bedded. This stratified material is glacial outwash, the drift caught up and modified, sorted, and distributed by streams of meltwater running away from the glacier. Some of the outwash forms small hills, **kames**, where sand and gravel accumulated by meltwater streams were dumped near or at the edge of the retreating ice. Other accumulations show the structure and shape of deltas built out into lakes. These deltas, called **delta kames**, are left behind as flat-topped hills that are frequently exploited as commercial sand and gravel pits.

The gravelly kames grade into the silts and clays that are characteristic of lake-bottom sediments. The silts and clays typically show the alternating coarse and fine layers that we call **varves** (see Fig. 2-17). Putting all of this together, we infer that a stream of glacial meltwater once discharged into a lake near the ice front, building a delta of sand and gravel. The varved sediments of the lake are the finer materials that were carried in suspension beyond the delta. In summer, only the coarser silt was laid down on the lake floor. In winter, when the lake froze over, the water became still, and even the finest clay settled out. This concept of an annual cycle of deposition, by which pairs of mud layers of different

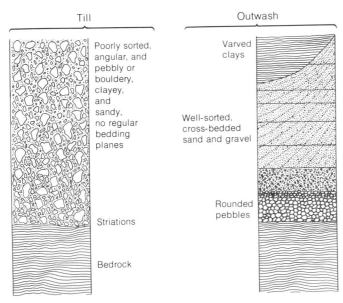

**Figure 11-20**
Comparison of generalized sections of glacial till and outwash.
Till is a mixed, poorly sorted accumulation of material of all sizes
left behind by melting ice. Outwash is a sorted, stratified deposit of
water-transported drift.

lithology were laid down, was used to tell how long a lake had
been in existence while the ice melted. By matching the varves
of many lakes of different ages in a glaciated region, geologists
were able to construct a chronology that revealed the total time
elapsed since the last glaciation.

What of the hollows, the undrained depressions, or **kettles**,
found in so many areas of both till and outwash? Many kettles
in outwash are steep-sided and occupied by lakes or ponds. Again
the clue came from observing the ablation of modern glaciers,
many of which melt back unevenly, sometimes leaving behind
immense, isolated blocks of ice, around which course streams of
meltwater carrying glacial sediment. Imagine what would happen
if a block of ice about a kilometer in diameter were left behind
by a retreating glacier. A mass of ice that big would take a long time
to melt—thirty years or more—and it might become partly buried
by outwash sand and gravel. By the time it could melt completely,
the main ice margin might have retreated so far that little outwash
would reach the hole left by the ice block. If the hole were deep
enough so that its bottom lay below the groundwater table, a lake
would form.

The territory behind a terminal moraine has its own kettles and
is rich in other forms created by ice movement and by meltwater.
**Ground moraine** is a mantle of till deposited underneath the ice.

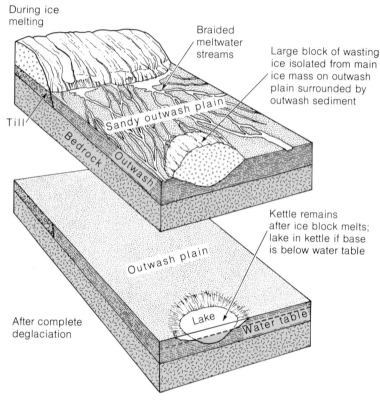

During ice
melting

Braided
meltwater
streams

Large block of wasting
ice isolated from main
ice mass on outwash
plain surrounded by
outwash sediment

Till

Sandy outwash plain

Bedrock

Outwash

Kettle remains
after ice block melts;
lake in kettle if base
is below water table

Outwash plain

After complete
deglaciation

Lake

Water table

**Figure 11-21**
Evolution of an outwash kettle. As a glacier retreats, it may leave behind large
blocks of wasting ice that are gradually buried by outwash from the receding ice
front. After the front has retreated far enough from the region, outwash sedimentation stops, the ice block melts, and a depression remains, filled with water
if it is deep enough to intersect the groundwater table.

It may be so thick that it completely buries bedrock everywhere,
or it may be a relatively thin cover that exposes bedrock knobs
and pavements in many places. **Eskers** are long, narrow ridges
of sand and gravel situated in the middle of ground moraine.
Varying in height and width from a few meters to tens of meters,
they wind for kilometers in a direction roughly parallel to the
direction in which the ice moved. The well-sorted materials of
eskers look like water-transported outwash, and the eskers themselves have the shape of stream channels; some even show "tributary" eskers. The deduction made from this geologic evidence
is that eskers form from meltwater streams flowing in tunnels
along the bottom of a melting glacier. The water seeps through
crevasses and cracks in the ice and gradually opens tunnels that
run downhill to the end of the glacier. Eskers, like kames, are
favorite sources of sand and gravel for road construction and
other uses.

**Figure 11-22**
Retreat of the ice margin has revealed an esker formed by a stream that ran under the glacier. The esker is the winding ridge of sand and gravel running from upper right to lower left. The grooves at right are in soft glacial sediment and are not bedrock striations. [Photo by B. Washburn.]

**Drumlins** are streamlined hills of till and bedrock that parallel the direction of ice movement. The shape of a drumlin is like that of an upside-down elongate spoon with the gentle slope in the down slope direction. They may be 25 to 50 meters (75 to 150 feet) high and a kilometer long. Some drumlins are all till; others are largely bedrock with a cover of till. Their shape is like that of rock knobs that have been smoothed and polished by the ice. Knobs, drumlins, and irregular elongate hills of ground moraine are some of the different shapes taken by materials of different strength and resistance to erosion when overridden by an ice sheet.

Identification of a glacial landform is a matter of perceiving its shape, which is best done from a hilltop, or a good topographic map and matching it with the nature of the material making it up. The contrast between the heterogeneous mixture of sizes and shapes of till and the rounded, evenly sorted gravel or sand of waterlaid, stratified drift is a good first guide. Just the presence of many lakes, swamps, and other undrained depressions is sufficient evidence to put one on the watch for other characteristic glacial landforms.

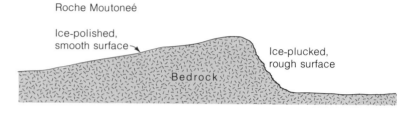

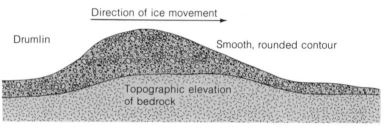

**Figure 11-23**
Two similarly shaped glacial landforms. (Top) Roche Moutoneé, a polished rock knob, is formed as the ice rides over a bedrock protuberance, smoothing the side facing the direction from which the ice moved and roughening the down-flow side by plucking cracked and jointed rock. (Bottom) Drumlins are composed of till; some have a bedrock base with a streamlined shape formed by ice movement. The steep side commonly faces the direction from which the ice came, but the reverse is also found.

**Glacial Topography and Culture** Three-quarters of a century ago, a geologist in Illinois said that he could predict voting habits by where people lived with respect to glacial drift. The northern part of the state is covered with fertile black soil formed on relatively unweathered glacial till deposited during the most recent glaciation, which ended about 10,000 years ago. There lay the voting strength of the prosperous Republican farmers. South of the terminal moraine that records that glaciation, the land is far less productive; gray clayey soil was formed on a till of an older glaciation. The much greater time elapsed since the older glaciation, more than 60,000 years, allowed deep weathering that reduced its fertility. Farmers were Democrats on that soil. Their voting habits have changed since then, but the effects of the glacial epoch cannot be underestimated as an influence on our lives. Much of the richest farmland in the northern tier of midwestern states is on glacial drift. Much of the abundant good water supply of the glaciated terrain is from aquifers of sand and gravel in the drift. Sand and gravel deposits are abundant. Lakes, one of our best recreational resources, are another of the gifts of the glacial epoch. But there are liabilities too. Any New England farmer can speak with feeling about all of the rocks in his fields of bouldery till, and geologists are sometimes frustrated by the drift that covers bedrock and prevents them from mapping geologic formations and mineral resources.

From an intensive study of landforms, geologists deduced that a great continental glaciation had effectively extended both polar ice caps far into what are now temperate climates. They also reasoned that the glaciation was recent, for everywhere the drift was soft, like freshly deposited sediment. It also lay at the surface everywhere and so was one of the last sediments to be deposited, covered only here and there by recent river and other deposits. Then, early in this century, matching of varved clays from glacial lakes gave a good absolute date: the most recent retreat of ice had taken place between 8,000 and 15,000 years ago. In the 1940's, the decay of radioactive carbon, $C^{14}$, was applied to the dating of materials younger than about 30,000 years, and the youngest glacial deposits of the Pleistocene Epoch, the ice ages, were opened to accurate dating.

In the course of mapping the glacial deposits, geologists found that there were actually several different layers of drift. Furthermore, the older layers had well-developed soils, including fragments of plants that botanists said were characteristic of warm climates. As the evidence grew, a new conclusion became inescapable: during the ice age, the great continental glaciers alternately advanced and retreated. Between glaciations, there were long interglacial periods, during which warm climates returned to a large part of the formerly glaciated lands. At least four major periods of advance and retreat could be deduced on the basis of landforms and stratigraphy of moraines. Some of the interglacial periods lasted longer than the glacial ones. In the past decade, we have learned from the study of deep-sea deposits, which give us some measure of ancient ocean temperatures, that there have been many shorter, more regular cycles of glaciation and deglaciation. This has been confirmed by more accurate dating of glacial sediments by radioactive decay of $C^{14}$ and other elements. Moreover, in recent years, volcanic materials that are found in some regions between older glacial layers on land have been dated by the potassium-argon method.

Some of the indirect effects of glaciation were explored as this picture emerged. The origin of the Pleistocene loess deposits, the thick layers of windblown dust, was traced to extensive flood-plain and outwash plain deposits of fine muds that came from rock flour ground up by the glacier (see Chapter 10 and Fig. 10-29). Many river valleys that carried the meltwaters of receding glaciers became choked and filled with that sediment. As the exploration of the seas started, oceanographic research vessels brought up samples of marine sediment of the glacial epoch. In the northern and southern parts of the oceans, much of the Pleistocene sediment was like stratified drift, the product of melting icebergs.

Some deposits of the early Pleistocene, too old for $C^{14}$ dating, have been dated in the same way as the more ancient rocks — that is, on the basis of the fossils they contain. Land deposits have been dated on the basis of snails, and marine sediments have been dated on the basis of foraminifera, small single-celled organisms that secrete shells of calcite.

Pleistocene foraminifera, the tiny calcareous animals that live in the ocean, bear testimony to the coldness of the water by the special ratio of the isotopes of oxygen in the calcium carbonate of their shells. The proportion of ordinary oxygen, $O^{16}$, and the heavy oxygen isotope, $O^{18}$, depends on the temperature of the water, and the shells from different layers of sediment show a warming and cooling of the sea that correlates with expansion and contraction of glaciers on land.

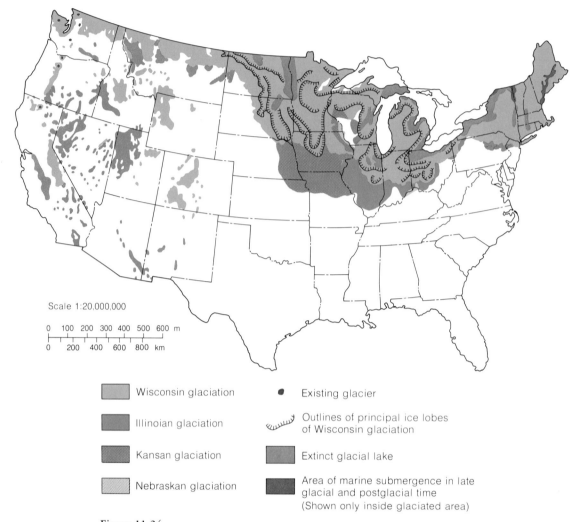

Scale 1:20,000,000

```
0  100 200 300 400 500 600 m
├──┼──┼──┼──┼──┼──┤

0     200   400   600   800  km
├──┼──┼──┼──┼──┤
```

| | | | |
|---|---|---|---|
| ▨ Wisconsin glaciation | | ● Existing glacier | |
| ▨ Illinoian glaciation | | ⌒⌒⌒ Outlines of principal ice lobes of Wisconsin glaciation | |
| ▨ Kansan glaciation | | ▨ Extinct glacial lake | |
| ▨ Nebraskan glaciation | | ▨ Area of marine submergence in late glacial and postglacial time (Shown only inside glaciated area) | |

**Figure 11-24**
Glacial geology of the conterminous United States. [Compiled by C. S. Denny, U.S. Geological Survey.]

**Sea-level Changes**   We know from the enormous volume of ice that occupied land during the Pleistocene and from the way the hydrologic cycle works that the sea is the likely source of the ice. Most of the water that precipitated as snow did evaporate from the ocean. The more water that was tied up in ice, the less the ocean could be replenished by river runoff to keep its steady level. As the ice accumulated, the level of the sea lowered everywhere in the world's oceans.

The total volume of ice on Earth today is a little more than 25 million cubic kilometers. Total ice volume during the height of the ice age can be estimated from the area covered by the ice sheets combined with calculations of the thickness of ice necessary to keep the glaciers moving so many hundreds of miles from their areas of accumulation. Startling as it may seem, the volume must have been about 70 million cubic kilometers. The extra 55 million cubic kilometers that came from the sea to make the additional ice lowered sea level by about 130 meters (425 feet)! Imagine the series of changes caused by that great a reduction in sea level. Much of the shallow sea areas adjacent to the shorelines of continents were left above sea level. In glaciated regions, the ice sheets extended far beyond present shorelines. Some of the newly emerged coastal plains in nonglaciated areas were more than 100 kilometers (62 miles) wide. Rivers extended new channels across them and created new deltas at the water's edge. Wide strips of beaches were formed. Mammoths, huge Pleistocene elephants, roamed those low coastal plains, and no doubt Pleistocene Man did too, for early prehistoric cultures were evolving in the land beyond the ice sheets. The relics of all of these happenings have been found by oceanographic research vessels exploring the shallow areas that became reflooded by the sea as the ice melted.

The figures for ice and sea-level changes can be looked at in another way. What if the 25 million cubic kilometers of water now tied up as ice were to melt? The change in sea level would be catastrophic, for the melting of all existing ice would raise the oceans by about 65 meters (210 feet). This would inundate many major cities of the world, such as London, New York, and Tokyo, plus the great bulk of the low-lying areas of the continents, where much of the world's population lives. Those who would tinker with the Earth's climate in ways that might cause the ice to melt are well advised to remember this simple calculation.

But what if it should happen naturally? What if we are really just living in an interglacial period like those during which the Pleistocene ice sheets withdrew? On the other hand, since the 25 million cubic kilometers of ice on Earth today is vastly more than existed during most times in the history of the planet, the Earth could be considered to be in a glacial age. It has been only 10,000 years since the last retreat of huge continental glaciers, and we have no reason to suppose that the ice will not advance again. It appears that there are some very real and possibly practical reasons, in addition to our natural scientific curiosity, why we ought to think about the causes of ice ages.

**Causes of Ice Ages**  For many years geologists have speculated, calculated, and theorized, but no complete agreement yet exists as to the causes of ice ages. The theories themselves have two different parts. One is an explanation of the cooling of the Earth's atmosphere that led to the glacial epoch several million years ago.

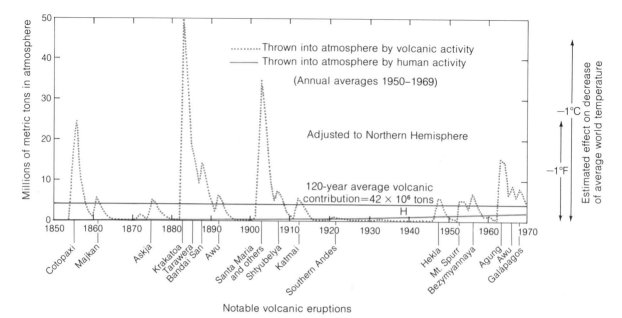

**Figure 11-25**
World-wide load of atmospheric dust thrown into the stratosphere by volcanic eruption since 1850 in comparison to that contributed by human activity. Peaks of volcanic activity may have an affect on world average temperature, as shown in the estimates on right side of graph. [After J. M. Mitchell, "Pollution as a Cause of the Global Temperature Fluctuation," in *Global Aspects of Pollution*. Copyright 1970 by Springer-Verlag New York, Inc.]

The other is an accounting for the oscillations during the Pleistocene — the growth and advance of ice during glaciations, and the shrinkage and retreat during the interglacial periods. We will cover just a few of the major ideas to give some notion of the kind of thinking that has gone on.

Most theories have concentrated on reasons why the Earth would cool, causing the onset of glaciation. Perhaps most naturally, some have sought an explanation in a possible variation in the heat radiated from the Sun. It has been suggested that there may be periodic changes in the rate at which the Sun burns its nuclear fuel, but no firm evidence has been found. It remains a speculation. According to another theory, injections of great quantities of volcanic ash into the atmosphere would block a portion of the Sun's rays and so cool the Earth. Though some cooling could be expected from this mechanism (see Fig. 11-25), no one has produced strong evidence of a period of volcanic activity in the late Cenozoic that was significantly greater than at other times.

A more attractive theory for general cooling relates to the positions of the continents relative to the poles. Today there are large land areas in the broad, cold regions surrounding both north and south poles. At certain times in the geologic past, one or the other pole was in the middle of the oceans, where currents and general movement of ocean waters allowed a sufficient exchange of heat with the atmosphere to make the distribution of temperature relatively uniform from poles to equator. Once the land areas were near or at the poles, the heat difference between poles and equator was increased, and the polar regions became very cold. If this theory is correct, the events leading to ice ages must have been set in motion about 20 million years ago, for that was when the continents approached their present positions relative to the poles. Indeed, there is now some evidence of small glaciers that started well before the main, or Pleistocene, ice ages.

The alternation between glacial and interglacial epochs has been explained as the result of astronomical cycles. The shape of the Earth's orbit around the Sun changes periodically, putting us sometimes slightly closer and sometimes slightly farther from the Sun. In addition, the Earth wobbles slightly on its axis of rotation, and this wobble also affects very slightly the heat received by the Earth at different latitudes. These motions have been worked out carefully and seem to fit with the durations of the glacial and interglacial periods. The astronomical theory of oscillations has to be superimposed on a theory of Earth cooling in order to arrive at a reasonable explanation as to why these oscillations took place during the Pleistocene.

The Pleistocene is not the only epoch for which there is a record of glaciation. Ancient glaciations are recorded by rocks that are the lithified equivalents of till. Why glacial periods come every so often in the history of the Earth remains uncertain. Nor is there any certainty about the effects that future glaciations may have on Man's existence on Earth. If there is another glacial episode, it will happen on a time scale of thousands of years, slow enough for all of us to ignore, but with implications of yet another change for the continuous record of human culture, which was well advanced when early Man was living in caves in southern France during the last ice age and changed greatly after deglaciation. Our first sensitive indicators would be the advance of mountain glaciers in Alaska and elsewhere. But we would have no way of knowing whether such an advance indicated any more than a temporary period of slight cooling that might last only a few years or tens of years and then be followed by another warm period.

1. Glaciers are masses of ice formed by the gradual transformation of snow by recrystallization and pressure.

2. The size of a glacier is determined by the rate at which snow accumulates and the rate of ablation by melting, evaporation, and iceberg calving.

3. Ice moves downhill partly by internal plastic flow, which is mainly the result of many small slips along planes of individual ice crystals, and partly by sliding along its base.

4. Surface ice breaks into crevasses as a result of the movement of ice masses over steep changes in slope and along valley walls. Glaciers move with a wide range of speeds; some glaciers move in fast surges.

5. Valley glaciers form in alpine mountains and erode U-shaped valleys, hanging valleys, and cirques.

6. Continental glaciers are of great size and exist today only in polar regions, where melting is almost nonexistent, as in Greenland and Antarctica.

7. Glaciers produce many erosional and depositional landforms. Ice abrades material from the walls and floors of valleys and transports it to ice fronts, where it accumulates as moraines.

8. Landforms of Pleistocene continental glaciers include moraines and various forms of till and drift, such as kettles, kames, varves, eskers, and drumlins.

9. There were several glaciations during the Pleistocene, each recorded by a till of a different age and by temperature fluctuations of the oceans. Sea level changed extensively as the ice sheets waxed and waned, at times leaving large areas of the continental shelves of the world exposed as land.

10. Causes of glaciation have been sought in astronomical theories to explain oscillations during colder periods and general cooling as a result of movements of continents with respect to the poles.

## EXERCISES

1. List the ways in which snow and sand are similar and glacial ice and marble are similar.

2. Mt. Washington is the highest peak of the White Mountains of New Hampshire, at 1920 meters (6,293 feet), and official weather records show winter temperatures as low as 60°C below zero. Discuss the reasons why it has no glacier and what changes in the current conditions might produce one.

3. One of the dangers of exploring glaciers is the possibility of falling into a crevasse. What topographic features of a valley glacier or its surrounding rocky terrain would you use to indicate that you are on a part of the glacier that might be badly crevassed?

4. What differences in the composition and texture of the glacial till would you expect between two glaciated areas, one on a geologically young series of coastal plain soft shales and loosely cemented sands, the other on an old granitic-metamorphic rock terrain.

5. A number of ancient rock formations have been the subject of controversy concerning their possible glacial origin. What characteristics of glacial till or glacial landforms might you expect to use to establish that a rock 270 million years old is a tillite?

6. You are assigned the job of mapping a terminal moraine of a late Pleistocene glaciation. What characteristics of topography or till composition would you use to distinguish it from ground moraine on the one hand and outwash on the other? How might you distinguish it from an esker?

7. If you were in charge of a global watch to monitor any indications of a resumption of extensive continental glaciations like those of 15,000 years ago, what kinds of observations would you direct your teams to make? How much warning in centuries, years, days, or hours might you expect to give the city of Chicago that another glaciation was imminent?

# BIBLIOGRAPHY

Field, W. O., "Glaciers." *Scientific American,* September, 1955. (Offprint No. 809.)

Flint, R. F., *Glacial and Quaternary Geology.* New York: John Wiley & Sons, 1971.

Knight, C. and N. Knight "Snow Crystals." *Scientific American,* January, 1973.

Paterson, W. J. B., *The Physics of Glaciers.* London: Pergamon Press, 1969.

Price, R. J., *Glacial and Fluvioglacial Landforms.* New York: Hafner Pub. Co., 1973.

Robin, G., "Ice of the Antarctic." *Scientific American,* September, 1962. (Offprint No. 861.)

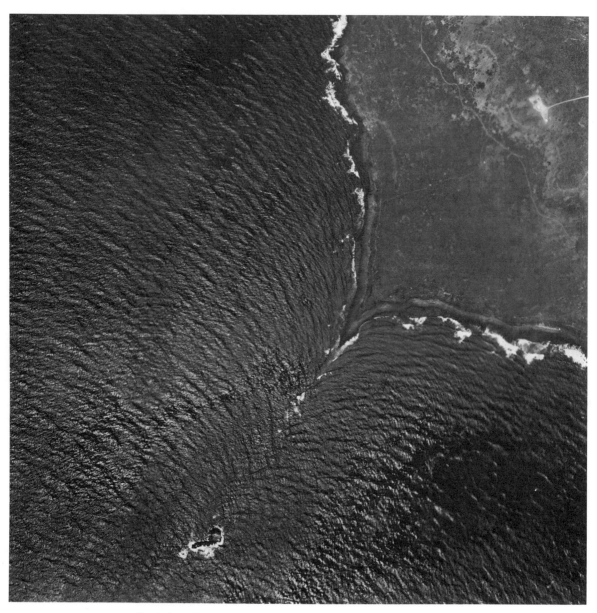

Vertical air photo showing waves approaching a point of land and being bent (refracted) around it. In the lee of the offshore rocks, the waves form an interference pattern. [Photo from U.S. Hydrographic Office Publication 234.]

# 12

## Ocean Processes

*The sea is the end of the line for the weathered debris of continental erosion. The material brought to the ocean by rivers and worn away from its coasts by wave erosion is distributed along its shores as beaches or transported by currents to the continental shelf and deeper parts of the ocean. Beaches are the expression of a balance between the supply of sand provided by erosion of the shore and the transport of sand by rivers, wave erosion, and longshore currents. The continental slopes, continental rises, and abyssal plains are shaped by turbidity currents that transport much material from the shallower to the deeper parts of the sea. The current patterns of the general circulation of the oceans influence the distribution of fine-grained pelagic sediment that mantles the topography of the deep oceans.*

About 100 years ago, in 1872, H.M.S. *Challenger,* a small wooden warship converted and fitted out for scientific study of the seas, set out from England on a four-year voyage over the world's oceans. From that first systematic study eventually flowed fifty thick volumes on the currents, the underwater topography, the bottom sediments, and the life of the deep sea, which opened the eyes of land-bound geologists. What had been the realm of groundless speculation had suddenly been transformed into a matter of serious interest: the acquisition of observations about the ocean and the building of theories to explain them. Early musings on "featureless plains" beneath the water were abandoned as the discoveries rolled in: the mid-Atlantic ridge and other high ranges of submarine mountains, deep troughs and trenches, vast areas of hilly topography, and, to be sure, some great flat plains. The rocks and sediments of the sea floor were found to be as

**Figure 12-1**
H. M. S. *Challenger,* as depicted in the official *Challenger* report. [Courtesy New York Public Library.]

diverse as the topography: fine-grained brownish clays, blue clays, sands, silts, and, in some places, even pebbles and cobbles. In many parts of the deep sea, the *Challenger* scientists found "oozes," sandy or muddy sediments dominated by the shells of tiny organisms, some of calcium carbonate, some of silica. Perhaps as surprising was the discovery of igneous rocks and unusual cobble-shaped lumps of heavy rocks: manganese nodules.

The broad outlines of the problems of marine geology were laid out by the voyage of the *Challenger.* What was the origin of the varied topography? What explained the different arrays of sediment and other rock types on the bottom? What role did ocean currents, tides, and waves play in forming the sea floor? Today, a hundred years after these questions first came to be asked, hundreds of oceanographic research vessels from different countries are exploring the sea, providing some answers to those questions and discovering new facts that raise new problems.

**Figure 12-2**
Pioneer and modern oceanographic research vessels. (a) The
Woods Hole Oceanographic
Institution's *Atlantis I*, the first
ship built specifically for a wide
range of oceanographic research.
It was delivered in 1931 and
retired in 1965. (b) Its "replacement," the *Atlantis II*, a modern
ship fully equipped for general
oceanographic research. [Cour-
tesy Woods Hole Oceanographic
Institution.]

More than any other endeavor, it was oceanographic research that has led to our current ideas of the dynamics of the Earth. A hundred years ago, the study of the sea was primarily either pure science (science for its own sake) or applied science in the service of the shipping trade. In the decades after World War II, much oceanographic research was financially supported by nations interested in the military advantages that might accrue to them from a knowledge of the sea, which was soon to become the home of the nuclear missile-launching submarine. Now oceanographic research is made more urgent, both by the possibility that the resources of the sea can be used to support the world's growing population and by the companion danger of pollution of the seas.

The study of the oceans—oceanography—has united geology, geophysics, and geochemistry with biology, physics, and meteorology to become a field of knowledge much too broad to cover in this book. In this chapter, we will discuss those aspects of oceanography that continue the story of the erosional, transportational, and sedimentational processes that shape the surface of the Earth's crust beneath the sea as well as on land. We will start with the most familiar part, the edge of the sea.

## THE EDGE OF THE SEA

The **coast**, the broad region that is the meeting place of land and sea, can be carved into many kinds of shapes: steep rocky cliffs, broad low beaches, crescents of small beaches in small bays alternating with rocky headlands, or wide, sweeping, sandy tidal flats. The forces that shape coasts are essentially the same as those that form other topographies: the destructive processes of erosion operating in conjunction with currents that transport and deposit debris and the tectonic forces that cause uplift or subsidence of the Earth's crust.

**Wave Erosion**   At the **shoreline**, the line along which the water meets the land, the major erosive agent is wave action. The force of high waves breaking against a rocky cliff during a storm can be awesome in its violence; you can feel the ground shake with the shock. Nearby sensitive seismometers, used to detect far-distant earthquakes, will register the small tremors of the solid earth caused by the sudden impact of tons of water on solid rock. Moderately high waves, about 3 meters (10 feet) high at the breaking point, have been measured to produce a force of more than 70 grams per square centimeter (30 pounds per square inch). So when you see waves break against a concrete seawall and throw water 5 meters (16 feet) into the air, you can estimate an impact force equivalent to dashing about 175 grams (0.4

**Figure 12-3**
A rocky headland (Heceta Head) on the Oregon coast. To the extreme right
is a small sandy pocket beach formed in a recess between headlands. [Photo
by R. Siever.]

**Figure 12-4**
Waves splashing against a rocky shore. Point Lobos, Monterey County, California.
[Photo by M. R. Campbell, U. S. Geological Survey.]

Henry David Thoreau visited Cape
Cod four times between 1849 and
1857 and kept a record of his
observations that was published as
*Cape Cod*. His description of the
beach and other parts of the Cape
remains one of the great examples of
the combination of natural history
with conversations with and
observations of people that was
common in the nineteenth century.
One good edition is the one put
out by W. W. Norton & Co. with
an introduction by Henry Beston.

pounds) against each square centimeter of rock surface. It is not surprising that such seawalls, built to protect homes just back of the shore, quickly start to crack and constantly have to be kept in repair.

The waves are even more destructive than the force of the water alone, for the waves catapult pebbles, cobbles, and, in intense storms, boulders against the shoreline rocks. The physical destructiveness of the waves is further enhanced by the chemical weathering action of sea water that is forced by wave pressure into every tiny crack and crevice. The chemical decay extends and widens the cracks, preparing them for the disintegrating physical beating by the waves.

Where waves break against bluffs of softer unconsolidated material, such as soils or recent glacial deposits, the rate of erosion may be extraordinary. The high sea cliffs of glacial materials facing the open Atlantic along the Cape Cod National Seashore are retreating at a rate of about a meter each year. Since Henry Thoreau first walked the length of those cliffs in the middle of the nineteenth century, about 6 square kilometers (2.3 square miles) of land area has been eaten by the ocean. At that rate, real-estate land (or "sea") values on Cape Cod will be worth little in a few thousand years!

**The Forces that Shape Shorelines**  Just as the surface of the land is the product of interplay of tectonic forces lifting up the land, erosion wearing it down, and sedimentation filling in the low

**Figure 12-5**
Sandy Cliffs of glacial deposits at Wellfleet, Cape Cod. Composed of easily erodible, unlithified clays and sands, these cliffs retreat about a meter per year under the attack of the waves. [Photo by R. Siever.]

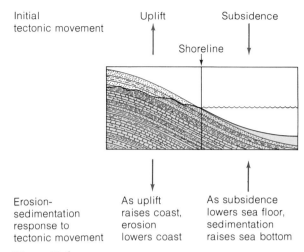

Initial tectonic movement    Uplift    Subsidence

Shoreline

Erosion-sedimentation response to tectonic movement

As uplift raises coast, erosion lowers coast

As subsidence lowers sea floor, sedimentation raises sea bottom

**Figure 12-6**
Shoreline profiles are a balance between tectonic forces and erosional and sedimentational processes that counter them. The greater the uplift of the land, the greater the erosion; the greater the subsidence of sea floor, the greater the tendency for sedimentation to build up the bottom. Similarly, subsidence of coastal lands decreases erosion, and uplift of sea floor inhibits sedimentation.

spots, a shoreline is a result of the same kinds of forces. The shore may be uplifted or it may subside with respect to sea level. If it is uplifted, erosion, principally the work of waves, will attack it by the same kind of negative-feedback process explored in Chapter 5 for landscape evolution: the more the uplift, the more intense the erosion that counters it. If the land subsides with respect to sea level, a different negative-feedback process operates: the low spot tends to get filled in by sedimentation. Uplift and subsidence may operate simultaneously in different parts of the same region. For example, some coasts are broken up by tectonic forces into a pattern of large blocks bounded by faults, which allows some blocks to be raised while others are depressed.

Tectonically uplifted coasts, where erosion is active, have a topography of prominent cliffs and headlands jutting out into the sea alternating with narrow inlets and irregular bays, some filled with small rocky beaches. Jagged, irregular shapes dominate in hard, rocky terrain. In softer sediments, the topography is characterized by the gentler slopes of straight bluffs, marked with occasional landslide and slump scars. The waves undercut the cliffs, causing huge blocks to fall into the water, where they gradually become worn away. Along such coasts, the retreat of the sea cliff may leave isolated remnants standing in the sea far from the shore. As the waves erode, they plane the surface just below sea level and create a **wave-cut terrace**.

**Figure 12-7**
Wave-cut terrace carved in rock in late Cenozoic time and now tectonically
elevated above sea level. At sea level, a new wave-cut terrace is being formed.
San Luis Obispo County, California. [Photo by G. W. Stose, U. S. Geological
Survey.]

Once the uplift stops, erosion wears away at the land, making
the shoreline move back from the sea while the land surface is
lowered. Given a very long period of complete stability, the coastline would be reduced to a low coastal plain. This hypothetical
state is not reached in nature, for the shoreline, not far from the
junction of oceanic and continental crust, is normally involved in
upward or downward tectonic movements, in many places in
relation to convergences or divergences of lithospheric plates.
The shorelines of virtually the entire Pacific Ocean are notably
unstable, as are those of the Mediterranean Sea.

Much of the eastern shore of North America has been a subsiding coast over the past 100 million years, as has much of the
shore of the Gulf of Mexico. The long-term sinking of the crust is
thought by some to be characteristic of a continental margin when
it coincides with the trailing edge of a lithospheric plate. Where
the land sinks, sediment tends to build up. These coasts have long,
wide beaches and wide, low-lying plains of relatively recently
deposited sediment behind the shoreline. Typical shoreline landforms along such coasts are sandbars (shallow, narrow ridges of
sand parallel to the beach just offshore); low-lying sandy islands;
and wide tidal flats that are either sandy or muddy. There is a
balance here between the subsidence that creates depressions and
the sedimentation of debris eroded from the land that tends to fill
up the low places. The relative intensity of the two processes
determines the result. A plentiful supply of sediment from the land
combined with only slight subsidence produces a shoreline that is

Figure 12-8
Aerial view of the coastal edge of the New Jersey coastal plain at Barnegat
Inlet, Ocean County, New Jersey. At the bottom of the photo is the shoreline,
in back of which are tidal inlets, sand bars, flats, and salt marshes that are
gradually being filled with sediment. [Photo by W. T. Lee, U. S. Geological
Survey.]

built up and advances seaward. Small amounts of sediment combined with deep subsidence results in deep basins, bays, and other
indentations of the coast.

**Pleistocene Sea-level Changes**   Thus far, we have covered long-
term local or regional movements of the solid crust with respect
to sea level, assuming that the general level of the oceans remains
constant. Studies of the Pleistocene glacial epoch have revealed
that there have been short-term rises and falls of sea level, some
more than 130 meters (425 feet), caused by the removal of water
from the sea to form huge continental glaciers (see Chapter 11).
These world-wide changes in sea level, called **eustatic** changes,

affected all coasts. During periods of lowered sea level, erosion dominated in a situation equivalent to tectonic uplift of the land. Valleys were deepened, and formerly submerged rocky terrains were attacked by waves. As sea level rose again when the ice melted, the deepened valleys were flooded far inland. Today, many of the rivers of the central Atlantic coast gradually widen and become mixed with sea water long before they reach the main shoreline. These long fingers of sea indenting the land are called **drowned valleys**. To some extent, all of the shores of the world were slightly drowned by the sea-level rise that accompanied the melting of glaciers 10,000 years ago. The shores of northern Europe and North America are now rising tectonically because the Earth's crust is still rebounding as a result of being unburdened of the great weight of glacial ice. This, of course, works counter to the sea-level rise. Recent studies show that the sea level from Maine to Virginia has been rising at a rate of 0.25–0.35 centimeters per year; the rate has been much faster than that for some short periods.

With this additional complexity, we now have four major forces shaping shorelines: tectonics, erosion, sedimentation, and eustatic changes in sea level. How erosion and sedimentation interact is most clearly shown by beaches.

## BEACHES

Beaches change constantly, and their dynamics can teach us much about the processes that operate at shorelines. The most obvious things about a beach are the materials of which it is made — sand and pebbles — and the action of the waves. Perceptive observation of these two gives a picture of the beach as a dynamic environment, not a fixed, static landform: the waves, in constant motion, move the material of the beach, acting sometimes to destroy the beach and sometimes to broaden and extend it. If we stop to consider the waves, we can begin to see how they work.

**Waves**  The best place to start looking at waves is from the rise in elevation that is behind every beach. Best is a high bluff, but a low sand dune will do. Once the first sensation of the whole sweep of the shore starts to fade, you can notice some of the elements of regularity about the waves. Waves appear to form some distance from the shore, build up, and break into splashing **breakers**, or **surf**, near the shore. The breakers, separated by troughs, succeed one another in an orderly way. Some distance from the shore, the waves can be distinguished only as low, broad, regular, rounded ridges, called **swell**. The swell becomes higher as it approaches the shore and assumes the familiar sharp-crested wave shape. The crest builds up even more as the wave rolls closer to the shore, finally to a steep high wall of water that then breaks forward in

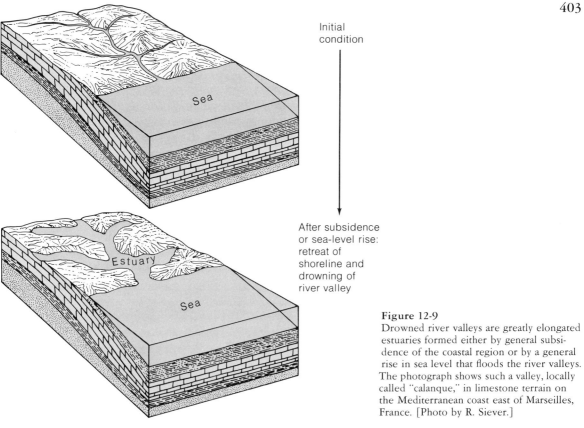

Initial condition

↓

After subsidence or sea-level rise: retreat of shoreline and drowning of river valley

**Figure 12-9**
Drowned river valleys are greatly elongated estuaries formed either by general subsidence of the coastal region or by a general rise in sea level that floods the river valleys. The photograph shows such a valley, locally called "calanque," in limestone terrain on the Mediterranean coast east of Marseilles, France. [Photo by R. Siever.]

**Figure 12-10**
Effect of wind of force 8 (Beaufort scale; see Table 10-1) on the surface of the sea. Wind speed, 65 km/hr (39 mi/hr); wave period, 6 sec; wave height, 5 m (17 ft). [From Atmospheric Environment Service — Environment Canada.]

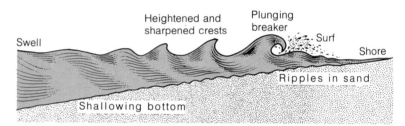

**Figure 12-11**
Formation of a plunging breaker as swell approaches a gently sloping beach. As low regular swell meets a shallowing bottom, wave crests become higher and steeper until the breaker plunges.

a collapse of splashing surf. The time interval between waves, the **period**, may be as short as a few seconds or as long as 15 or 20 seconds. A closer watch on the wave period may reveal more complex regularities, such as a few small waves with short periods alternating with larger waves with long periods. An experienced surfer learns the particular sequence of waves on a beach on any given day and knows that it changes from day to day.

Still looking at the shore from some height, you can distinguish a regular horizontal pattern to the crests of waves as they approach the shore. Far from shore, the lines of swell are parallel to each other but usually at some angle to the beach. As the waves approach the beach, no matter what the direction or angle, the parallel rows of waves start to bend gradually, so that they are approaching the shore in a direction that is more nearly at right angles to the shoreline; then they finally break into surf and sweep up on shore almost, but rarely exactly, at right angles (that is, the waves themselves lie nearly parallel to the shore when they break.

How do these observations fit together to give an explanation for wave patterns? First we must consider the movement of a piece of wood or other light material floating on the water. The piece moves a little forward on a wave crest and then a little backward as a trough between waves passes. The wave form moves steadily toward the beach, but the water itself just moves back and forth. Experiments in large laboratory tanks in which artificial waves are generated show that small floats at different depths in the water all move with a characteristic motion: approximately circular vertical orbits. The orbits have large radii at the top of the water, but they gradually decrease to zero at some depth below. Careful experiments of this kind show a relation between the **wavelength** of the wave pattern, which is the distance from crest to crest, and the orbital motion. At a depth of about one-half the wavelength, orbital motion essentially stops.

Now we can infer what happens as waves approach the shore across a shallowing bottom. At the distance from shore at which the bottom is only one-half the wavelength from the surface, the orbital motions of the lowest levels of the water start being restricted because the water can no longer move vertically:

**Figure 12-12**
Wave refraction along a long, straight, sandy beach on Cape Cod. The view is north along the north-south beach; the waves in deep water approach from the southeast and bend to the north as they come close to shore. [Photo by R. Siever.]

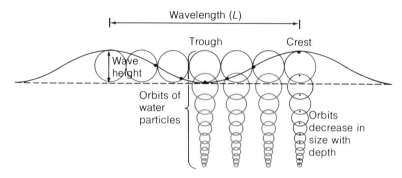

**Figure 12-13**
Wave forms are produced by orbital motions of water particles; each water particle continues orbiting about the same position while the wave form travels.

right next to the bottom, the water can only move back and forth horizontally. In the next higher level, the water can move vertically just a little, this motion combining with the horizontal motion to give a flat elliptical orbit. Higher in the water, the orbits become more circular the farther they are from the bottom.

It is the flattening of the orbits that makes waves "feel" the bottom, for the distortion of the orbits into ellipses causes the whole wave to decrease its velocity. While the wave slows down, its period remains the same—the waves keep coming in from deeper water at the same rate. As a consequence, the wavelength shortens, because all of these waves must follow the fundamental law of motion:

$$\text{Wave velocity} = \frac{\text{Wavelength}}{\text{Wave period}}$$

which is usually abbreviated to $V = L/T$. A further consequence of this change is that the wave grows higher and steeper and the wave crests become sharper. The wave **steepness**, which is defined as the ratio between its height and the wavelength, increases to the point at which the water can no longer support itself, at a ratio of about 1 to 7, and the wave breaks with a crash. The depth of the water at this point is about 1.3 times as great as the wave height. The distance from shore is highly variable, depending on how rapidly the bottom shallows. On a steeply sloping bottom, many small waves will break right at the shore. Sandbars and gently sloping bottoms will cause the waves to break farther out. Thus, the location of the **surf zone**, where the waves break, is a good guide to the depth of the water.

Now contrast this picture of regularity with what you see at the beach during a storm or a strong wind. Though the waves still break on shore with some semblance of regularity, the whole

The whitecaps seen in strong winds far from shore in deep water are not the same as breaking waves. They are caused by the wind blowing off the top parts of wave crests.

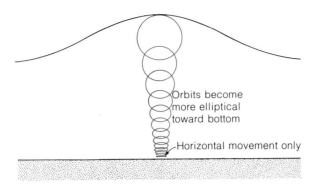

**Figure 12-14**
Orbits of water particles become elliptical as they approach a shallow bottom. At the bottom, particles move back and forth only.

scene is much more wild and confused. In deeper water, whitecaps are everywhere, the surface of the water is rough with ripples, or little waves on big ones, and no distinct wave pattern can be seen. We can find the key to all by watching a calm sea surface as still air is freshened by a breeze. Ripples, tiny waves less than a centimeter high, take shape as the breeze comes up. As the wind increases, the ripples grow into full-sized waves, the size depending on the speed of the wind, how long it blows in any one direction, and the distance that the wind blows over the water, which is called the **fetch** of the wind. The energy of motion of the wind is thus transferred to the water, much in the way that a child's energy is transferred to a toy automobile—the longer and harder he pushes it, the faster it goes.

Storms make irregular choppy patterns of large waves that radiate outward from the storm area, just as dropping a pebble into a still pond makes ripples that move out as ever widening circles. As the waves leave the storm area, they become more regular swell of lower height, which can travel hundreds of kilometers. Some such swells have been shown to cross the width of an ocean. Now we can begin to see why waves of different heights and periods break against the shore. Let's return to the pebbles dropped in a pond: Dropping two pebbles in different places in a pond will produce ripples that cross and interfere with each other in some places but that, in areas far from the source, will appear to merge into two different, approximately parallel sets. In a similar way, two storms will produce waves of different sizes and periods that interfinger as they approach the shore. Timing the periods and sizes of waves at the shore can give you a rough idea of the many storms near and far that generated the swell. The Pacific shores of North America are far better for surfers because waves from the many storms in various parts of

Storm waves can build to awesome heights. Sea captains have estimated wave heights greater than 25 m (80 ft) in hurricanes. One giant wave was estimated, by careful sighting by an officer of a U.S. Navy ship, to have been about 35 m (115 ft) high!

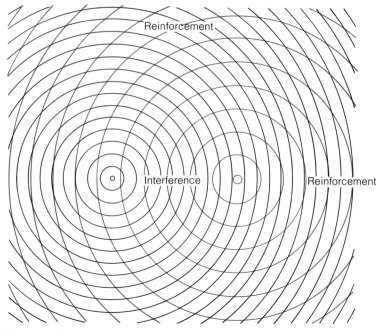

**Figure 12-15**
Waves from two storm centers interfere and reinforce each other, just like ripples from two pebbles tossed into a pond. At some distance from the storms, the waves will appear to come from the same general direction but with different wavelengths.

the great expanse of the Pacific Ocean are reinforced by prevailing westerly winds as they roll eastward. Surf along the Atlantic shore is smaller because the westerly winds blow against the advancing waves and decrease their effect. The best surfing on the Atlantic is on its eastern shore—along France's coasts, for example.

**Wave Refraction** There is one more piece of the story to be explained: those parallel lines of waves that bend toward the shore as they approach. As a wave approaches the shore at an angle, the parts closest to shore will feel the bottom first, the circular orbits becoming elliptical, and so that part will slow down; each successive segment along an individual wave crest meets the shallowing bottom and also slows. Meanwhile, the segment closest to shore has moved into even shallower water, and thus slows even more. In this way, not by separate segments but in a continuous transition along the wave crest, the line of the wave bends as it slows, in a process called **wave refraction** because of its similarity to the bending of light rays in optical refraction. The process can be compared with ranks of marching soldiers turning a corner, the ones closest to the corner marching most slowly and the farthest ones most quickly.

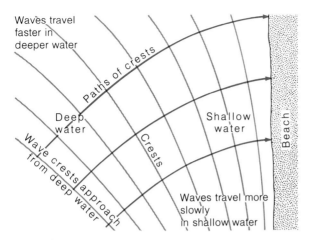

**Figure 12-16**
Wave refraction, the bending of wave crestlines as they
approach a shore from an angle. The part of the wave that
first encounters shallow water slows in speed while that
part of the wave still in deeper water continues to move
with its original speed, making the wave change angle, just
like a rank of marchers wheeling around a corner.

Wave refraction produces some special effects on an irregular
shoreline with indented bays and projecting headlands. Around
the headlands, the water shallows more quickly than the surrounding deeper water on either side, and the waves are refracted—
bent—towards the projecting part of the shore from both sides.
The waves converge around the point of land and expend propor-
tionately more of their energy breaking there than at other places
along the shore. Thus, erosion by waves is concentrated at headlands and tends to wear them away more quickly than along
straight sections of shoreline. Conversely, wave refraction operates
in a bay to make the waves diverge and expend less energy: the
waters there are deeper, so the waves are refracted to either side
into shallower water. The minimal wave energy expended along
the inner shores of bays makes the water quieter there and
which makes them secure places for mooring ships. It is in bays
that beaches form on an irregular shoreline of headlands and
indentations. To explore more of the interactions of waves and
beaches, we should take a closer look at how beaches are affected
by the waves.

**Longshore Drift and Longshore Currents** One of the more
effortless kinds of observations the student of beaches can make
is of the manner in which the waves roll up onto the sloping front
of the beach, the **swash**, and fall back down again, the **backwash**. The swash of a wave has enough current competence to
lift up sand grains and carry them along, in the same way that
river currents do. Strong waves are competent to lift up and move
large pebbles and cobbles. The backwash carries particles back

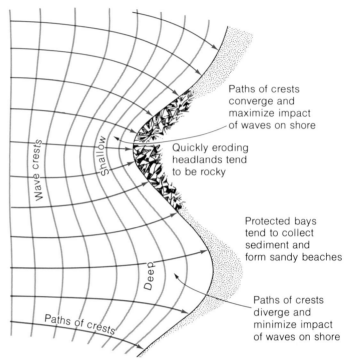

**Figure 12-17**
Wave refraction around a headland and bay. Because of the configuration of shallow and deep water around projections and indentations of the shoreline, wave energies are concentrated at headlands and dispersed at bays.

down again to about their original positions on the slope. Because, despite wave refraction, waves usually break at some small angle to the shoreline, the swash moves up the beach slope at a small angle. But backwash runs down the slope at right angles, down the steepest slope available to it. Thus, the sand grain in one swash-and-backwash cycle is displaced along the shoreline by a small amount. The net result is a zig-zag path in a direction along the shore that is determined by the wave direction. This **longshore drift** of sand on a beach, which is an imperceptibly slow movement to the observer of an hour or a day, has been demonstrated by experimenters using a tracer—slightly radioactive sand grains (though not radioactive enough to be dangerous to anyone).

Coupled to longshore drift is the **longshore current,** which is induced by waves approaching the shoreline at an angle: the water is transported along the beach, just as the sand grains are, by the combination of swash and backwash. The water transport creates the current, which is strong enough to carry sand grains along in shallow offshore zones. Longshore currents build up with increasing distance down the shore, and the water imperceptibly

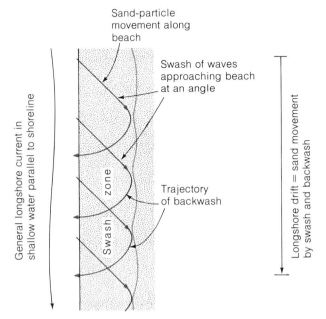

**Figure 12-18**
Longshore drift is the zig-zag movement of sand grains
thrown up on the shore by waves approaching at an angle
and washing back out at approximately right angles to the
beach, thus causing the sand to progress down the beach
in the direction toward which the waves are angled. Longshore current in the shallow water is created in the same
way, by the zig-zag movement of water.

piles up — the difference in height is too small to see or measure —
until a critical point is reached. There, the water breaks out to
sea at right angles to the shoreline, through oncoming waves, to
form a rapidly moving **rip current**. The combination of angular
waves, longshore current, and rip current makes a closed loop
along which the water continually moves. A long beach will
typically embrace many such loops.

The spacing, intensity, and precise location of the rip currents
are controlled by wave angle, wave height and period, and the
topography of the beach, but the relation between these factors
is so complex that exact predictions are very difficult to make.
Rip currents are dangerous to swimmers and surfers, however,
and a knowledge of their characteristics may help to save lives.
They can flow at a velocity more than 1 meter per second, too
strong a current for inexperienced swimmers or children to fight
against. The best way to get out of them was pointed out by Francis
Shepard, an American marine geologist. (Shepard was one of
the first to recognize the significance of rip currents and to stop
calling them by the colloquial name "rip-tides," for they have

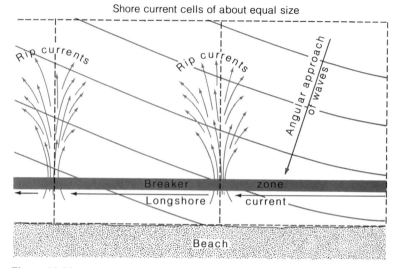

**Figure 12-19**
Rip currents flow out from the shore at certain points, usually regularly spaced, so that the shore can be divided into cells of water circulation along the beach. As a result of oblique wave attack, which forms the longshore current, the water tends to pile up, forming a rip current that breaks through the surf zone, fans out, and becomes dissipated.

nothing to do with tides.) Because they are narrow currents, the best way to avoid being carried out to sea is to swim parallel to the shore and not fight the current at all. In a short distance, the swimmer will be out of the current and can easily swim with the waves into shore.

It is best, of course, to avoid rip currents entirely. They can usually be detected by the lack of breaking waves, for the current both erodes a deeper channel and hinders the advancing waves. A good place to watch for them is at a bend or slight indentation of the beach where the water seems deeper than usual.

**The Beach Budget** If sand constantly leaves the beach by longshore drift and current, why doesn't the sand eventually disappear? Though this does not happen on natural beaches, it may very well happen when naïve engineers interfere with nature in certain ways. To understand why, we have to make an input–output budget for the beach, averaged over a significant time, such as a year. Because wind directions change from day to day, the drift may go one way one day and reverse itself the next as the waves approach from the opposite direction. On a few coasts, the net movement for the whole year will be zero; on most, however, there is a prevailing direction, and that will be the direction of net movement of sand. In a bay, where waves diverge because of refraction, the sand is transported from both sides to the central beach, where it accumulates. In contrast, any sand at a headland is soon transported away from it, and bare rocky shores are the rule in such places.

What happens at the end of a beach, as where a straight shoreline turns a sharp corner? Any sand in longshore transport will be carried off the end of the beach and into the deeper water beyond, where it gradually builds up a submerged bar. With continued growth, it rises above the surface and so extends the beach as a narrow **spit** of sand. From the shapes of spits and bars, a geologist can infer the direction of longshore transport; from that, he can predict the predominant directions of longshore currents and, thus, the average wind and wave directions. Favorite resorts in many parts of the world are on sand beaches that have been extended for great distances by longshore currents.

There is another output. Any beach will "leak" sand to the deeper water offshore, mainly because of rip currents and intense storms. Once there, it is rarely moved, for only the highest waves can stir up sand that lies more than 10 meters deep. This sand, therefore, is permanently lost by the beach. In shallow zones of the beach, sand may move in and out from the shoreline in a pattern related to wave activity. In times of intense storms and

**Figure 12-20**
The beach budget is a complex balance between erosion and sedimentation in which such forms as spits and bars grow as beach cliffs and other sources supply sand. If input is out of balance with output, the beach will tend to grow or erode.

| Input | Output |
|---|---|
| Cliffs back of shore eroded by waves | Sedimentation in deep water by rip currents and waves |
| Beach eroded by longshore drift and current | Growth of beaches (including spits and bars) downcurrent due to deposition of sand by longshore drift and current |
| Rivers upcurrent eroded by longshore drift and current | Growth of backshore dunes due to deposition of sand by offshore winds |

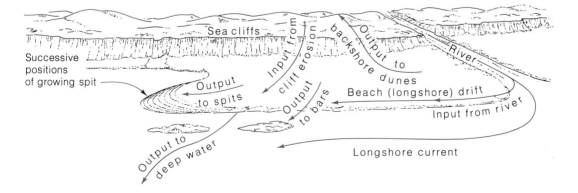

high waves, much of the beach above the normal high-water mark may be removed to the submerged nearshore area. In periods of relatively calm weather, that sand will gradually be shifted back up to the beach, and the beach will widen. A typical seasonal fluctuation of this kind is shown by some beaches that narrow during the stormy winter period and broaden during quiet summer weather.

If sand moves down the beach and extends it, there must be an input of sand to balance this output. Erosion of cliffs and headlands supplies some of the material, but much of it is brought to the shore by rivers. A river tends to drop its sediment to form a delta as it meets the ocean. Where longshore currents are strong, however, deltas never get a chance to form, because the sediment is quickly distributed to beaches, to bars, and offshore along the coast. Where sea cliffs are of sandy sediment, much replenishing sand is produced by erosion. Because cliffs of harder rock are broken into sand too slowly to keep pace with longshore transport, beaches do not readily form downdrift from them.

Beaches export sand in another direction: landward to form sand dunes. Sea breezes and winds blow sand inland to form the dune ridge common to most beaches. That sand may be blown many meters high and accumulate as thin dunes even on top of high bluffs. We can summarize a simplified budget of a beach, or any segment of it, as follows, neglecting short term fluctuations from beach to shallow water during storms and calm periods: If input exceeds output for some months or years, the beach will grow; if output consistently dominates over input, the beach will shrink, sometimes to the vanishing point.

In the natural state, beaches tend to maintain a balance between input and output. But most beaches in populous sections of the country are not in the natural state: people build cottages, pave beach parking lots, erect sea walls to "protect" the beach, and construct piers and breakwaters. The usual consequence, unless the engineers responsible are aware of the dynamics of the beach, is to make the beach shrink in one place and grow in another. The classic example is a narrow pier built out from shore at right angles to it. Much to the surprise of the builders, the sand, during the following months and years, disappears from one side and enlarges the beach greatly on the other. This is an expensive way to find out that there is a longshore current on that beach. The current and drift bring sand toward the pier from the upcurrent direction (the dominant wind direction, usually) and, stopped at the pier, dump the sand there. On the downcurrent side of the pier, the current and drift pick up again and erode the beach. On this side, however, there is no replenishment of sand by the current because it is blocked by the pier, so the beach shrinks and disappears. If the pier is removed, the beach will relax to its original state. It is important to learn that a beach is a dynamic form, and

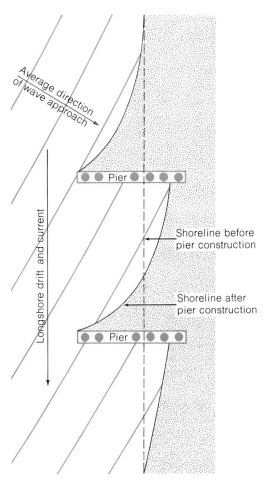

**Figure 12-21**
Construction of piers (frequently called "groins")
along a shore to control erosion of a beach may
produce unwanted changes if longshore current and
drift are not considered. A typical change induced
is erosion downcurrent of the pier.

any disturbance of its input–output balance, by altering either the
source of material or the transport mechanism, will affect it in
predictable ways. Thus, construction on (or other alteration of) the
dunes or bluffs at the back of a beach may strongly affect the width
of the beach by changing its supply of sand.

One of the consequences of oil spills from tankers offshore is
the fouling of beaches by heavy oils that gradually alter to tarry
or asphalt-like lumps and layers. The oil seeps into the pores of
the sand and binds it together. Temporarily, the beach may be
a disagreeable mess, but sooner or later, because a beach is constantly in motion, it will cleanse itself. The oil lumps are broken

**Figure 12-22**
Beach at West Falmouth, Massachusetts, a few days after an oil spill not far
offshore. A year later this beach had returned to its original state, but buried
tarry layers could be found at places on the upper beach. [Courtesy of Woods
Hole Oceanographic Institute.]

up, dispersed, and mixed with the immense volume of sand on
the beach. The problem is that the process may take many months
or years, which makes the beach's ability to cleanse itself a small
consolation to the animals and people that populate it in the
meantime.

## TIDES

One subject that philosophers and scientists of ancient Greece did
not cover was the tide, the daily rise and fall of the sea. Ancient
observers of the sea in places as widely separated as China and
Iceland, however, devoted much thought to the subject. The
Greeks were observant enough—they just happened to live on a
large inland sea with almost no tide, whereas the people of civilizations that developed along oceanic coasts were familiar with the
daily alternation of high and low water. By modern times, it had
long been known that there was a relation between the position
and phases of the Moon, the heights of the tides, and the time of
day at which the water would reach its highest level. It was not
until Isaac Newton formulated the laws of gravitation, however,
that the tides were understood to be the result of the pull of the
Moon and the Sun on the oceans.

The Earth and the Moon attract each other strongly with a gravitational pull that is slightly greater on the sides of the two planets closest to each other. That attractive force causes a slight bulge in the solid Earth, the oceans, and the atmosphere (and the Moon, too, of course). The deformation in the solid Earth is too small to be observed, except by sensitive instruments, but the much larger bulge in the water is easily seen as the tide. On the side away from the Moon, the gravitational attraction for the ocean water at the surface is at a minimum. There is a greater attraction for the solid Earth underneath, which, in turn, is closer to the Moon than the water. Thus, the solid part of the Earth is pulled more towards the Moon than the water on the far side, and the water appears to be pulled away from the Earth as another bulge. As the Earth rotates, the tides move around it, one always facing the Moon, the other directly opposite.

The Sun, though much farther away, has so much more mass that it, too, causes tides. The Sun tides are a little less than half the height of Moon tides. The two sets of tides are not synchronous, those related to the Sun coming every twenty-four hours, once each "solar" day. The time of rotation of the Earth with respect to the Moon is a little longer than the solar day: it is twenty-four hours and fifty minutes, because the Moon is moving around the Earth. In that twenty-four hours and fifty minutes, the "lunar" day, there are two high waters, with two low waters in between.

When the Moon, Earth, and Sun line up, the combined gravitational pull of Sun and Moon reinforce each other and produce

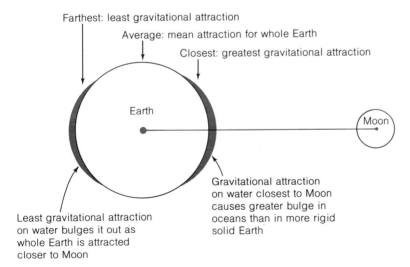

**Figure 12-23**
Moon's gravitational attraction causes two bulges of water on the Earth's oceans, one on the nearest side and one on the farthest side. As the Earth rotates, the bulges always face the Moon. Thus two high tides (bulge closest to Moon and bulge farthest from Moon) pass each point on Earth's surface each day.

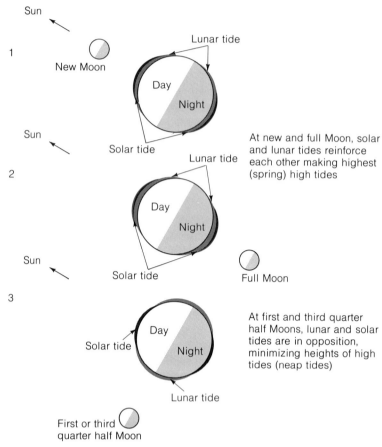

**Figure 12-24**
The relative positions of Earth, Moon, and Sun determine the heights of high tide during the lunar month. The highest high and lowest low tides (spring tides) come at new and full moons; the lowest high and highest low tides (neap tides) come at first- and third-quarter half moons.

very high tides, the **spring tides**. Such high tides come every two weeks at full and new Moon. The lowest tides, the **neap tides**, come between, at first- and third-quarter Moons, when the Moon and Sun are at right angles to each other with respect to the Earth. The heights of the tides are very different in various parts of the oceans. Because the oceans are of various shapes and sizes, the water of the tide responds in complex ways. It is a bit like connecting a great number of large and small pots and bowls in a complicated pattern and sloshing water back and forth between them. Thus, in Hawaii, the difference between low and high water is only about 0.5 meter; at Seattle, however, it is about 3 meters (10 feet). Extraordinary tides occur in a few places, such as the Bay of Fundy, where the tidal range can be more than 12 meters

(40 feet). Inland lakes, such as the Great Lakes, are virtually tideless. Tides are calculated for years ahead with great precision by computers for many of the world's coastlines, especially those with important commercial shipping and harbors.

Along shallow coasts, tidal movements give up energy through friction of the water with the sea floor—energy that must ultimately come from the rotation of the Earth and Moon. That frictional loss is enough to slow the rotation of the Earth by a very small amount. The average length of the day has been calculated to have lengthened by 0.001 second in the past 100 years. That may seem negligible, but over many millions of years of geologic time, that amount of slowing can mount up: The Earth must once have been rotating much faster, while the time of its revolution around the Sun was unaffected. The Moon's rate of revolution about the Earth would have been faster, and the Moon would have been closer to the Earth. That means that the tides would generally have been much higher, that there were many more days in a year, and that the days were shorter.

Striking support for this last idea came from the world of fossils. Corals, tiny marine animals that secrete calcium carbonate and make modern coral reefs, lay down a microscopically thin layer of calcium carbonate each day; the layers laid down in summer are thicker than those laid down in winter. Careful counting of these layers in fossil corals has convinced many paleontologists that, 400 million years ago, there were nearly 400 days in a year. Evidence of stronger tides is not so good, but some geologists who study tidal sediments of the past think they can see the effects of stronger **tidal currents**, the rapid movements of water in shallow parts of the oceans.

**Tidal Currents**  The movement of tidal waters near shorelines causes tidal currents that can reach speeds of a few kilometers per hour. As the tide rises, the water flows in toward the shore as a **flood tide**, moving into shallow coastal marshes and up small streams. As the tide passes the high stage and starts to fall, the **ebb tide** moves out, and low-lying coastal areas are exposed again. Such tidal currents cut channels through **tidal flats**, which are muddy or sandy areas that lie above low water but are flooded at high water. Tidal flats may be narrow strips seaward of the beach, or they may be extensive areas covering hundreds of square kilometers. When the tide advances on some wide tidal flats, it may move so rapidly that areas are flooded faster than a person can run. The tidal-flat beachcomber is well advised to consult a local tide table before wandering.

Tidal flats may be separated from the open sea by long sand bars or **barrier islands** parallel to the shoreline. The islands are bars of sand that have been built up above the high-water level and are stabilized by vegetation. The barrier islands or bars are broken

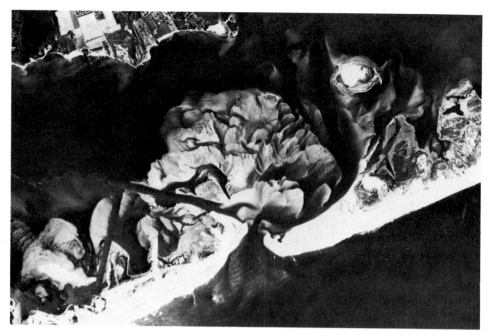

**Figure 12-25**
Aerial view of a barrier island (white) that was breached by a hurricane in September 1947, after which a lobate tidal delta formed between the island and the mainland. [Courtesy of M. M. Nichols, U.S. Dept. of Agriculture.]

by occasional tidal channels through which the water rushes in a strong current. Outside these channels, tidal currents may create tidal deltas of sand transported from the tidal flat during ebb tides.

## CONTINENTAL SHELVES

Bordering the North American coasts of the Atlantic and the Gulf of Mexico are broad, shallow sea-floor platforms that, though submerged, are clearly parts of the continental mass. These platforms, and similar ones around other continents, the **continental shelves**, extend from the edges of the continent to a depth of a few hundred meters, where they give way to steeper slopes that go down several thousands of meters to the main ocean floor. For most of the first century of ocean exploration, marine geologists were satisfied to say that continental shelves extended to depths of about 200 meters (650 feet), where, for the most part, the continental slopes started. Then, growing interest in the economic uses of the sea, from fishing to mining, provoked international legal discussions of the territorial rights of nations to the continental shelves. Marine geologists, finding that the old definition

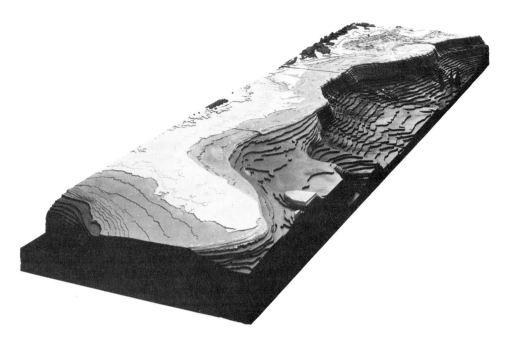

**Figure 12-26**
Two views of a three-dimensional model of the Atlantic continental shelf, slope, and rise of part of North America. (Top) View looking northwest; Florida is in the foreground, Nova Scotia and part of the Grand Banks at the far edge. (Bottom) Close-up of the section from Nova Scotia at right to New York and Long Island at left. Near the left margin is the Hudson submarine canyon, and in the middle foreground is the area of Georges Bank, off Cape Cod. [Courtesy of K. O. Emery and E. Uchupi, Woods Hole Oceanographic Institution.]

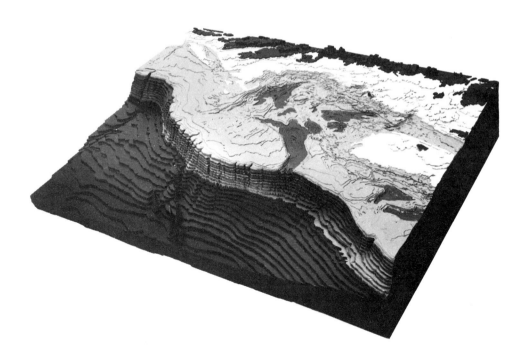

was too arbitrary, began to be more precise: in a United Nations' report of 1957, they recommended a 650-meter depth limit, one that would include all of the world's continental shelves. Legal questions sometimes impose themselves on scientific matters in this way.

Continental shelves vary widely in width, but they average about 65 kilometers; the general slope of the surface is slight, the angle of inclination averaging only 0°07'. Most of the surface is fairly uneven, with small hills and ridges alternating with basin-like depressions, broad valley-like troughs, and occasional narrow, steep-walled valleys called **submarine canyons**. Most areas of the continental shelves were above sea level during the Pleistocene glaciations and, as we noted in Chapter 11, much of erosional and sedimentational character of their surface was formed then. Waves and tidal currents acting on the shelves since the last glaciation have modified the surface, mainly by depositing sand near the shoreline and silts and muds in the deeper water farther from shore. Pleistocene relic sands may be found at shallow depths far out on the shelf, even near the shelf edge, as on Georges Bank southeast of Cape Cod, one of the famous fishing grounds of the Atlantic. These sands are now being dispersed, redistributed, and altered by modern processes. There are about 400 billion tons of this sand in the top 3 meters (10 feet) of sediment off the northeastern coast of the United States, enough to supply the needs of nearby areas for thousands of years, a valuable sea-bed resource.

Continental shelves may be narrow and broken by isolated deep basins and high ridges or escarpments, as off the coast of California, where water hundreds of meters deep may be within sight of land. This coast bears all of the topographic earmarks of a tectonically disturbed area, which is in keeping with the highly faulted and deformed coastal belt just above the water. Some narrow shelves, however, show little or no evidence of tectonic disturbance. One of the best known of these is off the east coast of Florida, where the powerful Gulf Stream, a rapidly flowing current of water, runs close to shore.

The structure of broad, gently sloping continental shelves was worked out mainly by geophysicists at the Lamont Geological Observatory of Columbia University during the 1950s. They sent strong sound waves down through the water, which penetrated the sediments and were reflected back to their oceanographic research ship (see p. 000, Chapter 5). From the way in which the sound waves were reflected by layers of sediment and rock, they were able to deduce the structure of the shelf: a pile of sediment several kilometers thick built out from the continent. The suspicion that the shelves had such a structure had early been stimulated by speculation on the narrowness of the Florida shelf near the Gulf Stream. There, it was thought, the current in the sea was

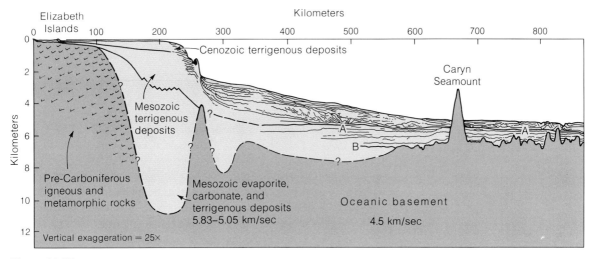

**Figure 12-27**
A profile of the Atlantic continental margin of North America off southern New England based on seismic data. Stratification, based on continuous seismic profiling, is shown diagrammatically for the younger sediments. [From K. O. Emery and E. Uchupi, *Atlantic Continental Margin of North America*. Copyright 1972 by American Association of Petroleum Geologists.]

so fast—up to about 10 kilometers per hour—that the sea floor was kept scoured clean of sediment. Farther north, where the Gulf Stream moved out to sea, the continental shelf widened and was discovered to consist of sedimentary layers.

In the last few years, the implications of plate-tectonic theory for continental-shelf development have been assuming increasing importance. According to some of these ideas, the continental shelves of the Atlantic Ocean type are thought to have originated as the trailing edges of continental margins on plates that broke apart and have been moving away from each other since about 180 million years ago, when the Atlantic Ocean started opening. As the edge of North America moved away from the newly formed mid-Atlantic ridge, it subsided below sea level, and erosional debris from the continent accumulated on its seaward edge. As the Atlantic widened, the shelf, slope, and rise gradually built out from the continent.

Narrow, tectonically deformed continental shelves, such as those of the Pacific, are characteristic of continental margins that coincide with the leading edges of lithospheric plates where they terminate at transform faults or subduction zones. In that tectonic environment, continental shelves accumulate sediments, but they are likely to be deformed as well as affected by volcanism and metamorphism. Much remains to be learned of these kinds of coasts, where sedimentation in deeper water plays a relatively more important role.

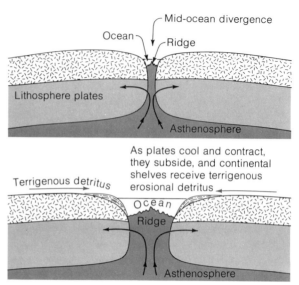

**Figure 12-28**
As two continental masses on lithospheric plates diverge
at a mid-ocean ridge, the plates cool and contract, and their
trailing edges subside, the newly formed continental shelves
receiving terrigenous erosional detritus from the continents.
See also Figure 21-13.

## CONTINENTAL SLOPE AND RISE

Continental shelves are abruptly terminated by the **continental
slopes**, which have an average inclination of about 4°—a drop of
about 70 meters in a horizontal distance of 1 kilometer. That is a
healthy slope, but not exactly a steep cliff, as one might gather
from diagrams (such as Fig. 12-27) that greatly exaggerate the
vertical scale. The sediments of most slopes are muds and silt,
derived from continental erosion, that have been carried across
the shelf and draped over the edge. Like the surface shelf sediments, much of the uppermost layers of mud of the slope was also
deposited when sea level was lower during the Pleistocene
glaciations. Geophysical exploration of the slopes below broad
continental shelves reveals inclined layers of sediment that have
been deformed a little by slumping and sliding down the incline.
In contrast, continental slopes of tectonic coasts may be steeper,
and older rock may be mantled with only a thin veneer of sediment. Rock outcrops are common on such slopes.

The lower parts of many continental slopes become gentler
and merge into a more gradually sloping apron of sediment extending into the main ocean basins. These aprons are the **continental rises**. Their average inclination is less than half a degree,
only about an eighth of that of the continental slope; but they
extend for hundreds of kilometers, whereas continental slopes

have an average width of only 20 kilometers. Rises are fairly smooth topographically, being broken only by an occasional seamount, submarine canyon, or channel. The rises are kilometers-thick piles of muddy, silty, and sandy sediments that have been carried from the continental shelf, down the slope, finally to come to rest on the deep-sea floor.

The continental slope is typically gullied by many valleys and interrupted by occasional submarine canyons. The appearance of the slope suggests erosion, as do the canyons, but we cannot draw again on the Pleistocene lowering of sea level for an explanation, because even the wildest estimate of sea level lowering would not envision the drop of thousands of meters that would be necessary to expose some of these slopes to land erosion processes. The erosion must have a marine explanation. The explanation came in a rush of developments around 1950, when a hypothesis about a previously unknown type of sedimentation was proposed. The hypothesis was largely accepted after stormy debate among sedimentary geologists that rivalled more recent debates about plate tectonics. The hypothesis proposed a new type of current, called **turbidity current**, and its characteristics ultimately helped to explain submarine canyons, the gullies on continental slopes, the formation of continental rises, and the flat plains that cover large areas of the deep-sea floor, the **abyssal plains**.

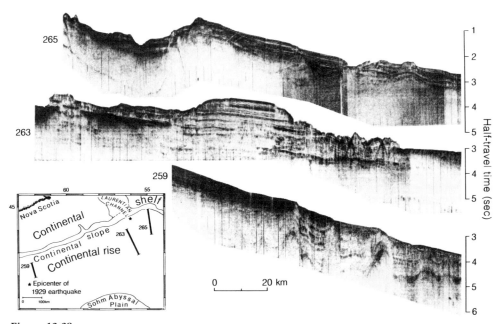

**Figure 12-29**
Bathymetric and seismic profiles of the continental rise off Nova Scotia, showing details of topography and structure that are characteristic of slumped and deformed areas where the cable breaks of the Grand Banks turbidity current occurred. [From K. O. Emery and E. Uchupi, *Continental Margins of North America.* Copyright 1972 by American Association of Petroleum Geologists.]

# TURBIDITY CURRENTS

Strictly speaking, such currents were not new to science in the 1950s. The idea was born in the latter part of the nineteenth century in Geneva, Switzerland, where the Rhone River enters Lake Geneva. The river water is cold and full of suspended clay from the meltwaters of alpine glaciers, and, at that time, when the lake was unpolluted and clear, the river water could be seen through the still lake water above, moving along a channel on the bottom as a distinct muddy current. The river water is denser than the overlying lake water, primarily because of its suspended sediment load and secondarily because it is colder. Being denser, it sinks to the bottom and runs downhill along it. The difference in density also acts as a barrier that slows down mixing of the current with the overlying waters, and so the current maintains itself for a long distance until it reaches the flat bottom of the lake. There it fans out in a thin sheet of muddy water and loses momentum, gradually coming to a stop, the suspended sediment slowly settling out on the bottom as a thin layer. This general kind of current, a **density current**, is known also from oceans, where they can arise because of differences in temperature or salinity alone — colder and saltier water is denser. The name turbidity current is applied to density currents that owe their density to suspended sediment.

**Figure 12-30**
A thick sequence of graded turbidite beds. The lighter, thicker units are sandstones that are coarsest at an abrupt base and grade upward through finer grain sizes into the darker shale beds. Each pair of sandstone-shale units represents one turbidity flow in a deep marine basin. Miocene of the Apennines. [Photo by Paul Edwin Potter. From F. J. Pettijohn, P. E. Potter, and R. Siever, *Sand and Sandstones.* Copyright 1972 by Springer-Verlag New York, Inc.]

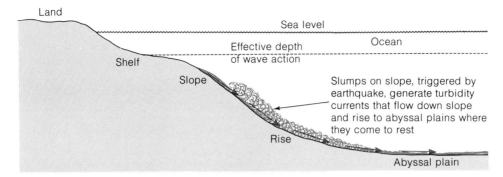

**Figure 12-31**
Typical formation of turbidite currents in the ocean. Slumps on the continental shelves generate turbidity currents that flow downslope to abyssal plains, where they gradually slow to a stop and deposit turbidite sediments.

Turbidity currents became a subject of mild interest in the 1930s as possible eroders of the submarine canyons that were then being extensively mapped. It was not until the late 1940s, however, that they became a focus of activity for marine and land geologists alike. The man who sparked much of that resurgence of interest was a Dutch geologist, *Philip Kuenen*, who combined his experience as an oceanographer and field geologist with a series of important laboratory experiments. Kuenen produced turbidity currents and studied their dynamics in his laboratory by pouring muddy water into the end of a flume with a sloping bottom. He showed how such currents can move swiftly, at many kilometers per hour, the speed depending on the steepness of the slope and the density of the current. Because of its speed and turbulence, a turbidity current was able to transport large quantities of sand introduced with the suspension or picked up by the current as it flowed over a sandy bottom. The sediment layers produced by settling after the current started to lose speed formed a **graded bed** — that is, a bed with coarse material at the bottom, grading upward to fine material at the top. Each turbid flow produced a single graded bed; using a series of flows, Kuenen produced a sequence of graded beds.

The brilliant next step was the recognition of such graded beds on land: thick marine sandstone and shale sequences produced long ago under water as the product of such currents. The final step was tying all this together with what was being learned about the sediments and topography of the sea floor to formulate a hypothesis: Turbidity currents are formed by large slumps of mud and sand draped over the edge of the continental shelf onto the continental slope. The sudden slump or slide, perhaps triggered by an earthquake, throws mud into suspension, creating a dense turbid layer, which then starts flowing down the slope, eroding and picking up sediment as it picks up speed. These currents gully the slope and excavate submarine canyons. As the turbidity current reaches the change of inclination where the slope meets

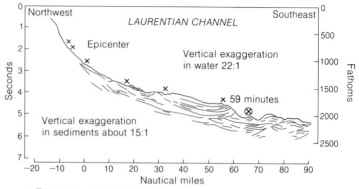

× Telegraph cables broken during earthquake
⊗ Telegraph cable broken 59 minutes later

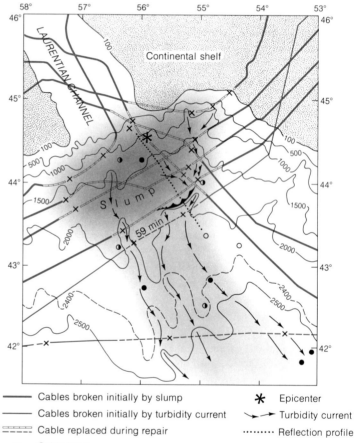

——— Cables broken initially by slump     ✳ Epicenter

——— Cables broken initially by turbidity current     �“→ Turbidity current

〓〓〓 Cable replaced during repair     ⋯⋯ Reflection profile

×   Cable break

Sediment cores containing:     ● Sands and silts
                                   ○ Undisturbed hemipelagic sediments
                                   ◑ Disturbed hemipelagic sediments

**Figure 12-32**
Map and seismic reflection profile of Grand Banks area where cable
breaks signalled the beginning of a slump and turbidity current. [From
B. Heezen and C. L. Drake, Copyright 1964 by American Association of
Petroleum Geologists.]

the continental rise, the current slows, and some of the sediment starts to settle. Many currents continue across the rise, cutting channels as they go, to reach the level bottom of the ocean basin, the abyssal plain, where they fan out and come to rest. The deposits of the rise and abyssal plain are graded beds of sand, silt, and clay; **turbidites** was the name coined for them.

Revolutionary as these ideas were, confirmation soon started rolling in. A series of sudden breaks in trans-Atlantic telephone and telegraph cables on the continental slope and rise off the Grand Banks of Newfoundland had taken place in 1929 following an earthquake. The locations of the breaks were plotted on a map and the times of breaking were noted. A pattern quickly became clear. There was a rapid breaking of cables high on the slope, followed by a sequential rupturing going down the slope and farther from the epicenter of the earthquake. But the breaks downslope occurred much too long after the earthquake to have been caused by earthquake waves. The turbidity-current hypothesis gave the only really plausible explanation: The earthquake provoked a slump that activated a turbidity current so fast and powerful that it snapped cables as it raced down the slope and rise. The current maintained itself for at least 700 kilometers and was calculated to have reached speeds of 40–55 kilometers (25–34 miles) per hour. Sediment cored from the deep sea near the path of cable breaks showed a thin graded bed at the surface, probably the material that settled from the current. Similar patterns of cable breaks on continental slopes in other parts of the world showed up as records of cable companies were searched.

The most compelling evidence of turbidity currents was the distribution of turbidite sediment in the oceans. No other mechanism could as simply or satisfactorily explain how coarse sand particles could be transported to deep waters beyond the reach of waves and tidal currents. How else could graded beds have filled in the abyssal plains, the lowest parts of the ocean basins? The turbidity-current hypothesis was especially attractive because it simultaneously explained the formation of the slope, rise, and plain.

Turbidity currents have not explained everything. Though much is known about the physics of density currents, we have still much to learn about the complex workings of turbidity currents. Turbidity currents may not be the sole agent responsible for excavating submarine canyons in solid, hard rock, and there is also some evidence for the existence of localized deep-ocean currents that can also transport sand. Like other major theories, the turbidity-current idea has become more complicated as it has been worked out in more detail. More recently, however, this important simplifying hypothesis, linked to the history of the oceans and continent edges, has become a vital part of plate tectonic theory and its applications. This has come about as a result of careful consideration of the topographic and geophysical profiles of the oceans.

## THE PROFILE OF THE OCEANS

The Atlantic Ocean is fairly symmetrical about the mid-Atlantic ridge, and is bounded on both sides by continental shelves, slopes, and rises. At the ridge, a high, rough topography of basalt is created by the upward movement of material from the partially molten asthenosphere to create new oceanic lithosphere. Away from the ridge, the topography becomes lower and less rugged, a region of **abyssal hills**, as the lithosphere subsides as a result of cooling. The slopes become thinly covered with the fine sediment that accumulates slowly over all of the ocean floor. As the lithosphere sinks more, it becomes swamped by turbidity-current deposits that bury the varied hilly topography and create a level abyssal plain. Closer to a continent, the plain becomes a continental

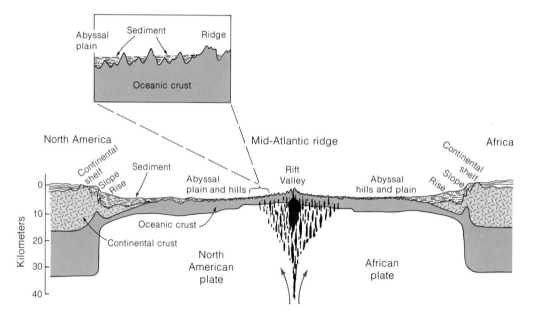

**Figure 12-33**
Topographic and geophysical profiles of the Atlantic Ocean. (Top) Geophysical and topographic cross section deduced from seismic measurements and echo-sounding profiles from New York to the northwest coast of Africa. [After B. C. Heezen, "Physiographic Diagram of the North Atlantic Ocean," p. 102, Figure 49. Copyright 1959 by Geologic Society of America.] (Bottom) Topographic profile of the floor of the Atlantic from Martha's Vineyard (Mass.) to Gibraltar. A portion of an abyssal plain extends from 750 miles to about 1000 miles off the east coast of the United States. The island at 2300 miles is one of the Azores. [From B. C. Heezen, "The Origin of Submarine Canyons." Copyright © 1956 by Scientific American, Inc. All rights reserved.]

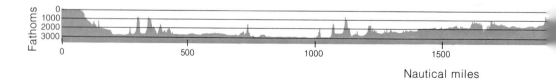

rise, with the original basalt topography buried under many kilometers of turbidite.

How this came to be is the subject of much speculation, a subject we can call "paleo–plate tectonics." It is the same picture we discussed on p. 423. As the Atlantic Ocean opened up, the first sediments deposited in the narrow rift-valley ocean, which was perhaps something like the Red Sea is today, were coarse turbidites flowing into deep waters at the steep, newly broken edges of the continental plates. As the plates separated farther, continental shelves were built up over the trailing edges of the plates as they subsided, and the rise sediments built up to make a thick sediment pile merging with the continental slope. With continued spreading of sea floor, the turbidites began to fill in the low spots on the rough, basaltic, oceanic crust, gradually building up to form the abyssal plains. This conception ties together tectonics and turbidite sedimentation in a consistent theory that explains much of what we know about the Atlantic Ocean today.

The Pacific Ocean is not so easily fitted into such a simple scheme. Continent edges there are narrow, and deep basins may be filled with turbidites very close to shore. Thick accumulations of turbidites are found as submarine fans or subsea deltas where submarine canyons that start near the shore empty onto the deep-sea floor. Turbidite sediments form in trenches, the subduction zones in which oceanic crust is consumed as it turns downward into the asthenosphere. Many of these turbidites are crumpled and folded, scraped off, and plastered against the overriding plate edge. Much of the sediment may not be carried down with the oceanic crust along the subduction zone because the sediment, saturated with water, is much less dense than basalt and tends to "float" at the surface of the lithosphere. Some, of course, may be dragged down, mixed with volcanics and other sediments, and metamorphosed by pressure and heat as it goes down.

We have given emphasis to turbidity currents because they are major sediment eroders and transporters, but they are not the only (or even the major) currents of the oceans, nor are turbidites the only significant sediment type of the deep sea. The general circulation of the oceans, and of the atmosphere above them, is responsible for the patterns of transport of much of the fine-grained sediment of the sea.

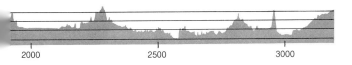

2000          2500          3000

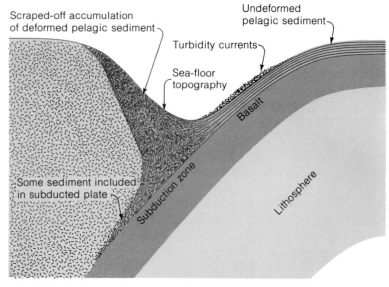

**Figure 12-34**
Sedimentation and deformation of sediment in an oceanic trench-subduction zone. Much of the pelagic sediment and turbidity current deposits in the trench are "scraped off" as the oceanic plate moves down the subduction zone, the sediments being of much lower density than the underlying lithospheric rocks. Some sediment may be carried along with the subducting plate.

Homer warned that the main danger in navigating the Strait of Messina was the whirlpool Charybdis, (now known as Garofalo). The whirlpool lies opposite Scylla (now Scilla), a rock near the Italian shore that, according to Homer, was inhabited by a creature whose name was also Scylla, a six-headed monster with an appetite for sailors. The expression "between Scylla and Charybdis," which is descriptive of situations in which the avoidance of one danger means exposure to another, has survived to this day.

## OCEANIC CIRCULATION AND TRANSPORT OF TERRIGENOUS SEDIMENT

For centuries, mariners have known that there are surface currents in the sea that can either speed a ship on its course or hinder its progress. At least as early as 800 B.C., the Phoenicians and Greeks knew the currents of much of the Mediterranean Sea. Homer's *Odyssey* mentions the treacherous waters of the Strait of Messina between Italy and Sicily. In North America, it was that versatile politician-publisher-scientist Benjamin Franklin who started the serious study of ocean currents by publishing the first chart of the Gulf Stream.

Early investigations of currents made use of drift bottles— floating bottles set free in the ocean with cards inside asking the finder (for a small reward) to return the card with information about the date and place of its recovery from the sea. Modern physical oceanographers have invented instruments to measure directly the speed and direction of currents. Current speed near the surface is frequently measured by rotating vanes attached to a meter that gives the number of rotations per minute. Current direction and current speed at depth are now most commonly determined by means of the Swallow float (named after its inventor, the English oceanographer John C. Swallow), a closed aluminum tube designed to float at any of various levels in the water. The

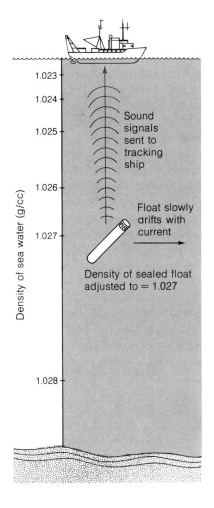

**Figure 12-35**
Devices such as the Swallow float are designed to measure deep currents in the ocean by sending out sound signals as they float with the current. Before the float is lowered, its density is adjusted to that of the water at the depth to be tracked. It then sinks to its equal-density level and begins to drift with the current. The density of ocean water varies from place to place depending on the salinity and temperature.

Swallow float takes advantage of the fact that the water of the oceans becomes denser as the pressure increases with depth: Because the float is denser than surface water, it sinks. At a certain depth, the density of the float, relatively unchanged by pressure, is exactly that of the surrounding water, and there it stays, floating. The operating part of the Swallow float is a sound source— usually called a "pinger" because of the sound it makes—that is picked up by sensitive listening microphones attached to the research ship. Thus, as the float moves in the current, the ship is able to follow its position.

The surface currents of the oceans can be simplified into a pattern of large closed loops, called **gyres** (Fig. 12-37). The main currents are the outside paths of the gyres; there is little movement at their centers. Two subtropical gyres lie on either side of the equator, the northern one rotating clockwise and the southern one counterclockwise. Between the westward-flowing equatorial currents that are parts of these gyres, there is an oppositely directed eastward flow, the Equatorial Countercurrent. North and south

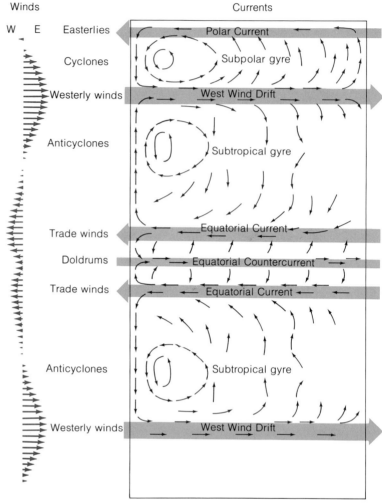

**Figure 12-36**
Idealized ocean, rectangular in shape and subject only to the horizontal wind
forces shown by the broad, gray arrows, would have the circulation patterns
indicated by the black arrows. Approximate relative velocities of surface winds
are indicated at left. [From W. Munk, "The Circulation of the Oceans". Copyright © 1955 by Scientific American, Inc. All rights reserved.]

of the subtropical gyres are smaller subpolar gyres; each of these
rotates in a direction opposite to that of its companion subtropical
gyre. The close correspondence between the gyres and the
directions of the prevailing winds over the oceans shows that the
winds and currents are linked: the winds constitute the primary
driving mechanism for the surface currents of the ocean.

The most famous ocean currents are the north–south sections of
the gyres. The north-flowing Gulf Stream is the strong western
part of the North Atlantic subtropical gyre. The south-flowing
California Current is the weak eastern section of the North

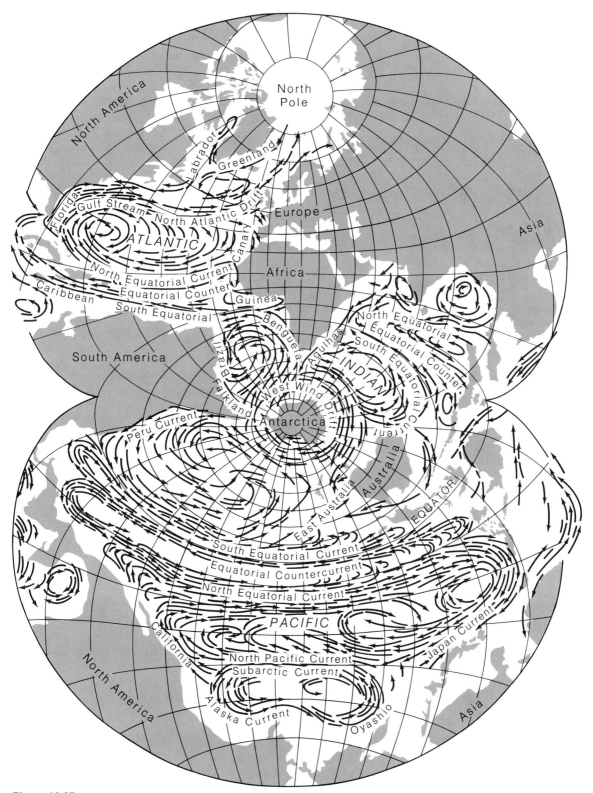

**Figure 12-37**
Major ocean currents shown on a map centered on the South Pole. The major gyres can be identified and compared with the idealized drawing of Figure 12-37. [From W. Munk, "The Circulation of the Oceans." Copyright © 1955 by Scientific American, Inc. All rights reserved.]

## Box 12-1   The Coriolis Effect

Ocean currents, propelled by the wind, are deflected to the right in the Northern Hemisphere and to the left in the Southern Hemisphere as a result of the **Coriolis effect** (also called the Coriolis acceleration). This effect is caused by the Earth's eastward rotation. If a giant cannon fired a ball from the equator toward a pole, the ball would also move eastward. At the equator, 0° latitude, the eastward movement of the surface of the Earth is about 1670 km/hr (about 1050 mi/hr), but this speed falls off as the circumference of the Earth gets progressively smaller with higher latitudes, so that, at 30° N latitude, the eastward speed of the Earth has dropped to 1450 km/hr (about 935 mi/hr). But, because the cannonball is not firmly attached to the Earth and is still moving at the higher rate of speed (if

there is no friction to slow it down), it will be moving eastward relative to the Earth's surface. Thus, an observer at 30° N latitude will perceive the cannonball as curving to the east. If an astronaut on the Moon were to be able to see the cannonball's path, however, he would perceive it as perfectly straight. This is the Coriolis effect. It is quantitatively important only when an object moving over a long distance at some angle to the equator is free to move with respect to the solid Earth. Thus, it does significantly affect ocean currents and atmospheric winds. The major ocean currents, such as the Gulf Stream, are deflected in accordance with the Coriolis effect, as are many more minor currents at and below the surface.

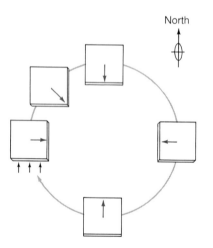

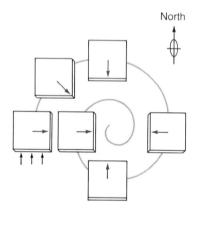

Coriolis acceleration, caused by the Earth's rotation, affects any object moving on the Earth. It is directed at a right angle to the direction of motion (to the right in the Northern Hemisphere). If a frictionless slab is set in motion toward the north by a single impulse (black arrows), the Coriolis effect (colored arrows) moves the slab in a circle. Presence of friction causes the slab to slow down, spiraling in toward the center of the circle in the right-hand part of the figure. A push to the north causes a spiral to the east. [From R. W. Stewart, "The Atmosphere and the Ocean." Copyright © 1969 by Scientific American, Inc. All rights reserved.]

Pacific subtropical gyre. The east–west equatorial currents flow at speeds of 3–6 kilometers (2–4 miles) per day; thus, they flow slowly enough that they warm up to adjust to the hot latitudes through which they flow. In contrast, the Gulf Stream flows more than an order of magnitude faster—40–120 kilometers (25–75 miles) per day—so that it does not have time to adjust quickly to its climatic surroundings. Thus, the Gulf Stream remains a warm current in the cold North Atlantic, only slowly transferring heat to the colder air above it. Such "rivers of the ocean," as the Gulf Stream has been called, contribute significantly to heat transfer from one part of the globe to another. Northern Europe would be almost uninhabitably cold, were it not for the mitigating effect on climate of the North Atlantic gyre.

The prevailing winds can cause vertical movements in the sea, sometimes causing deeper waters to move up to the surface—**upwelling**—or surface waters to move to lower depths—**sinking**.

A typical pattern of upwelling results from prevailing offshore winds, which tend to blow surface waters out to sea. When that happens, deeper waters move upward to replace the surface water, and a circulation loop is formed, with the return of surface water to replace the deep water completing the loop farther out to sea, in a usually diffuse general flow. Most deeper waters of the ocean are colder than those at the surface; therefore, when they reach the surface, they cool the air, which causes fog to form. This is what causes the foggy summer weather familiar to people who live along much of the Pacific coast of North America. Sinking of surface water can be induced by onshore winds. As the surface waters pile up against the coast (not something you can see with the naked eye), they sink. Upwelling and sinking near coasts, in most places, affects the water to a depth of about 100–200 meters (300–650 feet).

Upwelling is important for its biological side effects. Deeper waters tend to be rich in nutrients, because the sparse marine

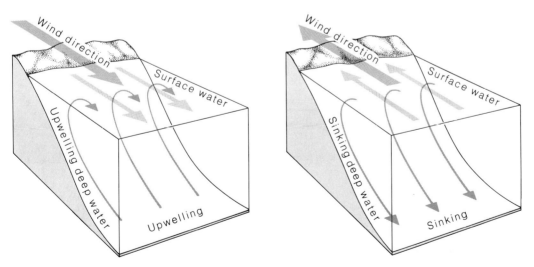

**Figure 12-38**
Upwelling and sinking are induced by offshore and onshore winds. Offshore winds drive surface waters away from shore, and deeper waters upwell to replace them. Onshore winds pile up surface waters along the coast; this causes a sinking and outflow of deeper waters to compensate.

life at depth is unable to utilize most of the phosphate, nitrate, and other dissolved matter in it. When that rich water reaches the sunlit surface, where organisms can thrive in abundance, the biological populations expand in response to the nutrients. Microscopic plant life blooms, and fish populations that live on it expand in proportion.

**Vertical Ocean Circulation** Imagine the Atlantic or Pacific Ocean as an elongate oval soup bowl stretched out in the north-south direction with the cold polar waters at opposite shallow edges and the equatorial waters in the middle. The surface waters are warm near the equator—around 25°C or warmer—while those in the Arctic and Antarctic are quite cold—about 0°C or a little above. Because cold water is denser than warm, the polar waters tend to sink and slide along the bottom toward the equator. As they do, they push in front of them the deeper waters, which tend to rise near the equator, being displaced from both directions. This simplified model explains how it happens that, just below the shallow, warm waters of the equatorial ocean, the deeper waters are cold. Because the dense, cold waters move slowly and mix with surrounding waters very slowly, they tend to retain their original temperature and salinity (saltiness of the water varies slightly from place to place in the ocean). Thus, the distinctive temperature and salinity of such waters allows them to be identified as large bodies or **water masses** and allows their place of origin to be determined. North Atlantic deep water can be traced from its polar origins, in this way, as it sinks and moves southward. Before it reaches the equator, it meets and rides over

**Figure 12-39**
Generalized vertical circulation of the Atlantic Ocean. (a) An ideal model of a North-South ocean that is cooled at the poles and warmed at the equator. Colder waters sink because they are denser, replacing warmer water, which rises. (b) Simplified model of Atlantic Ocean circulation, showing how Antarctic bottom water flows along the bottom more than 20° north of the equator, displacing the Arctic bottom water to an intermediate level.

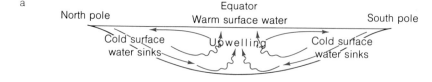

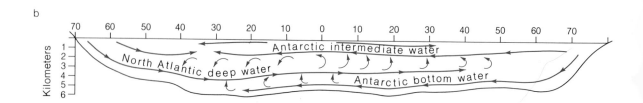

a mass of still colder and denser water, Antarctic bottom water, which has worked slowly northward from the cold Antarctic Sea.

The density-driven vertical circulation is much slower than the wind-driven horizontal circulation at the surface. Deep waters rise at the equator at speeds of only 2–5 meters (6–15 feet) per year. Polar waters take about 1000 years to circulate, a number that may be an important practical matter before very long. If pollutants are introduced into high-latitude bottom waters, they will not rise to the surface near the equator for hundreds of years. Thus, radioactive waste materials might lose a significant amount of their damaging radiation before they come back to the surface. On the other hand, once introduced to such a slow moving system, toxic materials may remain in circulation for a very long time.

The topography of the floor of the ocean affects deep water currents. They may be blocked by ocean ridges or they may flow through gaps or breaches in a ridge. The water masses are density-driven, so they follow topographic highs and lows just as turbidity currents do.

**Deep-sea Sedimentation**  The general circulation of the ocean is too slow to accomplish much erosion, except where such strong currents as the Gulf Stream travel along the shallower edges of the sea. Slow-moving currents do have enough mild turbulence, however, to keep fine-grained sedimentary particles suspended for a long time before they settle through the great depths of water to the bottom. Though sea water may appear clear and transparent, careful filtering of large volumes of deep water, such as samples of 100 liters (about 25 gallons), can extract a few thousandths of a gram of suspended mineral solids. The particles are smaller than 0.025 millimeter, and a great many of them are 0.001 millimeter or less. Identification of the particles shows that they are fragments of clay minerals, quartz, feldspar, and other common minerals produced by the weathering of rocks on the continents. The particles in the water are of the same general kind as those that constitute **pelagic sediments**, the gray and brownish muds that cover much of the varied topography of the deep-sea floor. There is little question that the muds on the bottom were formed by the slow, steady settling of land-derived erosional debris through the great depths of water.

Particles so small take a long time to settle through the water, for their rate of fall is slowed by the frictional resistance of the water. Even in perfectly still water, a particle 0.001 millimeter in diameter would take months to fall several kilometers to the deep-sea bottom. The settling is made even slower by the slow horizontal and vertical movements of the water which may lift the particle up again, extending its lifetime as a suspended particle for many years. Some oceanographers have estimated that it may take several hundred years for some of the smallest particles to finally reach the bottom.

Because the particles take so long to settle, they may be carried greater distances by horizontal ocean currents. A particle dropped into the Gulf Stream off North Carolina might end up on the bottom off the west coast of Europe. Patterns of bottom sediment have been mapped, revealing how distribution patterns are affected not only by strong currents like the Gulf Stream but the many lesser currents of the ocean.

The rates at which these particles are laid down as sediment are extremely slow, in most places about 1 millimeter every 1000 years. At that rate, an ocean bottom 4 kilometers deep would take 4 billion years, almost the whole age of the Earth, to fill up! Of course, the ocean bottom does not stay still for that long: as part of a lithospheric plate, it is moving horizontally at the rate of a few centimeters per year, eventually to be subducted into the mantle.

Though geologists have always known that wind transport plays some role in the deposition of pelagic sediments, it was only in the last decade that they began to realize how important that role is. That realization came as the result of oceanographic research ships taking air samples at sea, partly to test patterns of global air pollution, and was confirmed by melting down samples of ice from the Greenland and Antarctic glaciers and measuring the amount of airborne dust that had fallen there, far from the

**Figure 12-40**
Foraminiferal ooze dredged from a depth of 450 m off the coast of Central America. Enlargement about 15×. [Photo by Patsy J. Smith, U. S. Geological Survey.]

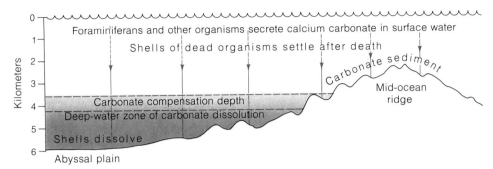

**Figure 12-41**
The calcium carbonate compensation depth is the level in an ocean below which the calcium carbonate of foraminifera and other shelled organisms that have settled from surface water will dissolve. The depth, which is a zone rather than a sharp boundary, varies some from ocean to ocean.

effects of human industrial pollution. This steady worldwide dust contribution forms about 10 percent of the pelagic sediment. At some times, particularly when there are violent volcanic eruptions, the proportion may be much greater.

Another important component of pelagic sediment is the calcite (calcium carbonate) shell material of pelagic foraminiferans, tiny single-celled animals that live in the surface waters of the sea. Though they are moved about by currents, most are large enough to fall relatively quickly to the bottom when they die. There they accumulate as **foraminiferal oozes.** Very early in the course of marine geologic exploration, it was discovered that oozes were abundant at depths less than about 4000 meters (about 13,000 feet) but rare in the deepest waters of the ocean. This could not be attributed to a lack of shells supplied by foraminifers, which live near the surface and, thus, are unaffected by the depth of the water. A closer look at many sediments from deeper water showed a few foraminifers, usually the larger and thicker-shelled ones.

At the same time, physical oceanographers were pointing out that the deeper waters of the oceans were colder, under higher pressure, and contained more dissolved carbon dioxide (from colder polar waters) than shallower waters. Each one of these factors contributed to making calcium carbonate more soluble at depth than at the surface. Then came the hypothesis: no matter how abundant the shelled organisms are in the surface water, as their shells fall into the deeper waters that are undersaturated with calcium carbonate, they will dissolve, either while settling or soon after they come to rest on the bottom.

An ingenious test for this hypothesis was made by an oceanographer at Scripps Institution of Oceanography, who reasoned that, with modern instrumentation, he ought to be able to observe directly any rate of solution. He took carefully machined, accurately weighed, perfect spheres of calcite, and lowered them over the side of a ship to various depths for various periods of time, and then brought them back up to weigh again. There was no change in the shallow depths, but he found a loss of weight in deeper waters, coupled with etched pits and irregularities, which confirmed that calcite was rapidly dissolved there.

The depth of water below which calcium carbonate starts to dissolve is called the calcium carbonate **compensation depth**. It has become important to an interpretation of the evolution of the ocean basins, using plate tectonics as a model, being developed by the National Science Foundation in its program of deep drilling of sea-bottom sediments. Drilled samples from some parts of the ocean that are now below the compensation depth show that, some millions of years ago, calcium carbonate was abundant in the sediments. Does this mean that the compensation depth of the oceans has changed with time? That would imply that there had been some drastic change in the circulation patterns of the ocean. The movements of oceanic lithospheric plates offered another alternative: As part of a plate was formed at a ridge or rise, it would have been topographically high, above the compensation depth and the resting place for much carbonate sediment, just as the mid-Atlantic ridge is today. As that portion of the plate moved

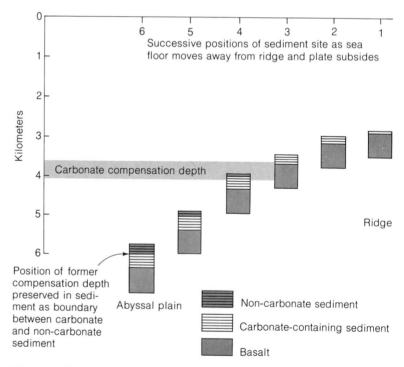

**Figure 12-42**
As a site of sedimentation is transported away from a mid-ocean ridge by plate movement, plate subsidence moves the site downward with respect to sea level, eventually below the calcium carbonate compensation depth. At that point, carbonate sediment is succeeded by noncarbonate deposits.

away from the ridge, it would have subsided gradually as it cooled, eventually passing below the compensation depth, after which time it would no longer have received much calcium carbonate sediment. This hypothesis is now being actively pursued with the aim of tracing the rates and paths of movement of oceanic plates in many parts of the oceans.

Calcium carbonate oozes are the most abundant of the biologically produced sediments on the sea floor. Oozes of silica—a noncrystalline form of silicon dioxide, like opal—form, in much the same way as calcium carbonate oozes, from the silica shells of diatoms, an abundant class of green, unicellular algae found in the surface waters of the oceans. These **diatom oozes** are found mostly in the Pacific and Antarctic oceans. Their distribution is governed partly by the abundance of nutrients at the surface. Even where the living organisms are abundant, however, diatom oozes are found only if there is little detritus from land erosion. In many places, such detritus forms the bulk of the sediment and so dilutes the slowly accumulating shells of diatoms and other silica-secreting organisms.

Some components of deep-sea sediment are formed by chemical reactions of sea water with the surface of the sediment. The most prominent of these are manganese nodules, those black lumpy accumulations first found by the *Challenger* expedition, which form at the surface of the sediment over many large areas of ocean bottom. Manganese nodules are estimated to cover as much as 20–50 percent of the Pacific Ocean floor. They form slowly as dissolved manganese and other metals, such as nickel, are precipitated both from the overlying sea water and from the sea water trapped in the pores of recently deposited sediment. The nature of the chemical reactions that form the nodules has long been a subject of scientific research. Recently, as geologists have become more concerned with conservation of natural resources and new sources of manganese ores, they have been estimating the amounts of manganese that could be recovered economically from the sea floor. At this time, the nodules are too expensive to recover, although they are relatively high-grade ores; but at some future time, the cost of mining land deposits of manganese may increase and the technology necessary to make undersea mining a practical matter may be developed. When that time comes, marine geologists and biologists will no doubt have to file some statements on the environmental impact of such mining activity on the water, the sediment, and the biological habitats of the deep sea. In order to do that, we will have to learn a great deal more about the deep sea than we now know.

The two concerns of practical importance—the utilization of the ocean for its mineral and food resources, and the pollution of the ocean by using it as the garbage can of the world—are intertwined. It is impossible to use the ocean's resources without

**Figure 12-43**
Manganese nodules on a ship's deck after having been freshly
dredged from the sea floor. Nodules vary from a few centimeters
to more than 10 cm in diameter. [From National Science
Foundation.]

interfering with it in some way. The imperative is to do so in such
a way that the alteration will not be damaging or irreversible. The
production of oil on continental shelves need not permanently
disrupt the biological population of the oceans nor destroy the
adjacent beaches, *if* there is sufficient geological knowledge and
rigorous enforcement of fail-safe engineering procedures for
the prevention of accidents. The oceans can be self-cleansing if
oil spills are few and far between. It is heedless dumping of wastes
on a global scale that will overload the capacity of the world ocean
to maintain its equilibrium.

## SUMMARY

1. Wave action is the major erosional agent operating along coasts. Rates of erosion by
   waves are fastest on shorelines of unconsolidated materials.

2. The shape of a shoreline is determined by wave erosion, modified by tectonic uplift
   or subsidence, sedimentation, and eustatic changes in sea level.

3. Waves are generated by wind, and are described by their wavelength, height, period,
   and velocity. Surf is the product of waves entering shallow water. Waves are radiated
   outward from storms at sea, their places of formation.

4. Wave refraction, the bending of waves as they approach shorelines, is controlled by the shape of the shoreline and the wind direction. Wave refraction affects erosion and sedimentation on beaches, and is also responsible for longshore currents and rip currents.

5. A beach is a dynamic form that is the result of a balance between material eroded from it and material transported to it along the shoreline. It can be described by a budget relating input and output. Man's interference can drastically alter the budget of a beach and, thus, its shape and size.

6. Tides originate from the gravitational pull of the Moon and Sun. Tidal currents are produced that are important distributors of sediments in tidal channels and tidal flats.

7. Continental shelves, slopes, and rises characterize the edges of continents. They are formed by sedimentation of erosional debris from the continents accumulating at the deep water margin.

8. Turbidity currents are density currents, formed by the suspension of mud in waters, that erode submarine canyons, gully the continental slopes, and deposit distinctive graded sediments, turbidites, on the continental rise and abyssal plain. Such currents are the agents of transport of coarser materials to the deep sea.

9. Plate tectonics explains the development of the profile of the oceans in terms of crust forming at the ridges, moving laterally away, and being covered with deep-sea sediment.

10. The horizontal circulation of the oceans is wind-driven into major and minor gyres. The Gulf Stream is the northward-flowing part of one of these, the North Atlantic subtropical gyre.

11. The vertical circulation of the oceans is a response to the greater density of masses of colder water formed in the polar oceans.

12. Pelagic sediments are combinations of fine-grained terrigenous clays, some water-borne and some transported by wind, combined with the calcareous or siliceous shells of small organisms, foraminifers and diatoms, that live in the uppermost layers of the oceans.

## EXERCISES

1. A certain stretch of shoreline includes a steep headland of sandstone with bays on both sides. Where would you expect to find the best beaches for swimming? How might you tell the direction of longshore currents from the distribution of beach sand?

2. Low submarine ridges of sand on the continental shelf off the northeastern coast of North America have been interpreted as former beaches. What sequence of geologic events can you hypothesize that would support this interpretation?

3. After a period of calm winds and no breakers along the northeastern shore of the United States, an intense storm with high winds passes over the area, followed by another period of calm. Describe the state of the surf along the beach during the storm and several days afterward.

4. What time(s) during one month that you have to spend at a rocky seashore would you pick to observe the maximum exposure of a wave-cut terrace? What time(s) to observe maximum erosion of a headland?

5. How might you account for the contrast between the broad continental shelf off the East Coast of North America and the narrow, almost nonexistent one off the West Coast?

6. Hoover Dam was built across the Colorado River in southern Nevada. When the river valley was flooded for many kilometers upstream, forming Lake Mead, engineers noted turbid muddy waters at the foot of the dam beneath clear waters above. How might this mud suspension have been produced?

7. Why would you expect thicknesses of pelagic sediments in depressions near the mid-Atlantic ridge to be much less than in depressions at the flanks, far distant from the Ridge?

8. Where would you be more likely to find foraminiferal oozes on the sea floor, on a mid-ocean rise or ridge or on an abyssal plain? In which of these two provinces would you be more likely to find turbidites? In what particular situation might you possibly find both kinds of sediment interlayered?

## BIBLIOGRAPHY

Bascom, W., *Waves and Beaches*. New York: Anchor Books (Doubleday and Company, Inc.), 1964.

Inman, D. L., and B. M. Brush, "The Coastal Challenge," *Science,* v. 181, pp. 20–32, 1973.

Gross, M. G., *Oceanography*. Englewood Cliffs, New Jersey: Prentice-Hall, Inc., 1972.

Heezen, B. C., and C. D. Hollister, *The Face of the Deep*. New York: Oxford University Press, 1971.

Menard, H. W., *Marine Geology of the Pacific*. New York: McGraw-Hill Book Company, 1964.

Moore, J. R. (editor), *Oceanography: Readings from Scientific American*. San Francisco: W. H. Freeman and Company, 1971.

Shepard, F. P., *Submarine Geology* (3rd ed.). New York: Harper and Row, 1973.

Sverdrup, H. U., M. W. Johnson, and R. H. Fleming, *The Oceans*. Englewood Cliffs, New Jersey: Prentice-Hall, Inc., 1942.

Turekian, K. K., *Oceans*. Englewood Cliffs, New Jersey: Prentice-Hall, Inc., 1968.

# 13

## Sedimentation
## and
## Sedimentary Rocks

*Sedimentation is the final stage of a process that begins with erosion and transportation of eroded materials to sites of deposition. Physical sedimentation is the deposition of such materials in the lowest places to which air and water currents can transport them. Chemical sedimentation is mainly the process by which sea water keeps its composition constant by depositing precipitates to balance the input of dissolved weathering products by rivers. Calcium carbonate makes up the largest volume of chemical sediment, and much of it is extracted from sea water by invertebrates and secreted as shells. Silica-rich sediments are also largely produced by organisms, mainly the diatoms. Bacteria play an important role in sulfide sedimentation in environments lacking oxygen. Gypsum and salt form from the evaporation of sea water in isolated basins. Sediments show the impress of the geomorphic environments in which they were deposited. Tectonism controls both subsidence in the depositional area and weathering at the source of the erosional debris. Chemical and physical changes after deposition result in lithification and many other alterations of composition and texture.*

**Sedimentation,** which means settling out of suspension, or deposition in a layer, is a word with many associations. The engineer designing a dam on a river for flood control worries about sedimentation filling the reservoir in back of the dam. The ship's captain navigating through sandy areas near shore is concerned that the shifting patterns of sedimentation may have built up shoals on which he may run aground. Geologists are concerned with all of the varied kinds of sedimentation on the surface of the Earth and how they play a role in the overall dynamics of the planet. Sediments include solid materials physically deposited by wind, water, or ice and dissolved substances chemically precipitated from oceans, lakes, or rivers. Physically deposited sediments have been discussed in relation to erosion and transportation in Chapters 8–12; here, we will consider them in relation to chemically deposited sediments and in the context of the accumulation of sediments in general.

## SEDIMENTATION AS A DOWNHILL PROCESS

Physical sedimentation starts where transportation stops. When the wind dies down, dust settles; when water currents decrease their speed, sand settles. On Earth, physical transportation and sedimentation follow a general downhill trend in response to gravity, from rock falls and mass movements downslope to river

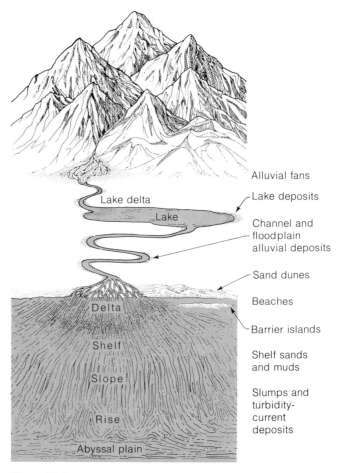

**Figure 13-1**
The downhill path of transportation and deposition takes material from the heights of mountains to the depths of the oceans, many different sedimentary environments being encountered along the way.

systems, and then down to the sea. In running water, sedimentation is a one-way street, each temporary stage of transport and sedimentation carrying the sediment farther toward the bottom of the deep sea. Much of it is permanently dropped along the way and never reaches the end of the line. Eolian sedimentation is different, for winds may blow material from low to high places and back again. But in the long run, eolian sedimentation is effectively a one-way street, too: Once windblown material drops to the ocean surface, it is trapped. It settles through the water and cannot be picked up again.

Chemical sedimentation is also a "downhill" process, but the driving force is chemical rather than gravitational. A major aspect of weathering and erosion is the chemical decay of rocks exposed to the water and carbon dioxide of the atmosphere. In the course of decay, ions from the rocks are dissolved, and rivers carry them to the sea (see Fig. 7-21). The ocean may be thought of as a huge chemical reservoir: Water continually evaporates from the surface and fresh river water runs in to replenish it. Although that keeps the amount of water fairly constant, it works also to enrich the sea in the dissolved ions: evaporation takes away only the water, the ions do not evaporate. Yet the sea maintains the same salinity. This is because of sedimentation of the dissolved material as chemical precipitates. Totalled over all of the oceans of the world, those precipitates must balance the total inflow of ions from weathering brought in by rivers. How and where the precipitates form are best illustrated by the most abundant chemically precipitated sediment, limestone. Limestone — $CaCO_3$ — and the related rock dolomite — $CaMg(CO_3)_2$ — are often called, from their chemical composition, **carbonate rocks**, or, more simply, **carbonates**.

Though the dissolved salt in sea water does not evaporate, the air above the ocean always contains some small crystals of sea salt that are left floating after droplets of sea spray from breaking waves have evaporated. This salty air causes automobiles near the ocean to rust much more than those inland, because the salt is corrosive to metal.

## CARBONATE DEPOSITION

The ocean is the scene of immensely varied kinds of carbonate sedimentation, from the formation of pelagic foraminiferal oozes in the deep sea (Chapter 12) to accumulations of sand and mud made up of calcium carbonate in shallow water. The chemical basis for carbonate sedimentation is the relative abundance of calcium and bicarbonate ions in sea water. The equation for carbonate sedimentation is:

calcium ion + bicarbonate ion $\longrightarrow$ calcium carbonate + hydrogen ion

$\qquad Ca^{++} \qquad\qquad HCO_3^- \qquad\qquad\quad CaCO_3 \qquad\qquad H^+$

At equilibrium (or saturation) there is, together with the solid calcium carbonate, a concentration of calcium and bicarbonate ions in solution whose (multiplication) product is constant. The precipitate may form if there is more calcium than bicarbonate, or vice versa, as long as their product exceeds the saturation value.

Two forms of calcium carbonate can precipitate, the hexagonal mineral calcite or the orthorhombic mineral aragonite (see Chapter 4). If the precipitation proceeds undisturbed and very slowly from slightly supersaturated solutions, calcite will form and remain stable indefinitely—a criterion for chemical equilibrium. In most precipitations, however, aragonite will form first, and then very slowly—sometimes over a period of many years—will transform to calcite. Most modern carbonate sediments are a mixture of the two forms, partly because the shells of organisms that contribute to carbonate sediments may be either or both. Almost all ancient limestones are calcite, because enough time has elapsed for all of the unstable aragonite to have changed to calcite.

In most of the oceans, sea water is fairly close to being saturated with calcium carbonate—that is, a chemist would expect that calcium carbonate would be precipitated if very many calcium or bicarbonate ions were added to it. The warm surface waters of many tropical areas are slightly supersaturated, whereas colder waters in high latitudes or at great depths are slightly undersaturated. This makes reasonable the notion that the ocean as a whole stays near the saturation point by a steady-state process. What would happen, then, if the inflow of river water stayed the same but its content of calcium and bicarbonate ions were to drop for some reason? Then the oceans would become undersaturated with calcium carbonate, and carbonate sedimentation might be expected to slow down or stop. Similarly, if the influx of those ions were greater, we would expect the ocean's composition to move towards supersaturation, thus stimulating more calcium carbonate sedimentation.

Thus, at first sight, the ocean seems to be a reasonable chemical system whose behavior is predictable. A closer look, however, shows that an important element has not been taken into account—the biological world.

**Biological Precipitation** Almost everywhere in the oceans, carbonate sediment is made up of the shells of organisms rather than of inorganic precipitates. Apparently, the conditions for precipitation are right, but it is accomplished by the organisms extracting calcium carbonate from the water for their shells rather than by an inorganic precipitate forming directly from the water as in the laboratory. Areas of extensive carbonate sedimentation are restricted to warm tropical seas, which shows the general correspondence in such places between the chemical condition of slight supersaturation and the high populations of organisms whose shells make the sediments. Shelled organisms live in colder undersaturated waters, but the shells tend to dissolve after the organisms die; the shells of pelagic foraminiferans, for example, settle into deeper waters below the calcium carbonate compensation depth, as we pointed out in Chapter 12. Carbonate-secreting organisms

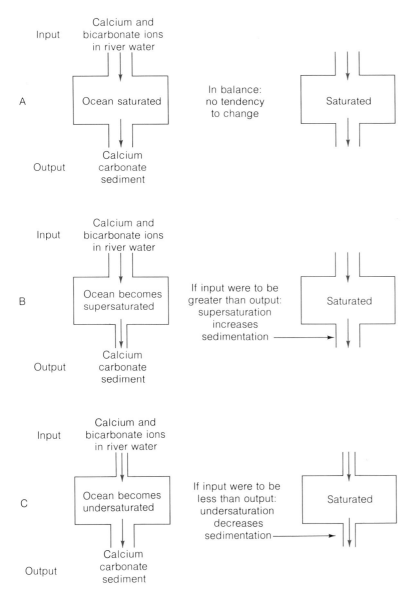

**Figure 13-2**
The feedback system that keeps the oceans approximately saturated with calcium carbonate. If river input increases (or decreases), the system will respond by increasing (or decreasing) the amount of sedimentation.

and carbonate sediments are also found in fresh-water lakes, some of which are saturated, and some of which are not.

An enormous array of organisms, from primitive one-celled animals to the common oysters, clams, and other highly evolved invertebrates, secrete some calcium carbonate. Carbonate-secreting plants range from primitive algae to some higher plants, such as

**Figure 13-3**
Underwater photograph of carbonate sediment formed from skeletal materials of algae and invertebrates in the Bahamas. A variety of algal species predominates in this view. [Photo by J. Swinchatt.]

some aquatic grasses. Vertebrates (and some invertebrates) do not secrete calcium carbonate but rather the harder calcium phosphate, to form bone and teeth. These are minerals containing small amounts of fluoride and hydroxyl ions that are variants of the mineral apatite. The physiology of carbonate and phosphate secretion varies greatly among the many groups of organisms, and much remains to be learned about it.

Carbonate-shelled organisms live in dependence on the sea (or lake) water around them, which provides nutrients and the dissolved materials from which they make their shells. They also live in a complex of intertwined relations with the other organisms in their habitat. The entire web of interdependence of organisms and their environment is the subject of the science of ecology. Because the ecological study of carbonate-secreting organisms and carbonate sediments in various environments has been carried on vigorously in the past two decades by geologists and paleontologists, we have learned how limestones are made. One sort of habitat, in particular, has always captured the attention of scientists—the coral reefs of tropical seas.

**Coral Reefs** From 1831 to 1836, Charles Darwin sailed on the *Beagle* as naturalist-scientist for a British surveying expedition, a voyage made famous because it allowed Darwin to observe and collect the great variety of organisms that laid the foundation for his theory of organic evolution. One of the products of the voyage was his analysis of Pacific coral reefs. Darwin was one of the first to explore in detail the relation of organisms to their geologic environment. Many of his ideas on coral reefs are still accepted today. More than a hundred years after the *Beagle* sailed, the coral reefs of the South Pacific became battlegrounds of World War II, and military and engineering needs sparked new observations. After the war, two coral reefs were used for atomic bomb testing, and the name of one of them became famous: Bikini (the other was Eniwetok). To geologists, those reefs became famous because of an extensive mapping and drilling program that was undertaken to determine the origin and dynamics of reefs so that the effects of bomb testing could be assessed.

The major physical characteristics of a coral island or **atoll** are: (1) an outer wave-resistant reef front, a steep slope facing the open ocean; (2) a flat reef platform in back of the reef and extending toward the island; (3) a shallow lagoon in back of the reef platform, protected from the waves by the reef; and (4) the island

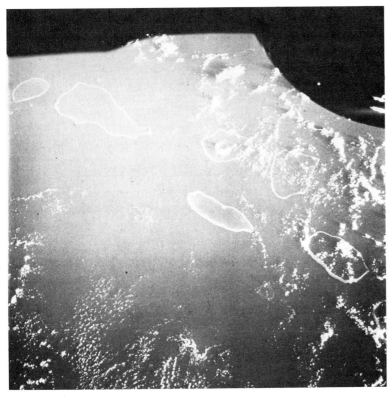

**Figure 13-4**
Tuamoto Islands, Pacific Ocean, taken from Gemini spacecraft. Shown in this photograph are eight atolls of the eighty-island Tuamoto archipelago, a part of French Polynesia. The coral reefs enclose more or less circular lagoons. [From National Aeronautics and Space Administration.]

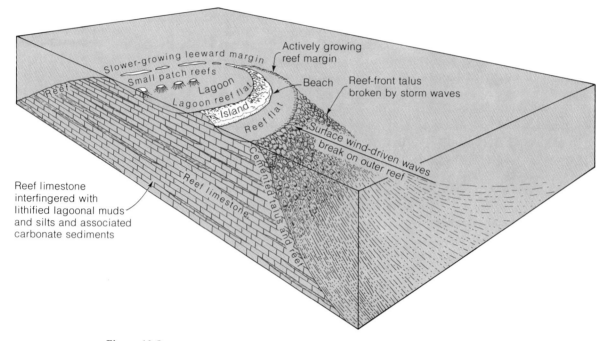

**Figure 13-5**
Diagrammatic drawing of a reef. The reef tends to grow more rapidly on the windward side, where waves bring fresh nutrients from deeper, open waters to the organisms.

itself. Some atolls have no central island, just a more or less circular reef, parts of which may be above water and forested, surrounding a central lagoon.

The main part of the reef tract is made up of actively growing coral, an organism that grows as colonies of great numbers of individuals joined to each other. The coral secretes its calcium carbonate as it grows, cementing itself to the carbonate remains of dead coral below, and the reef grows outward. The living stalks of coral get their food from tiny bits of organic matter brought to them by waves coming in from the open ocean. Reef-building corals need light to grow because they live in **symbiosis** (that is, in a mutually life-supporting relationship) with small green algae (zooxanthellae) that live in the translucent coral tissue and need light to live. The coral provides protection for the algae and the algae provides photosynthetic oxygen for the coral. The light requirement limits coral growth to shallow waters, less than about 20 meters (65 feet) because, as transparent as it may seem, sea water gradually filters out more and more light the deeper it gets.

Living in association with the coral are various kinds of coralline algae, which also secrete carbonate and thus help to cement the whole structure into a massive rock terrace. On the reef platform live many species of shelled and soft invertebrates. In the lagoon live many other kinds of organisms. The organisms feed upon each

**Figure 13-6**
An underwater photograph of a coral reef knoll in a few meters of water near North Rock, Bermuda.
The scale is 50 cm long. Below the scale is a sheet-like encrusting coral colony. At left center and
lower right are large heads of brain coral. The branching, tree-like forms are sea whips and sea fans.
[Photo supplied by Peter Garrett.]

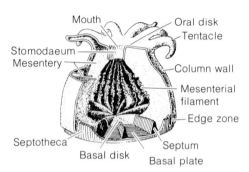

**Figure 13-7**
Cutaway drawing of a typical coral, showing
the relation of the soft body parts (mesentery)
to the skeleton parts of calcium carbonate:
basal plate and septum. [After *Treatise on
Invertebrate Paleontology,* Geol. Soc. Am. and
University of Kansas Press, 1956.]

other, and, in the process, the shells or other hard parts of the prey
organisms are broken up into sand-, silt-, or clay-sized particles.
These, together with the physical erosion of the reef front, which
produces fragments of coral and coralline algae, provide the carbonate sand of the island beaches and some of the fine carbonate
mud that floors parts of the lagoons. Some of the mud comes
from microscopic particles secreted by algae; some of the rest may
be produced by nonbiological precipitation from warm lagoon
waters that become supersaturated with calcium carbonate as
a result of evaporation.

A seasoned observer can deduce where samples of reef sediment
came from just by looking at the sediment and the kinds of shells
in it. Hard, cemented rock made up of coral and coralline algae
come from the reef front. Sediments from the main reef platform
may contain many kinds of clams and other shells, sometimes with
some carbonate sand from the erosion of the reef. The lagoon is
dominated by muds, sometimes built up into small mounds, occasionally spotted with little coral "heads" growing here and there.
The beaches are where the sand accumulates. The geologist uses
this information to infer the nature of ancient reefs formed in the
seas of the past—such as the reefs of the Silurian period, the
remains of which lie in a belt across what is now Indiana, Illinois,
and Wisconsin, or the great reefs of Permian age in west Texas.

**Origin of Coral Reefs** Ever since Darwin first studied reefs,
geologists have wondered about how they form. At first sight, there
appears to be a paradox: How can corals, which need light to live,
build islands that reach to the surface from the floor of the deep sea
thousands of meters below? There is no conceivable way that the
level of the sea could once have been low enough for coral to have

**Figure 13-8**
The Guadalupe Mountains of West Texas from the south. The drawing below is a reconstruction of
the sedimentary environment associated with a great reef that existed here in Permian times. [From
*Geology Illustrated* by J. S. Shelton. W. H. Freeman and Company. Copyright © 1966.]

formed so near the bottom, nor is there any evidence that much of the floor of the ocean was ever structurally elevated to sea level, which would have allowed the corals to begin growing. The clues Darwin saw were **fringing reefs**, reefs similar to atolls growing around the edges of volcanic islands. Other islands were mostly coral, but there was a remnant of volcanic rock. The theory he proposed over a century ago remains the most probable explanation:

The process starts with a volcano being built up from the sea floor. As the volcano stops its eruptive activity — temporarily or permanently — corals and coralline algae colonize the shore and build a fringing reef around the volcanic island. The central volcano may then be lowered almost to sea level by erosion. If such a volcanic island sinks below the waves as a result of some tectonic movement, actively growing coral and coralline algae may keep pace with the sinking and build up the reef. In this way, the volcanic

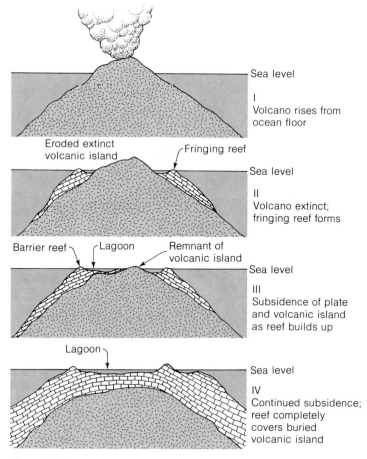

**Figure 13-9**
Evolution of a coral reef from a subsiding volcanic island. In most respects this chain of events was first proposed by Charles Darwin in the nineteenth century. The framework of plate tectonics provides an explanation for the subsidence of the islands.

center may disappear, to be replaced by an atoll with a central lagoon. One of the results of the deep drilling on Bikini and Eniwetok atolls was striking confirmation of this theory: volcanic rock was penetrated by the drills deep below the coral of the island. Oceanographic research parties have mapped many submerged flat-topped volcanoes that once must have stood above sea level and been eroded, but then subsided below sea level too fast for coral growth to keep up. More recently, as we will see in Chapter 21, plate-tectonic theory has proposed an explanation of the subsidence of volcanic islands on oceanic lithospheric plates as part of the process of sea-floor spreading. Of course, some coral islands have a much more complicated history, and sea-level changes during the Pleistocene affected all of the reefs then in existence.

Fascinating as coral-reef islands are, they account for a relatively small proportion of the carbonate sediment produced in the ocean. By far the largest bulk of carbonate sediments similar to the ancient limestones formed on the continents are formed on shallow-water platforms near or attached to the continents.

**Carbonate Platforms**   One of the best examples of a carbonate platform is the area of the Bahama Islands and the shallow banks surrounding them. This region of about 155,000 square kilometers

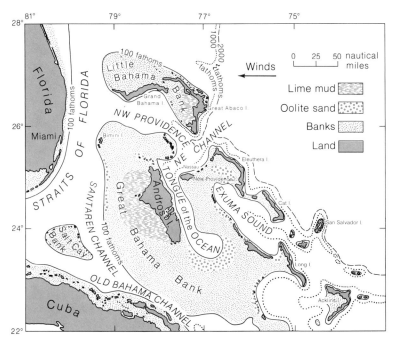

**Figure 13-10**
Map of carbonate sediments in the Bahama Islands. The reef fronts and islands are on the eastern margins of the banks, facing the westward-blowing trade winds. [After N. D. Newell and J. K. Rigby, "Geological Studies on the Great Bahama Bank," Soc. of Econ. Paleontologists and Mineralogists Spec. Paper No. 5, 1957.]

(about 60,000 square miles) is a flat-topped shallow platform, on the west separated from the mainland of Florida by the narrow, deep Straits of Florida and on the east dropping abruptly to the deep ocean floor, here more than 3600 meters (about 12,000 feet). Most of the platform is only a few meters below sea level, and it is dotted with many small islands. Coral and algal reefs are found on the eastern edges of the platform where the southwestward-blowing trade winds bring open-ocean water rich in nutrients to the organisms of the actively growing reef tracts. Much of the area of the shallow banks to the west is covered with carbonate sand and mud particles with essentially no land-derived detrital material, for supply of sediments from the continent is cut off by the Straits of Florida. The lack of terrigenous erosional debris is important because it allows life to be abundant on the platform: many of the organisms that live there might be rapidly "smothered" to death if there were a significant amount of noncarbonate clay and sand present.

A distinctive kind of carbonate sand, **oolite**, is formed where strong tidal currents flood the banks each day. Oolites are more or less spherical grains of aragonite that are made up of concentric layers surrounding a nucleus, like the structure of an onion. Most of the nuclei are fragments of shells. These nuclei get coated layers of aragonite as the tides bring in cold ocean waters that, warming and evaporating on the shallow banks, become supersaturated with calcium carbonate. The currents are strong enough to keep the grains rolling or suspended in the water, and, as a result, they get coated with precipitate on all sides. The limit to growth of the oolites is determined by the maximum grain size that can be moved by the current.

The variety of types of carbonate sediments found on the Bahamas is exemplified by a cross section of one of the islands, Andros, and the banks west of it (see Figure 13-12). The reefs on the east shore protect a narrow muddy lagoon, which lies in front of a sand beach on the island, like the lagoon of an atoll. The western side of the island leads down to a tidal flat. The tidal flat is covered with muds, sometimes finely rippled, interrupted by tidal channels, and in many places overgrown with **algal mats**, interwoven mucilaginous surfaces of blue-green algae. This sticky surface traps fine carbonate sediment and, as the algae grow around and upward through the carbonate, the sediment is incorporated in a layered mixture. In this way, the algae produce the layered structures called **stromatolites** that are found in many tidal-flat limestones. These structures have achieved great importance as evidences of algal life far back in the Precambrian. Also found on tidal flats are areas in which the calcite and aragonite of the original sediment have been transformed into the calcium-magnesium carbonate mineral dolomite, $CaMg(CO_3)_2$. In these areas, the sea water in the pores between the grains just below the surface has become

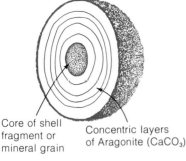

Core of shell fragment or mineral grain

Concentric layers of Aragonite ($CaCO_3$)

**Figure 13-11**
Cross section of an oolite, made up of concentric shells of calcium carbonate (aragonite) precipitated as the grain was rolled and suspended in current-agitated waters.

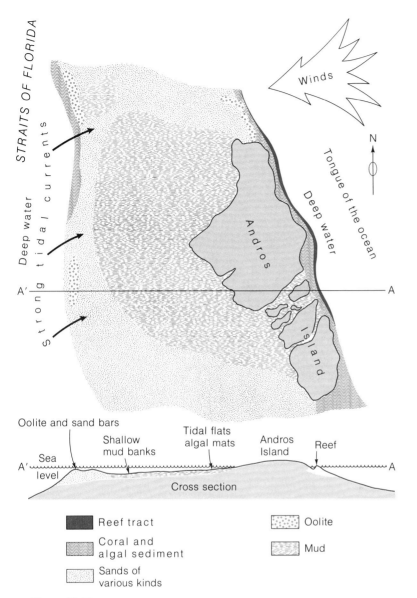

**Figure 13-12**
Map and cross section of Grand Bahama Bank west of Andros Island. [After various sources.]

enriched in magnesium ions owing to evaporation. That increase in magnesium promotes the exchange of some magnesium for calcium ions in the carbonate mineral, converting it to dolomite. Many ancient dolomites show evidence of this kind of tidal-flat origin.

Below the tidal flats west of Andros are extensive shallow banks, only a meter or two deep in many places, where the water is quiet and the bottom is carbonate mud. Some of the mud is in the form

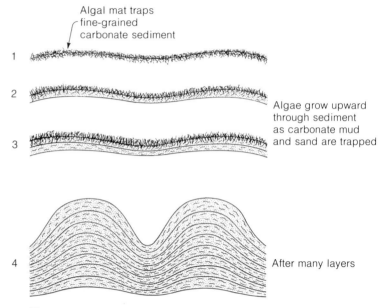

Algal mat traps
fine-grained
carbonate sediment

1

2

Algae grow upward
through sediment
as carbonate mud
and sand are trapped

3

4

After many layers

**Figure 13-13**
Stages in the growth of a stromatolite formed by the sediment-trapping
action of algae. As the sediment accumulates, the algae continue to grow
upward while the previously deposited layers become cemented and
lithified. The convex upward structures vary greatly in all dimensions
and in their regularity. A good example of one kind is shown in Figure
14-5.

of soft pellets, produced as fecal pellets by the many kinds of
organisms living on the bottom. Typical is a marine worm that
feeds by eating the mud, extracting for food the small amount of
organic matter it contains, and excreting the bulk of the carbonate
mud as pellets. In the mud area lives a variety of algae that secretes
microscopic needles of aragonite, about 0.003 millimeters long.
Similar needles occur abundantly in the mud. A small part of the
carbonate mud may also be the result of nonbiological precipitation from the warm, supersaturated water of the banks, which can
also produce aragonite needles. Most of the fine sediment is from
algal precipitation and the wearing down of coarser shell materials
by abrasion and animal ingestion.

On the far western side of the banks west of Andros, the movement of the water becomes greater and the currents winnow out
the fine material. Waves and tidal currents move sand consisting of
broken pieces of shells into submarine dunes and bars much like
those formed by currents in large rivers or those found on shallow
continental shelves in noncarbonate terrains.

Armed with the information about Pacific atolls and carbonate
platforms like the Bahamas, geologists have learned to recognize
many different kinds of ancient limestones that originated long ago
in similar situations. Thus, they have been able to make maps of

ancient reef tracts, extensive tidal-flat areas, sand banks, and the other kinds of areas in which animals, plants, and the physical environment interacted to produce distinctive carbonate deposits. A paleogeographic map of North America showing areas of carbonate deposition during one time period is shown in Figure 22-55.

Not all carbonate deposits form in the sea. Calcium carbonate is deposited as crusts and layers around hot springs, both by algae that grow in the hot water and by nonbiological precipitation. The deposits are **tufa**, which is porous and easily broken, and **travertine**, a denser variety that is familiar because of its wide use as a decorative stone for building facings and table tops. Carbonate is also found in **stalagmites** and **stalactites** in caves formed by groundwater in limestone formations. Evaporation of carbonate-saturated groundwater dripping from fractures in the cave roof form the needlelike stalactites. The stalagmites build up from the floor as the drips falls on that spot. Carbonate sediments can form in lakes from algal activity and nonbiological precipitation. They range from such salty lakes as the Great Salt Lake and the Dead Sea to the freshwater lakes of northern temperate regions, such as those of Minnesota and Wisconsin.

## THE VARIETY OF BIOLOGICAL SEDIMENTATION

Though calcium carbonate is the most abundant biological sediment, there is a great variety of other sediments whose formation is wholly or partially controlled by the biological world.

**Silica** The deposition of silica ($SiO_2$) has some similarities to the deposition of carbonates. Most of the silica of biological origin is secreted by single-celled organisms. The most important are small algae called **diatoms** that populate much of the surface layers of the ocean and freshwater lakes. These little organisms are extraordinarily efficient in removing silica from sea water to make their shells, which are an amorphous form of silica similar to opal in crystal structure. In contrast to the marine carbonate-secreting organisms, which live in waters that are nearly saturated with calcium carbonate, diatoms can extract silica from very undersaturated solutions. When diatoms growing in a laboratory aquarium use up all of the dissolved silica in the water, they will start to use dissolved silica etched from the glass walls of the aquarium, but how they do it is still a mystery. In addition to diatoms, there are small single-celled animals called **radiolarians** that are related to foraminifera and live much like them. Some kinds of sponges secrete silica, too. When diatoms and radiolarians die, they sink to the bottom of the sea and accumulate in the sediment. Under areas of high biological productivity, where silica-secreting organisms are abundant because of a high supply of silica and other nutrients in the water, the silica shells of dead organisms rain down and form silica-rich **diatom ooze** and **radiolarian ooze**. When these oozes become

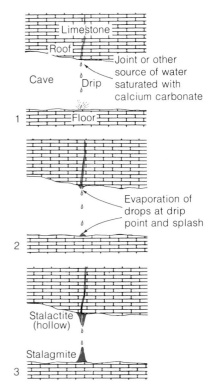

Figure 13-14
The formation of a stalactite and a stalagmite follows a drip of calcium-carbonate-saturated water from a cave ceiling.

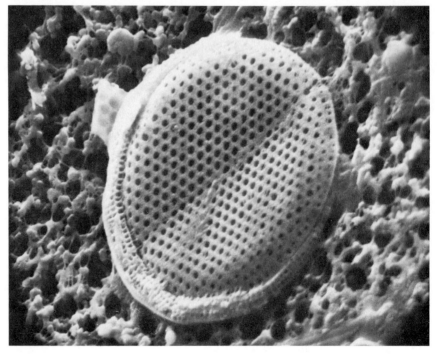

**Figure 13-15**
Two species of modern diatoms filtered from sea water. These photographs were taken with a scanning electron microscope at very high magnifications. The long dimension of each species is about 10 microns. [Photos by C. Stein and A. Eaton.]

cemented and hardened into rock, they are called **diatomite** and **radiolarite**. Diatomite is a useful rock: because of its high porosity, it has been used in filters and as fillers, and it is sometimes used as a mild abrasive. One of the best known diatomites is the Monterey Formation, which is exposed in many places along the coastal regions of central and southern California. Many ancient **cherts**, silica-rich rocks, originated in coastal waters rich in silica produced by the submarine or land weathering of volcanic rocks, which release much silica when they decay.

**Sulfide**   Organisms may control chemical sedimentation indirectly by changing the chemical conditions in the environment. One important example is the formation of ferrous sulfide ($FeS_2$), the mineral pyrite, by the indirect action of bacteria. The bacteria are in the same group as those that cause the noxious smell of rotten eggs. The smell is the gas hydrogen sulfide ($H_2S$). This gas is produced in sediments or sea water when these bacteria chemically change sulfur from its oxidized state, as the sulfate ($SO_4^{--}$) dissolved in sea water, to its reduced state, sulfide ($S^{--}$). These primitive bacteria, which can grow only in the absence of oxygen and so cannot respire as most organisms do, get their energy by converting the oxidized sulfur ion from its $+6$ state in sulfate, where the atom has lost 6 electrons, to the $-2$ state in sulfide, where the atom has gained 2 electrons. The hydrogen sulfide produced, itself a powerful reducing agent, changes ferric iron ($Fe^{+++}$) to ferrous iron ($Fe^{++}$) and precipitates the highly insoluble pyrite.

The presence of decaying organic matter is necessary for the $H_2S$ bacteria to thrive; this decay keeps the immediate environment free of oxygen. In most normal oxygenated or aerated environments, organic matter decays by using up oxygen, essentially as a slow combustion process similar to the way we respire:

$$\text{organic carbon} + \text{oxygen} \rightarrow \text{carbon dioxide}$$
$$C \qquad\qquad O_2 \qquad\qquad CO_2$$

As the oxygen is used, it is replenished by free mixing with the atmosphere, of which one-fifth is oxygen. In most shallow parts of the ocean, the mixing of the water by waves and currents keeps the water well aerated. In the deep ocean, the rate of replenishment is slower, taking as much as a thousand years; but, because there is relatively little organic matter in the deep sea, the oxygen does not completely disappear.

The oxygen *can* totally disappear in two situations: If there is a great amount of decaying organic matter in the sediment, such as in some biologically productive areas of tidal flats, the oxygen, slowly permeating into the sediment by diffusion, cannot keep up with the depletion of oxygen by biological activity, and the sediment becomes reducing—that is, lacking in oxygen. When that

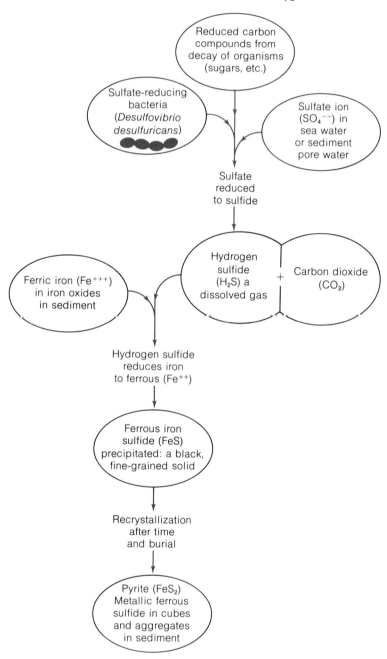

**Figure 13-16**
Pyrite, ferrous sulfide ($FeS_2$), is formed by a chain of reactions that starts with the reduction of sulfate ($SO_4^{--}$) from sea water by sulfate-reducing bacteria living in an environment free of oxygen.

happens, the bacteria that produce hydrogen sulfide become active, and pyrite is eventually formed.

In some areas of the sea, deep basins may be cut off from replenishment by aerated waters by a **sill**, a topographic ridge or barrier that prevents free circulation and mixing. In such a place, the waters of the basin do not mix well with oxygenated surface waters because the depth is too great for the deep waters to be affected by waves, and the greater density of those waters (caused by low temperatures) prevents general mixing by convection. Here again, as organic matter decays, oxygen is used up and replenishment by mixing with the aerated waters above is too slow; thus, bottom waters and sediment become reducing. Many fjords — drowned valleys scoured well below sea level by glaciers — have sills at the seaward end formed by terminal moraines or by the limit of the scouring action of the glacier. This causes them to be reducing below the level of the sill.

The deoxygenation of surface waters can also be brought about by some forms of pollution. When abundant phosphate, one of the important nutrients of organisms, pollutes a lake or river, the algae and many different aquatic plants will flourish, sometimes to the point of glutting the surface waters, a process called **eutrophication**. When these organisms die, their abundance is transferred to the bottom, where the extensive decay uses up oxygen faster than it can be replenished. The result is a great shift in animal populations, as fish and other organisms that need oxygen are cut off from much of the lake. The lake is not "dead," a term that many people have applied to Lake Erie; but its population has been drastically altered, and many species have disappeared, especially popular food and sport fishes, while "scummy" algae and weeds thrive.

Deoxygenation is the explanation for the black sands and muds just below the surface of many tidal-flat sediments. A quick scoop of the sediment will let you smell a whiff of hydrogen sulfide. The blackness comes from the first form of ferrous sulfide precipitated, a finely dispersed black substance that slowly transforms to crystalline pyrite.

Figure 13-17
Sea basins separated from the main body of open sea water by a sill may become deoxygenated as the water below the sill level stagnates from lack of circulation.

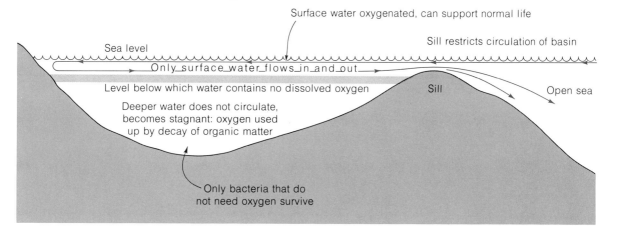

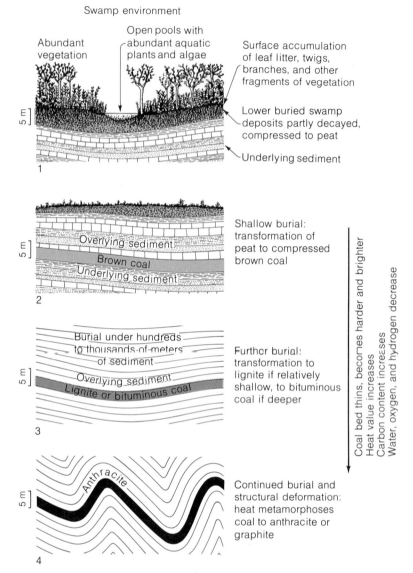

**Figure 13-18**
The process by which coal beds form begins with the deposition of vegetation.
Protected from complete decay and oxidation in a swamp environment, the
deposit is later buried and subjected to mild metamorphism, which transforms
it into lignite, bituminous coal, or anthracite, depending on depth of burial,
temperature, and amount of structural deformation.

**Coal** Lack of enough oxygen to completely decay vegetation on
land is responsible for the formation of one of our most valuable
sediments, **peat**, and its lithified equivalent, **coal**. The abundant
plant fossils in coal beds indicate that they originated in fresh-water
swamps; much as peat bogs form now. As the luxuriant plant growth
of such a swamp dies, it falls to the water-logged soil. The water

and rapid burial by falling leaves protect the twigs, branches, leaves, fruit, and seeds from oxidizing, just as thick leaf mats in some forests protect the lower layers from complete decay. The vegetation accumulates and gradually turns into peat, a porous brown mass of organic matter in which twigs, roots, and other plant parts can still be recognized. After more burial and chemical transformations of the aging organic matter, the peat becomes **lignite**, a very soft coallike material. Greater depth of burial, longer time, and higher temperature may ultimately metamorphose the lignite into **bituminous coal** or **soft coal**, and, in extreme cases, to **anthracite** or **hard coal**. The greater the metamorphism, the harder and brighter the coal and the higher its heat value. There is little doubt that, as our energy needs continue to grow and as oil and gas are less able to satisfy the demand for fuel, coal is becoming, once again, as it was up to about forty years ago, an extremely important chemical sedimentary rock.

## OTHER CHEMICAL SEDIMENTS

In an arid climate, the same general kind of barred basin that restricts circulation and causes deoxygenation can be the scene of a different kind of sedimentation — salt deposition.

**Evaporites** Sediments composed of the mineral halite or rock salt ($NaCl$), the calcium sulfate minerals gypsum ($CaSO_4 \cdot 2H_2O$) and anhydrite ($CaSO_4$), and many other less abundant salts formed by the evaporation of sea water are called **evaporites**. The thin crust of salt formed when a pond of sea water dries up is easy to observe along many shorelines, but not much salt forms in such a situation. The ratio of salt to water is such that, for every liter of sea water evaporated, about 35 grams (a little more than 1 ounce) of solid salt will crystallize. How much sea water had to evaporate then, to produce an evaporite formation 100 meters thick, 50 kilometers wide, and 100 kilometers long? The volume of the evaporite, about $5 \times 10^{11}$ cubic meters, at an average density of about 2.3 grams per centimeter, weighs about $1.15 \times 10^{18}$ grams, which represents the salt evaporated from about $3.3 \times 10^{16}$ liters of sea water (approximately $8.7 \times 10^{15}$ gallons)! A single pond containing that much water would occupy a volume well over 30,000 cubic kilometers. In other words, the water would have had to be more than 6 kilometers deep over the area covered by the formation, much more than the average depth of the oceans. This absurd conclusion alone led geologists to suspect that there must be some natural mechanism that constantly replenishes the sea water as it evaporates.

The study of the minerals formed by sea-water evaporation provided another lead. The first experiment of evaporating sea water was performed in 1849 by Usiglio (see Chapter 3), who showed how a sequence of minerals would be precipitated as evaporation proceeded, starting with calcium carbonate and calcium sulfate, progressing to sodium chloride, and then, in the final stages, to magnesium and potassium minerals. The important point, one that was amplified and carefully worked out in detail much later, was that a distinctive sequence involving a complex order of precipitation of many minerals would characterize complete evaporation. Some evaporites show those sequences and are interpreted as simple desiccation products. Yet a great many thick evaporite deposits consist only of gypsum, or anhydrite and halite, or a combination of the three. Did incomplete evaporation go only so far as to produce halite, in which case an even greater volume of water would have been required? Why are there so few evaporites that contain the last salts to precipitate just before complete drying?

A resolution of these difficulties came with the theory that typical evaporite sequences are deposited in relatively shallow arms of the sea, or in basins connected to the sea by narrow openings. As surface water evaporates from the basin, the dissolved materials become more concentrated, normally by limestone deposition, until the first evaporite precipitate, typically gypsum, starts to form. The concentrated solution formed at the surface, called a **brine**, is denser than the water below by virtue of its greater content of dissolved salt, and so tends to sink to the bottom. As it does, normal sea water flows in at the top, through the seaway from the open ocean. A continuous circulation pattern is set up by the denser brine flowing out through the bottom of the seaway. Thus, a steady-state system is created in which gypsum is continually removed as a precipitate while sea water circulates through the evaporite basin. Circulation of this sort occurs in the Mediterranean Sea and the Straits of Gibraltar. Atlantic Ocean water flows

**Figure 13-19**
Simplified diagram of circulation of the Mediterranean Sea through the Straits of Gibraltar. Because influx of fresh water from rivers into the Mediterranean does not balance evaporation from the surface, there is an inflow of water from the Atlantic Ocean to replace sinking surface water that has become more saline and thus denser as a result of evaporation.

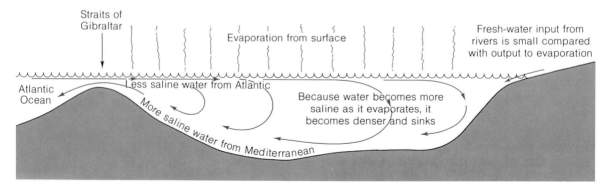

in at the top of the straits to the saltier Mediterranean, which has a high evaporation rate, and saltier Mediterranean bottom water flows out to the Atlantic through the bottom of the straits. The Mediterranean, however, does not evaporate sufficiently now to form gypsum. Alternative circulation patterns have been proposed, including ones in which the dense brines move down into and through permeable sediment below the basin and reappear along lower slopes of an adjacent ocean basin. This mechanism also works for evaporites formed on evaporating tidal flats, where the brine formed by evaporation sinks into the upper unsaturated groundwater zone, slowly precipitating gypsum as it moves down.

A recent hypothesis, based on Deep Sea Drilling Program holes in the Mediterranean that contain abundant evaporite, proposes that the Mediterranean was cut off from the Atlantic at Gibraltar about 6 million years ago and dried up completely to form a temporary salt-covered desert.

## CLASTIC SEDIMENTS

Though chemical sediments tell us much about the nature of chemical processes operating at the Earth's surface, they tell us little about transportation and deposition of solid particles, particularly in relation to the tectonic and erosional forces that produce detritus. **Clastic,** or **detrital,** sediments, those made up of particles broken and weathered from pre-existing rock, tell that story, much of which has been covered in earlier chapters on erosion and transportation. Clastic sediments—shales, sandstones, and conglomerates—are far more abundant than the chemical precipitates. They constitute well over three-quarters of the total mass of sediments in the Earth's crust, and, of that fraction, shale is by far the most abundant rock type, being about three times as common as the coarser clastics. Clastics dominate because mechanical erosion is constant and widespread over the Earth.

Sedimentation of particles is largely controlled by the strength of the currents that carry them: the larger the particle, the stronger the current must be in order to carry it. It was natural, then, for

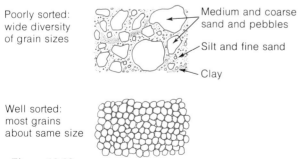

Poorly sorted: wide diversity of grain sizes

Medium and coarse sand and pebbles

Silt and fine sand

Clay

Well sorted: most grains about same size

**Figure 13-20**
Sorting of sedimentary particles produces sediments with a small size range of particles. Also see Figures 8-8 and 8-9.

Table 13-1
Classification of sand by grain size

| Classification | Grain size (mm) |
| --- | --- |
| Very coarse sand | 1.0–2.0 |
| Coarse sand | 0.5–1.0 |
| Medium sand | 0.25–0.5 |
| Fine sand | 0.125–0.25 |
| Very fine sand | 0.0625–0.125 |

geologists to distinguish the various clastic sediments on the basis of particle size. Such a division reflects the conditions of sedimentation: A gravel, or its lithified equivalent, conglomerate, calls to mind strong currents of rapidly flowing rivers in mountains, or high waves on a rocky beach. Mud, or its rock equivalent, shale, suggests quiet waters that allow the finest particles to settle. Sand and sandstone imply moderate currents, such as those of rivers and shores, as well as the strong winds that blow sand into dunes.

**Sand and Sandstone** Sandstone, more than any other clastic sediment, contains easily read information about its origin. Sand particles are large enough to be seen with the naked eye, and, under a microscope, the minerals can be identified by their optical characteristics and their textures can be studied. In contrast, most of the particles in shale are too fine for optical study, even with high-power microscopes. Sandstones are more abundant and widespread than conglomerates, and they tend to occur as erosion-resistant formations that make easily observed outcrops.

The range of sizes of sand and sandstone grains has been studied extensively as an aid in deciphering their origin. Table 13-1 shows how sands are divided on the basis of their average grain size. But average size does not tell the whole story. If all of the grains in a sample are close to the average size, the sand is considered well sorted; if there are many grains much larger and smaller than the average, the sand is poorly sorted. Because **sorting** is related to the kind of depositing current, we can distinguish between well-sorted beach or dune sands and poorly sorted turbidite sands.

The bedding of sand and sandstone gives even more useful information, for we can recognize crossbedding, ripple marks, and bedding-plane structures, such as flute casts, associated with turbidites (see Chapter 12). These help us to identify the sedimentary environment in which the deposit was laid down. Crossbedding and other current structures also are guides to the current direction. They can be measured in the field to make a **paleocurrent** map, showing the directions of sediment transport. If the sandstone were deposited by a river, the paleocurrent map

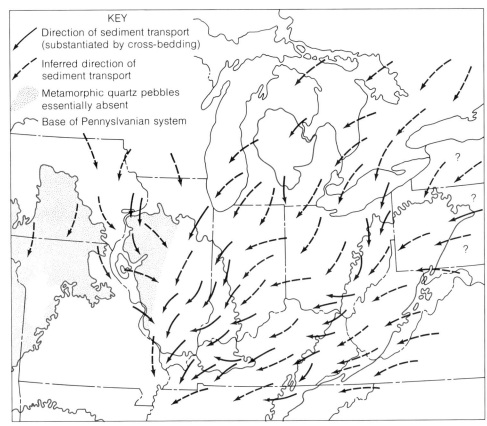

KEY

⟋ Direction of sediment transport
(substantiated by cross-bedding)

⟋ Inferred direction of
sediment transport

▒ Metamorphic quartz pebbles
essentially absent

⌒ Base of Pennyslvanian system

**Figure 13-21**
Map of inferred paleocurrent directions based on cross-bedding in the basal Pennsylvanian sandstones of part of the northeastern United States. General directions correspond to areas distinguished
by presence or absence of metamorphic quartz pebbles. [After P. E. Potter and R. Siever, "Sources
of Basal Pennsylvanian Sediments in the Eastern Interior Basin," *Jour. Geology*, v. 64, p. 242, 1956.
Copyright © 1956 by University of Chicago Press.]

can be viewed as a map of the general land slope down which the
rivers ran. If the bed structures of a turbidity-current are measured,
we can construct a paleocurrent map of the submarine slopes
down which the muddy suspension flowed. Thus, we can map
part of the route of sediment migration from eroding source to
depositional basin.

Paleocurrents, grain size and sorting, and many other characteristics of sandstones are studied intensively by petroleum geologists,
for much of the world's oil is found in the pore spaces of sandstones. Knowledge of how the sand was deposited guides geologists
in their exploration for new oil fields and assists them in the reservoir engineering required to extract the oil from the ground. Such
knowledge is also a necessity for the geologist seeking groundwater
supplies in porous sandstone beds.

**Gravel and Conglomerate** If sand grains are easy to study, the pebbles of a conglomerate are that much more so. The variety of pebbles is as great as the different kinds of outcrops from which they were eroded, and the information they provide is explicit. A granite pebble is sure evidence of the existence of a mass of exposed granite in the drainage basin of the rivers that brought the pebbles to the site of deposition. The number and size of the pebbles is directly related to the strength of the current that transported them. Because there are relatively few kinds of places in which currents strong enough to transport pebbles can flow, many times we can infer the origin of the pebbles with some confidence. Mountain streams carry gravels, and alluvial fans are among the important places in which they are dropped. Many river channels carry pebbles in their upper and middle reaches. Beach gravels are common along rapidly eroding coasts of hard rocks. Glacial outwash typically contains much gravel.

Because pebbles and cobbles abrade and become rounded very quickly in the course of transport, their shapes and the roundness of edges and corners are good guides to the distance they have traveled from their source area. Maps of average or maximum pebble size can be made that show a steady decrease in pebble diameter in a certain direction. In some studies, cross-bedding in the formation shows the same direction. Because there is a limit to the size of pebble any river can carry, generally about 25 centimeters in diameter, one can project upstream from measured sizes, knowing the rate of size increase from the map, to the position

**Figure 13-22**
Conglomerate and associated sandstone, Puente Hills, California. [From *Geology Illustrated* by J. S. Shelton. W. H. Freeman and Company. Copyright © 1966.]

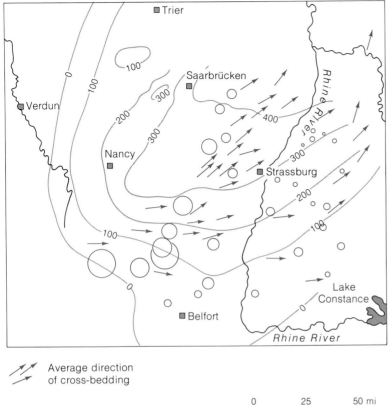

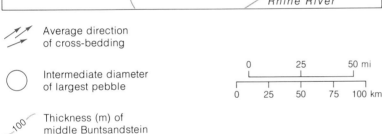

| | Average direction of cross-bedding |
| | Intermediate diameter of largest pebble |
| | Thickness (m) of middle Buntsandstein |

0    25    50 mi

0    25    50    75    100 km

**Figure 13-23**
Change in pebble size associated with cross-bedding directions in the Triassic Bunter Sandstone of France and Germany. Note the increase in the thickness of sandstone in the down-dip direction of cross-bedding; this increase is associated with aggradation of an alluvial plain. [From P. E. Potter and F. J. Pettijohn, *Paleocurrents and Basin Analysis,* Copyright 1963 by Springer-Verlag New York, Inc.]

where 25-centimeter pebbles would be found. That would be the source area. In this way, ancient mountain fronts have been deduced from sediment characteristics.

Conglomerates can be of local origin within the sedimentary environment. Strong storm waves can tear up previously deposited, partly compacted sediment and redeposit the irregular pebbles in a new matrix of sand or mud when the storm dies down. This happens most commonly with such fine-grained deposits as shales. The boulders and cobbles of a coral reef that accumulate along its margin as a talus slope may also form conglomerates. Owing to the nature of the erosional processes on land, gravel talus slopes against continental mountains are continually stripped away and are poorly preserved in the rock record.

**Mud and Shale**   The fine-grained sediments of the Earth, which are the most abundant and widespread, reveal the least about their formation. All too often, their outcrops are completely covered with soil or rubble. The material itself must mainly be studied by X-ray diffraction or chemical methods. Much of what we can determine from shales about their formation is a general deduction from their fine grain size: they are the result of slow settling from very gentle transporting currents.

On a river floodplain, muds and silts may be deposited from a waning flood or in an oxbow lake formed from a cut-off meander. Muds are left behind by ebbing tides along many tidal flats where wave action is not great. Below the depth of effective wave transport, mud settles over much of the deeper parts of continental shelves, slopes, and rises. And a slow, steady deposition of fine particles makes its contribution to the extensive deep-sea sediments over ridges, trenches, abyssal hills, and abyssal plains. Mud deposited by waning turbidity currents accounts for much of the sediment of abyssal plains.

Many muds and shales are mixed with chemical sediments; calcareous shale and shaly limestone are the most prominent types. Cherts, evaporites, and other chemically precipitated rocks may be mixed with mud, and some sediments grade from one pure rock type to the other with all the intermediate transitions. Frequently, it is this chemical component that gives the shale a distinctive character and, thus, a name. Black or organic shales are those that contain abundant organic matter, which is evidence of formation in a poorly oxygenated environment that allowed the organic compounds to escape oxidation and decay. Ferruginous or iron-rich shales are those rich in such iron minerals as hematite.

Perhaps the major significance of shales lies in their mineral composition. Most shales are at least one-third clay minerals; thus, they represent what we might call the dregs of weathering. Each shale bed is evidence of the weathering of a great volume of feldspar and other silicate minerals to produce kaolinite, illite, montmorillonite, and chlorite, the major groups of clay minerals.

Though shales may not provide a good indication of the geological conditions at their time of formation, they are economically very useful as the raw material for brick, tile, china, and pottery of all kinds. Mixed with limestone, shale becomes the raw material for Portland cement, the common cement used in making mortar and concrete. At this point in history, one kind of shale stands out as most valuable: oil shale. One of the world's major oil-shale reserves is in the western United States. The Green River shale owes its high oil content to the geological transformation of abundant algae and other organisms that grew in an enormous lake that covered parts of Wyoming, Colorado, and adjacent states. The

Green River shale has, in addition to its high organic content, a most unusual group of chemically precipitated minerals, the product of alkaline conditions in the lake.

## SEDIMENTARY ENVIRONMENTS

How does one make order out of the almost bewildering array of different kinds of sediments of all ages and take advantage of all the information they contain about what the earth was like when they were deposited? One way is to use the concept of the **sedimentary environment**—that is, all of the important physical and chemical factors or variables that control the formation of a particular sedimentary type.

We might describe a sedimentary environment in a limited way by making a statement about the average kinetic energy—that is, the energy of motion of its water; thus, we might thereby lump together deposits of point bars in rapidly flowing rivers, sandbars off a beach, and some turbidity-current deposits on the continental rise. Or we might characterize a chemical environment by saying that the water is relatively pure—that is, fresh. That would put together soils, sediments formed in rivers and lakes, and permeable rocks through which fresh groundwater percolates. These kinds of environmental definitions limited to one variable generally fail to be useful because they lump rather diverse sediment types and are not easily used for reconstructing the past. History is read in terms of a paleogeographic map, for example, which shows the distribution of land, sea, plains, and mountains, and where what kinds of sediments were being laid down, during some past time period (see Fig. 22-55).

A more useful general definition of a sedimentary environment is one that emphasizes the geomorphologic approach as a meaningful way to group the variables. Thus, we might describe a lagoonal environment as one of limited area, separated by some barrier from the open sea, in which the water is relatively calm and the depths shallow; some varieties may be further characterized by restricted circulation and lack of oxygen in bottom waters, and reef lagoons may be distinguished as a special type. This approach, which unites all the variables by the geometric shape of the place and its relation to adjacent environments and larger geographic and geologic elements of the Earth's surface, has proved most useful. Table 13-2 lists some major sedimentary environments and their characteristics. Many more could be listed, and many can be subdivided into smaller units, down to small areas that have been called "microenvironments." Environments may be mixed and not entirely sedimentary. For example, environments around volcanic islands may be characterized by mixtures of shallow- and deep-water sediments (derived by the erosion of the volcanic rocks on land) with submarine volcanic flows and pyroclastic deposits from violent eruption. What is important in any such

use of a particular environment is that a set of rock characters—
the only evidence a geologist sees—should be associated with a
geomorphologic environment in which certain well-defined
processes operate.

Table 13-2
Major sedimentary environments

| Environment | Sediment types |
|---|---|
| *Dominantly detrital* | |
| Alluvial | Moderate to poorly sorted point bar and channel sands and gravels interbedded with sandy shales; occasional coal or peat beds; very coarse debris in alluvial fans. |
| Deltaic | Well-sorted bar-finger sands and delta-front silts and clays interbedded with marine clays and coastal-swamp deposits. |
| Tidal | Moderately sorted tidal-channel sands, silts, and muds; abundant invertebrate faunas; pyrite in black deoxygenated layers below surface in some places. |
| Beach and bar | Well-sorted sands and occasional gravels with well-rounded and worn grains interbedded with marine and nonmarine deposits. |
| Shallow-water marine (continental shelf) | Sands, silts, and muds with variable marine fauna distributed in sheet-like geometry and interbedded frequently with some carbonate sediment. |
| Deep-sea turbidite (continental rise, abyssal plain, and trench) | Graded sequences of poorly sorted sand to fine clay-size particles interbedded with pelagic clays; some individual units can be traced laterally for long distances. |
| Deep-sea pelagic (abyssal hills, and abyssal ridges and rises) | Fine clay, some of windblown origin, very finely laminated, frequently with manganese nodules or crusts, interbedded with carbonate or siliceous oozes. |
| *Dominantly chemical* | |
| Reefs | Complex of carbonate rock types, from massive cemented reef rock of corals and/or algae to back-reef carbonate sands, lagoonal muds, and beach sands. |
| Carbonate platforms | Great variety of reef, lagoonal, tidal-flat, and shallow-bank deposits of oolitic and shell-fragment sand, and algal-mat and fine-grained carbonate muds. |
| Deoxygenated barred basins | Pyrite-rich and organic-rich muds with no bottom-dwelling marine fauna. |
| Restricted evaporite basins | Gypsum, anhydrite, halite, and other salts interbedded frequently with carbonates and some muds. |

# BURIAL AND ACCUMULATION OF SEDIMENT

A bar in a river may exist for only a few months, from the high-water stage that made it to the next, which destroys it. Of all the sediments that are made day by day, only a small fraction is preserved and gives us a record of geologic time. Sedimentary environments differ in the probability of preservation, and the differences are related, in part, to the rapidity of sedimentation and burial. In the deep sea, deposits slowly accumulate at very low rates, only a few millimeters every thousand years. Once deposited, marine sediments are unlikely to be eroded and redeposited, because bottom currents strong enough for the task are rare. That is the reason why deep-sea pelagic sediments offer the most complete historical records of organic evolution, of temperature changes during ice ages, and of other geologic patterns.

In contrast, the shallower waters of the continental shelf deposit sediments at much higher rates—many centimeters every thousand years, on the average. These sediments get buried faster, but waves and currents are much more likely to scour the bottom and rework it mechanically. Furthermore, the higher the sedimentary accumulation, the more the waves are likely to rework and redistribute it, so that there is a level of sedimentation above which accumulation stops. The importance of **subsidence** of the crust is that it allows sedimentation to continue and the accumulation to build up. This process is one of slight positive feedback, in the sense that tectonic subsidence is reinforced by the sedimentary load. The further sedimentation proceeds, the greater the weight on the crust, because each layer of sediment is much denser than the sea water it replaces. The effect of the weight is to push down

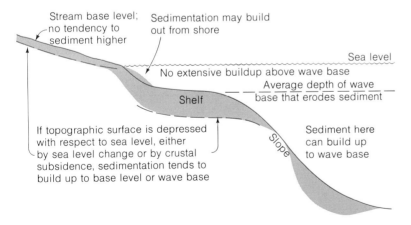

**Figure 13-24**
Sedimentation is closely dependent upon subsidence in those environments where the site of deposition is close to either base level of erosion on land or wave base in the sea.

the crust (see the discussion of isostasy in Chapter 20), allowing still more sedimentation to proceed. This process does not continue endlessly, nor can it account for all of the subsidence shown by most continental shelves. The reason for this is that, as subsidence continues and sediments get laid down, more and more of the crust at that point consists of sediments; because these are much less dense than average crust, the effect of additional sediment tends to become negligible. Subsidence initiated by tectonic mechanisms and enhanced by sediment weight is evidenced by great thicknesses of sediments.

**Geosynclines** The first theorizing about the evidence of sediments of great thickness did not come from modern oceanographic study of the continental shelves, but from the field work of the State Geologist of New York in the middle of the nineteenth century, *James Hall*. As a result of extensive mapping of sediments of the Appalachian Mountains, correlating them stratigraphically, and measuring their thicknesses, Hall concluded that there were many tens of thousands of feet of sediment that had accumulated in a long, relatively narrow trough bordering the continent. He envisioned it as having been something like a large downfold in the surface that steadily accumulated sediment as it subsided. Later, the name **geosyncline** was applied to the sediment-filled trough.

Geosynclines vary from a few tens of kilometers to hundreds of kilometers wide, and they may be thousands of kilometers long. The nature of the sediments they contain gives us a good picture of their origin, for some are turbidite sands formed in deep water and others are shallow carbonate platform limestones; in some places, alluvial deposits are also found. A great many of the sediments are shallow-water ones, including many of those near the bottom of the trough. The great thickness and the widespread distribution of such shallow-water deposits in geosynclines indicate that they were not deep trenches in the ocean that were gradually filled to the surface. On the contrary, as Hall reasoned, the evidence points to a gradually subsiding mobile belt that generally received sediment in proportion to its sinking, but whose surface was shallow most of the time.

Many geosynclines can be subdivided into an inner belt closer to the continental platform, the **miogeosyncline**, and an outer belt closer to an ocean basin, the **eugeosyncline**. The miogeosyncline tends to be filled with lesser thicknesses of limestones, shales, and alluvial sandstones, though other kinds of sediment may also be found. The greater thicknesses of the sediments in the eugeosyncline consist of turbidite sandstones and shales, pelagic limestones and shales mixed with volcanic rocks, submarine basalts, and ash falls and flows. The explanation of the origin of geosynclines in terms of plate-tectonic theory is outlined in

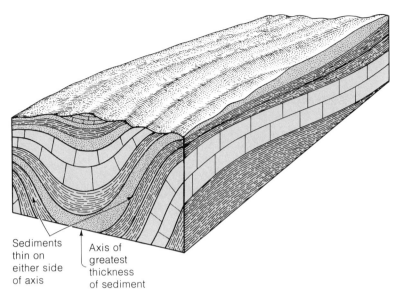

Sediments thin on either side of axis

Axis of greatest thickness of sediment

**Figure 13-25**
A geosyncline is a long, relatively narrow belt in which thick sediments accumulate. The general synclinal form is mainly a result of downwarping due to the
great thickness of sediment along the axis rather than a result of structural
deformation.

Chapter 12 and given in detail in Chapter 21.

Structurally deformed, metamorphosed, and injected with
igneous intrusions, geosynclines were the forerunners of the major
mountain chains of the world: the Alps, the Himalayas, the Appalachians, and the Western Cordilleran belt of the Americas. Only
in the youngest mountain belts are the youngest geosynclinal
beds preserved. As the mountains wear away, older and older
sediments are stripped off by erosion, in many places uncovering
igneous and metamorphic rocks below. As geologic time goes on,
the less likely it is that sediments uplifted into mountain ranges
will survive erosion. In general, less and less sediment is found
preserved as one moves back through the history of the Earth.

**Continental Platforms** Wide areas of the North American
continent between the Appalachians and the Rocky Mountains
are parts of the continent that have been relatively stable at least
since the Cambrian Period began. Sedimentary accumulations are
much thinner than in the geosynclines to the east and west, though
there are sedimentary basins, such as the Illinois Basin, that have
received as much as 3000 meters (10,000 feet) of sediment. Deep-
water formations are conspicuously absent, and many of the
sediments are shallow carbonate-platform sediments and alluvial
(or near-shore marine) sandstones and shales. Individual formations
more or less continuous with those in geosynclines are much
thinner on the platform. Subsidence has been much less on the

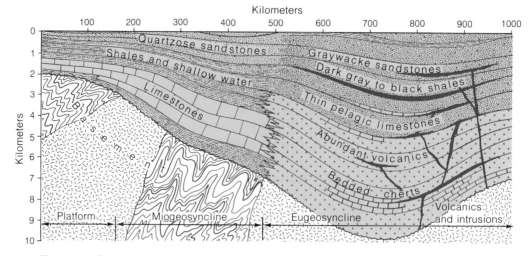

**Figure 13-26**
Diagrammatic cross section of a geosyncline and adjoining platform area. The platform and miogeosyncline differ largely in thickness of sediment, whereas miogeosyncline and eugeosyncline differ
also in kinds of sedimentary rocks.

platform, and no great mountain-making episodes have affected it. Conversely, erosion has not eaten so deeply into the surface sedimentary rocks, so that Precambrian igneous and metamorphic rocks, the basement, are exposed only in a few places, such as in the Ozark Mountains or the Black Hills. The key to platform sedimentation is not so much in sediment types and the environments in which they formed, for there is almost complete overlap with geosynclines in that regard, but in the thinness of the accumulation, a response to only mild and intermittent subsidence.

Subsidence, which influences environment and thickness of accumulation, is only one aspect of the tectonic control of sedimentation. As important is the indirect effect that tectonics has on the mineral composition of detrital sediments.

## TECTONICS AND SEDIMENT COMPOSITION

Contrasts in the mineral composition of clastic sediments can be startling. Compare two alluvial sandstones: an arkose, a sandstone rich in feldspar, with a pure-quartz sandstone. The arkose may contain as much as 30–40 percent feldspar, the remainder quartz and other mineral and rock fragments. The grains are coarse and angular, showing little sign of abrasion. The quartz sandstone may be more than 95 percent quartz with no feldspar. The grains are well-sorted and well-rounded. The contrast is not in the environment, for both are alluvial, but in the source of the erosional debris. The source of the arkose is likely a granitic igneous rock;

the rocks eroded to make the quartz sand might well have been another sandstone.

We can generalize further about feldspar, the most abundant mineral of the crust, as a detrital mineral in comparison to quartz. As shown in Chapter 6, feldspar chemically decomposes to clay minerals, given enough time for reaction with rain and soil water. Quartz is essentially unaffected by most chemical weathering. The survival of feldspar in a weathering terrain is dependent upon the rate of chemical decomposition in relation to the rate of mechanical erosion that breaks up the rock and transports it. The ratio of chemical to mechanical erosion is related to the topography of the terrain—the more elevated and rugged the mountains, the more rigorous the mechanical erosion. Thus, we may relate the abundance of feldspar to mountainous topography, which itself is the product of tectonic activity. Low-lying areas not subject to severe mechanical erosion produce detritus low in feldspar: there, much of the feldspar decomposes to clay before it can be transported.

This tectonic background to sediment composition is a major basis for the reconstruction of ancient tectonic events, of high mountain chains that once shed abundant detritus and are now gone. For example, the sandstones of Devonian age of the Catskill Mountains of New York are a very thick accumulation of alluvial and deltaic sands that were transported from a great mountain range to the east, roughly parallel to the present low hills of the Taconic Mountains east of the Hudson River. In the same way, in Canada, geologists have been able to reconstruct some fragmentary patterns of vanished mountain ranges, some more than 2 billion years old, from Precambrian sediments. Sediments are the hieroglyphics of geologic history.

## SEDIMENT INTO ROCK: DIAGENESIS

Once a sediment is deposited and buried by other sediments, it is not immune to change. One has only to compare freshly deposited muds and sands with ancient shales and sandstones to see the obvious differences in hardness, cohesion, and porosity. The many processes that produce the changes that take place in a rock's composition and texture after deposition are lumped together in the term **diagenesis**. Generally, they operate to harden the soft sediment into rock—that is, to **lithify** it. Diagenesis may also alter the mineral composition by dissolving some of the original minerals and precipitating new ones.

The major physical diagenetic change is **compaction**, a decrease in porosity caused by mineral grains being squeezed closer to each other by the weight of overlying sediment. Sands are fairly well packed as they are deposited, so they compact little, even when buried deeply. Newly deposited muds, however, are highly porous: they have a high water content when deposited, often well over 60 percent, and their layers thin drastically by compaction. The porosity of a sediment may also decrease greatly from the precipitation of diagenetic minerals in its pore spaces.

Chemical diagenetic changes are the result of two general tendencies. The first is a gradual approach toward chemical equilibrium of the nonequilibrium mixture of diverse minerals from many different kinds of rock that have been brought together as detritus in the sediment (and, perhaps, mixed with some chemical precipitate from the environment, such as calcite). Thus, sedimentation may mix minerals from two very different kinds of igneous rocks, minerals that would have been incompatible under the original conditions of formation of the igneous rocks, such as a sodium-rich plagioclase feldspar from a granite and a calcium-rich one from a basalt. Diagenetic chemical changes tend to dissolve the calcium-rich feldspars and precipitate sodium-rich feldspars, thus moving the rock toward chemical equilibrium—namely, a more homogeneous plagioclase composition. A different example of lack of equilibrium is a grain of aragonite in a carbonate sediment, which, given enough time, tends spontaneously to change to calcite, which is the form of calcium carbonate stable at low pressures and temperatures. This tendency to equilibrium results in many other chemical reactions between incompatible or unstable minerals, which result in the formation of new minerals and, thus, bring the rock to a new composition in equilibrium with its surroundings.

The second tendency is for a sediment to be buried more or less deeply in the crust. As a sediment is buried, it becomes subjected to increasingly high temperatures—on the average 1°C for each 30 meters (100 feet) of depth (see Chapter 15)—and high pressures—on the average about 1 atmosphere for each 4.4 meters of depth (1 pound per square inch for each foot of depth). As minerals and the surrounding groundwater in pore spaces are heated up and put under greater pressure, they tend to react chemically to form new minerals. This process, when carried far enough, becomes metamorphism, in which the entire character of the rock alters. The boundary between diagenesis and metamorphism is somewhat arbitrary, usually drawn at a temperature of about 300°C.

The list of specific diagenetic changes is great. Few of these are universal; their occurrence depends on the geologic situation. A few important examples include the change of unstable opaline silica shells of diatoms to quartz, the stable form; transformation of the unstable swelling clay, montmorillonite, to a stable mica type of clay, illite; the precipitation of calcite and quartz as pore-filling cement in sandstones; and the alteration of volcanic glass to clay minerals and zeolite minerals. All of these changes must be understood before we can deduce the nature of the original sediment that was laid down, which we need to do in order to interpret geologic history.

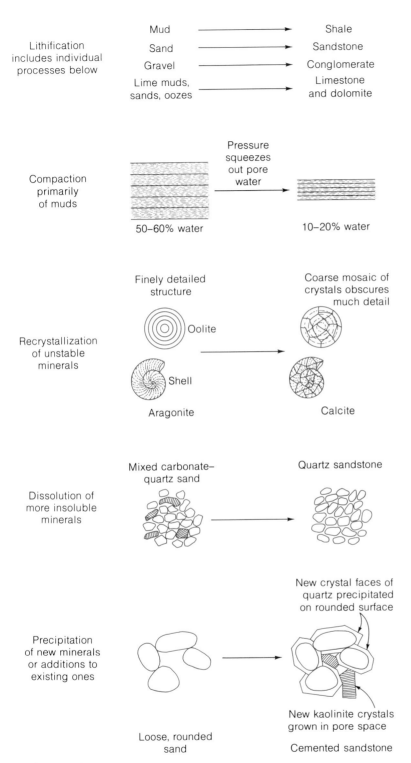

**Figure 13-27**
Some changes in composition and texture that are produced by diagenetic processes. Most of the changes tend to transform a loose, soft sediment into a hard, lithified sedimentary rock.

486

**The Skin of the Earth:**
**Surface Processes**

Diagenesis is of some practical value, because recovery of oil, gas, or groundwater from formations depends on the porosity of the rock, which may have been largely determined by diagenetic cementation. The formation of oil itself is a diagenetic process by which organic remains of many kinds of organisms are gradually altered to liquid petroleum or natural gas, or both. The transformation of peat through bituminous coal to anthracite is another diagenetic process. Understanding all of these processes better enables us to search more intelligently and successfully for new mineral and energy resources.

## SUMMARY

1. Physical sedimentation is controlled by gravity, and chemical sedimentation is dominated by precipitation from sea water, which keeps the ocean constant in composition.

2. Carbonate sediments form from supersaturated sea water in warm parts of the ocean's surface, primarily through the activities of organisms that secrete carbonate.

3. Coral reefs and atolls are built by large communities of carbonate-secreting corals and coralline algae. Their structure of reef front, lagoon, and island is made by the complex interaction of many kinds of organisms and the sea. Reefs form around originally volcanic islands, which may later erode and subside to be covered over by reef.

4. Carbonate platforms near or attached to continents in tropical seas account for most of the shallow-water carbonate sediments of the world. Oolite, carbonate sands, muds, and tidal-flat deposits, such as stromatolites, are characteristic of the various environments of the shallow bank areas of the Bahama Islands.

5. Silica is secreted by diatoms, radiolarians, and sponges. Siliceous oozes and cherts are formed by the sedimentation of the skeletons of these organisms on the sea floor.

6. Precipitation of sedimentary sulfides is accomplished by the action of sulfate-reducing bacteria on organic matter in such deoxygenated environments as silled basins or tidal flats.

7. Coal is formed by the partial decay and transformation of organic matter laid down in swampy environments. Vegetation is first changed into peat and then, in an ascending series of metamorphic stages, to lignite, bituminous coal, and anthracite.

8. Evaporites are deposited in restricted basins of the sea in which evaporation continues to produce precipitates of gypsum and salt as sea water is replenished by a connection to the open sea. Some evaporites form from complete desiccation. Others form by evaporation on tidal flats in arid environments.

9. Clastic sediments are divided into conglomerate, sandstone, and shale on the basis of grain size, which is a reflection of current strength.

10. A study of the grain size and bedding characteristics of sand and sandstone gives information on current type and direction. Paleocurrents are inferred from cross-bedding and other directional structures.

11. Gravel and conglomerate are formed by strong currents characteristic of rapidly flowing rivers and strong waves. Pebble size and roundness can be used to infer the distance of transport from the sediment source area.

12. Mud and shale are deposited from settling of fine material from slow or waning currents on river floodplains and in deeper marine environments. Many shales are mixed with chemical components. The clay minerals of shales come from the weathering of silicate minerals.

13. Sedimentary environments are described in terms of geomorphology, which links important sedimentary characteristics to particular geographic locations.

14. Sediment accumulates as a result of subsidence in the depositional area. Geosynclines, which are the locations of sedimentary accumulations of great thickness, can be subdivided into miogeosynclinal and eugeosynclinal tracts.

15. A continental platform is a large area of a continent over which subsidence has been slight for a long time. As a result, sedimentary accumulations on such platforms are thin.

16. Tectonic control of sediment composition is evidenced by the ratio of quartz, a stable mineral, to feldspar, an unstable mineral in weathering environments. Mountain building results in more rapid mechanical erosion of feldspar; thus, abundant feldspar in sediments indicates tectonism.

17. Diagenesis is the process by which sediments are changed physically and chemically and lithified, partly as a result of burial and increases in temperature and pressure.


# EXERCISES

1. A few years ago some marine biologists thought that dynamiting certain areas of a coral reef for engineering purposes was upsetting the ecological balance of the reef and would lead to the death of large populations of corals. Their fears have not been substantiated (so far), but if they had been right and all of the corals had died, what events might characterize the further evolution of the reef? Could you recognize these events from the sediment produced?

2. What reasons can you think of why the Pacific Ocean has so many coral atolls and the Atlantic so few?

3. Early in the Precambrian era there were no carbonate-secreting organisms. Assuming the oceans, with that exception, were much like those of today, where and what kind of carbonate sediment, if any, would you expect to have been deposited?

4. The Black Sea is known to lack oxygen in its deep waters. What does this suggest about the nature of the Bosporous and Dardanelles, the straits that connect the sea with the Mediterranean? What kinds of sediment might you expect to find at the bottom of the Black Sea?

5. The general oceanographic circulation conditions of some ancient evaporite basins seem to have been similar to those of some present-day deoxygenated basins, yet only some of the evaporite basins contain much organic matter and sulfides. Why might this be so?

6. A geologist has mapped a formation of a certain age as changing from a conglomerate on the western edge of his map area to interbedded sandstone and shale on the eastern edge. On the basis of this information alone, what deductions might be made about the depositional environment and paleogeography of the formation?

7. In the Mississippi delta country of Louisiana, oil wells have been drilled 7000 meters or more to penetrate Miocene formations of nearshore shallow-water sandstones and shales. How do you think this much sediment has accumulated?

8. How do you think the processes that have made mud into shale might generally differ from those that have made sand into sandstone?

## BIBLIOGRAPHY

Blatt, H., G. Middleton, and R. Murray, *Origin of Sedimentary Rocks*. Englewood Cliffs, New Jersey: Prentice-Hall, Inc., 1972.

Garrels, R. M., and F. T. Mackenzie, *Evolution of Sedimentary Rocks*. New York: W. W. Norton and Co., 1971.

Krumbein, W. C., and L. L. Sloss, *Stratigraphy and Sedimentation* (2nd ed.). San Francisco: W. H. Freeman and Company, 1963.

Kuenen, P. H., "Sand." *Scientific American*, April, 1960. (Offprint No. 803.)

Laporte, L. F., *Ancient Environments*. Englewood Cliffs, New Jersey: Prentice-Hall, Inc., 1968.

Selley, R. C., *Ancient Sedimentary Environments*. Ithaca, New York: Cornell University Press, 1970.

# 14

# Earth and Life

*Earth is probably only one of many billions of planets in the universe on which the conditions necessary for the development of life were met. The compositions of the crust, the atmosphere, and the oceans have not only determined the course of biological evolution but have been themselves affected by life processes. The primitive Earth had an oxygen-free atmosphere conducive to the origin of life. Some time after primitive life started, photosynthesis evolved and oxygen began to build up in the atmosphere, eventually to near present levels. Life depends on the environments in which it evolved and to which it has adapted. Human activities may affect those environments seriously by increasing the levels of carbon dioxide, which might affect the global climate, and by introducing large quantities of toxic metals into the biosphere.*

The Earth gives us permission to live. It provides the basics: the air we breathe, the water we drink, and the food we eat — the plants and animals that share the planet with us. For modern civilizations, it provides the mineral and energy resources necessary to innumerable systems that we take for granted, from the fertilizers required by modern high-productivity agriculture to the jet fuel required by our airliners. The surface environment of Earth plays a profound role in our lives: We can exist only in a narrow range of temperatures and pressures, and we are dependent on the abundance of many chemical compounds — water in particular. It has become abundantly clear, particularly in the past decade, that the relation between life and the environment is not one-way. Not only does the physical environment control life, the biological world has a profound effect on the environment. Rich, black soil—an environmental necessity for productive agriculture—is formed largely by the interaction of organisms with elements of the physical environment. The oxygen in the atmosphere—an element of the environment that is required by the respiratory mechanisms of all complex forms of life—has been manufactured by green plants in the process of converting sunlight to usable chemical energy by means of photosynthesis. The human population of the Earth is increasingly altering the global environment: We add more and more carbon dioxide to the atmosphere by burning coal and oil, for example, and such pesticides as DDT have been dispersed throughout the waters of the continents and oceans.

In this chapter, we will explore the interaction between Earth and life. Why is there life here and not on the Moon? How did life start and how did it evolve? Above all, how precarious is our existence on this planet, and what are the possible effects of our technology and growing population on our environment?

## LIFE BEGINS

Throughout most of human history, the subject of the origin of life was exclusively a mythological or a theological concern. Discussion of the matter dealt mainly with the precise meaning of the revealed truth of the Judeo-Christian scriptures or that of the sacred texts or oral traditions of other religions; all explanations were framed in terms of the actions of a creator. Although the rise of geology in the eighteenth and nineteenth centuries (as part of the "Age of Enlightenment") permitted the age of the Earth to be discussed in scientific terms rather than in terms of a literal interpretation of the Bible, it was not until the twentieth century that scientists seriously opened the subject of the origin of life to nontheological discussion. In the nineteenth century, Louis Pasteur waged a decisive battle against the idea of "spontaneous generation" of life, challenging the commonly held view that, by some such process, maggots, flies, and bacteria came to life on rotting garbage. Pasteur showed that neither bacteria nor any other kind of life could arise from any kind of dead organic matter if it were first sterilized by heat. Life originated from other life only, he argued. Right as he was about life as we now know it, he and others simply ignored the question of whether there was ever a beginning to the process.

In 1924 a young Russian biochemist, *A. I. Oparin,* opened up the whole matter by asserting that there had to have been a beginning, sometime in the early history of the Earth, and that it was possible to make some intelligent guesses about how it might have happened. He theorized that, in its early evolution, the Earth's atmosphere lacked oxygen but contained many reduced gases, such as ammonia ($NH_3$), hydrogen ($H_2$), and methane ($CH_4$).

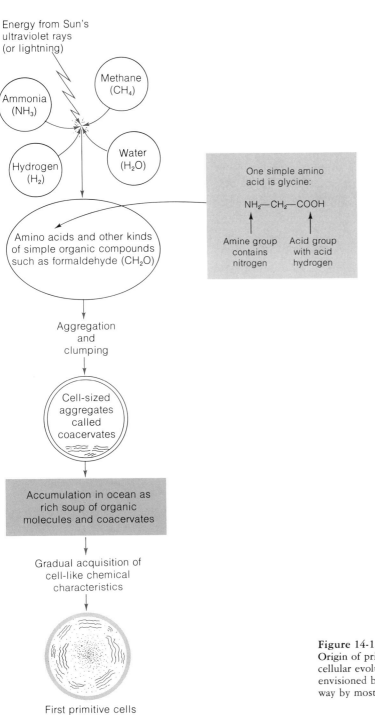

Energy from Sun's ultraviolet rays (or lightning)

Ammonia ($NH_3$)

Methane ($CH_4$)

Hydrogen ($H_2$)

Water ($H_2O$)

One simple amino acid is glycine:

$$NH_2-CH_2-COOH$$

Amine group contains nitrogen

Acid group with acid hydrogen

Amino acids and other kinds of simple organic compounds such as formaldehyde ($CH_2O$)

Aggregation and clumping

Cell-sized aggregates called coacervates

Accumulation in ocean as rich soup of organic molecules and coacervates

Gradual acquisition of cell-like chemical characteristics

First primitive cells

Higher evolution

Plant and animal kingdom

**Figure 14-1**
Origin of primitive organic compounds, followed by precellular evolution, leading to the first cells—a scheme first envisioned by A. I. Oparin and now accepted in a general way by most scientists studying the origin of life.

In that kind of atmosphere, energy from ultraviolet rays from the Sun, or lightning, might have created organic molecules similar to the basic building blocks of life, such as the amino acids, which are molecules that form proteins when chemically strung together in long chains. Once the building blocks were formed, they might have tended to clump together in larger and larger units, which would then have begun to attain some of the characteristics of a primitive cell. All of this gradual synthesis would have taken place in the early ocean, which he described as a "soup" of organic molecules. In that soup, the primitive cell-like aggregates would gradually have acquired more life-like attributes, a process culminating in the appearance of the genetic substance governing reproduction and inheritance. This stage would have marked the beginning of life. Oparin thus showed how a series of small, statistically probable chemical events could lead to the formation of life. A generation later, the DNA (desoxyribonucleic acid) molecule was shown to be the genetic substance, and its structure and mode of operation were worked out. Once DNA was formed, cells could divide, each keeping a strand of DNA, which would cause it to resemble its parent and, at the same time, would offer the opportunity for change and evolution through random alteration of the strand. (The changes in the organism produced by alterations of its DNA cause it to interact differently with the physical environment. Some changes are beneficial to the organism in its interaction with the environment and others are not; thus, some individual organisms survive and others die. This phenomenon, called natural selection by Darwin, was instrumental in directing the course of evolution.)

About thirty years later, a young American graduate student in chemistry, Stanley L. Miller, devised a crucial experiment that showed how it really might have happened. He built an apparatus in which he could simulate lightning bolts in a reducing atmosphere by setting off powerful electric sparks in a glass chamber filled with a mixture of water, ammonia, and methane. After sparking such a mixture for many hours, Miller was able to identify a number of amino acids. In the last decade, partly stimulated by the space program and the search for life on other planets, the study of the origin of life has become an active science linking biology with geology and astronomy.

The essence of the theory remains much as Oparin first suggested, the important condition being the oxygen-free atmosphere. The necessity for a reducing atmosphere arises because synthesis and preservation of a variety of large organic molecules takes place easily only in such surroundings. Many now believe that the major energy source required to link small molecules together to make larger ones is ultraviolet radiation. On today's Earth, most ultraviolet light is filtered out by a layer of ozone ($O_3$), a three-atom oxygen molecule formed in the stratosphere from the more

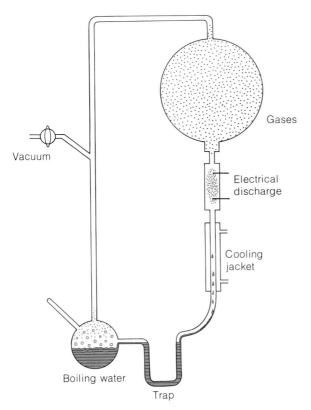

Vacuum

Gases

Electrical
discharge

Cooling
jacket

Boiling water

Trap

**Figure 14-2**
Experiment of S. L. Miller, in which amino acids were made
by circulating methane ($CH_4$), ammonia ($NH_3$), water vapor
($H_2O$), and hydrogen ($H_2$) past an electrical discharge. The
amino acids, collected at the bottom of the apparatus, were
detected by paper chromatography. [From "The Origin of
Life" by G. Wald. Copyright © 1954 by Scientific
American, Inc. All rights reserved.]

ordinary two-atom oxygen molecule ($O_2$). In the ozone layer, some
oxygen molecules are split by absorption of the energy of ultraviolet rays to form single oxygen atoms. These then combine with
oxygen molecules to form the ozone, which itself absorbs ultraviolet rays very effectively. Were it not for the ozone layer, we
would all be irradiated with powerful ultraviolet light, which
would induce such intense sunburn that life on the surface of the
land would be impossible. Strange that the rays that now tend to
destroy life worked then to build it.

The early ocean "soup" was fairly dilute, perhaps with a thin
scum of oily organic matter on its surface. Much current research
is concerned with the next step: the making of large aggregate
molecules of protein-like or DNA-like materials of the sort that
led to the evolution of the first cell. Experiments are now beginning
to show the nature of the many chemical steps involved in the
building of life-like molecules.

Some scientists believe that the
existence of the ozone layer would
be threatened by many supersonic
transport (SST) planes flying
through the stratosphere: the jet
gases would chemically react with
the ozone, converting it to oxygen,
thus leaving us unprotected from
ultraviolet radiation.

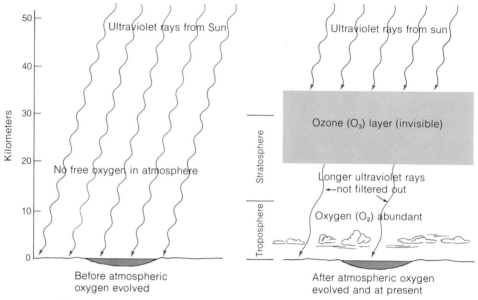

**Figure 14-3**
The evolution of oxygen in the atmosphere led to the formation of an ozone layer in the stratosphere as oxygen atoms and molecules combined to form the three-atom molecule $O_3$. This layer absorbs much of the shortwave ultraviolet rays from the sun and protects surface life from this cell-damaging radiation.

## THE OLDEST FOSSILS

When did all of this take place? Paleontology gives us an important date: In a dense chert in a sedimentary sequence in South Africa that is more than 3.2 billion years old, there are microscopic spherical and filamentous forms of carbonaceous organic matter that are probably fossils of primitive cells. Studies of the chemical composition of the carbon material support the microscopic evidence. If these forms are cells, then the whole process of life may have begun relatively early in the Earth's history, as early as 3.5 or even 4 billion years ago. Just as the search for the oldest rock—presumably igneous—continues, there is the possibility that very old sediments will be found that will allow geologists, with certainty, to push the origin of life even farther back than 3.2 billion years (the age of the oldest known sedimentary rock with evidence of life).

We are sure that, by about 2 billion years ago, well-organized algal life was thriving. A formation that old, exposed on the north shore of Lake Superior, the Gunflint chert, contains a diversity of forms that are undoubtedly biological. Formations of stromatolitic limestones of about the same age have been mapped in Canada and South Africa; the stromatolites are of the same sort as those formed by algal mats today. By one billion years ago, a higher form of life

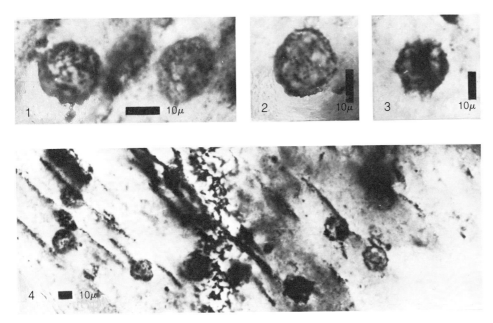

**Figure 14-4**
Organic, alga-like microscopic fossils in black chert of the early Precambrian Fig Tree Series near Barberton, South Africa. (1) Spheroidal and distorted fossils showing typical irregularly reticulate surface texture. (2) Section through one spheroid showing cell wall of variable thickness. (3) Spheroid containing "coal"-like organic material interpreted as the altered remnants of the original cell material. (4) Spheroidal fossils showing varying degrees of preservation. Diagonal layers are laminae of organic matter parallel to bedding. [Photos by E. S. Barghoorn. From J. W. Schopf and E. S. Barghoorn, "Alga-like Fossils from the Early Precambrian of South Africa," *Science,* v. 156, pp. 508–512. Copyright 1967 by the American Association for the Advancement of Science.]

**Figure 14-5**
Algal stromatolites from the 3-billion-year-old Bulawayan Limestone of Rhodesia. Typical stromatolitic, crescentic laminations are shown on a weathered rock surface. [Photo by Keith Kvenvolden.]

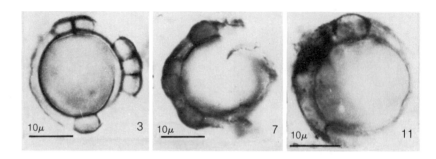

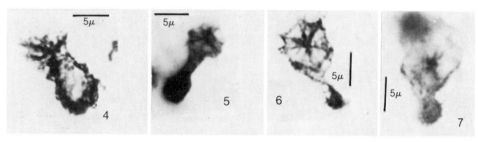

**Figure 14-6**
Microorganisms from the Precambrian Gunflint Chert of Ontario, about 2 billion years old. The top
row shows sections through three different individuals of the same alga-like organism *Eosphaera
tyleri*, thought to be a free-floating colonial photosynthetic organism. The bottom row shows four
different individuals of a peculiar primitive organism, *Kakabekia umbellata*, first discovered in the
Gunflint Chert. [Photos by E. S. Barghoorn. From E. S. Barghoorn and S. A. Tyler, "Microorganisms
from the Gunflint Chert," *Science*, v. 147, pp. 563–577. Copyright 1965 by the American Association
for the Advancement of Science.]

was established: the cell with a nucleus. (Cells of this sort are the ones that most of us are familiar with, but they are restricted to relatively advanced bacteria and higher organisms; the earliest life forms, such as primitive bacteria and algae, had only non-nucleated cells.) We know of these organisms from the Bitter Springs formation of central Australia. Those rocks have the same lithology as the other Precambrian sediments in which the oldest fossils are preserved: unmetamorphosed cherts in which there is excellent preservation of organic structures.

The first shelled organisms appeared at the beginning of the Cambrian Period. Cambrian fossils include an incredible diversity of species of invertebrates. Just before that time, in very late Precambrian sediments transitional to the Cambrian, higher forms of multicellular (or **metazoan**) life without shells had developed. Sometime between 600 million and one billion years ago, higher forms of life evolved so rapidly that the rate of evolutionary development, compared to the rates of earlier times, could be described as explosive. From a world populated by primitive algae and bacteria, there emerged many thousands of species of animals, from sponges and mollusks to worms and trilobites. From then

**Figure 14-7**
A late Precambrian fossil from rocks of the Ediacara
Hills, South Australia. This annelid worm is one of
many soft-bodied organisms whose development
preceded the evolution of the shelled invertebrates of
the Cambrian Period. [Photo by M. F. Glaessner.]

on came the march of evolution that paleontologists have mapped
so well: the appearance of the first vertebrates, land plants, flower-
ing plants, the dinosaurs, the mammals and finally, a few million
years ago, the genus *Homo.*

**Evolution of Atmospheric Oxygen**   There was a radical transformation from the first life that developed in a reducing atmosphere to the kinds of organisms of the Cambrian and later
periods that needed oxygen for respiration. The first organisms
must have lived without oxygen, like some primitive bacteria of
today, such as the sulfate-reducing species that are instrumental
in precipitating pyrite (Chapter 13). At some time around 3 billion
years ago, some of the primitive species must have developed
one of the great biological inventions: photosynthesis.

Photosynthesis is a complex chemical process by which plants
that contain chlorophyll are able to convert carbon dioxide and
water into carbohydrates, the sugars needed for biological energy.
In the equation below, we show the production of $CH_2O$, a simple
formula representing all of the many carbohydrates that actually
form.

$$\text{carbon dioxide} + \text{water} \xrightarrow[\text{sunlight and chlorophyll)}]{\text{(in the presence of}} \text{carbohydrates} + \text{oxygen}$$
$$\quad CO_2 \qquad\qquad H_2O \qquad\qquad\qquad\qquad CH_2O \qquad\qquad O_2$$

In this reaction, approximately 112,000 calories for each 30 grams
of carbohydrate produced is converted from sunlight to chemical
energy tied up in sugars, the energy that is the basis of life processes. At the same time, one molecule of oxygen is produced
for every atom of carbon incorporated into organic matter.

At the beginning of the evolution of photosynthesizing plants,
there were no animals to breathe oxygen, so much of the oxygen
produced by photosynthesis must have started to accumulate in
the atmosphere. This free oxygen made possible the beginning of
a new method of utilizing energy: respiration. Respiration is

Chemical reactions in the upper
atmosphere split a small number
of water molecules into hydrogen
and oxygen gas. A small amount
of atmospheric oxygen may have
originated in this way, but the rate
of production by these reactions is
very slow and, for practical
purposes, negligible.

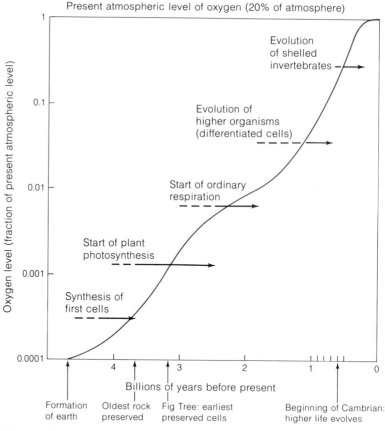

**Figure 14-8**
One hypothesis of the evolution of oxygen in the atmosphere in relation to the origin of life and evolution of higher organisms. There is as yet no general agreement on exactly when and to what levels oxygen accumulated in the Precambrian and how close the levels of oxygen in Phanerozoic time came to the present-day level.

oxidation of carbohydrate, which releases the energy gained from photosynthesis.

$$\text{carbohydrate} + \text{oxygen} \rightarrow \text{carbon dioxide} + \text{water}$$
$$\quad CH_2O \qquad\qquad O_2 \qquad\qquad CO_2 \qquad\quad H_2O$$

In this reaction, one molecule of oxygen is used up in the conversion of each atom of organic carbon to carbon dioxide. The first respirers were undoubtedly primitive bacteria that could live on small amounts of oxygen, perhaps as little as a hundredth of the amount now present in the atmosphere. This development may have taken place as early as 2.7 billion years ago. For a long time oxygen must have continued to accumulate slowly, as photosynthetic production exceeded its depletion by respiration. As a result, we can deduce that an equivalent number of organic carbon atoms were buried as organic matter in sediments. That organic matter can be looked upon as the buried excess of photosynthesis over respiration. In its simplest form the total budget since oxygen first started accumulating is:

$$\left(\begin{array}{c}\text{number of carbon atoms}\\ \text{in organic matter}\\ \text{produced by photosynthesis}\end{array}\right) = \left(\begin{array}{c}\text{number of molecules}\\ \text{of oxygen produced}\end{array}\right)$$

$$-\left(\begin{array}{c}\text{number of carbon atoms}\\ \text{in organic matter}\\ \text{oxidized by respiration}\end{array}\right) = -\left(\begin{array}{c}\text{number of molecules}\\ \text{of oxygen used up}\end{array}\right)$$

$$\left(\begin{array}{c}\text{number of carbon atoms}\\ \text{in organic matter}\\ \text{preserved in sediments}\end{array}\right) = \left(\begin{array}{c}\text{number of molecules of}\\ \text{oxygen accumulated}\\ \text{in atmosphere}\end{array}\right)$$

The photosynthesis-respiration couple that links organisms to the oxygen and carbon dioxide in the atmosphere and oceans.

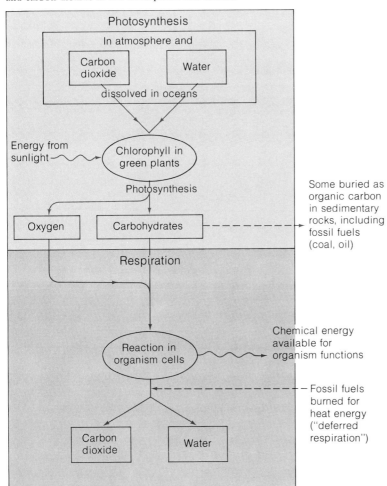

The total amount of carbon in organisms living today is so small in comparison to the total for all the organisms that have lived in the past that it can be neglected.

At a point some time just before the Cambrian, atmospheric oxygen reached levels sufficiently close to today's to allow for the rapid evolution of the higher invertebrates. For the rest of geologic time, the oxygen in the atmosphere has been maintained by the photosynthesis of the green plants of the world, much of it by green algae in the surface waters of the ocean. This story is one of the most dramatic examples of how far-reaching the consequences of the origin and evolution of life were for the surface of the Earth.

## OUR OXYGEN SUPPLY

In the late 1960s, geochemists were called upon to demonstrate that our oxygen supply was vast and that pollution would not deplete it even in the distant future. At that time, some biologists

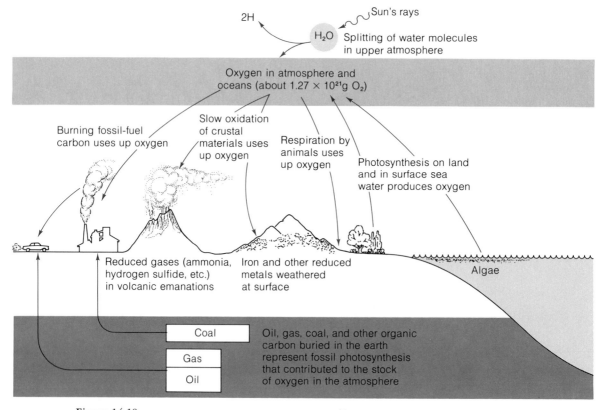

**Figure 14-10**
Input and output of oxygen—atmosphere and oceans. Photosynthesis and respiration are in approximate balance and far outweigh the other inputs and outputs. The amount of oxygen in the atmosphere is so great in proportion to the output rate that even if all photosynthesis were to stop, our oxygen supply would last well over 2000 years.

concerned with environmental quality voiced concern that pollution was endangering our oxygen supply and posed the possibility that we might inadvertently suffocate ourselves. The concern came from reports that DDT was interfering with photosynthesis of some species of green algae. The reasoning followed that, if photosynthesis stopped all over the world, we would "soon" use up all the atmosphere's oxygen by respiration. The concern was shown to be groundless, for there is such an immense amount of oxygen in our atmosphere and oceans (about $10^{21}$ grams) that, even if all photosynthesis stopped tomorrow and all other respiring life went on as before, it would be several thousand years at least before oxygen would be significantly depleted by respiration and all of the other reactions in which oxygen is used up. In addition, only some species of algae seem to be vulnerable to DDT, and there is no evidence that the Earth's major photosynthesizers, oceanic algae, are detectably affected. Some concern about the environment is justified, but there need be no fear of running out of oxygen.

## EXTRATERRESTRIAL LIFE

The evolution of life on Earth was an expectable consequence of the evolution of the planet. Earth was large enough for its gravitational attraction to hold an atmosphere. The temperature of most of the surface was low enough that water was liquid. All of the chemical elements needed for our varied biological processes were present. Given this conjunction of circumstances, was it not inevitable that life forms based on the organic chemistry of carbon and water would arise? If so, then would it also not be inevitable that life is not a strange accident on a unique planet but an event that we would expect to be repeated in various places and at various times in the universe? This was the way Harlow Shapley, one of the great American astronomers of the past generation, started to think about the subject. He then made a simple calculation. How many stars had about the same composition and quality of luminosity as our Sun, and thus could heat and light a planet such as ours? From his knowledge of our galaxy and the rest of the universe, he estimated that there were about $10^{20}$ stars with the proper characteristics. Then, how many of those stars might have a series of planets like the ones in our solar system? One in a thousand is a conservative estimate. How many of those planets might be placed, as the Earth is, just the right distance from their stars for them to have the right temperature? One in a thousand, to continue being conservative. How many of those planets might be big enough to hold atmospheres but not so big as to hold all of the hydrogen too? Again, one in a thousand. After all these and other estimates, he ended with a most conservative estimate of the number of planets in the universe on which life might be expected to develop: 100,000,000. With such a large number, he

Some atmospheric oxygen is used up by the oxidation of metals that occurs in the weathering of rocks, but the amount per year is very small compared with the huge quantities used by respiration and provided by photosynthesis. Accelerated burning of fossil fuels also uses up oxygen, but this amount is also too small to affect the overall picture seriously. So far, we have used up only 7 out of every 10,000 oxygen molecules by burning coal, oil, and gas, and if we were to keep burning fuel at an increasing rate until the year 2000, we would still have used up only 20 out of every 10,000 oxygen molecules.

Since Shapley's estimates, others have modified the estimates of the likelihood of life existing elsewhere in the universe. Some have modified the estimates upward on the supposition that the number of planets like ours is larger; others have decreased the estimates by arguing that the probability of formation of cells and reproductive life, even given favorable conditions, is much smaller.

thought it exceedingly probable that life does indeed exist out there somewhere—perhaps too far ever to reach even by radio communication, but there, nevertheless. Thus was born a new kind of scientific enterprise: **exobiology**.

Part of the excitement about the first trips to the Moon was the thought of confirming our ideas about the origins of life. The Moon was lacking an atmosphere; it was too hot on the day side and too cold on the night side; and there was no free water detectable. We did not expect to find life there. And we didn't find it. The most careful analysis of the first Moon rocks failed to reveal any signs of life, neither cells of organisms nor telltale organic molecules that could only have been formed by biological processes. So we have important negative evidence that gives us some confidence in our reasoning.

Mars will be another story. Because Mars has a thin atmosphere (mostly carbon dioxide) and there is evidence of water (the geomorphology of some terrain that resembles river valleys), the chances that some form of life exists are better than on the Moon. That in itself does not make the probability high, for certainly any form of even very primitive life, at least as we know it on Earth, could not survive on Mars. At this point, we have recourse to observation, and that is the object of one series of experiments planned for an instrument landing. One of the instruments that will be landed on the Martian surface is a device that can detect small quantities of amino acids in soil. If such biologically important compounds exist there, it will be a signal for us to look further.

Life in outer space? It's doubtful; but astronomers have detected small quantities of about twenty simple organic compounds in the nearly empty space between stars, one of which is formaldehyde ($CH_2O$), the simplest building block of carbohydrates. Some of these compounds are also found in meteorites. Thus, it is apparent that the basic units of the compounds of life can be formed in space and transported by meteorites. Now we must indeed recognize the probability that life exists throughout the universe.

Back here on Earth, the importance of all of these ideas is that they make us focus on why we are here and what the conditions are for our continued existence on this planet. How stable is our environment? Can we, the members of one species, so alter the Earth that we make life more difficult for ourselves?

## HAZARDS TO THE ENVIRONMENT

A visit to the "environment and ecology" section of a public library or paperback bookstore may frighten you. Titles are made up of words like "crisis," "survival," "threat," and "danger." Perhaps more revealing is the use of such words as "frail" and "fragile" to describe the Earth. How do we reconcile the geological evidence that the Earth is a dynamic global system that is re-markably stable in its steady-state equilibrium with the doomsday view? To what extent have people suddenly wakened to the real possibility that we are so efficient at controlling our own environment that we may foolishly be tipping "the balance of nature" on a global scale, and to what extent have they simply become aware of what the Earth has been like all the time?

Environmental hazards divide themselves into local and regional versus global problems. Landslides and earthquakes are natural and local phenomena. We describe, in Chapter 8, how poor engineering and construction practices can precipitate landslides and, in Chapter 19, how minor earthquakes can be provoked by pumping fluids at high pressures into subsurface formations. At the opposite extreme are the global hazards, such as the world-wide increase of carbon dioxide in the atmosphere, that might influence the Earth's climate.

Many concerns for specific regions or types of environment are purely esthetic—the preservation of natural beauty, wilderness areas, and, in general, "unspoiled" landscape. To the extent that such concerns are based on uniqueness of the landscape, a knowledge of geology may contribute some. We know that river gorges may be common in a variety of terrains, but there is little question that there is only one of the magnificence of the Grand Canyon of the Colorado.

**Medical Geology** The environmental hazards that are of a different sort of importance are those to public health. The link between geology and health has spawned a new field, **medical geology,** and its questions are new: Are the levels of lead in the atmosphere much above the natural level, and, if so, are they potentially a medical danger? How widespread is mercury pollution, and how toxic are a few parts per billion in drinking water? In such matters, the job of the scientist is dual: One is to measure the spread of a pollutant through the Earth-surface system of atmosphere, surface waters, biological populations, and sediments, and to determine how much man's activities are altering that spread. The other is to answer a series of medical questions: What bodily damage may result from long continued exposure to low levels of toxic metals or other pollutants? What are levels at which obvious disease symptoms appear? At the extreme, what are the lethal doses? We must also consider the ill effects of deficiencies of some metals in our diet.

**Toxic Metals** The toxic metals are some of the most serious potential offenders. Mercury, for example, which is known to be toxic, first hit the headlines a few years ago when a Canadian graduate student discovered high levels of the metal in the tissues of lake fish. We know that mercury occurs in small amounts in many rocks—about 0.2 part per million in granite and less than half that amount in the average crustal rock. The mercury in rocks is steadily contributed in small amounts to natural waters by ordinary chemical weathering processes. Most natural waters contain only a few parts per billion, and thus are harmless. Some part of the mercury in water is naturally converted to an organic form, methyl mercury, which is the form most harmful to organisms. The medical data indicate that chronic mercury poisoning may arise from high levels (many parts per million) of the metal dissolved in water, much of it as methyl mercury. At these levels, mercury affects the nervous system in hidden ways, with few well-defined symptoms shown for long periods of time after the exposure. Acute mercury poisoning has occurred in people living around several bays in Japan, where shellfish, severely contaminated by mercury-laden industrial wastes, were eaten as a large portion of the diet. For these reasons, the World Health Organization has recommended the maximum permissible mercury level for human food, including fish, at 0.5 part per million.

Enough work has been done on the circulation of mercury at the Earth's surface to show how the interaction between the biological world and mercury in the physical environment works. In the past fifty years, many thousands of tons of mercury have been mined for use in electrical equipment, chemical processing plants, and in pesticides. We can look on this as an accelerated weathering process, by which much more mercury than normal is being released from rocks. Though some of the mercury in some chemical processes is reused, a great deal of it escapes into natural waters or is vaporized into the atmosphere. From there, it is distributed to lakes, the ocean, and various sedimentary environments. A fraction of the mercury, converted to methyl mercury by bacteria, is ingested by organisms and accumulates in their tissues. As larger animals eat smaller ones, more of the metal accumulates in the larger ones, so that very large fish, such as tuna or swordfish, may contain relatively large amounts, perhaps a few hundredths of a part per million. In waters that are polluted by industrial waste, the levels may be higher. Eventually, this material is absorbed by sedimentary particles, particularly the clays, and is buried out of reach of the biological system.

The serious question about mercury is, just how much does the mercury in our waters exceed natural levels, and how widespread are such high levels? It is by no means clear that, in most places in the world, natural levels have been exceeded greatly. We know of a few places where industrial wastes have been uncontrolled and the pollution has reached dangerous levels; but we have only recently begun to pay much attention to these elements, and what we need to do is monitor unpolluted and polluted natural waters and their biological populations much more carefully. In company with other toxic metals, mercury has been in Man's diet at very low levels since he evolved several million years ago. Exactly what the range of those levels is—and by how

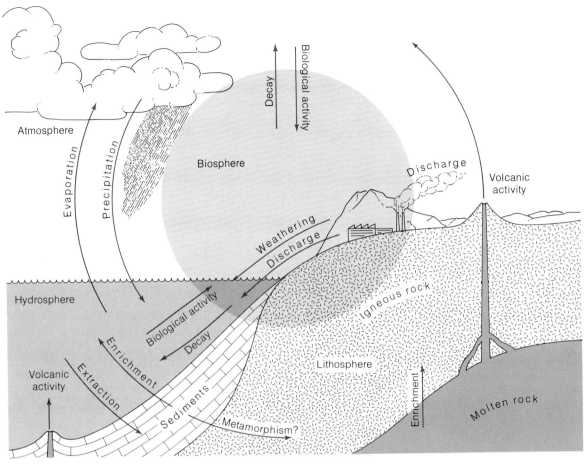

**Figure 14-11**
Mercury cycle disperses the metal throughout the lithosphere, hydrosphere, and atmosphere and through the biosphere, which interpenetrates all three. Mercury is present in all spheres in trace amounts, but it tends to be concentrated by biological processes. Man's activities — in particular, certain industrial processes — may now present a threat by significantly redistributing the metal. [From "Mercury in the Environment" by L. J. Goldwater. Copyright © 1971 by Scientific American, Inc. All rights reserved.]

much they can be exceeded without danger — remains to be determined. There is also good evidence that minimum levels of some metals are required for good health. If, as a result of our fear of absorbing "poisonous" substances, we were to try to use only distilled water for drinking and food preparation, we might produce more harmful consequences than those produced by most impurities in water. We are, after all, adapted to a natural habitat in which small amounts of almost every element are found.

Other metals have the same general cycles of behavior. The most important, from the medical point of view, are lead, cadmium, arsenic, chromium, and nickel. Lead is particularly important, both because it is very toxic and because it is largely dispersed through the atmosphere from automobile and industrial emissions. Two aspects of the problem are apparent: Locally, in inner cities or near major highways, the lead levels from emissions may be very high, compared to those in suburban or rural regions; and globally, there is evidence that the entire atmosphere is being charged with more and more lead. Just how rapidly that is happening and how long the lead stays in the atmosphere before settling out are facts we will have to know before we can assess the environmental dangers to public health.

All environments are self-cleaning by sedimentation. Sooner or later, the contaminants we worry about will settle out of the atmosphere, lakes, and oceans. That is small consolation to most of us, because rates of sedimentation are so slow that, even if sources of excessive pollution are cut off, dangerously high levels may remain for a hundred years. In a particular environment, any pollutant tends toward a steady-state level that is determined by the balance between input rate from natural and human activity, its dispersal through the environment, and the sedimentation rates in that environment. Historical records show whether the system is out of balance, whether the levels are increasing because inputs are greater than outputs. Monitoring in the future may also show

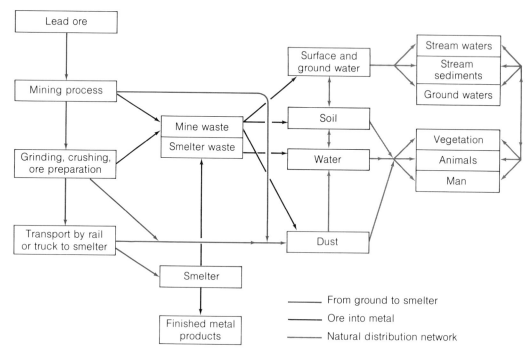

**Figure 14-12**
Movement of lead through surface environments, beginning with the mining of lead ore. At each stage of processing or transport, lead-rich material may get into the air or into surface waters and soil, some of it eventually to be ingested by plants and animals. [Modified from B. G. Wixson, 1973.]

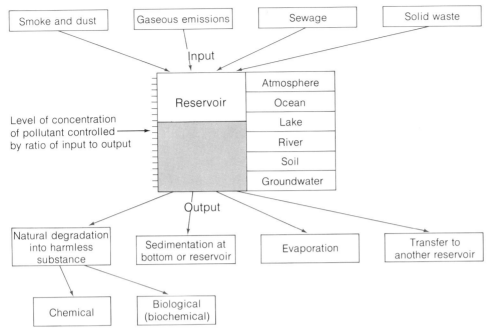

**Figure 14-13**
Inputs and outputs of pollutants in relation to concentration levels in reservoirs. The level is controlled by the input-output ratio. The rate of change of level depends on the rates of the processes involved.

how levels decrease as artificial inputs are lowered and the natural output continues. Input-output relations are shown well by the carbon dioxide system.

## CARBON DIOXIDE AND CLIMATE

The small amount of carbon dioxide in the atmosphere (a little over 320 parts per million) has a profound effect on our climate. The atmosphere is relatively transparent to the incoming visible rays of the Sun. Much of that radiation is changed by reflection at the Earth's surface to infrared, invisible long-wave rays that radiate back away from the surface. The atmosphere, however, is relatively opaque and impermeable to infrared rays because of the combined effect of clouds and carbon dioxide, which strongly absorbs the radiation instead of allowing it to escape into space. This absorbed radiation then heats the atmosphere, which radiates heat back to the Earth's surface. This is called the "greenhouse effect," by analogy to the warming of greenhouses, whose glass is the barrier to heat loss. Any process that alters the amount of carbon dioxide in the atmosphere may conceivably affect the Earth's climate.

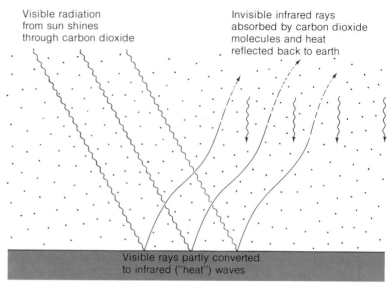

**Figure 14-14**
How the greenhouse effect works. Just as the glass of a greenhouse transmits
light rays but holds in heat, the carbon dioxide of the atmosphere transmits
visible radiation from the sun but absorbs and reflects back to Earth the infrared
rays from the surface.

Since the start of the industrial revolution, about the beginning
of the nineteenth century, we have been pumping carbon dioxide
into the atmosphere at an accelerating rate by our burning of coal,
oil, and gas. The carbon dioxide level of the atmosphere has been
increasing, as shown by systematic measurements in various places
in the world. In addition, calculations have been made of the
amounts of carbon dioxide added to the atmosphere from the
figures for fuel consumption. There is a pronounced discrepancy:
the increased level in the atmosphere is not as high as would have
been predicted by the additional supply. This suggests some loss
or output from the atmosphere, something absorbing the extra
carbon dioxide, moderating the effect of the increased input.

The carbon dioxide that is "missing" from the atmosphere has
been mixed with the oceans. Gas molecules of carbon dioxide
in the air are in equilibrium with dissolved gas molecules in the
water. As the concentration of gas in the air increases, there is a
tendency toward re-establishing equilibrium: the water dissolves
more gas, taking some of the excess from the air. In this way, the
oceans are absorbing some of the carbon dioxide produced by
the industrial revolution (about 50 percent) and keeping the
atmosphere much closer to its natural levels. Nevertheless, in
spite of the ocean's moderating effect, carbon dioxide levels are
expected to reach about 375 parts per million by the year 2000,
a significant increase over 320 parts per million in 1970 and 295

parts per million in the middle of the nineteenth century. Yet it is doubtful that this has seriously affected our climate and that it will so affect it in the next decades.

The possible increase in temperature is not great, partly because atmospheric carbon dioxide is already absorbing about as much radiation as it can and the rest will pass through unabsorbed, and partly because any rise in temperature will produce more cloudiness, which will tend to lower the radiation coming in from the Sun. The best guess now is that, in the next thirty years, the carbon dioxide effect alone will increase the global average temperature by about 0.5°C, but that the countering effect of cloud cover may moderate this almost entirely. (It is doubtful that the small increase in cloudiness would be noticed by the average person.)

The issue is important because an increase in the average temperature of only a few degrees could lead to the melting of glacial ice, a rise in sea level, and flooding of coastal regions. Too important to be left to calculations that depend upon some current uncertainties, carbon dioxide levels continue to be monitored by research laboratories so that we may better predict what will happen in the future.

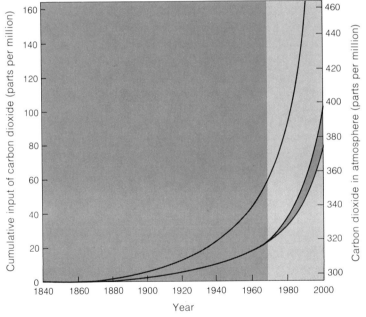

Figure 14-15
The lower curve shows the increase in atmospheric carbon dioxide since 1860, with a projection to the year 2000. The upper curve shows the cumulative input of carbon dioxide. The difference between the two curves represents the amount of carbon dioxide removed by the ocean or by additions to the total biomass of vegetation on land. [From "The Carbon Cycle" by B. Bolin. Copyright © 1970 by Scientific American Inc. All rights reserved.]

## USING LAND WITH GEOLOGICAL FORESIGHT

Life is not all hazards. Part of our concern for the environment stems from feelings that we may not be making the most sensible decisions about our use of land. A geologist can foresee the unhappy consequences of scalping a hillslope of soft, potentially waterlogged formations and building many small homes on the resulting unstable tract. The location of highways, dams and reservoirs, strip and underground mines, large-scale tract housing, and many other enterprises is best not left to chance development, for the costs of unwise choices are too often borne by the community as a whole. Though zoning regulations on building are traditional in many urban areas, they rarely take geological realities into account.

Any construction has to be firm on its foundations, but that is only one aspect of the problem. Builders have to take into account what will happen to the foundations if water from many septic tanks waterlogs the soil. Underground piping of gas, water, and sewage may be affected by slumping, landslides, or earthquakes. Discharge of sewage from septic tanks may find its way into subsurface water supplies, contaminating them with dissolved chemical substances. Garbage disposal is an increasing problem as communities move against open dumps and uncontained burning of trash. One response, spreading the garbage and covering it with soil as sanitary landfill, may also result in groundwater contamination if the proper site is not chosen. The formations below the landfill should be sufficiently impermeable to prevent downward seepage into the water table.

Land-use planning may have to take larger regions of the land into account. Is it wise, for example, to plan for the urbanization of an area known to be the recharge of an underground aquifer that supplies drinking water to the population? Another question is one of surface-water storage. There is always the temptation to drain marshes and swamps and build on the reclaimed land, but there is usually little realization that such places are important storage depots for water (see Chapter 7). Once the wetlands are drained and paved, the area is much more subject to flash flooding: because there is no opportunity for the water to soak into the ground, it runs off immediately.

There are many other choices to be made that involve evaluations of the suitability of land for agriculture, recreation, or transportation. Once the suitability of a given piece of land is determined, we must still face the priority problem: we must determine whether one use of the land is more important than another. Those priorities, such as designating land for recreation rather than building development, are socially and politically determined, but the choices among the alternatives have to be illuminated by geological knowledge. If we are to live with development and utilization of land and resources for different purposes, it ought to be based on planning that is as intelligent as possible. The geological contribution is an understanding of how all contemporary surface processes are related, and how the crust of the Earth beneath has been patterned by past geological events.

## SUMMARY

1. The origin of life on Earth came early in its history as a response to the reducing composition of the primitive atmosphere. Energy, probably as ultraviolet radiation, produced small to medium size organic molecules, such as amino acids, which then grew to larger sizes. Ultimately, the molecules became extremely complex and evolved into true cells.

2. The oldest fossils are primitive cells of Precambrian bacteria and algae found in chert beds. The oldest known are 3.2 billion years. By 2 billion years ago, algal life was well established. By the beginning of the Cambrian period, higher forms of life had evolved and shelled organisms started to provide a good geologic record of evolution.

3. Atmospheric oxygen started to evolve as a result of the development of photosynthesis by early single-celled plants; during the Precambrian, it gradually accumulated to levels close to that of today. Our oxygen supply is now so great that there is no cause for fear that it will be depleted.

4. It is likely that life exists elsewhere in the universe. No life has been found on the Moon, but Mars remains a possibility.

5. Environmental hazards—local or global—may be primarily esthetic, or they may be related to problems of health or safety. Medical geology relates levels of potentially harmful substances, such as lead or mercury, to sources of pollution, to dispersal in the environment, and to rates of removal of pollutants by sedimentation.

6. Global environmental problems are illustrated by carbon dioxide, which has been increasing in the atmosphere as a result of burning fossil fuels. Much of the increase has been absorbed by the oceans. The small increase in temperature expected to be produced by the carbon dioxide increase will probably be balanced by increased cloudiness, so the world's climate will not be significantly affected.

7. Intelligent decisions on the use of land in urban and rural areas for various purposes depend on a knowledge of geological processes and the local geological situation.

## EXERCISES

1. Construct a table showing the ranges of temperature, pressure, and atmospheric composition that make life possible in the form that we know it on Earth. Which, if any, of the other planets might satisfy all of the conditions?

2. Suggest the probable changes in the composition of the Earth's atmosphere in the next several thousand years if all photosynthesis were to stop suddenly today.

3. If you were part of a team of scientists planning a hypothetical manned landing on Mars, what sorts of things would you ask the astronauts to look for as evidence of life of any kind on the surface? What kinds of organic molecules would you presume to be evidence of life?

4. An unsubstantiated report suggested that very high amounts of aluminum in the diet might be harmful to health. From what you know of geochemical and environmental cycles, as well as your knowledge of common aluminum products, suggest possible routes by which people might be subject to unusually high aluminum intake.

5. If you were the mayor of a coastal city, how might you react to a report that some engineers wanted to melt the polar ice caps to produce large quantities of fresh water?

6. A land developer would like to build many houses on a shoreline strip of low-lying coastal marsh below bluffs of relatively unconsolidated shales and sandstones 100 meters high. What geological reasons can you think of that would make you argue against such a development?

## BIBLIOGRAPHY

Laporte, L. F., *The Earth and Human Affairs*, San Francisco, Canfield Press, 1972.

Goldwater, L. J., "Mercury in the Environment." *Scientific American,* May, 1971. (Offprint No. 1221.)

McHarg, I. L., *Design with Nature.* New York: Natural History Press, 1969. (Also issued in paperback by Doubleday/Natural History Press.)

Oparin, A. I., *The Origin of Life.* New York: Macmillan, 1938. (Reprinted in Paperback by Dover Publications, Inc., New York, 1953.)

Shklovski, I. S., and C. Sagan, *Intelligent Life in the Universe.* New York: Holden-Day, Inc. 1966. (Reprinted in paperback by Dell Pub. Co. (Delta), New York.)

Singer, S. F., *Global Effects of Environmental Pollution.* New York: Springer-Verlag, 1970.

Wald, G., "The Origin of Life." *Scientific American,* August, 1944. (Offprint No. 47.)

# Part III

## THE BODY OF THE EARTH: INTERNAL PROCESSES

The materials that make up the atmosphere, the ocean, and the rocky skin of the Earth, and the forces that deform the outermost layers, have their origin in the deeper interior. The internal heat of the Earth provided the energy that led to the formation of plates and their embedded continents, and it provides the energy that keeps the plates moving. These deep-seated forces are so powerful as to cause earthquakes and volcanic eruptions and to form deep-sea trenches, geosynclines, and mountain belts, mainly in association with plate boundaries. The geologist can only infer the properties of the deep interior, using as clues the nature of plate motions, the materials brought up by volcanoes, and physical measurements of such phenomena as heat flow, variations in the magnetic field and gravity, and the travel times of seismic waves. Man's first explorations of other planets will enable him to test his theories of Earth's evolution in the perspective of other bodies that have evolved differently.

# 15

# The Internal Heat of the Earth

*Earth's internal engine is driven by radioactively generated heat and heat left over from the differentiation that took place early in the planet's history. The work done by the engine registers on the surface in the form of plate motions, earthquakes, mountain-making, and volcanic eruptions. In fact, almost all of geology is ultimately related to Earth's internal heat.*

Three hundred years ago the English physicist and chemist Robert Boyle wrote on the nature of terrestrial heat with a prescience that was truly remarkable. In a recent historical review the following quote from Boyle's book *Of the Temperature of the Subterraneal Regions as to Heat and Cold* was reprinted:

> ". . . it seems probable to me, that in these yet inpenetrated Bowells of the Earth, there are great store-houses of either actual Fires, or places considerably Hot, or, (in some Regions) of both; from which Reconditories (if I may so call them) or magazines of hypogeall heat, that quality is communicated, especially by Subterraneall Channells, Clefts, Fibres or other Conveyances, to the less deep parts of Earth, either by a propagation of heat through the substance of the interposed part of the Soil . . . Or else, (which is perhaps the most usuall way,) by sending upwards hot minerall Exhalations and Steams."

Boyle anticipated many of the ideas we will take up in this chapter—for example, the temperature increase with depth caused by internal heat sources and the transfer of heat to the surface by conduction and the motion of hot matter.

The evidence of Earth's internal thermal energy is everywhere. Boyle cited volcanoes, hot springs, and the elevated temperatures in mines and boreholes to document the flow of heat from the interior to the surface. Today we would add such examples as global plate motions, earthquake activity, and the uplift of mountains, all of which represent mechanical work whose ultimate source of energy is the heat in the interior.

Despite these ubiquitous manifestations of internal heat, we cannot positively reconstruct Earth's thermal history by working backward in time from present-day observational data. It would be like trying to write a history of civilization on the basis of what is in today's newspaper. An infinite number of thermal histories could have caused our planet to evolve to its present state. Nevertheless, we can make plausible models by invoking as always our general knowledge of the Earth and the other planets and making the best use of such data as that gained from studies of heat flow, radioactivity, the ability of rocks to transmit heat, rheological (flowage) properties of heated rock, and travel times of seismic waves. We have no guarantee that these models correspond to anything that happened, but they do provide insight about the present thermal state of Earth as well as its possible history.

Before we consider the source of Earth's internal heat and how this heat is transferred to the surface, we give a brief overview of the thermal history of the Earth, summarized from Chapter 1.

**Figure 15-1**
Geothermal power plant at the Geysers, Sonoma County, California. Heat from the interior converts water to steam, which is piped through turbines to generate 81,000 kilowatts of electrical power. [U.S. Geological Survey.]

**Heat Sources** Radioactivity and the conversion of gravitational to thermal energy are thought to be the major sources of internal heat. The processes of planetary accretion and adiabatic compression warmed the interior, and Earth started on its evolutionary course about 4.7 billion years ago with an initial temperature which may have been somewhere near 1000°C. Radioactivity then took over, and the internal temperature began to rise. Core formation was triggered perhaps 4 billion years ago, or earlier, when the temperature rose to the melting point of iron. The sinking of vast drops of iron to the core would have liberated some $2 \times 10^{37}$ ergs of gravitational energy in the form of heat. This source of heat was great enough to produce extensive melting and reorganization, resulting in a differentiated Earth, zoned into core, mantle, and crust.

The principle types of rocks that make up the Earth's crust are granites and basalts. Seismologists and petrologists tell us (Chapters 16, 17, and 19) that peridotite is an excellent candidate for the major constituent of the mantle. The heat produced in these rocks by the radioactive disintegration of uranium, thorium, and the radioactive species of potassium can be determined in the laboratory. Table 15-1 summarizes the results. Because the radioactive elements like uranium have become concentrated in granite as a by-product of Earth's early differentiation, this common rock leads in radioactive heat production.

Within each gram of granite about 300 ergs of thermal energy is produced per year. It is easy to see by multiplying this number by $2.7 \times 10^{25}$ grams—the number of grams of granite in a hypothetical spherical shell 20 kilometers thick and of the same diameter as Earth—that some $10^{28}$ ergs of thermal energy can be produced by an outer layer of granite only 20 kilometers thick. This is equal to the total heat reaching the Earth's surface each year as determined by measurements. It is about 1000 times more than the energy released each year in earthquakes, and it is about 250,000 times more than the energy of a 1-megaton nuclear explosion! This simple calculation demonstrates not only the importance of radioactivity as a present-day heat-producing agent but also

One calorie (cal) of heat will raise the temperature of one gram of water by 1°C. One cal = $4.18 \times 10^7$ ergs. One $\mu$cal = $10^{-6}$ cal. A megaton nuclear explosion releases $4 \times 10^{22}$ ergs.

Uranium is one of the heaviest elements, yet it became concentrated in the outermost layers of the Earth instead of sinking to the core. This is an example of how chemical affinity can be important as well as gravity in determining where elements end up. Uranium has a strong attraction for oxygen, which is most abundant in the crust. Oxygen migrates upward because it forms lightweight and easily meltable compounds with calcium, sodium, potassium, aluminum, and silicon.

Table 15-1
Radioactive heat production in common igneous rocks.

| Kind of rock | Amount of radioactive element in rock (parts per million) | | | Amount of heat produced (ergs/gram/year) |
|---|---|---|---|---|
| | Uranium | Thorium | Potassium | |
| Granite | 4 | 13 | 4 | 300 |
| Basalt | 0.5 | 2 | 1.5 | 50 |
| Peridotite | 0.02 | 0.06 | 0.02 | 1 |

gives some measure of the total volume of granite in the Earth's crust, for a layer much more than 10 to 20 kilometers (0 to 12 miles) thick would generate more heat than that observed. We will see that much of the heat flowing out of the continents originates in the radioactivity of near-surface granitic rocks, whereas heat flow from the sea floor, where there is no granite, has a deeper source.

**Heat Flow by Conduction**    Heat is energy in transit. It may seem obvious that heat flows from regions of high temperature to regions of low temperature — or in the example of our planet, from the interior to the surface. Not so obvious is the mechanism of heat flow or the rate at which it occurs. Heat energy in a solid exists as the vibration of atoms. The intensity of the vibrations determines **temperature**. Heat is **conducted** when the thermally agitated atoms and molecules jostle one another, thus mechanically transferring the vibrational motion from the hot region to the cool one (Fig. 15-2).

The quantity of heat transferred per unit time between two points is proportional to the *temperature difference* per unit distance and to the property called **thermal conductivity**, which differs for each substance and is a measure of its ability to conduct heat. Rock is a very poor heat conductor: it has a very small thermal conductivity. To transfer heat through only 10 meters (30 feet) of rock takes about three years, which is why underground pipes don't freeze and why underground vaults maintain nearly constant temperature despite large seasonal temperature fluctuations at the surface. A lava flow 100 meters (300 feet) thick would take about 300 years to cool. Heat entering one side of a plate of rock 400 kilometers (250 miles) thick would take about 5 billion years to flow out the other side. In other words, if the Earth were to have cooled by conduction only, heat from depths greater than about 400 kilometers would not yet have reached the surface!

Beer does not freeze in Moscow "in cellars that were not above 12 or 14 foot deep" (Robert Boyle, 1671).

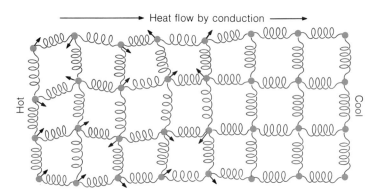

Figure 15-2
Schematic representation of heat flow by conduction through a solid. Interatomic forces which bind atoms to each other are indicated by springs. Heat applied at the left induces thermal agitation of the atoms. Heat is conducted when the vibrations gradually spread to the right.

When a substance is so hot that it begins to "glow," like a red-hot poker, heat can be transferred by radiation, most of which is emitted in the near infra-red and visible region of the spectrum. This **radiative transfer** can be a more efficient process than conduction in some materials, and at one time it was thought to be an important factor in making heat transfer more efficient at depths of 100 kilometers (60 miles) and below. A recent experimental study has shown, however, that silicate minerals are more opaque than first estimated, reducing the efficiency of heat loss by this means so that it is probably less important than conduction. There is no question that heat conduction must be an important agent of heat transfer in the interior. But it may be overshadowed by convection.

**Heat Flow by Convection**   The phenomenon called **convection** (Fig. 15-3) is rather common. One can see it happening in a cooling cup of coffee (Figure 15-4) or in a rapidly heating kettle of water. Because liquids conduct heat poorly, a kettle of water would take a long time to heat to the boiling point if convection did not distribute the heat rapidly. Convection is at work when a chimney draws, or when warm tobacco smoke rises, or when cumulus clouds form on a hot day. All of these examples of convection are governed by the fact that a heated fluid, either liquid or gas, expands and rises because it becomes less dense than the surrounding cooler and heavier material. Thus instead of heat being transferred slowly by conduction, it is transported more rapidly upward by the moving heated material. Colder material flows in to take the place of the rising fluid, is in turn heated, and rises to continue the cycle. The regular flow circuit of rising warm fluid and sinking cold fluid shown in Figure 15-3 is called a **convection cell**.

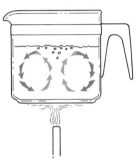

Figure 15-3
A familiar example of convection is seen when water is heated in a coffee pot.

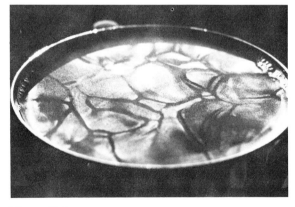

Figure 15-4
Convection in a coffee cup. A thin layer of white dust on the surface shows a cellular pattern that marks regions of rising columns of hot fluid. Dark lines mark regions where the spreading, cooled liquid descends into the body of the fluid. [Photo by V. J. Schaeffer, State University of New York, Albany.]

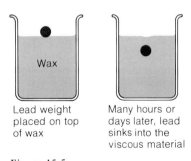

Lead weight placed on top of wax

Many hours or days later, lead sinks into the viscous material

Figure 15-5
Materials can behave as solids over short times and as viscous fluids over long intervals. Cold wax is a solid, but a lead ball placed on top of the wax slowly sinks into the interior in a few hours or days, as if the wax were a viscous fluid. Earth's mantle is supposed to show this dual behavior, permitting slow convections to occur.

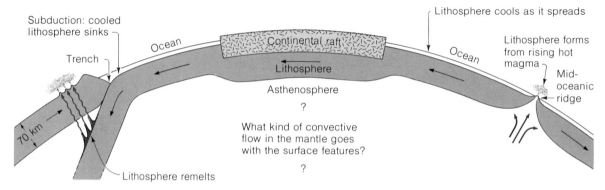

**Figure 15-6**
The motion of plates, spreading from mid-ocean ridges and sinking in subduction zones, is the surface manifestation of convection currents in the interior. The nature of the flow in the interior is uncertain.

Lord Rayleigh, one of the great English physicists of the last century, studied the conditions under which convection would take place. He found that a fluid placed between a hot lower surface and a cool upper surface, as in Figure 15-3, would convect when certain conditions were met. Convection is fostered by a large temperature difference and a high **coefficient of thermal expansion** — a measure of how much a material expands when its temperature is raised. This enhances the density contrast between the warm and cool regions. Increasing the distance between the hot and cold boundaries also encourages convection. Convection is inhibited by **viscosity** — the measure of a fluid's resistance to flow — and by a high thermal conductivity, which would make heat transfer by conduction more efficient, and convection, in a sense, less "necessary."

Up to now we have been discussing convection as a phenomenon of fluids, but under certain conditions solids can also "flow" (Fig. 15-5). Over short terms, like seconds to years, the Earth's mantle behaves as a rigid solid, transmitting seismic waves efficiently and responding elastically to the tidal pull of the Moon. But when stresses are applied over millions of years, the mantle is weak, and we will see direct evidence for this when we discuss "floating" mountains in Chapters 19 and 20. Thus under long-enduring conditions of high pressure and temperature, the mantle "creeps" and behaves as an extremely viscous substance, so that convection is indeed a possibility.

Now for the key question — one that had been debated by geophysicists for years. Is convection an important process of heat transfer within the Earth? Is convection occurring at the present time or has it occurred at any time in the past? The debate hinged on the numerical values of the factors that inhibit and foster convection — for example, the temperature differences, viscosity, conductivity, coefficient of expansion, and the size of the cells. Since most of these quantities were unknown for the conditions

The silicone compound "Silly Putty" illustrates how a solid can flow over long periods. One can bounce it like a ball or break it with a sudden blow, but a ball of Silly Putty will flow into a pancake shape under its own weight if left on a table overnight.

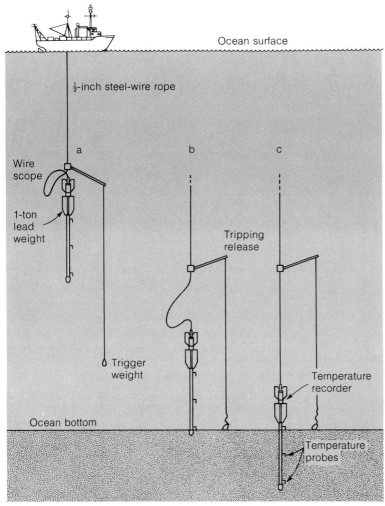

Ocean surface

½-inch steel-wire rope

Wire
scope

a        b        c

1-ton
lead
weight

Tripping
release

Trigger
weight

Temperature
recorder

Ocean bottom

Temperature
probes

**Figure 15-7**
Heat flowing out of the sea floor is measured by plunging a core tube into the sediments. Thermometers on the side of the tube record the temperature increase with depth, and the thermal conductivity of the sediments is measured when the core is retrieved. The product gives the heat flow.

of the mantle and could only be estimated indirectly, it was difficult to resolve the debate on this basis. With the discovery of sea-floor spreading and plate-tectonics the surface manifestations of convection have, in a sense, been directly detected, although the flow pattern in the mantle is still unknown (Fig. 15-6). The rising hot matter under mid-ocean ridges builds new lithosphere, which cools as it spreads away, eventually to sink back into the mantle where it is resorbed. This *is* convection transport, in that heat is delivered from the interior to the surface by the motion of matter. The process is enormously more complicated than is indicated by the simple convection cells shown in Figure 15-3.

It involves melting, upwelling, solidification, horizontal movement, sinking and resorption. The motion may be jerky as a result of sticking and slipping at plate boundaries.

For most geophysicists the debate over convection no longer concerns its existence as much as its scale and periodicity. Does convection involve the entire mantle or only the upper few hundred kilometers? Is it continuous—that is, repetitive, cycle after cycle—or transient in nature, occurring once every few hundred million years when the internal temperature builds up to a critical value? We don't know the answers, but some possible models will be discussed later in this chapter. In any case, convection in connection with the creation and spreading of the sea floor is an important mechanism for the transfer of heat from the mantle under the ocean.

The British geologist Arthur Holmes was among the first to propose convection as the driving mechanism of continental drift. When he did so in the 1930's Holmes was 30 years ahead of his time, because corroboration had to wait for the extensive exploration of the sea floor that began after World War II and the massive gathering of data that led to the concept of sea-floor spreading.

## THERMAL STATE OF THE INTERIOR

**Heat Flow Observations**  Except for the heat received from the Sun, the heat flow from the interior is the most important terrestrial energy source. About 2 times $10^{20}$ calories, or $10^{28}$ ergs, of energy per year reaches the surface from the interior. This averages out to 1.5 microcalories ($\mu$cal) per square centimeter of surface area per second. This exceeds by a factor of about 1000 the annual energy release of earthquakes and by about 10 all of the energy used by man. It is more than a thousand times the energy required to lift the Rocky Mountains by 1 centimeter (0.4 inch). Although internal heat energy is more than enough to raise mountains and make earthquakes, it is a puny amount compared to the energy received from the Sun. So far as controlling climate is concerned it is the Sun, delivering five thousand times the energy from the interior, that is the significant factor. Ultimately both solar and internal heat are radiated into space. Solar heat has its geological consequences, however, in "driving" the atmosphere and hydrosphere, the chief agents of erosion. In a real sense, the Earth's internal heat engine builds mountains, and the Sun's external heat engine destroys them.

Several thousand measurements of heat flow have been made on land and on the sea floor. One obtains a heat flow value by first measuring the rate of increase of temperature with depth in the Earth (the temperature gradient). This multiplied by the thermal conductivity of the rock gives the outward heat flow per

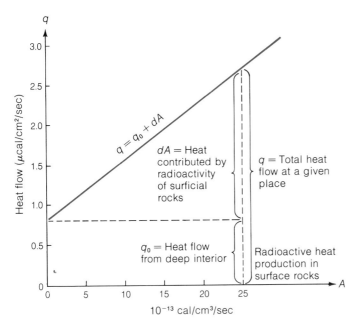

**Figure 15-8**
According to R. F. Roy, D. D. Blackwell, and F. Birch, heat reaching
the surface includes a contribution from the higher radioactivity of
surface rocks added to the heat flowing out of the deeper interior. By
plotting the heat flow $q$, measured at many places in a region, against
the radioactive heat production $A$ of surficial rocks at each place of
measurement, the two contributions can be disentangled. The intercept
of the curve gives $q_0$ the amount of "deep heat," and the slope of the
line gives $d$, the thickness of the surficial zone of high radioactivity.

unit area and per unit time. Thermometers inserted in boreholes
measure the temperature increase with depth. Sea-floor measurements of heat flow are more plentiful, not only because the oceans
cover more area, but also because they are easier to perform.
Because temperatures at the land surface undergo both diurnal
(daily) and seasonal variations, deep holes are needed to remove
the perturbations introduced by these effects and by flowing
groundwater, which may cool or heat the rocks abnormally. The
ocean shields the sea floor from these effects, and this makes it
possible to obtain accurate temperature gradients in holes only
a few meters deep. Drilling is unnecessary, since the soft deep-
sea mud can be easily penetrated by the temperature probe as
it falls into the sea bottom under its own weight (see Fig. 15-7).

**Continental Heat Flow**  The continental crust is mostly granite
in the uppermost portion, and granite is radioactively the "hot-
test" rock, as Table 15-1 shows. We should expect, therefore,
that some of the heat flowing from the continents originates in
the granitic layer, but not all of it. The main problem for the
specialists in the field is to determine the heat budget—that is, to

An example of a heat-flow measurement: At one place in the Rocky Mountains, the temperature increases with
depth by 25°C in 1 km ($10^5$ cm). The
conductivity of the rock, measured by
bringing a sample to the laboratory,
was 0.008 cal/cm·sec·deg. The heat
flow is given by the product

$$\frac{25°C}{10^5 \text{ cm}} \times \frac{0.008 \text{ cal}}{\text{cm·sec·deg}}$$

$$= 2 \times 10^{-6} \text{ cal/cm}^2/\text{sec}$$

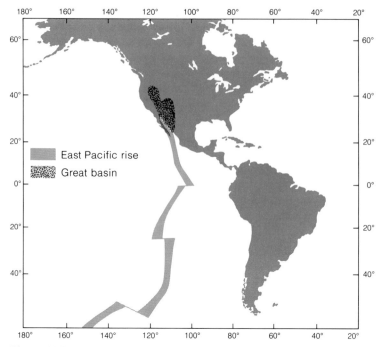

**Figure 15-9**
According to H. W. Menard, the divergence zone marked by the East Pacific rise continues into Arizona, New Mexico, Utah, and Nevada, accounting for the high heat flow, recent volcanism, and elevated topography of this region.

assess how much heat originates in the granite, how much comes from the deeper mantle, and how this all relates to the geologic history of a region.

Geophysicists have recently developed a remarkably simple way to unscramble the diverse factors. They took all of the North American continental data, separated the heat-flow values according to geological province, and for each province plotted heat flow against the radioactive heat production of the surface rocks measured in microcalories per cubic centimeter per second. For each province, the data showed a straight-line fit, and all the straight lines were nearly parallel but had different intercepts (Fig. 15-8). The straight lines represent the equation

surface heat = deep heat + depth of × granitic heat
flow                              granite     production

($\mu$cal/cm²/sec) ($\mu$cal/cm²/sec)   (cm)   ($\mu$cal/cm³/sec)

or

$$q = q_0 + d \times A$$

The intercept of the straight line $q_0$ is the heat originating in the deeper crust and underlying mantle. The slope of the line is $d$, the thickness of the radioactively "hot" surficial granitic layer. The value of $d$ measured from the slope of the line turns out to be about 10 kilometers (6 miles), which shows the remarkable extent to which differentiation has concentrated these radioactive elements at the top of the crust. The value of $q_0$ for the east-central United States is about 0.7 $\mu$cal/cm²/sec, which is to be compared with the average heat flow reaching the surface in this region of 1.0 $\mu$cal/cm²/sec. We see that 30 percent of the heat flux originates in the granitic layer of this geologically old province and 70 percent comes from below. Because of the low radioactivity of the basalts and peridotites underlying the granitic layer in the lithosphere, most of this deep heat must be introduced from below the lithosphere — that is, from regions deeper than about 100 kilometers. The quantity $q_0$ could include heat delivered by convection in the mantle.

The average surface heat flow, $q$, for the Basin and Range province (Arizona, New Mexico, Utah, and Nevada) is about 2 $\mu$cal/cm²/sec, of which 1.4 $\mu$cal/cm/sec is contributed by deep heat, $q_0$. Here too, in this geologically young and active region, some 70 percent of the heat has a deep source. But twice as much deep heat is being delivered here than in the geologically inactive east-central part of the country. Consequently, some geologists believe that the Basin and Range province represents a continuation of the East Pacific rise divergence zone into the continent (Fig. 15-9). This implies that the rising plume of a convection current lies under the Basin and Range province, which would account for such observations as the excess heat flux, the thin crust, which shows evidence of recent fracturing, the recent volcanism, and the earthquake activity. Does this mean that California will one day split from the rest of the country, with an ocean basin opening somewhere in Utah and New Mexico? At the rate of divergence — a few centimeters per year — we need not worry about the political consequences.

The Sierra Nevada, near the California-Nevada boundary, shows a surprisingly small deep-heat flux, $q_0 = 0.4$ $\mu$cal/cm²/sec, about half the total heat flow for this mountain region. Some geophysicists have speculated that the Sierra Nevada block, because of an ancient plate collision, is underthrust by an old and still cool plate, reducing the flow of deep heat (Figure 15-10).

The first two examples seem typical of the rest of this planet: low heat flow ($\sim$1 $\mu$cal/cm²/sec) in geologically old and inactive areas, such as Precambrian shields; high heat flow ($\sim$2 $\mu$cal/cm²/sec) in regions of more recent mountain-building or volcanism, such as the Alps or the western United States. Special cases like the Sierra Nevada, however, require unique explanations.* In each region the highly radioactive rocks at the top of the continental

---

*Heat flow is also low in regions of rapid sedimentation, such as along the Gulf Coast of the United States. The sediments blanket the heat from the interior, in effect temporarily insulating the surface from the heat sources below.

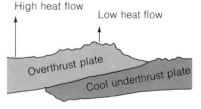

Figure 15-10
To account for the small heat flow found in the Sierra Nevada Mountains, K. E. Torrance and D. L. Turcotte propose that because of an ancient plate collision this block rests on an old and still cool plate, reducing the heat flow from the deep interior.

crust account for less than 50 percent of the total outward flow of heat. The mantle contributes the remainder, the amount being about twice as much in regions of recent orogeny (mountain-building) and volcanism than in geologically old and stable provinces. Combining all of the regions, the average heat flow for the continents is about 1.5 $\mu$cal/cm$^2$/sec.

The separation of continental heat flow into contributions from near-surface granites and deeper mantle sources and the identification of different geological provinces as "hot" or "cold" are new approaches to the understanding of intracontinental geology that are just beginning to be used.

**Heat Flow from the Sea Floor**   The continents may be likened to rafts embedded in large plates. The rafts have grown through geological time, and the oldest rocks found on Earth have been preserved on continents for nearly 4 billion years. Continents are difficult to destroy; they may be deformed, but they survive plate convergence because they are light and keep floating. In marked contrast, the sea floor is created at mid-ocean ridges and destroyed in subduction zones in a time scale of 100 to 200 million years. Furthermore, sea-floor crust and lithosphere are made of rocks like basalts and peridotites, which show far less radioactivity than the continental granites. From these considerations we should expect the sea floor and continents to show completely different heat-flow characteristics. Yet the average heat flow from the sea floor does not differ significantly from the average heat flow from the continent. When the first results emerged in the early 1950's, they produced quite a shock in geophysical circles. Could this puzzle be explained?

Like the heat-flow observations made on land, the oceanic observations show a pattern, and from them a correlation with the geology of the sea floor has been found. Oceanographers tell us that although there is some variability in the observations, the average heat flow for the major oceanic provinces is as follows: ocean ridges, >2; ocean basins, ~1.3; ocean trenches, <1.0 $\mu$cal-/cm$^2$/sec. The reader should be able to anticipate the explanation of the decrease in heat flow from ridge crest to ocean basin to deep-sea trench. Mid-ocean ridges sit atop the rising plume of hot magma that carries heat from the deeper mantle. The magma cools, solidifies, and becomes attached to the oceanic lithosphere. As the lithospheric plate spreads away from the ridge, it loses its heat by conduction to the sea floor and gradually cools (Figure 15-11). For this reason, heat flow should decrease with the age of the sea floor, hence with distance from a mid-ocean ridge. The oldest parts of the sea floor should have the lowest heat flow, and these parts should be found farthest from ridges and closest to deep-sea trenches, where the cold slabs sink back into the mantle.

Oceanic heat flow is thus dominated by the process of cooling of the recently created oceanic lithosphere. Geophysicists believe

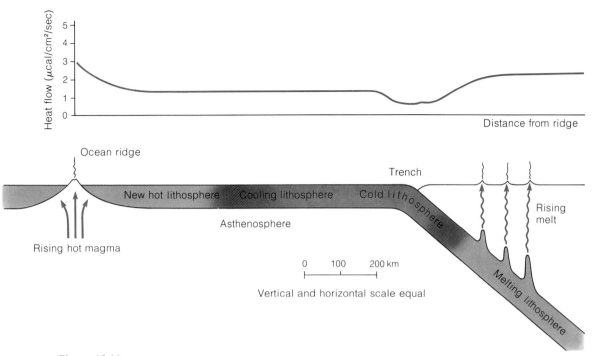

**Figure 15-11**
The pattern of heat flowing out of the sea floor. High values are observed over mid-ocean ridges. As the sea floor spreads and cools, less heat reaches the surface. On the side of deep-sea trenches where the lithosphere plunges into the asthenosphere, rising magma from the melting lithosphere produces higher heat-flow values.

that this process may account for as much as 45 percent of the total heat flow from the Earth, and that this may represent a major mode by which the Earth has cooled. Because of the convective formation and destruction of the sub-oceanic lithosphere, the time scale for decrease in heat flow is about ten times more rapid on the sea floor than on continents. Regions that have "cooled down" to low heat-flow values of 1 $\mu$cal/cm$^2$/sec are 1 billion years old or more on land, but only 100 million years old on the sea floor. The sea floor cools ten times more rapidly than the continents because the ocean bottom is in a sense the top of a convection cell, and conduction, the process that governs cooling on the continents, takes place more slowly.

We return to the puzzle of the equality of average heat flow for the continents and oceans despite the entirely different thermal regimes. Remember that on the continents some 30 percent of the heat flow originates in the radioactively "hot" granitic crust, which is absent on the sea floor. An early explanation was offered by Sir Edward Bullard, who proposed that the subcrustal rocks of the sea floor contain more radioactivity than their subcontinental counterparts, in order to compensate for the absence of a heat-producing granite layer on the sea floor. Until the uranium content of such mantle rocks as peridotites, can be studied in more detail after being dredged or drilled from the sea floor, this hypothesis remains in the running. Alternatively, we could take the easy way out and explain the phenomenon as a coincidence or as part of some grand scheme of things that still eludes us.

**Temperatures in the Earth** The average increase of temperature with depth, as measured in bore holes or mines, is about 2 or 3°C per 100 meters (300 feet). How can we infer temperatures in the Earth at depths greater than those we can reach with a thermometer—that is, below about 8 kilometers (5 miles)? The problem is not an easy one, and the experts need all the help they can get, especially for the lower mantle and core.

Temperatures in the continental crust are fairly well known from measurements of surface heat flow, which can be separated into heat contributions from the radioactive surficial rocks and from the underlying region by means of graphs like the one in Figure 15-8. A knowledge of these two sources of heat is sufficient to fix the temperatures with some assurance for the continental crust. **Geotherms**, or temperature-depth profiles, typical of such geologically old and stable regions as the eastern United States and of the young and active regions of continents, exemplified by the Basin and Range province, are shown in Figure 15-12. The higher heat flow in the tectonically active region shows up in the more rapid increase of temperature with depth. Forty kilometers below the surface, temperature rises to almost 1000°C in the active region. This is close to the point of initial melting of such deep crustal and mantle rocks as basalt and peridotite. In contrast, temperature under the stable region reaches the relatively low value of about 500°C at a depth of 40 kilometers (25 miles). The association of tectonic activity and volcanism with subcrustal temperatures near the melting point, and of long-term geological stability with relatively cool regions of the crust, is an important discovery. For example, it supports the speculative notion that the Basin and Range province may overlie the rising plume of a convection current.

Temperatures in the suboceanic lithosphere are dominated by the process of cooling down from the high temperature of the magma that builds the slab in the first place. Referring again to Figure 15-11, we could make a good guess that the temperature under the ocean would be close to 0°C (the temperature of the bottom water) at the top of the slab, and close to 1200°C (the melting point) at the bottom of the slab, where it is in contact with the partially molten asthenosphere. Where the slab plunges deep into the mantle, we would expect it to warm up gradually to the high temperatures of the material surrounding it.

One of the important tasks of the next generation of geophysicists is to find a method for directly determining the temperature in the deep interior. Until this is done we can only speculate on

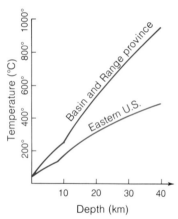

**Figure 15-12**
Temperature variation with depth in different regions, according to the ideas of R. F. Roy, D. D. Blackwell, and F. Birch. The higher heat flow in the Basin and Range province implies higher temperatures at depth in this young and active region. The temperature rises to almost 1000°C at 40 km. In the geologically older and more stable eastern United States, the temperature reaches only 500°C at this depth.

the nature of convection within the Earth using indirect evidence. We know that we can't simply extrapolate the temperature curves for the crust (Fig. 15-12) by straight-line extension all the way to the depth of the Earth's core, at 2900 kilometers (1800 miles), for that would give a temperature of at least 25,000°C, which would mean that most of the Earth would have to be molten. This impossible situation shows once again that the important radioactive elements are concentrated in the outermost layers, and that the rate of temperature increase with depth (temperature gradient) must lessen below this radioactively "hot zone." More efficient heat transfer in the deeper layers will also decrease the temperature gradient so as not to give impossibly high values.

At present all we can do is sketch a possible geotherm, showing temperatures as they might occur all the way down to the center of the Earth (Figure 15-13). In order to "anchor" the geotherm

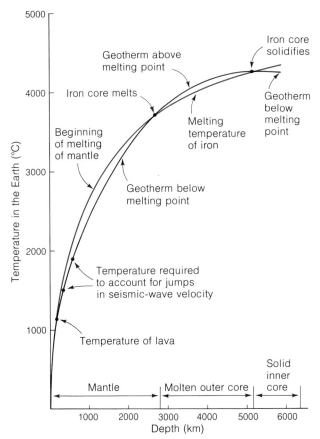

**Figure 15-13**
A geotherm showing how temperature may increase with depth from the surface to the center of the Earth. Temperatures in the central region may reach 4000° to 4500°C. The geotherm exceeds the melting point in the fluid iron core and in the partially molten asthenosphere, which is a source of lava. [Data from D. L. Anderson, G. Kennedy, and G. Higgins.]

at a few points to keep it from drifting too far from reality, the following data are used:

1. Seismology tells us that the asthenosphere is partially molten. We therefore "fix" the temperature at a depth of 100 km (60 miles) at the **solidus**, or point of incipient melting, which laboratory geologists tell us is about 1100 to 1200°C. This is just about the temperature at which lavas issue from oceanic volcanoes, which provides a nice check.

2. At depths greater than 300 kilometers (200 miles) in the mantle, the temperatures must be below the solidus, since seismic shear waves cannot pass through a partially molten region without being severely attenuated, and this is not observed below the asthenosphere. We use an estimate of the solidus recently inferred from laboratory studies as an upper limit for mantle temperatures below the asthenosphere.

3. The two rapid increases in seismic wave velocities at depths of about 400 and 700 kilometers (250 and 450 miles) are important clues, which we will discuss in Chapter 19 on Seismology. They are caused by a more dense packing of the atoms of a major mineral constituent of the mantle — olivine. The sudden densification, or **phase change**, occurs when critical pressures and temperatures are reached. A few years ago the phase change at 400 kilometers (250 miles) was actually verified in the laboratory when rocks were squeezed and heated to duplicate the pressure and temperatures at these depths. Therefore, we know the pressure and temperature at which the phase change takes place. The conditions for the phase change at 700 km can be estimated theoretically. Geophysicists recently used seismological data and new laboratory data to fix the temperatures at about 1500°C and 1900°C for the depths 400 and 700 kilometers (250 and 450 miles), respectively.

4. We will see in Chapter 19 that the core is primarily iron, molten in the outer part and solid in the inner part. Recently a new melting-point curve for iron (Fig. 15-13) has been used by geochemists to fix temperatures at the high pressures in the core. The temperature at the mantle-core boundary must exceed the melting point of iron to account for the liquid core, and stay below that of the mantle to account for its solidity. It is an interesting and useful coincidence that these upper and lower bounds on temperature are close together at the mantle-core boundary, fixing the temperature at about 3700°C. Moreover, the geotherm must cross over and fall below the iron melting-point curve at 5100 kilometers (3200 miles) in order to provide for the solid inner core insisted upon by the seismologists. This gives a temperature of about 4300°C. It seems unlikely that the temperature at Earth's center could be much above this value.

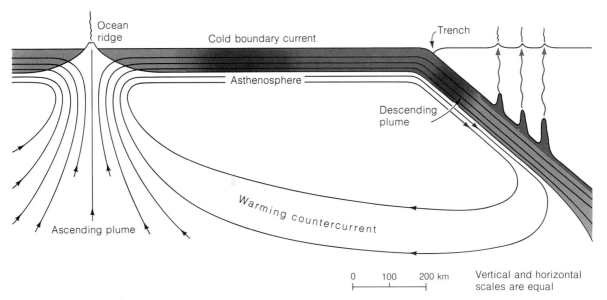

**Figure 15-14**
Possible convection flow in the upper mantle, according to D. H. Turcotte and E. R. Oxburgh. An ascending hot plume spreads laterally under the ocean ridges; cooling by conduction to the ocean, it solidifies to form a boundary layer, the cold, rigid lithosphere. The descending cold plume coincides with the sinking lithosphere of the subduction zone.

**The Tectonic Engine** It is widely accepted that tectonic phenomena — that is, plate motions, mountain-building, and earthquakes — are accounted for by internal convection currents. Many schemes for convective flow have been advanced. They vary somewhat in scale, some involving the entire mantle and others limiting the flow to the upper few hundred kilometers.

Some fascinating calculations have led to a proposed scheme of convection in which the flow of matter takes place in the outer several hundred kilometers of the Earth. The model explains many of the features summarized earlier in Figure 15-11; these features are repeated in Figure 15-14 with the convective flow lines superimposed. An ascending hot plume spreads laterally, cools by conduction to the ocean and solidifies to form a cold, brittle boundary layer — the lithosphere. A descending cold plume, the subducted lithosphere, begins the return flow; the cycle of convection is completed by a counter-current several hundred kilometers below the surface, which heats up and rises again. Buoyancy forces of the light, hot ascending plume and of the heavy, cold descending plume drive the flow. The energy source is radioactivity plus perhaps some of the original heat left over from the early years of the planet. The velocity of the flow, which is calculated for the model, agrees with the sea-floor spreading rates, and the heat coming out of the surface matches the observational data quite well. The results encourage us to believe that the men who suggested these ideas may be on the right track in the search for the mechanism that makes the tectonic engine run.

The question of the driving force behind plate tectonics has become so important that it is the basis of a new international scientific program called The Geodynamics Project. Many of the questions posed in this chapter will be attacked in a concerted, multinational research effort over the next decade.

## SUMMARY

1. The internal heat of the Earth is derived from the radioactivity of uranium, thorium, and potassium. Heat left over from the early years of the planet may be an additional source.

2. Much of the internal heat reaches the surface by convective flow in the mantle; the large-scale surface manifestation of this process is the spreading sea floor. New lithosphere created from rising magma cools as it spreads, eventually to sink back into the mantle where it remelts. Measurements of high heat flow over mid-ocean ridges and low heat flow over deep-sea trenches are consistent with this picture. The creation of new sea floor is an important mechanism by which the Earth gets rid of its internal heat.

3. Heat flow on the continents correlates with regional geology—low values in old stable regions and high values in regions of recent rifting, orogeny, and volcanism.

4. Temperatures increase rapidly with depth in the outermost 100 kilometers (60 miles) of the Earth, reaching the solidus in the asthenosphere (100 to 300 kilometers or 60 to 190 miles). Below 300 kilometers it must rise less rapidly or the Earth would be extensively molten. It is everywhere below the solidus until the core is reached. Here the temperature is about 3500°C—above the melting point of iron, as required by the liquid iron outer core. The temperature at the center is about 4000 to 4500°C, according to current views.

## EXERCISES

1. Give some "kitchen physics" or common experience examples of the following concepts. *Example*: Hot tea in a porcelain cup is easier to handle than in a metallic cup because porcelain is a poorer conductor of heat and will less likely burn the hand or the lip.
   a) uses of low-conductivity and high-conductivity materials,
   b) convection,
   c) materials of different viscosity,
   d) work done by heat.

2. Enlarge upon the statement "in a real sense Earth's internal heat builds mountains, and external heat from the Sun destroys them."

3. Almost twice as much heat flows out of the ground in Utah as in Massachusetts. Why?

4. More heat flows out of the crust along mid-ocean ridges than older ocean basins. Why?

5. Although the geologist has direct access only to the upper few kilometers of the Earth, he can infer temperatures at greater depths. How?

6. It is widely believed that convection currents in the mantle provide the driving forces for plate motions. What would be the connection between rising, warm convection plumes, sinking cold plumes, and sea-floor spreading and subduction?

# BIBLIOGRAPHY

American Geophysical Union, *Terrestrial Heat Flow.* Geophysical Monograph Number 8, American Geophysical Union, 1965.

Sclater, J. G., and Jean Francheteau, "The Implications of Terrestrial Heat Flow Observations on Current Tectonic and Geochemical Models of the Crust and Upper Mantle of the Earth." *Geophysical Journal of the Royal Astronomical Society,* v. 20, pp. 509–542, 1970.

Turcotte, D. L., and E. R. Oxburgh, "Continental Drift." *Physics Today,* v. 22, pp. 30–39, 1969.

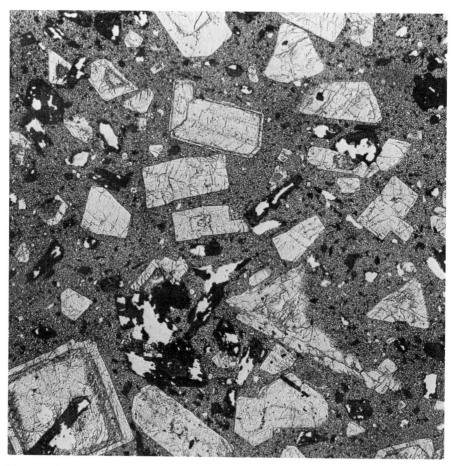

Microscopic view of a thin section (transparent slice) of diorite porphyry. The white crystals are feldspar; the large, dark, rectangular ones are hornblende. The large crystals were all once floating in a melt, which is now represented by the fine-grained groundmass. Width of field 2.5 mm. [Photo courtesy of C. B. Hunt.]

# 16

## Formation of Igneous Rocks

*Localized heat and pressure deep in the Earth cause deep crustal and mantle materials to melt, forming magmas. Magmas crystallize to form igneous rocks of diverse composition as they intrude higher portions of the crust or erupt at the surface. Experimental studies of the crystallization behavior of melts indicate that the compositional diversity of igneous rocks is a result of partial melting of the original material and chemical differentiation. In this way, silicic magmas can be produced from originally mafic melts. Pressure and the presence of water in the melt greatly affect melting temperatures and the formation of partial melts.*

The connection between the interior heat of the Earth, volcanoes, and the formation of igneous rocks (those crystallized from a molten state) seems to be straightforward. Simple observation of many volcanoes reveals how cooling and congealing of molten lava produce the rock we call basalt. More than 200 years ago, geologists drew the conclusion that ancient basalts were also of volcanic origin, even though, in many cases, all traces of the volcanoes that produced them had long since vanished. Other volcanoes produce lavas that crystallize to form rocks of different composition, ranging from andesite to rhyolite. Somehow, the composition of the melt varies from place to place, and this determines the composition of the rock (see Fig. 4-32).

## GRANITE

It is not as easy to demonstrate that granite, one of the commonest of coarsely crystalline materials, is a rock formed by crystallization from a hot, molten mass. It is found in high mountain chains, such as the Sierra Nevada and the Alps, and in the cores of deeply eroded ancient mountains, such as those of the Scottish highlands, but this in itself is no great clue to its origin. The details of its relations to other rocks, however, were most important clues to James Hutton as he worked in the field in Scotland. He saw cross-cutting relations, in which sedimentary rocks were somehow fractured and invaded by the granite, as if it had been forced into them as a liquid. Even more significant were the mineralogical and textural changes in the sediments bordering the granite, which he inferred were the results of great heat.

Hutton guessed that the granite was formed by molten material rising from great depths. As it rose, it intruded the rocks of the upper part of the crust and cooled slowly to form crystals of quartz, feldspar, mica, and other minerals. The texture, a mosaic of interlocking crystals, could not be confused with the texture of a sandstone consisting of grains of the same minerals. Many of the minerals in granite show sharp crystal faces, and the crystals are interlocked and held together like the pieces of a jig-saw puzzle. Sandstone, in contrast, consists of grains rounded by abrasion and

**Figure 16-1**
Cross-cutting contact of granite (below) with shales and tuffs (above). Bedding dips gently to the right. A small, nearly vertical, light-colored dike is at the right. Shasta County, California. [Photo by J. P. Albers, U. S. Geological Survey.]

**Figure 16-2**
Photomicrograph of a thin section of granite in polarized light. The large light crystal at lower left and the dark one above it are microcline, a variety of potassium feldspar. The more homogeneous light- and medium-gray crystals at center and upper right are quartz. This interlocking mosaic of crystals results from crystallization of a melt. [Photo by A. H. Koschman, U. S. Geological Survey.]

bound together by a mineral cement between the grains. The minerals of granite are so insoluble in water that it would be hard to conceive of the rock having been precipitated from sea water, like an evaporite, which may have the same kind of crystalline texture. Thus granite, just as basalt, came to be recognized as the product of crystallization of a **magma**, a large body of very hot molten rock.

In the second half of the nineteenth century, after the polarizing microscope had been invented and had been used for mineral identification of small crystals in rocks, geologists described the multitude of igneous rocks of the granite family, discovering in them the same range of variations as they saw in the extrusive rocks formed by volcanism. Techniques of chemical analysis, which were well advanced, were joined with microscopic examination and field study to produce a classification scheme of the sort we still use (see Figs. 4-28 and 4-32). The major division of igneous rocks into fine-grained (as typified by basalt) and coarse-grained (as typified by granite) was based on their differing origins. Volcanism gave rise to extrusives, which quickly cooled, at the

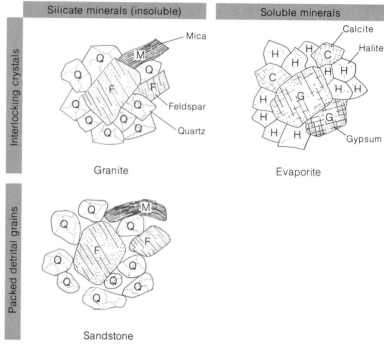

**Figure 16-3**
Contrasts between the textures of two rocks composed of the same silicate minerals—one igneous (granite), the other sedimentary (sandstone)—and between the composition of two rocks having the same interlocking crystalline texture—one igneous (granite), the other sedimentary (evaporite).

relatively low temperature of the Earth's surface, to form glasses or fine crystals. Plutonism—the formation of a magma at depth, where temperatures and pressures were higher—gave rise to intrusives, which slowly cooled to form coarse crystals.

The explanations for the variety of compositions of the granite series, from granite to gabbro, were more elusive. Complicating the picture further were the ultramafic intrusives composed of olivine and pyroxene, which have few corresponding extrusives. What accounts for the formation of magmas of such a variety of mineralogical compositions? What temperatures were required for the formation of melts of different kinds? If these temperatures could be estimated, then we might have some idea of how deep in the Earth the magmas formed and crystallized. The effect of pressure might also prove to be important.

## CRYSTALLIZATION FROM MELTS

By the early twentieth century, as all the geological data on igneous rocks accumulated, there were still no good answers to detailed questions about the origin of such rocks, either from field studies

Table 16-1
Chemical analysis of a granite
from Cumberland, England.

| Element | Formula* | Relative proportion (%) |
|---|---|---|
| Silicon | $SiO_2$ | 68.55 |
| Aluminum | $Al_2O_3$ | 16.21 |
| Ferric iron | $Fe_2O_3$ | 2.26 |
| Magnesium | MgO | 1.04 |
| Calcium | CaO | 2.40 |
| Sodium | $Na_2O$ | 4.08 |
| Potassium | $K_2O$ | 4.14 |

*Formulae are conventionally shown as oxides, though the elements are not actually present in this form.

Table 16-2
Mineralogical analysis of a granite
from Cumberland, England.

| Mineral | Formula | Relative proportion (%) |
|---|---|---|
| Quartz | $SiO_2$ | 24.4 |
| Orthoclase | $KAlSi_3O_8$ | 36.2 |
| Plagioclase | $\left\{ \begin{array}{l} NaAlSi_3O_8 \\ CaAl_2Si_2O_8 \end{array} \right\}$ | 33.6 |
| Biotite | $K(Mg,Fe)AlSi_3O_{10}(OH)_2$ | 5.8 |

or from chemical analyses. These questions about where and how magmas were formed arose at a time when the young science of physical chemistry was maturing. Thermodynamics, the science of heat and energy, was then being applied to problems of crystallization from solutions and melts; by the turn of the century, it began to be applied to geological problems. The new way to approach the problem was by experimentation. Scientists mixed chemicals in the proportions in which they were known to occur in igneous rocks, melted them in high-temperature furnaces, and allowed them to crystallize, observing carefully the temperatures of melting and crystallization and the compositions of the minerals that were formed.

Igneous rocks contain many elements. In addition to the abundant elements—oxygen, silicon, aluminum, iron, sodium, potassium, magnesium, and calcium—there are a host of minor elements. The experimentalists wisely chose not to tackle the terribly complicated natural mixtures immediately but to take a few very simple compositions and to work them out first. One of those is the plagioclase feldspar system, a group of feldspars ranging from albite, a sodium aluminosilicate ($NaAlSi_3O_8$), to anorthite, a calcium aluminosilicate ($CaAl_2Si_2O_8$). Plagioclase is a major mineral component of igneous rocks, and its systematic change of composition is basic to the classification of the range of the intrusives, from granite to gabbro, and the range of extrusives, from rhyolite to basalt. The results of the experiments with plagioclase were most instructive.

The experiments showed that melts of various plagioclase compositions, if allowed to crystallize very slowly, would form crystals through a complicated path of crystallization (see Box 16-1). The first crystals to form started out with more of the calcium component, anorthite, and slightly depleted the calcium present so that the melt became relatively richer in sodium. As crystallization

The results of many crystallization experiments with melts of compositions corresponding to those of various members of the plagioclase series are shown in the accompanying figure. The chemical composition is shown on the horizontal axis of the graph as the percentage of the two components, albite and anorthite, in a simple mixture of the two. Natural plagioclases are a solid-solution series — that is, there is a continuous range of compositions from one side to the other, the basic crystal structure remaining the same throughout (see Table 4-5). Just as the solid minerals are, the melts of albite and anorthite are completely miscible in all proportions.

Because we are interested in temperatures of crystallization, we have plotted temperature, in degrees centigrade, on the vertical axis. The graph tells us what to expect if a melt of composition X (about 30 percent anorthite), heated to about 1500°C, is allowed to cool slowly. There is no change as the liquid cools until a temperature of about 1370° is reached, at which point the first crystals of plagioclase form. These crystals are very different in composition from the liquid; they turn out to be a little over 70 percent anorthite. As the first crystals form, because the proportion of anorthite to albite being taken out of the melt is much higher than the original composi-

tion at X, the liquid becomes a little depleted in anorthite and starts to move down the upper curve.

As more crystals grow, they form from a liquid of slightly changed composition, and so are a little less rich in anorthite. In the meantime, the first-formed crystals are no longer in equilibrium with the liquid, because the melt is now more albitic. If crystallization is very slow, these first crystals will continuously react with the changing liquid; and, as new crystals form, the composition of all of the crystals will travel down the lower curve. This process continues — the liquid moving down the upper curve and the solid moving down the lower curve — until, at a temperature of about 1190°C, the last drop of liquid (by now containing only about 5 percent anorthite) forms crystals of the original composition, X. After that, the entire crystalline mass continues to cool without further change. The melt has crystallized into a plagioclase of the same composition as the original melt, but by a complicated route. When a solid plagioclase is melted, it follows exactly the same route in reverse. The first liquid formed is the same as the last liquid drop of the crystallization process, and liquid and crystals move up their respective curves until the last crystal is melted, when the liquid attains the X composition.

proceeded, the earlier crystals were gradually transformed to crystals containing relatively more and more sodium, until the final homogeneous mass had the same composition as the original melt. The key to the process was the continuous transformation of the composition of earlier crystals, by a reaction with the remaining melt, to new compositions, so that, at any point in the course of crystallization, all of crystals present would have the same composition.

## MAGMATIC DIFFERENTIATION

When *N. L. Bowen* published his pathfinding experimental work on feldspars in 1913, he had natural geologic processes very much in mind. He focused on the course of crystallization in situations in which the feldspars did *not* continuously react with the liquid to

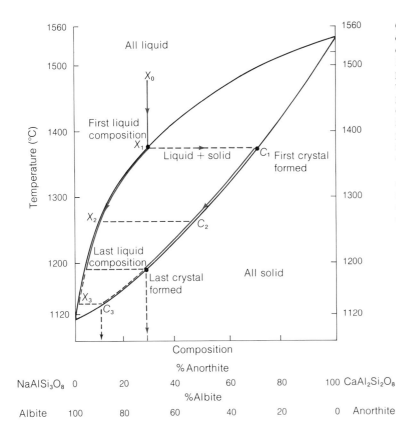

Crystallization diagram for the cooling of a plagioclase feldspar melt. The original melt at composition $X_0$ (about 30 percent anorthite) cools to a temperature of about 1370°C ($X_1$), at which point the first crystal of composition $C_1$ is formed. As successive crystals are formed, the liquid changes in composition to that indicated by point $X_2$, at which time all crystals have reacted to form crystals of composition $C_2$, the last liquid forming a crystal with the composition of the original melt. If crystals of composition $C_2$ are withdrawn at that point, the liquid will change to composition $X_3$ and the final crystal will be of composition $C_3$.

change their compositions. This would occur, for example, if a magma cooled so rapidly that there was time only for the outer surface of the earlier-formed crystals to react with the changing liquid, so that only the outer rim of each crystal would change composition during the crystallization, each successive layer being covered by another layer more rich in albite. The result of such a process is a mass of **zoned crystals**, the interiors of which are anorthite-rich and that gradually grade to albite-rich material toward their exteriors. The zoning of plagioclase has another implication: If the anorthite-rich cores of the growing crystals are unable to react with the liquid, the liquid will be more albite-enriched than it would be in a slow, continuous reaction. If, for example, at any stage of crystallization, the crystals already formed were to be withdrawn somehow, the solution at that point, already albite-enriched, would behave as though it had just started to crystallize, and the next crystals to form from it would be more albitic.

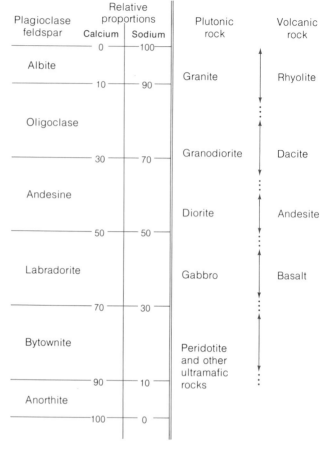

**Figure 16-4**
Igneous rock types are classified according to the relative proportions of the feldspars (sodium plagioclase) and anorthite (calcium-plagioclase). The boundaries between rock types are gradual; the divisions between the feldspars are arbitrarily defined.

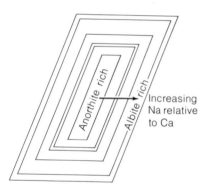

**Figure 16-5**
Drawing of a zoned plagioclase crystal formed during the fractional crystallization of a silicate melt in which crystallization was too rapid to permit continuous reaction of already formed crystals with the changing melt composition. As the melt congeals, the layers of the crystal become more albite rich.

Then, if there were a slow, continuous reaction thereafter, the remaining liquid would ultimately form crystals with an anorthite content much lower than the original liquid would have, had it cooled slowly in a continuous reaction from the outset.

Bowen proposed a theory of **chemical differentiation** based on this idea. If the early crystals formed deep in the crust in a large opening filled with melt, a **magma chamber**, the crystals might settle to the bottom and thus be withdrawn from further reaction. Or structural deformation midway in the crystallization process might squeeze the remaining liquid from the crystal mush into other places, where it would continue to crystallize. Whatever the exact details of the process, **fractional crystallization** of the melt—that is, crystallization in which crystals do not continuously react with the melt—could account for albite-rich plagioclases

a
Fractional
crystallization

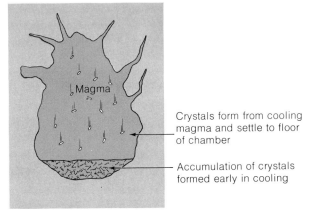

Crystals form from cooling
magma and settle to floor
of chamber

Accumulation of crystals
formed early in cooling

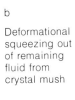

b

Deformational
squeezing out
of remaining
fluid from
crystal mush

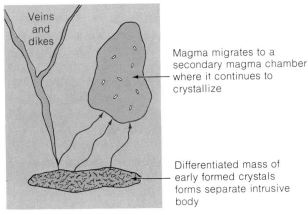

Magma migrates to a
secondary magma chamber
where it continues to
crystallize

Differentiated mass of
early formed crystals
forms separate intrusive
body

**Figure 16-6**
Two stages in the evolution of a differentiated magma. In the first stage (a), crystals formed early settle to the floor of the magma chamber. During cooling, structural deformation may squeeze remaining liquid from the chamber and separate the early-formed crystals as a distinct intrusive body (b), while the liquid migrates elsewhere to form veins, dikes, and other magma chambers, where it continues to crystallize.

forming from an originally anorthite-rich melt. Basaltic magmas are anorthite-rich melts, and granites contain albite-rich plagioclases. Magmatic differentiation relates the two in the sense that the basaltic magma is the starting material that gradually differentiates to a more silicic melt by a fractional crystallization process that, if carried far enough, forms granite.

**Mafic Minerals** Magmas do not just crystallize to form plagioclase feldspars; they also produce such mafic minerals as olivine, pyroxene, amphibole, and black mica (biotite). Melting–crystallization relations among those minerals are a little more complex, but their order of crystallization can produce differentiation, too. In a process in which slow cooling allows all crystals to react

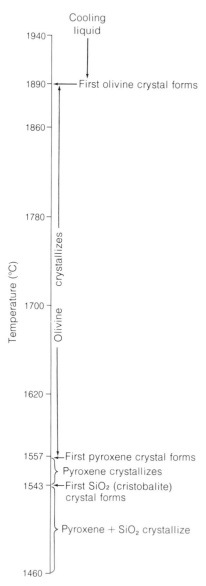

Cooling liquid

1940 —

1890 — First olivine crystal forms

1860 —

1780 —

Olivine crystallizes

1700 —

1620 —

1557 — First pyroxene crystal forms
Pyroxene crystallizes

1543 — First SiO₂ (cristobalite) crystal forms

Pyroxene + SiO₂ crystallize

1460 —

**Figure 16-7**
Sequence of events in the crystallization of a cooling liquid of magnesium and silica. This sequence forms one part of the discontinuous reaction series by which different mafic minerals are successively crystallized from a melt.

completely with the liquid, olivines crystallize first; then, when the liquid reaches a certain point, pyroxene starts to form and all the olivine is converted to pyroxene. In such a **discontinuous reaction** series, reactions take place between minerals of two definite compositions at a particular temperature (a little over 1550°C for magnesium olivine and pyroxene), rather than over a continuous range of compositions and temperatures, as with the plagioclase series. Just as in the continuous reactions of plagioclase formation, the first-formed crystals of olivine may settle out or become coated by later-formed pyroxene crystals, and so be withdrawn from further reaction. Then the melt becomes enriched in silica and finally crystallizes to form the high-temperature silica mineral cristobalite (or, at lower temperatures, tridymite or quartz).

**The Reaction Series** Bowen combined the continuous and discontinuous fractional crystallizations of the major minerals of igneous rocks into a general scheme for magmatic differentiation that is now called the **Bowen reaction series**. A high-temperature magma of basaltic composition starting to cool and crystallize will gradually differentiate by fractional crystallization along two paths—the continuous plagioclase-feldspar series and the discontinuous mafic-mineral series. The melt converges toward a final low-temperature granitic composition of orthoclase and albite feldspar, muscovite mica, and quartz. After the early crystallization and separation of olivine crystals, the magma corresponds to a diorite intrusive in composition—or to its fine-grained relatives, andesite volcanics, if the magma reaches the surface at that stage. At later stages, the magma becomes more silicic, ending with granite and its extrusive equivalent, rhyolite. By the time this has happened, so many crystals have formed that only about 10 percent of the original liquid is left.

The Bowen differentiation theory accounts for many field relations and for the range of compositions of igneous rocks, and it is consistent with the known temperatures of magmas of different compositions. A number of layered intrusive bodies are known in which the bottom layers are olivine and successively higher layers contain lower-temperature minerals arranged in order of the reaction series. Rhyolite volcanics erupt at times from large basaltic volcanoes toward the end of an eruption episode. Zoned plagioclase crystals are common in granites and granodiorites, as one would expect. In addition, there are very few extrusive equivalents of the most mafic intrusives of olivine and pyroxene, which suggests that most mafic plutonic bodies are not formed from a magmatic liquid—which might be expected to erupt somewhere—but are accumulations of crystals in a magma chamber. As discussed in Chapter 17, the temperatures of basaltic lavas are much higher than those of more silicic compositions and are approximately in harmony with the experimental results.

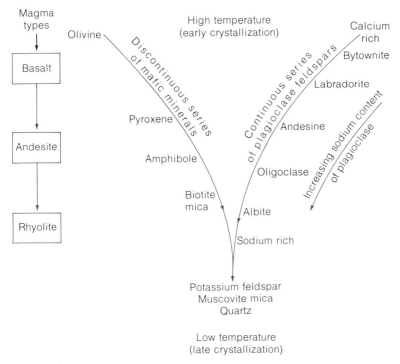

**Figure 16-8**
Bowen's reaction series, showing how the sequence of fractional crystallization of a melt leads to the formation of differentiated magmas.

Bowen's theory has been elaborated and extended by much work in the field and laboratory. As more elements and minerals were added to the scheme and various crystallization histories of magmas were worked out, it proved to be a workable and encompassing theory that explained the diverse facts of the occurrences of many igneous rocks, particularly those of the continents. More than any earlier theory, it integrated field observation, experimental evidence, and sound physical chemical theory and so set the stage for further work on the origin of igneous rocks.

Many of the facts, however, are difficult to reconcile with simple differentiation as the major mechanism for silicic magma generation. Huge batholiths—enormous bodies of granite, such as those found in California or Idaho—seem to require the existence of vastly more voluminous bodies of basaltic magma from which they would have differentiated. Hence, many believe that these granites did not originate by differentiation from basalt. Oceanic basalts, typically those formed by upward movement of the asthenosphere at mid-ocean ridges, do not show differentiation. Great areas of granite in regionally metamorphosed areas merge into gneissic metamorphosed bodies of the same composition; there, much of

One of the other phenomena associated with the reaction series is the tendency for the structures of the discontinuous series to become increasingly complex as temperature drops. Olivines are made of isolated silica tetrahedra, pyroxenes of single chains of tetrahedra, amphiboles of double chains, and micas of continuous sheets (see Chapter 4). Fractionation of minor chemical elements also accompanies the series.

the granite seems to be of metamorphic origin (see Chapter 18). Large granitic bodies associated with plate-convergence zones are now thought by many to have been formed by melting of some combination of sediments, igneous rocks, and metamorphic rocks that would give a melt of granitic composition. We will consider further the origin of the many kinds of granite after we turn to the origin of the magmas themselves.

## ORIGIN OF MAGMA

Up to this point, we have taken for granted that magma occurs deep in the Earth, but we have not asked where, when, and how. We must reject the older notion that all of the interior of the Earth is molten, because the data we have used to infer its nature tell us that it is solid for thousands of kilometers down to the boundary of the core (see Chapter 19). Magma chambers under active volcanoes have been surveyed by geophysical methods, and it is clear that magma occurs as discrete liquid segregations in the otherwise solid interior.

Mid-ocean ridges and rises are major zones of basaltic magma generation: **flood basalts** flow intermittently from their rifts and fissures. The same kind of flood basalts are spread extensively over continental tectonic belts and over some stable shield areas as well. Granite, granodiorites, and other silicic intrusives are essentially restricted to the continents, accompanied in places by silicic extrusives. The differences in the distribution of rock types are reflected in the more silicic composition of average continental crust (including sedimentary, granitic, and basaltic layers) as contrasted with the more mafic composition of average oceanic crust.

The differences in composition can be related to melting temperatures. All magmas must form deep in the Earth where temperatures are high. Basaltic magmas form at temperatures well over 1000°C. In contrast, granitic melts may form at temperatures several hundred degrees lower. Some idea of the depths of formation of these magmas can be obtained from the geothermal gradient in Figure 15-13. To get a better picture, we must consider other

Table 16-3
Compositions of average oceanic crust and average continental crust.
Averages include compositions of sedimentary, granitic, and basaltic layers.

| Crust | Composition (%) | | | | | | | |
|---|---|---|---|---|---|---|---|---|
| | $SiO_2$ | $Al_2O_3$ | $Fe_2O_3$ | $FeO$ | $MgO$ | $CaO$ | $Na_2O$ | $K_2O$ |
| Continental | 60.2 | 15.2 | 2.5 | 3.8 | 3.1 | 5.5 | 3.0 | 2.9 |
| Oceanic | 48.7 | 16.5 | 2.3 | 6.2 | 6.8 | 12.3 | 2.6 | 0.4 |

Source: Data from Ronov and Yaroshevsky, 1969.

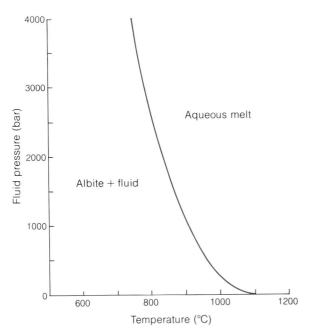

**Figure 16-9**
The melting curve for albite in the presence of water, as measured by fluid pressure in bars (1 bar equals 0.987 atmospheres). As fluid pressure increases, the melting temperature is lowered. [After *Earth Materials* by W. G. Ernst. Copyright © 1969 by Prentice-Hall, Inc.]

factors that influence melting and crystallization, including pressure and the presence of water in the melt.

At atmospheric pressure, pure albite in the presence of some water remains solid up to temperatures a little over 1000°C. Above that temperature, the albite melts to form a liquid containing much water, an aqueous melt. If more water is present, the melting temperature of the albite is lower (see Fig. 16-9, in which the relative amount of water present is indicated by the pressure of water vapor).

In the same way, the melting temperatures of all of the feldspars drop considerably if there are large amounts of water present. Thus, melting can occur in some places in the lower parts of the crust, both because temperature and pressure are higher there than in the upper parts of the crust and because there is more water present than in the relatively dry mantle below. Pressure alone has an important effect: Figure 16-10 shows the melting of dry basalt as a function of pressure and temperature, plotted together with geothermal gradients. The effect of increased pressure is to raise the melting points of basalt and the silicate minerals. At a depth of several hundred kilometers, we can expect some melting of relatively dry mantle material. This is consistent with the location of

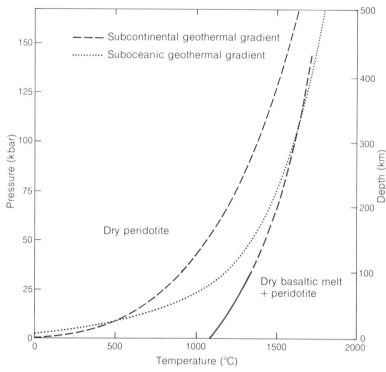

**Figure 16-10**
Calculated geothermal gradients under continents and ocean basins (somewhat speculative). The melting of basalt in the absence of water has been experimentally investigated only in the pressure–temperature region indicated by the solid line. [After *Earth Materials* by W. G. Ernst. Copyright © 1969 by Prentice-Hall, Inc.]

the asthenosphere, the partially molten zone found at these depths by seismological methods. This partial melting would account for the widespread occurrence of basaltic magma.

Geologic field evidence long ago provided evidence that the role of water in granitic melts is important. Numerous **hydro-thermal veins** branch off from the sides and tops of many intrusive bodies. Such veins are filled with minerals that contain much chemically bound water, which, together with a variety of other evidence, shows that abundant water was associated with the magma. Hydrothermal ("hot water" in Greek) veins are of great importance in the geology of many metal ores, which are found as rich deposits in such veins. Some of the most valuable deposits of mercury, gold, silver, lead, and copper are mined from them (see Chapter 24). The hot-water solutions percolating upward from the intrusives carried these metals with them, and the metals were precipitated from them along the walls of joints and cracks as they rose and cooled.

Water-rich silicic magmas, such as those that produce hydrothermal veins, cannot be produced by simple differentiation of deep mantle material that has been melted into basalt. Such magmas are likely produced by partial melting of a mixture of quartz-rich, water-rich portions of the crust—sedimentary rocks, in particular. Sediments are richer in silica and aluminum than the parent rocks from which they were eroded, and have lost their mafic minerals by chemical weathering.

This leads to the conclusion, amply suggested by field evidence, that many large granite intrusions originate as silicic melts deep in the crust. At depths of 35–40 kilometers, temperatures of about 600°C, and water pressures of about 10,000 atmospheres, such melts would form from mixtures of sedimentary and metamorphic rocks that are compositionally equivalent to granodiorite. This seems likely to happen in deeply subsiding geosynclines, in which sediments, partly metamorphosed by heat and pressure, move down into the melting region.

In the 1960's, experimental studies of the melting behavior of mantle materials at extraordinarily high pressures (well over 30,000 atmospheres) were made possible by newly developed high-pressure equipment. The results led many geologists to propose an additional mechanism for the origin of batholiths—that of partial melting of mantle materials. This mechanism also explained many of the facts about the voluminous andesite volcanics that occur in the same active tectonic belts, such as the western North American Cordillera; the average composition of the batholiths in that tectonic belt come close to that of andesite. A silicic melt forming in the deep crust, however well it may explain some granite batholiths, fails to provide a mechanism for the extensive andesite volcanics, which must have come from a magma of that composition; but partial melting of such mantle materials as eclogite and peridotite (pyroxene–olivine–garnet rocks), if it occurs in the presence of some water, does give a reasonable explanation for andesite magmas. The attractiveness of the notion of partial melting of mantle rocks is that it may explain both granitic batholiths and andesite volcanics.

With the development of the idea of plate tectonics, partial melting in the upper mantle has assumed new significance: it has been linked to the downward movement of lithospheric plates at subduction zones associated with volcanic island arcs. As a lithospheric plate moves down, carrying with it the oceanic crust, some sediment, and water, partial melting occurs as it sinks into the asthenosphere. This melting produces magmas of various compositions in or above the descending plate at various depths, depending on the amount of water, the temperature, and the pressure. These magmas are then intruded or extruded in the mountain belts associated with crustal deformation near the subduction zone. This theme will be further explored in the next chapter and integrated with a general view of magmatism and volcanism in relation to plate tectonics.

## SUMMARY

1. Intrusive igneous rocks are formed by emplacement of silicic magmas in the crust.

2. Slow crystallization of plagioclase feldspars involves continuous reaction of already formed crystals with the surrounding melt until complete crystallization results in all of the crystals having the same composition as the original melt.

3. Magmatic differentiation may occur as a result of fractional crystallization, in which earlier formed crystals do not continuously react with the melt. This differentiation can produce silicic magmas from originally mafic ones. The Bowen reaction series describes how the feldspars and mafic minerals successively crystallize during differentiation.

4. Magmatic differentiation can explain some occurrences of granite and silicic volcanics, but it is probable that many granites are formed by metamorphism or by partial melting of a mixture of sediments and earlier granitic crust, either in the crust or in a subduction zone in the upper mantle.

5. Differences in the composition of melts and the places in which they originate in the crust or mantle are accounted for by the effects of pressure and water content on the melting temperatures of basalt and the silicate minerals.

## EXERCISES

1. Compare those aspects of the mineralogical composition of granites that suggest an origin from a melt with evidence from field relations.

2. What evidence could you use to decide which of two intrusions formed at different temperatures might have been crystallized from the hotter magma and which from the cooler?

3. What kinds of intrusive igneous rocks would you expect on a hypothetical planet of the same crustal composition as Earth, on which all magmas cooled extremely slowly and always maintained equilibrium between crystals and remaining melt?

4. Why would you not generally expect basaltic magmas to be generated in the upper 10 kilometers of continental crust?

5. What is the evidence for appreciable quantities of water being included in some magmas?

## BIBLIOGRAPHY

Barth, T. F. W., *Theoretical Petrology* (2nd ed.) New York: John Wiley and Sons, Inc., 1962.

Bowen, N. L., *The Evolution of the Igneous Rocks.* Princeton, N. J.: Princeton University Press, 1928. (Reprinted by Dover Publications, Inc., 1928.)

Ernst, W. G., *Earth Materials.* Englewood Cliffs, New Jersey: Prentice-Hall, Inc., 1969. (Chapters 2 and 5.)

Wyllie, P. J., "Experimental Petrology: An Indoor Approach to an Outdoor Subject." *Journal of Geological Education,* v. 14, no. 3, June, 1966.

Wyllie, P. J., *The Dynamic Earth: Textbook in Geosciences.* New York: John Wiley and Sons, Inc., 1971. (Chapters 7 and 8.)

# 17

## Volcanism

*Volcanoes can be beautiful, informative, and both beneficial and dangerous. They serve as windows through which we can dimly perceive the interior. Volcanism occurs in many different styles, principally in the vicinity of plate boundaries.*

For decades geologists have argued about the source of magma, the hot, mobile material that is generated within the Earth and which solidifies into igneous rock. There is now little doubt, so the seismologists tell us, that the asthenosphere, which extends from about 75 to 250 km in depth, is the one large region in the mantle that is partially molten (see Chapter 19). It is reasonable, therefore, to identify this layer as a major source of magma. Remelting of sections of the lithosphere may provide another source of magma. At certain places, perhaps where the lithosphere is fractured or otherwise weakened, the magma rises, squeezed up by the weight of the overlying crust, some of it eventually reaching the surface where it erupts as **lava**. Lava differs from the parent magma to the extent that the volatile constituents (gases) are partially lost to the atmosphere or ocean and other chemical components are gained or lost en route to the surface. Despite these differences, lava provides important clues to the chemical composition and physical state of the upper mantle.

**Figure 17-1**
Cerro Negro, photographed in 1968. This volcano, near Managua, Nicaragua, is a cinder cone built on an older terrane of lava flows. [Photo by Mark Hurd Aerial Surveys.]

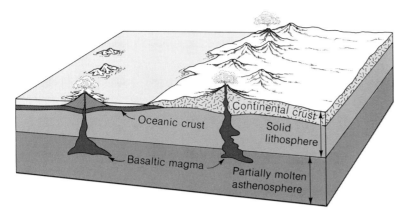

**Figure 17-2**
Basaltic magma, originating in the partially molten asthenosphere, rises through the lithosphere to erupt as basaltic lava on the surface.

Volcanic activity is not randomly distributed, but shows a definite pattern that correlates strongly, though not exclusively, with the positions of plate boundaries, where lithosphere is created or destroyed (Fig. 17-3). Not long ago our views of volcanism were for the most part colored by the grandeur and the spectacular nature of visible eruptions. We now think that the creation of the sea floor is volumetrically the most important volcanic process, one that is hidden almost entirely from direct observation by the overlying ocean. At the end of this chapter, we will return to the question of the global pattern of volcanism: the different compositions and eruptive styles of lavas and their association with plate boundaries and other geological features.

## VOLCANIC DEPOSITS

Volcanic rocks are important not only for what they reveal about the interior but also because the chemical-mineralogical composition of lavas affects eruptive styles and the kinds of landforms they build upon freezing. To understand the variety of shapes of volcanoes—why some eruptions are explosive and others gentle—we must begin with the rocks themselves.

Figure 17-4 gives a brief review of volcanic rock types. In Chapter 4, igneous rocks were classified into four major groups on the basis of silica content and mineralogy: felsic, intermediate, mafic, and ultramafic. The major minerals associated with these rocks were listed (see Fig. 4-30). Texture is also important in rock classification. Very slow cooling gives rise to large-sized mineral grains; faster cooling results in a fine-grained rock, and very rapid cooling results in glass, since no time is available for crystal growth. The coarse-grained, or intrusive, analogs of lava, which will be important in discussing the origin of magma, are: rhyolite-granite,

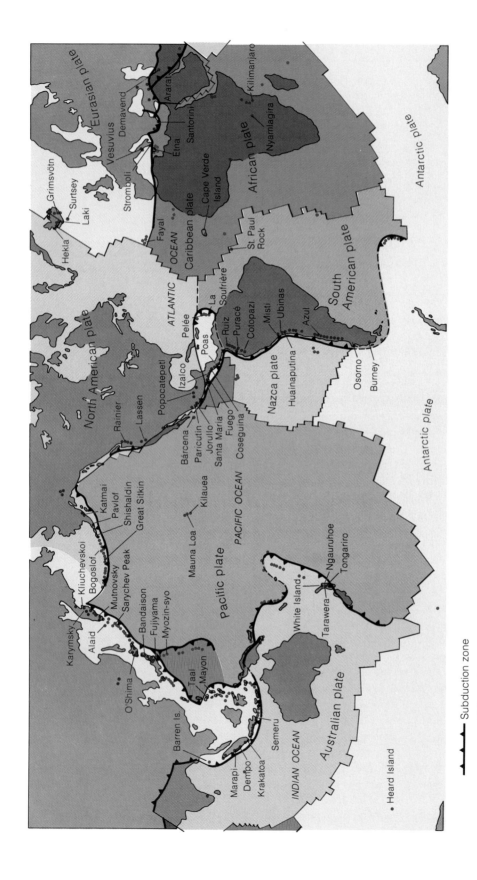

— Subduction zone

— Ridge axis

— Transform

∘∘ Volcano

**Figure 17-3**
The principal active volcanoes of the world are not distributed randomly on Earth's surface: they tend to be associated with plate boundaries.

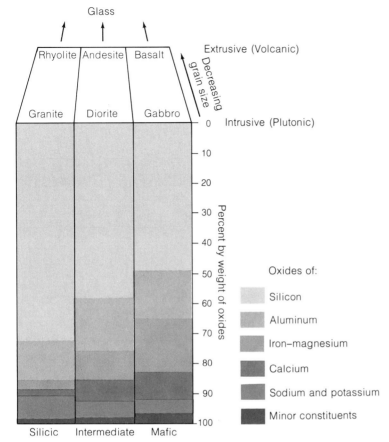

**Figure 17-4**
Average composition of the major types of lava (basalt, andesite, and
rhyolite) and their intrusive counterparts (gabbro, diorite, and granite).
Note the systematic increase in silica and the decrease in iron-magnesium
oxides in going from basalt to andesite to rhyolite.

andesite-diorite, and basalt-gabbro. Ultramafic rocks almost never
reach the surface through volcanism. The principal lavas are basalt
and andesite, the mafic and intermediate types, respectively.

During eruptions, magma moves in response to pressures
within the volcano, and is effusively deposited on the surface as
lava flows or violently ejected as ash. Rapid cooling follows and
indeed we find that most volcanic rocks are fine grained, glassy
or pyroclastic. We recall that pyroclastics are fragmented deposits
formed from material ejected or exploded into the air from a
volcanic vent.

**Lava Flows**  As do all liquids, lavas flow downhill. Basaltic melts
are highly fluid and are known to maintain their fluidity longer
than the more silicic lavas; basalt can flow fast and far (Figs. 17-5,
17-22). Flow velocities as high as 100 kilometers/hour (60 miles/

**Figure 17-5**
Basaltic lava flow from the 1970 eruption of Kilauea, Hawaii. [Photo by
T. McGetchin, M.I.T.]

hour) have actually been observed, but velocities of a few kilometer/hour are more common. Lava streams extending more than 50 kilometers (30 miles) have been witnessed in historic times. On flat terrain, basaltic lava spreads out in thin sheets, successive flows often piling into immense lava plateaus. The great Columbia Plateau of Oregon and Washington is one example; the lunar maria* may have also formed in this way. Such sheets can sometimes be confused with sedimentary strata. In one sense they are like sediments, for they obey the laws of original horizontality and superposition. Basalt erupts with temperatures in the range 1000 to 1200°C.

Rhyolite, the most silicic lava, has a lower melting point and erupts with temperatures of 800 to 1000°C. It is much more viscous than basalt and flows very slowly, forming thick, bulbous deposits. Andesite, with intermediate silica content, shows intermediate properties.

Lava flows can be distinguished by a variety of interesting and sometimes bizarre surface and internal features. The basaltic flows on Hawaii fall into two categories according to their surface forms: **pahoehoe** (pronounced pahoyhoy) and **aa** (ah ah). These euphonic terms are among the few Hawaiian words in the scientific vocabulary. Pahoehoe is a highly fluid lava that spreads in sheets. A thin, glassy elastic skin that forms at the surface is dragged into ropy, filamented folds as flow continues below the surface (Fig. 17-6). Aa is a slower-moving lava flow whose thick skin is broken

*Pahoehoe* in Hawaiian means "ropy"; *aa* is what it says—or what a bare-footed Hawaiian says while walking on it: "ah . . . ah . . . oh . . . ah."

---

*Plural of *mare,* from the Latin for "sea"; pertaining to the large dark areas on the Moon.

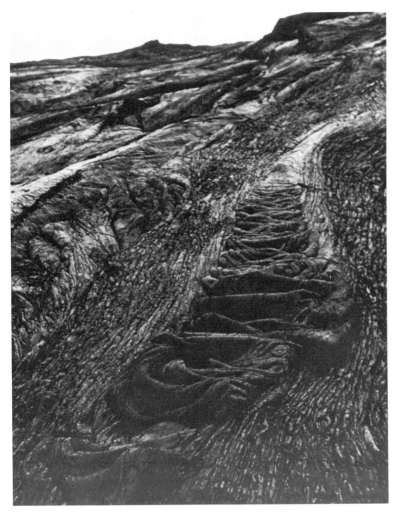

**Figure 17-6**
Pahoehoe lava from the 1970 eruption of Kilauea, Hawaii. [Photo by T. McGetchin, M.I.T.]

into an indescribably rough, jagged, clinkery surface (Fig. 17-7). The blocks ride on the viscous, massive interior, advancing as a steep front of angular boulders in the manner of a tractor tread. Many is the mainland haole (non-native) who, "site" unseen, has paid dearly for an acre of Hawaiian aa. His shoes would be cut to ribbons if he tried to walk on his own land. Aa may differ in behavior from pahoehoe because it has lost its volatiles and consequently its fluidity. Single flows commonly grade from pahoehoe near their source into blocky aa "downstream." The scale of the blockiness varies from small, football-sized blocks (as at Mt. Etna, Italy) to house-sized blocks (at Hekla, Iceland). The blocks, regardless of scale, are loose and unstable, and their surfaces are covered with gas cavities, and hence are jagged. The local topographic relief on blocky aa flows is commonly 10 feet or more.

**Figure 17-7**
An Aa lava flow in the saddle between Mauna Kea and Mauna Loa, Hawaii.
[Photo by T. McGetchin, M.I.T.]

**Figure 17-8**
Pillow lava, characteristic of underwater volcanic eruptions,
photographed at an ocean depth of 3300 m off Hawaii. Area
of photograph approximately 1.25 by 1.25 m. [Photo by
J. G. Moore, U.S. Geological Survey.]

**Figure 17-9**
A chunk of vesicular lava, Newberry Caldera, Oregon.
[From *Volcanic Land Forms and Surface Features* by J. Green
and N. M. Short, Springer-Verlag, New York. Photo by
J. Green.]

They are truly treacherous to cross. A good pair of boots has a mean lifetime of about a week, and the traveler or geologist can count on cut knees and elbows. The instability of house-sized blocks makes some flows downright hostile and sufficiently dangerous to be treated with respect, like rattlesnakes!

**Pillow lavas**, piles of ellipsoidal, sack-like blocks about a meter in dimension, are characteristic of underwater eruptions of basalt and andesite (Fig. 17-8). Apparently the lava surface, chilled by contact with water, forms a tough, plastic skin, convex upwards, filled with liquid. On cooling, the skin becomes glassy and the interior crystalline, and often radially cracked. The identification of pillow structure in a geological section is a sure indication of underwater deposition. It has also been a valuable guide as to which is top and bottom in deformed sequences.

Dr. James Moore of the U.S. Geological Survey and some colleagues performed the remarkable feat of swimming up to an advancing underwater lava flow and observing the formation of pillows. This is their report:

"A team of five divers from the U.S. Geological Survey, the University of Hawaii, and the Lockheed Aircraft Company dived on the lava flow from Kilauea Volcano on the south coast of the Island of Hawaii on April 19th (1971). This is the lava flow that has been going on for six weeks and is presently flowing into the sea. The flow is advancing under water as a wall of rubble which is some 300 to 500 feet seaward of the new sea cliff. The front of the rubble wall is at the angle of repose of about 45°, and the base of it is covering over the old ocean floor to a depth of about 100 feet. Tongues of lava, circular in cross section, extend down the front of this rubble slope. Some of these are as long as 200 feet and they are 3 to 4 feet in diameter, generally. Budding off of them are typical pillows. We could approach one of these elongated lava tongues, in which lava was actually flowing inside and whose surface was too hot to touch on the outside. Generally, the surface was dark, but periodically it would crack and bright incandescence could be seen inside for a few seconds until the crack solidified and then it would crack in some other place. There was a continuous loud mixture of noises: hissing, cracking, small explosions, and rumbling noise as the lava went down through these tubes. General water temperature was slightly elevated around the flowing lava tongues from a normal of 76°F to about 80°F, but in cracks and around behind these lava tubes the water was boiling, and many bubbles were coming up out of cracks in the tubes. One surprising feature was the concentration of marine life including fish, lobsters, and eels which were crowded ahead of the advancing rubble flow at the base of the rubble, apparently having been displaced from their normal habitats along the coast. Underwater visibility was better than expected."

Lavas show many other features. They can be glassy, fine grained, or even crystalline, depending on the rate at which they cooled. When water vapor or other gases are released, ovoid gas cavities, or **vesicles**, can form (Fig. 17-9). **Pumice**, an extremely vesicular, generally rhyolitic lava, occasionally has so much void space that it is light enough to float. **Xenoliths**, foreign inclusions

**Figure 17-10**
Giants Causeway, an example of columnar jointing in lava. [Courtesy of Northern Ireland Tourist Board.]

picked up by a magma, provide important clues about the source region and also the rock formations through which lava passed en route to the surface. When lava cools and contracts, shrinkage cracks (**joints**) often form. Some flows develop impressive **columnar jointing** perpendicular to the cooling surfaces (Fig. 17-10). Where flows are thin and cool rapidly, the joints may be highly irregular. On Hawaii one can walk into **lava caves**, or **tubes**, and sometimes find lava stalactites and stalagmites (Fig. 17-11). Tunnels form where the source of an enclosed stream of lava was cut

**Figure 17-11**
Lava tube in Hawaii, with stalactites formed from lava dripping down from the roof of the chamber. [From Geophysical Laboratory of the Carnegie Institution of Washington, Washington, D.C.]

**Figure 17-12**
Lava fountain building a cinder and spatter cone during the 1960 eruption of Kilauea, Hawaii. [Photo by T. McGetchin, M.I.T.]

off and the remaining lava drained from the channel. Some geologists think the deep valley named Hadley Rille, where Apollo 15 landed, was formed by the collapsed roof of a lunar lava tube. **Spatter cones** are steep-sided, conical hills built from the spatter of **lava fountains** spewed out of vents (Fig. 17-12).

**Pyroclastic Deposits** Water and dissolved gases are important constituents of magma; before eruption, the confining pressure of the overlying rock keeps these volatiles from escaping. When the magma rises close to the surface and the pressure drops, the volatiles may be released with explosive violence, shattering the lava into fragments of various sizes, shapes, and textures. This is more common in the highly silicic, viscous rhyolitic and andesitic lavas than in basalts, which are more fluid and release volatiles more quickly.

**Pyroclasts** are the fragmentary volcanic rock materials that are ejected into the air. The particles, whether rocks, minerals, or glass, are classified according to size; the finest are called **dust**, fragments up to about half a centimeter, are called **ash**, and chunks larger than several centimeters are called **bombs** (Fig. 17-13). Bombs have several characteristic shapes: ellipsoidal, discoidal, and irregularly rounded; typical dimensions range from baseball to basketball size. Some volcanic bombs, however, can be block-busters. Blocks as heavy as 100 tons are known to have been thrown more than 10 kilometers in violent eruptions. Volcanic ash, and especially dust, can be carried great distances. Dust from the great Krakatoa eruption of 1883 reached the upper levels of the atmosphere and was dispersed throughout the world. More recently, in 1965, the eruption of Agung on Bali gave rise to large amounts of dust, which meteorologists measured from the violet color the particles produced in the stratosphere (Fig. 17-14).

Some scientists believe that long periods of volcanic activity can induce worldwide climatic changes, even trigger ice ages, by ejecting so much dust as to reduce the amount of incoming solar radiation. Meteorologists think that volcanism has a much greater effect in blocking the Sun's rays than does man's pollution.

**Figure 17-13**
A field of volcanic bombs and other ejecta, Haleakala volcano, Maui, Hawaii. Source cone is in the background. [From *Volcanic Land Forms and Surface Features* by J. Green and N. M. Short, Springer-Verlag, New York. Photo by J. Green.]

**Figure 17-14**
A gigantic explosion of the volcano Bezymianny in central Kamchatka, Siberia, destroyed the southeastern slope of the volcano and produced an ash cloud that reached a height of some 35 km. [G. S. Gorshkov, Soviet Geophysical Committee.]

**Figure 17-15**
Volcanic breccia, Absaroka Mountains, Wyoming. [Photo by Robert Decker, Dartmouth College.]

Sooner or later pyroclasts must fall, mostly near the source, building up deposits. When the fragments become cemented together (lithified), the rocks formed from the smaller fragments are called **volcanic tuffs**; those formed from the larger ones are called **volcanic breccias** (Fig. 17-15). There is as much variety in volcanic deposits as there is in sedimentary ones, and there are many similarities between the two. Some volcanic deposits show graded bedding, not unlike that of certain sedimentary rocks, because the coarse fragments settle first, followed by the finer debris. Volcanic beds often show lateral sorting, a gradation from coarser material near the vent to finer material with increasing distance from the vent. Even dune structures and cross-bedding are found, having been formed by winds blowing at the time of fallout and also by the vigorous radial (sidewise) expansion of the volcanic eruption cloud. Similar bedded and dune-like deposits were observed around some of the sites at which atomic bombs were exploded; from details of the forms of those deposits, it is possible to estimate the wind velocities produced by the blasts.

A spectacular and often devastating form of eruption occurs when hot ash and dust fragments and gases are ejected in a glowing cloud that rolls downhill with amazing speed. The solid particles are actually buoyed up by the hot gases, so that there is little frictional resistance to this incandescent, fluidized avalanche, or **nuée ardente** (Fig. 17-16). In 1902 a nuée ardente with an internal temperature of 800°C flowed down from Mont Pelée on Martinique at a speed of 100 kilometers per hour (60 miles per hour). In one minute, and with hardly a sound, it enveloped the town of St. Pierre and killed 28,000 people. Deposits left behind by a nuée ardente are poorly sorted, without much indication of bedding, as would be expected. Typically, the fragments are still hot,

**Figure 17-16**
Nuées ardentes (glowing avalanches, or ash flows) plummeting down the slopes of Mount Mayon in the Philippines during the eruption of 1968. [Photo by W. Melson, Smithsonian Institution.]

**Figure 17-17**
Hand specimen of welded tuff from an ash-flow sheet. The dark lenses are
collapsed pumice fragments enclosed in a light-colored matrix composed mostly
of welded glassy shards. Laboratory experiments indicate that welded tuffs form
at temperatures in the range 700°–900°C. [Photo from R. G. Schmidt and
R. L. Smith, U.S. Geological Survey.]

hence soft, when deposited and become compacted and stuck
together to form **welded tuffs** or **ignimbrites** (Fig. 17-17). The
welding agent that makes the fragments stick together is glass.
Vast ignimbrite sheets, also called ash flow deposits, more than
100 meters thick, covering tens of thousands of square kilometers
are known on the continents. These pyroclastic deposits represent the high-silica counterpart of the plateau basalts.

## ERUPTIVE STYLES

Now that we have discussed something of the several types of
volcanic deposits that have poured out or been blown out from
the interior, it is natural to consider the different ways in which
eruptions occur, and the different results produced. Eruptions
do not always produce the majestically symmetrical cone of a
Fujiyama, which has become a symbol to a nation. The hundreds
of thousands of square kilometers of monotonous **flood basalt**,
such as that of the Columbia Plateau, represent another variant.
Eruptions such as those of Hawaii's Kilauea have been heralded

**Figure 17-18**
A phreatic explosion on the volcano Surtsey in the Atlantic Ocean off the coast of Iceland. [Copyright by Solarfilma, Reykjavik, Iceland.]

by warning events that were followed by relatively quiet outpourings of lava, which moves slowly and predictably, destroying property but not lives. Pyroclastic eruptions, such as the one at Mont Pelée, or **phreatic explosions** (violent steam blasts), such as that of Krakatoa in 1883, can wipe out entire populations (Fig. 17-18). Teams of archeologists and marine geologists have pieced together the story of the volcano Thera (formerly called Santorin), situated in the Aegean Sea. It appears that the eruption of Thera in the fourteenth century B.C. was an event of super-Krakatoan proportions. The volcanic debris and sea waves that were produced caused destruction in coastal settlements over a large part of the eastern Mediterranean. The mysterious disappearance of the Minoan civilization has been attributed to this cataclysm, and it is thought that the legend of the lost continent of Atlantis may have its origin in this event.

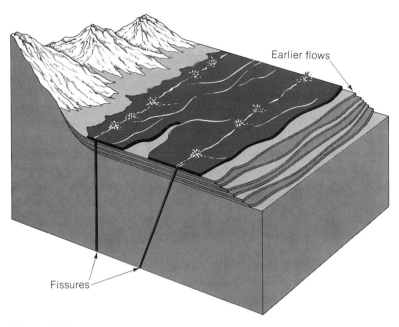

**Figure** 17-19
Schematic diagram of a fissure eruption of highly fluid basalt. Lava rapidly flows away from fissures and forms widespread layers, rather than building up into volcanic mountains. [After R. S. Fiske, U.S. Geological Survey.]

**Fissure Eruptions** Either lava or pyroclastic materials may emanate from long, narrow fissures or a group of such fissures (Fig. 17-19). The lavas are usually basaltic; the pyroclastics originate from more silicic magmas. The rather fluid basaltic lavas flow away from the fissures rather than build volcanoes, and often flood extensive areas, building lava plains or piling up into lava plateaus—hence the names "flood basalts" and "plateau basalts" (Fig. 17-20). The basalt flows that made the Columbia Plateau buried 130,000 square kilometers (50,000 square miles) of the pre-existing topography, which had a relief of about 1.5 kilometers (1 mile). Some individual flows were more than 100 meters (300 feet) thick, and some were so fluid that they spread more than 60 kilometers (40 miles) from their source. The total volume of basalt poured out in this Miocene volcanic episode has been estimated at more than 100,000 cubic kilometers (25,000 cubic miles)! An entirely new landscape with new river valleys has since been created on top of the old surface. Other famous examples of continental plateau or flood basalts are the Deccan of India, which has an areal extent of more than 500,000 square kilometers (200,000 square miles), and the Parana of Brazil and Paraguay, covering more than 750,000 square kilometers (300,000 square miles).

When rocks are dredged from a mid-ocean ridge or samples are collected from such occasional oceanic ridge peaks as the Azores, which stick out above water, they always seem to turn out

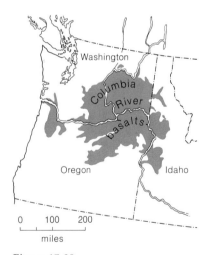

**Figure** 17-20
Map showing area covered by the Columbia River flood basalts. [After R. S. Fiske, U.S. Geological Survey.]

**Figure** 17-21
View across Snake River Canyon and part of the Columbia Plateau in Idaho and Oregon. Flood basalts form the layered rocks on the upper part of the land surface. [Photo from U.S. Geological Survey.]

to be basalt. The basalt originates in fissure eruptions in connection with sea-floor spreading. The process is seemingly unspectacular because it is mostly unseen, but volumetrically it dwarfs all of the continental outpourings combined. The fissures are coincident with the ridge crests and often are marked by median **rift valleys** and seismic belts (see Fig. 1-18). The global oceanic ridge-fissure system can be traced over distances totaling some 50,000 kilometers (30,000 miles) (Fig. 1-17). Enormous amounts of basalt have poured out of this world-encircling sequence of cracks, enough to build the crust of all of the present sea floor in the past 200 million years.

Figure 17-22
The volcano Surtsey was born in the sea south of Iceland in 1963. Some 300 million m³ of lava and 600 million m³ of ash poured out of the sea floor to build this new island. [Copyright by Solarfilma, Reykjavik, Iceland.]

Iceland, an exposed segment of the mid-Atlantic ridge, provides an unexcelled opportunity to view the process of fissure eruption and sea-floor spreading directly. The island is composed mostly of basalt, with small amounts of the more felsic lavas. Repeated geodetic surveys show that Iceland is in a state of tension, literally being pulled apart, one half moving eastward with the Eurasian plate, the other westward with the American plate. Tensile cracks develp, magma flows in from below, and overflows onto the surface. Evidence of basalt flooding from great fissures has been found in the geologic record at many places, but actual flooding has been witnessed by man only on Iceland. In a single eruption in 1783, a crack 32 kilometers (20 miles) long, called the Laki fissure, opened and spewed out some 12 cubic kilometers (3 cubic miles) of basalt. At the conclusion of such an episode the lava solidifies to form a nearly vertical **dike** in the fissure and near horizontal beds on the adjacent surface. With the next episode of lateral spreading, a new crack forms, and another flow pours out over the old one. This is the way Iceland grows, by repeated fissure eruptions, as sketched in Figure 17-23. Although the details may differ under water, it is reasonable to expect the sea-floor crust to grow in a similar fashion. The upwelling basalt frequently produces ridges centered on the deep valleys of the fissures rather than Icelandic plateaus. The ridges gradually subside into ocean basins as they move away from the zone of spreading, to be replaced by fresh ridges. This is discussed further in Chapter 21, which summarizes what we know about plate tectonics.

Fissure eruptions of pyroclastic materials rather than lavas are more likely when the parent magma is more silicic than basalt.

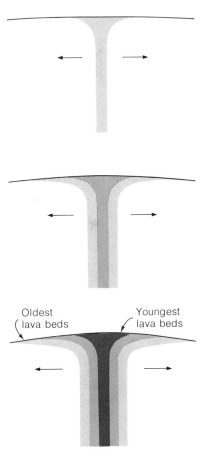

Figure 17-23
Iceland is an exposed part of the mid-Atlantic ridge. This schematic diagram shows the mechanism of its growth, by repeated fissure eruptions and lateral spreading.

**Figure 17-24**
Ignimbrite, or ash-flow sheets, are deposited when intensely hot, gas-charged volcanic dust, ash, and pumice spread swiftly over the surface to settle and cool as flat sheets. The example shown is on the Pajarito Plateau, near Los Alamos, New Mexico. Source of the ash lies beyond the mountains in the background. [Photo from R. G. Schmidt and R. L. Smith, U.S. Geological Survey.]

Such eruptions have produced extensive ignimbrite sheets, although man has never witnessed one of these spectacular events. The early Tertiary ignimbrites of the Great Basin in Nevada and adjacent states cover an area of about 200,000 square kilometers, and are as much as 2500 meters thick in some places. In Yellowstone National Park, successions of forests were buried by broad ignimbrite sheets.

**Central Eruptions** Unlike the linear sources of fissure eruptions, central eruptions are much like point sources of magma. The extrusive materials issue from a **central vent**, or **pipe**, and give rise to the most familiar of all volcanic features—the **cone**.

Lava cones are built by successive lava flows. Basaltic lava is relatively free flowing and spreads widely. If there is a large supply, a broad **shield-shaped** volcano can be built up, many tens of kilometers in circumference and more than one or two kilometers in height. The slopes are relatively gentle, 6 to 12°. Mauna Loa, on Hawaii (Fig. 17-26), is the classic example of a **shield volcano**. Although it rises 4 kilometers (2.5 miles) above sea level, its actual size is hidden by the ocean. Measured from the sea floor, Mauna Loa rises 10 kilometers (6 miles) and has a basal diameter of 100 kilometers (60 miles)! It grew to this enormous size by the accumulation of individual lava flows, each only a few meters thick. The island of Hawaii is actually formed by the overlapping of a group of shield volcanoes. When the lava supply is less plentiful, a smaller lava cone is the result.

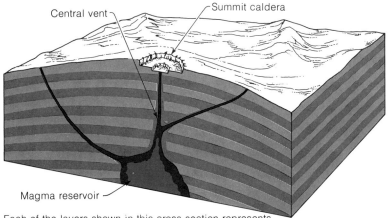

Central vent ⌐ Summit caldera

Magma reservoir

Each of the layers shown in this cross section represents
the accumulation of many hundreds of thin lava flows

**Figure 17-25**
Schematic diagram showing how a shield volcano is built up from the accumulation
of thousands of thin basaltic lava flows that spread widely and cool as gently
sloping sheets. [After R. G. Schmidt, U.S. Geological Survey.]

**Figure 17-26**
The shield volcano Mauna Loa on Hawaii, the giant among the active volcanoes of the world. Smaller
craters lead upward to Mokuaweoweo caldera, the large depression at the summit. [U.S. Air
Force photo.]

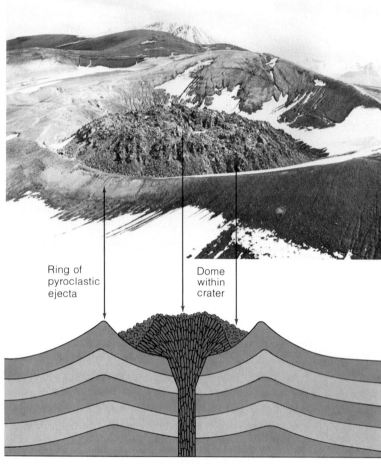

**Figure 17-27**
Volcanic domes are bulbous masses of lava. The lava masses are so viscous that they do not flow, but pile up over the vent. Shown here is Novarupta Dome in the Katmai region of Alaska. [After A. Chidester and R. G. Schmidt, U.S. Geological Survey.]

Silicic lavas are so viscous that they can just barely flow, as is shown by the shapes of the **volcanic domes** they produce (Fig. 17-27). The domes give every appearance that the lava had been squeezed out of a vent like cold toothpaste, with very little lateral spreading, to form a bulbous steep-sided mass. Some silicic lavas are so viscous that no spreading occurs; instead, a cylindrical, or spine-like, shape is retained.

**Cinder cones** are built of fragmental material, explosively ejected from a vent (Fig. 17-28). The profile of the cone is determined by the **angle of repose**—that is, the maximum angle at which the debris remains stable and does not slide downhill (Figs. 17-29 and 17-1). The larger fragments, which fall near the summit,

**Figure 17-28**
Ejecta building a cinder cone on the flank of Mt. Etna, Sicily, during the 1969
eruption. Fragments with dimensions of 0.1 to 1 m³ were thrown out of the
crater with a velocity of about 50 m/sec. [Bernard Chouet, M.I.T.]

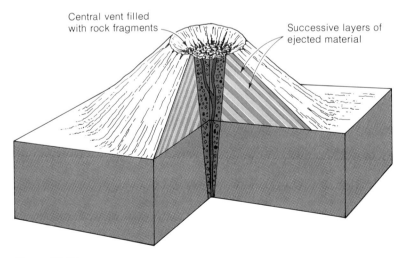

**Figure 17-29**
Diagrammatic representation of the structure of a cinder cone. Ejected material
is deposited as layers that dip away from the crater at the summit. The vent
beneath the crater is filled with fragmental debris.

**Figure 17-30**
Paricutín in eruption. This volcano was born in the field of a Mexican farmer in 1945. It is now a dormant (temporarily inactive) volcano 400 m high. [Photo by Tad Nichols, Tucson, Arizona.]

can form stable slopes exceeding 30°. Finer particles are carried farther from the vent and form gentle slopes of about 10° at the base of the cone. The classic, aesthetically pleasing, concave-shaped, volcanic cone reflects this variation in the angle of repose. Parícutin in Mexico erupted in a farm field in 1943 and built a cinder cone 150 meters high in its first six days of activity. It is now a dormant (sleeping) volcano about 400 meters (1300 feet) high (Fig. 17-30).

When a volcano emits lava as well as pyroclasts, a **composite cone** or **strato-volcano** is built of alternating lava flows and beds of pyroclasts (Fig. 17-31). This is the most common form of the large continental volcanoes, represented by such famous examples as Fujiyama, Vesuvius, Etna, and Stromboli (Fig. 17-32).

If the central vent of a volcano grows too high or becomes plugged, the lava may find an easier route by opening a subsidiary vent and building a small cone on the flanks of the main volcano.

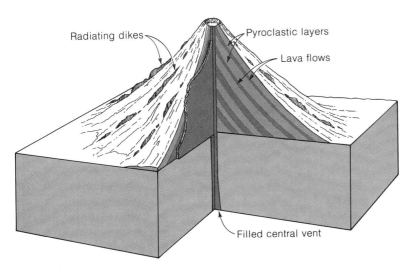

**Figure 17-31**
Structure of a composite volcano built up of layers of pyroclastic material and lava flows. Lava, solidified in fissures, forms rib-like dikes that strengthen the cone. [After R. G. Schmidt, U.S Geological Survey.]

**Figure 17-32**
Mt. Fuji, symbol of many things to the Japanese nation and one of the most majestic mountains of the world, is a composite volcano. [Photo by S. Aramaki, The University of Tokyo.]

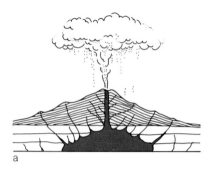

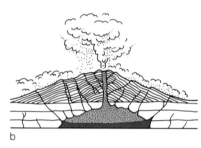

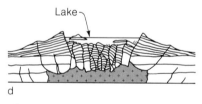

**Figure 17-33**
The evolution of the caldera at Crater
Lake, Oregon. (a and b) Eruptions from
the summit of the vanished, former
Mount Mazama deplete the magma
chamber below. (c) Caldera results from
collapse of the mountain summit into
the empty magma chamber. (d) Lake
forms in the Caldera; minor eruptions
into the lake continue until the magma
below the surface finally solidifies.
[Modified from H. Williams. From
*Principles of Geology* by J. Gilluly,
A. C. Waters, and A. O. Woodford,
W. H. Freeman and Company. Copyright
© 1968.]

Hundreds of subsidiary cones pimple the slopes of Etna, Mauna
Loa, and Mt. Newberry (central Oregon), to cite a few examples.
Flank eruptions can also originate from fissures.

**Craters** form the summits of most volcanoes, and are usually
centered over the vent. During the eruption of a lava volcano, the
upwelling lava overflows the crater walls; when eruption ceases,
the lava that remains in the crater sinks back into the vent. When
pyroclastic cones erupt, the material is literally blasted out of the
crater, which later becomes partially filled by debris falling back
into it. Because crater walls are steep, they may cave in or retreat
by erosion after an eruption, becoming enlarged to several times
the vent diameter. Etna, for example, has a central vent 300 meters
(1000 feet) in diameter and at least 850 meters (2800 feet) deep.

**Calderas** are large, basin-shaped depressions of volcanic origin.
Most calderas are collapse features that form after a magma chamber becomes drained and no longer supports part of the overlying
cone. The subsidence can occur catastrophically or gradually,
leaving a steep-walled, flat-floored depression. Later activity can
lead to the growth of new cones in the caldera floor. Figure 17-33
diagrams the evolution of a caldera. Some famous examples are
the still-active Kilauea (Hawaii), the dormant Crater Lake (Oregon),
and the newly active Krakatoa (Indonesia) (Fig. 17-34). Perhaps
the largest and most spectacular caldera known is the one on
Olympus Mons, a huge volcano photographed on Mars from the
Mariner 9 spacecraft (Fig. 17-35).

At one time it was thought that a caldera was formed by a huge
explosion, as if the volcano had "blown its top." Geologic mapping of the debris around volcanoes, however, has shown that
more volcanic rock subsided than was blasted out. The pattern
of down-faulting was consistent with a different explanation—
that "the roof caved in."

This is not to say that volcanic explosions have not left their
mark on the Earth's surface. Hot, gas-charged magma encountering
groundwater or sea water results in a highly explosive mixture.
Vast quantities of superheated steam are generated with disastrous consequences, in the form of phreatic explosions. The
destructive eruptions of Krakatoa and Taal (Philippines) are two
of the many examples of such cataclysmic events. Another is
Bandai-san (Japan), which blew out in 1888 and left an amphi-
theater 2.5 kilometers (1.5 miles) across with walls 350 meters
(1200 feet) high.

A somewhat rare but especially interesting type of volcanic
feature is the **maar**. Maars are shallow, generally round, pit-like
craters commonly about a kilometer in diameter and surrounded
by a thin, low rim, or **ejecta blanket**, composed of both pyroclastic ejecta and nonvolcanic debris ripped from the vent walls.
The name "maar" is the German word for lake; the features are
named for the beautiful lake-filled craters in the Eifel district of
Germany. Maars in deserts, however, are dry. These shallow
craters are believed to have formed by violent eruptions, driven

**Figure 17-34**
Crater Lake, Oregon, fills the caldera in the stump of the vanished Mt. Mazama. [From *Geology Illustrated* by J. S. Shelton, W. H. Freeman and Company. Copyright © 1966.]

by a flow process not unlike the expansion of gases in a rocket engine. One can imagine a maar in eruption to look much like the exhaust jet of a Saturn 5 rocket placed upside down in the ground, blowing rocks and gas into the air. Such an eruption was observed by a Chilean Air Force pilot when the maar Nilahue erupted in 1955. The jets lasted for about 30 minutes, each blast followed by periods of quiescence of about the same duration. Activity stopped after about a month, presumably after the volcano literally ran out of gas.

**Diatreme** is the term given to the filled volcanic vent, or pipe, left behind when an extinct maar has been completely removed by erosion. Ship Rock, which towers 515 meters (1700 feet) over the surrounding, flat-lying sedimentary rocks of New Mexico, is an excellent example. It has been viewed by thousands of transcontinental air travelers, and looks like a gigantic black skyscraper in the red desert (Fig. 17-37). The eruptive mechanism that produces maars and diatremes has been pieced together in much detail from the geologic record to be found in the ejecta and the diatremes. Rock and mineral data tell us that some maars are

Diamond, the hardest substance known, is formed when carbon is subjected to high pressures, corresponding to depths of about 125 km in the Earth's mantle. Gem-quality diamonds only occur naturally, whereas small industrial diamonds are now produced synthetically.

produced by the upward movement of materials from great depths —well within the upper mantle. The evidence indicates that they are formed by gas-charged magmas that melt their way to the surface, finally ejecting gases, lava fragments from the vent walls, and fragments from the deep crust and mantle—all with explosive energy and sometimes with supersonic velocity.

The underground workings of the fabled Kimberley diamond mines of South Africa penetrate diatremes, which contain mantle materials (including diamonds) and fragments of rock encountered en route to the surface. Geologists view this pipe in much the same way they would a 300-kilometer (180-mile) long drill core in which the rocks were scrambled by eruption. Careful field and laboratory work on such scrambles of rock has enabled geologists to unravel their mysteries and deduce something of the layering of the Earth's crust and mantle well beyond the reach of the deepest drill hole.

**Figure 17-35**
Caldera of the gigantic volcano Olympus Mons, photographed on Mars during the Mariner 9 mission. The caldera is 65 km in diameter, and the volcano is twice as broad as the most massive volcanic pile on Earth. [From National Aeronautics and Space Administration.]

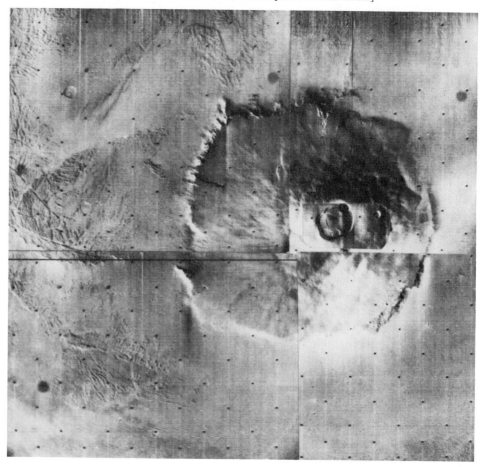

**Figure 17-36**
Maar on Nunivak Island, off the west coast of Alaska. [Photo by T. McGetchin, M.I.T.]

**Figure 17-37**
Shiprock is a diatreme, or volcanic pipe, exposed by erosion of its enclosing rock. It towers 515 m above the surrounding, flat-lying sediments of New Mexico. [Photo from U.S. Geological Survey.]

**Other Volcanic Phenomena** When a nuée ardente meets a river, it can be transformed into a landslide or mudflow of volcanic debris called a **lahar**. Lahars can also be produced when a crater lake is breached, or when glacial ice is melted by lava flows or nuées ardentes. One formation in the Sierra Nevada contains 8000 cubic kilometers (2000 cubic miles) of material of lahar

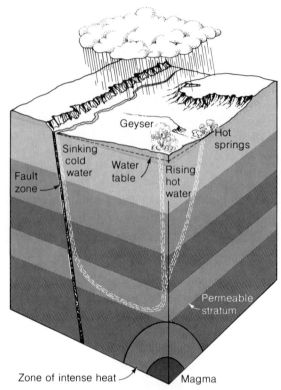

**Figure 17-38**
Surface water percolating down along fractures
occasionally reaches the vicinity of a magma chamber.
The heated fluid rises through other channels to erupt as
steam and hot water, in geysers and hot springs. [After
D. E. White, U.S. Geological Survey.]

origin. Lahars have been known to carry huge boulders for tens
of kilometers. When Kelut (Java) erupted in 1919, the crater lake
containing 380,000 cubic meters (14 million cubic feet) of water
was displaced, and the resultant lahar rushed down the flanks and
killed some 5500 people.

Volcanic gases have been studied by courageous and occasionally foolhardy volcanologists. Water vapor is the main constituent
(70 to 95 percent), followed by carbon dioxide, sulfur dioxide, and
traces of nitrogen, hydrogen, carbon monoxide, sulfur, and chlorine. Enormous amounts of volatiles are released. Parícutin emitted some 18,000 tons of water in a single day.

Both the nature and origin of volcanic gases are of considerable
interest and importance because they formed the oceans and the
atmosphere. Some volcanic gas may be derived from deep within
the Earth and may be making its way to the surface for the first
time (**juvenile gas**); some may be recycled groundwater and ocean
water or atmospheric gas or gases that had been trapped in rocks.

Emissions of gas and vapor that are unaccompanied by the eruption of lava or pyroclastic ejecta often mark the late stages of

In 1938 two Russian volcanologists,
V. F. Popkov and I. Z. Ivanov, made
measurements of lava temperatures
and collected gas while floating down
a river of lava on a raft of colder lava
for a distance of 2 kilometers. The
surface temperature of their raft was
300°C, and the lava river showed a
temperature of 870°C! The occasion
was the eruption of Biliukai on Kamchatka. Pliny the Elder may have been
the first scholar to lose his life studying a volcano. He was killed in the
great eruption of Vesuvius, which
destroyed Pompeii in A.D. 79.

volcanic activity. Circulating groundwater can reach and be heated by buried magma (which retains heat for a long time), producing **hot springs** and **geysers**. A geyser is a hot-water fountain that spouts intermittently with great force, frequently accompanied by a thunderous roar. The best known example is *Old Faithful* in Yellowstone Park, which erupts irregularly, about every 63 minutes, sending a jet of hot water as much as 60 meters (200 feet) high. **Fumaroles** are vents that emit gas fumes or steam. All of these volcanic emanations contain dissolved materials that precipitate as the water evaporates and cools, forming various sorts of encrusting deposits (for example, travertine), some of which are valuable mineral deposits.

**Figure 17-39**
Old Faithful Geyser, Yellowstone National Park. Hot water and steam are violently ejected about once every 63 min. In between eruptions, hot water presumably fills underground cavities. Further heating converts some of the water into steam, generating the pressure that forces the discharge. [Photo from U.S. Geological Survey.]

**Figure 17-40**
Geologists collecting gas samples from the hot, newly formed cinder cone of Kilauea Iki, Hawaii, during the eruption of 1959. [Photo from U.S. Geological Survey.]

## CASE HISTORIES OF VOLCANOES

The stories of individual volcanoes make fascinating reading, not only for their phenomenological content but also for their human interest value. Some volcanoes have changed the course of ancient history—for example, Thera and Vesuvius.

**Kilauea—An Instrumented Volcano**   The giant shield volcano Mauna Loa and the smaller Kilauea on its eastern flank make up the southern half of the island of Hawaii. Because the U.S. Geological Survey operates a volcano observatory on the rim of Kilauea caldera, this volcano is perhaps the best studied in the world, and what has been learned from its study has influenced our notions of volcanic processes profoundly. A modern network of seismographs has been installed to provide information on the internal structure and activity of Kilauea. Tiltmeters (instruments that measure tilting of the ground) were emplaced to observe the swelling and deflation of the volcano due to the accumulation and underground movement of magma. A geochemical laboratory was established to study the systematic variations in the chemistry and petrology of the lavas and the chemistry of the emitted gases.

The U.S. Geological Survey used these new techniques to monitor the Kilauea eruption of 1959–1960. The following material is taken from one report. Between the 14th and 19th of August, 1959, the seismographs detected a swarm of small earthquakes at a depth of 55 kilometers beneath Kilauea caldera. This marked the entrance of magma into the conduits leading from the asthenosphere, through the lithosphere, to the surface. The upward migration of the magma could be traced by weak seismic disturbances

**Figure 17-41**
Core drilling into crust of lava lake of Kilauea Iki, Hawaii, in 1967. At a depth of about 30 m, the drill penetrated melt. Note the geologist wearing asbestos gloves in order to handle the retrieved core barrel, which gets heated to about 300°C. Volcanoes are studied to increase our understanding of their origin and development and our knowledge of the deep interior. [Photo from U.S. Geological Survey.]

originating at depths of 5 to 15 kilometers. Between August and October, the tiltmeters showed that the volcano was beginning to swell as the magma rose and filled a shallow reservoir below the caldera.

The first sign of an outbreak occurred in September, when seismographs picked up a series of small earthquakes near the northeast rim of Halemaumau, a deep pit within the caldera. By November this swarm of tremors exceeded a thousand per day. At this time the summit of Kilauea was swelling three times more rapidly than in prior months. Then on November 14 the quakes increased tenfold in number and intensity, signifying that the eruption fissure was splitting its way to the surface. That evening the lava broke out of a kilometer-long fissure in the wall of Kilauea Iki, a crater just east of the caldera. The earthquakes stopped, but the seismographs recorded an almost continuous tremor that typically accompanies lava outpouring. In the next seven days about 30 million cubic meters (1 billion cubic feet) of lava poured into Kilauea Iki, accompanied by spectacular lava fountains. The tiltmeters showed the volcano to be deflating as the reservoir emptied its contents. There were repeated cycles of swelling and eruption until December 21, when Kilauea Iki became dormant. The Geological Survey scientists expected more trouble, however, because the volcano began to inflate even more than it had in November. More lava was in the shallow reservoir below the caldera than when the eruption began. The pressure would have

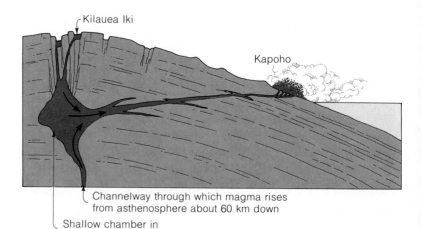

**Figure 17-42**
Schematic diagram of the 1960 eruption on the flank of Kilauea, which destroyed the village of Kapoho on Hawaii. In an earlier stage, lava filled Kilauea Iki crater on the summit, forming a lake 125 m deep, but more than half drained back into the vent. Following the cessation of activity at Kilauea Iki, the flank eruption at Kapoho, 45 km away, covered about 10 km² with lava. [After J. P. Eaton and R. G. Schmidt, U.S. Geological Survey.]

to force open new rifts, and dikes of lava would split their way to the surface, signalled by earthquakes. In early January of 1960, a new swarm of earthquakes was detected, this time not far from the village of Kapoho, about 40 kilometers (25 miles) east of the caldera. Then on January 13, as expected, a flank eruption broke out (Fig. 17-42). It came through a crack about a kilometer long, a few hundred meters north of the village. Within the next four weeks, more than 100 million cubic meters (3.5 billion cubic feet) of lava poured out of the fissure, destroying Kapoho, but causing no casualties. A new landscape was created as the lava flowed to the sea. Twenty-foot walls were built in a futile attempt to divert the lava and save a seashore community. When it was all over,

Box 17-1   Earth Scientists' Hotline

Scientists who study short-lived natural phenomena such as volcanic eruptions, earthquakes, meteorite falls, and tsunami waves require immediate notification that an event has occurred in order not to lose valuable data. Such a service is rendered by the Smithsonian Institution's Center for Short-Lived Phenomena. The Figure shows an event notification report issued in connection with the Kilauea eruption of 1972.

Specialists are notified by telephone or cable of specific events of particular significance to them so that expeditions can reach the site in the crucial early hours. Interested individuals can subscribe and receive these event cards routinely by writing to the Smithsonian at the address shown. A fee is charged to cover mailing and preparation costs, but these cards constitute the best "newspaper" of current natural history events now available.

| EVENT | 9-72 | KILAUEA VOLCANIC ERUPTION-1972 | 14 FEBRUARY 1972 | 1354. |
|---|---|---|---|---|

"The Mauna Ulu vent on Kilauea's east rift is erupting again after more than 3 months of inactivity. A new lava lake was discovered in Mauna Ulu's summit crater about 0900, February 5. It may be that the eruption began sometime between February 2 and 4, as shown by a slight increase in tremor on a nearby seismograph, but a severe storm and lack of visibility prevented earlier inspection of the area. The level of the lava lake progressively rose from the bottom of the crater, formerly about 150 m deep, and when discovered it was 80 m below the crater rim. It continued to rise, and on February 7, at a level 25 m below the rim, lava spilled over a notch at the east end of the crater and flowed into a collapsed trench about 450 m long that extends down Mauna Ulu's east flank. At the east end of the trench, lava drains into a tube, travels SSE about 250 m, and emerges at the north edge of the subsidence bowl that marks the site of old Alae crater. Thin flows have spread across about a quarter of the floor of the depression. The surface of the lava lake in Mauna Ulu crater is presently characterized by vigorous fountains that reach as much as 25 m in height. The entire surface is continually agitated by fountains and bursts of escaping gas. A prolonged period of tumescence of Kilauea's summit preceded this eruption, which is 11 km ESE of the summit. The outbreak is mild, and only slight deflation has occurred to the present. Harmonic tremor is weak, and only a few small earthquakes, none strong enough to be felt, have accompanied the eruption."

EVENT  NOTIFICATION  REPORT

| TYPE OF EVENT | GEOPHYSICAL |
|---|---|
| DATE OF OCCURRENCE | FEBRUARY 5, 1972 AND CONTINUING |
| LOCATION OF EVENT | ISLAND OF HAWAII |
| | U.S.A. |
| REPORTING SOURCE | U. S. GEOLOGICAL |
| | SURVEY |
| SOURCE CONTACT | DR. DONALD W. |

PETERSON, HAWAIIAN VOLCANO OBSERVATORY

HAWAII NATIONAL PARK, HI   96718, USA

SMITHSONIAN INSTITUTION
CENTER FOR SHORT-LIVED PHENOMENA
60 Garden Street
CAMBRIDGE, MASSACHUSETTS 02138
UNITED STATES OF AMERICA
CABLE  SATELLITES NEW YORK
TELEPHONE  (617)-864-7911

sao-slp-5

the tiltmeters showed that the reservoir under the caldera had been drained in supplying lava for the Kapoho eruption. Confirmation was given by the fact that the amount by which the summit subsided was volumetrically equal to the lava erupted.

Geologists believe that the basaltic magma in the partially molten asthenosphere is lighter than the rock in the overlying lithosphere. The magma could therefore be squeezed to the surface by the weight of the lithosphere, if an escape route were available. A fracture under Hawaii taps this source of magma, and the lava rises to fill the shallow chamber under the caldera. The rest of the story is as given above.

**Mont Pelée**   The term **peléan** is an adjective that is taken from the name of this volcano, and is used to describe violent eruptions and explosions of gas-charged and very viscous magmas. It is an aptly chosen word, for on May 8, 1902 at 8:02 A.M. the Caribbean coastal town of St. Pierre, below Mont Pelée on Martinique, was destroyed by a nuée ardente, and all but two of its 28,000 inhabitants perished within minutes. Once the nuée ardente exploded laterally out of the mountain with very little warning, nothing could be done to save the panicked victims. The glowing volcanic avalanche plunged down the slopes at a speed of about 100 kilometers per hour and engulfed the town with a searing 800°C gas-glass-dust emulsion (Fig. 17-43). The gas was mostly $CO_2$. It is sobering to scientists who render advice to others to recall the statement of Professor Landes of St. Pierre's College, issued the day before the cataclysm: "The Montagne Pelée presents no more danger to the inhabitants of Saint Pierre than does Vesuvius to those of Naples." Professor Landes perished with the others.

The following is an account of an officer aboard a ship in the harbor, quoted by K. Wilcoxon:

"I saw St. Pierre destroyed. The city was blotted out by one great flash of fire. Nearly 40,000 people were killed at once. Of eighteen vessels lying in the roads, only one, the British steamship *Roddam* escaped and she, I hear, lost more than half of those on board. It was a dying crew that took her out. Our boat, the *Roraima*, arrived at St. Pierre early Thursday morning. For hours before entering the roadstead we could see flames and smoke rising from Mt. Pelée. No one on board had any idea of danger. Captain G. T. Muggah was on the bridge, and all hands got on deck to see the show. The spectacle was magnificent. As we approached St. Pierre we could distinguish the rolling and leaping of red flames that belched from the mountain in huge volumes and gushed into the sky. Enormous clouds of black smoke hung over the volcano. . . . There was a constant muffled roar. It was like the biggest oil refinery in the world burning up on the mountain top. There was a tremendous explosion about 7:45, soon after we got in. The mountain was blown to pieces. There was no warning. The side of the volcano was ripped out and there was hurled straight toward us a solid wall of flame. It sounded like a thousand cannon.

**Figure 17-43**
View of St. Pierre toward steaming Mont Pelée, after it was destroyed by a nuée ardente in the eruption of May 8, 1902. [Copyright 1902 by Underwood and Underwood. Library of Congress.]

"The wave of fire was on us and over us like a flash of lightning. It was like a hurricane of fire. I saw it strike the cable steamship *Grappler* broadside on, and capsize her. From end to end she burst into flames and then sank. The fire rolled in mass straight down upon St. Pierre and the shipping. The town vanished before our eyes.

"The air grew stifling hot and we were in the thick of it. Wherever the mass of fire struck the sea, the water boiled and sent up vast columns of steam. The sea was torn into huge whirlpools that careened toward the open sea. One of these horrible, hot whirlpools swung under the *Roraima* and pulled her down on her beam end with the suction. She careened way over to port, and then the fire hurricane from the volcano smashed her, and over she went on the opposite side. The fire wave swept off the masts and smokestacks as if they were cut by a knife.

"Captain Muggah was the only one on the deck not killed outright. He was caught by the fire wave and was terribly burned. He yelled to get up the anchor, but before two fathoms were heaved in, the *Roraima* was almost upset by the boiling whirlpool and the fire wave had thrown her down on her beam ends to starboard. Captain Muggah was overcome by the flames. He fell unconscious from the bridge and overboard. The blast of fire from the volcano lasted only a few minutes. It shrivelled and set fire to everything it touched. Thousands of casks of rum were stored in St. Pierre, and these were exploded by the terrific heat. The burning rum ran in streams down every street and out into the sea. This blazing rum set fire to the *Roraima* several times.

"Before the volcano burst, the landings of St. Pierre were covered with people. After the explosion, not one living soul was seen on the land. Only twenty-five of those on board were left after the first blast.

"The French cruiser *Suchet* came in and took us off at 2 p.m. She remained near by, helping all she could, until 5 o'clock, then went to Fort de France with all the people she had rescued. At the time it looked as if the entire north end of the island was on fire."

**Krakatoa** The 1883 explosion of Krakatoa, in the strait between Java and Sumatra, was one of the greatest witnessed by man. Krakatoa was an island formed from nested volcanic cones in an ancient caldera. The caldera, 6 kilometers (4 miles) across, was a remnant of a prehistoric andesitic stratovolcano. On August 27, following numerous smaller explosions, Krakatoa blew its top in a phreatic explosion with a TNT equivalent of about 100 megatons (million tons) ($2 \times 10^{11}$ pounds). The sound of the explosion was heard in Australia, nearly 2000 kilometers away. The atmospheric pressure waves were recorded on weather barographs on the other side of the Earth. Nearly 20 cubic kilometers (5 cubic miles) of debris was discharged into the air, much of it falling as ash over an area of some 700,000 square kilometers (300,000 square miles). Almost total darkness settled on Jakarta, 150 kilometers (100 miles) away when the dust blotted out the Sun. Fine dust rose to the stratosphere and drifted around the Earth, producing brilliant red sunsets for several years and lowering Earth's mean annual temperature by a few degrees by partially blocking solar radiation. The explosion generated a giant sea wave that reached a height of almost 40 meters (130 feet) and drowned 36,000 people in nearby coastal towns. The sea waves were recorded on tide gauges as far away as the English Channel. After the eruption most of Krakatoa had disappeared, leaving in its place a water covered basin 300 meters (1000 feet) deep. It is believed that much of the energy was provided by the violent expansion of hot steam after the walls of the volcano were initially ruptured, permitting sea water to pour into the magma chamber. The result can be viewed as one of the biggest steam-boiler explosions in recorded history.

Following forty-four years of quiescence, another cycle of volcanic activity began in 1927. A new island, *Anak Krakatoa* (Child of Krakatoa), is now building up by repeated eruptions and has reached a height of more than 100 meters (330 feet).

## THE GLOBAL PATTERN OF VOLCANISM

At the outset of this chapter, we mentioned the correlation of active volcanism with present-day plate margins. As can be seen from Figure 17-3 most volcanoes occur in connection with belts of tension, where plates diverge, and belts of compression, where plates converge. To a lesser extent, volcanoes are also found within plates. Examples are Hawaii, which lies within the Pacific plate, and the eastern rift zone, which lies within the African plate. The global pattern of volcanism in association with plate boundaries also extends to the types of lavas that erupt and consequently to the styles of eruptions. Most lavas that issue from vents in oceanic divergence zones and from intraoceanic volcanoes are basaltic. Where ocean plate collides with ocean plate, basalts and andesites predominate (see Fig. 17-4) as a reminder of the classification of lavas). Adjacent to the zone along which an ocean plate converges against a continental margin, rhyolitic ignimbrites are found as well as basalt and andesite. If we are allowed to present heuristic arguments (reasonable ones, but without proof), make some broad generalizations, and not try to account for all details at this early stage of the theory of plate tectonics, this worldwide pattern can be understood in a consistent and insightful manner. But first we must say something about mafic, intermediate, and silicic magmas and their possible modes of origin.

**Magmas—Where Do They Come From?** Our current views of the principal zones of the upper few hundred kilometers of the Earth are summarized in Figure 1-16. The continents float like rafts on the rigid plates of the lithosphere. The continental crust is made up mainly of granite and other felsic rocks. Such rocks are almost entirely missing from the sea floor, the oceanic crust consisting mostly of the mafics basalt and gabbro. The crust is separated from the rest of the lithosphere by an abrupt change in chemical composition called **M discontinuity**. Beneath this discontinuity the lithosphere consists of ultramafic rock (even less silica and more ferromagnesium minerals than mafic), underlain by the weak, partially molten asthenosphere, which is also ultramafic. We will give more details about these zones in later chapters.

We surmise that basalt must be the molten component of the asthenosphere for several reasons. Recalling the case history of Kilauea on Hawaii, we know from seismic evidence that the magma taps the asthenosphere and that basalt is the eruptive result. Peridotite is our best guess for the ultramafic rock that makes up most of the upper mantle. When peridotite is heated in the laboratory to the temperature at which melting begins, basalt is the first liquid to form.*

The felsic lavas at the silicic end of the scale of volcanic rocks, such as rhyolite, must have another origin. The compositional contrast with basalt is so great that the location and nature of the parent rocks or the evolution of the magma must be different. That rhyolitic ignimbrites are found on continents and not in the oceanic environment (where a felsic crust is absent) is an important clue. One way to make rhyolite is obvious — by completely remelting granitic continental crustal rocks to produce silicic magma. Alternatively, sediments derived from these rocks and deposited on the sea floor adjacent to continents could, if reheated, yield silicic magma. One favorable environment for the formation of silicic magma would therefore be a convergence zone where the leading edge of one of the colliding plates is a continental margin. All the ingredients for "cooking up" silicic magmas are present: heat, deformation, and an abundance of parent materials of the proper composition (see Fig. 1-21).

Andesitic lava — intermediate to the felsic and mafic end members — can be derived in several ways. Suppose a basaltic magma cools in a magma chamber until crystals begin to form. The first mineral to appear is olivine, which sinks to the bottom, leaving a more silicic magma. This process of gravitational settling of heavy crystals can continue as long as new minerals crystallize; at each stage the residual liquid, or differentiate, would have a different composition, depending on which minerals were formed and separated from it. In principle, the differentiation series represented by the transformation from basalt to andesite to rhyolite is possible. It is doubtful whether rhyolite, the ultimate differentiate, commonly originates from a parent basaltic magma, but an intermediate differentiate, andesite, has been found in an oceanic environment — in Iceland, where it undoubtedly formed as a differentiate from the much more abundant basalt. In fact, Icelandic geologists have shown that the longer the period of repose between eruptions, the more silicic is the eruptive, since there is more time for fractional crystallization and settling to occur (see Chapter 16 for a discussion of fractional crystallization).

As was mentioned in the preceding chapter, andesite might also be derived from the partial melting of mantle materials in the presence of water.

Because andesites are roughly intermediate in composition between basalt and rhyolite, they can also result from a mixture of mafic and silicic magmas or by assimilation of surrounding

---

*Basalts are highly variable, a subject we will not go into except to say that composition can depend on such factors as local composition of the parent rock, depth of melting, amount of water present, degree of partial melting, and assimilation of other rocks en route to the surface.

granitic rock by a mafic magma. Nevertheless, the occurrence of large volumes of andesite in conjunction with subduction zones argues for still another source, perhaps the most important one. In recent years it has been suggested that a downgoing, cold, oceanic plate begins to remelt when it plunges diagonally into the asthenosphere. Heat flowing in from the hotter asthenosphere and the heat generated by friction along the contact between the upper surface of the descending plate and the overriding plate would supply the thermal energy needed for melting. Thus the basaltic crust of the old sea floor, chemically enriched with silica by its clinging veneer of wet ocean-bottom sediments, could melt into andesitic magma, which would rise buoyantly to erupt from volcanoes in the overlying volcanic arc. We are now in a position to summarize some of the major features of volcanism on a global scale (Fig. 17-44).

**Ocean-Ridge Volcanism** A worldwide system of fissures, along which basalt erupts, coincides with the ridge-rifts depicted in Figure 1-17 and in the schematic section of Figure 17-44. The fracture between the separating plates extends to the asthenosphere. A mush of peridotite and basaltic magma rises buoyantly into the fracture to form new lithosphere. As the hot peridotite from the asthenosphere rises in the gap between the separating plates, the reduced pressures induce further melting, adding to the supply of basalt. The overflow forms ocean ridges, volcanoes, and the basaltic sea floor crust. Basaltic plateaus such as Iceland can also build up.

**Figure 17-44**
Schematic representation of volcanism on a global scale, showing the association with plate boundaries, particularly the regions of formation and destruction of the lithosphere.

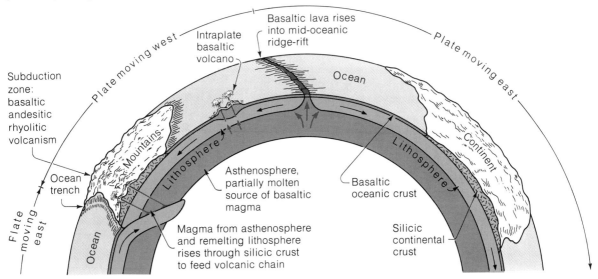

**Intraocean Volcanism**  Hawaii is the type example of an active volcano that rises from the sea floor but is removed from a plate boundary. A channelway penetrates the lithosphere and taps the basaltic melt of the asthenosphere, as shown in Figure 17-44. The magma rises buoyantly and builds immense shield volcanoes superimposed on the oceanic crust. Why such conduits exist within plates and the mechanism that maintains them are as yet unresolved questions.

**Guyots,** mentioned in Chapter 5, are extinct, truncated volcanic cones of basalt that rise from the sea floor to within 1 or 2 kilometers of sea level. The flat tops indicate that the cones were truncated by erosion at an earlier stage, when they projected to sea level, and that truncation was followed by subsidence. One view has it that guyots originated as volcanoes on or near ridge crests, became extinct, then were truncated by ocean-wave erosion and carried away by the spreading sea floor. As the oceanic lithosphere spread, it cooled and contracted, causing the sea floor to subside, dropping the flat-topped volcano below sea level (Fig. 17-45).

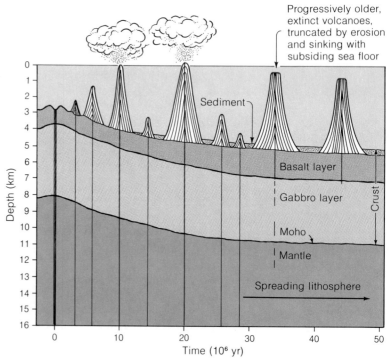

Progressively older, extinct volcanoes, truncated by erosion and sinking with subsiding sea floor

**Figure 17-45**
Guyots probably began as undersea volcanoes that grew near spreading centers. Many projected above sea level, became extinct, and were eroded down to flat-topped cones by wave action. The volcanoes rode along on the moving plate as they grew, then were truncated, finally to sink beneath the sea surface as the plate that carried them subsided. [After "The Deep Ocean Floor" by H. W. Menard. Copyright © 1969 by Scientific American, Inc. All rights reserved.]

**Intracontinental Volcanism** Fissure eruptions of basalt, like those in the Pacific Northwest of the United States, imply that a fracture penetrated the continental lithosphere and that the basalt spurted rapidly to the surface without much contamination from the silicic crust. Perhaps tensile fractures developed in the early stages of an abortive spreading episode. Eruptions of basalt, marking the initial stages of continental rifting, or breakup, can be documented in several parts of the world. For example, basalt flows found today in fault-bounded troughs in eastern North America erupted at the beginning of the breakup of Pangaea some 200 million years ago. Basalt is found in association with the rift valleys of East Africa—a feature that some geologists interpret as representing an abortive stage in the breakup of that continent.

The vast ignimbrite sheets, which are the rhyolitic counterpart of flood basalts, have no easy explanation. Perhaps magma rising slowly from the upper mantle assimilates large quantities of the granitic or sedimentary crust and in this way is transformed into a rhyolitic magma.

**Convergence-zone Volcanism** The many phenomena that occur where plates converge are now being sorted out by Earth scientists. One of the main features of a region of plate collision is the chain of volcanoes that stands subparallel to the adjacent deep-sea trench and marks the edge of the overriding plate. Where two oceanic plates converge, an island arc builds up from the sea floor, typically by the extrusion of basalts and andesites. The basalts probably derive from the asthenosphere above the descending plate, and the andesites might come from the partial melting of the basaltic crust and the ocean-bottom sediments attached to the descending plate, as was mentioned earlier. The Philippine arc and the Marianas Arc are examples of this type of convergence zone. The growing island arc contributes erosion products and volcanic debris to the offshore region and beyond into the trench. In a later stage this pile can be compressed, heated, intruded, and uplifted to form new continental crust and mountains. Japan is a prototype of this process, and Fujiyama is an example of an andesitic cone rising within a growing arc. Mt. Pelée on Martinique, which squeezes out viscous silicic lavas and explosive nuées ardentes, which are deposited as ignimbrites, is another example.

When an oceanic plate is overridden by a plate carrying a continent on its leading edge, as in Figure 17-44, an arcuate mountain chain is thrown up in the zone of compression near the continental margin. The Chilean trench and the adjacent Andes Mountains are examples of such arcuate mountains, and the magmatic activity is typically manifested by the ejection of huge quantites of ash and the eruption of andesitic and basaltic lavas. Basalt would occur if magma from the asthenosphere could reach the surface without being contaminated by sediments or the intervening felsic crust. But the more felsic eruptives seem to imply either contamination by sediments or actual melting of continental crust. Parícutin (Mexico), Cosequina (Nicaraugua), Irazu (Costa Rica), Cotopaxi (Equador), and Calbuco (Chile) are examples (see Fig. 17-3).

**The Growth of Continents**  It is basic to the theory of sea-floor spreading that the sea floor is ephemeral: it is created by spreading at ridge-rifts and destroyed by subduction. Furthermore, continents and island arcs can be eroded but not destroyed, since they are too light to sink back into the mantle. Continents grow on their leading edge by sedimentation and magmatism of the kind just described. Island arcs build up from the sea floor. According to the theory, ocean basins eventually close up; continents run into continents, trapping arcs and sediment-filled troughs between them. Perhaps this also contributes to continental growth. Even the erosion products may eventually be recycled back to the continents after spending some time on the sea floor (see Fig. 1-21).

## VOLCANISM AND HUMAN AFFAIRS

Can volcanic eruptions be predicted? This question is of real concern. Some 200,000 people have been killed by volcanoes in the past 500 years. With the increased concentration of population, the toll could go higher. The geologic record shows that ignimbrite sheets have been deposited on a vast scale—something man has not yet witnessed. If the Katmai (Alaska) eruption of 1912 had been centered in New York (chosen only to show the scale of the event), the city would have been destroyed, one foot of ash would have covered Philadelphia, and fumes would have reached Denver.

Volcanoes kill and destroy in several ways: lava flows, nuées ardentes, and phreatic explosions and their consequences, such as tsunamis, ash falls, and lahars. A shield volcano like Kilauea, whose eruptive style is the release of basaltic lava, provides the best chance of successful prediction. The premonitory phenomena are many: earthquakes, ground tilt, seismic tremor. There may be no way to prevent Kilauea from damaging property, but it need not be a killer. Erupting basaltic lava usually flows slowly enough for people to get out of the way, and in some situations the lava streams might be diverted from property.

Volcanoes erupting gas-charged, silicic (hence viscous) magma are more dangerous; they explode, and the instant of the explosion is almost impossible to predict. The beginning of a period of dangerous activity might be foretold by seismographs, tiltmeters, temperature monitors, gas detectors, or changes in gravity and magnetic fields. At the first indication of activity, a local population might at least be alerted for possible evacuation. This, however, would be more like long-range or climatological forecasting than a weather report for the next day. Phreatic explosions must be treated similarly, for the danger zone can extend over hundreds of kilometers, especially if the volcano has a history of generating tsunamis. Krakatoa is the prime example.

Lahars occur in conjunction with eruptions near ice fields or bodies of water. The paths of lahars should be predictable, both from topography and recent history. Even though some minutes or hours of warning might be possible after an eruption has occurred, it seems better to prevent habitation of possible lahar paths below volcanoes.

Although it will be difficult, it may soon be possible to predict future eruptions of known active or temporarily dormant volcanoes. But what of the long-extinct volcanoes? Do they come to life suddenly—as Vesuvius did, or as Mt. Shasta might? An even more difficult problem of prediction is posed by Paricutin, which rose up from a flat farm field. Finding out how to sense the movements of deep lava in relation to possible new outlets to the surface will be a real challenge.

**Beneficial Aspects of Volcanism** We have seen something of the beauty of volcanoes and also something of their destructiveness. Volcanoes contribute to man's well being in numerous ways. The atmospheres and the oceans originated in volcanic episodes of the distant past. Soils derived from volcanic materials are exceptionally fertile. Emissions of volcanic rock, gas, and steam are

**Figure 17-46**
Chemicals such as boric acid and ammonia are recovered from natural steam generated by volcanic heat. Larderello, Italy. [Photo by F. M. Bullard, University of Texas.]

also sources of important industrial materials and chemicals, such as pumice, boric acid, ammonia, carbon dioxide. The thermal energy of volcanism is being harnessed more and more. Most of the houses in Reykjavik, Iceland, are heated by hot water tapped from volcanic springs. Geothermal steam is exploited as a source of energy for the production of electricity in Italy, New Zealand, the United States, Mexico, Japan, and the U.S.S.R. Exploration for new geothermal energy concentrations is actively underway in the United States and Mexico.

## SUMMARY

1. Volcanism occurs because molten rock inside the Earth rises to the surface, squeezed up by the weight of the overlying layers.

2. The classification of lavas into felsic, intermediate, and mafic is based on the decrease in silica content and the increase in iron-magnesium mineral content. The chemical composition and gas content of lava are important in determining the form an eruption takes.

3. Basalt can be highly fluid and erupt in sheets from fissures, often building lava plateaus. Shield volcanoes grow from repeated eruption of basalt from vents. Silicic magma is more viscous and, when charged with gas, tends to erupt explosively. The pyroclastic debris piles up into cinder cones or covers extensive areas with ignimbrites or ash-flow sheets. Stratovolcanoes are built of alternating layers of lava flows and pyroclastic deposits.

4. Oceanic volcanoes are almost entirely basaltic. The formation of the sea floor from basaltic magma rising from the asthenosphere into fissures of the ocean ridge-rift system is volumetrically the most significant form of volcanism. The more intermediate and silicic eruptives tend to occur on islands and continental margins, in regions where plates collide. They may originate by remelting of the sinking lithosphere or by assimilation of silicic sea-bottom sediments or silicic continental crust into rising magma.

## EXERCISES

1. It seems reasonable to identify the asthenosphere as a major source of magma. Why? What forces magma to rise to the surface?

2. What is the difference between magma and lava? Give some examples of types of lava and their coarse-grained, intrusive counterparts.

3. Describe the principal styles of volcanic eruptions and the deposits and landforms they produce.

4. What is the association between plate boundaries and volcanism? Can the eruptive style and composition of volcanic deposits be correlated with plate boundaries?

Green, Jack, and N. M. Short, *Volcanic Landforms and Surface Features.* New York: Springer-Verlag, 1972.

Herbert, Don, and Fulvio Bardossi, *Kilauea: Case History of a Volcano.* New York: Harper and Row, 1968.

U.S. Geological Survey, *Atlas of Volcanic Phenomena.* U.S. Geological Survey, Washington, D.C. 20242

U.S. Geological Survey, *Volcanoes of the United States* (a pamphlet), U.S. Geological Survey, Washington, D.C. 20242.

Wilcoxon, K. H., *Chains of Fire — The Story of Volcanoes.* Philadelphia: Chilton Books, 1966.

Split Mountain, at the crest of the Sierra Nevada batholith, near Bishop, California. The dark rock above has been metamorphosed and deformed by the intrusion of the granite below. [Photo by John S. Shelton.]

# 18

# Plutonism and Metamorphism

*Evidence of the processes of igneous intrusion and metamorphism by heat and pressure is found in the structure and form of plutonic bodies and the metamorphic rocks that are associated with them. Plutons take a variety of shapes and sizes, depending on the kinds of rocks intruded, the volume of magma, and the deformation accompanying emplacement: dikes, sills, laccoliths, lopoliths, and batholiths are all geological expressions of intrusion. Metamorphic rocks, associated with mountain belts and igneous activity, are of three main kinds. Cataclastic rocks are mechanically broken and ground up rocks produced by high deformational pressures. Contact metamorphic rocks are produced in baked zones of high temperature bordering intrusions. Regionally metamorphosed rocks are made by the combination of pressure and temperature operating over wide areas. As is true in igneous rocks, texture and mineralogy are the pressure gages and thermometers of metamorphism.*

Much is known of the mechanisms by which extrusive rocks are formed, because the eruption of volcanoes and the ejection of lava have been open to direct study in many places at the Earth's surface. This is not true, however, of the mechanisms by which igneous and metamorphic rocks are formed at depth; indirect evidence has to be relied upon in the study of their formation, long after the rocks have cooled. The forms that molten igneous masses take when they intrude the crust are deduced from direct evidence gained through geological field work done millions or billions of years after the rocks were emplaced. Similarly, metamorphism, the geological effects of regional increase of heat and pressure, has to be studied in the same way, long after the fact. The biologist dissects his specimen to see what goes on inside, but the geologist must, in a sense, wait for Earth to dissect itself. Field study of igneous and metamorphic rocks can only proceed where the crust has been uplifted and subjected to deep erosion. Thus the study of these kinds of rocks is indissolubly linked with structural geology and the history of mountain belts.

Holes have been drilled into the Earth's crust in various places where heat flow and seismic activity indicate active igneous and metamorphic processes are now operating. One of these is under the Salton Sea, in Southern California, where the North American lithospheric plate abuts the Pacific plate along the San Andreas transform fault (see Fig. 22-61). The temperatures in the drill hole are so high that superheated steam and water are produced, along with a great quantity of unusual chemical substances, including many metallic elements, all of which are dissolved in water. From the chemical composition of the waters, geochemists have inferred that a magmatic body at depth is metamorphosing the sediments around it. Drill holes and mine workings are of chief importance, however, not in regions of current igneous activity, but in regions where they extend the eyes of geologists to older rocks that are now cooled but still buried. In such places, the details of structure and form are mapped as a guide to further exploration for metal ores, for many of the great economic mineral resources of the Earth are formed by igneous and metamorphic activity.

## PLUTONS

Igneous bodies that have congealed from magma underground are called **plutons**. When uplift and erosion expose them to study, or when they are examined in mines and boreholes, they are found to be variable in size and shape and in their relationship to the **country rock** — that is, the invaded rock surrounding them. Plutons include **laccoliths, lopoliths, stocks,** and **batholiths.** Figure 18-1 depicts these different kinds of plutons as well as **dikes** and **sills**.

Magma rising through the crust makes space for itself in several ways. It can eat its way upward by breaking off blocks of the invaded rock, which sink and are assimilated at depth (**stoping,** the name of the process, is borrowed from the mining term for the removal of ore by working upwards); it can heat the invaded rock enough to make it flow plastically out of the way; and it can forcibly split rocks apart or bow them up. The structure of overlying rocks often gives the clue to the process of intrusion.

**Sills, Laccoliths, and Dikes**   There are probably few open voids or cavities at depths much greater than 8 or 10 kilometers. The pressure due to the weight of the overlying rock would tend to close them. Since there are no vacant spaces for sills, laccoliths,

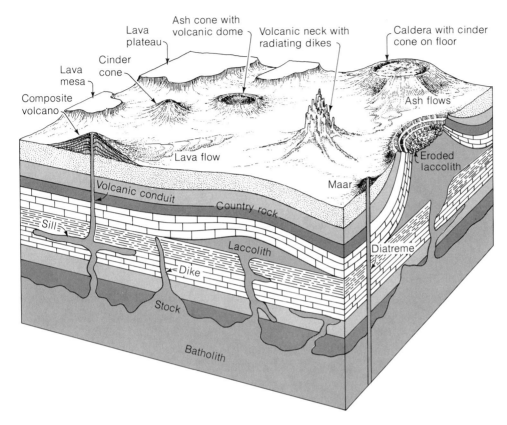

**Figure 18-1**
Block diagram of plutonic and volcanic structures. [After R. G. Schmidt and H. R. Shaw, U. S. Geological Survey.]

and dikes to fill, these intrusions must make their own space by wedging open and penetrating cracks, joints, or planes of weakness, the force coming from the pressure that drives the magma. A **sill** is a tabular pluton that has been formed by injection of magma between beds of layered rock in a **concordant** fashion; that is, its boundaries are parallel to the layering. Sills range in thickness from a mere centimeter to as much as hundreds of meters, and can extend laterally over considerable areas (Figs. 18-2 and 18-3). The Whin Sill in northern England is the type example. It averages about 25 meters (80 feet) in thickness and has an area of some 400 square kilometers (1550 square miles). The 300-meter (1000-foot) thick Palisades Sill, which overlooks the Hudson River in New York and New Jersey, is another example.

Sills can be distinguished from sheets of extrusive rock because they lack the ropy, blocky, vesicular or pillow structure of lava flows. Moreover, sills are typically coarser grained, and the rocks above and below them show thermal effects, such as baking or bleaching.

602

**Figure 18-2**
Laccolith and sill intruding sandstone and shales in the Henry Mountains, Utah. [From *Physical Processes in Geology* by A. M. Johnson. Freeman, Cooper & Company. Copyright © 1972.]

**Figure 18-3**
Diabase sills in Victoria Land, Antarctica. The concordant sill in the center of the mountain face sent discordant tongues into its floor and turned abruptly into a thick dike toward the left side. [Photo by W. B. Hamilton, U. S. Geological Survey.]

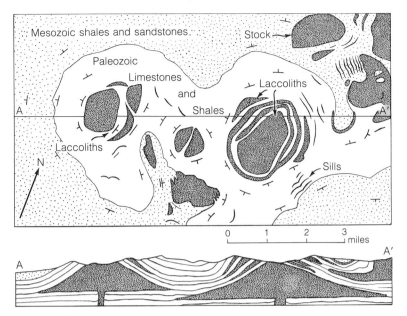

**Figure 18-4**
Map and cross section of laccoliths and stocks in the Judith Mountains, Montana.
[After W. H. Weed and L. U. Pirsson, U. S. Geological Survey.]

Like sills, **laccoliths** are formed by the injection of magma along bedding planes of flat, layered rocks, but instead of being tabular they typically have a mushroom shape (Figs. 18-2 and 18-4). Unlike sills, laccoliths dome the overlying layers upward. Sills and laccoliths are both injected at high enough pressures to overcome the weight of the overlying rocks, which are uplifted to accommodate the magma.

**Dikes** are tabular plutons that are discordant; that is, they cut across the layering of the country rock. Dikes can force open pre-existing fractures, but more often their channels follow cracks opened by the pressure of magmatic injection. Some individual dikes can be traced across country for tens of kilometers. Widths vary from centimeters to many meters (Fig. 18-5). When country rock is deformed by intrusion, many cracks are formed, all of which are possible channels for the invading magma. For this reason, dikes rarely occur alone; more typically, large numbers, or **swarms**, of hundreds or thousands of dikes are found in a region.

Some dikes crop out roughly in the form of circles or ellipses. Called **ring dikes**, these features are thought to be the erosional remnants of intrusions that filled cylindrically shaped fractures. Supposedly, a fracture of this sort forms when a crustal block with roughly circular section sinks into a depleted magma chamber. The resulting ring dike bounds a circular area of subsidence. Ring dikes have been found with diameters ranging up to 25 kilometers.

**Figure 18-5**
Dikes of dark igneous rock cutting across biotite gneiss. Gunnison County,
Colorado. [Photo by W. R. Hansen, U. S. Geological Survey.]

**Lopoliths and Batholiths**   These are more massive intrusions,
generally formed deeper in the crust than the plutons discussed
in the preceding section. A **lopolith** is a large, floored intrusive
whose center has sagged downward, both roof and floor, to form
a bowl-shaped body. The Duluth lopolith, the type example, is
a huge intrusion of gabbro. It crops out on both sides of Lake
Superior's western end, and is inferred to continue beneath
the lake. It is estimated to be 250 kilometers across, 15 kilometers
thick, and to have a volume of about 200,000 cubic kilometers.
Geophysical data suggest that, buried beneath younger rocks,
it extends all the way to Kansas. If so, it is enormously more
voluminous than estimated above.

Batholiths, the largest of plutons, are discordant intrusives
that are, by definition, at least 100 square kilometers in area.
Similar, but smaller, discordant plutons are called **stocks**. The
bottoms of batholiths are unplumbed; whether they remain constant or increase in cross section with depth is unknown, but they
are estimated to extend perhaps 10 to 30 kilometers into the
crust. These vast bodies, composed primarily of granite or granodiorite, are visible today because they have been uplifted and
exposed by erosion, which has stripped away some of the overlying country rock (Figs. 18-6 and 18-7).

Lopoliths and batholiths are generally coarse-grained—a
consequence of the slow cooling that deeply buried bodies
experience. Most lopoliths are differentiated, showing conspicuous

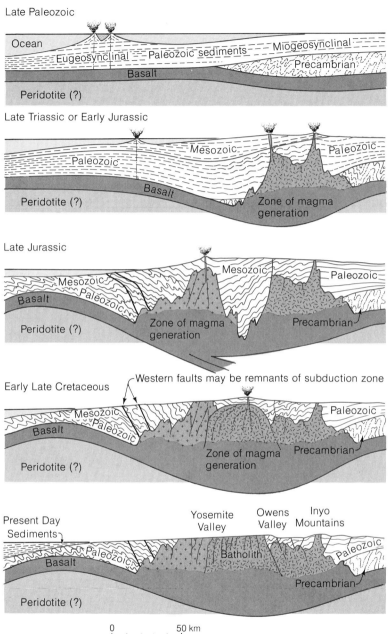

**Figure 18-6**
Generalized sections across Sierra Nevada, showing existing structure of this
huge batholith (bottom) and a model for its evolution (top four sections). According to a plate-tectonics interpretation, magma generation, intrusion, and
deformation would be associated with subduction in late Triassic or early Jurassic
time, which continued actively until the late Cretaceous about 80 million years
ago. [After P. C. Bateman and J. P. Eaton, *Science,* v. 158, p. 1407, 1967. Copyright 1967 by the American Association for the Advancement of Science.]

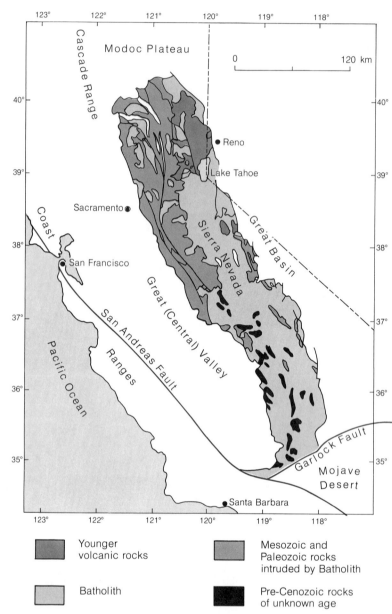

**Figure 18-7**
Generalized geologic map of the Sierra Nevada and adjacent areas. [After
P. C. Bateman and J. P. Eaton, *Science,* v. 158, p. 1407, 1967. Copyright 1967
by the American Association for the Advancement of Science.]

zones or bands of contrasting minerals. During the slow process
of crystallization that takes place in these igneous masses, crystals
become segregated; the heaviest and first to form sink through the
melt under gravity to a level governed by their density (Fig. 18-8;
see also Chapter 16). Convection currents in the cooling magma

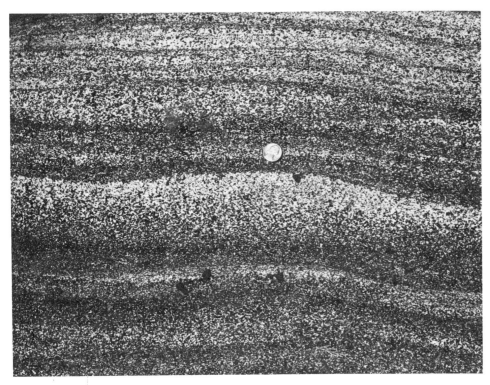

**Figure 18-8**
Banded structure or layering in granitic intrusion, possibly due to gravity settling. Sierra Nevada batholith. [From P. C. Bateman, L. D. Clark, N. King Huber, J. G. Moore, and D. Rinehart, U.S. Geological Survey.]

might also influence the mineral stratification by sorting out groups of minerals.

Some batholiths have sharp contacts with country rock, and show flow structures in which elongated, needle-shaped crystals are arranged in parallel alignment, like logs aligned parallel to the flow of a river. These features imply that the batholiths were formed by injection of an igneous magma. Other batholiths grade into country rock without sharp contacts, and exhibit structures that vaguely resemble those of sedimentary rocks. Such features imply that pre-existing sediments were **granitized** — converted in place to granite by hot solutions and gases percolating up from great depths (Fig. 18-9). The origin of batholiths, linked with the origin of granite (Chapter 16), is one of the most durable of controversies; it is still a hotly debated subject. Geologists today subscribe to various combinations of two views: that plutons are the result of igneous intrusions, and that they are produced by granitization of pre-existing rocks in place.

As we will see, batholiths form in the cores of many mountain ranges. Moreover, immense regions of exposed Precambrian rocks

**Figure 18-9**
Contact between country rocks (dark) and granitic intrusion (light rock) exposed over a vertical distance of nearly 2 km in Pine Canyon, Sierra Nevada. [From P. C. Bateman, L. D. Clark, N. King Huber, J. G. Moore, and D. Rinehart, U. S. Geological Survey.]

in continental interiors are made up of successions of overlapping batholiths, many of which show the characteristics of roots of ancient mountains long since removed by erosion. All of this implies a connection between large-scale plutonism and the mountain-making process—a connection that will be further explored in Chapter 22.

What happens to a rock when it is baked and squeezed in the Earth? The change in mineral composition and texture may be ever so slight or it may be so great that the rock is completely altered in appearance. The degree of change depends mainly on the pressures and temperatures to which the rock was subjected. One good place to see the effect of heat alone is in an outcrop where a shale has been intruded by a dike or sill. At some distance from the intrusion, the shale begins to show a loss of bedded structure, and its clay minerals have recrystallized to micas. At the contact with the igneous rock, the shale has lost all of its original texture, and is now a coarsely grained, crystalline, **contact metamorphic** rock containing pyroxenes and anhydrous (lacking water) aluminosilicate minerals like andalusite ($Al_2SiO_5$). The conclusion is inescapable that the high temperature of the intrusion caused the disappearance of clay minerals, which were

**Figure 18-10**
Contact metamorphism displayed by a limestone intruded by a sill. The hammer lies along the contact of the sill, the dark gray rock below. The limestone above the sill is bleached to a light and moderate gray in contrast to the darker unmetamorphosed limestone above. Gallatin County, Montana. [Photo by I. J. Witkind, U. S. Geological Survey.]

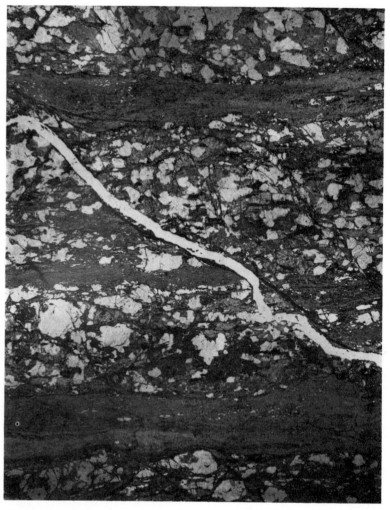

**Figure 18-11**
Photomicrograph in polarized light of a cataclastically deformed rock, a gneiss taken from a fault zone. Bands of coarse, granulated gneiss with large, light crystals alternate with darker bands of finely granulated mylonite. Gilpin County, Colorado. [Photo by W. A. Braddock, U. S. Geological Survey.]

originally formed by low-temperature weathering processes, and the appearance within the rock of new minerals, such as pyroxenes, that are characteristic of high temperatures.

The transforming effect of pressure and tectonic movement alone on rocks is obvious along fault planes, where rock is sheared and ground up to a paste. The mineralogy may be unaffected even though the rock has become pulverized. These **cataclastic** rocks, the mechanically fragmented metamorphic products of intense folding and faulting, are associated with a variety of tectonic structures.

**Figure 18-12**
Structurally deformed and complexly metamorphosed Precambrian basement rocks in northeastern Tennessee. [Photo by W. B. Hamilton, U. S. Geological Survey.]

The great bulk of metamorphic rocks are those that have been subjected to both high temperature and pressure over large regions, with the consequent destruction of original igneous or sedimentary structures and the growth of new minerals. These rocks are the products of **regional metamorphism** rather than the products of localized reactions along fault planes or the borders of igneous intrusions, with which cataclastic and contact metamorphic rocks are associated. Regionally metamorphosed rocks are found in the cores of deeply eroded, folded, and intruded mountain belts in company with granitic batholiths. The formation of new minerals goes hand in hand with structural deformation, and both produce characteristic textures.

## METAMORPHIC TEXTURES

The ways in which the crystals in a metamorphic rock have been rearranged are guides to the processes that produced the new rock from its original sedimentary or igneous parent. The grain and bedding textures in detrital sediments are mainly the products of fluid flow regimes, and the crystalline textures of igneous rocks and some chemical sediments are the result of primary crystallization from a melt or solution; in contrast, the textures of metamorphic rocks are the result of recrystallization or conversion of one mineral to another in the solid state.

**Foliation and Fracture Cleavage**  Most conspicuous in a large variety of deformed metamorphic rocks are parallel planes that generally cut the rocks at an angle to the bedding of the original sediment, though it may coincide with it in some places. This **foliation** may be so weak as to be barely seen, or it may so dominate that it is the only visible structure. Many shales part fairly easily along bedding planes. As many such shales are traced into

**Figure 18-13**
Slaty cleavage in a schist. [Photo by W. B. Hamilton, U. S. Geological Survey.]

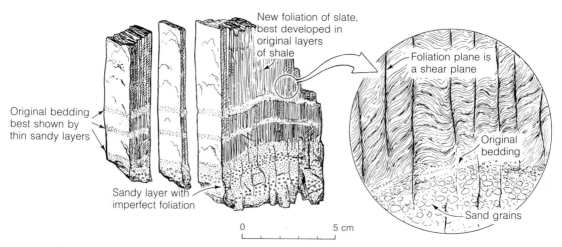

**Figure 18-14**
Fragments of slate (left) showing foliation (vertical lines) and relics of original bedding. The enlargement shows small offsets of the bedding along the surfaces of foliation of the cleavage. [From *Principles of Geology* (3rd ed.) by J. Gilluly, A. C. Waters, and A. O. Woodford. W. H. Freeman and Company. Copyright © 1968.]

**Figure 18-15**
Intricately folded veinlet of quartz cutting across planes of schistosity in a mica schist. [Photo by J. C. Reed, U. S. Geological Survey.]

structurally deformed mountain belts, they become transformed into **slates,** harder rocks that typically show a **fracture cleavage**. This variety of foliation is named from a tendency to cleave or break into moderately thin sheets at more or less regular intervals. The cleavage is not only at an angle to the bedding but sometimes can be seen to offset it by a small amount at each fracture. The cross-cutting relationship is evidence of the deformational origin of the foliation. The almost perfect cleavage of some slates makes them ideal for use as roofing tile. Others that cleave into thicker sheets are more suitable for flagstones. The cleavage is sound textural evidence of metamorphism.

If the slate can be traced into an even more deformed region, the foliation is found to become finer and more all pervasive, the rock showing the lineation of **flow cleavage,** characterized by

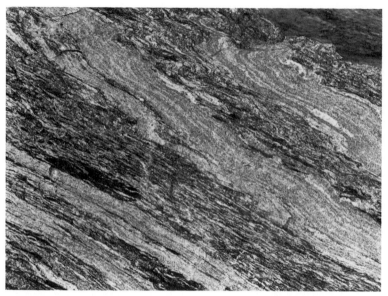

**Figure 18-16**
A banded gneiss of metamorphosed sandstone and schist. The darker beds are a mixture of flattened lenses of dark biotite schist and lighter gray quartz-feldspar rock similar to the lighter metamorphosed sandstone layers. Great Smoky Mountains, South Carolina. [Photo by J. B. Hadley, U. S. Geological Survey.]

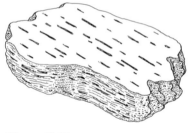

**Figure 18-17**
Lineation is a metamorphic texture formed by parallel orientation of prismatic or needle-shaped crystals of such minerals as pyroxenes or amphiboles. Viewed end on, the needles appear as dots.

very thin sheets along which platy minerals, such as mica, have grown parallel to the foliation planes. Further growth of crystals to larger size, visible to the naked eye, accompanied by some segregation of minerals into lighter and darker bands produces **schistosity**, the coarse foliation of **schists.**

The coarsest foliation is shown by **gneisses**, in which coarse bands of segregated light and dark minerals are prominent throughout the rock. There is little or no tendency to part or split along these bands, and there are relatively few platy minerals parallel to the bands.

Amphiboles form needle-like crystals in many metamorphic rocks. In some the needles show a marked **lineation**, or parallel alignment. In schists this texture is oriented parallel to foliation. Where foliation is not well developed, the lineation may be the most prominent textural feature. Good foliation is most common in the mica-bearing slates and schists formed from shales. Good lineation is best developed in rocks of mafic composition, which metamorphose to form large amounts of elongate amphibole crystals.

Foliation, lineation, and other metamorphic textures are the product of **preferred orientation** of crystals, the pronounced tendency toward parallel alignment of crystals. The orientation is related to the directions of the compressional forces of defor-

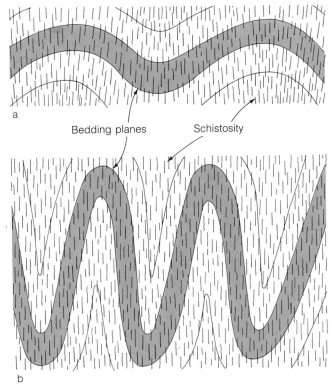

Bedding planes          Schistosity

**Figure 18-18**
(a) In moderately folded bedded rocks, the angle between
schistosity and bedding planes varies from the limbs to the apex of
a fold. (b) In tightly folded rocks the schistosity is essentially
parallel to bedding over most of the fold.

mation responsible for crystallization or recrystallization. In
simplest relation to folded structures, mica flakes grow with their
shortest dimension (the crystallographic $c$-axis) perpendicular
to the major compressional force. The plane of the flake or sheet
is parallel to (and is the main cause of) the cleavage or foliation,
which is itself parallel to the plane of the folding, as shown in
Figure 18-18 (see also the section "Axial Plane" in Chapter 22).
Preferred orientation probably results from a combination of
crystal growth in the preferred direction, some plastic deformation
or flowage of an existing crystal, and the physical rotation of a
crystal during deformation.

**Mineral Textures**  Because most contact metamorphic rocks
have undergone little or no deformation, they lack foliated structures. Their texture is nondirectional; platy or elongate crystals
are randomly oriented. Contact rocks of this type are called **hornfels**, and the texture is **hornfelsic**.

**Figure 18-19**
A hornfels breccia. Darker fragments of hornfels
were broken during the intrusion and recemented
with a lighter hornfels. Both kinds show a granular
texture and absence of foliation. [Photo by H. C.
Granger, U. S. Geological Survey.]

In both contact and regionally metamorphosed rocks, new minerals may grow into large crystals surrounded by a much finer-
grained matrix of other minerals. These **porphyroblasts** form in
the solid state from the chemical ingredients of the rock, and
they grow at the expense of the matrix surrounding it. Porphyroblasts look something like phenocrysts, the large crystals in predominantly fine-grained igneous rocks, but the textural relations
are the opposite. Prophyroblasts grew after the main rock matrix
was formed, whereas phenocrysts formed first, after which the
matrix crystallized around them.

**Figure 18-20**
Feldspar porphyroblasts and thin quartz veins in biotite schist. The light-colored feldspar crystals have grown large at the expense of the surrounding fine-grained matrix. [Photo by J. E. G. W. Greenwood, U. S. Geological Survey.]

Some metamorphic rocks have developed an interlocking mosaic of crystals with no outstanding preferred orientation or foliation. Marbles, metamorphosed limestones, may develop this texture together with some dark mineral banding. Similarly altered silicate rocks are called **granulites**.

**Cataclastic Textures** The shearing and grinding associated with faulting and intense folding produce a broken, granulated texture in which pre-existing crystals are pulverized and strung out in bands or streaks. Coarse-grained rocks of this type, **friction breccias**, consist of particles ranging from a millimeter to a meter in diameter. The finer-grained, more pulverized variety, **mylonite**, is composed of grains 0.01 to 0.1 millimeter in diameter. A distinctive texture, **augen** (from the German for "eye"), is formed by the shearing and abrasion of large porphyroblasts. They give the name to strongly banded, sheared **augen gneiss**.

All of these textural details serve as evidence of the pressure changes that took place during or after heating and of the rapidity of heating during metamorphism. Though our knowledge of the mechanisms by which all of the many textures are formed is still far from complete, the textures themselves nevertheless constitute some of the best information that the field geologist has for interpreting the extent to which pressures and/or temperatures were increased and in what sequence. They are also indispensable tools of the trade for the structural geologist as he unravels the patterns of folding and faulting.

**Figure 18-21**
Augen gneiss. The augen (eyes) are large white or light gray potassium feldspar
crystals that grew as porphyroblasts and were then sheared and abraded as the
rock was structurally deformed. [Photo by H. L. Foster, U. S. Geological Survey.]

## REGIONAL METAMORPHISM

As was mentioned in Chapter 3, geologists have learned to use
minerals as pressure gages and thermometers. This has been well
developed in the study of metamorphism. By mapping characteristic **index minerals** in the field, geologists were able to define
broad zones or belts of metamorphism, ranging from least to most
intense. On a typical traverse from weakly to strongly metamorphosed shales, slates, and schists, one may go from detrital micas
and other unchanged sedimentary minerals in the shale toward
the metamorphic rocks and encounter the first new mineral,
chlorite, in the slates. Moving in the direction of increasingly
intense metamorphism, one may successively meet biotite and
garnet in the schists, then staurolite, kyanite, and finally silli-
manite, the schists becoming progressively more coarsely foliated.
Lines on a map connecting points where an index mineral first
appears are **isograds**, signifying a change in metamorphic grade.
The isograds are rough measures of pressure and temperature
conditions. Metamorphic minerals do not everywhere appear in
exactly the same sequence as metamorphic grade increases, because pressure and temperature may not increase at the same rate
as metamorphism becomes more intense. Thus, in some areas,
high pressure may have built up with relatively little increase in
temperature; in other places, temperature may increase greatly
with relatively little pressure increase. Most common are the
regionally metamorphosed rocks that have formed by a combina-

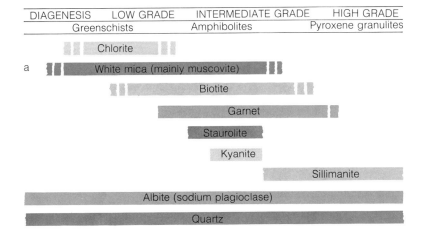

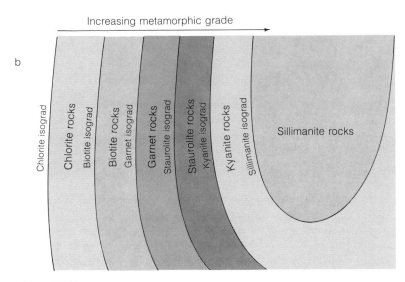

**Figure 18-22**
(a) Changes in mineral composition of shales metamorphosed under conditions of intermediate pressure and temperature. (b) Idealized map view of a regionally metamorphosed terrane in which shales have been metamorphosed under the same conditions. The isograd lines mark the first appearance of the index mineral.

tion of moderately high temperatures and pressures. This kind of metamorphism is a more severe extension of the mild changes caused in sediments by diagenesis after burial. The dividing line between diagenesis and metamorphism is thus more or less arbitrarily drawn at low temperatures and pressures (see Chapter 13). Metamorphic rocks formed at high temperature and pressure may later be subjected to another set of metamorphic conditions, perhaps as a part of a second orogeny. If the second episode is at a lower temperature and pressure, the rocks will be lowered in metamorphic grade, a process called **retrograde** metamorphism.

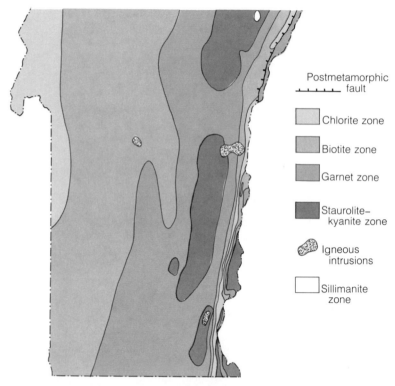

**Figure 18-23**
Metamorphic zones in the southern part of Vermont. The trend of these zones generally parallels the structural lineaments of the terrain. [Simplified from "Centennial Geologic Map of Vermont." Copyright © 1961, State of Vermont.]

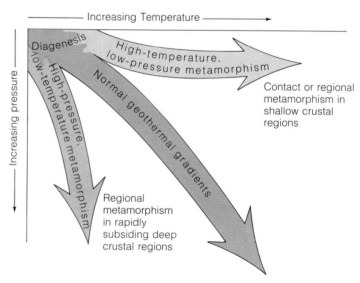

**Figure 18-24**
Different metamorphic routes of pressure and temperature increase may be induced by combinations of tectonic and igneous activity that produce different groups of metamorphic rock types.

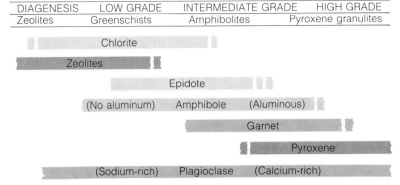

| DIAGENESIS | LOW GRADE | INTERMEDIATE GRADE | HIGH GRADE |
| Zeolites | Greenschists | Amphibolites | Pyroxene granulites |

**Figure 18-25**
Changes in mineral composition of basalts and the other mafic rocks metamorphosed under conditions of intermediate pressure and temperature. Compare with mineral assemblages of shales metamorphosed under the same conditions (Fig. 18-22) to see the effect of original composition on mineralogy.

The course of metamorphism is partly determined by the chemical (and thus mineralogical) composition of the original rock. The examples given above were drawn from shales. Volcanic rocks rich in mafic minerals show a different progression of associated minerals with increasing metamorphic grade. The lowest-grade rocks contain a variety of **zeolite** minerals, complex hydrous aluminosilicates formed by the alteration of mafic volcanic glasses. The next higher grade, called **greenschist**, contains chlorite (a mica), epidote (an aluminosilicate), and actinolite (an amphibole)—all minerals containing much iron, magnesium, and calcium—plus albite (sodium plagioclase feldspar). Following the greenschists in grade are the **amphibolites**, characterized by hornblende (an amphibole), sodium-calcium plagioclase feldspar, and garnet. The highest metamorphic grade of mafic volcanics includes the **pyroxene granulites**, containing pyroxenes and anorthitic plagioclase. This metamorphic mineral assemblage does not differ greatly from the mineral composition of gabbro, the intrusive equivalent of basalt, though the textures are different.

If a mafic volcanic is subjected to high temperatures but relatively low pressures, pyroxene granulites are the common product. In metamorphic belts where pressures are high and temperatures low, the typical metamorphic rock is a **blueschist**, characterized by glaucophane, a blue amphibole. At extremely high pressures and variable moderate to high temperatures **eclogites** form. These rocks, which many geologists think are characteristic of many parts of the earth's mantle, are pyroxene-garnet rocks.

Many of the same mineral groups are represented in all of these varied metamorphic rocks. The amphiboles, pyroxenes, and garnets are divided into many species of variable chemical composition, each more or less characteristic of a different rock type. It is this range of compositions that is so useful in applying the re-

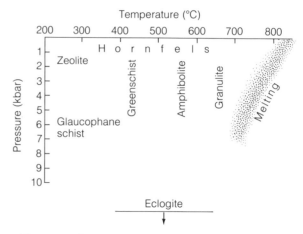

**Figure 18-26**
Generalized and simplified metamorphic facies diagram, showing the distribution of metamorphic rock types in relation to the temperature and pressure fields in which they are formed. There are no sharp boundaries between any of these facies. [Modified and simplified from various sources.]

**Figure 18-27**
A veined gneiss of migmatite origin. The darker layers are strongly metamorphosed sedimentary rocks. The lighter layers are quartz-feldspar rock introduced by invasion of partly molten material. [Photo by J. B. Hadley, U. S. Geological Survey.]

sults of experimental synthesis of metamorphic rocks at controlled temperatures and pressures, for the mineral compositions are strongly affected by pressure and temperature as well as by the total chemical composition of the rock. The experimental work gives the basic pressure-temperature data. The metamorphic geologist first obtains samples of rocks in the field, noting carefully where they were collected; then takes them to the laboratory, where he analyzes the exact compositions of the minerals by a number of methods. The most powerful technique makes use of the electron microprobe to give the chemical analysis of a small area of a single crystal only a few thousandths of a millimeter in diameter. From this field and laboratory information, the geologist is able to plot pressure-temperature lines on a map to show the direction in which the intensity of metamorphism increased or decreased.

All of this work on pressures and temperatures of metamorphic rock formation resulted from a concept enunciated in 1920 by a great Finnish metamorphic geologist, *Pentti Eskola.* He drew a chart on which the different metamorphic rocks could be related to pressure and temperature; a modified version of his chart is shown in Figure 18-26. There are no exact boundaries, because the rock types depend partly on the total, or bulk, chemical composition of the rock. This idea has been one of the most durable in metamorphic geology, and remains the cornerstone of modern interpretation.

**Extreme Metamorphism** The most far-reaching effects of metamorphic transformation can be seen where erosion has cut deeply into the crust in such ancient mountains as the Adirondacks of New York. The rocks are badly deformed, stretched, and contorted, and show much evidence of having flowed as plastic solids. In addition, the rocks are shot through with veins and intrusions of igneous rock. In such country it is difficult to say which rocks are metamorphic and which are igneous, particularly for a kind of veined gneiss frequently called **migmatite.** This is used for rocks of mixed origin. Such a rock is partly metamorphic in having been converted to its present condition largely as a solid, but it is also partly igneous in that parts of it have at times been molten. Arguments have raged over the origin of these rock types for 50 years, for many of them look remarkably like intrusive granites.

Some veined gneisses are truly intrusive, formed by the injection of magma in veins that show contact metamorphic margins or evidence of having forced the surrounding rock apart. Others seem to have been transformed at high temperatures in the solid state by the diffusion of elements, an exchange process in which individual atoms or ions migrate slowly through the crystalline material. In this process elements like sodium and potassium replace others like calcium and magnesium and convert the rock to a granitic composition. The rocks may exhibit original textures, and there may be no evidence that they have disrupted the surrounding rock, which suggests that they are products of recrystallization rather than melting. More problematic gneisses seem to have formed by partial melting, and others seem to be extreme examples of the kind of mineral segregation into bands that is typical of lower-grade schists and gneisses. Many geologists believe that large volumes of batholithic granites are migmatites rather than true intrusives. Careful, detailed mapping coupled with analysis of the pressures and temperatures of mineral formation is needed to evaluate the origin of these rocks in any particular area.

## CONTACT METAMORPHISM

Igneous intrusions are surrounded by margins of altered rock called **aureoles**. The width and nature of an aureole depends on the temperature of the intrusive magma and on the depth in the crust at which the intrusion took place, which determines the pressure and temperature of the rocks being intruded. Aureoles are most prominent around plutons injected at temperatures about 1000°C in shallow crustal rocks only a few kilometers from the surface. In such a situation, there is a strong change in temperature away from the magma contact. Magmas intruded deep in the crust are not that much hotter than the surrounding rocks, and aureoles are obscured by general regional metamorphism.

Contact aureoles display sequential zones of index minerals in the same way as regional metamorphic isograds. The particular minerals formed by recrystallization are strongly influenced by both temperature and the bulk chemical composition of the rock. Thus limestones show patterns that differ greatly from those of

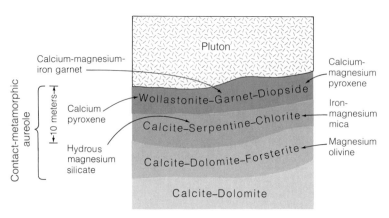

**Figure 18-28**
Skarns are banded rocks produced by contact metamorphism of limestones and dolomites. Grading from unaltered rock toward the contact, they change progressively from pure carbonate marble to bands composed of different calcium-magnesium silicate minerals, finally to a carbonate-free silicate rock.

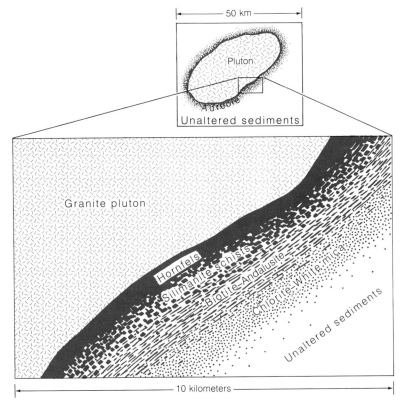

**Figure 18-29**
A contact aureole in sandstones and shales intruded by a granite pluton is a series of zones characterized by mineral assemblages formed at progressively lower temperatures with greater distance from the intrusive.

shales, because their compositions are so different. Pressures are generally low to moderate. The thickness of the zones may be as little as a few centimeters adjacent to small dikes and sills and as much as several kilometers wide bordering a large granite pluton.

The metamorphism of limestone to marble in aureoles leads to a number of changes in the mineralogy of the rocks, many of them involving the clay or silica impurities in the calcite or dolomite of the sediment. One of the minerals formed by such metamorphism is wollastonite (calcium pyroxene), a product of the reaction between silica impurities (in the form of quartz silt or chert) and the calcium carbonate of the limestone:

$$\text{calcite} + \text{silica} \rightarrow \text{wollastonite} + \text{carbon dioxide}$$
$$CaCO_2 \quad SiO_2 \qquad CaSiO_3 \qquad CO_2$$

This reaction produces carbon dioxide gas, which generally escapes. Though silica and calcium carbonate have no tendency to react at low temperatures, we know from experiments at near-surface pressures that at around 500°C they will combine. As is true of many other reactions, the temperature at which these substances begin to react rises as the pressure increases.

If the rock contains dolomite, the mineral diopside (calcium-magnesium pyroxene) will form in the same way as wollastonite. If diopside forms near the contact, it gives way farther out to tremolite (calcium-magnesium amphibole), a mineral that contains the same elements but which crystallizes at lower temperatures. Different patterns of zones form from the interplay of compositional and temperature factors. A common pattern shows a gradation from the unaltered dolomite to a tremolite zone a few hundred meters wide, then to a somewhat narrower diopside zone, and finally to an inner contact zone of wollastonite that may be only a few meters wide.

Contact metamorphism of silicate rocks produces similar patterns but with different minerals. Outer zones contain micas, amphiboles, and calcite. Inner zones are characterized by pyroxenes, including wollastonite and the alumino-silicate andalusite ($Al_2SiO_5$). The general nature of contact aureole patterns illustrates the tendency for the minerals that contain **volatiles**, the easily escaping elements or compounds (for example, carbon dioxide bound chemically in calcite), to be found in outer zones and for dry, gas-free minerals to be found in inner, hotter zones.


## METAMORPHIC CHEMICAL CHANGES

The mineral transformations of regional and contact metamorphism illustrate the general tendency for systematic chemical changes that accompany increased temperature and pressure. One of the major processes affecting sediments is the loss of volatiles, mainly water and carbon dioxide. The opposite is true of the metamorphism of some continental mafic volcanics, many of which may erupt as relatively dry, volatile-free lavas and become hydrated and carbonated (combination with carbon dioxide to form carbonate minerals). Organic materials containing carbon and hydrogen (such as the hydrocarbons in petroleum) lose their volatile gas, hydrogen.

The high solubility of silica and two of the alkali metal elements, sodium and potassium, in hot aqueous fluids associated with metamorphism can lead to solution and transportation of these elements from deeper to shallower zones in the crust, forming the granite-like veined gneisses called migmatites.

Even though these chemical transformation processes are known to take place, a great deal of regional and contact metamorphism seems to have taken place with little or no bulk chemical change — excluding losses of water and carbon dioxide. It is this fact that enables metamorphic geologists to "look through" the metamorphic mineral assemblage to reconstruct the original mineralogy of the sediment or igneous rock that was transformed. Knowing the total mineral composition of the rock, he can recast it, using his knowledge of experiments on mineral reactions, such as the conversion of calcite and silica to wollastonite, to infer the former presence of a mineral assemblage that would have been stable at lower temperatures and pressures. In this way, former rock patterns, including sedimentary facies, can be reconstructed even though almost all direct evidence has been obliterated. The study of metamorphism illustrates again how knowledge of process—how things happen—is inextricably tied to the history of the Earth, how things *did* happen.

It is the link between rock formation and the history of the Earth, combined with questions about the ultimate nature of the forces causing regional metamorphism and their association with mountain-building, that is causing metamorphic geologists to use plate-tectonic theory as a new conceptual framework for their ideas. Greenstones, metamorphosed basalts, have been dredged from such zones of plate divergence as the mid-Atlantic ridge. At these active zones of basalt upwelling, localized heating transforms newly extruded basalts into low-grade metamorphic rocks. Blueschists, metamorphosed sediments whose minerals indicate that they were formed under very high pressures but relatively low temperatures, were recognized for years as the result of rapid and deep subsidence of geosynclinal sediments into the crust, with little heating. Metamorphism at a subduction zone is now seen as the explanation. The subducted plate, carrying deep-water marine sediments, such as turbidites and cherts, moves down so rapidly that its great mass heats up very slowly while the pressure increases rapidly. Just the opposite pair of conditions, high temperatures and low pressures, characterizes the metamorphic rocks of volcanic island arcs at plate convergences. Here the igneous rocks that are generated by melting at plate margins come close to the surface and transform shallowly buried volcanics and sediments. Recognition of these broad metamorphic zones along ancient continental margins can thus be used to reconstruct something of the pattern of ancient plate motions. We discuss the full story of the relationships between the theory of plate tectonics and igneous and metamorphic processes in Chapter 21, after the stage has been set by exploring how seismological and other geophysical investigations led to the theory.

1. The forms of igneous intrusions, such as plutons, dikes, and sills, are a guide to their method of emplacement at depth. Such bodies may intrude by stoping, by heating surrounding rock so that it gives way by plastic flow, or by fracturing or bowing up.

2. Sills and laccoliths are concordant; dikes are discordant.

3. Lopoliths and batholiths are very large plutons. Some batholiths appear to be granitized sediments.

4. Metamorphic rocks are of three kinds: contact, cataclastic, and regional. Contact metamorphism is produced by heat at the borders of an igneous intrusion. Cataclastic rocks are mechanically fragmented by folding and faulting. Regionally metamorphosed rocks are the product of generalized heat and pressure associated with orogeny.

5. Metamorphic textures include foliation, the general term for parallel planes, fracture cleavage, flow cleavage, schistosity, and lineation. These textures are formed by the preferred orientation of platy and needle-shaped crystals of new minerals grown during metamorphism. Hornfelsic textures form in contact metamorphic rocks. Porphyroblasts are large crystals formed at the expense of a fine-grained matrix. Granulites have little or no preferred orientation.

6. Cataclastic rocks include friction breccias, mylonites, and augen gneiss.

7. Regional metamorphism produces a series of mineral assemblages that may be used to map isograds that are indications of pressure and temperature conditions. Metamorphic facies are a grouping of rock types by their mineral assemblages in relation to pressure and temperature of formation.

8. Extreme metamorphism can produce migmatites, which may be a major component of some batholiths.

9. Contact metamorphic aureoles are altered margins around plutons in which zones of index minerals appear in definite sequences.

10. Systematic chemical changes normally accompany regional metamorphism, mainly the loss of volatiles, carbon dioxide, and water.

11. Metamorphic belts are now being interpreted in the framework of plate tectonics.


## EXERCISES

1. What differences in the modes of formation of sills and lava flows account for their differing texture and general appearance?

2. What properties of some granitic batholiths are difficult to reconcile with formation by the simple emplacement of a large body of differentiated magma?

3. Which kind of pluton would you expect to have produced the most intense contact metamorphism of surrounding rocks, and why — a granite intrusion very deep in the crust or a gabbro intrusion at moderate depths?

4. What properties of a group of metamorphic rocks of medium and high grade would you use to infer the nature of the rocks before metamorphism?

5. Would you choose foliation or chemical composition as the better property to use in determining metamorphic grade? Why?

6. Why might you characterize weathering and metamorphism as opposites with respect to volatile elements or compounds?

## BIBLIOGRAPHY

Billings, M. P., *Structural Geology* (3rd ed.). Englewood Cliffs, New Jersey: Prentice-Hall, Inc., 1972. Chapters 16, 17, 18, and 19.

Ernst, G., *Earth Materials.* Englewood Cliffs, New Jersey: Prentice-Hall, Inc., 1969.

Harker, A., *Metamorphism.* London: Methuen, 1932.

Turner, F. J., *Metamorphic Petrology.* New York: McGraw-Hill Book Company, 1968.

Turner, F. J., and J. Verhoogen, *Igneous and Metamorphic Petrology.* New York: McGraw-Hill Book Company, 1960.

Winkler, H. G. F., *Petrogenesis of Metamorphic Rocks* (2nd ed.). New York: Springer-Verlag, 1967.

Part of the San Andreas fault, cutting across the Carrizo Plains in eastern San Luis Obispo County, California. [Photo by Robert E. Wallace, U.S. Geological Survey.]

# 19

# Seismology and the Earth's Interior

*The basic causes of earthquakes are strains induced by plate motions. In fact, by their locations and the nature of the ruptures they produce, earthquakes define the plate boundaries. Analysis of seismic waves, together with laboratory studies of rocks, helps us to infer the composition and state of the Earth's interior.*

Seismology deals primarily with earthquakes and seismic waves. Because earthquakes are among the scourges of mankind, seismologists are concerned with ameliorating their destructiveness. They do this by assessing seismic risks in different geographic regions, so that sensible building and zoning codes can be written, and by researching the problems of **tsunamis**\* and earthquake prediction, and even earthquake control. But this is not the only job of seismologists.

By studying the pattern of earthquake occurrence, seismologists have provided one of the essential clues to the development of the concept of plate tectonics: earthquake belts demark plate boundaries, the zones along which plates collide, diverge, or slide past one another (Fig. 19-1). The modern seismograph, which records the waves generated by earthquakes and explosions, provides the most important means of probing the deep interior, a region that would otherwise be closed off from observation.

---

\***Tsunamis** are potentially destructive sea waves generated by earthquakes and volcanic explosions. They cross the ocean in a matter of hours, traveling with a velocity of 700 to 800 km/hr (440 to 500 mi/hr), and can reach heights of 10 to 20 m (30 to 60 ft) when they impact on a coast.

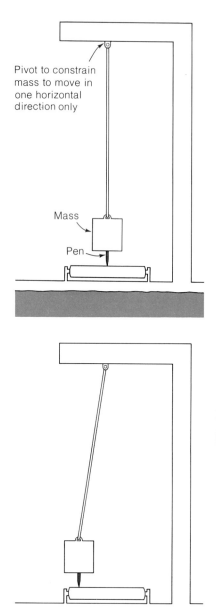

**Figure 19-1**
The San Andreas fault—the boundary between the American and Pacific plates. The block in the background is being displaced to the right relative to the block in the foreground, as evidenced by the stream offset. Carrizo Plains, California [Photo by R. E. Wallace, U.S. Geological Survey.]

## SEISMOGRAPHS

The seismograph is to the Earth scientist what the telescope is to the astronomer—a tool for peering into inaccessible regions. The ideal seismograph would be a "skyhook," a device fixed to a frame outside of the Earth, so that when the ground shakes, the seismic vibrations could be measured by the changing distance between the fixed device and the ground. Because this is impossible to achieve, a compromise is struck by loosely coupling a mass to the Earth so that the ground can vibrate without causing appreciable motion of the mass. The mass is coupled to the Earth by

**Figure 19-2**
Schematic diagram illustrating the principle of the seismograph. Because of its inertia, the mass does not keep up with the motion of the ground. The pen traces the difference in motion between the mass and the ground, in this way recording vibrations that accompany seismic waves. The instrument schematized here records horizontal ground motion.

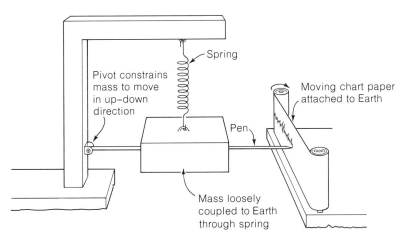

**Figure 19-3**
Schematic diagram depicting a seismograph that records vertical ground motion.

means of a pendulum (Fig. 19-2) or by suspending it from a spring (Fig. 19-3). When the ground moves, the mass tends to remain stationary because of its inertia, and the displacement of the Earth relative to the stationary mass is used to sense the ground movement. The most advanced electronic technology* has been employed to magnify this motion, so that modern instruments can detect ground displacements as small as $10^{-8}$ centimeter (1/4 inch $\times\ 10^{-8}$) — an astounding feat, considering that such small displacements are of atomic size. Actually this is more sensitivity than can even be used on Earth, because the ground moves in a continual state of unrest, shaken by the action of the winds, ocean waves, and man's machinery: in most places such sensitive seismographs would be driven off scale by Earth's "noise." The Moon is the place for man's most advanced seismograph; there, no winds, no ocean sounds, no mechanical vibrations can overload it. The lunar seismographs left behind by the astronauts can detect the seismic waves generated by a 1-kilogram (2.2 pound) meteor striking anywhere on the Moon's surface. A modern seismograph is shown in Figure 19-4.

The strain seismograph shown schematically in Figure 19-5 was developed by H. Benioff, a leader in the design of electronic musical instruments and seismographs. By electronic measurement of the change in distance between two concrete piers placed about 30 meters (100 feet) apart, this instrument is capable of detecting extensional and compressional responses of the ground dur-

---

*For the electronics hobbyist, we might mention that the magnification of a seismograph is achieved by means of moving-coil and variable-reluctance transducers, like those on phonograph pickups, and low-noise, ultra-high-gain, solid-state amplifiers. The output is recorded photogalvanometrically on film, on paper with direct-recording pen motors, on AM, FM, or digital magnetic tape.

ing seismic vibrations and the pull of the Moon on the solid Earth (body tides, just like those to which the ocean responds, but much smaller). A photograph of an actual installation is shown in Figure 19-6. A record made by a strain seismograph is shown in Figure 19-7.

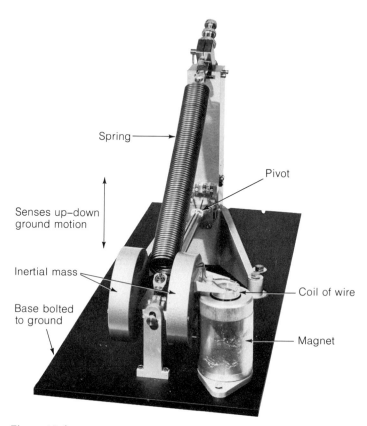

**Figure 19-4**
The Press-Ewing seismograph, an example of a modern instrument. Airtight cover, electronics, and recording system not shown.

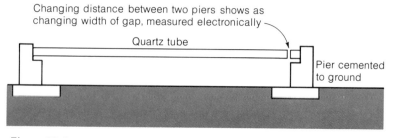

**Figure 19-5**
Schematic diagram of the Benioff strain seismograph. The distance between two piers attached to the ground changes due to tides in the solid Earth and the passage of seismic waves. These variations are measured by placing an electronic motion detector in the gap.

**Figure 19-6**
Example of an actual strain seismograph installation in an underground tunnel. This system is so sensitive that it could detect a change of 1 mm in the distance between New York and California.

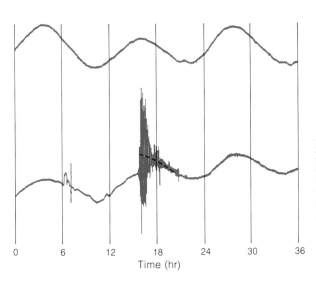

Time (hr)

**Figure 19-7**
Record made by a strain seismograph. The slow periodic movements are the Earth tides; the more rapid vibrations are the seismic waves from an earthquake. [From "Resonant Vibrations of the Earth" by Frank Press. Copyright © 1965 by Scientific American, Inc. All rights reserved.]

# EARTHQUAKES

**What Is an Earthquake?** The rupture of the San Andreas fault, which resulted in the devastation of San Francisco in 1906, provided the evidence that led H. F. Reid, one of the official investigators of this catastrophe, to advance his **elastic rebound theory** of earthquakes. Earthquakes are associated with large fractures, or faults, in the Earth's crust and upper mantle. Consider the fault between the two hypothetical crustal blocks in Figure 19-8. Suppose that surveyors had located lines running perpendicular to the fault from block $L$ to block $R$, as shown in part a of the diagram. Blocks $L$ and $R$ are moving in opposite directions, but because they are pressed together by the weight of the overlying rock, friction locks them together, just as a brake can lock the wheel of a car if enough force is applied. Instead of slipping taking place along the fault, the blocks are deformed in the vicinity of the fault, and the surveyors' lines are bent as shown in Figure 19-8*b*. As the rock is strained, elastic energy is stored in it in the same way that it is stored in a wound-up watch spring. The movement continues, the strain builds up until the frictional bond that locks the fault can no longer hold at some point on the fault, and it breaks, as in Figure 19-8*c*. The blocks suddenly slip at this point, which is the **focus**, or **hypocenter**, of the earthquake. Once the rupture is initiated it will travel at a speed of about 3.5 kilometers per second (2 miles/second), continuing for as much as 1000 kilometers (600 miles). In great earthquakes, the **slip**, or offset of the two blocks can be as large as 15 meters (50 feet). Figure 19-8*d* shows the two blocks after the earthquake, displaced by the amount of slip. Once the frictional bond is broken, the elastic strain energy, which had been slowly stored over tens or hundreds of years, is suddenly

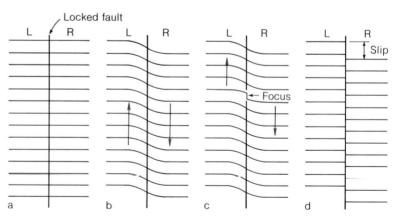

**Figure 19-8**
Sketch illustrating the elastic rebound theory of an earthquake. The two simulated crustal blocks $L$ and $R$ are being forced to slide past one another (a). Friction along the fault prevents slip (b), but the deformation builds up until the "frictional lock" is broken (c) and an earthquake slip occurs (d).

**Figure 19-9**
The earthquake of 1906 was caused by slip along the San Andreas fault. The
offset fence shown here shows a slip of nearly 3 m. Scene is near Bolinas,
California. [Photo by G. K. Gilbert; courtesy of R. E. Wallace, U. S. Geological
Survey.]

released in the form of intense seismic vibrations—which constitute the earthquake. The seismic waves are propagated large
distances in all directions away from the fault. Near the focus
the waves can have large, destructive amplitudes. The process
may be likened to storing elastic energy by slowly drawing out
the rubber band of a sling shot, and then releasing it suddenly
to propel a pebble.

Strictly speaking, the elastic rebound theory is an incomplete
one. The reason for this is that the pressure holding the blocks
together is so great that the frictional bond is actually stronger
than the rock itself. In other words, the block would prefer to
fracture elsewhere rather than slip along the fault. Yet faults do
exist, and movement occurs along them periodically. To complete
the theory, we need a means of "lubricating" the fault or reducing
the locking pressure. Geologists working in the field of rock
mechanics are currently trying to remove some of the mystery
of the mechanics of faulting.

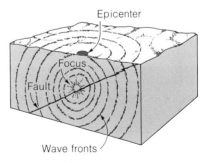

**Figure 19-10**
Diagram showing the focus and epicenter of an earthquake. The focus is
the site of initial slip on the fault. Also
shown are seismic waves radiating from
the focus. [After *Principles of Geology*
(3rd ed.) by J. Gilluly, A. C. Waters, and
A. O. Woodford. W. H. Freeman and
Company. Copyright © 1968.]

**Earthquakes—How Big and How Many?** The time between great earthquakes is about 50 to 100 years in California and somewhat less in more active seismic regions, such as Japan or the Aleutians. Thus the time required to build up the elastic strain energy in the rocks adjacent to a fault is enormous compared with the time that elapses during the release of stored energy, for earthquakes last only a few minutes. The amount of stored energy can be gauged in several ways: perhaps the two most common methods are measuring the distortion of surveyed lines, as in the example just discussed, or measuring the energy of the released seismic waves. About $10^9$ ergs (a unit of energy) of elastic strain energy is released from each cubic meter (1.3 cubic yards) of rock at the time of an earthquake—the equivalent of a fire cracker per cubic meter. This may not seem very impressive until one adds up the cubic meters affected by a great earthquake: suppose the fault is 1000 kilometers (600 miles) long, extends 100 kilometers downward, and distorts surveyed lines as far as 50 kilometers on either side of the fault. This amounts to a strained volume of $10^{16}$ cubic meters, each cubic meter contributing $10^9$ ergs, which gives a total of $10^{25}$ ergs. This is one big fire cracker indeed—about the equivalent of 1000 nuclear explosions, each with a strength of 1 megaton (1 million tons) of TNT!

Energy release is the most precise way of measuring the size of an earthquake, but it is a long, complicated process to determine the fault dimensions, the slip, and the other factors needed to compute it. Seismologists have therefore adopted the **Richter magnitude scale**, which is based on the amplitude of seismic waves recorded by seismographs. Actually, magnitude is based on the logarithm of the maximum amplitude plus an empirical factor that takes into account the weakening of seismic waves as they spread away from the focus. Thus seismologists all over the world can study their records and in a few minutes come up with nearly the same value for the magnitude of an earthquake. Seismographs are sensitive enough to easily detect earthquakes of magnitude less than 1. The largest earthquakes yet recorded show Richter magnitudes of about 8.5. Because these magnitudes are based on a logarithmic scale, a change in magnitude of 1 unit corresponds to a change in the amplitude of seismic waves by a factor of 10. Figure 19-11 shows how the magnitude of an earthquake is determined in actual practice.

Table 19-1 shows how magnitudes and energy can be related in a rough way to earthquake effects, and also indicates the number of earthquakes each year. The table demonstrates the fortunate fact that most earthquakes are small. Each year, 800,000 little tremors are recorded by instruments, but not felt by humans. Great earthquakes, with magnitudes exceeding 8, occur about once every 5 to 10 years. Damage begins at magnitude 5 and increases to nearly total destruction in nearby settlements for earthquakes

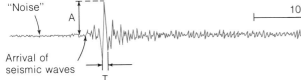

"Noise"

A

10 seconds

Arrival of
seismic waves

T

**Figure 19-11**
Determination of earthquake magnitude from a seismograph recording.
Dividing $A$, the maximum trace motion, by the magnification of the seismograph
gives the maximum ground motion $a$, measured in microns ($10^{-6}$ meters). $T$ is
the duration of one oscillation, or the period of the seismic wave in seconds.
Magnitude $m = \log (a/T) + B$, where $B$ is a factor that allows for the weakening
of seismic waves with increasing distance from the earthquake. *Example:* An
earthquake 10,000 km away ($B = 6.8$ from table of data) produced a ground
motion $a = 10$ microns with period $T = 1$ sec. Thus $m = \log 10 + 6.8 = 7.8$. The
correction factor $B$ is found empirically, so that a seismograph located any place
in the world would give the proper magnitude of an earthquake, regardless of
distance to it.

with $m > 8$ (> means "greater than"). The San Fernando (Los
Angeles) earthquake of 1971 was only of magnitude 6.6, yet the
damage bill amounted to a billion dollars. It is no wonder that
Californians worry about the great shock that should visit them
about every 50 to 100 years!

The report in Figure 19-12 from the Smithsonian Institution's
Center for Short-Lived Phenomena was issued a few days after
the San Fernando earthquake. The description is in the dry language of the specialist, with no hint of the fear or personal tragedy
that accompanies these catastrophes. As damaging as the earthquake was, the seismic energy release was about a thousand times
smaller than some truly great earthquakes, such as those in San
Francisco (1906), Tokyo (1923), Chile (1960), and Alaska (1964).

**Table 19-1**
Earthquake magnitudes, energies, effects, and statistics.

| Characteristic effects of shallow shocks in populated areas | Approximate magnitude | Number of earthquakes per year | Energy (ergs) |
|---|---|---|---|
| Damage nearly total | ≥8.0 | 0.1–0.2 | $>10^{25}$ |
| Great damage | ≥7.4 | 4 | $\geq .4 \times 10^{24}$ |
| Serious damage, rails bent | 7.0–7.3 | 15 | $0.04–0.2 \times 10^{24}$ |
| Considerable damage to buildings | 6.2–6.9 | 100 | $0.5–23 \times 10^{21}$ |
| Slight damage to buildings | 5.5–6.1 | 500 | $1–27 \times 10^{19}$ |
| Felt by all | 4.9–5.4 | 1,400 | $3.6–57 \times 10^{17}$ |
| Felt by many | 4.3–4.8 | 4,800 | $1.3–27 \times 10^{16}$ |
| Felt by some | 3.5–4.2 | 30,000 | $1.6–76 \times 10^{15}$ |
| Not felt but recorded | 2.0–3.4 | 800,000 | $4 \times 10^{10}–9 \times 10^{13}$ |

*Source*: Data from B. Gutenberg.

| EVENT | 17-71 | SAN FERNANDO EARTHQUAKE | | 26 FEBRUARY 1971 | 1139. |

"The San Fernando earthquake of 6:01 a.m., Feb. 9, 1971 was responsible for 62 deaths and an estimated property damage of one billion dollars. The main shock was located 14 kilometers east of Newhall; 42 aftershocks were located in the region between the main shock and the zone of surface faulting in the northernmost San Fernando Valley. The earthquake was associated with oblique-slip, reverse faulting along the southern front of the San Gabriel Mountains, thrusting the mountains relatively to the southwest over the northern San Fernando Valley. The aftershock pattern, the focal plane solution and the surface ruptures in the San Fernando Valley are all consistent with faulting on a plane striking east-west and dipping at 50-60° beneath the San Gabriel Mountains.

"The extensive damage to buildings in San Fernando is closely linked to the location of the zone of ground breakage. Preliminary estimates of the length of faulting at the surface give at least 11km with horizontal compression of 1 m and locally, left lateral slip of.as much as 1.7m.

"The main shock had a magnitude of 6.6 on the Richter scale with the largest aftershock having a magnitude of 5. Over 1000 aftershocks greater than Richter magnitude 2 were recorded in the first 3 days following the main shock. Faulting at the time of the earthquake was either on a plane that (i) strikes east-west with dip 60° N, or (ii) strikes N152°E with dip 50°W. For the most probable fault plane, the E-W north-dipping nodal plane, slip is in a southwesterly direction. The slip has approximately equal components of reverse and left lateral displacement on this nodal plane."

EVENT INFORMATION REPORT : 4

| TYPE OF EVENT | GEOPHYSICAL |
| DATE OF OCCURRENCE | 9 FEBRUARY 1971 |
| LOCATION OF EVENT | LOS ANGELES |
| COUNTY, CALIFORNIA, U.S.A. | |
| REPORTING SOURCE | SEISMOLOGICAL LAB |
| CALIFORNIA INSTITUTE OF TECHNOLOGY, PASADENA, CALIFORNIA, 91109, U.S.A. | |
| SOURCE CONTACT | NATIONAL CENTER |
| FOR EARTHQUAKE RESEARCH, U. S. GEOLOGICAL SURVEY | |
| MENLO PARK, CALIFORNIA, USA. | |

This report is based on notifications received from the Center's correspondents and is disseminated for information purposes only. The Smithsonian Institution bears no responsibility for its accuracy.

SMITHSONIAN INSTITUTION
CENTER FOR SHORT-LIVED PHENOMENA
60 Garden Street
CAMBRIDGE, MASSACHUSETTS 02138
UNITED STATES OF AMERICA

CABLE : SATELLITES NEW YORK
TELEPHONE: (617)-864-7911

sao-slp-2

| EVENT | 17-71 | SAN FERNANDO EARTHQUAKE | | 26 FEBRUARY 1971 | 1140. |

SURFACE FAULTING
"The most intense surface faulting associated with the San Fernando earthquake occurred in an E-W trending zone about 130 meters wide. The zone trends eastward to the hills from central San Fernando turning southward from about 500 meters south of Lopez dam. The fault zone runs south to the San Fernando Industrial Park and the Foothill Nursing Home and thence eastward along the base of the hills north of the town of Tujunga for about 8 kilometers.

"The deformation is characterized by compression normal to the surface trace. Asphalt, concrete and sod are buckled into compression ridges 30 cm or more in height and horizontal shortening normal to the zone locally exceeds 1m. In the western part of the zone, linear features such as curbs, white lines, etc., are locally offset as much as 1.7 meters in a left-lateral sense. The zone of faulting is closely correlated with one major zone of damage. In another major zone of extensive structural damage near the Olive View Sanitarium and Veterans' Hospital careful investigation revealed no tectonic rupture but whether tectonic deformation other than surface faulting has taken place is being investigated.

FREQUENCY OF EARTHQUAKES
"During the first three days after the main shock, thousands of aftershocks occurred. We counted more than 1,000 events which were clearly recorded at our SBLG station ($\Delta$=70km) with local magnitude greater than 1.5 (or possibly 2). The counted earthquakes were plotted on a histogram of two-hour intervals. Within three days, aftershocks decreased from about 40 per hour to less than 5 per hour. Such exponential decrease of the number of aftershocks is normal."

EVENT INFORMATION REPORT : 5

| TYPE OF EVENT | GEOPHYSICAL |
| DATE OF OCCURRENCE | 9 FEBRUARY 1971 |
| LOCATION OF EVENT | LOS ANGELES |
| COUNTY, CALIFORNIA, U.S.A. | |
| REPORTING SOURCE | SEISMOLOGICAL LAB |
| CALIFORNIA INSTITUTE OF TECHNOLOGY, PASADENA, CALIFORNIA, 91109, U.S.A. | |
| SOURCE CONTACT | NATIONAL CENTER |
| FOR EARTHQUAKE RESEARCH, U. S. GEOLOGICAL SURVEY | |
| MENLO PARK, CALIFORNIA, U.S.A. | |

This report is based on notifications received from the Center's correspondents and is disseminated for information purposes only. The Smithsonian Institution bears no responsibility for its accuracy.

SMITHSONIAN INSTITUTION
CENTER FOR SHORT-LIVED PHENOMENA
60 Garden Street
CAMBRIDGE, MASSACHUSETTS 02138
UNITED STATES OF AMERICA

CABLE : SATELLITES NEW YORK
TELEPHONE: (617)-864-7911

sao-slp-2

**Figure 19-12**
Event information report issued on the occasion of the destructive San Fernando, California, earthquake of 1971. [From Center for Short-Lived Phenomena, Smithsonian Institution.]

Table 19-1 also points up the interesting fact that the few large earthquakes each year release more seismic energy than the hundreds of thousands of small shocks combined. This should put to rest the notion that small earthquakes act as a safety valve, gradually releasing strain in harmlessly small amounts and thus forestalling a big shock. About $10^{25}$ ergs of seismic energy are released each year. This is about 0.1 percent of the heat energy reaching the Earth's surface from the interior which is equivalent to about 2 percent of the electric energy produced by man.

**Earthquakes—Where Do They Occur?** A **seismicity** chart showing the map locations, or **epicenters**, of almost 30,000 earthquakes that occurred between 1961 and 1967 is reproduced in Figure 19-13. The seismologists who prepared the map placed all of the epicenters in the memory of a computer so that statistical studies on distribution and timing could be made and also to make use of automatic plotting machines linked to the computer. Seismologists have known for decades that earthquakes tend to occur in belts—for example, the "ring of fire" surrounding the Pacific Ocean. In recent years, however, it has become possible to detect the more numerous, smaller earthquakes and to improve methods of locating epicenters, so that seismic belts can now be defined more accurately. Interestingly, the increase in the number of seismic observatories, and the use of computers that made this possible, was stimulated by research done in the 1960's, during the negotiations for a nuclear test-ban treaty; the purpose of the research was to determine whether small underground nuclear explosions could be detected and distinguished seismically from earthquakes.

The high-quality seismicity maps showed that narrow belts of epicenters coincide almost exactly with the crest of the mid-Atlantic, the east Pacific, and other oceanic ridges, where plates separate. Earthquake epicenters are also aligned along transform faults, where plates slide past each other. But earthquakes that originate at depths greater than about 100 kilometers (60 miles) typically occur near margins, where plates collide (Fig. 19-14). The foci of these earthquakes are distributed on well-defined planes that dip into the mantle, and these occur in close association with deep-sea trenches, island arcs, young mountains, and volcanoes. It is a basic tenet of the theory of plate tectonics that these deep earthquakes actually define the positions of subducted plates which are plunging back into the mantle beneath an overriding plate (Fig. 19-15). This global correlation between topography, geology, and seismicity provided the essential data for defining the boundaries of the lithospheric plates. It may seem like a simple matter now to draw a line through the seismic belts and so define the plates depicted in Figure 1-17, but this important advance could not have been made without the "knowledge

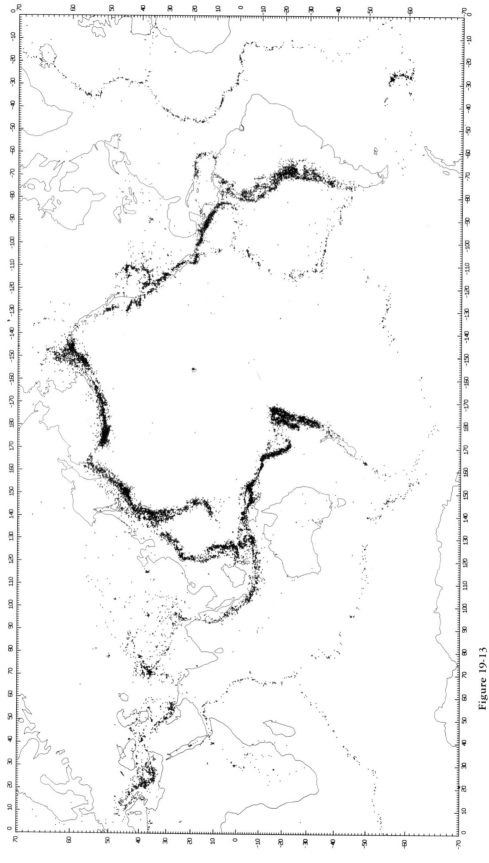

**Figure 19-13**
Epicenters of some 30,000 earthquakes recorded in the years 1961–1967, with focal depths between 0 and 700 km. [Epicenters by U. S. Coast and Geodetic Survey. Computer plot by M. Barazangi and J. Dorman, Columbia University.]

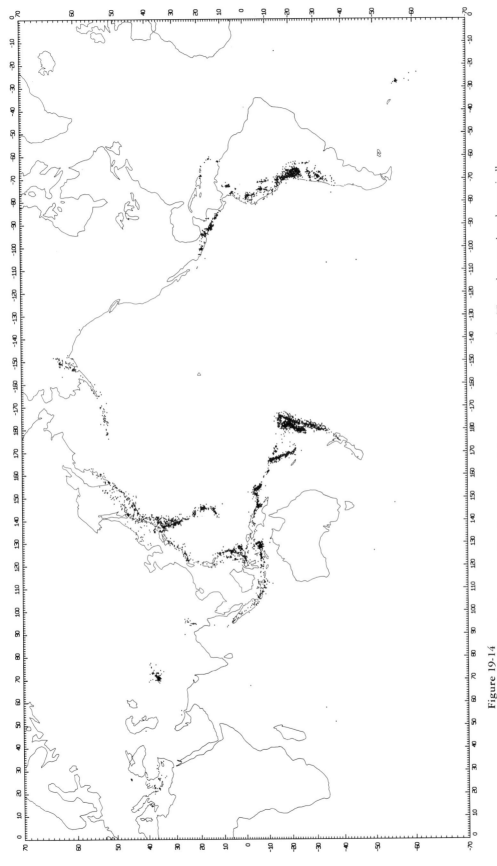

**Figure 19-14**
Subset of earthquakes from preceding figure with focal depths greater than 100 km. These deep earthquakes typically originate near margins where plates collide, and thus serve to identify such places. [Computer plot by M. Barazangi and J. Dorman, Columbia University.]

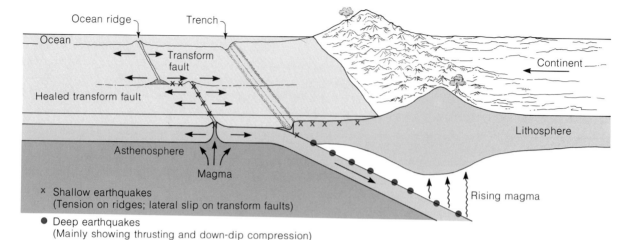

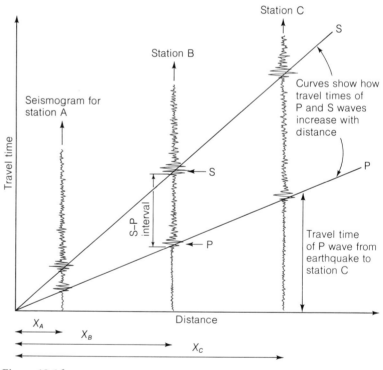

**Figure 19-15**
Diagram illustrating the association of earthquakes with plate boundaries: ocean ridges, transform faults, and trenches. [After L. Sykes, J. Oliver, B. Isacks, and P. Molnar.]

**Figure 19-16**
The time interval between the seismic S- and P- waves increases with the distance from the epicenter. The time interval observed at a given station is matched against the travel-time curves for P- and S-waves until the distance is found at which the separation between the curves agrees with the observed S–P time difference. Knowing the distance from the three stations A, B, and C, one can locate the epicenter.

explosion" in seismic data and the imaginative, uninhibited, and synthesizing minds of about ten workers who sifted through those data in the years following 1967.

Although most earthquakes are recorded at plate boundaries, the seismicity map shows that a small percentage originate within plates. Some of these have been quite destructive, as is indicated by these examples; New Madrid, Missouri (1812), Charleston, South Carolina (1886), Boston, Massachusetts (1755). Apparently, stresses can exceed the strength of rock within lithospheric plates, albeit infrequently.

A brief digression on how earthquake epicenters are located is in order. The principle is quite similar to deducing the distance to a lightning bolt from the time interval between the flash and the sound. Later in this chapter we will describe the two types of seismic waves — the first-arriving $P$ and the slower travelling $S$ waves, which take almost twice as long to reach the seismograph. The lightning flash may be likened to the $P$ waves of earthquakes and the thunder to the $S$ waves. The time interval between the arrival of $P$ and $S$ waves therefore increases with the distance traveled by the waves, and for each $S$-$P$ time interval there is associated a definite distance to the epicenter. This is indicated on the travel-time chart for $P$ and $S$ waves in Figure 19-16, which shows diagrammatically how the $P$ and $S$ wave travel times depend on distance and how the $S - P$ interval increases with distance. To get an approximation of the epicenter, the seismologist simply reads off the $S - P$ interval on the seimogram from a given station and uses a nomogram* like that in Figure 19-16 or a table to get the distance to the epicenter. Knowing the distances from three or more stations enables him to pinpoint the epicenter (Fig. 19-17). He can also deduce the time of the shock at the epicenter because he knows the arrival time of the $P$ waves at each station, and from a nomogram or a table he can tell about how long the waves took to reach the station. Once an approximation has been made, the exact location can be found by making refinements. The estimated location of the epicenter is used to predict the arrival time of $P$ waves at each station. These predicted times are compared with the observed values, and the discrepancies are used as a guide to improving the estimated distance to epicenter and origin time. For example, if all of the predicted times exceed the observed arrival time for $P$ waves, then the estimated time of the earthquake is too late. If all of the stations on the east show earlier predicted times and the stations on the west are late, then the estimated epicenter is moved west. All of this is done in a computer, and the procedure is iterated — that is, repeated with corrections — until the discrepancies are reduced to small values, signifying that a precise location has been found.

*A nomogram is a graph used to determine an unknown quantity from two or more known quantities, saving the time to solve an equation or look up the answer in a table.

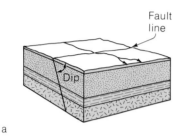

Fault
line

Dip

a

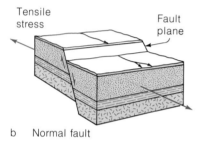

Tensile
stress

Fault
plane

b   Normal fault

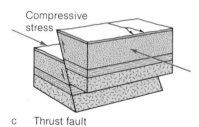

Compressive
stress

c   Thrust fault

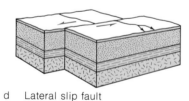

d   Lateral slip fault

**Figure 19-18**
Summary of types of fault movement
and the stresses that cause them: (a)
situation before movement takes
place; (b) normal fault due to tensile
stress; (c) thrust (or reverse) fault due
to compressive stress; (d) lateral slip
(or strike slip) fault due to shearing
stress.

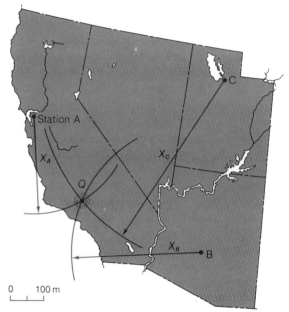

Station A

$X_A$

$X_C$

Q

$X_B$

B

C

0    100 m

**Figure 19-17**
Knowing the distance, say $X_A$, of an earthquake from a
given station, as by the method of the preceding
figure, one can only say that the earthquake lies on a
circle of radius $A$, centered on the station. If, however,
one also knows the distances from two additional
stations $B$ and $C$, the three circles centered on the
three stations, with radii $X_A$, $X_B$, $X_C$, intersect uniquely
at the point $Q$, the epicenter.

**Obtaining Stress Patterns**   When an earthquake occurs, one
block slips relative to an adjacent one along a **fault plane** (Fig.
19-18*a*), as has already been indicated. The orientation of the
fault plane and the slip direction are of great interest because
they provide information about what happens at plate boundaries.

If the concept of plate tectonics is correct and seismicity is
primarily associated with boundaries along which plates separate,
collide, or slide past each other, then the fault orientations and
slip directions should be different for each type of plate junction.
Earthquakes in divergence zones should occur in response to
tension, as if the plates are being pulled apart, and **normal faults,**
in which the overlying block moves down the dip of the fault
plane, should characterize the earthquake mechanism, as in Figure
19-18*b*. Many earthquakes in convergence zones, where plates
collide, should show a compressive mechanism—for example,
**thrust faulting,** in which the overlying block moves up the dip
of the fault plane (Figure 19-18*c*). Where plates slide past each
other along transform faults, the earthquakes mechanism should
be simple lateral (sideways) slip along nearly vertical planes (Figure

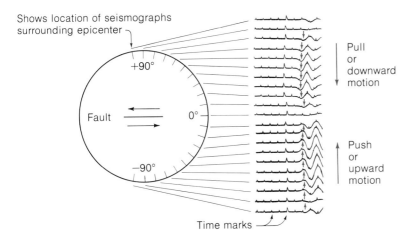

Shows location of seismographs surrounding epicenter

Pull or downward motion

Push or upward motion

Time marks

**Figure 19-19**
The initial motion of seismic waves shows a characteristic movement, a push away from the source or a pull toward the source, depending on the orientation of the fault, the direction of slip along the fault, and the direction of the seismograph from the earthquake. In this example, taken from a laboratory simulation, the stations located between 0° and +90° show a pull toward the source, because the slip along the fault is in a direction away from these stations. The fault orientation can be obtained from the direction 0°, which separates the stations into two groups, one showing an initial push (0 to −90°) and another showing an initial pull (0° to +90°). Because most earthquakes originate below the surface, seismologists must use this principle in analyzing their records in order to find the orientation of the fault and the direction of slip.

19-18*d*). Faults will be discussed further when we take up structural geology in Chapter 22.

Seismologists have discovered how to deduce these earthquake mechanisms from their seismograms. This is especially convenient, since very few earthquake faults break through to the surface, where the slip direction and fault orientation can be observed directly, and many small faults are not always clearly and unambiguously displayed in the field (contrary to our beautiful illustrations). How seismologists do this is shown in Figure 19-19, taken from a laboratory experiment in which an "earthquake" was induced under controlled conditions so that the pattern of seismic waves radiated from the fault could be studied. Although the scale is reduced by about a millionfold from that of nature, the simulation is valid. The experiment models a vertical fault plane along which the slip is lateral. The "seismographs" are positioned on a circle surrounding the fault; the motion recorded after the "laboratory earthquake" is shown at 10° intervals from −100° to +100°. Note that in the quadrant 0 to −90°, the **first motion** on the seismic trace is upward, which is consistent with the slip on that side of the fault being a push to the right. The first motion from 0 to +90° is downward, indicating a push to the left. Thus these positive and negative first motions help define the orientation of the fault plane and the directions of slip.* Although the model simplifies nature somewhat and real seismograph stations are not situated in simple circles surrounding faults, seismologists know how to allow for natural complexities and can uncover the true source mechanism of an earthquake from the radiation pattern of $P$ waves. This is what they find.

When the topography of mid-ocean ridges is examined in detail, the ridges are often found to be segmented, the segments being offset by transform faults (Fig. 19-15). Earthquake epicenters coincide with the ridge crests and with the transform faults between the offset ridge segments, as shown in the figure. In 1967, seismologists at Columbia University found that the pattern of $P$-wave motions radiated from earthquakes on ridge crests indicated that they originate in normal faults that run parallel to the ridge crest. This means that the axis of tension is perpendicular to the trend of the fault, as we have seen in Figure 19-18, just as predicted by the plate-tectonics concept, which specifies that the ridge crests mark the boundary between separating plates. Furthermore, it was found that earthquakes in transform faults between ridge crests show lateral slips, just as would be expected for a region where plates slide past each other in opposite directions. What elegant support seismology gives to the notion that plates spread from ridge crests! Outside the region between the ridge crests, the transform fault becomes **aseismic** — that is, produces no earthquakes. This is to be expected, since there is no differential slip in this region, where the plates move in the same direction on both sides of the fault. In a sense the fault has healed, and is evidenced only by topography usually by a scarp (cliff or steep slope), as shown in Figure 19-15.

So much for the receding edges: what about the leading edges on the other side of the moving plates, where collisions occur? Seismologists have found that many of the deep earthquakes that originate along leading edges show the predicted compressive mechanism, in the direction of the dip of the downgoing plate (Fig. 19-15). What a beautiful verification of the plate-tectonics concept!

**Earthquake Destructiveness — Can It Be Controlled?** Earthquakes cause destruction in several ways. Ground vibrations can shake structures and stress them to the point of failure and collapse. The ground accelerations caused by great earthquakes can approach and even exceed that of gravity near the epicenter, and

---

*Actually, a fault perpendicular to the one shown in the figure would produce the same pattern of up and down motions if the slip directions were ↑↓ rather than ⇆. Fortunately, however, the correct orientation of the fault plane can be deduced either from field evidence, if there is any, or from the aftershocks that follow the main event, for they originate in the fault plane, and their positions define the true orientation of the fault.

**Figure 19-20**
Oblique airview of structures that collapsed during the San Fernando, California, earthquake of 1971. [Photo by R. E. Wallace, U. S. Geological Survey.]

very few man-made structures can survive without severe damage. Certain kinds of soil lose their rigidity and "liquefy" when subjected to repeated seismic shocks. The ground simply slides away, taking man's creations with it (Figs. 19-21, 19-22). As was mentioned earlier, coastal earthquakes on occasion generate the awesome waves called "tsunamis," which can form walls of water as much as fifty feet high and sweep over low-lying coastal areas (Figs. 19-23, and 19-24). Avalanches, mudflows, and fire may accompany earthquakes and take their toll (Fig. 19-25). Of the 99,000 fatalities in the Tokyo earthquake of 1923, 38,000 were due to fire. Table 19-2 lists the human losses of historical earthquakes. A seismic-risk map of the United States, based on the earthquake history of the country, is shown in Figure 19-26. Many readers may be surprised to find that they live in a zone where there is risk of earthquake damage.

**Figure 19-21**
Foundation failure as a result of soil liquefaction caused these buildings in Niigata, Japan to topple during the earthquake of 1964. The structures themselves were built to withstand earthquakes; they toppled intact. [G. Housner, California Institute of Technology; National Science Foundation.]

**Figure 19-22**
Destruction of homes in Anchorage, Alaska, due to soil liquefaction during the earthquake of 1964. Geologists had mapped this area as unstable ground before these residences were constructed. [Photo by F. Press.]

Table 19-2
Some of the world's worst earthquakes (in lives lost).

| Year | Place | Deaths (est.) |
|------|-------|---------------|
| 856 | Corinth, Greece | 45,000 |
| 1038 | Shansi, China | 23,000 |
| 1057 | Chihli, China | 25,000 |
| 1170 | Sicily | 15,000 |
| 1268 | Silicia, Asia Minor | 60,000 |
| 1290 | Chihli, China | 100,000 |
| 1293 | Kamakura, Japan | 30,000 |
| 1456 | Naples, Italy | 60,000 |
| 1531 | Lisbon, Portugal | 30,000 |
| 1556 | Shen-shu, China | 830,000 |
| 1667 | Shemaka, Caucasia | 80,000 |
| 1693 | Catania, Italy | 60,000 |
| 1693 | Naples, Italy | 93,000 |
| 1731 | Peking, China | 100,000 |
| 1737 | Calcutta, India | 300,000 |
| 1755 | Northern Persia | 40,000 |
| 1755 | Lisbon, Portugal | 30,000–60,000 |
| 1783 | Calabria, Italy | 50,000 |
| 1797 | Quito, Ecuador | 41,000 |
| 1819 | Cutch, India | 1,500 |
| 1822 | Aleppo, Asia Minor | 22,000 |
| 1828 | Echigo (Honshu), Japan | 30,000 |
| 1847 | Zenkoji, Japan | 34,000 |
| 1868 | Peru and Ecuador | 25,000 |
| 1875 | Venezuela and Colombia | 16,000 |
| 1896 | Sanriku, Japan | 27,000 |
| 1897 | Assam, India | 1,500 |
| 1898 | Japan | 22,000 |
| 1906 | Valparaiso, Chile | 1,500 |
| 1906 | San Francisco | 500 |
| 1907 | Kingston, Jamaica | 1,400 |
| 1908 | Messina, Italy | 160,000 |
| 1915 | Avezzano, Italy | 30,000 |
| 1920 | Kansu, China | 180,000 |
| 1923 | Tokyo, Japan | 99,000 |
| 1930 | Apennine Mountains, Italy | 1,500 |
| 1932 | Kansu, China | 70,000 |
| 1935 | Quetta, Baluchistan | 60,000 |
| 1939 | Chile | 30,000 |
| 1939 | Erzincan, Turkey | 40,000 |
| 1948 | Fukui, Japan | 5,000 |
| 1949 | Ecuador | 6,000 |
| 1949 | Hait, U.S.S.R. | |
| 1950 | Assam, India | 1,500 |
| 1953 | Northwestern Turkey | 1,200 |
| 1954 | Northern Algeria | 1,600 |
| 1956 | Kabul, Afghanistan | 2,000 |
| 1957 | Northern Iran | 2,500 |
| 1957 | Western Iran | 1,400 |
| 1957 | Outer Mongolia | 1,200 |

Table 19-2 (*continued*)

| Year | Place | Deaths (est.) |
|---|---|---|
| 1960 | Southern Chile | 5,700 |
| 1960 | Agadir, Morocco | 12,000 |
| 1962 | Northwestern Iran | 12,000 |
| 1963 | Skopje, Yugoslavia | 1,000 |
| 1968 | Dasht-e Bayaz, Iran | 11,600 |
| 1970 | Peru | 20,000 |
| 1972 | Managua, Nicaragua | 10,000 |

Can anything be done about reducing earthquake hazards? Seismologists in the United States, Japan, the Soviet Union, and China are working hard to find answers to this question. Even though no major scientific breakthroughs have been made so far, damage and loss of life can be mitigated by encouraging sound building practices. Figure 19-27 shows something that should not be allowed—a residential area built in an active fault zone, the San Andreas, the most dangerous in the United States. Construction on unstable soils or in avalanche-prone areas should be prohibited. Engineers know how to design anti-seismic structures that will withstand most earthquakes, and building codes should require that this be done in high-risk areas.

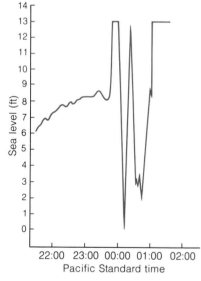

**Figure 19-24**
The tsunami from the great Alaskan earthquake of 1964, reached California, where several people perished and some coastal damage resulted from the waves. A tide gage at Crescent City, California, which usually serves to record ocean tides, made this record of the tsunami, showing sea level changes 6 ft more than normal.

**Figure 19-23**
Destruction at Seward, Alaska, caused by the tsunami generated during the earthquake of 1964. Although a warning system exists to alert people on distant coasts to the danger of tsunamis, it cannot yet function rapidly enough to help residents in the epicentral region. [Photo by F. Press.]

**Figure 19-25**
Monument standing atop an avalanche which buried the town of Khait in the
Tadzhik Republic of the USSR following the earthquake of 1949; 12,000
people were killed. The slide, more than 30 m thick, moved over the town with
a velocity of 100 m/sec. No sprinter could have outrun it. Pamir Mountains in
distance. [Photo by L. Knopoff, University of California.]

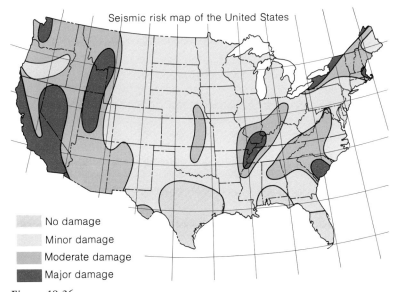

**Figure 19-26**
Seismic risk map of the United States, showing the relative danger in different
parts of the country. The map gives a rough idea of long-term hazard; zone 3
corresponds to potential for major damage; zone 0, no damage. Although no
frequency of activity is implied, major earthquakes in zone 3 on the West
Coast would occur more often than in zone 3 on the East Coast or in the
mid-continent. [After U.S. Coast and Geodetic Survey.]

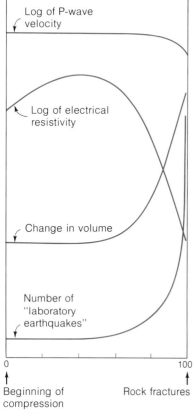

Log of P-wave
velocity

Log of electrical
resistivity

Change in volume

Number of
"laboratory
earthquakes"

0                                    100

Beginning of                    Rock fractures
compression

Percent of stress at faulting

**Figure 19-28**
Graphical summary of laboratory
studies made by W. F. Brace in his
search for a method of earthquake
prediction. A sample of granite is
compressed until it fractures. Several
physical properties are studied to see
if they change before the rock ruptures
and could thereby provide a warning.
The results are encouraging in that the
number of small "earthquakes"
(suddenly appearing, small cracks)
increases, and this also shows as an
increase in volume and as a decrease in
electrical resistance and P-wave
velocity of the rock prior to rupture.
These changes are now being looked for
along actual faults to see if they can
be used to predict real earthquakes.

**Figure 19-27**
Housing tracts constructed within the San Andreas fault zone, San Francisco
peninsula. The solid line indicates the approximate fault trace, along which
ground ruptured and slipped about 2 m during the earthquake of 1906. [Photo
by R. E. Wallace, U. S. Geological Survey.]

Ten years ago only astrologers, mystics, and religious zealots
were concerned with earthquake prediction. Today some of the
most respected scholars in seismology are actively working on this
problem. Increased knowledge of the earthquake mechanism has
encouraged seismologists to believe that earthquakes are preceded
by events that signal the coming of an earthquake within hours
or days or years. The challenge comes in learning how to recognize
them. Some encouraging leads are currently being pursued, such
as anomalous ground tilt or strain changes that precede earthquakes; the bunching of small foreshocks, all indicating the same
slip direction along a fault plane just before rupture; and changes
in such physical properties as porosity, electrical conductivity,
and elastic velocity in the hypocentral region just before faulting.
Some preliminary results of use in earthquake prediction are shown
in Figures 19-28 and 19-29. Scientists in the Soviet Union will
soon be using some of these methods to predict earthquakes several weeks in advance in one highly seismic region. An American
research program in earthquake prediction is just getting underway.

If earthquake prediction has just become respectable, earthquake control continues to boggle the imagination. Yet some

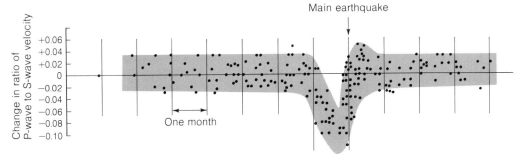

**Figure 19-29**
A possible method of earthquake prediction discovered by scientists in the USSR. Each dot is an observation of the ratio of the velocities of P- and S-waves from a small foreshock. Some months before a large earthquake, this ratio drops, and the earthquake can be predicted within days, signalled by the subsequent increase in the ratio.

chance discoveries have opened this intriguing possibility. In 1966, a dramatic correlation was found between the rate of high-pressure injection of waste fluids into a deep well and the frequency of earthquakes in the vicinity of Denver, Colorado (Fig. 19-30). Apparently the earthquakes were triggered by reduction of frictional resistance to faulting. The pressure exerted by the injected fluids "unlocked" a pre-existing fault and strains that had built up earlier were released. Perhaps someday earthquake-control wells will be spotted every few miles along the San Andreas fault, and fluid injected so as to cause the fault to creep continuously and slip frequently in controlled earthquakes rather than allowing strain to build up over periods of 50 to 100 years and be released in a large, damaging shock. Much research will be needed to achieve this important (and perhaps impossible) goal; the first steps have already been taken.

## EXPLORING THE INTERIOR WITH SEISMIC WAVES

To appreciate the importance of seismic waves in revealing the properties of the interior, one need only reflect on what the state of knowledge would be in the absence of this key tool. We might surmise the existence of a crust from the observation that most surface rocks are generally light and felsic or mafic compared to the more dense ultramafic intrusions that seem to invade the surface layers from below, but we could only speculate on its thickness. The sea-floor crust would of course be terra incognita. We probably would have guessed the mantle to be composed of ultramafic rocks, but could only wonder about its physical state, structure, and thickness. From the clues provided by iron meteorites, the large relative abundance of iron in the cosmos, and from our efforts to explain the Earth's density and magnetic field, we

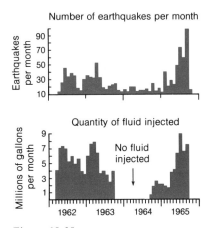

**Figure 19-30**
Correlation made by D. M. Evans between quantity of waste water injected in a deep well and number of earthquakes per month in the vicinity of Denver, Colorado. This unplanned "experiment" opens the distant possibility of earthquake control by fluid injection.

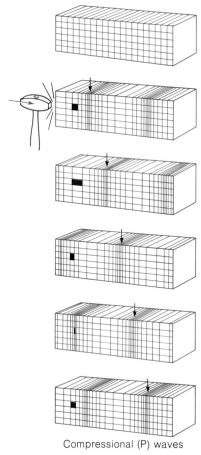

Compressional (P) waves

**Figure 19-31**
Stages in the deformation of a block
of material with the passage of
compressional, or P-waves, through
it. The undeformed block is shown
at the left. In the sequence from top
to the bottom, a crest of compression,
marked by an arrow, moves through
the block with the P-wave velocity.
It is followed by an expansion, and
any small piece of matter, like the
marked square, shakes back and forth
in response to alternating compressions
and expansions as the wave train moves
through. A sudden push (or pull) in
the direction of wave propagation,
indicated by the hammer blow, would
set up P-waves. [After O. M. Phillips,
*The Heart of the Earth,* Freeman,
Cooper & Co.]

might have been led to hypothesize the existence of a molten iron
core, but this would have been argued extensively. Its depth would
be uncertain, and the inner solid core would be unknown. Continental drift and sea-floor spreading would be debated, but the
overall concept of plate-tectonics — especially the fate of plates
in subduction zones — would probably have escaped us.

**Types of Seismic Waves**  As early as the 1800's, mathematicians
proved with pencil and paper the existence of compressional and
shear waves in elastic bodies. Not until the close of the nineteenth
century, however, did seismologists devise instruments sensitive
enough to detect such waves in the Earth — the $P$ waves and $S$
waves generated by sudden slip along a fault. Figure 19-31 shows
the faster-traveling $P$ wave as the propagation of a volume change
— a squeezing and unsqueezing of the medium; the individual
particles vibrate to and fro in the direction of wave propagation.
In Figure 19-32, the $S$ wave is shown as a traveling shearing disturbance, the material distorting in shape rather than changing in
volume; the particles vibrate back and forth at right angles to the
direction of propagation.

Figure 19-33 depicts the trajectory of $P$ waves as they travel
from the source of an earthquake or explosion into the interior,
emerging again at distant points. These wave paths and their travel
times have been determined empirically from the seismographic
records of earthquakes all over the world. Note particularly the
**shadow zone** and its geometrical relationship to the focus and the
core. The Earth's core deflects the waves and in effect casts a
shadow where very little $P$-wave energy reaches the surface. The
existence of shadow zones suggests that the core is molten, because
compressional waves decrease sharply in velocity when they pass
from a solid into a liquid of the same composition. The suggestion
became a firm pronouncement when seismologists found that
shear waves could not penetrate this region. Liquids transmit $P$
waves but not $S$ waves, since the fluids elastically resist and recover
from squeezing, but do not resist shearing.

When $P$ and $S$ waves encounter a boundary such as that between the core and the mantle, they are in general reflected back
as well as transmitted across it, just as light may be partly reflected
and partly transmitted at a water surface. If in the new medium
the wave velocity is different, the waves are bent, or **refracted**.
Because of all of these possibilities of reflection, transmission,
and refraction, $P$ and $S$ waves break up into several types as they
travel through the Earth, as shown in Figure 19-34. Follow the
wave $PcP$ in the figure as it bounces in radarlike fashion from the
Earth's core and yields its depth from the round trip time. The
wave $PKP$, which penetrates the core, is useful for exploring that
region. Many of the $P$-wave trajectories and travel times are
sketched in Figure 19-33.

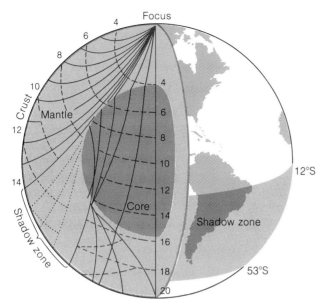

**Figure 19-33**
Cutout showing the pattern of *P*-wave paths through the Earth's interior. The numbers show the travel time in minutes for the waves to reach the associated broken line. Note the shadow zone, a region not reached by *P*-waves (for this hypothetical earthquake at the north pole) because they are deflected by the Earth's core. [After *Internal Constitution of the Earth* by B. Gutenberg (ed.), Dover Publications, 1951.]

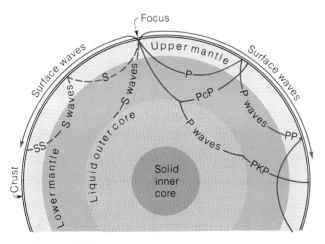

**Figure 19-34**
*P* and *S*-waves radiate from an earthquake focus in many different directions. Waves reflected from the Earth's surface are called *PP* or *SS*. *PcP* is a wave that bounces off the core, and *PKP* is a *P*-wave transmitted through the liquid core. *S*-waves cannot travel in a liquid.

**Figure 19-32**
Stages in the deformation of a block of material with the passage of shear, or *S*-waves, through it. A wave crest, marked by an arrow, moves through the block with the *S*-wave velocity as vertical planes shake up and down. Any small piece of matter, like the marked one, shakes up and down and experiences a shearing deformation (from a square to a parallelogram in the figure) as the shear wave passes through. A sudden shear displacement, indicated by the hammer blow at right angles to the direction of wave propagation, would set up *S*-waves. [After O. M. Phillips, *The Heart of the Earth,* Freeman, Cooper & Co.]

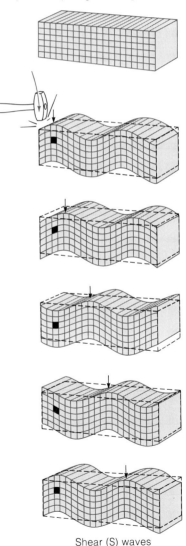

Shear (S) waves

**Box 19-1**

The use of seismic waves in exploring for oil is an important application of seismology. In off-shore prospecting a ship tows a sound source and underwater phones. Sound waves are generated by a pneumatic device that works like a balloon burst. The sounds bounce off rock layers below the sea floor and are picked up by the phones. In this way subsurface sedimentary structures that trap oil, such as faults, folds, and domes, are "mapped" by the reflected waves. This technique is used extensively to explore the submerged continental shelves and shallow seas for oil and gas deposits. Oceanographers use this method to study the sedimentary layers on the continental slope and rise and on the floor of the deep sea.

A section of the Gulf of Mexico, 30 km long and 10 km deep, in which folded sedimentary layers are revealed by reflected seismic waves. [From Petty Geophysical Engineering Co.]

Seismic method of prospecting for oil and gas offshore. [After U.S. Dept. of Interior.]

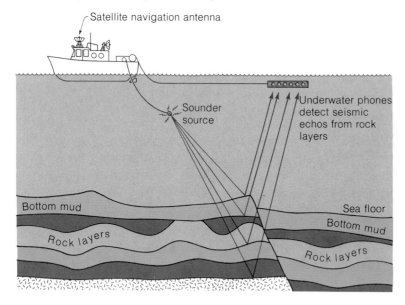

Satellite navigation antenna

Sounder source

Underwater phones detect seismic echos from rock layers

Bottom mud

Rock layers

Sea floor

Bottom mud

Rock layers

In addition to *P* and *S* waves, another category of seismic wave is the surface wave, guided in its propagation by the Earth's surface. These waves travel only through the outer layers of the Earth, just as the motion of ocean waves is mostly surficial. Figure 19-35 shows seismograms in which many of these seismic waves are labeled.

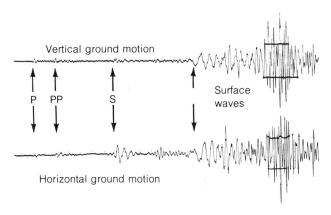

Vertical ground motion

Surface
waves

P PP S

Horizontal ground motion

**Figure 19-35**
Seismograph recording of *P,S,* and surface waves
from a distant earthquake.

Pluck a violin string and a tone is emitted; strike a bell with a hammer and it rings. The Earth also rings when it is disturbed by a great earthquake that causes the entire globe to vibrate like a bell for as long as several weeks. The tones of Earth's vibrations are pitched too low for the human ear to hear, but modern seismographs are sensitive enough to detect these low-frequency oscillations. The Earth can vibrate in different modes, actually an infinite number of them. Some are shown in Figure 19-36. The mode with the lowest pitch is the "football," or **spheroidal**, mode, which takes 53 minutes to execute one vibration. For our musician-readers, we add that in terms of what one might call music of the spheres, that vibration corresponds to E flat in the twentieth octave below middle C. The "balloon," or **radial**, mode has a frequency of one vibration in 20 minutes, and the twisting, or **torsional**, mode—vibrates once in 44 minutes.

**Finding Earth Models from Travel Times and Vibration Frequencies** The basic experiment with seismic waves is to measure precisely the travel times of *P, S,* and surface waves and the frequencies of the vibrations of the Earth. The tools used to make these measurements are thousands of sensitive seismographs and highly accurate clocks distributed throughout the world. From these measurements we can plot travel-time curves for the different kinds of seismic waves (Fig. 19-37). The travel times depend on how the velocities of compressional and shear waves change as they pass through materials having different elastic properties. The vibration frequencies of the Earth depend on the velocities of these waves as well as the density of different parts of the interior, just as the tone of a bell depends on its elasticity and density. Once all the seismological data are accumulated, the next step is to find Earth models whose *P*-wave and *S*-wave velocities and internal densities explain the data.

In the decade of the 1960's, seismologists worked to develop a method to distinguish underground nuclear explosions from earthquakes. They found that explosions excite surface waves with less efficiency than do earthquakes—a discovery that may stimulate statesmen to agree to an underground nuclear test-ban treaty.

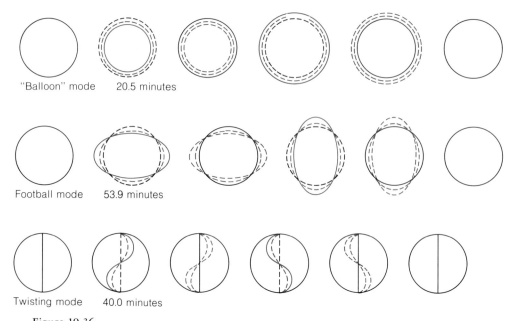

"Balloon" mode     20.5 minutes

Football mode     53.9 minutes

Twisting mode     40.0 minutes

**Figure 19-36**
Three of Earth's vibrational modes. The schematic illustration shows how the planet changes shape during these vibrations, but the actual movements are only a small fraction of a millimeter.

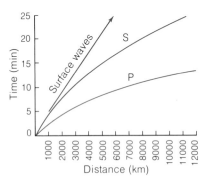

**Figure 19-37**
The time it takes $P$, $S$, and surface waves to travel a given distance can be represented by curves on a graph of travel time against distance.
Seismologists use these curves to locate earthquake epicenters and to determine the Earth's internal structure.

One method used to find Earth models has some similarity to gambling with dice. In fact, it is called the Monte Carlo method, named after the gambling casino. Imagine throwing dice to obtain a value for the density at some depth in the Earth. If the number 5 came up, it would be the density assigned to that depth in a trial model. Velocities of $P$ and $S$ waves can be randomly selected and tested in the same way. Actually, all this selecting and testing is done in a computer, which generates millions of random numbers instead of throwing dice. Follow the flow diagram of the computer program shown in Figure 19-38 to see the steps. The computer first randomly selects a velocity model of the Earth. The travel times of the $P$ and $S$ waves for this model are compared with the observed travel times from nuclear explosions. If the model fails it is discarded and the program loops back to the beginning; this happens repeatedly until a successful model is found. A density model is then sought until one is found that fits the known mass of the Earth. If this test is passed, the so-far successful velocity and density model is used to compute vibration frequencies for this trial Earth, and these are compared with about 100 observed ones for the real Earth. If this last test fails, the model is rejected, and the search begins again from the start. Millions of models are created and rejected in the search for a successful one. This may seem wasteful, but modern computers are so fast that a million Earth models can be created and tested every hour. Scientists, after all, are human, and despite their efforts to keep an open mind, they may be prejudiced in favor of certain ideas about what the Earth is like inside. The Monte Carlo method, being a random one, helps us guard against bias and preconceived notions in building Earth models. A computer that has been honestly programmed should not throw loaded dice.

If a model passes all the tests—that is, if it fits the observed data—we have a possible Earth model. Unfortunately, it is not necessarily a description of the real Earth, because the data currently available are somewhat incomplete and imprecise, and therefore many possible models fit the same set of data. Thus a model that uniquely describes the interior of the real Earth in all details still eludes us; we need more complete and more precise data for its specifications. Nevertheless, the close similarities shown by the various possible models accepted by the computer clearly indicate that modern seismology has made considerable progress: densities and $P$-wave and $S$-wave velocities can now be specified to within several percent of the true value for most of the Earth's interior.

A computer plot of the ranges of internal densities and $P$-wave and $S$-wave velocities for all possible Earth models is given in Figure 19-39. It is quite likely that within the rather narrow bands shown by these data there is a model that would accurately describe the real Earth. As more complete and more precise data

**Figure 19-38**
A computer must be told what to do, step by step. It makes up for this weakness by having an enormous memory and an ability to follow instructions at fantastic speeds. This illustration shows a "flow diagram" of instructions to a computer programmed to find models of the Earth that fit the observed travel times of seismic waves, the Earth's vibration frequencies, and mass.

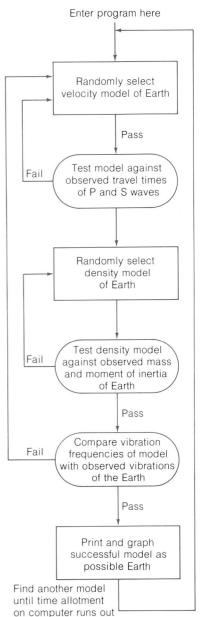

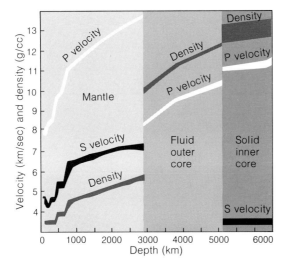

**Figure 19-39**
The variation of density and $P$- and $S$-wave velocity in the Earth's mantle and core probably lies within the narrow bands shown in the illustration. As data become more precise and plentiful, the bands of uncertainty will become narrower. Nevertheless, even today we can specify the density throughout most of the interior with an uncertainty of only a few percent of the actual value.

become available, these bands should become narrower, perhaps someday converging to give us a perfect model of the real Earth. But what can we say *now* about the interior?

**Composition, Structure, and State of the Interior** To know these properties of the interior is the main goal; velocity and density models represent only a means to this end. We use laboratory experiments to make the connection between seismology on the one hand and petrology and geochemistry on the other. High-pressure equipment and shock waves generated by explosives are employed, as described in Chapter 3, to learn how velocity and density would vary in different rocks, either in the solid or in a partially molten state, and either near the surface or in the deep interior. With this information the band of Earth models shown in Figure 19-39 can be interpreted so that we can make some statements about the materials and their state.

The major divisions into crust, mantle, and core (Fig. 19-34) were discovered from the analysis of reflected and refracted *P* and *S* waves, and have been known for more than 60 years. The boundary between the crust and the mantle is called the **Mohorvičić discontinuity** (M, or **Moho**, for short) after the Yugoslavian seismologist who discovered it in 1909. It separates rocks in which *P*-waves have velocities of about 6 to 7 kilometers per second (3.8 to 4.4 miles per second) from underlying mantle rocks, in which *P*-waves have a velocity of about 8 kilometers per second (5 miles per second). The field method of measuring these velocities is described in Box 19-2. From geological sampling to find all possible crustal and mantle materials and from laboratory measurements of the properties of these materials, we have learned to associate *P*-wave velocities with composition, as indicated in Table 19-3. We conclude from these measurements that the continental crust consists mostly of granitic rocks, with gabbro appearing near the bottom, and that no granite occurs on the floor of the deep ocean, the crust there being entirely basalt and gabbro. The mantle below the M discontinuity is almost certain to consist of the dense ultramafic rock peridotite or dunite. The crust is a distillate of the mantle, and therefore differs chemically from its parent. In this sense, the Moho is a chemical boundary located by seismic waves.

Table 19-3
Correspondence between composition and *P*-wave velocity in igneous rocks.

| Composition | P-wave velocities (km/sec) |
|---|---|
| Silicic (granitic) | 6 |
| Mafic (gabbro) | 7 |
| Ultramafic (dunite, peridotite) | 8 |

**Box 19-2    Seismological Sounding of the Earth's Crust and Upper Mantle**

Seismologists have developed a field procedure for measuring the thickness of the crust and the velocity of $P$ waves in the crust and at the top of the mantle. Small seismographs are placed on the surface in a line extending away from a "shot point," where an explosion is set off to generate $P$ waves. The waves leave the explosion in all directions—some traveling along the surface, others along the top of the mantle, as shown in the figure. A travel-time curve is plotted, each point representing the travel time required for the waves to reach the seismograph. The waves that travel along the surface plot on a straight line that passes

through the origin of the graph and has a slope of $1/V_1$. The slope of the line is measured to obtain the speed $V_1$ of $P$ waves in the crust. The waves traversing the mantle plot on a line with slope $1/V_2$ and with intercept $T$. Values of $V_1, V_2, T$ obtained from the graph are used to calculate the thickness from the formula

$$D = \frac{T}{2} \frac{V_1 V_0}{\sqrt{V_1^2 - V_0^2}}$$

A student familiar with trigonometry and Snell's law might try his hand at deriving this simple but important equation.

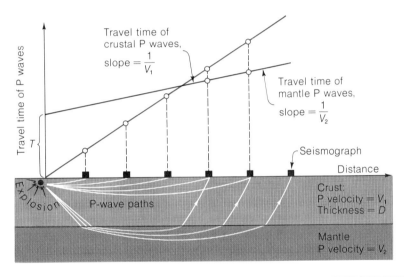

Nowadays seismologists are excited by the finer details, the variations within the crust, mantle, and core. Consider the lateral variations in crustal thickness in a section extending from continent to ocean, as shown in Figure 19-40; it is one of the most important recent seismological results. The thickness of the crust varies from about 35 kilometers to 10 kilometers in a section extending from continent to ocean. Under a high mountain the crust thickens to as much as 65 kilometers. If Figure 19-40 suggests to the reader that the continental crust floats on the denser mantle like an iceberg on the ocean, he has made a good observation. Icebergs float because they are less dense than sea water; flotation comes from the large volume of ice that lies below the sea surface. When

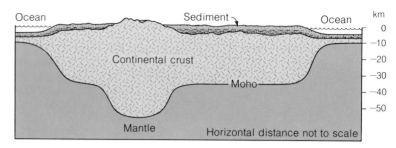

**Figure 19-40**
The lithosphere is topped by a relatively lightweight crust. Seismology reveals that the crust varies in thickness; it is thin under oceans, thicker under continents, and thickest under high mountains.

Archimedes' principle of buoyancy is applied to the flotation of continents and mountains, it becomes the **principle of isostasy**, which holds that the relatively light continents float on a more dense mantle; most of a continent's volume lies below the sea floor for the same reason that most of an iceberg lies below the ocean surface. Nature has contrived that large topographic loads like mountains and continents are compensated — that is, supported primarily by buoyancy rather than by the strength of the crust. Rocks, which we know to be solid and strong over the short term (seconds or years), are weak and flow like a viscous fluid when loaded over the long term (thousands to millions of years). When continents grow or mountains are pushed up, a supporting root must develop as part of the process to provide buoyancy and keep the new load from sinking. One variant should be mentioned. If for some reason — for example, regional heating — a part of the upper mantle becomes less dense than the adjacent mantle, it will also exert a buoyant force that can support elevated topography above it without the need for a crustal root. In a sense, the lower-density mantle serves as a root. This mode of isostatic compensation seems to be operating in the Basin and Range province (Utah, Arizona, Nevada) of the United States. The fact that this is also a region of high heat flow is consistent with the explanation. We shall see later that isostatic compensation does not operate in many places on the Moon where the outer layers seem to be colder and for this reason stronger and able to support loads. Isostasy was actually discovered from variations in gravity, as we shall see in the next chapter, where we take up the subject again. Seismological studies of crustal thickness have provided quantitative corroboration for the mechanism of isostasy.

In the years 1965 to 1970, geologists and geophysicists the world over concentrated their research efforts on the upper thousand kilometers of the Earth as part of the International Upper Mantle project. This concerted attack led to many exciting discoveries about a region that had previously been poorly known. We

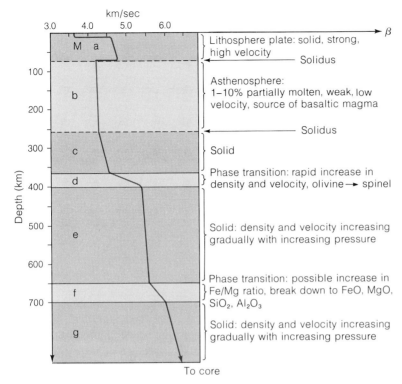

**Figure 19-41**
A modern view of the structure of the outermost 700 km of the Earth is
illustrated by a plot of *S*-wave velocity against depth. Note how velocity changes
mark the important zones: lithosphere, partially molten asthenosphere,
transitions to more dense molecular structures.

can illustrate the more important of these by discussing one of the
shear-wave velocity models found in the Monte Carlo search. The
model is divided on the basis of shifts in velocity into zones *a* to
*g,* as shown in Figure 19-41.

Zone *a* is the **lithosphere,** a slab about 70 kilometers (45 miles)
thick in which the continents are embedded. Crust forms the
uppermost part of this outer shell of the Earth. Its lower boundary
is marked by an abrupt decrease in shear-wave velocity. The lithosphere is characterized by high velocity and efficient propagation
of seismic waves, both of which imply solidity and strength. We
have already discussed in several places how the Earth's lithosphere
is made up geographically of about ten distinct plates, created
along mid-ocean ridges and destroyed in subduction zones.

Zone *b* is the **asthenosphere,** or zone of weakness. It is also
called the **low-velocity zone** for the obvious reason that the shear-
wave velocity takes on low values. Seismic waves are attenuated
more strongly in the asthenosphere than anywhere else in the
Earth. Because laboratory experiments show that seismic waves

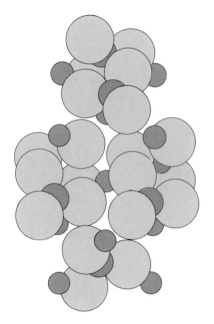

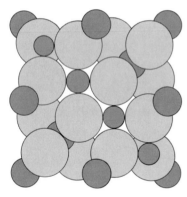

**Figure 19-42**
Olivine ($Mg_2SiO_4$ or $Fe_2SiO_4$) is a major mineral in the Earth's mantle. The figure at the top shows olivine in its low-pressure form. The large pale-brown atoms are oxygen, the medium-brown atoms are silicon, and the dark-brown ones are magnesium. When the pressure reaches a critical value, corresponding to a depth of about 400 km in the Earth, the molecule collapses into a more dense form (lower figure). Note the decrease in void space in the high-pressure form, in which the oxygen atoms are more closely packed. Seismologists have found where this transition occurs in the Earth, and petrologists have observed the transition in high-pressure laboratory experiments.

are slowed and absorbed in a crystalline-liquid mixture, most geophysicists and petrologists think that these features of the asthenosphere indicate partial melting, perhaps 1 to 10 percent melt. Velocity and density in both the lithosphere and asthenosphere would fit a peridotitic composition. These two zones therefore do not differ so much chemically as they do in physical state; the boundary at 70 kilometers marks the solidus (where melting begins), separating the solid lithosphere from the not-so-solid asthenosphere. In Chapter 17 it was proposed that the melt of the asthenosphere is the primary source of basaltic magma, which fits the picture nicely. A solid slab underlain by a weak layer might be more easily moveable, and perhaps this accounts for the mobility of the lithospheric plates.

The asthenosphere ends at a depth of about 250 kilometers (155 miles), and the rocks become solid again in zone *c*. The velocity increases slightly with depth in this region because of the effect of increasing pressure.

Zone *d,* about 400 kilometers (250 miles) below the surface, is thin but very important. Note the rapid increase in velocity, which correlates with the rapid increase in density in Figure 19-39. This **transition** is too rapid to be accounted for by a composition change or any other mechanism. A change of phase — that is, a rearrangement on the atomic level is required. The theoretical explanation was beautifully checked in 1969 when E. A. Ringwood and S. Akimoto squeezed olivine in their laboratories and found that at critical pressures and temperatures its atoms take up a more compact arrangement, changing into the spinel structure (Fig. 19-42). Olivine is the principal mineral in peridotite, and at a depth of about 400 kilometers in the Earth, conditions are just right for it to change phase. This is an excellent example of how the collaboration of specialists with different backgrounds (seismologists and petrologists) can, little by little, remove some of the mystery of Earth's interior.

Zone *e* is one of small change with depth, but zone *f,* near 700 kilometers, shows another rapid transition. This one has only been verified in the laboratory by shock-wave experiments, but since the sample is destroyed the transition has not been identified. Geophysicist F. Birch believes that minerals like olivine and pyroxene break down into dense, simple oxides like $FeO$, $SiO_2$, and $MgO$ at this depth. The entire region from 400 to 700 kilometers, containing zones *d, e,* and *f,* is sometimes called the **transition zone.**

The lower mantle, zone *g,* extending from 700 kilometers (435 miles) to the core at a depth of 2898 kilometers (1800 miles) is a region that changes little in composition and phase with depth. Density and velocity increase gradually, again due to increasing pressure.

The Earth's core is far away, but not out of the reach of seismic waves. We know that its outer region is fluid and its inner one solid

(Fig. 19-34). To obtain its composition, we use the same approach that has already proved so useful — comparison of laboratory experiments and seismological data. Look at Figure 19-43 to see how this is done. A band of Monte Carlo solutions for the density in the core is plotted. Also shown are the densities of nickel, iron, and a mixture of iron and silicon, determined by "shocking" these materials in the laboratory with explosives, as was described in Chapter 3. We see that nickel is too dense, iron is better but needs to be lightened by adding perhaps 15 percent silicon. Would other shocked elements fit the data? Perhaps, but our choice is limited by the relative abundance of elements; the core accounts for one-third the mass of the Earth, and must therefore contain relatively abundant elements. Iron is the only abundant element that approaches the required density under the pressure of millions of atmospheres at these great depths. It is a little too dense, as Figure 19-43 shows, so a plentiful "lightening" element like silicon must be added. Oxygen or sulfur might also be possible lightening elements, but these have not yet been "shocked" with iron to see if they fit.

In this way seismological observations and laboratory measurements of the properties of materials combine to give an incomplete but nevertheless good approximation of the Earth's interior. A zoned, differentiated Earth is found in which the major components are a metallic iron core and a rocky mantle consisting primarily of iron-magnesium silicates. The mantle includes a transition zone in which atoms are forced into closer packing, a partially molten asthenosphere, and most of the outer lithosphere. A thin, lightweight crust — the end-product of the differentiation process — caps the mantle.

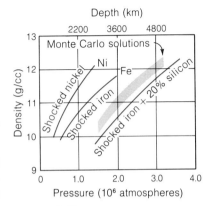

**Figure 19-43**
Density in the Earth's fluid core plotted against depth below the surface and against pressure. The true density is not precisely known but probably lies in the colored band. Despite this uncertainty, comparison with the densities for iron, nickel, and iron-silicon mixtures found in laboratory studies enables seismologists to conclude that the core is mostly iron, but slightly less dense than pure iron, as if a small amount of a "lightening" element like silicon were present.

## SUMMARY

1. Most earthquakes originate in the vicinity of plate boundaries. The mechanism of earthquakes is governed by the kind of plate boundary: fracture under tensile stress at boundaries of divergence, fracture under compressive stress boundaries of convergence, and lateral slip along transform faults.

2. Great earthquakes release in a few minutes huge amounts of elastic strain energy that had been slowly stored in the rocks of the fault zone over tens or hundreds of years. The source of this strain is plate motions.

3. Richter magnitudes are determined from the size of the ground motions, as measured when seismic waves are recorded on seismographs. The three types of seismic waves are $P$ waves, $S$ waves, and surface waves. The entire Earth can be set into global vibration by great earthquakes.

4. From a study of the travel times of seismic waves and the frequency, or pitch, of the global oscillations, seismologists have found that the earth is divided into shells — that is, it is a zoned, differentiated planet:

1. A slab-like, mostly ultramafic lithosphere, broken into large, mobile plates.

2. A partially molten asthenosphere, the primary source of basaltic magma, as evidenced by reduced velocity and high absorption of seismic waves.

3. A transition zone, where atoms are forced into a closer packing due to the high pressures.

4. A lower mantle, predominantly iron-magnesium silicate.

5. A fluid core, mostly iron but with one or more "lightening" elements.

6. A solid iron central core.

The continents with their lightweight silicic crusts — the end-products of the differentiation process — are embedded in the lithosphere.


## EXERCISES

1. What is an earthquake? How is its magnitude measured? How many earthquakes cause serious damage each year?

2. How does the distribution of earthquake foci correlate with the three types of plate boundaries?

3. Seismograph stations report the following data for an earthquake: Dallas, $S - P = 3$ minutes; Los Angeles, $S - P = 2$ minutes; San Francisco, $S - P = 2$ minutes. Use a map of the U.S. and travel time curves (Fig. 19-37) to obtain a rough epicenter.

4. Taking into account mass hysteria, false alarms, reduction in casualties and other possible consequences, do you think the objective of predicting earthquakes is a worthwhile goal?

5. Draw a cross section of the Earth, showing, to scale, the crust, lithosphere, asthenosphere, transition zone, core-mantle boundary, fluid outer core, solid inner core. What are the major characteristics of each region?


## BIBLIOGRAPHY

Anderson, D. L., "The San Andreas Fault," *Scientific American*, pp. 53–68, November 1971. (Offprint No. 896.)

Bolt, B. A., "The Fine Structure of the Earth's Interior," *Scientific American*, pp. 24–33, March, 1973.

Division of Earth Sciences (NRS-NAC), *Seismology, Responsibilities and Requirements of a Growing Science, Part II.* Division of Earth Sciences, National Research Council, National Academy of Sciences, Washington, D.C., 1969.

Isacks, B., J. Oliver, and L. R. Sykes, "Seismology and the New Global Tectonics," *Journal of Geophysical Research*, vol. 73, pp. 5855–5899, 1968.

Richter, C. F., *Elementary Seismology.* San Francisco: W. H. Freeman and Company, 1958.

Tectonophysics, "Forerunners of Strong Earthquakes," *Tectonophysics*, vol. 14, pp. 177–344, 1972.

# 20

## The Earth's Magnetism and Gravity

*Earth's magnetic field originates in the fluid iron core. Many rocks "remember" the direction of the magnetic field that existed at the time of their formation, and this remanent magnetization provides not only a history of the magnetic field but also a history of the movements of the lithosphere. Variations in the gravity field over the surface correlate with subsurface geological changes. Mountain roots, an example of isostatic compensation or buoyant support of mountains, were inferred from the gravity anomalies they produced.*

A body, such as the Earth, is surrounded by a gravitational field, which attracts other bodies through interaction with their gravitational fields. Just why this occurs is unknown, yet, since Newton's time, we have been able to calculate the effects of gravity—such as the acceleration of an apple falling from a tree, the orbits of planets, the trajectories of spacecraft, the gravitational attraction of a mountain, and so on. Similarly, without knowing its basis, we observe and use the fact that magnetism is associated with moving electric charges, such as an electric current in a wire, or electrons spinning around the nucleus of an atom, as depicted in Figure 20-3. In this chapter, we will see how a planet's magnetic and gravity fields can be used to infer conditions and processes in its interior. For instance, just a few years ago, the Earth's magnetic field provided the key that unlocked the secret of sea-floor spreading.

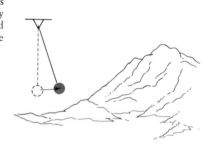

**Figure 20-1**
Three examples of gravitational attraction: an object falling in Earth's gravity field, a planet held in orbit by the Sun's gravity, a pendulum deflected by the gravitational attraction of the mass of a mountain.

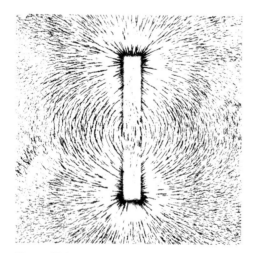

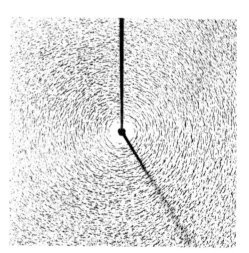

**Figure 20-2**
Magnetic fields of a bar magnet and of a wire carrying an electric current made evident by the alignment of iron filings on paper. The filings are aligned parallel to the direction of magnetic field lines. [From *PSSC Physics,* 3rd edition, D. C. Heath and Co. Reprinted by permission of the publisher.]

# EARTH'S MAGNETIC FIELD

**The Earth as One Big Magnet** Twenty-five hundred years ago, the Greeks discovered the magnetic properties of lodestone, one of the naturally occurring ores of iron, the mineral called magnetite. It took about another thousand years for the Chinese to invent the first crude magnetic compass, which was simply a piece of lodestone suspended on a string. Travelers of the fourteenth century brought news of this discovery back to Europe, where it was adopted as a navigational aid, making possible the great feats of such men as Columbus and Magellan during the Age of Exploration. It remained for William Gilbert, physician to Queen Elizabeth I, to explain in his book *De Magnete,* published in 1600, how the magnetic compass works. He offered the proposition that "the whole Earth is a big magnet" whose field acts on the small magnet of the compass needle to align it in the north–south direction.

The Earth's magnetic field can be fairly well described by postulating a small but powerful permanent bar magnet located near the center of the Earth and inclined about 11 1/2° from the geographic axis, as shown in Figure 20-4. The "lines of force" of the magnetic field, shown in the figure, imply the presence of a magnetic force

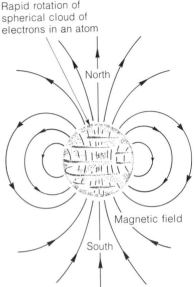

Figure 20-3
Moving electrons are equivalent to electric currents and create magnetic fields as they whirl around the nucleus of an atom. [After *Magnets* by F. Bitter, Doubleday Anchor Books. Copyright © 1959 by Educational Services, Inc.]

Rapid rotation of spherical cloud of electrons in an atom

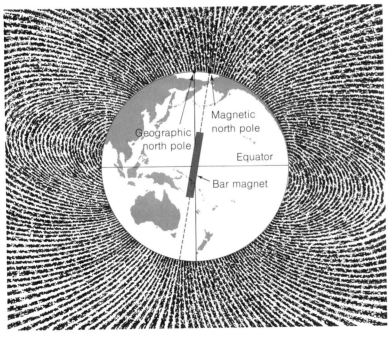

Figure 20-4
Earth's magnetic field is much like the field that would be produced if a giant bar magnet were placed at the Earth's center and slightly inclined (11°) from the axis of rotation.

at each point in space. A magnetic needle that is free to swing rotates under the influence of this magnetic force into a position parallel to the local line of force, approximately in the north–south direction. The north-seeking end of the needle points to the **magnetic north pole**. A compass needle that is free to swing in a horizontal plane does not point to true geographic north, but deviates slightly, east or west, depending on where the observer is, as Figure 20-4 implies; the angle of east or west deviation from true north is called the **declination**. In New York, the compass points about 10° east of geographic north; in California, the declination is about 20° west. The declination would be zero everywhere, making a navigator's life simple, if the magnetic north pole coincided with the geographic north pole in Figure 20-4 — that is, if the hypothetical bar magnet at the center of the Earth lined up perfectly with the axis of rotation.

A magnetized needle free to pivot in a vertical plane (called a **dip needle**) measures the **inclination** of the magnetic field — that is, its angle of deviation from the horizontal. At the magnetic poles, the dip needle swings into a vertical position, as shown in Figure 20-5, which is how some early explorers located these two points on the Earth's surface. Near the equator, where the inclination is zero, the dip needle takes a horizontal position.

The compass and dip needle together give the direction of the geomagnetic field. To characterize the field fully, one needs to also measure its intensity or strength with a device called a **magnetometer**, which, in some versions, simply measures the force exerted by the Earth's field on a small, standard magnet. The unit of magnetic-field strength is the gauss (named after the great mathematician Karl Friedrich Gauss). Ordinary horseshoe magnets have fields of about 10 gauss. The Earth's field is about 0.5 gauss near the surface.

**The Puzzle of the Earth's Magnetism** Unfortunately, though a good description of the field can be given assuming a permanent magnet at the center of the Earth, this idea has a fatal defect — heat destroys magnetism and magnetic materials lose their permanent magnetism when the temperature exceeds a certain value called the **Curie point** (named after Pierre Curie). For most magnetic materials, the Curie point is about 500°C, a temperature exceeded below depths of about 20 or 30 kilometers in the Earth. In other words, the Earth cannot be permanently magnetized below this depth, so the notion that there is something like a bar magnet near the center of the Earth is eliminated, even though it represents the field nicely.

Another way to make magnetic fields is with electric currents as the sketch in Figure 20-2 shows. About a billion amperes of current would be needed to produce the Earth's magnetic field, and

**Figure 20-5**
A suspended magnet needle aligns itself in the direction of the Earth's magnetic field. This diagram shows that the inclination of the needle would vary from vertical over the pole to horizontal at the equator. [After *Continental Drift* by D. and M. Tarling, Doubleday Anchor Books. Copyright © 1971 by G. Bell and Sons, Ltd.]

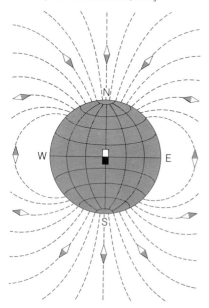

this is nearly as much electric current as the total amount generated by man. Dynamos in power plants make electricity using an electrical conductor, propelled by steam or falling water, rotating in a magnetic field. Where inside the Earth is there a dynamo with the capacity to generate this much current?

The obvious place to look is the Earth's fluid iron core, because fluids can move readily and iron is a good conductor. Walter Elsasser and Edward Bullard proposed that a dynamo actually exists in the Earth's core—a **self-exciting dynamo**, which generates its own magnetic field. Figure 20-6 shows the principle of such a dynamo. No one believes that a rotating disc like the one in the figure actually exists in the core; what is proposed is that the fluid iron is stirred into convective motion by heat generated from residual radioactivity in the core. A small, stray magnetic field would interact with the moving fluid iron to produce electric currents, which would then create their own magnetic field, starting up a self-exciting dynamo. This sounds vague because it is: the dynamo theory for the core is exceedingly difficult to spell out in detail—in fact, the mathematical equations have never been completely solved. We do know enough, however, to show: (1) that the mechanism is feasible; (2) that the magnetic pole and the geographic pole (about which the Earth rotates) must nearly coincide, as they do today; (3) that the polarity of the magnetic field is a matter of chance—that is, that the north magnetic pole could just as easily be a south magnetic pole. Large-scale fluid motions in the core seem to offer the best possibility for explaining the Earth's magnetic field.

**Changes in the Earth's Magnetic Field**  The fact that the magnetic compass was a key factor in navigation led governments to subsidize the science of geomagnetism. As a result, magnetic observations have been made since the sixteenth century. This historic documentation of the geomagnetic field over the last 400 years shows that declination, inclination, and field strength vary gradually the world over—a phenomenon called **secular variation**. For example, the compass needle in London has swung from 11° east of true north in 1580, to due north in 1660, to 24° west of north in 1820, and back to 7° west of north in 1970, as Figure 20-7 shows. This amounts to a drift of about 0.1° per year, a rapid rate of change compared to most geological processes. The solid parts of the Earth, even when they creep plastically, change much more slowly. Because fluid motions can occur this fast, however, again we resort to the fluid core as the likely source of the secular variation. Thus fluid motions in the core not only account for the origin of the main field, but also for its fluctuations, which originate perhaps in small eddies superposed on the large-scale convective motion.

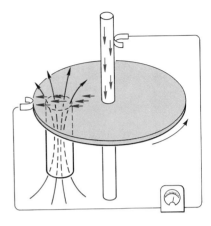

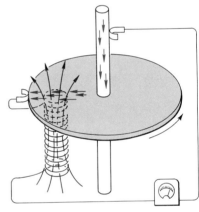

Figure 20-6
Simple dynamo generates an electric current (colored arrows) when a copper disc is turned through the magnetic field of a bar magnet. If the bar magnet is replaced by a coil of wire, the same electric current creates a magnetic field in the coil, which keeps the system going. As long as the disk keeps rotating, the current will flow, and the magnetic field will remain in this "self-exciting dynamo." The Earth's magnetic field is thought to originate in a self-exciting dynamo, but one that is enormously more complicated. [From "The Earth as a Dynamo" by W. M. Elsasser. Copyright © 1958 by Scientific American, Inc. All rights reserved.]

1580
11° east of north

1660
Due north

1820
24° west of north

1970
7° west of north

**Figure 20-7**
A compass needle does not remain
fixed in direction for all time, but
drifts gradually in response to slow
changes in the Earth's magnetic field.
The example above shows how a
compass in London varied in direction
between 1580 and 1970.

**Fossil Magnetism** Not too long ago, an Australian graduate
student found a fireplace in an ancient campsite where the aborig-
ines cooked their meals. He carefully removed several stones that
had been baked by the fires, first noting their orientation. Then
he measured the direction of magnetization of the stones, and he
found that they were magnetized exactly opposite to the present
geomagnetic field. He proposed to his disbelieving professor that,
as recently as 30,000 years ago, when the campsite was occupied,
the geomagnetic field was reversed from the present one—that
is, that magnetic north was south.

Surprisingly enough, Earth scientists have discovered how to
find out what the geomagnetic field was like in the past, not just
thousands but millions of years before man developed instruments
and started keeping records. The Curie point was mentioned earlier as the temperature at which permanent magnetism is lost
when a magnet is heated. It is also a property of many very hot
magnetizable materials that, in cooling below the Curie point,
they become magnetized in the direction of the surrounding magnetic field. This is **thermoremanent** magnetization, magnetization
"remembered" by the rock long after the magnetizing field has
disappeared. Thus, the Australian student was able to determine
the direction of the field when, after the last meal was cooked in
that ancient fireplace some 30,000 years ago, the stones cooled
down and took on the magnetization of the surrounding geomagnetic field. One might also think of an ancient volcano in eruption,
say 100 million years ago. When the lava solidified and cooled
below the Curie point, it became magnetized, leaving us with a
permanent record of the geomagnetic field in mid-Cretaceous
time, just as a fossil leaves a record of ancient life (Fig. 20-9).

Some sedimentary rocks can also take on a remanent magnetization. Recall that marine sedimentary rocks are formed when
sedimentary particles that have settled through the ocean to the
sea floor become lithified. Magnetic grains among the particles
—chips of the mineral magnetite, for example—would become
aligned in the direction of the Earth's field while falling through
the water, and this orientation would be incorporated into the
rock with lithification. The **depositional remanent magnetization** of a sedimentary rock would then be due to the parallel
alignment of all of these tiny magnets, as if they were compasses
pointing in the direction of the field prevailing at the time of
deposition (Fig. 20-10).

With this important tool of **fossil magnetism** or **paleomagnetism** for exploring the ancient magnetism of the Earth, scientists
roamed the surface of the planet, in the 1950s and 1960s, collecting
old rocks on every continent and determining their magnetism and
ages in order to reconstruct the history of the geomagnetic field.
This worldwide effort led to the discovery of three startling phenomena: (1) **polar wandering**, (2) **reversals of the geomagnetic
field**, and (3) **the sea floor as a magnetic "tape recorder."**

a    Above the Curie point, atoms take random directions

b    Below the Curie point, atoms become parallel within domains

c    Below the Curie point, in the presence of an external magnetic field, domains line up

**Figure 20-8**
Magnetic properties of a magnetizable material. (a) At temperatures above the Curie point, heat "agitates" the atoms; they take random directions, and their randomly oriented magnetic fields cancel. (b) Below the Curie point, domains form in which the atoms line up, giving each domain a definite magnetization, represented by the larger arrows. (c) If the material cools below the Curie point, in the presence of an external magnetic field, the domains line up parallel to the external field, and the material becomes permanently magnetized.

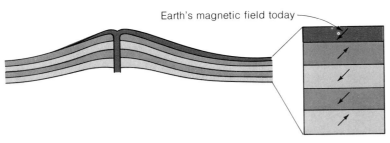

Earth's magnetic field today

**Figure 20-9**
Schematic illustration of how lava beds preserve the record of reversals of the Earth's magnetic field. The modern flow at the top shows the direction of the field today. Earlier flows record the directions of ancient fields that existed when the lavas cooled.

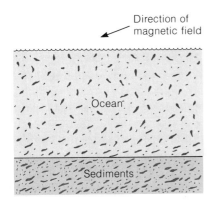

Direction of magnetic field

Ocean

Sediments

**Figure 20-10**
Schematic illustration of how a sedimentary rock can take on a magnetic field. Magnetic mineral grains, transported to the ocean with other erosion products, become aligned with the Earth's magnetic field while settling through the water. The initial orientation, is preserved in the lithified rock, which thus "remembers" the field that existed at the time of deposition.

*Polar Wandering.* Once the direction of remanent magnetism in a rock is measured, the position of the magnetic pole at the time the rock was magnetized can be obtained. For example, suppose a formation 600 million years old is found whose remanent magnetic inclination is horizontal—that is, 0°—and whose remanent declination is in a northerly direction. This means that, regardless of the present location of the land mass in which the formation is situated, it was close to the equator 600 million years ago, as Figure 20-5 shows, and if the land mass hasn't rotated, the magnetic pole was located 90° to the north. It is assumed, of course, that the formation hasn't been folded or magnetically disturbed, and that the geomagnetic field was always of the simple magnetic configuration shown in Figure 20-5. These assumptions are reasonable ones, but they are by no means readily proven. Ancient pole positions can be plotted for remanent inclinations and declinations found in rock formations using these assumptions, and the movement of the magnetic pole for hundreds of millions of years can be charted as in Figure 20-11. The magnetic pole seems to have wandered extensively in the past. Some 600 million years ago, it was in the mid-Pacific equatorial region. It moved to its present position gradually, following the path shown in the figure. Most geophysicists believe that the geomagnetic and geographic poles must maintain the close proximity that they do now, so it is assumed that the mid-Pacific equatorial region was also near the geographic north pole 600 million years ago. In the absence of any external influences, the direction of the axis of rotation of the Earth (which defines the geographic pole) remains fixed in space for all time. This is a requirement of the law of conservation of angular momentum, which was mentioned in Chapter 1. The only way out of this dilemma is for the "geography" to move relative to the axis of rotation. To an observer in space, the geographic pole would remain fixed and polar wandering would show as a movement of the surface features of the Earth over the globe. The continents would appear to "wander" about 90° to the north in some 600 million years, as Figure 20-11 implies. The sites of Denver and Paris would have been south of the equator 600 million years ago. Actually, it is the plates of the lithosphere that "wander" with respect to the axis of rotation, "carrying the geography along." Paleoclimatic data—that is, the rock record of ancient climates—such as the occurrence of coal beds (which indicate humid climates) in polar regions and glacial deposits near the equator, may be taken as additional evidence that polar wandering occurred (Fig. 20-12).

Polar wandering is really a manifestation of the mobility of the plates of the lithosphere. An observer in space would not simply see the map slip over the globe, he would also see it distorted, with the continents colliding and drifting apart and the oceans closing and opening.

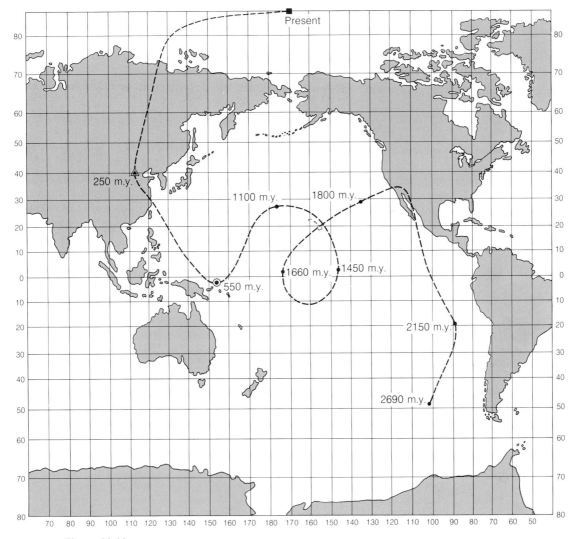

**Figure 20-11**
Positions of the magnetic field from the present to 2.69 billion years ago, according to T. M. Gates. Most geophysicists believe that the pole didn't actually wander, as the figure suggests; it remained fixed, close to the axis of rotation, while the "map"—that is, the outer layers of the Earth—drifted over the interior.

*Reversals of the Geomagnetic Field.* Imagine the state of shock and consternation if navigators and hikers the world over were to see their compass needles flip around with the north-seeking pole pointing south. A reversal of the Earth's magnetic field will take place in the next few hundred thousand years, except not as rapidly as the opening sentence of this paragraph implies. Erratically, but roughly every half-million years, the Earth's magnetic field changes polarity, taking perhaps a few thousand years to reverse its direction.

**Figure 20-12**
Coal outcrop below sandstone, Victoria Land, Antarctica. The presence of coal
implies tropical or subtropical conditions, suggesting that the Antarctic
continent was closer to the equator at one time. [Photo by W. E. Long, Ohio
State University.]

The reversals are clearly indicated in the fossil magnetic record
of layered lava flows, as shown in Figure 20-9. In the sequence of
rocks, each layer represents a progressively earlier period of geological time whose age can be determined by radiometric methods.
The direction of remanent magnetism can be obtained for each
layer and, in this way, the time sequence of flip-flops of the field —
that is, the **magnetic stratigraphy** — can be deduced. The history
of reversals going back almost 5 million years has been worked
out in this way, as shown in Figure 20-13. It has been found that
about half of all rocks studied are magnetized opposite to the
Earth's present magnetic field, which implies that the field has

flipped frequently over geologic time and that normal and reverse fields are equally likely. Although normal and reverse **magnetic epochs** (each of which is named after a famous magnetician) seem to last on the order of one-half million years, superimposed on the major epochs are transient, short-lived flips of the field, known as **magnetic events**, which may last anywhere from several thousand to 200,000 years. The Australian graduate student mentioned earlier apparently found a new reversal event within the present Bruhnes normal epoch.

Just why the field reverses is unknown. We are even uncertain whether the field dies down and builds up again in the opposite direction or simply tilts over. The simple self-exciting dynamo can have either polarity, and perhaps this indicates that the Earth's internal dynamo is easily changeable.

Some specialists on the microfossils of the sea floor have reported a curious correlation between extinctions of certain faunal types and reversal of the field. It is a matter of some concern if the disappearance of certain organisms is related to polarity changes. One explanation that has been advanced is the penetration to the Earth of lethal cosmic rays may be increased when the geomagnetic field is near zero. However, most cosmic rays are absorbed by the atmosphere rather than deflected by the magnetic field, so the effect on life is likely to be small. Perhaps the constituents of the microfauna somehow depend on the magnetic field in their internal biochemical reactions. Perhaps an external common cause, like a large meteorite impact, affects both the fauna and the field. Perhaps the correlation is spurious.

Although the cause of reversals remains for future scientists to explain, their occurrence has made possible an important discovery about the sea floor.

*The Sea Floor as a Magnetic "Tape Recorder."* During World War II, extremely sensitive airborne magnetometers were developed to detect submarines by their magnetic fields. With slight modification, these same instruments were adopted by oceanographers for towing behind their research ships. Used in this way, the magnetometers measure two things: the main planetary or geomagnetic field (see Fig. 20-5) and the local magnetic disturbance, or **magnetic anomaly**, due to magnetized rocks on the sea floor. The general, broad geomagnetic field could easily be subtracted, leaving a record of a highly variable magnetic anomaly originating in the magnetized rocks beneath the sea. Rocks strongly magnetized in the normal direction would show as a positive anomaly; reversely magnetized rocks would show as a negative anomaly.

Steaming back and forth across the ocean, sea-going scientists discovered amazing magnetic anomaly patterns like the one shown in Figure 20-14. In many areas, the bands of positive and negative

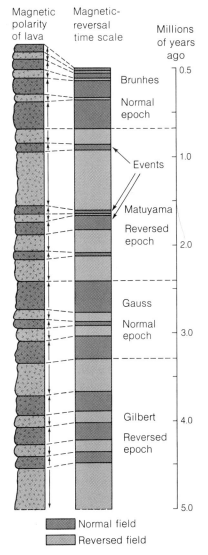

**Figure 20-13**
Schematic illustration of how magnetic polarities of lava flows are used to construct the time scale of magnetic reversals over the past 5 million years. In no one place is the entire sequence found; the sequence is worked out by patching together the ages and polarities from lava beds all over the world.

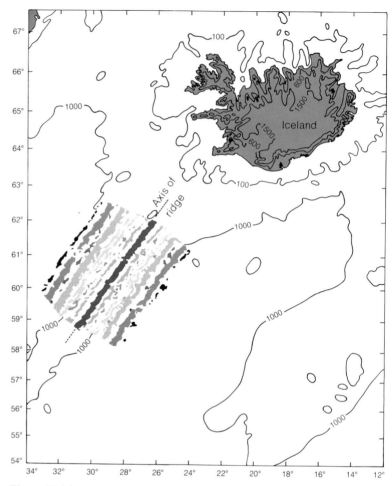

**Figure 20-14**
Magnetic anomaly pattern associated with the Reykjanes Ridge, a part of the
mid-Atlantic ridge southwest of Iceland. The white spaces correspond to rock
formations on the sea floor that are reversely magnetized. The strips shown in
tones of brown or gray correspond to normal magnetization—that is, similar
to the present-day direction. The almost perfect symmetry with respect to the
ridge axis is emphasized by the matching tones on opposite sides; the youngest
rock is brown and the oldest black, in accordance with the hypothesis of sea-
floor spreading. [After "The Origin of the Oceanic Ridges," by E. Orowan.
Copyright © 1969, by Scientific American, Inc. All rights reserved.]

magnetic anomalies are linear over hundreds of miles and show
an almost perfect symmetry with respect to the central magnetic
anomaly. The axis of symmetry coincides with the crest of the mid-
ocean ridges. If one folds a map of such anomalies along the ridge
axis, he will find that the magnetic bands on one side fall nearly
on top of those on the other side, almost like the ink-blot pattern

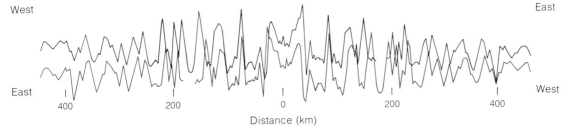

**Figure 20-15**
Symmetry of the magnetic anomaly of sea floor rocks on both sides of the ridge axis is demonstrated by reversing a record covering about 900 km (brown) and superposing it on the original (black). [After "Sea Floor Spreading" by J. R. Heirtzler. Copyright © 1968 by Scientific American, Inc. All rights reserved.]

of a Rorschach test. Figure 20-15 shows a magnetic profile across a ridge, which illustrates these features in greater detail. Note the amazing correspondence of the sequence of peaks and troughs of the magnetic-anomaly curve on both sides of the ridge.

This peculiar magnetic pattern puzzled scientists for several years until 1963, when two Englishmen, F. J. Vine and D. H. Mathews, and, independently, two Canadians, L. Morley and A. Larochelle, made a startling proposal: They reasoned that the positive and negative magnetic zones correspond to bands of rock on the sea floor that were magnetized during ancient episodes of normal and reversed magnetism of the Earth's field, and that the magnetic stripes could be used as evidence in support of the theory of sea-floor spreading (Fig. 20-16). They argued that the ocean progressively widens as new sea floor is created along a crack that follows the crest of mid-ocean ridges. Lava flowing up from the interior solidifies in the crack and becomes magnetized with either normal or reversed magnetization, depending on the direction of the Earth's field at the time. As the sea floor spreads away from the ridge, approximately half the newly magnetized material moves to one side and half to the other, forming two symmetrical, magnetized bands. Newer material fills the crack, continuing the process. In this way, the sea floor acts like a tape recorder that encodes, by magnetic imprinting, the history of the opening of the ocean in terms of the history of reversals of the geomagnetic field. By discovering how to "replay the tape" — that is, by dating the magnetic bands on the sea floor using the magnetic stratigraphy of Figure 20-13 worked out from lavas on land — geophysicists showed that the magnetic data could indicate how fast the ocean opened up. We would expect the rocks on the crest of the ridge to be normally magnetized, because they were extruded during the present normal magnetic epoch. Reversely magnetized rocks corresponding to a magnetic epoch of about one million years ago have been displaced

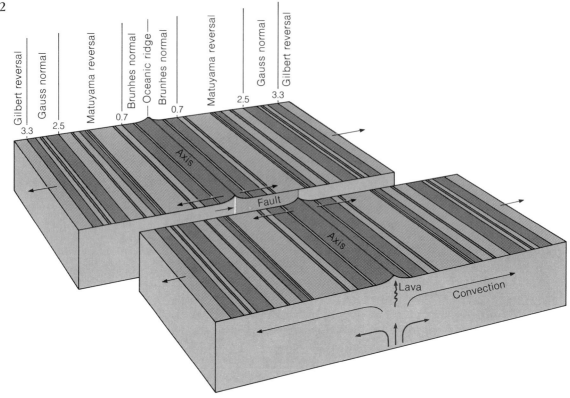

**Figure 20-16**
The succession of normal and reversed magnetic anomalies on the sea floor provides strong support for the concept of sea-floor spreading. Rocks of normal, or present-day, polarity are shown in shades of brown; rocks of reversed polarity are in shades of gray. The symmetry of the magnetic stripes and the correlation with the time scale of reversals worked out from lava flows on land suggests that molten rock upwelling along the ridge axis became magnetized as it cooled, was pushed out on both sides, and gradually moved outward with the separating plates. The separation of the two blocks represents a transform fault. The diagram is based on the studies of many scientists. [After "The Confirmation of Continental Drift" by P. M. Hurley. Copyright © 1968, by Scientific American, Inc. All rights reserved.]

some distance from the ridge—say, about 15 kilometers on each side of the ridge crest if the spreading rate is 1.5 centimeters per year. The reversal time scale can actually be followed through many oscillations of the Earth's field, and the corresponding magnetic bands extend out to many hundreds of kilometers from the ridge crest (Fig. 20-17).

The magnetized strips of sea floor provided a powerful tool to extend the history of magnetic-field reversals back in time some 80 million years into the Cretaceous period. Precise dating of reversals (and, therefore, of spreading rates) goes back only about 5 million years because the radiometric method of dating lava flows on land loses accuracy beyond this time. By simply assuming the

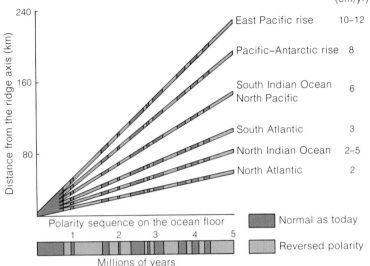

**Figure 20-17**
The rate of ocean spreading from magnetic anomaly patterns. The same magnetic polarity zone may be encountered at various distances from different ridges in different oceans, depending on the rate of spreading. Since the ages of the polarity zones are known from the magnetic-reversal time scale, the spreading rates can be calculated. The spreading rate of the Pacific Ocean along the East Pacific rise is the most rapid, amounting to 10–12 cm/yr. The North Atlantic shows the slowest rate, 2 cm/yr. [After *Continental Drift* by D. and M. Tarling, Doubleday Anchor Books. Copyright © 1971 by G. Bell and Sons, Ltd.]

spreading rate determined for the last 5 million years to be representative of a much longer period, the older normal and reversed magnetic epochs could be correlated over all the oceans and assigned dates. Figure 20-18 shows the sequence of 171 reversals, extending back 76 million years to the Cretaceous period, as they were worked out by this method. The magnetized strips on the sea floor are shown schematically below the observed magnetic anomalies. Note that the spacing of the strips differs in the several oceans because the sequence of reversals can be compressed or stretched out according to slower or faster spreading.

After geologists study a region, they often present the sequence of rock layers extending back in geological time as a stratigraphic section. Magnetic anomalies on the sea floor enable geophysicists to plot a "magnetic-stratigraphic" section, as in Figure 20-19, showing the historical sequence of worldwide magnetic reversals back in geological time to the Cretaceous period.

The power and convenience of the new magnetic stratigraphy in working out the history of ocean basins cannot be overemphasized. Simply by steaming back and forth, measuring the magnetic field of the magnetized rocks of the sea floor, and correlating the

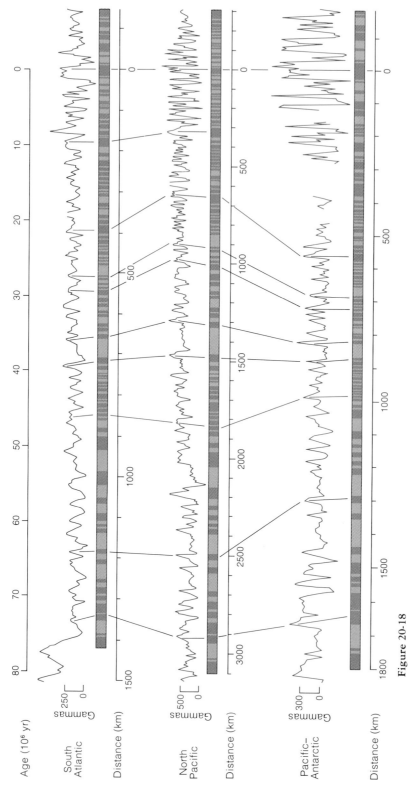

**Figure 20-18**

Magnetic anomalies recorded over the oceans reveal the same succession of magnetized lava formations on the sea floor, depicted by the dark and light stripes. The spacing may differ because the spreading rates vary, but each ocean shows the same sequence of some 171 reversals over the past 76 million years. The illustration shows the correlation of the anomalies from ocean to ocean. [After "Sea Floor Spreading" by J. R. Heirtzler. Copyright © 1968 by Scientific American, Inc. All rights reserved.]

pattern of reversals with the sequence shown in the preceding two figures, ages can be assigned to different regions of the sea floor without even examining rock samples! All one needs is good magnetic records, and these have already been obtained and interpreted for large sections of the world's oceans.

**Isochrons** (contours of the age of the sea floor) obtained in this way are shown in Figure 20-20. These contours show the time that has elapsed and the amount of spreading that has occurred since the magnetized rocks were injected as lava into a mid-ocean rift. Note how the isochrons show progressively older sea floor on both sides of the major ocean-ridge rifts, the more widely spaced isochrons of the east Pacific signifying a faster spreading rate than those of the Atlantic. We will see in the next chapter that these "magnetic" ages were verified when the deep-sea drilling project brought back rock samples from the sea floor that could be dated in the laboratory using fossils and radioactivity methods. What a coup for the scientists who discovered this tool!

**Magnetized Moon Rocks — A Puzzle**  Unlike the Earth, the Moon has no planet-wide magnetic field. There is no question about this. Soviet and American spacecraft have failed in several efforts to detect such a field. Yet magnetized rocks have been found lying on the lunar surface. Discordant data are the stuff of great discoveries, and planetary scientists are vying to explain these seemingly contradictory results. The leading hypothesis at this time proposes that the Moon rocks, in their remanent magnetism, "remember" an earlier period of lunar history some 3–4 billion years ago (the age range of the rocks), when the Moon did have a planetary magnetic field. This implies also the existence, at this early time, of a small liquid iron core that has since cooled and solidified. Is there a better way to manifest the power of modern geological and geophysical methods than to return a rock from the lunar surface, date it, measure its magnetic field, and then describe the physical state at the center of the Moon billions of years earlier?

# EXPLORING THE EARTH WITH GRAVITY

**The Indian Puzzle**  Some 150 years ago, during the great land survey of India, a curious discrepancy was uncovered by the British surveyors. The distance between Kaliana, some 100 kilometers (60 miles) south of the Himalaya range, and Kalianpur, 600 kilometers (375 miles) farther south, was determined in two precise ways—measurement over the surface and by reference to astronomical observations—and the results disagreed by some 150 meters (500 feet) in 600 kilometers. This may seem like a small

[685]

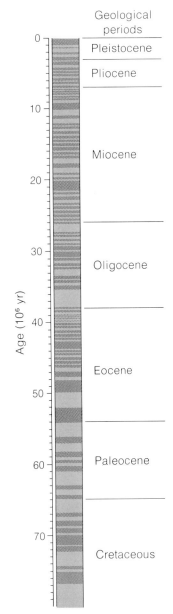

Figure 20-19
The time scale of geomagnetic field reversals from the present back to Cretaceous time, some 76 million years ago. This chronology was worked out using the dates established for recent reversals and the assumption that spreading rates found for the past 5 million years have remained nearly constant since the Cretaceous. [After "Sea Floor Spreading" by J. R. Heirtzler. Copyright © 1968 by Scientific American Inc. All rights reserved.]

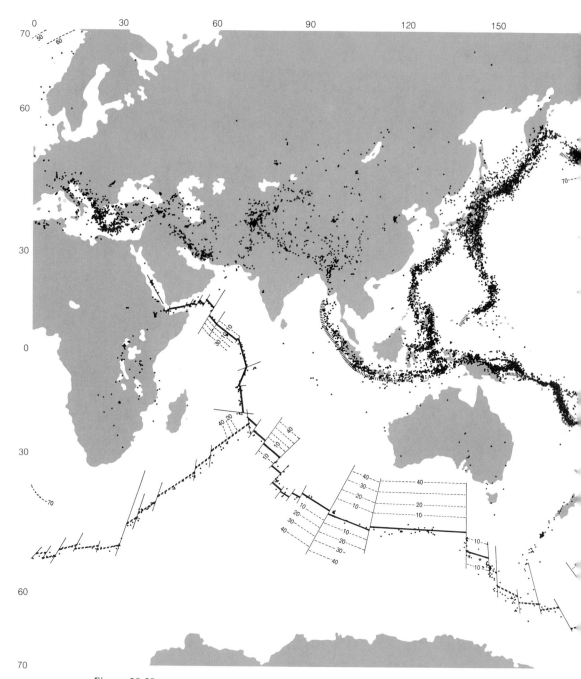

**Figure 20-20**
Worldwide pattern of sea-floor spreading is revealed when magnetic and earthquake data are combined. Mid-ocean ridges (heavy black lines) offset by transform faults (thin lines) show as a worldwide system along which new sea floor is continuously being extruded. The dating of magnetic anomaly stripes makes it possible to establish isochrons (thin, broken lines parallel to ridges) that give the age of the sea floor in millions of years since creation at the ridges. Hatched bands mark

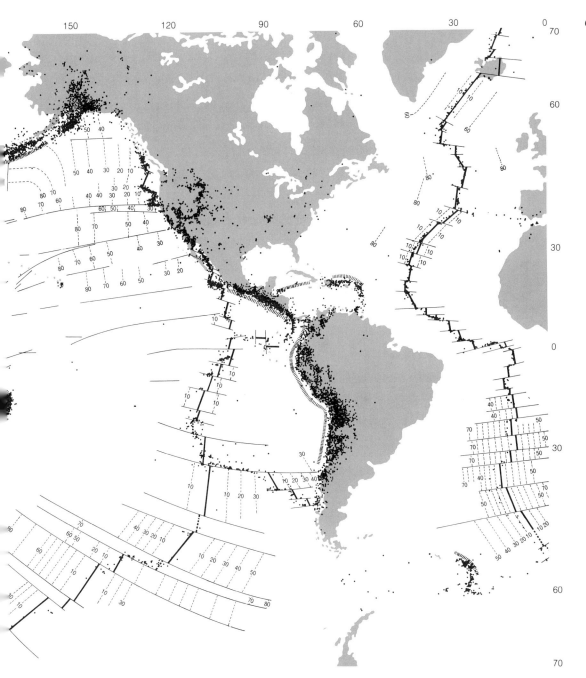

the deep-sea trenches in which sea floor is destroyed by sinking back into the mantle. Earthquakes (black dots) originate mostly at these plate boundaries, where spreading and subduction are underway. The magnetic data imply that the present oceans were created in perhaps 100 to 200 million years. [After "Sea Floor Spreading" by J. R. Heirtzler. Copyright © 1968 by Scientific American, Inc. All rights reserved.]

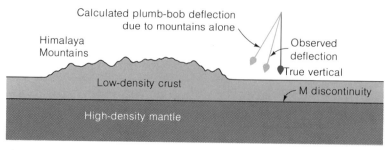

**Figure 20-21**
A plumb line ordinarily occupies a vertical position. Near a mountain system, we would expect the plumb bob to be deflected toward the mountains because of the gravitational attraction of their mass. The observed deflection is typically less than expected, a discrepancy whose explanation led to an important discovery. The diagram exaggerates the amount of deflection, which is small but readily measurable.

Examples of buoyancy

Iceberg

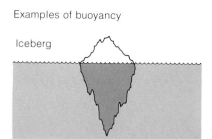

Boat

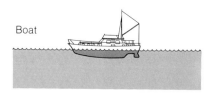

Mountain

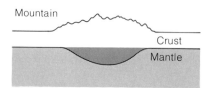

**Figure 20-22**
Examples of buoyancy. Icebergs and ship hulls float because the volume submerged is lighter than the volume of water displaced. Similarly, the volume of relatively light crustal rock projecting into the denser mantle provides a buoyant force that supports the mountain mass above.

amount, but it was an intolerable surveying error even by nineteenth-century standards. The astronomical method of measuring distance uses the angles of stars with respect to the vertical, which is defined by a plumb line (a weight suspended on a string). To account for the difference, it was proposed that the plumb line was tilted towards the Himalayas because of the gravitational attraction of the mountains on the plumb bob, causing an error in the distance measurement. When the calculation was actually made, it was found that the mountain should have introduced an even larger error—one of about 450 meters (1500 feet)—thus compounding the puzzle (Fig. 20-21).

In 1865, no less a figure than the Astronomer Royal, Sir George Airy, came forward with an explanation for this descrepancy that contained the basis of the principle of isostasy (already discussed in Chapter 19). Airy proposed that the enormously heavy mountains are not supported by a strong, rigid crust below, but that they "float" in a "sea" of denser rock. Stated otherwise, the excess mass of the mountains above sea level is **compensated** by a deficiency of mass in an underlying root. This root provides the buoyant support, in the manner of all floating bodies, just as a ship with a deep hull is buoyed up (Fig. 20-22). The plumb bob "feels" both the excess mass on top and the deficiency of mass below, hence the reduced deflection (Fig. 20-23). The resolution of the "Indian puzzle" not only led to the concept of isostasy but also introduced **gravity surveying** as a method for detecting mass variations in the interior by their corresponding gravity variations.

**Gravimeters**   The local value of gravity, *g*, can be obtained from the period* of a swinging pendulum or the acceleration of a falling

---

*Time required for one complete oscillation.

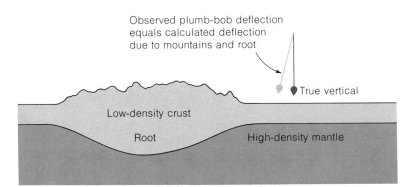

Observed plumb-bob deflection
equals calculated deflection
due to mountains and root

True vertical

Low-density crust

Root        High-density mantle

The Earth's Magnetism
and Gravity

**Figure 20-23**
The discrepancy between the observed and expected deflection of the plumb
bob in Figure 20-21 can be reconciled if the excess mass of the mountain is
compensated by a deficiency of mass in a "light" crustal root below. The
root provides buoyant support for the mountain, which otherwise would sink
into the mantle.

weight—an experiment performed in every elementary physics
course. Pendulums have been used in gravity surveys, but they
have mostly been supplanted by the modern gravimeter. This is
a device no more complex than a weight on a spring that stretches
or contracts as gravity increases or decreases from place to place
(Fig. 20-24). Although the principle is simple, the engineering of
the gravimeter is most elegant: This device, not much larger than
a thermos bottle, can detect gravity variations as small as 0.00001
centimeter per second per second. Thus, because the average
value of gravity is 980 centimeters per second per second, this
instrument can measure variations as small as $10^{-8}g$! The standard
unit of acceleration used in gravity surveys is the milligal, which is
0.001 centimeter per second per second. A modern gravimeter
can easily measure the difference in gravity between a table top
and the floor, even though the table top is only 1 meter farther
from the center of the Earth!

**Gravity Surveying**   The millions of dollars required to develop
the sensitivity achieved by the modern gravimeter was underwrit-
ten by the oil industry. This was the direct result of the recognition,
made some forty years ago, that the buried geologic structures in
which oil is trapped (such as folds, faults, and salt domes) often
produce **gravity anomalies** that are detectable by sensitive instruments. A gravity anomaly is a perturbation in the normal gravity
field caused by a change in subsurface mass due to a mountain
root, a salt dome, or any other lateral geological change. The idea
is to find and describe the mass anomaly from its gravity effect
and, thus, to explore the subsurface geology by making gravity
surveys. An example of a gravity anomaly is given in Figure 20-25.
Before such a profile can be produced, however, three important

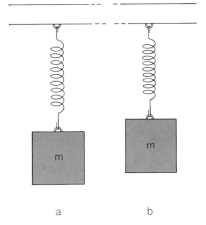

a          b

**Figure 20-24**
The gravimeter is simple in concept but
elegant in implementation. A mass
attached to a spring experiences a
larger or smaller pull as gravity varies.
The corresponding extensions or
compressions of the spring are measured
very precisely, so that small changes
in gravity can be observed.

Gravity anomaly

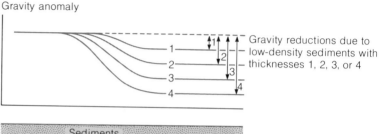

Gravity reductions due to low-density sediments with thicknesses 1, 2, 3, or 4

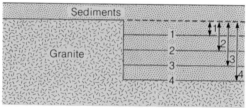

**Figure 20-25**
Schematic illustration of a gravity anomaly. The value of gravity changes across the structure shown because the less-dense sediments contain less mass than an equal volume of granite. The thicker the sedimentary deposit, the greater the decrease in gravity, as the curves show.

corrections must be applied to the value of $g$ from each station at which the gravimeter is read; otherwise, the most interesting anomalies would be obscured.

1. If the Earth were spherical and nonrotating, the gravitational attraction of the planetary body would be the same everywhere on the surface. Because of centrifugal force, however, things tend to fly outward from a rotating body; hence, gravity is less at the equator, or things weigh less there than at the poles. The same centrifugal force makes the Earth bulge outwards at the equator and flatten at the poles. Remember, from Chapter 3, that $g$ is proportional to $1/R^2$, in which $R$ is the distance to the center of the planet. $R$ decreases by 21 kilometers going from the equator to the pole of our flattened or spheroidal planet; hence, $g$ increases with latitude. An International Formula has been adopted that best describes $g$ everywhere on the Earth's surface, taking into account the Earth's shape and rotation.* This formula value is subtracted from the gravimeter reading in the search for anomalies, which could otherwise be masked by the big effects of rotation and flattening.

2. Because of topographic variations, gravity stations generally differ in elevation—that is, in distance $R$ from the center of the Earth. In our search for anomalous subsurface masses, we must

*By international agreement, $g = 978{,}049 \, (1 + 0.0052884 \, \sin^2\phi - 0.0000059 \, \sin^2 2\phi)$ milligals, in which $\phi$ is the latitude. Thus, gravity is about 0.5 percent stronger at the north pole where $\phi = 90°$ and $\sin\phi = 1$, than at the equator, where $\phi = 0°$ and $\sin\phi = 0$.

remove the obvious effect of elevation on the local gravity value (using the fact that $g$ varies as $1/R^2$ to get the correction), and all readings are corrected as if they were made at sea level. This is called a **free-air correction**, and it amounts to adding 0.31 milligals for each meter of elevation above sea level.

3. Finally, in order to highlight subsurface effects, it is important to account for all obvious near-surface masses that affect gravity. The free-air correction reduces the reading to sea-level only partially, because it allows for the distance to the center of the Earth but not for the attraction of the mass of rock between the station and sea level. The correction that completes the subtraction of the local topography is called the **Bouguer correction**. It amounts to subtracting about 0.1 milligals for each meter of rock above sea level. For gravity surveys on the ocean, where $g$ is measured at sea level, the Bouguer correction corrects for the low density and, therefore, the low gravitational attraction of water by increasing the gravimeter reading by an amount necessary to "convert" the ocean to rock. In this way, the obvious gravity deficiency of the ocean is removed in order to emphasize anomalous suboceanic masses. The free-air and Bouguer corrections are depicted diagramatically in Figure 20-26.

If there were no lateral variations in mass in the interior, the sum of all these corrections and the gravimeter reading would be close to zero—there would be no gravity anomaly, because everything would have been taken into account. However, if the sum were to differ from zero, we would have found an anomalous mass —that is, something more or less dense then the average rock: an ore body, an intrusion, a sedimentary basin, or a mountain root. The shape and magnitude of such a gravity anomaly can help us define the dimensions and density of the anomalous rock mass.

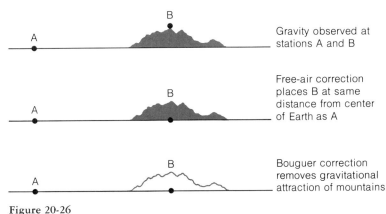

**Figure 20-26**
Gravity is measured at $A$ and $B$ to see if there is a difference in subsurface mass. To emphasize subsurface effects, corrections are made to the value of gravity at $B$, as if to bring $B$ to the same elevation as $A$, and also to remove the obvious gravitational attraction of the mountain. Any remaining gravity difference between $A$ and $B$ is ascribed to a change in subsurface geology.

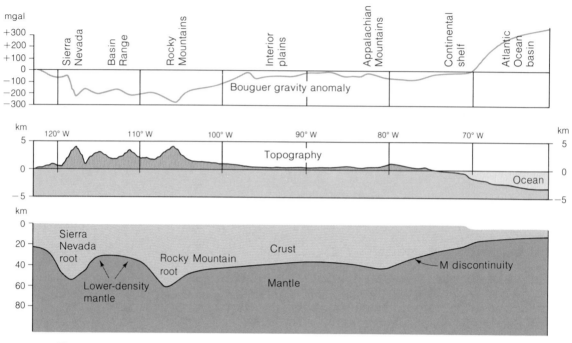

**Figure 20-27**
Transcontinental gravity survey from the Pacific Ocean to the Atlantic Ocean. The negative gravity anomalies over the mountainous regions, the near zero values at low elevations, and the positive values over deep oceans, mirroring the topography, demonstrates the role of isostatic compensation in shaping Earth's surface features.

Imagine driving across the United States, stopping every mile to measure gravity. You would thereby produce a gravimetric profile of the country crossing many geological provinces, and it might be interesting to see how the gravimeter can aid us in interpreting these contrasting features. This has actually been done, and the resultant gravity-anomaly profile is shown in Figure 20-27, which includes an extension into the Atlantic Ocean made with a shipboard gravimeter. The top and middle sections of the figure show the gravity profile and the topography and identify the major geological provinces. All the corrections have been applied to the gravity curve—that is, all the land mass above sea level and all of the *lack* of land mass in the oceans have been accounted for—so this wiggly curve should reflect subsurface mass variations. The bottom section of the figure gives an interpretation of the gravity-anomaly profile in terms of changes in the crust and mantle. Note the negative values—that is, the deficiency in gravity across the Sierra Nevada, the Rocky Mountains, and the Basin and Range province. The anomaly is close to zero in the low plains, becomes slightly negative under the Appalachian Mountains, approaches zero again in the coastal plains and continental shelf, and zooms to large positive values over the deep Atlantic.

The striking feature of the gravity-anomaly interpretation shown in the bottom section of Figure 20-27 is the way the Mohorovičić discontinuity mirrors the topography (except in the Basin and Range province). The gravity anomaly is strongly negative where the crust thickens to provide buoyant support for mountains. What causes the negative anomaly in these topographically high places is the mass deficiency due to the less dense crustal root displacing denser mantle. The high positive gravity values over the ocean basin signify the presence of excess mass: dense mantle rock is much closer to the surface here. This feature has been called an **antiroot**, and it mirrors the "negative" topography (water instead of rock) of the ocean basin. The Appalachians show a modest negative anomaly, which indicates that they have a shallow root. This is appropriate for an old mountain system: its root (and $g$ anomaly) is disappearing as its topography erodes away. One might have expected the structurally high Basin and Range province, with its average elevation of about one kilometer, to have a slightly thickened crust to go along with its negative anomaly. Actually, many geophysicists thought this to be the case until experiments with seismic waves revealed the thin crust. Because the seismic information reveals that there is no crustal root for the Basin and Range province, it is necessary to assume that the mass deficiency there is a result of the area being underlain by mantle material of relatively low density, as shown in the figure.

Many of these results were anticipated in Chapter 19, in which we discussed isostasy in the context of the structure of the crust and upper mantle as determined by seismic methods. Actually, the concept of isostatic compensation with its notions of "floating" continents and still higher floating mountains was discovered from gravity observations such as these. However, the seismological data mentioned in Chapter 19 contributed to our understanding by clearing up such questions as where the mass deficiencies were located and whether compensation involving crustal roots or compensation via low-density mantle was the isostatic mechanism responsible for them. Low-density mantle seems to go with a tectonic setting that includes recent volcanism, high heat flow, and low seismic velocities—which implies, perhaps, a partially molten mantle directly below the Moho. Some geologists suggest that these features, as they occur in the Basin and Range province, indicate that tension-producing forces, perhaps due to a spreading or divergence zone, are active within a continent. Compensation involving crustal roots is the predominant isostatic mechanism for continents as a whole, as well as for high mountains.

In this way, gravimetry and seismology combine to reveal the importance of isostasy to the study of geology on a regional and continental scale. The kinds of insights that emerge are given by the following examples: Mountains eroding down will be pushed up by the excess buoyancy of the root, until both root and mountain range nearly disappear. Such loads as ice caps or sediments

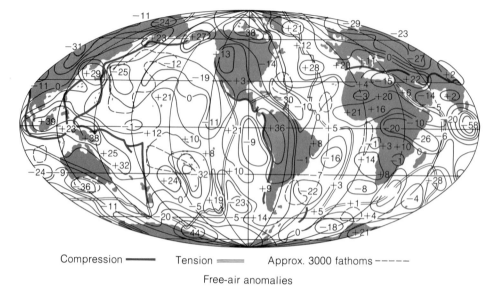

Compression ——— Tension ═══ Approx. 3000 fathoms ─ ─ ─ ─

Free-air anomalies

**Figure 20-28**
Worldwide pattern of gravity anomalies obtained by observing the fluctuations in the orbits of
satellites circling the Earth. Geophysicists are still pondering the meaning of these variations.
Heavy black line represents boundaries of colliding plates; double line indicates spreading zones.
[After W. M. Kaula, *Science,* vol. 169, pp. 982–985, Sept. 4, 1970. Copyright 1970 by the
American Association for the Advancement of Science.]

filling a basin can depress the crust. Where gravity studies indicate
that a positive or negative load (a depression, for example) is not
compensated, some force must be present that helps support it
and keeps it from subsiding or rising. Continents are unlikely
to be destroyed by subduction; their ability to float allows them
to "ride out" repeated episodes of splitting and collision.

Variations in gravity perturb the orbits of satellites as they circle
the Earth. Because these spacecraft can be tracked with high precision, the orbital fluctuations can be analyzed to determine the
gravity anomalies that cause them. A satellite gravity map of the
Earth developed in 1971 is shown in Figure 20-28. Broad gravity
anomalies are revealed that imply broad mass variation within the
mantle. Geophysicists are still pondering the meaning of these
new data. Do they represent differences in density due to convection in the mantle, or are they indicative of the mantle's strength
and its ability to support mass differences nonisostatically deep
within itself?

**The Fennoscandian Uplift — Nature's Experiment with Isos-
tacy** If you depress a cork floating in water with your finger,
and then release it, it pops up almost instantaneously. A cork
floating in molasses would rise more slowly; the drag of the viscous
fluid would slow down the process. If we could perform a similar
experiment on the Earth, much could be learned about the viscosity of the mantle and how it affects rates of uplift and subsidence. How convenient it would be if we could push the crust

down somewhere, remove the force, and then sit back and watch it rise.

Nature has been good enough to do this experiment for us. The load is a continental glacier—an ice sheet 2–3 kilometers thick that can appear, with the onset of an ice age, in the geologically short period of a few thousand years. The crust is depressed by the ice load, and a downward bulge develops on its underside to provide buoyant support. With the onset of a warming trend, the glacier melts rapidly. With the removal of the load, uplift of the depressed crust begins. The rate of uplift can be documented by dating ancient beaches that are now well above sea level (see Fig. 2-4). Such raised beaches can tell us how long ago a particular stretch of land was at sea level. The presence of a negative gravity anomaly can tell us how much of a relic root remains and how much more uplift will occur before the root disappears.

Such depression and uplift has occurred in Norway, Sweden, and Finland, as well as elsewhere in glaciated regions. The ice cap disappeared some 10,000 years ago and the land has been rising since. Figure 20-29 shows, for example, how much upwarping

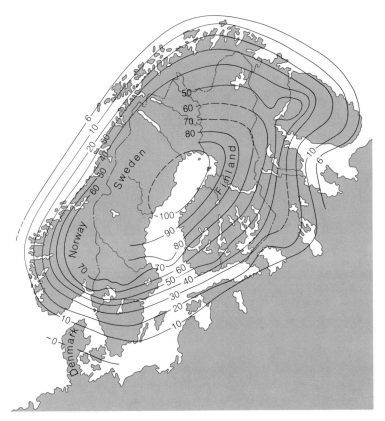

Figure 20-29
Map showing the amount of uplift in meters in Fennoscandia in the past 5,000 years, according to M. Sauramo. The crust, depressed by the weight of the ice cap of the last "ice age," is still rebounding some 10,000 years after the ice disappeared.

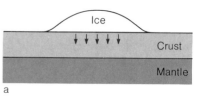

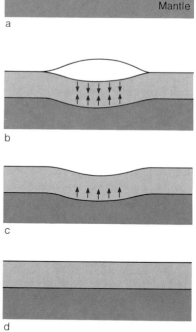

**Figure 20-30**
Schematic diagram of the mechanism of postglacial uplift. (a) Continental glacier grows, loading the crust. (b) Crust sags, and root develops to support ice load isostatically. (c) Glacier disappears, but root remains because of viscosity of mantle. Root evidenced by negative gravity anomaly. (d) Buoyancy of root leads to slow uplift. Root disappears, surface assumes original level, and gravity anomaly disappears. Arrows depict direction of forces due to ice load and root. Fennoscandia today is between stages c and d. Figure not to scale. Crust is about 40 km thick; a 3-km thick glacier would produce a root about 1 km thick—less if the strength of the crust supported part of the weight.

took place in the last 5000 years. The most intense upwarping has occurred near west-central Sweden, which is believed to have been overlain by the thickest ice. Some 200 meters of uplift has occurred in 10,000 years, indicating an average rate of 2 centimeters per year. There is a remaining negative gravity anomaly of −50 milligals, which implies that part of the root still remains and that about 200 meters more uplift must occur before isostatic compensation is complete. The region shows minor seismicity —additional evidence, perhaps, that stresses due to too much buoyancy are still present. Geophysicists have used these data to show that the weak zone (which "flows" so that the crustal root could develop) coincides with the asthenosphere, the partially molten zone discussed in earlier chapters. This region is not only an important factor in the mobility of the lithospheric plates and as the source of basaltic magma; its ability to "yield" makes it the key factor of isostatic compensation.

Man could not have designed a better experiment to demonstrate the isostatic mechanism than nature's demonstration with the Fennoscandian postglacial uplift.

1. Motions in the fluid iron core somehow set up a dynamo action that generates the Earth's magnetic field. The field can be fairly well described by an hypothetical bar magnet located near the center of the Earth and approximately aligned with its axis of rotation.

2. Many rocks become magnetized in the direction of the geomagnetic field that prevailed when they were formed. If the rocks are dated radiometrically, the history of the magnetic field can be recovered from this remanent magnetization.

3. The remanent magnetization of old rocks shows that the Earth's magnetic pole occupied different positions in the past. Actually, the magnetic pole probably did not wander, but stayed fairly close to the geographic pole. Thus, polar wandering is probably an indication that the lithospheric plates have been moving, changing the geography of the surface. Paleoclimatic evidence, such as coal in Antarctica and glacial till near the equator, support this idea.

4. Remanent magnetization has also led to the discovery of reversals in the magnetic field, and the history of reversals since Cretaceous time has been worked out. Although unexplained, these reversals have become a very important tool in dating the sea floor. When the sea-floor crust is formed at mid-ocean rifts, it becomes magnetized. This magnetic imprinting stays with the crust as it spreads away from the rift. The sequence of reversals shows as positive and negative magnetic anomalies, which a surveying ship can readily detect. Using the history of reversals, we can determine the age of the underlying sea floor and the rate of sea-floor spreading.

5. Gravity changes over the surface of the Earth are due to the planet's oblateness, its rotation, its topography, and differences in its subsurface mass. The first three factors can be allowed for, so that the remaining gravitational anomalies indicate subsurface geological differences.

6. Gravity anomalies associated with continents, oceans, and mountains show that the Earth's crust is not strong enough to support topographic loads over long periods. Roots, or downward bulges of the crust, develop and provide buoyant support. This is an example of isostatic adjustment. Another example is the depression of the Fennoscandian crust by the weight of a continental glacier. Although the ice cap disappeared some 10,000 years ago, uplift is still continuing in that area and will continue until the relic root disappears.

## EXERCISES

1. What evidence supports the hypothesis that the Earth's magnetic field originates in a fluid iron core?

2. Describe two mechanisms whereby rocks can take on remanent magnetism. Some meteorites show remanent magnetism, which has yet to be explained. Would you hazard a guess as to its origin?

3. You are an astronaut exploring another planet. You find rocks with remanent magnetism, yet there is no planetary magnetic field. What would you conclude? (Such rocks have been found on the Moon.)

4. What is the connection between the sequences of magnetic reversals worked out on land and the bands of positive and negative magnetic anomalies found on the sea floor?

5. Would the gravity anomaly (with Bouguer correction) show large negative values, values near zero, or large positive values at each of the following places?
   a) Rocky Mountains
   b) east coast of U.S.
   c) middle of an ocean basin
   Explain your answers.

## BIBLIOGRAPHY

Garland, G. D., *The Earth's Shape and Gravity*. New York: Pergamon Press, 1965.

Heirtzler, J. R., "Sea-Floor Spreading," *Scientific American*, vol. 219, pp. 60–70, December 1968. (Offprint No. 875.)

Strangway, D. W., *History of the Earth's Magnetic Field*. New York: McGraw Hill Book Company, 1970.

Takeuchi, S., S. Uyeda, and H. Kanamori, *Debate about the Earth* (revised ed.). San Francisco: Freeman, Cooper and Company, 1970.

# 21

# Global Plate Tectonics—The Unifying Model

*Geologists are gradually rejecting the notion of a rigid Earth with fixed continents and ocean basins. Most now believe that the Earth's lithosphere is broken into about a dozen plates, which for reasons not fully understood move over the interior. Plates are created along the crest of mid-ocean ridges and are pushed down into the mantle near deep-sea trenches. Continents, embedded in the lithosphere, drift along with the moving plates. Plate tectonics gives new life to the old ideas of continental drift and explains the distribution of many large-scale geological features and zones of activity—narrow belts of mountains, volcanic and seismic activity—in terms of their association with plate margins.*

Not long ago, on the occasion of an international geophysical meeting in Moscow, an interesting exchange occurred between two western participants—a younger man who had achieved prominence because of his work on plate tectonics and a well-known older scientist. The setting was a party in the apartment of a Soviet geophysicist, and the conversation was well lubricated by vodka. The din of cocktail party chatter stopped suddenly when the younger man called out to his older colleague, "Dr. _____, everyone tells me how brilliant you were in your younger days. If that's the case why didn't you discover sea-floor spreading and plate tectonics twenty years ago?" The explosive response of the older man needn't be recorded, but the question, properly generalized, is indeed thought provoking. Why did this concept, which unifies so much of geological thought, "arrive" so late in the history of the subject?

Actually, a key element of the concept—large-scale displacement of continents—had been around for a long time. The jigsaw-puzzle fit of the coasts on both sides of the Atlantic did not escape the attention of early natural philosophers. Francis Bacon remarked on the parallelism of the facing shores of the Atlantic in 1620. Antonio Snider published maps in France depicting continental drift as early as 1858. At the close of the nineteenth century, the Austrian geologist *Eduard Suess* put some of the pieces of the puzzle together and postulated the former existence of a single giant continent—Gondwanaland, made up of the combined present-day southern continents. Early in this century, *Alfred Wegener*, a German meteorologist, cited as further evidence of continental drift the remarkable similarity of rocks, geological structures, and fossils on opposite sides of the Atlantic. In the years following, Wegener continued to build up the case for continental drift, wherein he postulated that a supercontinent called Pangaea, once made up of all of the present continents, began to break up some 200 million years ago, with ocean filling the widening gaps. Figure 21-1 shows what Pangaea may have looked like and how it may have fragmented.

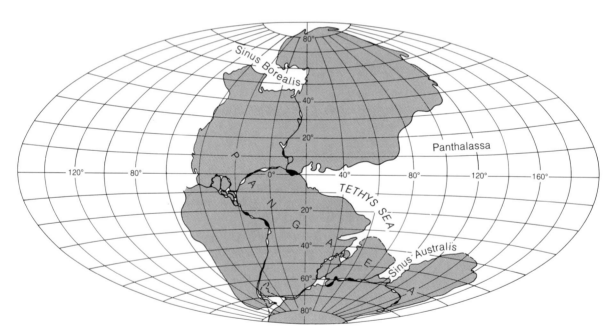

**Figure 21-1**
The ancient landmass Pangaea may have looked like this some 200 million years ago. Panthalassa evolved into the present Pacific Ocean, and the present Mediterranean Sea is a remnant of the Tethys. [After "The Breakup of Pangaea" by R. S. Dietz and J. C. Holden. Copyright © 1970 by Scientific American, Inc. All rights reserved.]

Although the theory received serious attention for about a decade, aside from a few vocal geologists in Europe and South Africa, "continental drift" never caught on. The proponents could not come up with a plausible driving force. More important, drift advocates buttressed their speculation with special pleading, selecting evidence patently favorable to their views, evidence that was far from incontrovertible. Aside from the geometrical matching of continents, their main arguments were based on the evolution of vertebrates and land plants, which showed similarities in development on different continents up to the supposed breakup time, after which they showed divergent evolutionary paths. "It has always happened that after several distinguished palaeontologists have presented evidence favourable to continental drift, some other equally distinguished ones have proceeded to point out other facts that are made more difficult to explain"—so argued *Sir Harold Jeffreys* in his influential book *The Earth*. Independent, diverse, corroborative evidence would be needed before the scientific establishment would abandon prevailing ideas and elevate an unorthodox speculation to the level of a generally accepted theory.

In 1928, *Arthur Holmes,* a widely respected British geologist, wrote an article invoking the mechanism of thermal convection in the mantle as the driving force. Holmes proposed that subcrustal convection currents "dragged the two halves of the original continent apart, with consequent mountain building in the front where the currents are descending, and ocean floor development on the site of the gap, where the currents are ascending." Holmes came close to expressing the modern notions of plates, divergence, and subduction when he speculated that a subcrustal basaltic layer serves as a conveyor belt that carries a continent along to the place where the belt turns downward into the mantle, leaving the continent resting on top. Figure 21-2, which depicts his concept, contains many of the ingredients of the theory of sea-floor spreading as we know it today. Nevertheless, Holmes called attention to the tenuous nature of his views when he wrote that "purely speculative ideas of this kind, specially invented to match the requirements, can have no scientific value until they acquire support from independent evidence."

Convincing evidence began to emerge as a result of extensive exploration of the sea floor during the years following World War II. In particular, the mapping of the mid-Atlantic Ridge and the discovery of the deep, crack-like valley or rift running down its center line sparked much speculation. In the early 1960's Harry Hess of Princeton University suggested that sea floors separate along the rifts in mid-ocean ridges, and that new sea floor forms by upwelling of mantle materials in these cracks, followed by lateral spreading (see Figs. 15-5, 17-44). The work of Vine and Mathews, mentioned in the previous chapter, showed how the

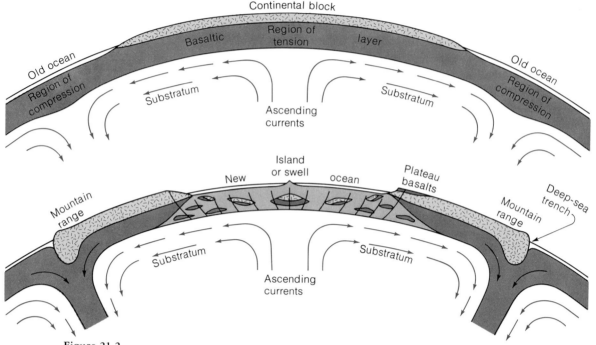

**Figure 21-2**
This early convection model, which remained unknown to many geologists for 30 years, was proposed by Arthur Holmes around 1930. It shows a continent being pulled apart by rising mantle currents, with new ocean developing from the growing rift. In the vicinity of a descending current, a mountain range and bordering deep-sea trench develop. Holmes's theory, which has been called "sea-floor stretching," is a forerunner of the modern theory of "sea-floor spreading." The figure is modified from an illustration published by Holmes in 1929 in the *Transactions of the Glasgow Geological Society.*

oceanic magnetic patterns could be explained by Hess's concept. Thus was born the theory of sea-floor spreading. Within a few years abundant confirmation would be available from the study of such diverse evidence as that provided by worldwide magnetic-anomaly surveys, the observation of earthquake mechanisms, the measurement of heat flow, and the determination of the thickness and age of the sedimentary layers of the sea floor.

It remained for a younger generation of geophysicists to broaden the concepts of continental drift and sea-floor spreading into the more general theory of plate tectonics. Beginning about 1967, they extended the ideas of Hess and Canadian geophysicist J. T. Wilson about the mobility of the lithosphere by identifying the separate lithospheric plates and discussing the geometry of their relative motions and the phenomena that occur at their boundaries. By the end of the 1960's the evidence became so persuasive that most Earth scientists, except for a few prominent holdouts, embraced these concepts. Textbooks were revised, and specialists began to think of the implications that the new discoveries held for their own fields.

Let us return to the question raised earlier about why these new concepts became generally accepted so late in the history of geology. There are different styles among scientists. Some scientists —those with particularly inquiring, uninhibited, and synthesizing minds—perceive great truths in advance of others. Although their perceptions may frequently turn out to be false, these failures usually go unrecorded. Most scientists, however, proceed more cautiously and wait out the slow process of gathering supporting evidence. The concepts of continental drift and sea-floor spreading were slow to be accepted simply because the audacious ideas came so far ahead of the firm evidence. The oceans had to be explored, a new world-wide network of seismographs had to be installed and used, the magnetic stratigraphy had to be painstakingly worked out, and the deep sea had to be drilled before the majority could be convinced. In a well-known European laboratory, a list is being assembled (in good humor) to record the names of Earth scientists in the order of the date of their acceptance of sea-floor spreading as a confirmed phenomenon. It is interesting that the names of scientists of distinction appear at both the top and the bottom of the list.

## PLATE TECTONICS—A REVIEW AND SUMMARY

Plate tectonics forms the conceptual framework of this book, and we have already introduced the basic ideas in earlier chapters. In this chapter we draw together and review the diverse lines of evidence that support the theory of plate tectonics, using illustrations primarily rather than words for material repeated from earlier chapters. The geometry of plate motions is discussed, as well as the geological phenomena associated with plate boundaries. We will begin a discussion of rock associations and orogeny within the framework of plate tectonics. The fragmentation of Pangaea since the Jurassic will be reviewed together with some speculation about continental drift and extinct plates in the pre-Jurassic. The chapter will close with some brief remarks on the driving mechanism of plate tectonics, but the reader shouldn't expect more than vague speculations, for the subject is just beginning to receive serious study.

**The Mosaic of Plates**    According to the theory of plate tectonics, the lithosphere is broken into a number of moderately rigid plates whose outlines are shown in Figure 1-17. The plates move continually, and their relative directions of motion are shown in the figure. According to the relative motions of adjacent plates, we can define three kinds of plate boundaries, or marginal zones: (1) zones of divergence or spreading, typically ocean ridges; (2) fracture zones, or transform faults; and (3) zones of convergence (Fig. 19-15).

Zones of divergence are boundaries along which plates separate; in the process of plate separation, partially molten mantle material upwells along linear ocean ridges, and new lithosphere is created along the trailing edges of the diverging plates. Such boundaries are characterized by active basaltic volcanism, shallow-focus earthquakes caused by tensile (stretching) stresses, and high rates of heat flow. The outpouring of magma along ocean ridges and the building of the oceanic lithosphere is volumetrically the most significant form of volcanism. Figures 1-18, 1-22, 15-11, 15-14, 17-23, and 17-44 emphasize the different aspects of divergence zones.

Transform faults are boundaries along which plates slide past one another, with neither creation nor destruction of lithosphere. Occasionally marked by scarps, transform faults are characterized by shallow-focus earthquakes with horizontal slips. Examples were given in Figures 1-22, 19-15, and 20-16.

Zones of convergence are boundaries along which the leading edge of one plate overrides another, the overridden plate being subducted, or thrust into the mantle, where lithosphere is resorbed. The thrusting mechanisms that operate along these collision boundaries tend to produce deep-sea trenches, shallow- and deep-focus earthquakes, adjacent mountain ranges of folded rocks, and both basaltic and andesitic volcanism. Convergence zones were illustrated in earlier chapters (Figs. 1-21, 15-11, 15-14, and 17-44).

Each plate is bounded by some combination of these three kinds of zones, as can be seen in Figure 1-17. For example, the Nazca Plate in the Pacific is bounded on three sides by zones of divergence, along which new lithosphere forms, and on one side by the Peru-Chile trench, where lithosphere is consumed. Continental margins may or may not coincide with plate boundaries. If they do, the continents tend to remain "floating" because of their lower density and are not readily subducted. Where two plates with continents at their leading edges converge, the crust thickens to form great mountain ranges like the Himalayas.

The global sum of plate creation and consumption is approximately zero. The Earth would otherwise change size in order to accommodate the new sea floor, and this doesn't seem to be happening; instead, the plates form and disappear and change in size and shape as they move.

**The Structure of Plates**  Figure 21-3 depicts some of the structural details of a rigid lithospheric slab, or what we call a plate, from its region of generation at a ridge axis to its region of subduction, where it is resorbed. Both oceanic and continental crust cap the plate; the continent, embedded in the moving plates, is carried along passively by it. Thus in a real sense, continental drift is simply a consequence of plate movements. Underneath is the plastic, partially molten asthenosphere—source of the raw materials that build new lithosphere. Once heated and partially

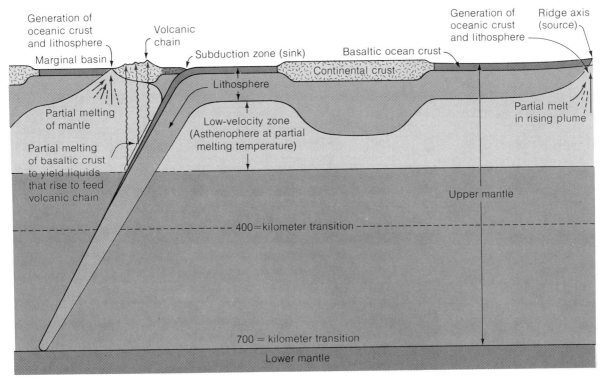

**Figure 21-3**
Cross section of the upper mantle. The lithosphere is a plate of solidified rock that "rides" on the partially molten asthenosphere. It is approximately 70 km thick under oceans and perhaps 100 to 150 km thick under continents. The continent is embedded in the plate and moves along with it. The lithosphere forms at mid-ocean ridges from a rising plume of partially molten rock; it sinks back into the mantle in subduction zones, where it remelts. The marginal basin, a small ocean like the Sea of Japan, is situated behind an island-arc volcanic chain, and separates the chain from a continent. The olivine-spinel transition at 370 to 400 km and the deeper transition at about 700 km mark regions where mantle material becomes more dense due to a closer packing of atoms. [After "Plate Tectonics" by J. F. Dewey. Copyright © 1972 by Scientific American. All rights reserved.]

melted, subducted lithosphere becomes a source of magma, which rises to feed the overlying volcanic chain. A generalized heat-flow profile (see Fig. 15-11) shows a large amount of heat coming out along the ridge axis, a lesser amount from the older, cooled slab, and higher heat flow from the volcanic chain of the subduction zone and the marginal basin behind it.

**Rates of Plate Motion** The velocities of moving plates are measured by dating ocean-floor magnetic anomalies (using the time scale of magnetic stratigraphy) and dividing the age of each anomaly into the distance between it and the ridge axis. The procedure was outlined graphically in the preceding chapter (Figs. 20-16, 20-17, and 20-20).

The worldwide pattern of sea-floor spreading is being worked out by using a combination of magnetic, seismic, and bathymetric

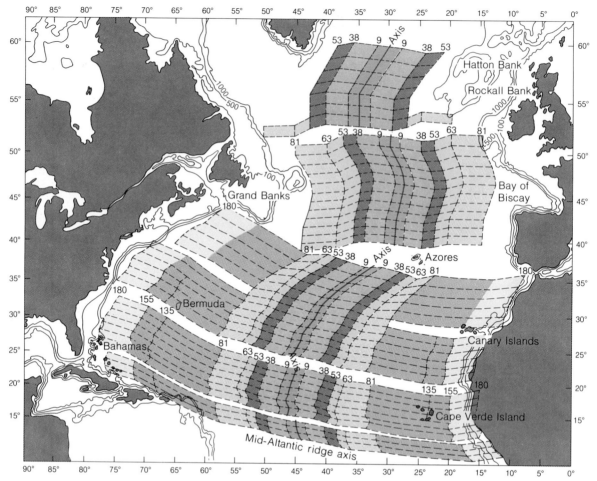

**Figure 21-4**
Isochron chart of the North Atlantic Ocean. Sea floor is divided into age zones by the isochron lines, which indicate the time elapsed, in millions of years, since the underlying rock was formed at the mid-Atlantic ridge. [After W. C. Pitman III and M. Talwani, *Bull. Geol. Soc. America*, v. 83, p. 619, 1972.]

data. The charts used earlier (Figs. 1-17 and 20-20) map the world's zones of spreading, subduction, and fracture; their geographic locations were obtained from the positions of ocean ridges, deep-sea trenches, earthquake epicenters, and other indications of activity. On the basis of spreading rates determined from magnetic data, isochrons (contours that connect points of the same age) were drawn to show the age of the sea floor in millions of years. The distance from a ridge axis to a 50-million-year isochron, for example, indicates the extent of new ocean floor created in the period of time represented by the isochron. In Figure 20-20, note how the isochrons are closely spaced in the Atlantic and widely spaced in the Pacific, where the spreading rate is higher. The fracture zones offset the isochrons, so that upon crossing one of these great faults there is an abrupt change in the age of the sea floor. This chart was prepared in 1968. To show how rapidly this field of study has developed, we include an isochron chart of the Atlantic made four years later (Fig. 21-4). The age of the Atlantic sea floor, and therefore the detailed spreading history of this ocean, has been recovered almost completely.

It was a great triumph for the magneticians, who worked out spreading rates, when the first results of the deep-sea drilling project were announced. This joint project of the major oceanographic institutions and the National Science Foundation had as its primary goal to drill through the sediments of the sea floor at many places in the world's oceans. By studying the sedimentary cores, it is possible that the history of the ocean basin can be worked out directly, in contrast to the indirect methods of magnetic anomalies. Since sedimentation begins as soon as an ocean exists, the age of the oldest sediment in the core, adjacent to the basaltic bedrock, dates the ocean floor at that spot. The age is obtained from the fossils found in the cores. Sediments older than about 150 million years have not been found, attesting to the "youth" of the sea floor. The sediments become older with increasing distance from mid-ocean ridges, confirming the prediction of the sea-floor-spreading hypothesis. Figure 21-5 is a plot of the ages determined from drill cores from the Atlantic and Pacific Oceans against ages predicted from the magnetic data. It is remarkable how close the experimental points approach the straight line, with slope of 1, which is theoretical and represents perfect agreement. In our opinion, one that is generally but not universally shared, this agreement clinches the concept of magnetic stratigraphy and the hypothesis of sea-floor spreading.

As an interesting aside, we have included a photograph of the drilling vessel Glomar Challenger (Fig. 21-6). It is 400 feet long, and amidships it carries a drilling derrick 140 feet high. The only ship of its kind in the world, it has the capability of lowering drill pipe several kilometers to the sea floor and of drilling thousands of meters into the sediments and underlying volcanic rock. Before the ship could accomplish such a feat, a technological breakthrough had to be made. A means had to be devised to hold the ship stationary, regardless of current, wind, or waves, during drilling. Otherwise, the drill pipe would break off. The problem was solved by developing a positioning device that uses sound waves from acoustic beacons planted on the sea floor. Any change in the ship's position is sensed by a computer that monitors the time of arrival of the sound pulses. The same computer controls bow and stern side thrusters and the ship's main propulsion to keep the vessel on station. The Glomar Challenger was the answer to those who said when lunar exploration started, "It's better to explore the ocean's bottom than the backside of the Moon." We ended up doing both.

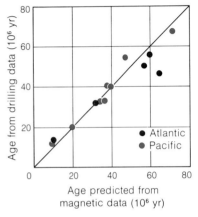

Figure 21-5
A comparison of ages of igneous rocks at different distances from mid-ocean ridges with ages obtained from fossils in the sediments immediately above the igneous rock. The igneous rocks were dated from their magnetic anomaly pattern. The sediments were recovered by deep-sea drilling operations. The 45° line is a theoretical one implying perfect agreement between these two methods of dating the sea floor. The substantiation of the "magnetic" ages by deep-sea drilling, shown by the close fit of the experimental points to the theoretical line, lends strong support to the concept of sea-floor spreading. [After C. L. Drake.]

**Figure 21-6**
The deep-sea drilling vessel Glomar Challenger, a unique facility capable of
recovering cores of sediment and underlying igneous rock from the floor of the
deepest oceans. The deep-sea drilling program was originally an American one;
it is now jointly supported and operated by the United States, the U.S.S.R. and
other countries. [From National Science Foundation.]

**Geometry of Plate Motion**  If the individual plates behave as
rigid bodies, which seems to be a reasonable first assumption,
several interesting and useful geometric consequences follow. By
"rigid" we simply mean that the distance between three points on
the same plate—say, New York, Miami, and Bermuda—do not
change, no matter how the plate moves. But the distance between
New York and Lisbon will of course increase because the two
cities are on different plates, which are being separated along a
narrow zone of spreading on the mid-Atlantic ridge. Listed here
are some geometric principles, mostly self-evident, that govern
the sliding of plates on a plane:

1. Along transform faults, surface area is conserved; that is,
it is neither created nor destroyed. Stated simply, no overlap,
buckling, or separation occurs at such boundaries; the two plates
merely slide past one another. Look for a transform fault if you
want to deduce the direction of plate motions, because the orien-

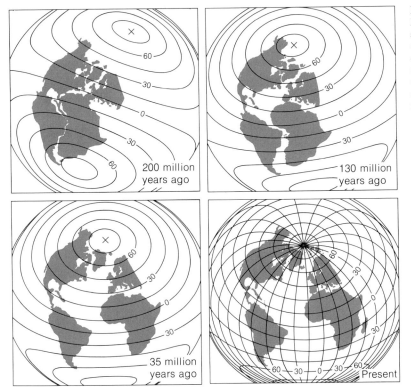

× = Ancient geographic pole

**Figure 21-7**
Magnetic and deep-sea drilling data are used to chart the northward drift of the continents and the opening of the Atlantic Ocean over the last 200 million years. The Central Atlantic, the Caribbean, and the Gulf of Mexico began to form about 200 million years ago in Triassic time when Africa and South America drifted away from North America. The South Atlantic opened about 150 million years ago with the separation of South America from Africa. As the continents drifted apart, so did they migrate in a northerly direction to their present positions. Note that the equator passed through the southern parts of the United States and Europe in Triassic time. [From J. D. Phillips and D. Forsyth, *Bull. Geol. Soc. America*, v. 83, p. 1579, 1972.]

tation of the fault *is* the direction of relative sliding of two plates, as Figures 1-22 and 19-15 show. Surface area is obviously not conserved at zones of convergence or divergence. The plates can move perpendicularly or obliquely to the trend of convergence boundaries, which are therefore not as reliable indicators of directions of movement as transform faults or divergence zones.

2. Magnetic anomaly stripes and isochrons are roughly parallel and symmetrical with respect to the ridge axis along which they were "created." Look at Figure 20-16 to see why this must be so. Since each magnetic strip or isochron marks the edge of an earlier plate margin, isochrons that are of the same age, but on opposite sides of an ocean ridge, can be brought together to show the positions of the plates and the configuration of the continents as they were in that earlier time. By this means we can reconstruct, for example, the opening of the Atlantic Ocean, as shown in Figure 21-7.

3. The point at which three plates meet is called a **triple junction**. Figure 21-8 shows an example of a point at which a spreading zone, a subduction zone, and a transform fault meet. If the relative motion between two pairs of plates is known, we can solve for the third by using a simple equation (see Box 21-1).

**Box 21-1    Solving for the Relative Motions of Plates**

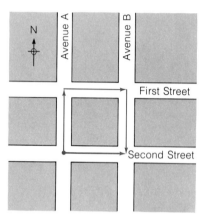

Velocity is a **vector** quantity, meaning that it has both direction and magnitude. A simple example of how vectors are added is shown in the upper figure. If a man walks single blocks north on Avenue A, east on 1st Street, south on Avenue B, he ends up at a place he could have reached directly by walking one block east on 2nd Street. In vector addition this direct route is equal to the sum of the segments that make up the long route.

In the lower figure (part a), let the velocities of the three plates meeting at a triple junction be $V_1, V_2, V_3$. The velocity of plate 2 relative to plate 1 is $V_2 - V_1$; $V_2 - V_3$ and $V_3 - V_1$ are the other possible relative plate velocities. It is clear that the sum of the relative velocities of the plates, taken in order around the triple junction, must be zero, since $(V_2 - V_1) + (V_3 - V_2) + (V_1 - V_3) = 0$, or $(V_2 - V_1) = (V_3 - V_1) + (V_2 - V_3)$. In the lower figure

(part a), the transform fault and the spreading zone show their usefulness in prescribing the directions of relative motions between plates 2 and 3 and 3 and 1, and presumably magnetic stripes give the velocity of motion. This enables us to solve for $V_2 - V_1$, the direction and amount of relative motion across the subduction zone. Part b of the lower figure shows the solution for $V_2 - V_1$ in terms of vector addition.

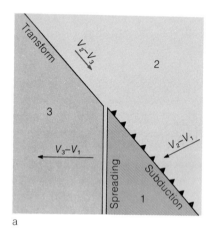

$$(V_2-V_1) = (V_3-V_1) + (V_2-V_3)$$

b

An example of an actual triple junction is the point where the Pacific, Cocos, and Nazca plates meet (Fig. 1-17). Three spreading zones meet at this junction, as shown in the enlarged part of Figure 21-9. The unknown motion that was found by vector addition was that between the Nazca and Pacific plates, the motions between the Pacific-Cocos and Cocos-Nazca plates having been worked out from transform faults and magnetic anomaly stripes. The arrows show the resultant plate movements. Note also how the isochrons bend to become parallel to the spreading centers, where they originated, and how they are offset by the transform

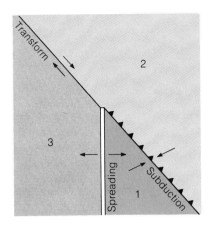

**Figure 21-8**
Illustration of a triple junction. Plates 1, 2, and 3 meet at the intersection of a spreading zone, a subduction zone, and a transform fault. An example would be the intersection of the Pacific plate, the North American plate, and the Cocos plate below the mouth of the Gulf of California. Three spreading zones intersect to form a triple junction where the Pacific plate, the Nazca plate, and the Cocos plate join (see Fig. 1-17 and 21-9). The arrows depict relative motion between adjacent plates.

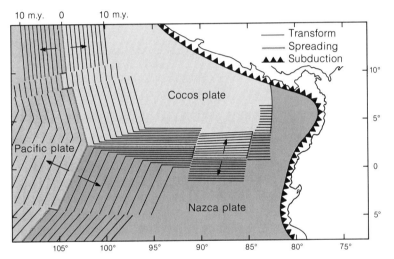

**Figure 21-9**
Triple junction formed by the intersection of three spreading zones in the southeast Pacific Ocean. The schematic isochrons are parallel to the ridge axes from which they migrate as the sea floor ages and spreads. The spacing between isochrons is a measure of the spreading velocities. [After R. N. Hey, K. S. Deffeyes, G. L. Johnson, and A. Lowrie, "The Galapagos Triple Junction," *Nature,* v. 237, p. 20, 1972.]

faults. The spacing of the isochrons reflects the spreading rates, which are largest for the Pacific-Nazca plates and least for the Cocos-Nazca plates.

Up to this point we have considered plates sliding on a plane. Although much can be learned about plate motions by making this simplification, plates do actually move on the Earth's spherical surface. A theorem of spherical geometry worked out two hundred years ago by Euler has been resurrected by geophysicists to handle the real situation. Box 21-2 explains how plate movements on a sphere can be described.

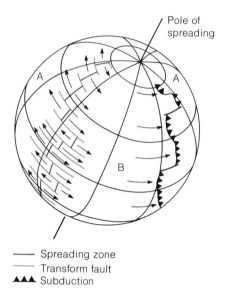

Pole of
spreading

—— Spreading zone
----- Transform fault
▲▲▲ Subduction

## Box 21-2  Plate Motions on a Spherical Earth

The above figure shows how the separation of two plates on a sphere can be described. The motion of plate $B$ relative to plate $A$ can be completely represented by a rotation of $B$ about some pole. (For completeness a schematic subduction zone is shown where plate $B$ sinks into the mantle.) Since surface area is neither created nor destroyed in slip along a transform fault, the fault must lie on what is called a *small circle* centered on this pole of rotation, which is also called a pole of spreading. If a tennis ball were sliced in two (not through the center) and put back together, one could rotate the two parts to see that this is so. The cut, which would describe a small circle, would also represent a transform fault. To locate the pole of spreading, one need only draw a great circle (any circle whose center is also the center of the ball) perpendicular to the cut at two or more positions. The intersection of the great circles is the pole of spreading. On a model of the Earth, if great circles are drawn perpendicular to transform faults between a pair of plates, their intersection locates the pole of spreading, which together with the spreading rate completely describes the relative motion of the two plates. The spreading rate is zero at the pole of spreading and increases to a maximum 90° away at the equator of spreading, as the figure indicates. This maximum equatorial value is frequently cited as the spreading rate between plates.

To see how a pole of spreading is located in practice, refer back to Figure 1-17, which shows the zone of spreading and the transform faults that are separating the African and American plates. Great circles perpendicular to the transform faults are drawn in the figure below. They intersect near the point 58°N, 36°W, off the southeast coast of Greenland. This is the pole of spreading of these two great plates. Don't bother going there, for there is nothing to be seen. The pole of spreading has no physical significance. It serves only as a construction point, a convenience for describing the relative motion of plates merely by giving the latitude and longitude of this point.

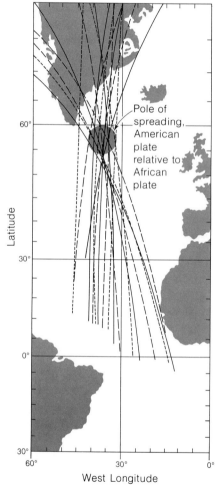

[After W. J. Morgan, "Rises, Trenches, Great Faults, and Crustal Blocks," *J. Geophysical Research*, v. 73, pp. 1959–1982, 1968.]

With the application of these geometric principles to get spreading directions, and magnetic anomalies to deduce spreading rates, the motions of the lithospheric plates are being worked out on a worldwide basis. Some results have already been pictured in Figures 1-17, 20-20, 21-4, and 21-9.

## SEA-FLOOR SPREADING AND CONTINENTAL DRIFT—RETHINKING EARTH HISTORY

One of us (F. P.) once wrote a paper dealing with the permanence of ocean basins. If he were allowed to expunge from the scientific record the one contribution he regrets the most, this would be it. The notion of the stability of global geographic features was not only a main tenet of the old geology but seems to be firmly rooted in the human psyche. We now know that the sea floor is ephemeral on the geological time scale. The present ocean basins are being created by spreading and recycled by subduction on a time scale of about 200 million years, which is less than 4 percent of the age of the Earth. There is little likelihood of finding extensive older remnants of sea floor. Continents, on the other hand, are mobile but permanent features. They are too buoyant to be subducted. They may be fragmented, moved, reassembled, deformed, and eroded at their surfaces, but their bulk does not seem to be much diminished. Old terranes with ages of around 3.5 to 3.7 billion years can still be found. Continents grow with time by the gradual accumulation of materials along their margins. New continental strips can therefore be added on in different places at different times, depending on the history of fragmentation, movement, and reassembly.

With the emergence of these revolutionary ideas, geologists are rethinking Earth history. The appearance of new developments in nearly every issue of the geological journals shows that the subject has definitely been revitalized. Rock associations, volcanism, metamorphism, the evolution of mountain chains—all are being re-examined in the framework of plate tectonics. Some of the new interpretations that we describe in this chapter may not stand the test of time. In this connection, the next edition of this book may show some changes, not so much in the big picture of plate tectonics as in the details of fitting regional geology into the overall framework. The student (as well as the authors of this book) should be cautioned against calling on plate tectonics for facile explanations of everything geological. It is not clear, for example, that the origin of such structures as the Ozarks, the Black Hills, or such intracontinental, sediment-filled depressions as the Michigan basin are related to plate movement. In what follows we draw in large part on the recent work of W. R. Dickinson of Stanford University, J. F. Dewey and J. B. Bird of the State University of New York, and R. S. Dietz, now at the National Oceanographic and Atmospheric Agency. In all fairness it should be mentioned that we are emphasizing one point of view, and that a diminishing number of well-intentioned geologists doubt some of the concepts that follow.

**Rock Assemblages and Plate Tectonics** The only record we have of past geologic events is the incomplete one found in the rocks that have survived erosion or subduction. In a number of previous chapters, some of the methods used to read the rock record were described. Here we explore the nature of the rock assemblages that would characterize different plate-tectonic regimes, as a first step in unraveling the history of past plate motions. Our motivation is to find ways to reconstruct the process of continent fragmentation and ocean development, to locate the sites of vanished oceans, and to recognize the sutures that mark ancient plate collisions.

Of the three kinds of plate boundaries, we might expect distinct suites (assemblages) of rocks to be associated with zones of divergence and convergence. At transform faults no distinct or characteristic rock assemblages are to be expected. Discontinuities across the fault are found, however, since rock formations formed and altered elsewhere have slipped past one another, and once-continuous formations or structural features are displaced. Within plates, where conditions are generally tectonically quiet, the stable offshore borderland between a continent and a deep ocean should receive an orderly sequence of undeformed sediments—the products of erosion of the adjacent land mass.

Think of everything that happens at a zone of divergence and you can predict the suite of rocks that would characterize the place and the process. Because there is extensive undersea volcanism, one would expect to find submarine basaltic lava, perhaps pillow lavas. Suboceanic crust and mantle is created here; dredge hauls and geophysical data show these layers to consist of mafic rocks, such as gabbro and peridotite, often showing evidence of alteration in a water environment (hydrous metamorphism). A carpet of deep-sea sediments would cover all of this. From Chapters 12 and 13 we remember that these deposits are recognized by thin layers of shale, limestone, and chert,* often intercalated with thin, discontinuous turbidites† and containing the fossil remains of pelagic (open ocean) marine organisms. A combination of deep-sea sediments, submarine lavas, and mafic igneous intrusions like that shown in idealized section in Figure 21-10, is called an **ophiolite suite.** The presence of narrow ophiolite zones in convergence features like the Alpine-Himalayan belt and the Ural and Appalachian belts may indicate that slices of oceanic crust and mantle

---

*A reminder: Chert is a siliceous rock; those derived from the siliceous ooze of the sea floor are made up largely of radiolarian (marine protozoans) or diatom (algae) skeletons.

†A turbidite is a turbidity-current deposit—the result of a sudden underwater flow of a mix of sediments and fluids, plunging downslope under gravity, settling finally on the deep-sea floor.

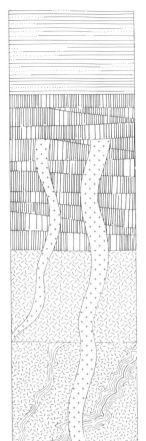

Deep-sea sediments:
shales, limestones,
cherts, turbidites,
fossils of pelagic
marine organisms

Basaltic pillow lava

Gabbro, evidence of
hydrous metamorphism

Peridotites and
other ultramafic
rocks, often showing
hydrous metamorphism

**Figure 21-10**
Idealized section of an ophiolite suite. The combination of deep-sea sediments, submarine lavas, and mafic igneous intrusions indicates a deep-sea origin. W. R. Dickinson and other geologists believe ophiolites to be fragments of the oceanic lithosphere emplaced on the continent as a result of plate collisions.

were thrust onto land when an ancient ocean finally disappeared as two continents converged. Some geologists believe that the Appalachians, for example, mark the site at which the ancestral Atlantic Ocean closed when North America and Africa converged about 375 million years ago. They also propose that the Atlantic reopened a few hundred kilometers east of this old suture, about 200 million years ago, in a spreading episode that is still underway.

Since the events that take place in a convergence zone are different from divergence-zone phenomena, so the rock assemblages have different characteristics. The main features of ocean-ocean or ocean-continent collision are shown in transverse section in Figure 21-11. Thick marine sediments, mostly turbidites, eroded from the continent or the island arc, rapidly fill the adjacent, elongate, marginal depressions. In descending, the cold oceanic slab stuffs the region below the inner wall of the trench with these sediments as well as with deep-sea materials brought in with the incoming plate. Regions of this sort are enormously complex and highly variable, as they include turbidites and ophiolitic shreds scraped off the downgoing slab by the edge of the overriding plate — all highly folded, intricately sliced and metamorphosed. They are

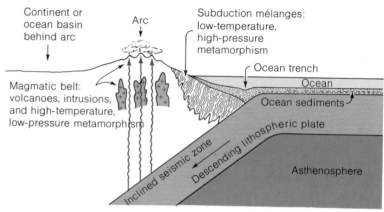

**Figure 21-11**
Geologic features and activities associated with plate collisions and subduction:
ocean trenches, mélange deposits, magmatic belts, metamorphism, volcanism.
Not to scale. Thickness of lithosphere is about 70 km, depth of ocean trench
10 km, distance from trench to arc 300 to 400 km. [After W. R. Dickinson,
"Plate Tectonics in Geologic History," *Science*, v. 174, p. 107. Copyright 1971 by
the American Association for the Advancement of Science.]

difficult to map in detail but recognizable by their distinctive mix
of materials and structural features; such a chaotic mess has been
called a **mélange**. The metamorphism is characteristically the kind
that takes place under high pressure and low temperature, because
the material may be carried relatively rapidly to depths as great as
30 kilometers, where recrystallization occurs in the environment
of the cold slab. Somehow, perhaps by buoyancy and mountain
building, the material rises back to the surface much later. Find a
mélange and you can't be too far from the place of downturn of
an ancient plate, long since consumed, but leaving this relic of
its former existence.

Refer again to Figure 21-11. Parallel to the mélange is a magmatic belt, coincident with the island arc or a mountain chain
near the continental margin as the case may be. Here the conditions are dominated by the rise of magma from the descending
plate. At the interface, where the descending plate slides past the
overriding one, perhaps friction is great enough to cause the upper
part of the downturned slab to melt. The liquids rise buoyantly
from depths of 100–200 kilometers to erupt and build the volcanic
chains on the leading edges of plates. The characteristic igneous
rocks produced are andesitic lavas and granitic intrusives. Island
arcs, built up from the sea floor, may contain larger amounts of
basalt; continental margins typically erupt rhyolitic ignimbrite and
are intruded by granitic batholiths below. These differences were
discussed in Chapter 17 on volcanism. In contrast to that in a
mélange, the metamorphism in the magmatic belts is typically the
result of recrystallization under conditions of high temperatures
and low pressures. This is because the hot fluids rise close to the

surface, delivering much heat to a low-pressure environment.

Paired belts of mélange and magmatism shown in Figure 21-11 are the signature of subduction. The details may differ from place to place, but the essential elements of these features of collision have been found in many places in the geologic record. One can see mélange in the Franciscan Formation of the California Coast Ranges and magmatism in the parallel belt of the Sierra Nevada to the east (Fig. 21-12). This paired belt marks the Mesozoic boundary between the colliding Pacific and American plates, and even shows

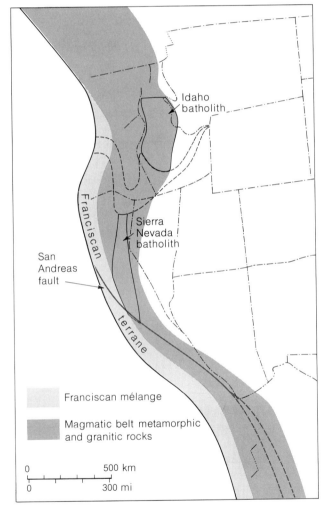

**Figure 21-12**
Paleogeologic map of the western United States shows the geology of the region as it was at the beginning of Tertiary time. The paired mélange and magmatic belts indicate a collision of the Pacific plate and the American plate in Mesozoic time, the Pacific plate being the subducted one. [After W. Hamilton and W. B. Myers, "Cenozoic Tectonics," *Reviews of Geophysics,* v. 4, p. 541, 1966.]

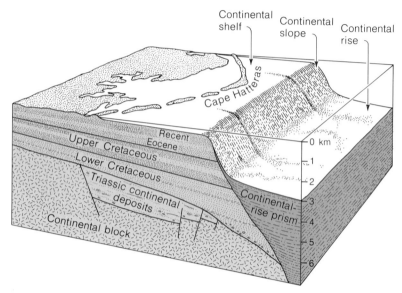

**Figure 21-13**
Geosynclines off the Atlantic coast of the United States. The sediments beneath the continental shelf are shallow-water, miogeosynclinal deposits that have been accumulating over the past 150 million years as the continental margin gradually subsided following the opening of the Atlantic Ocean. Under the continental rise are thick, deep-sea, eugeosynclinal deposits. Important accumulations of oil are believed to exist under the Atlantic continental shelf. [After "Geosynclines, Mountains and Continent-Building" by R. S. Dietz. Copyright © 1972 by Scientific American, Inc. All rights reserved.]

the polarity of the convergence by the spatial order of mélange on the west and magmatism on the east: the Pacific plate was the subducted one. Other paired belts—for example, in Japan—can be found along the continental margins framing the Pacific basin. The central Alps, a European example, were produced by the convergence of a Mediterranean plate with the European continent.

**Continental shelf deposits** are sedimentary rock assemblages that are laid down under tectonically quiet conditions in a geosyncline situated at a continental margin within a plate. Figure 21-13 shows the orderly sequence of deposits in what Dietz called the "living" geosyncline, which is still forming off the Atlantic coast of the United States. The continental margin there was formed when the American plate separated from the European plate. Resting on the offshore shelf is a wedge-shaped body composed of sediments that were eroded from the continent and deposited in shallow water. Because the unsupported trailing edge of the receding continent slowly sinks, the geosyncline continues to receive sediments for a long time. In addition, the load of the growing mass of sediment further depresses the crust isostatically, so that the geosyncline can receive still more materials from land. For every three feet of sediments received, the crust sinks two feet. The result of these two effects is that the geosynclinal deposits can accumulate in an orderly fashion to thicknesses of 10 kilometers or more. At the same time, the supply of sediments is sufficient to maintain the shallow-water environment of the geosyncline, or miogeosyncline as we called it in Chapter 13.

The deposits exhibit all of the indicators of shallow-water conditions, which were discussed in Chapter 13. At the bottom of the entire sequence are rift valleys containing basaltic lavas and nonmarine deposits formed during the early stages of continental fissuring. Subsequently, in the early stages of shelf deposition, sandy materials started to fill the depression. Much was dropped on the continental slope, only to be moved later to the eugeosyncline on the continental rise by turbidity currents. Very thick deposits can be built up in deep water in this way. As the shelf miogeosyncline builds up, deposition becomes dominated by shales and carbonate platform deposits—indicators of a decrease in the supply of detritus from the continent.

Think of what might happen to these geosynclines if the orderly, sequentially layered, gently dipping sediments were to become the leading edge of a plate in collision. In the following section we describe some of the many possibilities.

**Orogeny and Plate Tectonics**  Orogeny means mountain-making, particularly by folding and thrusting of rock layers. In the framework of plate tectonics, orogeny occurs primarily at the boundaries of colliding plates, where marginal sedimentary deposits are crumpled and magmatism and volcanism are initiated.

Consider first some scenarios of plate convergence. In Figure 21-14,*a*, a plate with a continent at the leading edge collides with another plate carrying a continent. In the early stage, during which the convergence is between continent and subducted oceanic lithosphere, a magmatic belt, folded mountains, and mélange deposits may be features of the overriding continental boundary. A "living" example may be found today along the Pacific coast of South America, where the American and Nazca plates are colliding. Refer back to Figure 1-17 to see the setting of the plates. The Andes, from which the name of the volcanic rock andesite is derived, lie in the magmatic belt; subduction is taking place under the Peru-Chile trench.

In a later stage, continent may meet continent, as shown in Figure 21-14,*b*. Since continental crust is too light to be carried down, the plate motions could be brought to a halt. Another possibility, the one depicted in the figure, is that the plate motions continue, with subduction ceasing at the continent-continent suture but starting up anew elsewhere. Cold and dense as the descending slab is, chunks of it may conceivably break off, fall freely into the mantle, and be resorbed. As Figure 21-14,*c*, shows, the suture will be marked by a mountain range made up of either folded or thrusted rocks, or both, coincident with or adjacent to the magmatic belt, and by a much-thickened continental crust. A prime

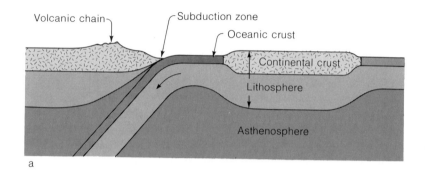

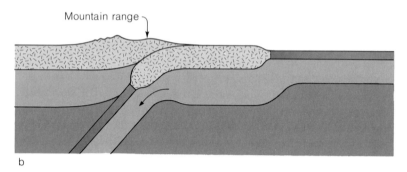

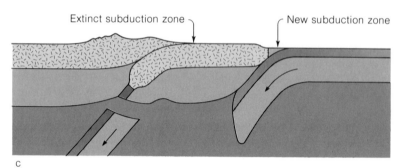

**Figure 21-14**
Schematic illustration of possible stages in plate collisions. (a) Convergence between plates with continental and oceanic lithosphere at leading edges. Magmatic belt, folded mountains, and mélange deposits are features of the overriding continental boundary. (b) Collision of continents, producing a mountain range, magmatic belt, and thickened continental crust. Since the continent is too buoyant to be carried down into the mantle, plate motions may be brought to a halt. (c) Alternatively, the plate may break off and a new subduction zone be started elsewhere. An extinct subduction zone may show as an old mountain belt within a continent. [After "Plate Tectonics" by J. F. Dewey. Copyright © 1972 by Scientific American, Inc. All rights reserved.]

example of continent-continent collision is the Himalayas, which began forming some 25 million years ago when a plate carrying India ran into the Asiatic plate (the uplift is still going on). This may be the way in which the root underlying the Himalayas originated (see Chapter 20).

**The Appalachian Mountains—A Case History** Some daring geologists are now attempting to reconstruct the events of a collision that began some 500 million years ago and which may have resulted in the formation of the Appalachians, an old and eroded mountain belt. Although still controversial, this avant-garde concept is presented here to illustrate how plate tectonics may be applied to geological problems in the years to come.

There are four chapters to the Appalachian story, as interpreted in the language of plate tectonics. The first chapter, summarized in Figures 21-15,*a* and *b*, begins more than 600 million years ago in late Precambrian time, when a continuous North American-African continent split apart, and an ocean, the ancestral Atlantic, opened. As North America receded from Africa and Europe, the newly formed North American continental shelf received sediments from the northwest under stable, intraplate conditions, and eventually developed into a miogeosyncline filled with shallow-water deposits. A thick sequence of deep-sea deposits grew in the eugeosyncline along the continental rise beyond the shelf, perhaps fed as well from Africa. The second chapter of the story opens in Early Ordovician times some 500 million years ago, when the ancestral Atlantic stopped opening and began to close again in an episode of contraction that was to last through Devonian time. The lithosphere broke along the edge of the North American continental margin, and plate loss began in a subduction zone adjacent to the continental rise (Figure 21-15,*c*). During the third chapter (Figs. 21-15,*c* and *d*), the compressive stresses of convergence collapsed the eugeosyncline of the continental rise, and the deep-sea sediments, altered by pressure, magmatic intrusions, and volcanism, were thrust up to form a lofty mountain range—the ancient Appalachians. The continental shelf deposits in the miogeosyncline to the west were subsequently folded into the series of ridges situated between the present-day Blue Ridge of Virginia and the Allegheny front of Pennsylvania. Sheets of rock were thrust onto the shelf as a result of the intense lateral compression of the continental-rise deposits. By middle Devonian times, about 375 million years ago, the ancestral Atlantic closed as North America collided with Europe and Africa, forming once again a single continuous continent—Pangaea (Fig. 21-15,*d*). The final chapter begins about 200 million years ago with the fissuring of Pangaea east of the old suture line. The modern Atlantic opened, and geosynclines of the continental shelf and rise developed anew (Figure 21-15,*e*). The modern shelf (Figs. 21-15,*f* and 21-16) occupies the site of the late Precambrian and early Paleozoic continental rise and also the site of the lofty mountain ranges of Ordovician and Silurian times, now almost completely eroded down to their roots east of the Piedmont. In this sense, North America may have grown as a result of the Appalachian orogeny, having added strips from the earlier episodes. How many times such strips may have been added in the

722

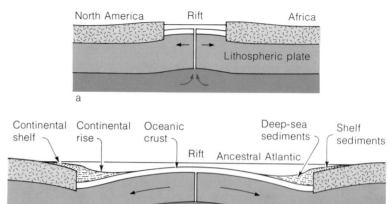

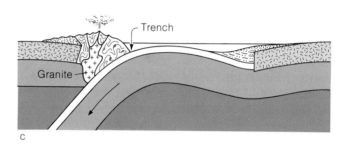

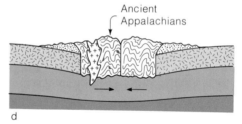

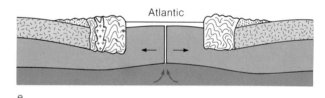

**Figure 21-15**
Stages in the formation of the Appalachian Mountains. According to the theory of continental drift, North America and Africa split apart more than 600 million years ago, an ancestral Atlantic opened, and sediments were deposited on the margins of the receding continents (a and b). About 500 million years ago, the Atlantic began to close, subduction was initiated, and the geosynclinal deposits were deformed, intruded, and uplifted (c). About 375 million years ago, the Atlantic closed as North America collided with Africa and Europe (d). The modern Atlantic opened about 200 million years ago with the fissuring of Pangaea, and marginal geosynclines began to develop anew (e and f) [After "Geosynclines, Mountains and Continent-Building" by R. S. Dietz. Copyright © 1972 by Scientific American, Inc. All rights reserved.]

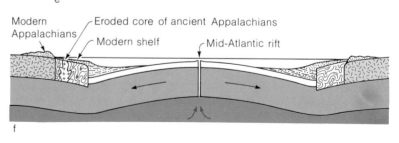

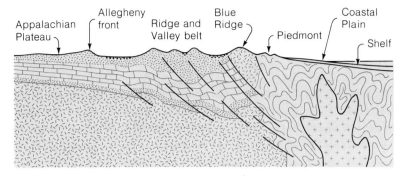

**Figure 21-16**
A section of the Appalachian Mountain belt. According to a plate-tectonics interpretation, the present-day basement east of the Blue Ridge is the eroded crystalline core of Ordovician and Silurian mountain ranges that were formed from the crumpling of a Precambrian and early Paleozoic eugeosyncline (Fig. 21-15, c and d.). The ancient miogeosyncline was folded into the series of ridges and valleys between the Allegheny Front and the Blue Ridge. The positions of the remnants of the ancient geosynclines to the west of the modern ones are indicative of the growth of the continent as a result of the Appalachian orogeny. [After "Geosynclines, Mountains and Continent-Building" by R. S. Dietz. Copyright © 1972 by Scientific American, Inc. All rights reserved.]

Precambrian is hard to know, but it may have been many.

How are geologists able to reconstruct, or hypothesize, such a complex sequence of events? They know how to recognize the deposits of shallow continental shelf and deep-sea deposits that are the trademarks of the expansion phase of ocean basins. They can recognize the imprints of ocean contractions and plate convergence by the fold belts, the thrust sheets, the magmatism and metamorphism, and the scraped-up ophiolites that such plate movements produce. Once these crustal features are recognized for what they are, the events that produced them can be pieced together to tell the story of the expansion, contraction, and re-expansion of an ocean basin—a story that spans more than 600 million years.

**The Grand Reconstruction**   As we have seen, there is evidence, albeit controversial, that before the present expansion of the Atlantic and contraction of the Pacific commenced, Pangaea had undergone an earlier episode of fragmentation and reassembly. Old mountain belts like the Appalachians and the Urals seem to suggest lines along which continental fragments collided *before* the most recent breakup of Pangaea, some 200 million years ago. In time geologists may be able to sort out more details of this complex jigsaw puzzle, whose individual pieces change shape over geological time. But even the reconstruction of events over the past 200 million years still involves some guesswork, because the subject is so new and evidence is still being gathered.

The sequence of Figures 21-1 and 21-17 to 21-20 summarizes one notion of the most recent breakup of Pangaea. Figure 21-1 shows the world as it may have looked in Permian times, a little more than 200 million years ago. Pangaea was an irregularly shaped land mass surrounded by a universal ocean called Panthalassa, the ancestral Pacific. The Tethys Sea, between Africa and Eurasia, was the ancestor of part of the Mediterranean. The fit of North and South America with Europe and Africa is probably correct in large details — at least it is the fit for which we have the firmest evidence. The positions of central America, India, Australia, and Antarctica are less certain.

The breakup of Pangaea was signaled by the opening of rifts from which basalt poured. Relics of this great event can be found today in the Triassic basalt flows of the east coast of the United States. Dating of these flows by radioactivity provides the estimate of about 200 million years for the beginning of drift.

The geography of the world after 20 million years of drift — at the end of the Triassic some 180 million years ago — is sketched in Figure 21-17. The Atlantic has opened, the Tethys has contracted, and the northern continents (Laurasia) have all but split away from the southern continents (Gondwana). New ocean floor has also separated Antarctica-Australia from Africa-South America. India is off on a trip to the north.

By the end of the Jurassic period, 135 million years ago, drift had been underway for 65 million years. The big event at this time was the splitting of South America from Africa, which signaled the birth of the South Atlantic (Fig. 21-18). The North Atlantic and Indian oceans are enlarged, but the Tethys Sea continues to close. India continues its northward journey.

The close of the Cretaceous period 65 million years ago sees a widened South Atlantic, the splitting of Madagascar from Africa, and the close of the Tethys to form an inland sea, the Mediterranean (Fig. 21-19). After 135 million years of drift, the modern configuration of continents becomes discernible.

The modern world, produced over the past 65 million years, is shown in Figure 21-20. India has collided with Asia, bringing its trip to an end. Australia has separated from Antarctica. Nearly half of the present-day ocean floor was created in this period. Figure 21-21 shows several schematic sections that summarize modern plate, ocean, continent, and island-arc relationships for the American, African, Eurasian, and Indian plates.

If the pattern of modern plate motions continues for another 50 million years, the continents of the world may have the configuration shown in Figure 21-22, but this is sheer speculation. The Atlantic and Indian oceans have expanded at the expense of the contracting Pacific. Australia is beginning to collide with Asia. A new ocean appears in its beginning stages in east Africa. Los Angeles has by-passed San Francisco's position on its northward journey to destruction in the Aleutian trench.

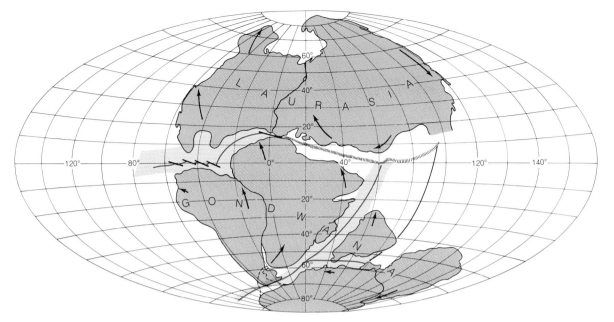

**Figure 21-17**
One view of world geography at the end of the Triassic Period, 180 million years ago, after some 20 million years of drift. New ocean floor shows as areas tinted in color. Spreading zones are represented by heavy brown lines, transform faults by black lines, and subduction zones by hatched lines. Arrows depict motions of continents since drift began. [After "The Breakup of Pangaea" by R. S. Dietz and J. C. Holden. Copyright © 1970 by Scientific American, Inc. All rights reserved.]

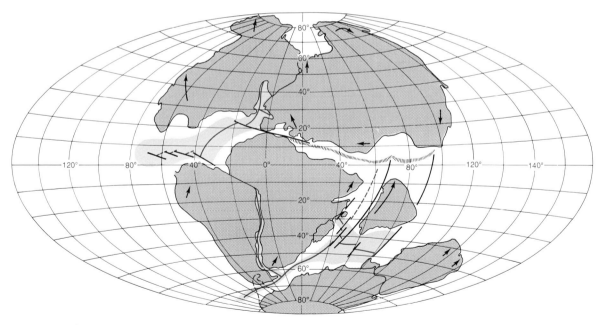

**Figure 21-18**
World geography at the end of the Jurassic Period, 135 million years ago, after some 65 million years of drift. [After "The Breakup of Pangaea" by R. S. Dietz and J. C. Holden. Copyright © 1970 by Scientific American, Inc. All rights reserved.]

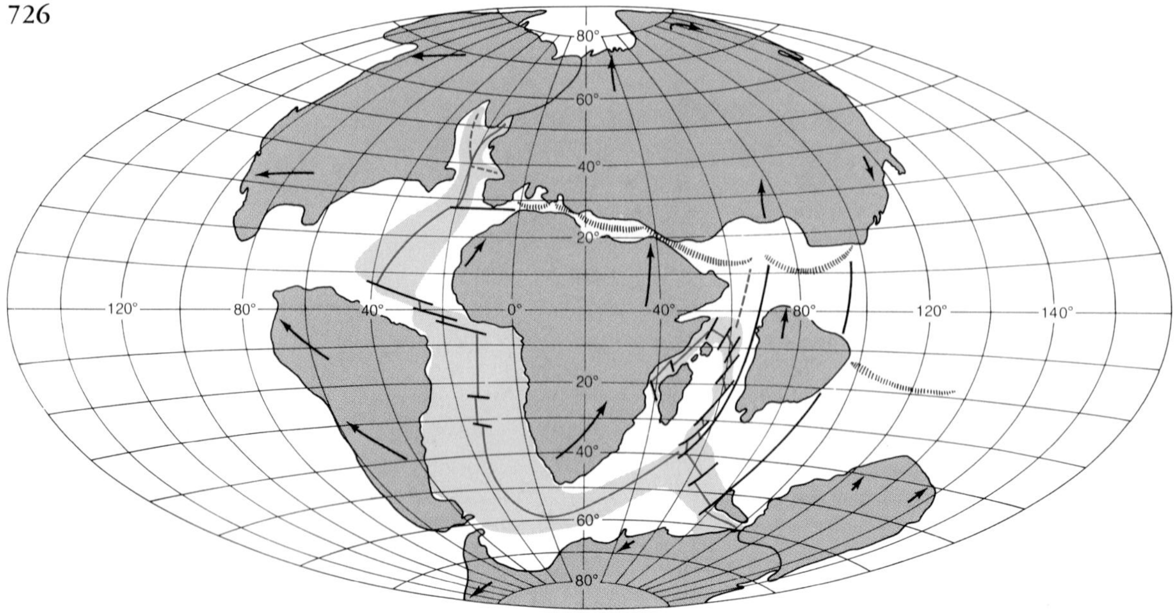

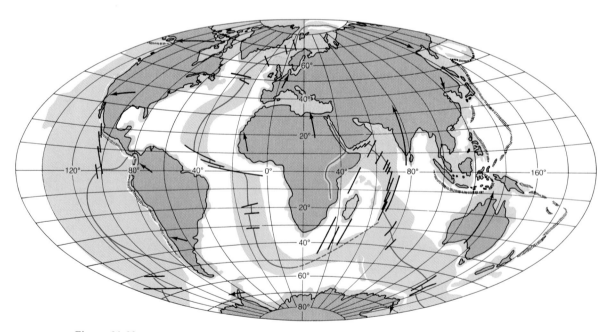

**Figure 21-19**
World geography at the end of the Cretaceous Period, 65 million years ago, after some 135 million years of drift. [After "The Breakup of Pangaea" by R. S. Dietz and J. C. Holden. Copyright © 1970, by Scientific America, Inc. All rights reserved.]

**Figure 21-20**
World geography today, showing sea floor produced during the past 65 million years, in the Cenozoic Period. [After "The Breakup of Pangaea" by R. S. Dietz and J. C. Holden. Copyright © 1970 by Scientific American, Inc. All rights reserved.]

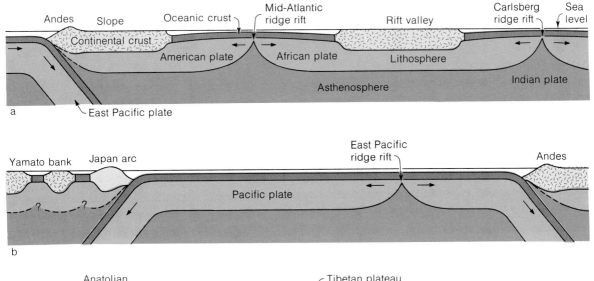

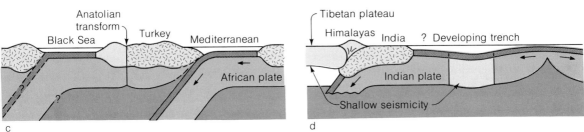

**Figure 21-21**
Schematic sections showing modern plate, ocean, continent, and island-arc relationships. [After
J. F. Dewey and J. M. Bird, "Mountain Belts and New Global Tectonics" *J. Geophysical Research,*
v. 75, pp. 2625–2647, 1970.]

There is not a branch of geology that is untouched by this
grand reconstruction of the continents, except perhaps crystallography. Economic geologists are using the fit of the continents
to find mineral and oil deposits by correlating the formations
in which they occur on one continent with their pre-drift continu-
ations on another continent. Paleontologists are rethinking some
aspects of evolution in the light of continental drift. For example,
during most of the age of reptiles, the continents we know today
were grouped together in two supercontinents, Laurasia and
Gondwanaland. These continents were fragmented during most
of the age of mammals, with faunas developing on the daughter
continents under conditions of isolation from one another. Is
this the reason why mammals diversified into so many more orders
than the reptiles, and in a much shorter period of time? Structural geologists and petrologists are extending their sights from
regional mapping to the world picture, for the concept of plate
tectonics provides the means of interpreting such geological
processes as sedimentation and orogeny in global terms. For

728

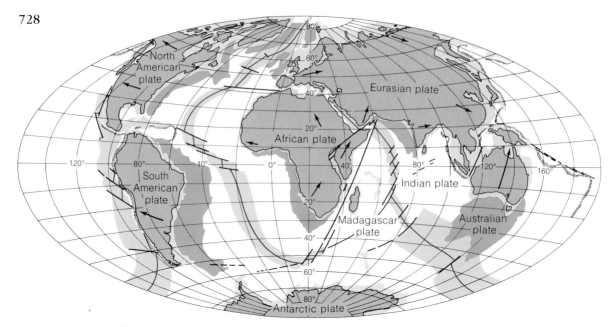

**Figure 21-22**
World geography as it may look some 50 million years from now if present-day plate movements continue. [After "The Breakup of Pangaea" by R. S. Dietz and J. C. Holden. Copyright © 1970, by Scientific American, Inc. All rights reserved.]

example, the Caledonian mountain belt that runs along the northwest margin of Europe is the predrift continuation of the Appalachian belt, and the trend of the Andes may be followed into Antarctica and Australia, as Figure 21-23 shows.

Oceanographers are reconstructing currents as they might have existed in the ancestral oceans, to understand better the modern circulation and to account for the variations in deep-sea sediments. Paleoclimatologists are "forecasting backwards" in time to describe termperature, winds, the extent of continental glaciers, and the level of the sea as they were in predrift times. What better testimony to the triumph of this once-outrageous hypothesis than its ability to revitalize and shed light on so many diverse topics!

## THE DRIVING MECHANISM OF PLATE TECTONICS

Up to this point everything we have discussed might be categorized as descriptive plate tectonics. The geometry and rates of plate motions, the consequences of plate separation and collision have been described. But what drives it all? Full understanding of the entire phenomenon awaits the answer to this question. The International Geodynamics Project will enlist the efforts of thousands of scientists to help find the underlying cause of plate motions.

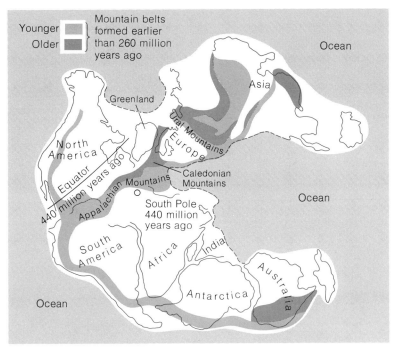

**Figure 21-23**
When the ancient continent of Pangaea is reconstructed, the Caledonian
Mountains of Europe and the Appalachian Mountains form a continuous belt,
indicative of a plate collision between continents antedating Pangaea. The
continuity of the mountain belt that extends from the Americas across Antarctica
and western Australia supports the reconstruction. The Urals and other old
mountains contain ophiolite zones, marking these sutures as the sites of
vanished oceans. [After "Plate Tectonics" by J. F. Dewey. Copyright © 1972
by Scientific American, Inc. All rights reserved.]

 

As is generally the case when there is an abundance of data in
search of a theory, many hypotheses have been advanced. Some
would have plates pushed by the weight of the ridges at the zones
of spreading or pulled by the heavy downgoing slab at subduction
zones. Others hold that the plates are dragged along by currents
flowing in the underlying asthenosphere. Figure 21-24 shows some
of these ideas. In line with the discussion in Chapter 15, we agree
with those who view the process not in piecemeal but as a highly
complex convective flow, involving rising, hot, partially molten
materials and sinking, cool, solid materials, under a variety of
conditions ranging from melting to solidification and remelting.
A significant part of the mantle must be involved, since slabs are
known to penetrate to depths of some 700 kilometers before being
completely resorbed. Figure 21-24,*d*, which was discussed in Chapter 15, shows one of the first computer models of the process
— one that neglects many of the effects just mentioned, but which
nevertheless accounts for many of the observations. A rising plume
of hot material, heated from below reaches the surface at a center

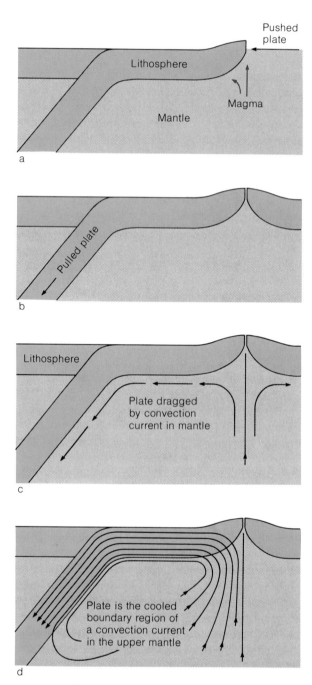

**Figure 21-24**
Schematic illustration of possible driving mechanism of plate
tectonics. (a) Plate is pushed by the weight of the ridges at
centers of spreading. (b) Plate is pulled by heavy downgoing
slab of subduction zone. (c) Plate is dragged by convection
current in mantle. (d) Plate is the cooled, brittle, boundary
region of a convection current involving the hot, low-
viscosity upper mantle.

of spreading. It moves away from the center, cools near the surface, and the cooled boundary becomes solid, strong lithosphere. Finally becoming heavier after it has cooled, the lithospheric slab sinks back into the mantle in a subduction zone, where it is reassimilated, to be heated and to rise again in the future. Among the problems left to the next generation of Earth scientists are the incorporation of such important details as the shapes of plates and the history of their movements, and the formation and growth of continents, into an explanation of the distribution of convective cells in time and space.

## SUMMARY

1. According to the theory of plate tectonics, the lithosphere is broken into about a dozen rigid, moving plates. Three types of plate boundaries are defined by the relative motion between plates: boundaries of divergence and convergence and transform faults.

2. In addition to earthquake belts, many large-scale geological features are associated with plate boundaries, such as narrow mountain belts and chains of volcanoes. Boundaries of convergence are recognized by deep-sea trenches, inclined earthquake belts, mountains and volcanoes, and paired belts of mélange and magmatism. The Andes Mountains and the trenches of the west coast of South America are modern examples. Divergence boundaries (for example, the mid-Atlantic ridge) typically show as seismic, volcanic, mid-ocean ridges. A characteristic deposit of this environment is the ophiolite suite. Transform faults, along which plates slide past one another, can be recognized by their topography, seismicity, and offsets in magnetic anomaly bands. Ancient convergences may show as old mountain belts, such as the Appalachians.

3. Age of the sea floor can be measured by means of magnetic anomaly bands, and the "stratigraphy" of magnetic reversals worked out on land. The procedure has been verified and extended by deep-sea drilling. Isochrons can now be drawn for most of the Atlantic and large sections of the Pacific, enabling geologists to reconstruct the history of opening and closing of these oceans. Using this method and drawing on geological and paleomagnetic data, the fragmentation of Pangaea over the last 200 million years can be sketched.

4. Although plate motions can now be described in some detail, the driving mechanism is still a puzzle. An attractive hypothesis proposes that the upper mantle is in a state of convection with hot material rising under divergences and cool material sinking in subduction zones. The plates, according to this notion, would be the cooled, upper boundary region of the convection cell.

## EXERCISES

1. Summarize the principle geological features of subduction zones and divergence zones.

2. Explain the following in the context of plate tectonics:
   a) Iceland.
   b) San Andreas fault of California.
   c) Orogeny.
   d) Aleutian trench
   e) "Jigsaw-puzzle" fit of the Americas and Europe-Africa.
   f) Andes Mountains.

3. How do we know that spreading along the East Pacific rise is faster than along the mid-Atlantic ridge?

4. What would an astronaut look for on Mars to find out if plate tectonics is an active process on the planet?

5. How would one recognize the boundaries between ancient plates no longer in existence?

## BIBLIOGRAPHY

Cox, A., (editor), *Plate Tectonics and Geomagnetic Reversals.* San Francisco: W. H. Freeman and Company, 1973.

Dewey, J. F., "Plate Tectonics," *Scientific American,* May, 1972. (Offprint No. 900).

Meyerhoff, A. A., and H. A. Meyerhoff, "The New Global Tectonics: Major Inconsistencies." *American Association of Petroleum Geologists Bulletin,* v. 56, pp. 269–336, 1972.

Tarling, Don and Maureen, *Continental Drift.* Garden City, New York: Doubleday and Company, 1971.

Wilson, J. Tuzo (compiler), *Continents Adrift — Readings from Scientific American.* San Francisco: W. H. Freeman and Company, 1970.

# 22

## Deformation of the Earth's Crust

*Forces in the crust have been active throughout geologic time. Rocks respond to these forces by folding and fracturing. In a typical mountain-making (orogenic) episode, a thick section of sediments, deposited in a geosyncline along the margin of a stable block, is crumpled. Mountains are raised as the belt is deformed and the sediments are thrown into a series of folds and faults. Intrusion and metamorphism are typical. After the highlands are eroded down, a final stage of rejuvenation by uplifting often occurs. The cycle of geosyncline formation and orogeny can be integrated with the plate-tectonics concept of opening and closing oceans in a simple and natural way.*

What a spectacular movie could be made if a camera could be fixed to film the Earth's surface by snapping one frame every thousand years. Replayed at standard speed, a movie made this way would condense almost 50 million years of geologic time to an hour-long feature. Geologic movements that normally are invisibly slow could be viewed and really appreciated in such a fast-motion replay. One could see the drift of continents accompanying the opening of some oceans and the closing of others. In some regions, uplift, tilting, faulting, or folding would give rise to plateaus, mountains, rift valleys, and other geologic structures, and the continuous interplay between tectonic activity and erosion would be evident. Subsidence in other places would lower portions of a continent beneath the sea. The idea is not as far-fetched as it sounds, because movies of this sort have been made in the manner of animated cartoons: the frames of such films are photographs of sketches based on interpretations of the geologic record.

When a geologist examines the geologic record and makes a geologic cross section showing beds in various attitudes—flat-lying, tilted, folded, faulted, perhaps intruded by igneous rocks, metamorphosed, or covered by lava flows—he does so to unravel the geologic structure of the region and then to deduce the sequence of events that we call geologic history (Fig. 22-2). He knows how to recognize those features (such as lithology, bedding, subaqueous slump, ripple marks, or mud cracks) that are connected

**Figure 22-1**
Apollo 16 photograph of western North America, showing the major deformational features of the Cordillera: the Pacific Coast Ranges, the Great Valley of California, the Sierra Nevada, the Basin and Range province, the Colorado Plateau, and the Rocky Mountains. The Great Plains of the stable interior are evident in the distance. Also shown is the Gulf of California, a region of active sea-floor spreading, with Baja California rifting away from the American plate. [From National Aeronautics and Space Administration.]

with the original formation of the sedimentary beds, using the criteria described in Chapters 2 and 13. Deformation structures, such as folds and faults, which are imposed by events subsequent to formation, can then be interpreted in terms of the sequence in which they occurred. This chapter takes up the subject of geologic structures—particularly the changes ranging in extent from meters to hundreds of kilometers that take place in rocks due to deformation of the Earth's crust.

## HOW ROCKS DEFORM

Folding and faulting are the most common forms of deformation of the rocks that make up the outer layers of the Earth's crust. For years, geologists were baffled by the problem of how rocks, which seem strong and rigid, could be distorted by crustal forces into folds or broken along faults (Figure 22-3). The question still does not have a complete answer, but much understanding has come from laboratory experiments in which rocks were squeezed under the conditions of high pressure and temperature that are known to occur in the crust.

Figure 22-4 shows what happens to a small cylinder of marble, for example, when it is squeezed by applying **stress** at the ends. Stress is a force divided by the area over which it is applied. (For example, a weight of 4 kilograms applied over an area of 2 square centimeters exerts a stress of 2 kilograms per square centimeter. Pressure is also measured as force per unit area. One kilogram per square centimeter is approximately equivalent to atmospheric pressure.) The experiment consists of applying stress by pushing down on one end of a sample of rock with a piston, while at the same time applying pressure to all sides of the sample to simulate the confining pressure to which materials deep in the crust are subjected because of the weight of the overlying rock. Temperature can also be raised to realistic, deep Earth values, usually to several hundred degrees Centigrade. The figure shows an unstressed cylindrical sample and two samples that have been **strained** by 20 percent—that is, changed in length (in this case), shortened by 20 percent owing to the applied stress. Low confining pressures, which simulate shallow depths in the crust, allowed the middle sample to fracture; high confining pressures, which are equivalent to conditions at greater depths in the crust, made the third sample deform **plastically** under stress—that is, made it flow, without any cracking, into its shortened, bulging shape. The investigators concluded that, if a bed of this particular marble were stressed near the surface, it would tend to deform by faulting; but, if it were deeper than a few kilometers, it would behave plastically and fold. Experiments also show that, if rocks are hot when they are stressed, plastic flow will occur more readily.

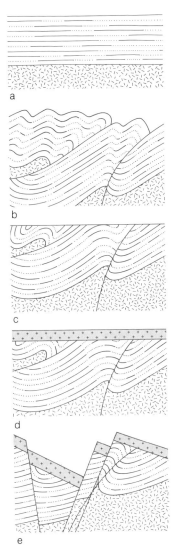

**Figure 22-2**
Stages in the development of the Basin and Range province, as postulated by W. M. Davis. (a) Deposition of stratified sediments. (b) Folding and thrusting. (c) Development of an erosional surface. (d) Sheets of lava cover the eroded surface. (e) Block faults develop, breaking up the earlier features. A geologist sees only the last stage and attempts to reconstruct from the structural features all of the earlier stages in the history of a region. [After *The Evolution of North America* by P. B. King.] Copyright © 1959 by Princeton University Press. [Redrawn with permission of Princeton University Press.]

**Figure 22-3**
Small-scale folds in interbedded shales and cherts of the Franciscan Formation in Central California. [From A. M. Johnson, *Physical Processes in Geology*, Freeman, Cooper & Company. Copyright ©
1972.]

Rocks, like most solids, can be classed as **brittle** or **ductile**, according to how much plastic flow they show. A brittle material ruptures, rather than flows, when stressed beyond a critical value. Ductile substances are capable of considerable plastic flow before they break. Glass, near room temperature, is a familiar example of a brittle material, and modelling clay is obviously ductile. The marble discussed in the previous paragraph would be brittle at shallow depths, but it would respond to stress as a ductile substance deeper in the crust.

The graph in Figure 22-5 summarizes the deformation behavior of many rocks under stress. The ordinate shows the applied stress and the abscissa shows the corresponding deformation. When the stresses are low, the strains follow **Hooke's elastic law**: strain changes in the same proportion as the applied stress. The graph shows this elastic behavior as a straight line starting from zero stress and strain. In this region of low stress and elasticity, the

**Figure 22-4**
Marble cylinder deformed in a laboratory by pushing down at one end. From left to right: (a) undeformed; (b) 20 percent strain (shortened by 20 percent of original length), 270 atmospheres confining pressure; (c) 20 percent strain, 445 atmospheres confining pressure. Note how the sample deforms plastically at higher confining pressure and fails by fracture at lower confining pressure. [Photo by M. S. Paterson, Australian National University.]

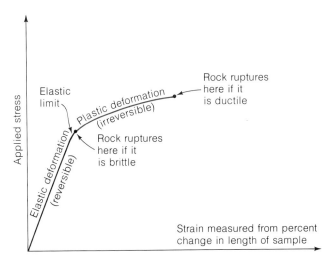

**Figure 22-5**
Schematic graph showing how a rock deforms when stress is applied. Initially, the deformation is elastic and reversible, with stress proportional to strain. If stress is increased beyond the elastic limit, the rock undergoes a permanent, plastic deformation. Brittle rocks rupture with little or no plastic deformation. Ductile rocks undergo much plastic deformation before they break.

deformation is reversible — that is, the sample returns to its original size and shape when the stress is removed, like the spring in a scale when a weight is removed. If the stress is increased, the **elastic limit** is reached, beyond which the rock begins to deform by plastic flow. On the stress–strain diagram, the curve bends over where this occurs, which means that small increases in stress lead to larger deformations. Plastic flow is a permanent distortion — the original shape of the material is not restored if the stress is removed. If the stress increases further, the material finally breaks when its **rupture strength** is reached. For brittle materials, the region of plastic deformation is small or absent and the material breaks with little additional stress after the elastic limit is exceeded. Naturally, the shape of the curve varies for different materials. Wet clay would show little elasticity and much plasticity, whereas a rock like basalt would show no plasticity at near-surface conditions.

Nature is more complex, of course, than the conditions under which such simple experiments are performed. For example, stresses connected with crustal deformation may continue for thousands or millions of years, whereas laboratory geologists have never squeezed individual rock samples for much longer than a few years. Rocks may flow under stresses much lower than those required in the laboratory, if the stresses are applied for long periods and would flow more readily if wet or hot. Nevertheless, the experiments shed some light on how rocks respond to stresses, and they give us more confidence in our interpretations of field evidence. When we see folds and faults in the field, we can remember that some rocks are brittle, that others are ductile, that the same rock can be brittle at shallow depths and ductile deep in the crust, and that every rock flows plastically if stressed beyond its elastic limit, or ruptures if stressed beyond its strength. Other things being equal, laboratory experiments teach us that we should expect to find that granite is stronger than sandstone, that crystalline basement rocks are more brittle than the plastically deformable young sediments that may cover them, that sedimentary rocks increase in ductility from limestone and sandstone through clay and shale to salt and gypsum.

## REGIONAL MOVEMENTS OF THE EARTH'S CRUST

**Orogenic movements**   A striking feature of continents is that they are marked by long, relatively narrow mountain chains made up of strongly folded and faulted rocks. The movements that produce these deformed belts are termed **orogenic**, and they are generally thought to be the result of horizontal compression. The Appalachian Mountains provide a prime example, and the case history given in the previous chapter told the story of this orogenic belt in terms of the compression of a belt of geosynclinal sediments

**Figure 22-6**
Rocks respond differently to forces in the Earth's crust. In this tilted section, sandstone beds were able to withstand deformation, but shales show evidence of flowage and changes in original thickness. Zumaja, Spain. [Photo by R. Siever.]

accompanied by episodes of intrusions and metamorphism. The Alps, the Urals, and the Rockies are other examples. We will say more about orogenic belts later in this chapter, following a discussion of folds and faults.

**Epeirogenic Movements**    Sedimentary rock sequences all over the world record the same kind of history of downward and upward movements of the crust that we can see in the Grand Canyon (Chapter 2). **Epeirogeny** is the term used to describe these movements. Great thicknesses of marine deposits of shallow-water origin, once buried hundreds or thousands of meters below the sea floor, are now found well above sea level. These deposits give evidence of continued slow subsidence during sedimentation. In many cases, later elevation to their present position above the sea is due to simple uplift. Gaps in the rock sequence represented by unconformities (Fig. 22-7) tell of uplifts of large regions that stopped sedimentation and started stripping away previously deposited rocks. Fossil trees and other plants embedded in coal deposits now mined deep in the Earth tell of earlier times when they grew in the sunlight of the surface. Raised beaches and marine shells now found at elevations of 250 meters above sea level substantiate the postglacial uplift of Fennoscandia. Other examples

**Figure 22-7**
Unconformity in which Precambrian sandstone is deposited on an erosional surface of gneiss. Loch Assynt, Scotland. [Photo by John Haller, Harvard University.]

that indicate vertical movements are incised meanders, coral reefs raised above sea level, and drowned river valleys (see Figs. 2-4, 22-8, 22-9 and Chapters 5 and 9).

Along seacoasts, the geologist is careful to distinguish between movements in which one portion of the crust went up or down relative to another part of the crust and the worldwide sea-level changes that accompanied Pleistocene glacial and interglacial periods. Whether the land goes up or the sea goes down (or vice versa) makes no difference in shaping such landforms of seacoasts as terraces or eroded headlands.

Though many of the vertical movements are connected with orogeny, epeirogenic movements commonly affect large regions without extensive folding or faulting. A typical product of this kind of downward movement, which is usually slow and intermittent, is a **basin**, a bowl- or spoon-shaped depression that gets filled up with sediment. An example of gradual and intermittent

**Figure 22-8**
Marine planation surfaces. The various foreground surfaces and the several surfaces of the distant island were all formed by wave erosion. Subsequent vertical movements raised the surfaces above sea level. San Benito Islands, Baja California. [Photo by W. B. Hamilton, U.S. Geological Survey.]

**Figure 22-9**
Regional uplift, evidenced by a delta raised about 30 m above sea level. An old delta is being dissected by streams as a new delta is being deposited. Antarctic Sund, E. Greenland. [Lauge Koch Expedition; courtesy of John Haller, Harvard University.]

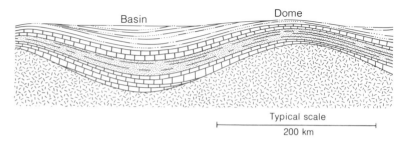

**Figure 22-10**
Idealized section of a dome and basin, evidence of vertical movements in the relatively undeformed interior of the United States.

downward movement of the crust in one part of the relatively undeformed or stable broad region between the Appalachians and the Rockies is the Michigan Basin. Covering much of the lower peninsula of Michigan, this is a circular area of about 500,000 square kilometers that subsided throughout much of Paleozoic time and received sediments more than 3 kilometers thick in its central, deepest part. The structure of the formations in this basin has been likened to a pile of saucers (Fig. 22-10).

Upward movement with little or only moderate faulting or folding can produce broad uplands and plateaus. The Colorado Plateau, the Adirondack Mountains, and the Black Hills of South Dakota are examples of general upward movements. The Black Hills structure is an oval-shaped **dome** — an area of uplift sloping off more or less uniformly in all directions from the highest point — that rises more than 2 kilometers above the Great Plains surrounding it. The larger area in which the Black Hills are located was once lower, and it subsided to receive a blanket of Paleozoic and Mesozoic sediment more than 2 kilometers thick. Sometime between the late Cretaceous and Oligocene, the whole area of about 50,000 square kilometers was uplifted, as if pushed up from below by a piston, without being extensively crumpled or broken. Subsequent erosion has stripped away the sediments overlying the central part of the uplift, exposing the Precambrian igneous and metamorphic rocks below. Here is where the famous Homestake Gold Mine was discovered in the Precambrian rocks near Lead, South Dakota (Fig. 22-11). Some geologists speculate that the doming of the Black Hills was caused by an upwelling of magma from deep within the crust.

Geologists have no ready mechanisms to account for most of these slow and broad epeirogenic movements, although hypotheses exist for some of them. The Fennoscandian uplift, as mentioned in Chapter 20, represents the slow upward recovery of the crust following the removal of the glacial load that had depressed it. The sediment-filled downwarp off the east coast of the United States may be evidence of the foundering of the edge of the continent after it drifted apart from North Africa. The subsidence of

**Figure 22-11**
The Black Hills of South Dakota, a domal uplift in the Great Plains. Erosion
of the central part exposes the core of Precambrian igneous and metamorphic
rocks and the succession of Paleozoic and Mesozoic sediments that slope away
from the center. [After *Geology of Soils* by C. B. Hunt. W. H. Freeman and
Company. Copyright © 1972.]

the sea floor as it moves away from the mid-ocean ridge may
represent contraction of the newly formed lithosphere as it cools
in the spreading process, as mentioned in Chapter 21 (Fig. 22-12).
Some of the broad uplifts, such as the Tibetan Plateau, may be
connected with plate collisions; in general, however, warping within
the continents and far from plate margins is still a puzzle.

**Recent Movements**  Up and down movements, or **warping**, of
large regions are not restricted only to the geologic past, but are
measurable as they occur in our own time. For example, the city
of Venice is slowly sinking into the Adriatic Sea at a rate of 4 millimeters per year. The process is mostly one of coastal downwarping,
though man-made subsidence due to the withdrawal of water and
natural gas from the underlying sediments is, unfortunately, hasten-
ing this beautiful city's demise (Fig. 22-13).

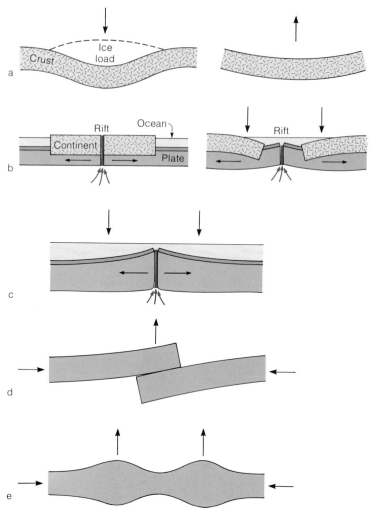

**Figure 22-12**
Some proposed mechanisms for vertical movements (not to scale). (a) Glacial
ice load buckles the crust; slow uplift follows removal of ice. (b) Rifting and
separation of two segments of a continent; receding margins sink. (c) Formation
of lithosphere at mid-ocean ridges. Spreading sea floor subsides as plate cools
and contracts. (d) Two plates collide; overriding plate is uplifted. (e) Crust
deforms by thickening due to horizontal forces.

Precision surveys across the United States have revealed the
pattern of vertical movements shown in Figure 22-14. Large regions
are currently sinking and rising at rates of 1–15 millimeters per
year. If the pattern persists for the geologically short time of only
1 million years, much of New England and the Gulf Coast will have
sunk to the sea floor and plateaus a few kilometers in height will
have grown in the Midwest. These seemingly slow rates of displacement, unnoticed by the inhabitants, are all that it takes to
produce highlands and basins.

**Figure 22-13**
Venice is slowly subsiding into the Adriatic Sea. Water level now reaches
the base of the columns of this old building. The raised sidewalk, of recent
construction, will be awash in several decades if subsidence continues unimpeded.
[Photo by Roberto Frassetto, National Research Council of Italy.]

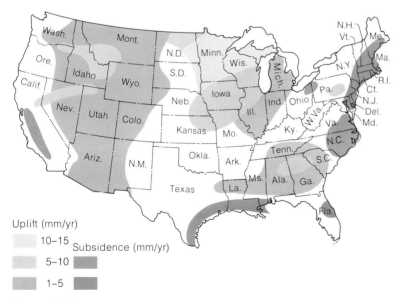

**Figure 22-14**
Crustal-movement map showing areas of present-day uplift and subsidence in
the United States. [Photo by S. P. Hand, National Oceanic and Atmospheric
Administration.]

There is one type of crustal movement that the inhabitants of an affected area do take notice of: the sudden and large displacements along faults that produce large earthquakes. Integrated over thousands or millions of years, the accumulated displacements due to earthquake movements may be of the same order as the gradual displacements depicted in Figure 22-14. The catastrophic nature and the large size of each impulse of displacement is what sets earthquake movements apart in the minds of men.

As discussed in Chapter 19, earthquake displacements can be horizontal and vertical and, in the case of a truly great shock, can amount to as much as 10 or 15 meters. Horizontal displacements along the San Andreas fault at the time of the San Francisco earthquake of 1906 amounted to as much as 5 meters (Fig. 19-9). The earthquake of 1872 that devastated the Owens Valley just east of the Sierra Nevada in California produced a vertical displacement of 4 meters. At the rate of one such earthquake every few thousand years, this lofty mountain range could have been raised in but a few million years, even allowing for erosion.

The permanent displacements accompanying large earthquakes are not restricted to the vicinity of the fault. They can affect immense areas. The great Alaskan earthquake of 1964 is a well-documented example. Figure 22-15 shows the amount of uplift and subsidence that took place in an area of more than 200,000 square kilometers: The maximum uplift amounted to 13 meters and the greatest subsidence was 2 meters. All of this occurred in a few minutes as the Pacific plate slipped a few meters deeper into the mantle beneath southern Alaska. It was not very difficult to document the pattern of vertical displacements along the Alaskan coast. Barnacles and certain other marine organisms that live permanently attached to the rocks of the shore are limited in their vertical distribution by the reach of the tides, such that they grow in horizontal bands along the shore whose upper limits may correspond with, say, the mean high-water level. So close are such correspondences that, following the Alaskan earthquake, uplift could be measured by comparing the height of the band of dead barnacles with that of the band of new, living barnacles (Figure 22-16). Similarly, subsidence could be determined from the height to which high tide inundated land plants. Differences in the levels of pre- and post-earthquake beaches, tide-gage readings, and inland surveys of elevation along roads and railroad routes also provided data on permanent vertical displacement. The displacements accompanying earthquakes, integrated over thousands or millions of years, can be a major factor in regional deformation.

The broad, gradual uplifts described in the preceding pages almost always entail some slight undulating deformation of the rocks. Rocks that look flat-lying and horizontal prove to be slightly tilted and gently folded when careful surveys are made. Within the continent, in apparently flat terrain, there are also small faults along which very small amounts of movement occurred. Thus, folds and

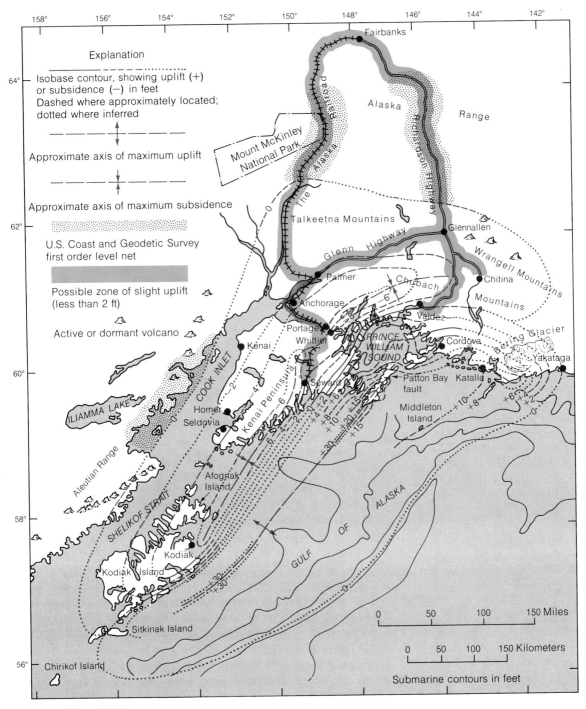

**Figure 22-15**
Permanent uplift and subsidence caused by the great Alaskan earthquake of 1964. Contours show displacements in feet—uplift to the southeast and subsidence to the northwest of a hinge line running from the vicinity of Kodiak to that of Valdez. [After G. Plafker, U.S. Geological Survey.]

**Figure 22-16**
Geologist measuring distance from sea level to the upper limit of barnacle growth on a sea cliff (white band). Barnacles normally grow below the level of high tide, but uplift due to the Alaskan earthquake of 1964 has raised them out of the sea. By measuring the distance from the present level of high tide to the upper growth limit on the cliff, the amount of uplift can be obtained. [Photo by G. Plafker, U. S. Geological Survey.]

**Figure 22-17**
Former sea floor of Montague Island, Alaska, exposed by 8 meters of uplift that accompanied the Alaskan earthquake of 1964. Whitish coating on rocky surface consists mainly of dessicated remains of sea-bottom animals and plants. [Photo by G. Plafker, U. S. Geological Survey.]

minor faults are widely distributed in most parts of the crust. In orogenic belts, strong folding and faulting are the major structural features. Folds and faults are the details of the patterns of deformation, strong and weak, that geologists map in the field as the clues to the larger panorama of tectonics. We shall go on to treat the geometry of these folds and faults as they are exhibited best in strongly deformed terrains.

## FOLDS

The term "fold" implies that a structure that originally was planar, like a sedimentary bed, has been bent. The deformation may be produced by horizontal or vertical forces in the crust. One type of deformation can result when large rock masses glide down an inclined bedding plane, fault plane, or unconformity under the force of gravity.

Folding is the most common form of deformation of layered rocks, and its most typical manifestation is in mountain belts. In young mountain systems, where erosion has not yet erased them, majestic, sweeping folds can be traced, some of them with dimensions of many kilometers (Fig. 22-18). On a much smaller scale,

**Figure 22-18**
Large-scale folds in lower Paleozoic graywacke, northeast Victoria Land, Antarctica. [Photo by W. B. Hamilton, U. S. Geological Survey.]

very thin beds can be crumpled into folds a few centimeters long (Fig. 22-19). Folds can be gentle or the bending can be severe, depending on the magnitude of the applied forces and the ability of the beds to resist deformation. Ancient mountains, long since eroded away, have left a partial record that enables geologists to reconstruct their former grandeur — the remnants of intense folding found today in the stable Precambrian terrains of continental interiors (Fig. 22-20).

In order to characterize and interpret a deformed rock layer, a geologist must first describe it. Ideally, it would be easiest to visualize a deformed stratum if the elevation of its surface could be mapped and contoured, just like a topographic map (Chapter 5)

**Figure 22-19**
Complex fold patterns in strongly deformed marble, northern Norway. [Photo by J. Haller, Harvard University.]

**Figure 22-20**
Folds in Precambrian iron formation, Beresford Lake area, Manitoba. This is a
relic of an ancient mountain-making episode in the continental interior, a region
that has remained relatively stable for approximately the past billion years. [From
Geological Survey of Canada, Ottawa.]

charts the Earth's surface in terms of contours. Sometimes, particularly when a search for oil or minerals is involved, there is
sufficient information to do just this, using such evidence about
the depth of the layer as data obtained from mines, boreholes,
outcrops in valleys, or perhaps the radarlike echoes of seismic
waves (see Box 19-1).

More often, the geologist has to reconstruct the geometry of
folds and faults from sparse surface information only (Fig. 22-21).
He begins with a topographic map showing the contours of the
Earth's surface. From his field observations, which involve much
walking and surveying, he superimposes on the topographic map
the pattern of outcrops of the different formations. The geologist
is ever hopeful the soil cover doesn't hide all of the outcrops. If

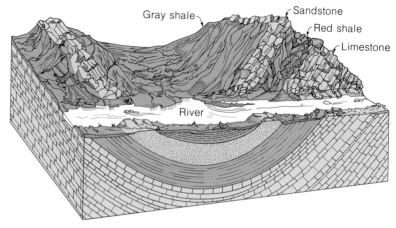

**Figure 22-21**
Geologists typically work from surface outcrops of rock formations to reconstruct subsurface structures. This block diagram shows schematically one example of the surface expression of a fold in a sequence of sedimentary rocks.

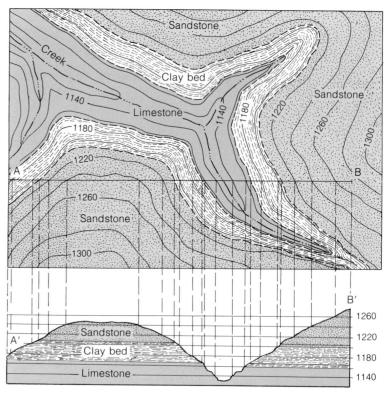

**Figure 22-22**
Construction of a geologic section from a geologic map. In this example the formations are horizontal and therefore crop out at the same topographic level—that is, between the same elevation contours. [After *Principles of Geology* (3rd ed.) by J. Gilluly, A. C. Waters, A. O. Woodford W. H. Freeman and Company. Copyright © 1968.]

there are hills and valleys and the beds are horizontal, the outcrop of a given stratum will always fall at the same topographic level — that is, on the same contour line (Figure 22-22). More often, the beds are tilted, folded, or faulted, and the same formation will outcrop at many different elevations. Using such criteria as relative age (indicated by fossils or radioactive dating) and lithology, the geologist keeps track of the sequence of layers. In order to figure out the structure of the deformed beds, he must also keep track of the geometric attitude or orientation of the beds at outcrops. He does this by marking on his map the **dip** and **strike** of the beds wherever they outcrop. The dip is the angle of inclination of the bed from the horizontal, in the direction of steepest descent: water flows in the down-dip direction. The strike is at right angles to the dip direction. It is the intersection of the plane of the bed with the horizontal plane. Figures 22-23 and 22-24 show how dip and strike are measured in the field and plotted on maps; topographic and geological maps are described in more detail in Appendix V.

**Figure 22-23**
Illustration of the strike and dip of an inclined bed.

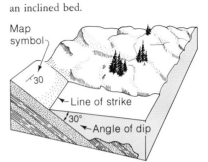

**Figure 22-24**
Geologists measuring the dip of inclined beds using a clinometer. The dip is given by the angle between the horizontal and the edge of the instrument, which is aligned parallel to the bedding planes. The horizontal is found by rotating a level bubble until the bubble is centered. [From U. S. Geological Survey.]

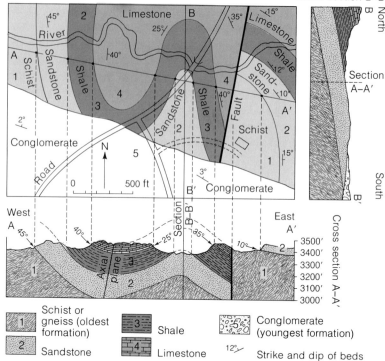

**Figure 22-25**
Geological map with formations, strikes, and dips indicated. Geologic sections
A-A' and B-B' are inferred from the map. Note how the surface features
indicated on the map along the lines A-A' and B-B' are transferred to the
corresponding cross sections, and how the subsurface reconstruction is consistent
with the surface observations. [After U. S. Geological Survey.]

Once the formations are mapped and the dips and strikes are
recorded from many places, one can attempt to reconstruct the
deformed shape of the bed, even when erosion has removed
sections of it. It is like putting together a three-dimensional jig-saw
puzzle with missing pieces. Figure 22-25 shows a geological map
with formations, strikes, and dips indicated, and below it is a cross
section inferred from the map, with eroded portions indicated by
dashed lines. The up folds and down folds of the beds show in the
section as they would appear if a vertical cut could be made through
the Earth in the direction indicated on the map, providing the
geologist was correct in his reconstruction.

Upfolds or arches of layered rocks are called **anticlines**, and
downfolds or troughs are **synclines** (Fig. 22-26). A steplike bend
in otherwise gently dipping or horizontal beds is a **monocline**
(Fig. 22-27).

In order to discuss the different types of folding with an economy
of language, we need to define the parts of a fold: The two sides
of a fold are its **limbs**. The **axial plane** is a surface that divides a

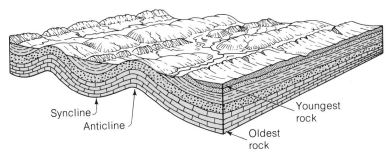

**Figure 22-26**
Upfolds, or anticlines, and downfolds, or synclines. Erosion of the folds
produces parallel ridges where resistant rocks crop out and parallel valleys
on the easily eroded formations. [After U. S. Geological Survey.]

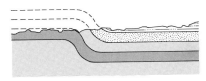

**Figure 22-27**
Schematic illustration of a monocline.

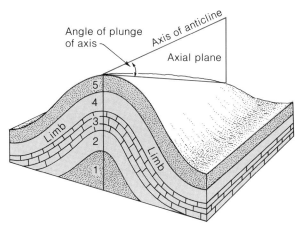

**Figure 22-28**
Diagrammatic illustration of parts of a fold.

fold as symmetrically as possible. The line made by the inter-
sections of the axial plane with the beds is the **axis** of the fold. If
the axis of a fold is not horizontal, the fold is said to **plunge** by an
amount given by the angle the axis makes with the horizontal.
Parts of a fold are illustrated in Figure 22-28.

It is easy to see (Fig. 22-25) that, for a folded sequence of layers,
the oldest beds would be found at depth in the core (or central
axis) of the anticline and the youngest rocks on the surface over
the axis of the syncline. On a geological map or cross section,
eroded anticlines would be recognized by an axial core of older
rocks, bordered on both sides by younger rocks dipping away
from the axis. In the core of synclines, the map would show
younger rocks on the axis with older rocks dipping inward on both
sides, as in Figures 22-25 and 22-26.

**Types of Folds** We should not expect every fold to have a vertical axial plane with limbs dipping symmetrically away from the axis. When a horizontal force is applied to layers that have an initial dip or that change laterally, say, in thickness, strength, or

Symmetrical folds

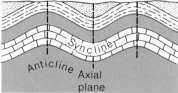

Axial plane is vertical

Overturned folds

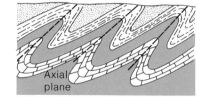

Upper limb of syncline and lower limb of anticline, tilted beyond vertical limbs of folds, dip in same direction

Asymmetrical folds

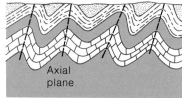

Beds in one limb dip more steeply than those in the other

Recumbent folds

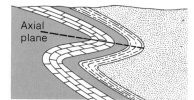

Beds in lower limb of anticline and upper limb of syncline are upside down; axial plane is nearly horizontal

**Figure 22-29**
Diagrammatic illustration of symmetrical, asymmetrical, overturned, and recumbent folds.

**Figure 22-30**
Overturned syncline in Cretaceous limestone, Lake Lucerne, Switzerland. [Photo by W. B. Hamilton, U. S. Geological Survey.]

ductility, or when the force itself is uneven, the folds can be thrown into **asymmetrical** shapes, with one limb dipping more steeply than the other. This is a common situation. When the deformation is intense, the fold can be **overturned**, with the lower limb of an anticline or the upper limb of a syncline tilted more than 90° from its original attitude. Both limbs of an overturned fold dip in the same direction, as in Figure 22-29, and the overturned limb shows an inverted sequence in an outcrop, with older beds on top of younger ones. In **recumbent** folds, the axial plane is horizontal, or nearly so, and one limb has been rotated 180° into a completely upside-down sequence (Fig. 22-31). Figure 22-32 shows how an

Figure 22-31
A recumbent fold, north of Grandjeans Fjord, Greenland. From top of cliff to valley bottom, the difference in altitude is about 800 meters. [Lauge Koch Expedition; courtesy of J. Haller, Harvard University.]

overturned fold might evolve. The spectacular recumbent folds of the Alps, some of the first ones to be mapped in detail, are particularly famous among geologists.

As mentioned earlier, domes and basins are folds in which beds dip radially away from, or toward, a point, respectively (Fig. 22-11). The outcrops of such formations on the surfaces would tend to be circular or elliptical (Fig. 22-33). Domes are very important in oil geology, because oil tends to migrate upward and become trapped against impervious rocks at the high point of a dome.

Follow the axis of any fold, and sooner or later the fold will die out. When an anticline disappears, for example, its axis descends and the folds get smaller and smaller, as in Figure 22-34. The axis of a syncline ascends as the fold disappears. The outcrops of eroded remnants of a plunging fold show a characteristic pattern like the bow of a canoe, because the limbs converge where the fold disappears (Fig. 22-34). Successions of (eroded) plunging anticlines and synclines show on geological maps or air photographs as a zig-zag pattern of outcrops (Figs. 22-25 and 22-35).

Folds tend to occur not in isolation but in elongated groups. When a geologist finds a **folded belt,** he infers that the region at one time was compressed by horizontal crustal forces. The Appalachians were mentioned earlier as an excellent example of a folded mountain belt (Fig. 22-36), and one current hypothesis, discussed in Chapter 21, invokes a plate collision to account for the wrinkling of the once flat sheets of Paleozoic sediments.

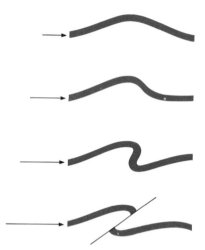

**Figure 22-32**
Diagrammatic illustration of how an overturned fold might evolve as a result of an increasing horizontal force. In the bottom sketch the formation has actually ruptured.

**Figure 22-33**
Circular ridges of resistant formations reveal the shape of an eroded dome at Sinclair, Wyoming. [Aerial photo by J. R. Balsley, U. S. Geological Survey.]

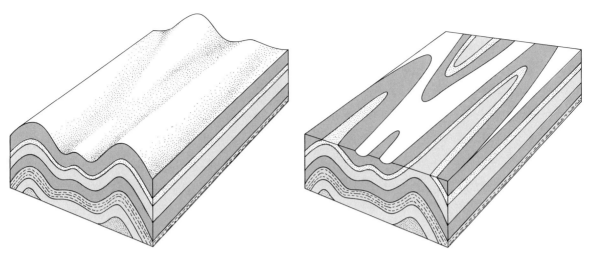

**Figure 22-34**
(Left) Schematic block diagram of plunging folds. (Right) the eroded remnants of plunging folds show a characteristic pattern in which the limbs converge like the bow of a canoe.

## FRACTURES

Rocks yield to deforming forces in several ways: Some layers will crumple into folds, some will break, some will fold first and then break if the applied stresses build up beyond the strength of the rock.

Fractures may be divided into two categories, **joints** and **faults**. A joint is a crack along which no appreciable movement has occurred. If there is displacement of the rocks on both sides of a fracture and parallel to it, the fracture is a fault.

**Joints**　These are structural features found in almost every outcrop. In some places, they are randomly irregular. They are particularly interesting to the geologist if they show a regional pattern —for example, if all the cracks in a particular area are approximately parallel. Very often there are two or more sets of joints that intersect, breaking the rocks into large rectangular blocks or parallelepipeds. Regularity in a joint system often implies uniformity in the stress system that produced them. For example, the compression of sedimentary layers produced by the weight of overlying beds can produce one set of joints and the decompression of erosional unloading can establish another set. Regional stresses—compressional, tensional, or shear—that have long since vanished, can leave their imprint in the form of distinct sets of joints. The contraction of layers of cooling lava can result in columnar jointing (see Fig. 17-10).

As noted in Chapter 6, joints provide channels for the flow of water through rock. Because of the greater ratio of surface area to volume, the irregular blocks formed by intersecting joint systems

**Figure 22-35**
The erosional remnants of plunging folds show a zig-zag pattern in this view taken in the Valley and Ridge belt of the Appalachian Mountains, 30 miles northwest of Harrisburg, Pennsylvania. In the drawing below, the imaginary trench reveals the subsurface structure. [From *Geology Illustrated* by J. S. Shelton. W. H. Freeman and Company. Copyright © 1966.]

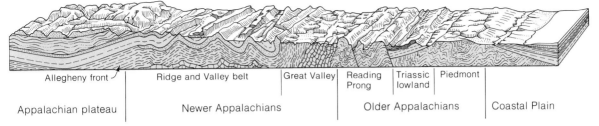

**Figure 22-36**
Block diagram of a section of the Appalachian Mountains, an example of the remnants of a folded mountain belt. [After *Stream Sculpture on the Atlantic Slope* by D. Johnson. Columbia University Press. Copyright © 1931.]

Figure 22-37
Intersecting joints form rectangular blocks in volcanic rocks near Braintree, Mass. [Photo by J. Haller, Harvard University.]

are more readily attacked by air and water than are massive, unjointed formations; thus, the weathering process is accelerated by jointing (Fig. 22-37). Joints also can provide underground channels for the flow of magma, often leaving in evidence a swarm of parallel dikes, or cracks filled with igneous rock. Magma injected under high pressure can enlarge existing joints or open new ones.

Block before faulting

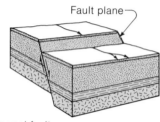

Normal fault

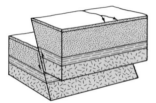

Reverse fault

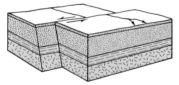

Strike-slip fault

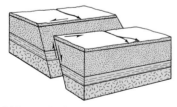

Oblique-slip fault

**Figure 22-38**
Types of faults.

**Faults** Faults have already been discussed in the chapters on seismology and plate tectonics (Chapters 19 and 21), so the review here will be brief. Like folds, faults are a common feature of mountain belts, particularly where the deformation is intense. As discussed in Chapter 21, margins where plates collide, pull apart, or drift past each other are the sites of subduction zones, rift valleys, or transforms, all of which involve faults. Some transform faults—the San Andreas fault of California, for example—show such large displacements that the offset of the two plates may amount to hundreds of kilometers.

The different categories of faulting are distinguished by the direction of motion along the fracture plane (Fig. 22-38). A **dip-slip** fault involves displacements up or down the dip of the plane. A **strike-slip** fault is one in which the movement is horizontal, parallel to the strike of the fault plane. A combination of dip-slip and strike-slip movements would describe an **oblique-slip** fault.

Faults need a further characterization, since the movement can be up or down, or right or left. A **normal fault** is one in which the rocks above the fault plane move down relative to the rocks below. A **reverse fault**, then, is one in which the rocks above the fault plane move up relative to the rocks below. A reverse fault in which the dip is small so that the overlying block is pushed predominantly horizontally is a **thrust fault**. Finally, if, as one faces a strike-slip fault, the block on the other side is displaced to the right, then the fault is a **right-lateral** fault; **left-lateral** faults are displaced in the opposite direction (see Figs. 1-23, 19-9, and 22-39 for some photographs of faults).

Low-angle thrusts or overthrusts with displacements of many kilometers are often found in intensely deformed mountain belts. These thrusts are an expression of compressive forces, which shorten the crust. The surficial layers accommodate the crustal shortening, in this case, by breaking, with one sheet overriding another. An example of thrusting from California and southern Nevada is shown in Figure 22-40.

**Grabens**—a term sometimes used synonymously with rift valleys—are long, narrow troughs bounded by one or more parallel normal faults. Tensional crustal forces, literally pulling the crust apart, are responsible for these down-dropped fault blocks. The East African rift valleys, the mid-ocean ridge rifts, and the Rhine River Valley are famous examples. A **horst** is the opposite, a ridge formed by parallel reverse or, more commonly, normal faults (Figs. 22-41, 22-42, and 22-43).

Because a fault implies that dislocation has occurred along a fracture, it can be recognized in the field by the disruption of formations on either side. In some places, the two segments of an offset formation can be found, and the amount of relative displacement can be measured directly (Fig. 22-39), but this becomes increasingly difficult with large offsets. In establishing the time of

**Figure 22-39**
Thrust fault in limestone, Gusta, Norrland, Sweden. Offset of once-continuous bed, indicated by line, is about 0.8 m [Photo by J. Haller, Harvard University.]

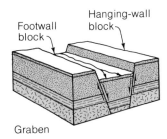

Graben

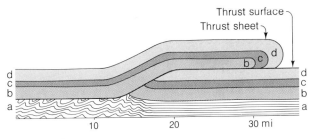

**Figure 22-40**
Schematic diagram of large-scale thrust sheet, modeled by C. K. Longwell after examples in California and southern Nevada.

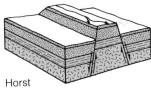

Horst

**Figure 22-41**
Block diagrams of a graben and a horst.

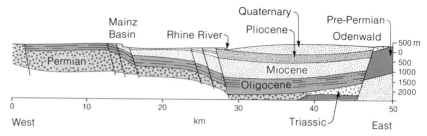

**Figure 22-42**
Geologic cross section of the Rhine graben. [After P. Dorn.]

**Figure 22-43**
The Red Sea bifurcates to form the Gulf of Suez on the left and the Gulf of Aquaba on the right. The Arabian Peninsula on the right, splitting away from Africa on the left, has opened these great rifts, now flooded by the sea. [Photo by Apollo 7 astronauts; from National Aeronautics and Space Administration.]

**Figure 22-44**
Slickensides on rhyolite, Braintree, Mass. [Photo by J. Haller, Harvard
University.]

faulting, geologists use the simple rule that a fault must be younger
than the youngest rocks it cuts and older than the oldest undis-
rupted formation that covers it (see Fig. 22-2).

Crushed or ground up rock is often found in a fault zone. In
some places, polished and striated surfaces result from friction
along a fault plane. These surfaces are called **slickensides** (Fig.
22-44); because they align in the direction of relative movement
of the fault blocks, they tell the geologist what this direction was,
though not with complete reliability. If movement has occurred
recently, the effect on topography, drainage patterns, vegetation,
and man's artifacts can often be recognized (see Figs. 19-1
and 19-9).

Several cross sections of folded and faulted belts are shown in
Figures 22-36, 22-45, 22-46, and 22-47 to illustrate some of the
ways rocks deform in response to crustal forces.

## TOPOGRAPHIC EXPRESSION OF DEFORMATION

Needless to say, deformation—in the form of mountain belts with
their folds and faults, plateaus, grabens, and strike-slip faults—
leaves its mark on the configuration of the Earth's surface. These

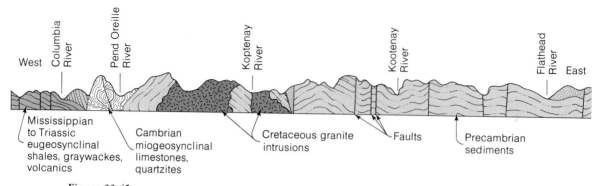

**Figure 22-45**
Geologic cross section from the Cascade Mountain front of northeast Washington to the Rocky
Mountains of northwest Montana. Deformed eugeosynclinal and miogeosynclinal deposits and
granitic intrusions characterize much of this region. [After R. G. Yates, G. E. Becraft, A. B. Campbell,
and R. C. Pearson, Canadian Institute of Mining and Metallurgy.]

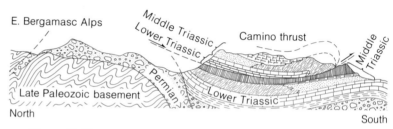

**Figure 22-46**
Unconformities, folds, and thrust sheets in a section of the southern Alps.
[After *Structural Geology* by L. U. De Sitter. McGraw-Hill Book Company.
Copyright © 1964.]

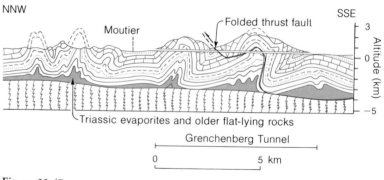

**Figure 22-47**
Folds and faults in a section of the Juras, an example of a folded mountain belt.
According to one hypothesis, the beds overlying the Triassic were sheared off
and deformed independently of the older rocks below.

**Figure 22-48**
The Jura Mountains, in a view looking northeast over the town of Moutier, Switzerland. The near and distant ridges are anticlines of upper Jurassic limestone. The crest of the anticline can be seen in the water gap in the distant ridge. The town is in a synclinal valley. See previous figure. [Photo by J. Haller, Harvard University.]

topographic expressions are often a guide to the structures that control them. Even with such relatively small-scale features as the shapes of hills and valleys and the courses of streams, the controlling factors are the structural elements in a complex interaction with erosion.

It should also be apparent that, the older the structure, the more likely it is that erosion has erased it and the less evident is its physiographic expression. Thus, present-day surface relief is largely due to movements that occurred during and since the Tertiary Period. These movements and earlier erosion have tended to obscure Mesozoic and Paleozoic structures. Precambrian deformation no longer shows as mountains but only as relic remnants

**Figure 22-49**
Rugged topography of a young mountain belt. View over Bedretto Valley, Switzerland, to the
front of the Pennine Alps on the left. The formations behind the front have glided a long distance
from their original place of deposition to a position over the beds in the foreground. [Swissair.]

of folds and faults in the basement rocks of the continental interior.
Examples of Tertiary deformation with marked physiographic
expression are the Alpine and Himalayan belts, the island arcs and
deep-sea trenches of the Pacific, the great rift valleys of Africa,
the rejuvenated Late Paleozoic Appalachians, the Rocky Mountains, and the Pacific Coast Ranges (Fig. 22-49).

Although most features of topography are relatable to some
particular structural element, one should not expect that the crests
of anticlines always form ridges and that the troughs of synclines
always become valleys. When stratified rocks are deformed, the
resistance of the individual beds to weathering and erosion are
important factors in controlling landform, as well as whether the
layers are flat-lying, tilted, folded, or faulted. In the block diagram
of the Appalachians (Fig. 22-36), one can see all combinations of
anticlinal and synclinal hills and valleys.

Up to this point, we have discussed the slow, up and down movements of the crust, the deformation of layers by folding and faulting due to horizontal crustal forces, and the emplacement of plutons. These structural elements, in conjunction with the kinds of rocks involved—in short, the sequence in space and time of magmatic, sedimentary, and deformational episodes—determine the geological character of a region. In this section, we will discuss the broad regional characteristics of continents, using North America as an example.

The geological fabric of a continent is not random. Similar structures tend to occur in association so as to make up distinctive regions—such as regions of folded mountain belts, upwarped mountains, fault-block mountains, stable interior platforms, marginal depressions, and so on. Moreover, the spatial relations of these regions seem to have a pattern, as we can see from Figures 22-1, 22-50, and 22-51, in which the major structural divisions of

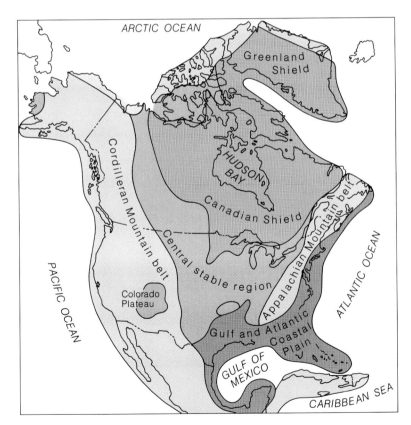

Figure 22-50
Major tectonic features of North America.

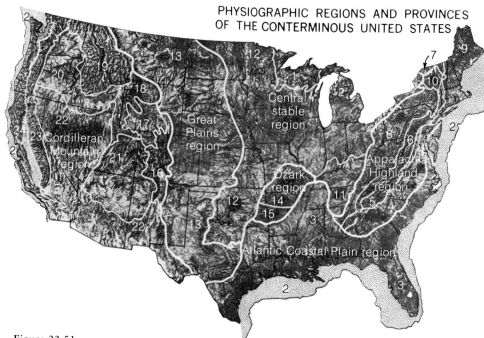

PHYSIOGRAPHIC REGIONS AND PROVINCES
OF THE CONTERMINOUS UNITED STATES

**Figure 22-51**

Physiographic regions and provinces of the conterminous United States. [After U.S. Geological Survey.]

1. *Superior Upland.*   Hilly area of erosional topography on ancient crystalline rocks.
2. *Continental Shelf*   Shallow, sloping submarine plain of sedimentation.
3. *Coastal Plain*   Low, hilly to nearly flat terraced plains on soft sediments.
4. *Piedmont province.*   Gentle to rough, hilly terrain on belted crystalline rocks becoming more hilly toward mountains.
5. *Blue Ridge province*   Mountains of crystalline rock 3,000 to 6,000 feet high, mostly rounded summits.
6. *Valley and Ridge province.*   Long mountain ridges and valleys eroded into strong and weak folded rock strata.
7. *St. Lawrence Valley.*   Rolling lowland with local rock hills.
8. *Appalachian Plateaus.*   Generally steep-sided plateaus on sandstone bedrock, 3,000 to 5,000 feet high on the east side, declining gradually to the west.
9. *New England province.*   Rolling, hilly, erosional topography on crystalline rocks in the southeastern part, changing to high mountainous country in the central and northern parts.
10. *Adirondack province.*   Subdued mountains on ancient crystalline rocks rising to more than 5,000 feet.
11. *Interior low plateaus.*   Low plateaus on stratified rocks.
12. *Central Lowland.*   Mostly low, rolling landscape and nearly level plains. Most of area covered by a veneer of glacial deposits, including ancient lake beds and hilly, lake-dotted moraines.
13. *Great Plains.*   Broad river plains and low plateaus on weak stratified sedimentary rocks. Rises toward Rocky Mountains, reaching altitudes above 6,000 feet at some places.
14. *Ozark Plateaus.*   High, hilly landscape on stratified rocks.
15. *Ouachita province.*   Ridges and valleys eroded on upturned folded strata.
16. *Southern Rocky Mountains.*   Complex mountains rising to more than 14,000 feet.
17. *Wyoming Basin.*   Elevated plains and plateaus on sedimentary strata.
18. *Middle Rocky Mountains.*   Complex mountains with many intermontane basins and plains.
19. *Northern Rocky Mountains.*   Rugged mountains with narrow intermontane basins.
20. *Columbia Plateau.*   High rolling plateaus underlain by extensive lava flows; trenched by canyons.
21. *Colorado Plateau.*   High plateaus on stratified rocks cut by deep canyons.
22. *Basin and Range province.*   Mostly isolated ranges separated by wide desert plains. Many lakes, ancient lake beds, and alluvial fans.
23. *Cascade-Sierra Nevada Mountains.*   The Sierra Nevada, in the southern part of the province, are high mountains eroded from crystalline rocks. The Cascades, in the northern part of the province, are high volcanic mountains.
24. *Pacific Border province.*   Mostly very young, steep mountains; includes the extensive river plains in California.

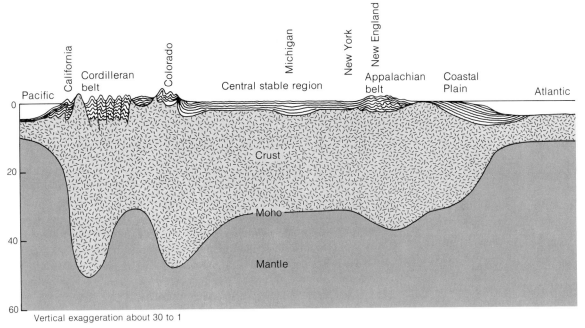

**Figure 22-52**
Schematic section across the United States, depicting major structural regions and variations in crustal thickness.

North America are indicated. A transcontinental section showing the major structural divisions schematically is shown in Figure 22-52.

**Shields** The Precambrian Canadian **shield** is the great stable portion of the continent. It has been relatively undisturbed since Precambrian time, except for gentle warping. It is dominated by granitic and high-grade metamorphic rocks, such as gneisses — which, together largely with highly deformed and metamorphosed sediments and volcanic rocks, imply a series of intense mountain-making episodes in Precambrian time before the stable conditions set in (Fig. 22-53). The shield also includes some very old sediments that were hardly touched by deformation and metamorphism. This primitive block contains one of the oldest and most complex records of geological history, much of it still unknown in detail, and it is famous for major deposits of iron, gold, copper, and nickel. Other examples of shields can be found in Fennoscandia, Siberia, central Africa, Brazil, and Australia.

South of the Canadian shield (see Figs. 22-50 and 22-52) is another stable region that, in a sense, is a subsurface continuation of the shield, inasmuch as it contains Precambrian basement rocks covered by a veneer of Paleozoic sediments. The sediments, which were laid down under quiet conditions, have remained unmetamorphosed and only slightly deformed to this day. There is abundant

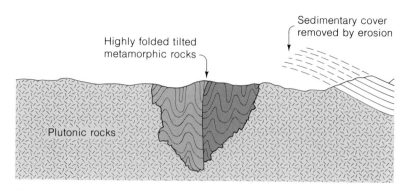

**Figure 22-53**
Idealized section in the Canadian Shield. The highly deformed and metamorphosed rocks in this region indicate that an intense orogenic episode took place in Precambrian times, before stable conditions set in.

evidence for up and down movements—basins, domes, marine transgressions and regressions, overlaps, and unconformities. The deformation is mostly in gentle tilts, folds, and faults with small displacements. The Russian platform, which includes much of European Russia, is an analogous structure. Figure 22-10 shows a representative section on a stable block.

**Orogenic Belts**  Referring to Figures 22-50 and 22-51, we see that, surrounding the great stable interior of the continent, there are **orogenic** belts, regions that were deformed by folding and faulting and were subjected to plutonism and metamorphism at various times in the Paleozoic, Mesozoic, and Cenozoic eras.

Orogenic (or folded) mountain belts usually show certain general features that enable us to say something of how they were formed. Figure 22-54 reviews this important concept, which was discussed in Chapter 21 in terms of plate-tectonic theory. These deformed belts occur typically on the margins of stable shields in close connection with geosynclines. A geosyncline is a subsiding linear trough that earlier received a thick succession of layered sediments, sometimes more than 10,000 meters (30,000 feet) thick—which contrasts strongly with the thin sedimentary veneer on the shield. Typically, two geosynclines are involved, an inner miogeosyncline with sandstone and limestone shelf deposits and an outer eugeosyncline with a great thickness of such deep-water sediments as turbidites, shales, and pelagic limestones, and volcanic rock (see Chapter 13). Folded mountain chains almost always seem to be the result of a wave of deformation that begins in the eugeosyncline and later extends to the miogeosyncline. Lateral forces develop that intensely compress the crust in some places by 20 percent or even more, as if it were caught in the jaws of a vise. In response to this squeezing, the overlying sedimentary

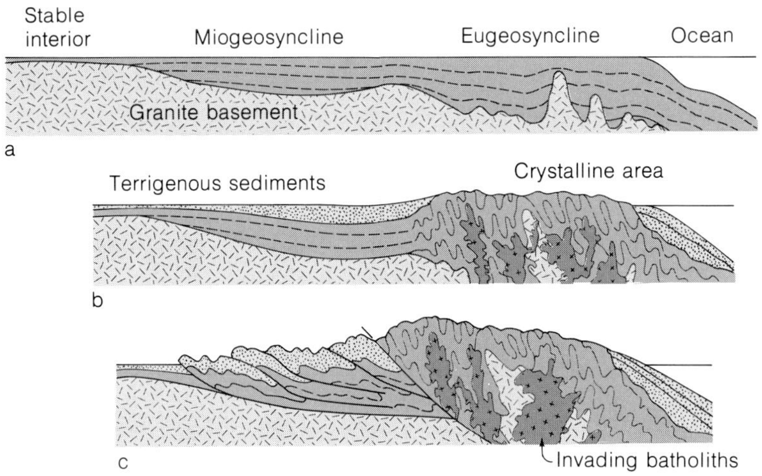

**Figure 22-54**
General features of an orogenic belt, showing its inferred development from
geosyncline to deformed belt. (a) Geosynclinal stage. (b) After deformation of
eugeosyncline and deposition of terrigenous sediments eroded from the newly
formed mountains. (c) After final deformation of miogeosyncline. According to
the plate tectonics hypothesis, stage a would follow plate separation and the
opening of an ocean basin, with deposition on the receding margin of a continent.
Stages b and c would occur with plate collision and the closing of an ocean basin.
[After *The Evolution of North America* by P. B. King. Copyright © 1959 by
Princeton University Press. Redrawn with permission of Princeton University
Press.]

cover in the eugeosyncline is deformed into great folds and thrust
faults, generally trending parallel to the edge of the shield. Intrusions work their way into this compressed and contorted belt.
New mountains are thrown up, reversing submarine slopes, so
that erosion leads to episodes of rapid deposition of terrigenous
sediments towards the continent on one side, both in arms of the
sea and on land, as well as towards deeper water on the other side,
as Figure 22-54 shows.

In a final stage, the miogeosyncline is deformed. The deforming
forces are typically directed from the ocean basins towards the
interior shield. This can be seen in the frequent transition from
gentle folds adjacent to the shield to severe folding and thrusting
and increasing metamorphism and magmatism with increasing
distance away from the shield. Also, the deformed belts are
asymmetrical in other ways—the folds lean or are overturned towards the shield, and the thrust sheets have been pushed in the
same direction.

Batholiths form the cores of many continental mountain ranges,
which implies that they have some connection with the mountain-
making process. For example, the great mountain ranges that
extend for thousands of kilometers along the western margins of
the North American and South American plates have cores of
contiguous batholiths. Huge areas of exposed Precambrian terrain

in the now stable areas within continents are made up of successions of batholiths, and many of these seem to be the roots of mountains that have long since been eroded. In Chapter 18, we discussed the idea that the formation of some batholiths was connected with the subduction process at the margin of colliding plates. It was hypothesized that reheating and melting of the descending plate could provide rising, hot fluids as raw materials for batholith formation along the margins of the overriding plates (see Fig. 21-11).

Very often, following the culmination of major folding and after erosion wears the mountains down, the belt is rejuvenated by later upwarping, and the old structures still remaining are exposed. Rejuvenated fold belts are exemplified by the Alps, the Urals, and the Appalachians.

All of these features are well developed in the Appalachian belt, which bounds the stable interior on the east in a zone running northeast from Alabama to Newfoundland. This classical region which served as the model for Figure 22-54, was discussed in terms of plate tectonics in Chapter 21. In the block diagram of the Appalachians of Figure 22-36, the remains of the deformed eugeosynclinal belt are depicted as the "older Appalachians." The folded miogeosyncline now shows as the Ridge and Valley belt in the "newer Appalachians."

Bounding the central stable region of North America on the south is a fold-and-thrust fault belt, not unlike the inner zone of the Appalachians. Its surface expression can be seen in the Ouachita Mountains of Arkansas and eastern Oklahoma, and the Marathon Mountains of western Texas. These mountains seem to be a continuation of the Appalachian belt, which curves west through Alabama, Mississippi, and Arkansas and then southwest to the Big Bend region of Texas.

On the west, the central stable region is bounded by a complex of orogenic zones of several types (refer to Fig. 22-51). This is the region of the North American Cordillera, a mountain belt containing some of the highest peaks on the continent. Across its middle section between San Francisco and Denver, the Cordilleran system is about 1600 kilometers (1000 miles) wide and includes several contrasting physiographic provinces: Region 24 of Figure 22-51, the Coast Ranges along the Pacific Ocean; the Great Valley of California; Region 23, the lofty Sierra Nevada; Region 22, the Basin and Range province (a region of faulted and tilted blocks showing as numerous narrow mountain ranges and valleys extending from the California–Nevada border to western Utah); Region 21, the high table land of the Colorado Plateau; Region 16, the rugged Rocky Mountains, which end abruptly at the edge of the Great Plains on the stable interior (Region 13). The Apollo 16 photograph of western North America (Fig. 22-1) clearly shows these features.

The Cordilleran system is topographically higher and more extensive than the Appalachians because its main orogeny was more recent, having occurred in the last half of Mesozoic and early Tertiary time. However, the form and height of the Cordillera we see today are not due to this major episode of folding, faulting, plutonism, and metamorphism. Rather, they are manifestations of more recent events that took place in Tertiary and Quaternary time and rejuvenated the mountains. At that time, for example, the Colorado Plateau and the Central and Southern Rockies attained much of their present height as a result of a broad regional uplift. Stream erosion was accelerated, the mountain topography sharpened, and the canyons deepened. Block faulting, which broke the crust into mountains and valleys in the Basin and Range province and tilted the Sierra Nevada block, occurred in Tertiary and Quaternary time. At that time, too, volcanism spread sheets of ignimbrite widely over the southern parts of the Basin and Range province and covered large parts of the Pacific Northwest with lava. This culminating stage of uplift, block faulting, and volcanism served as a model for Figure 22-2. The geologist knows how to read through the confusion of these recent events and reconstruct the orogenic history of the region. The entire story of the Cordillera is a complicated one, and we tell it here in highly simplified form. It is a story that begins with geosynclines.

From the nature and distribution of rocks, geologists can reconstruct the distribution of lands and seas of the past. A map of North America as it was in the Orodovician period, some 475 million years ago, is shown in Figure 22-55. The main belt of the Cordillera existed at this time as two undersea troughs, trending in a north–south direction and separated in places by islands. These geosynclines received voluminous deposits of Paleozoic and earlier Mesozoic rocks—primarily a marine section of volcanics, graywackes, and cherts in the outermost western eugeosynclinal belt, and a limestone, sandstone, and shale accumulation in the inner eastern miogeosynclinal belt. A restored section of these Ordovician geosynclines, one with all the affects of subsequent geological events removed (so that it looks as it did in Ordovician time), is shown in Figure 22-56.

Deformation accompanied by metamorphism and plutonism took place at various times in the Paleozoic and Mesozoic along much of the length of the eugeosyncline, but the climactic episode of deformation that ended geosynclinal conditions and refashioned much of the region into a land area began early in the Jurassic and extended well into Cretaceous time (Fig. 22-57). Intense folding and thrusting, followed by batholithic intrusions, deformed and uplifted this outer belt. These batholiths, and the deformed and metamorphosed sediments they intruded, now form large parts of the Klamath Mountains of southwestern Oregon and northern California and the Sierra Nevada of eastern California.

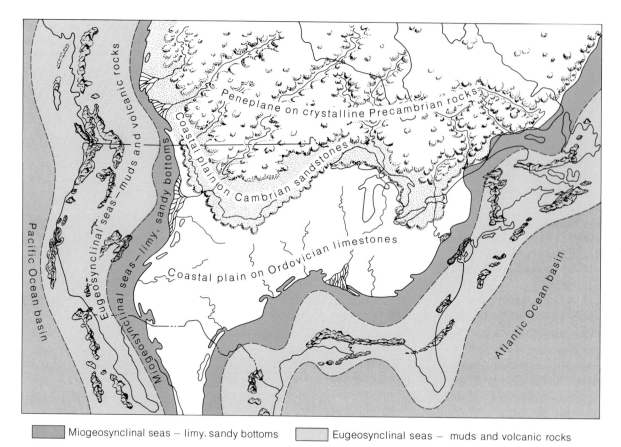

Miogeosynclinal seas – limy, sandy bottoms          Eugeosynclinal seas – muds and volcanic rocks

**Figure 22-55**
Geography of North America in early Ordovician time, about 475 million years ago, reconstructed from the evidence of today's rocks. Off the coasts are the miogeosynclinal and eugeosynclinal seas. [After "The Origin of Continents" by M. Kay. Copyright © 1955 by Scientific American, Inc. All rights reserved.]

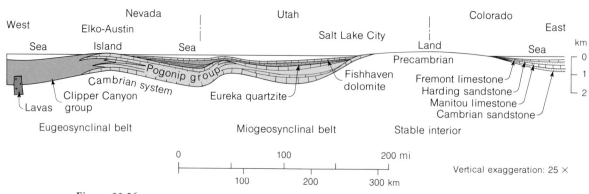

**Figure 22-56**
Restored section of Ordovician rocks from central Nevada to central Colorado, with subsiding geosynclinal belts in the west and stable interior to the east. [After *Stratigraphy and Life History*, edited by M. Kay and E. H. Colbert. Copyright © 1965 John Wiley & Sons.]

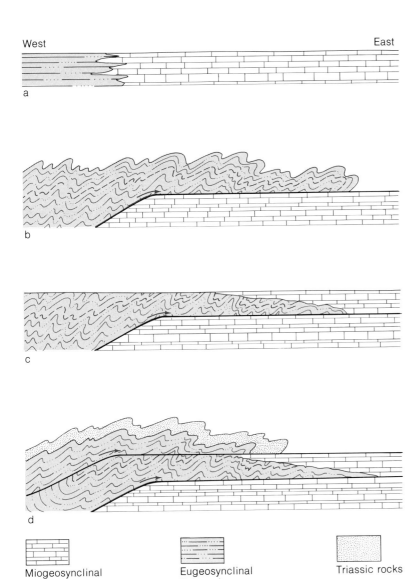

West                                                                East

a

b

c

d

Miogeosynclinal
Paleozoic
rocks

Eugeosynclinal
Paleozoic
rocks

Triassic rocks

**Figure 22-57**
Structural history of north-central Nevada, illustrated diagrammatically.
(a) Deposition of eugeosynclinal and miogeosynclinal formations during early
to middle Paleozoic time. (b) Eastward thrusting of eugeosynclinal over
miogeosynclinal deposits during late Mississippian or early Pennsylvanian time.
(c) Renewed miogeosynclinal deposition over thrust sheets during Pennsylvanian
and Permian time. (d) Renewal of eastward thrusting during mid-Mesozoic time.
[After *The Evolution of North America* by P. B. King. Copyright © 1959 by
Princeton University Press. Redrawn with permission of Princeton University
Press.]

As in the Appalachians, the orogeny reached its climax earlier in the outermost, seaward belt. Thus, the deformation of the Cordillera progressed in a complex wave to the east. It culminated in late Cretaceous and early Tertiary time, as the miogeosyncline and the shelf deposits along the adjacent margin of the stable continental platform further to the east were deformed by an episode of intense faulting and thrusting. This phase of the orogeny deformed the geosynclinal deposits of the miogeosyncline from Central Nevada eastward to the edge of the continental platform area in Wyoming, central Colorado, and New Mexico, along what is now the eastern edge of the Rocky Mountains. Here, the deformation involved the relatively thin sediments of the continental platform in high, broad anticlines that have since been eroded to expose the basement granitic rocks. The Precambrian Pike's Peak granite of the Colorado Front Range is one of these. The deformations of the geosynclines and the intrusions of batholiths accompanying this great orogeny are preserved in the record of the rocks still remaining, as in Figure 22-45, a section across the Cordillera near the Canadian border.

Well after the compression of the two great troughs, and mostly in late Tertiary time, the spectacular episode of **block faulting**, mentioned earlier, occurred in a region extending southeast from southern Oregon to Mexico and including Nevada, western Utah, and parts of eastern California, Arizona, New Mexico and western Texas. This latest structural imprinting, superposed on the earlier-deformed geosynclinal rocks, is responsible for the present-day features of the Basin and Range province. Thousands of high-angle faults sliced the crust into innumerable upheaved and down-dropped blocks, forming hundreds of discontinuous, narrow mountain ranges and intervening, alluvium-filled basins separated by normal faults. Successive movements along the faults over millions of years resulted in the large vertical displacements evident today (Figs. 2-22 and 22-58,$c$). Some of the faults have remained active through the Cenozoic (Fig. 22-59). Unlike folded mountains (Fig. 22-58,$b$), such as those of the Appalachians, which uplift primarily unmetamorphosed sedimentary geosynclinal deposits, the kinds of rocks and structures exposed in **fault-block mountains** vary greatly. Plutons, lavas and tuffs, metamorphic rocks, and folded geosynclinal deposits can be found in the different ranges. The intervening lowlands, once formed, accumulate sediments eroded from the adjacent mountains of upthrown blocks.

The Sierra Nevada fault block consists of granitic batholiths and tightly folded and metamorphosed beds, structures formed during the Late Jurassic–Early Cretaceous orogeny of the western trough previously mentioned. The mountains of this earlier episode were worn down, leaving a deeply eroded remnant after the Mesozoic. About 10 million years ago, in Pliocene time, the block was pushed up on the east and tilted to the west. In 1872, the

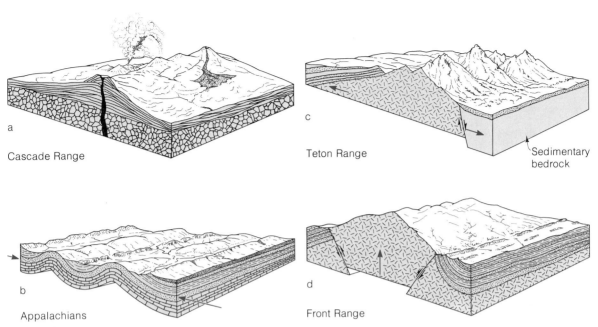

a

Cascade Range

c

Teton Range

Sedimentary
bedrock

b

Appalachians

d

Front Range

**Figure 22-58**
Mountains vary in form and origin. (a) Mountains formed by volcanic action. (b) Mountains resulting
from folded layers of rock. (c) Mountains formed from fault blocks. (d) Mountains originating in
vertical uplift. [After U. S. Geological Survey.]

**Figure 22-59**
Normal faulting in the Basin and Range province. The fault scarp formed during
the Dixie Valley-Fairview Peak (Nevada) earthquake of December 16, 1954.
Dip-slip displacement of 2.1 m and strike slip of 3.3 m occurred at this location.
Faulting over a distance of about 100 km accompanied this magnitude 7.1
earthquake.

**Figure 22-60**
Fault-block mountains. The Sierra Nevada (skyline) viewed from the Panamint Mountains, California. Sierra fault scarp in distance. In the middle ground is the Argus range, a late-Quaternary upfaulting of Plio-Pleistocene basalts. A small fault valley, side up, cuts the alluvium of Panamint Valley, below the center of the picture. [Photo by W. B. Hamilton, U. S. Geological Survey.]

bounding fault on the east was the site of one of the greatest earthquakes in United States history. Mt. Whitney is one of many peaks of the Sierra Nevada standing more than 4000 meters above sea level. (We used the Sierra Nevada as an example in a discussion of plutonism in Chapter 18; see Fig. 18-6.)

Some other examples of fault-block mountains are the Wasatch Range of Utah, the Teton Range of Wyoming, and the tilted edges of the rift valleys of east Africa and the Dead Sea–Jordan Valley rift of Israel.

As mentioned earlier, the modern mountainous topography of the Rocky Mountains is due to epeirogeny — a broad upwarping that took place in Cenozoic time, after the earlier episodes of folding and deep erosion. The region was raised 1500–2100 meters (5000–7000 feet) in the last 15 or 20 million years, pushing Precambrian basement rocks and their veneer of later-deformed sediments above the level of their surroundings (Fig. 22-58,*d*). Other examples of **upwarped mountains** are the Adirondacks, the Black Hills, the Labrador Highlands, and the mountains of Fennoscandia.

The Colorado Plateau, the classic showplace of geology discussed in Chapter 2, seems to be an island of the central stable region, cut off from the interior by the Rocky Mountain orogenic belt. Since the late Precambrian, it has been a stable shelf area — with no thick geosynclinal deposits and no major orogeny, with movements mainly up and down, and with slight deformation in places into broad folds and basins. Volcanics and sedimentary debris from adjacent mountains filled the lowlands of the plateau in Cenozoic times.

The Coast Ranges of California lie between the Great Valley and the Pacific Ocean. These mountains contain rocks younger than those to the east. Great thicknesses of clastic rock, graywackes, shales, cherts, and pillow basalts accumulated in a geosyncline that formed seaward of the troughs discussed earlier, mainly after the climax of the orogenies to the east. These deposits show strong deformation in the form of a series of long fault slices trending generally northwestward, some of which were in existence as far back as mid-Tertiary and others of which, such as the famous San Andreas fault, are still active today.

Just how the spectacular Cordilleran orogenies just described can be fitted into the framework of plate tectonics is not entirely clear, and many geologists are now beginning to work on this important question.

The Atlantic **coastal plain** and the **continental shelf** that is its offshore extension (see Chapters 5 and 21) are features that began developing after the close of the Appalachian orogeny. The outer metamorphic zone of the Appalachians were block-faulted in Triassic time, and a series of long, narrow, down-dropped grabens developed along a belt running from South Carolina to Nova

---

### Box 22-1  The Alps

The Alps, another great example of an orogenic belt, are a system of ranges extending for some 1000 kilometers (600 miles) in an arcuate course running from the French–Italian Mediterranean coast through Italy, Switzerland, and Austria. The highest peak is Mont Blanc in southeastern France, with an elevation of 4810 meters (15,781 feet). In the development of the Alps, a whole series of geosynclines separated by anticlinal ridges received Mesozoic deposits. The southern Triassic geosyncline was strongly disturbed by faulting and folding—but without a major magmatic phase—and developed into the Lombardy Alps (see Fig. 22-46). The central geosyncline, a Jurassic one, now the Pennine Alps (see Fig. 22-49), was subjected to extreme folding, with recumbent folds piled up one atop the other; it also experienced magmatic and strong metamorphic episodes. The northern geosyncline, which is now the Helvetides, is of Cretaceous age; it shows no magmatic phase,

but strong folding with thrust sheets developing from asymmetric anticlines, the sheets seemingly piled upon one another. In a later stage of compression and uplift, huge masses of sediments were detached from the rising belt and glided down the flanks of ranges, pulled by gravity and intricately folded as they slid into neighboring troughs in which Tertiary sediments were accumulating. The Alps have been termed unique in the grandeur of their scale and the complexity of their folding and overthrusting.

By Pliocene time, these complex structures were eroded to a low terrain. The present grandiose topography of the region reflects rather recent epeirogeny, with most of the uplift and erosion having occurred prior to the Pleistocene glaciation.

Movement of the African block against Europe has been proposed as the source of the Alpine orogeny, which has been developing since early Mesozoic time.

**Figure 22-61**
Salton Sea and Imperial Valley of California, looking toward Arizona. High heat flow, recent
volcanism, active faults, and progressive widening suggest that the trough originates as part of the
spreading process that is opening the Gulf of California. The region has high potential as a source of
geothermal energy. [National Aeronautics and Space Administration, *Science,* and W. A. Elders,
R. W. Rex, T. Meidav, P. T. Robinson, and S. Biehler, University of California at Riverside.]

Scotia. These basins trapped a thick series of red, clastic deposits,
which were later intruded by basaltic sills and dikes. The Connecticut River valley and the Bay of Fundy are examples (Fig. 22-62).

Following this, in early Cretaceous time, the deeply eroded and
beveled coastal plain and continental shelf began to subside and
to receive sediments from the continent. Cretaceous and Tertiary

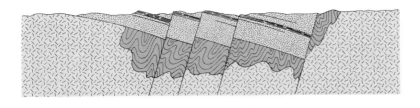

0    10 km

Granitic intrusives
Pre-Triassic

Deformed and
metamorphosed sediments
Pre-Triassic

Triassic sandstones
and shales with
interbedded basaltic
sills, surface flows,
dikes

Figure 22-62
Diagrammatic section of Triassic basins of Connecticut. Nonmarine sediments
were trapped in basins formed by tilted fault blocks. Basaltic flows intruded
and covered these deposits.

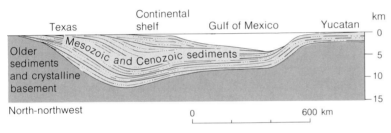

Figure 22-63
Schematic section across the Gulf of Mexico from Texas to Yucatan, showing
inferred thicknesses of Mesozoic and Cenozoic sediments. Maximum thickness
near the Texas coast may exceed 10 km. [After *The Evolution of North America*
by P. B. King. Copyright © 1959 by Princeton University Press. Redrawn with
permission of Princeton University Press.]

sediments up to 5 kilometers thick filled the slowly subsiding
trough, and even more material was dumped into the deeper water
of the continental rise (Fig. 21-13). This is the living geosyncline
referred to in Chapter 21; it is possibly a forerunner of some
future orogeny.

The Gulf coastal plain and shelf are continuous extensions of
the Atlantic ones, though interrupted in a way by Florida. The
Mississippi, Rio Grande, and other rivers draining the interior of
the continent have delivered sediments to fill a trough some
10–15 kilometers (40,000–50,000 feet) deep running parallel to
the coast (Fig. 22-63). The Gulf coastal plain and shelf are a rich
reservoir of petroleum and natural gas. The Atlantic shelf is now
being actively explored for these and other resources.

## DEFORMATION FORCES

Ever since deformation of the crust was first recognized, geologists have speculated about the origin of the forces that shape the Earth's surface features. At one time, orthodox thought had it that the Earth was supposed to be shrinking, and the crust was likened to a wrinkled skin like a dried apple. For a short time, recently, a few proposed that the Earth was expanding, in order to account for the fragmentation of continents and the opening of ocean basins. Convection currents in the mantle have been invoked at various times in the past generation—to move continents, to erode mountain roots, to downbuckle the crust.

In Chapter 21, a hypothesis was described that was startling in its simplicity: that at least some geosynclinal–orogenic cycles were controlled by plate tectonics. It was proposed that the geosynclinal phase may be a tensional feature that forms when a continent fragments and the trailing edges of the separating plates founder. The Triassic graben and the geosyncline now filling off the Atlantic Coast would then be consequences of the opening of the Atlantic Ocean, which began about 180 million years ago. To generalize, the continental shelves surrounding other young oceans, such as the Indian, Antarctic, and Arctic oceans, would be of the Atlantic type, formed with the breaking and separation of pre-existing larger continents in Mesozoic and Cenozoic times (see Figs. 21-15 and 21-17). When a continental margin coincides with the leading edge of a plate in collision with another plate, the hypothesis predicts folding, faulting, earthquakes, magmatism, and volcanism. This can be seen in the circum-Pacific and Alpine–Himalayan belts, which are active today. It would follow that the orogenic belts discussed in the previous section were formed when blocks of continental crust collided with other crustal plates and crumpled marginal geosynclines during episodes of closing of the ancient Atlantic and Pacific oceans. The asymmetric structure of an orogenic belt is accounted for by the basic asymmetry of an oceanic plate plunging under an overriding continental plate along the collision boundary (see Figs. 21-14 and 21-15).

The force that drives plate tectonics must still be specified. As we mentioned in previous chapters, some form of mantle convection is usually invoked at this early stage in the development of the hypothesis. Upwarping, block faulting, and slow vertical movements of the stable interior are among the elements that have yet to be integrated with the plate-tectonics concept. This integration may have to await the next generation of geologists and geophysicists.

1. When rocks are subjected to stresses, they deform. Laboratory studies show that rocks vary in strength. Some deform by plastic flow, and some fracture—depending on the kind of rock, the temperature, the surrounding pressure, the magnitude of the stress, and how fast it is applied.

2. Geologic structures in rock formations that result from deformation include folds, domes, basins, joints, and faults. The geologist begins to deduce the deformational history of a region by fixing the ages of the formations using radioactive isotopes or fossils. He then tries to bracket the age of the deformation by finding a younger undeformed formation lying unconformably on an older deformed bed. Sediments that accumulate as a result of deformational movements also provide clues.

3. Forces acting within the crust can deform large regions. In some cases, the regional movements are simple up-and-down displacements (epeirogeny) without much deformation of the rock formations. In other cases, horizontal forces can produce extensive and complex folding and faulting (orogeny).

4. A continent typically is divided into large regions that have had different deformational histories. North America, for example, contains a large central stable region that has been relatively undisturbed since Precambrian time, except for gentle vertical movements. Surrounding the stable interior are orogenic belts—the mountainous Cordillera and Appalachians, which were deformed by folding and faulting and were subjected to plutonism and metamorphism at various times in the Paleozoic, Mesozoic, and Cenozoic eras.

5. A typical orogenic episode is preceded by subsidence of marginal troughs in which sediments then accumulate. These geosynclines are subsequently deformed. Great folds and thrusts are generated, along with metamorphism and intrusion by batholiths. The resulting belt is then uplifted and eroded. A culminating stage of uplift or block folding again raises the region, which accounts for many of the mountainous features we see today. An explanation of orogeny in plate-tectonics terms would have the deformation phase associated with plate collisions.

## EXERCISES

1. Evidence of vertical crustal movements is frequently found in the geological record. Give some examples of such evidence.

2. In the field, you find a geological section consisting of horizontal beds of Silurian limestone overlying tilted Cambrian slates. What is the geological story that this sequence tells?

3. Draw a geological cross section that tells the following story: A series of marine sediments were deformed into folds and thrust faults. This was followed by erosion. Volcanic activity ensued, and lava flows covered the eroded surface. A final stage of block faulting broke the crust in numerous places.

4. Anticlines are upfolds and synclines are downfolds. Nevertheless, one often finds synclinal ridges and anticlinal valleys. Explain.

5. Summarize the stages of a typical orogenic episode in a series of sketches. Provide legends for the sketches.

6. Summarize the main features of the major structural regions of North America.

## BIBLIOGRAPHY

Billings, M. P., *Structural Geology.* Englewood Cliffs, New Jersey: Prentice-Hall, Inc., 1973.

Dott, R. H., Jr., and R. L. Batten, *Evolution of the Earth.* New York: McGraw-Hill Book Company, 1971.

Hills, E. S., *Elements of Structural Geology.* New York: John Wiley & Sons, 1963.

King, P. B., *The Evolution of North America.* Princeton: Princeton Univ. Press, 1959.

Wilson, J. Tuzo (compiler), *Continents Adrift—Readings from Scientific American.* San Francisco: W. H. Freeman and Company, 1970.

# 23

# The Planets—A Summary of Current Knowledge

*The planets fall into two groups—the gaseous, giant, outer planets, which have retained their Sun-like composition, and the small, stony, Earth-like, inner planets. Space exploration has opened up Earth's neighbors—the Moon, Mars, and Venus—for geological study, and they show surprising diversity. The Moon was geologically active in its early history, but it has been dead for three billion years. Mars, like Earth, is geologically youthful, and shows evidence of recent volcanism, of tectonism, and, surprisingly, of erosion by water and wind, despite the absence of liquid water and the presence of only a very thin atmosphere. despite the absence of liquid water and the presence of only a very thin atmosphere. Venus, Earth's twin in many respects, is blanketed by a thick, heavy atmosphere of $CO_2$, which causes oven-hot temperatures to occur at the surface. Only when we know why the planets followed different evolutionary paths will we understand Earth's workings and environment and why life could originate here.*

Man's interest in the planets dates back to prehistoric times, when the ancients noticed that a few of the brightest stars seemed to drift across the sky, not following the pattern of daily and annual motion of the fixed stars. It was the Greeks who called them *planetai*, or wanderers, and first mapped their paths, and it was the Romans who named them after their gods: Mercury, Venus, Mars, Jupiter, and Saturn.

The invention of the telescope—first used to explore the skies by Galileo in 1609—opened the possibility for describing distinctive planetary features. Viewed through the telescope, the planets appeared to be discs, rather than points of light like the stars. They had individual features—Saturn with its rings; Jupiter

**Figure 23-1**
Haystack Observatory with its 120-ft diameter radio "dish" antenna. Radar waves from this facility are a key tool for the exploration of the surface features of the Moon, Mars, and Venus. [From M.I.T. Lincoln Laboratory.]

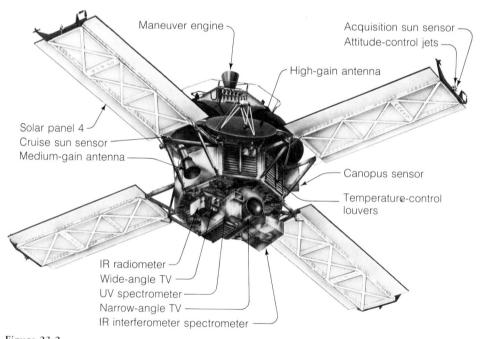

**Figure 23-2**
The Mariner 1971 spacecraft returned TV pictures and other information that altered our views about Mars. [Jet Propulsion Laboratory and National Aeronautics and Space Administration.]

with its alternating bands of yellow, blue, and brown clouds and its coterie of moons; Mars with its seasonally changing ice caps and its supposed canals; and, of course, our cratered Moon with its highlands and "seas." The asteroid belt and the outermost planets Uranus, Neptune, and, finally, Pluto were discovered in later centuries by astronomers searching the skies with even more powerful telescopes. As early as the nineteenth century, some geologists turned their attention to the Moon, seeking to explain its surface features in terms of Earth experience.

After World War II, new ideas of the origin of our solar system, and of the origin of the Moon in particular, started a revival of research in solar-system astronomy.

Space exploration since the first successful satellite in 1957 has quickened interest in the planets. Recent years have seen the birth of a new science, **planetology**, the comparative study of Earth and the other planets. Manned exploration of the Moon, automated and instrument-laden planetary satellites and probes, radar and radio telescopes exploring with radio waves rather than with light, optical telescopes fitted with electronic devices to sense faint infrared and other invisible radiation in addition to light—these are the powerful new methods of this infant science. Figures 23-1 and 23-2 show some of these new tools.

In this chapter, we will review some of the recent discoveries of planetology. Perhaps the most intriguing aspect of the study of other planets is that we may one day arrive at a general theory of planetology that will account for both the origin of our solar system and the evolution of the individual planets with their differing features. Only in this way will we really come to understand our own Earth, its workings, its environment, and the reasons why life could originate here.

## SOME VITAL STATISTICS OF PLANETS

Our solar system is made up of the Sun, 9 planets, 32 satellites, and countless asteroids and comets inhabiting a region some eight billion miles across, yet this is a minuscule domain by cosmic standards. To bring planetary sizes and distances into perspective, one might think of the Sun as a baseball. A grain of sand about 30 feet away, then, would represent Earth; Jupiter would be a pea 150 feet away; Pluto, the outermost planet, would be another sand grain about 1200 feet distant; and the nearest star would be another baseball 1500 miles away. The layout of the planets, held in elliptical, approximately regularly spaced orbits by the sun's gravitational attraction, is shown in Figure 1-3. Except for Pluto and Mercury, the planetary courses around the Sun lie approximately in the **ecliptic**, which is the name for Earth's orbital plane. Mercury's orbit is tilted 7°, and Pluto's is tilted an anomalous

Table 23-1
Vital statistics of the planets

| Planet | Diameter (km) | (mi) | Mass (Earth = 1) | Density (Water = 1) | Gravity (Earth = 1) | Number of satellites | Time for one rotation on axis (Earth hours or days) | Time for one revolution around Sun (Earth years) | Distance from Sun (10⁶ km) | (10⁶ mi) |
|---|---|---|---|---|---|---|---|---|---|---|
| *Terrestrial planets* | | | | | | | | | | |
| Mercury | 4,835 | 3,005 | 0.055 | 5.69 | 0.38 | 0 | 59 days | 0.241 | 57.7 | 36.8 |
| Venus | 12,194 | 7,579 | 0.815 | 5.16 | 0.89 | 0 | 243 days | 0.616 | 107 | 66.9 |
| Earth | 12,756 | 7,920 | 1.00 | 5.52 | 1.00 | 1 | 1.00 days | 1.00 | 149 | 92.6 |
| Mars | 6,760 | 4,201 | 0.108 | 3.89 | 0.38 | 2 | 1.03 days | 1.88 | 226 | 141 |
| *Giant planets* | | | | | | | | | | |
| Jupiter | 141,600 | 88,000 | 318. | 1.25 | 2.64 | 12 | 9.83 hr | 11.99 | 775 | 482 |
| Saturn | 120,800 | 75,000 | 95.1 | 0.62 | 1.17 | 10 | 10.23 hr | 29.5 | 1421 | 883 |
| Uranus | 47,100 | 29,300 | 14.5 | 1.60 | 1.03 | 5 | 10.80 hr | 84.0 | 2861 | 1777 |
| Neptune | 44,600 | 27,700 | 17.0 | 2.21 | 1.50 | 2 | 15.00 hr | 165. | 4485 | 2787 |
| Pluto | 14,000? | 8,700? | 0.8? | 4.2? | ? | 0 | 6.39 days | 248. | 5886 | 3658 |

17° from the ecliptic, as the figure shows. It has recently been discovered that Pluto's orbit comes quite close to Neptune at times, and that it actually lies inside Neptune's orbit over a portion of its path. This new result suggests that Pluto may be an escaped satellite of Neptune. The asteroid belt fills a conspicuous gap between Mars and Jupiter; it contains perhaps 100,000 small bodies in orbit around the Sun, an estimate obtained by viewing this region through large telescopes.

Table 23-1 lists the vital statistics of each planet—its dimensions, its mass and density, the lengths of its day and year, the pull of its gravity, and so on. Vital is an appropriate word: in this age of space exploration, planetary masses, dimensions, and positions are essential elements in planning the navigation of spacecraft. Planetary diameters are measured using telescopic observations or radar. To determine a planet's mass, one must see how its gravitational attraction affects the motion of celestial bodies that pass nearby, such as other planets, satellites, comets, asteroids, and, nowadays, spacecraft. A planet's density, which is simply its mass divided by its volume, is a key factor in determining its composition because the elements present, and the compounds they form, have characteristic densities, and the mean density of the planet depends on the details of the mix.

The planets—except for Pluto, about which little is known—can be separated into two distinct groups: the inner, small, high-density terrestrial planets and the outer, large, low-density giant planets, as Table 23-1 indicates. The terrestrial planets resemble the Earth in density, and it is reasonable to expect a similar composition—namely, a rocky ball mainly of magnesium silicates, possibly admixed with iron. The giant planets are more like the Sun in density and, therefore, in composition; they are light because, like the Sun, they are largely composed of hydrogen and helium, the lightest elements. In this way, the giant planets seem to have preserved much of the composition of the original solar nebula. Saturn is actually less dense than water—which means that it would float, if an ocean big enough could be found.

The Earth-like planets somehow lost the lighter elements or never had them. Perhaps, because of their smaller mass and weaker gravitational pull, they couldn't hold on to these light gases. More likely, these elements were expelled from the inner reaches of the solar system before the terrestrial planets formed. We can only guess at the reason. Perhaps it was magnetic spinup, as described in Chapter 1, or the newly formed Sun may have experienced flareups that blasted these gases away.

The Earth, "with its atmosphere and oceans, its complex biosphere, its crust of relatively oxidized, silica-rich, sedimentary, igneous, and metamorphic rocks overlying [a magnesium silicate mantle and a core] of metallic iron, with its ice caps, deserts, forests, tundra, jungles, grasslands, fresh water lakes, coal beds,

**Figure 23-3**
View of Earth from the Mediterranean Sea to the Antarctic polar ice cap. Almost the entire
continent of Africa and the Arabian peninsula are visible. The Red Sea and Gulf of Aden between
Arabia and Africa are two of the newest oceanic regions created by sea-floor spreading. [From
National Aeronautics and Space Administration.]

oil deposits, volcanoes, fumaroles, factories, automobiles, plants,
animals, magnetic field, ionosphere, mid-ocean ridges, convecting
mantle, . . . is a system of stunning complexity." This quotation,
essentially a one-paragraph summary of much in this book, is from
a popular article by geochemist John S. Lewis. It is given here in

order to set the scene for the discussion of the other, vastly different planets. It is essential to remember Earth's history when we come to consider her neighbors in the solar system.

793
The Planets—
A Summary of
Current Knowledge

## THE MOON

Earth's satellite is exceptional, in that it is so large compared to its parent planet. For this reason, it is often regarded as a planet—a partner to the Earth—rather than a mere satellite. The Moon is the best-known planet next to Earth, thanks to its proximity and the programs of manned and unmanned lunar exploration that have been conducted by the United States and the Soviet Union.

**Project Apollo**   The signal to proceed with Project Apollo, the American program of manned lunar exploration, was given by President John F. Kennedy. It was basically a politically motivated decision made at a time when the superpowers were competing for world leadership. In those days, it was supposed that a demonstration of superior space technology was also a demonstration of a superior political-economic system. Regardless of its origins as a nonscientific enterprise, the Apollo missions have paid off handsomely in terms of scientific discoveries. One might even say that the explorations of the sea floor and the Moon shared in opening the renaissance of major discoveries and new theories in the Earth sciences.

Lunar exploration provided photography of unprecedented resolution; data from a network of emplaced observatories containing seismographs, magnetometers, and other geophysical tools; and, most important, Moon rocks (Figs. 23-15, 23-18, and 23-19). Although much (if not all) of this could have been achieved eventually with unmanned, automatic spacecraft, the value of having reasoning, trained human beings collecting samples and emplacing instruments should not be underestimated.

**Lunar Provinces**   The view of the Moon in Figure 23-4 shows the major provinces of the Moon, together with locations of the landing sites of the exploratory missions. This side always faces the Earth because, for some not fully understood reason, the Moon rotates only once on its axis for each revolution around the Earth. Although the far side of the Moon has been photographed from lunar-orbiting spacecraft (Fig. 23-5), current technology prevents us from landing there. The lighter-colored, highly cratered highlands stand out from the large, dark basins, first called maria (singular: mare), or seas, by Galileo because they resembled oceans in the blurred view through his rudimentary telescope.

The Apollo 17 mission included Dr. Harrison H. Schmitt, a petrologist trained at Caltech and Harvard. He was the first geologist to walk on the lunar surface. Earlier astronauts, who were flyers by profession, were nevertheless well versed in the rudiments of geology.

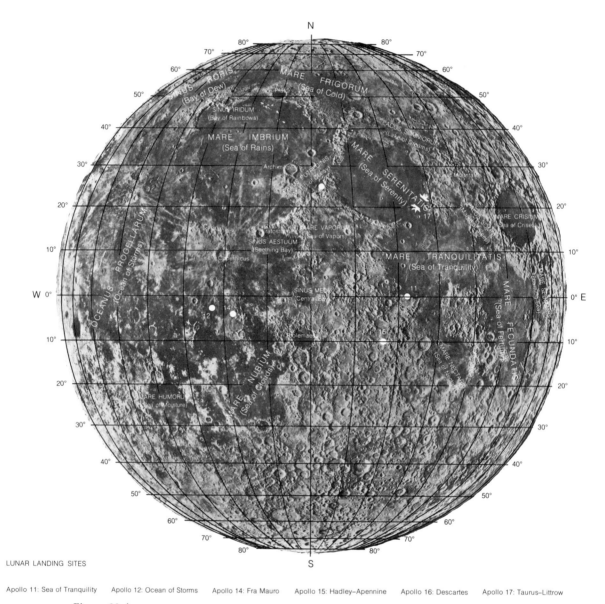

LUNAR LANDING SITES

Apollo 11: Sea of Tranquility    Apollo 12: Ocean of Storms    Apollo 14: Fra Mauro    Apollo 15: Hadley–Apennine    Apollo 16: Descartes    Apollo 17: Taurus–Littrow

**Figure 23-4**
View of the lunar surface, identifying major features and showing Apollo landing sites. [U.S. Air Force Aeronautical and Chart Information Center and National Aeronautics and Space Administration.]

**Surface Processes on the Moon** The Moon's gravity is only one-sixth that of Earth. It lacks the gravitational pull to retain an atmosphere as Earth does, and it lacks water. For these reasons, and also because it is tectonically almost dead, the surface features of the Moon are profoundly different from those of the Earth. As we have seen in earlier chapters, Earth's surface is molded by two opposing natural forces: air and water erode rocks and remove the debris to the oceans, whereas plate motions, along with associated mountain making and volcanism, continually renew the surface. As a result of this ongoing process of creation and destruction, most of the Earth's surface features are fresh ones — younger than 200 million years, that is, formed in the last few

**Figure 23-5**
Apollo 8 view of the Moon over east limb. Right half of photograph covers regions not visible from Earth. Bright rayed crater at top right is Bruno. For comparison with previous figure, irregular mare in left center is Fecunditatis. Mare Crisium is above it. [From National Aeronautics and Space Administration.]

**Figure 23-6**
Alsep, the automatic observatory left behind on the surface of the Moon by the Apollo astronauts. Data from seismographs, magnetometers, heat-flow probes, gas detectors, and other instruments will be transmitted back to the Earth for several years from four of the landing sites. Photograph provides a good view of the lunar "soil," or regolith, showing small craters, ejecta, and astronauts' footprints. [From National Aeronautics and Space Administration.]

percent of geological time. To be sure, many older relics, both rocks and structures, that date back 3–3.7 billion years can be found on continents, but these rocks cover only a small fraction of the Earth's surface.

Imagine the excitement among geologists when the first lunar rocks picked up by the Apollo astronauts proved to be 3.6 billion years old! This was no lucky find. All Moon rocks examined so far were formed or recrystallized 3.0–4.5 billion years ago. (These dates are derived from radioactive dating in Earth laboratories.) Consistent with these indications that the Moon has been geologically inactive for more than 3 billion years, the seismographs installed by the astronauts have indicated that the rate of seismic energy released by moonquakes is one million times less than that of earthquakes (see Box 23-1). This negligible amount of moonquake activity clinches the idea that the Moon is now tectonically almost dead. We might also have reached the same

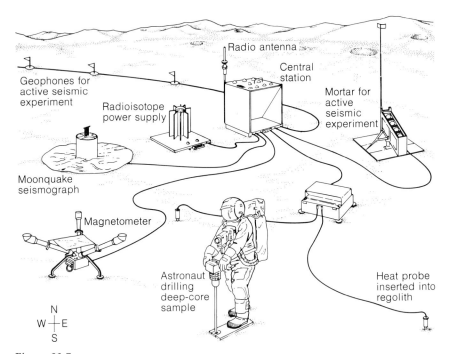

**Figure 23-7**
Schematic illustration of the Alsep observatory, showing several of the experiments deployed on the lunar surface. In the active seismic experiment, a mortar fires an explosive charge on signal from Earth to generate seismic waves in the lunar crust, which are picked up by the geophones. [From National Aeronautics and Space Administration.]

conclusion from photographs that show none of the trademarks of plate tectonics—no globe-circling volcanic ridge-rifts, no transform faults, no extensive trenches with adjacent mountains—in short, no moving plates. Not only is the Moon "dead," when it was "alive" it had a different life style.

In the absence of erosion and orogeny as we know them on Earth, what does shape the lunar surface? What accounts for the principal morphological feature on the lunar surface—the nearly circular craters ranging in diameter from inches to hundreds of miles? The craters themselves, rimmed with rubble, often with rays (which are narrow, bright streaks of debris radiating outward), provide the clues (Figs. 23-5 and 23-9). The Moon's surface, unprotected by a blanket of air, has been riddled by meteorite impacts for the last 4.6 billion years. Coming in with velocities of as much as 150,000 kilometers per hour (roughly 100,000 miles per hour), these celestial projectiles punch into the crust, vaporize, and explode when the energy of motion is almost instantaneously converted into heat. The blast excavates a crater, throwing out debris locally in the form of a surrounding rim, but also hurling ejecta to distant points. To give some idea of what an impact can do, a meteorite 3 meters (roughly 10 feet) in diameter and weighing 500 tons explodes with the energy of 100

## Box 23-1 Moonquakes

Seismologists were surprised to discover that the weak and infrequent moonquakes picked up by their instruments were concentrated around the time when the Moon was closest to the Earth—that is, when the tides on the Moon, pulled by the Earth's gravity, were at a maximum. Apparently, moonquakes are triggered by tidal strains. Earthquake prediction would be so much simpler if this also happened on our own planet. Meteorites impacting on the Moon also produce seismic tremors, as if they were explosions, and seismologists had to learn to distinguish these events from moonquakes. The seismologists also tell us that moonquakes occur at a depth of about 1000 kilometers near the boundary between a thick, strong lunar lithosphere and an underlying "soft" asthenosphere.

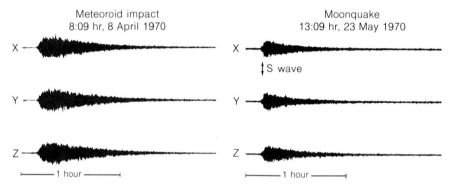

Recordings were made from seismographs on the moon. Traces *X* and *Y* were made by instruments that detect horizontal ground motion; trace *Z* indicates vertical ground motion. Signals on the left are seismic waves generated by a meteorite impact. Signals on the right are from a moonquake. The sharp *S*-waves on the moonquake recording helps distinguish these events from impacts. [From National Aeronautics and Space Administration.]

tons of TNT and blasts out a crater 150 meters (roughly 500 feet) in diameter. A large meteorite weighing $10^6$ tons explodes with as much energy as is released each year by earthquakes.

If the incoming body is big enough, a mare hundreds of miles in diameter can be excavated (see Figs. 23-4 and 23-5). In the smaller gravity field of the Moon, unimpeded by air, huge amounts of ejecta can be hurled for hundreds of kilometers, scarring the surface and littering it with soil and blocks of rock. Some of the ejecta fall back to the surface with high enough speed to punch out secondary craters at some distance from the primary crater.

A giant collision can throw up immense piles of rock—enough to form mountains. The most famous example is the great collision site known as Mare Imbrium, the largest circular basin on the moon. The massive Apennine Mountains, which tower some 5 kilometers (3 miles) above the floor of the basin on its southeast margin, were formed in this cataclysmic event, as were piles

**Figure 23-8**
Apollo 15 photograph of cratered, undulating surface showing fine-grained regolith with widely scattered pebbles and cobbles. Large block in front of astronaut was ejected from a distant impact. [From National Aeronautics and Space Administration.]

of rubble and numerous grooves, ridges, and secondary craters radiating from the collision area. This region was the landing site for the Apollo 15 mission (landing site 15 in Fig. 23-4). The astronauts drove to the front of the massive Apennines in their lunar rover, collecting rocks and photographing the terrane. An example of their work is shown in Figure 23-10.

Craters from meteorite impacts are known on Earth. They are rare because the atmosphere slows all but the largest bodies and literally burns up (oxidizes) the smaller ones, almost all of them completely by the time they hit Earth. The bright light of shooting stars is produced by that burning. The evidence of impact of the larger objects that manage to get through is obliterated as a result

**Figure 23-9**
Apollo 16 mission to Descartes. The photograph was taken from the rim of a crater about 1 km in diameter and 230 m deep. The rim, 50 m high, is a flap of lunar formations blown out and over-turned by the impact explosion. A meteorite about the size of a football field would have produced a crater this big. The rocks in the foreground are breccias. [U.S. Geological Survey and National Aeronautics and Space Administration.]

of erosion by wind and water and burial by sediments. Craters and old buried structures believed to have been caused by impacts are called **astroblemes**, which means "star wounds" (Fig. 23-11). It has been estimated that one crater with a diameter greater than one kilometer is punched out by a meteorite every 50,000 years on Earth.

Craters on the Moon can be preserved for fantastically long times. The only processes working to erase them are a sort of sand blasting by small, high-velocity meteorite grains, called **micro-meteorites**, blanketing by ejecta from other impacts, a direct hit

**Figure 23-10**
Apollo 15 lunar rover parked near Hadley Rille (right center background). The boulder in the foreground is basalt, similar to that in formations that crop out along the sides of the rille. [From National Aeronautics and Space Administration.]

**Figure 23-11**
Meteor Crater near Flagstaff, Arizona is about the same size as the lunar crater in Figure 23-9. It was caused by the impact of a large meteorite in prehistoric times. Thousands of fragments of the meteorite have been found nearby. [From U.S. Geological Survey.]

by another meteorite, or burial by a lava flow. The contrast between a young and old crater side by side in Figure 23-12 shows that the eroding effect of micrometeorites tends to round off the sharp edges and to fill the bottoms of craters. The pitted surface of a Moon rock exposed to micrometeorite bombardment can be seen in Figure 23-13. The process of wearing down is extremely slow—perhaps 10–20 feet of abrasion in the 4.6 billion years of the lifetime of the Moon. A crater is more likely to be obliterated by burial under ejecta from nearby craters than by micrometeorite erosion. The larger the crater, however, the better its chances of survival. Craters larger than a few hundred feet might survive for billions of years.

**Figure 23-12**
Lunar landscape, showing contrast between old and young craters. Old craters have rounded rims and filled bottoms as result of erosion by micrometeorite bombardment. Young craters have sharp edges and bright rays emanating from them. [U.S. Geological Survey and National Aeronautics and Space Administration.]

**Figure 23-13**
This rock, collected during the Apollo 12 mission, is covered by small "zap" pits caused by micrometeorite impacts. Many of the pits are glass lined. [U.S. Geological Survey and National Aeronautics and Space Administration.]

**The Regolith** The Moon has a soil of sorts that is called the **regolith**. It is not a soil in the conventional sense of the word, as is the Earth's soil (see Chapter 6), which results from the breakdown of rocks mainly by the chemical and mechanical actions of water. The lunar regolith is simply the accumulated fragmental debris of billions of years of exposure to bombardment by meteorites, cosmic rays, and particles blown out from the Sun (the solar wind), as depicted in Figure 23-14. Its thickness generally ranges from a few meters to a few tens of meters. It accumulates at the remarkably slow rate of about one millimeter per million years. Contrast this to the sediments of the deep sea of the Earth, which is laid down 5000 times more rapidly. Geochemists have also found a way to determine that the rate of turnover or "gardening" of the lunar soil due to the continuous bombardment is about 1 meter in 0.5 billion years. The lunar soil is particularly important because it is a mixture of materials thrown in from many different areas. One sample of soil, therefore, helps us to figure out what the rocks are like in other places on the Moon from which samples might not otherwise be collected. Figures 23-6, 23-8, 23-9, and 23-10 show what the surface of the regolith looks like. Close inspection shows that, in addition to rock fragments, glass spherules

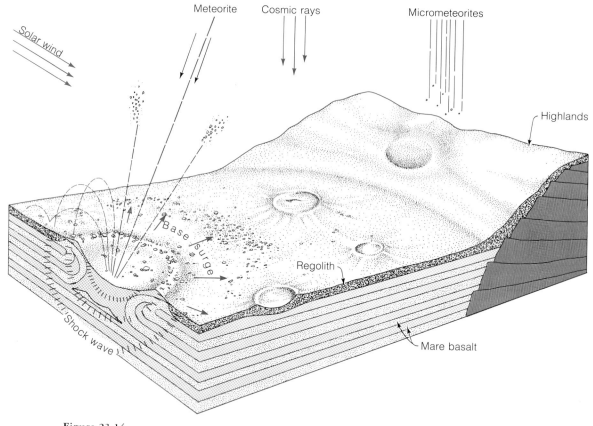

**Figure 23-14**
The original crust of the moon, still present under the highlands, was broken up by large impacts, following which the basins were flooded by basaltic lava. Erosion of the surface, extremely slow compared to the rate of the process on Earth, removes only about a millimeter in a million years. Bombardment by micrometeorites is believed to be the main cause. Larger meteorite impacts occur rarely, but they create craters and scatter ejecta over large distances. The regolith, up to a few tens of meters thick, is the accumulated, fragmental debris of billions of years of exposure to meteorite bombardment, cosmic rays, and the solar wind. [After "The Carbon Chemistry of the Moon" by G. Eglinton, J. R. Maxwell, and C. T. Pillinger. Copyright © 1973 by Scientific American, Inc. All rights reserved.]

are an important component. Presumably, these are formed at the sites of impacts, where melting is induced by the shock and the liquid rock is ejected as a fine spray. The droplets form glassy spherules if they freeze in flight, or they splatter rocks with a glassy crust if they remain molten until they land (Fig. 23-15).

**Lunar Stratigraphy  Photogeology** is mapping geology from photographs. Lunar geologists have used photographs to get relative ages of different regions of the Moon by counting the density of craters. The oldest terrane should show the highest

**Figure 23-15**
Glass-coated lunar rock. Heat, generated by impact, melts surrounding material, which splatters over the surface and covers rocks with a glassy crust (dark area). [From National Aeronautics and Space Administration.]

crater count simply because it has been bombarded for the longest time. The densely cratered highlands, for example, in the south-central portion of the Moon (Fig. 23-4), are the oldest regions. In many places, the highlands are covered by ejecta blankets thrown out from the circular maria. This cover is punctured by more craters than the lava surface of the maria. Hence, the impact that produced the ejecta blanket must be older than the volcanism that filled the maria with lava. The reader may wish to establish relative ages of the different features in Figures 23-5 and 23-16 by using this simple method of crater counting, as well as by looking for other evidence of "freshness," such as craters with sharp-edged features, or bright rays that have not yet disappeared due to erosion or other aging effects.

Stratigraphic mapping on the Moon—that is, working out the chronological relations of the many lunar surface features—is quite different from similar procedures on the Earth. There are no fossils, and the events that define periods are quite different

**Figure 23-16**
Apollo 8 photograph of craters in Mare Fecunditatis. [From National Aeronautics and Space Administration.]

in nature and time scale than those on the Earth. The principal
formations are the original crust, lava flows and ejecta blankets of
varying ages, and the regolith. One proposal for defining time
periods on the Moon is as follows:

| Period | Event |
|---|---|
| Pre-Imbrian | Formation of highlands; period of most intense meteorite bombardment |
| Imbrian | Formation of maria |
| Eratosthenian | Formation of craters whose rays have been obliterated |
| Copernican | Formation of craters with rays still visible |

The periods are named after the Mare Imbrium and the craters
Eratosthenes and Copernicus (Fig 23-4).

To be sure, the law of superposition holds on the Moon, but
sometimes it must be inverted. Consider the case of material
ejected from a crater, as in Figure 23-9 or 23-16. The deepest
rocks encountered by the projectile are concentrated at the top
of the ejecta that forms the rim. The stratigraphy is inverted,
as if the formation were a flap laid over on itself. The shallowest
ejecta are found on the surface, but farther out on the ejecta
blanket. These features have been discovered on Earth in connection with underground nuclear explosions that were shallow
enough to "blow out."

**The Maria** Controversy raged for years, prior to the Apollo
program, about the nature of the material in the maria. The lunar
basins are not filled with sediments deposited in ancient seas, nor
are they filled with dust abraded from the highlands, as some had
supposed. Analysis of the returned samples from three maria
(Tranquillitatis, Fecunditatis, and Procellarum) leaves no doubt
about the answer: They are flooded by layer upon layer of basaltic
lava, not unlike the basaltic flows of Iceland or the Columbia
River Plateau discussed in Chapter 17. Apparently, when the
maria basins were plowed out by the impact of large meteorites
early in the Moon's history, the partially molten interior was
exposed and lava subsequently flowed out to fill the depressions
(Fig. 23-17). Mare Crisium and Mare Fecunditatis, both good
examples of this, can be seen in the northwestern and western
portions, respectively, of Figure 23-5. The Apollo 15 astronauts
visited Hadley Rille, a long, narrow channel incised in Mare
Imbrium, running parallel with the Apennine front (Fig. 23-10).
The astronauts could see and photograph ledges in the canyon
wall that were some 60 meters thick. These are individual lava

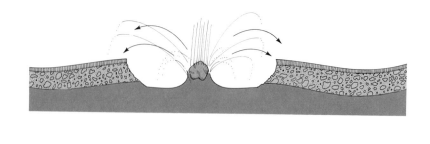

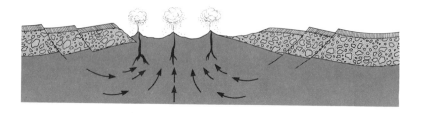

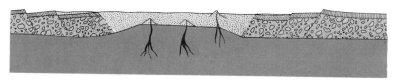

**Figure 23-17**
Model for the origin of the lunar mare. (a) The impact of a large meteorite
excavates a basin, ejecting material to great distance. (b) The high temperatures
generated by the impact causes volcanic activity. Alternatively, volcanism may be
caused by fractures that reach partially molten rock beneath the lunar crust.
(c) The basin fills with lava from the interior. [After National Aeronautics and
Space Administration.]

flows, layers that flowed one on top of the other in the process of
filling the basin. Hadley Rille is thought by some to be the result
of the collapse of an underground lava tunnel whose roof caved in
after the lava ceased flowing.

**Lunar Rocks—What They Are and What They Reveal**  Many
hundreds of pounds of rocks have been returned to the Earth
from the Moon. This precious cargo has been disseminated to
hundreds of specialists all over the world. The United States and
the Soviet Union have recently agreed to exchange lunar material
and to cooperate in other ways in lunar exploration. Many investigators were given new laboratories by NASA, with funds to
develop analytical equipment of unprecedented precision, to
ensure that all possible information be gleaned from the lunar

samples. In this way, lunar exploration has inspired international cooperative and spurred innovations that might otherwise have been delayed.

The Moon's crust turns out to be a relatively simple assemblage of rocks. The highlands are formed of anorthosite, an igneous rock consisting predominantly of calcic plagioclase feldspar. This material may represent the first melting episode on the Moon, perhaps dating back 4.5 billion years, nearly to the beginning of the solar system. Because anorthosite is rich in aluminum, a light element, and poor in iron, one of the heaviest elements, it is relatively light; its crystals floated up to form a primitive crust. This crust was later broken up by large impacts, following which the basins were flooded by iron-rich basalt.

The lunar basalt flowed out beginning perhaps 4.1 billion years ago, with the last flooding occurring about 3.1 billion years ago. It is startling that the Hawaiian volcano Kilauea is erupting on Earth at this very moment, yet the youngest lava on the Moon is

**Figure 23-18**
Vesicular basalt collected by Apollo 15 astronauts. The holes, or vesicles, were gas bubbles "frozen" into the lava when it solidified. [From National Aeronautics and Space Administration.]

3 billion years old! For some as yet unexplained reason, internal activity on the Moon mostly ceased at this point, as if the internal heat engine were shut off. The Moon simply stopped evolving (in the sense that its surface feature have since changed relatively little) somewhat more than 3 billion years ago, leaving a "half-baked" planet with its ancient record still preserved. Four billion years ago, Earth's crust may have had a similar composition; but Earth continued on its evolutionary path to this very day, producing a complex and rich variety of rocks, an ocean, and an atmosphere, and initiating organic evolution. (Not a shred of evidence of biological life has been found on the lunar surface.)

In general terms, the Moon's basalt is similar to the commonest volcanic rock on Earth, yet lunar rocks differ in several essential features from their Earth counterparts. Materials that vaporize at relatively low temperatures—such as water, carbon, nitrogen, sulfur, chlorine, mercury, zinc, and lead—are depleted in Moon rocks. On the other hand, Moon rocks are enriched in certain refractory elements—that is, elements that are not easily melted, such as titanium and zirconium. The extreme dryness of the Moon rocks is striking. Even the driest Earth rocks show the presence of some water bound in various physical and chemical ways, but

**Figure 23-19**
Apollo 15 rock sample. The vugs (mineral-lined cavities) in this basalt contain pyroxene crystals. [From National Aeronautics and Space Administration.]

water is absent from Moon rock. Astronauts can forget about trying to find water on the Moon to sustain supposed future settlements.

The big question is, at what stage did the Moon undergo this gain and loss of materials? Some specialists believe that these depletions and enrichments existed in the original matter that formed the Moon and that the Moon started off as chemically different from the Earth. The Moon may have formed in another part of the solar system in a high-temperature environment, which resulted in the loss of the volatile elements; subsequently, it was "captured" by the Earth. Alternatively, the Moon may have formed near the Earth but, because of its smaller size, may simply have captured different chemical elements than the Earth did. Other hypotheses exist, evidence that the debate about the Moon's origin will continue for some time.

A photograph of the Moon taken a few hundred million years ago would show essentially the same features as a modern picture— Figure 23-5, for example. Astronaut David Scott demonstrated this dramatically on the Apollo 15 mission when he picked up a rock at the foot of the Apennines that, as it turned out, had been lying there unmoved for 400 million years! A picture of the Earth taken 200 million years ago would show a surface completely different from the one the Earth shows today. The Atlantic would not yet have been formed and the map would be dominated by the supercontinent Pangaea. Incidentally, the youth of the Earth's surface features dispells the old speculation that the Moon was torn out of the Earth leaving the Pacific Ocean as a scar. The Pacific is a recent feature of this rapidly changing planet, and it couldn't have existed in its present shape or form 4.6 billion years ago.

One additional rock type should be mentioned because, although it has the same name as an Earth rock of similar appearance, it is peculiar to the Moon—the **lunar breccia**. Rocks of this type are cohesive agglomerates or clusters of rock fragments and fine regolith, apparently stuck together by the heat and pressure generated at the sites of meteorite impacts (Fig. 23-20). Some volcanic breccias on Earth, though of different origin, resemble them.

**Magnetized Rocks and Mascons** Two additional discoveries, seemingly contradictory, should be mentioned in our discussion of the Moon. Lunar rocks were found to be magnetized. This remanent magnetization in the returned samples cannot be isolated examples because a magnetometer aboard a satellite launched into lunar orbit by the Apollo 15 astronauts shows that the whole lunar crust has a small residual magnetism. The simplest explanation is that the Moon once had its own planetary magnetic field. If so, it existed at the times of formation of the magnetized rocks—that is,

**Figure 23-20**
Lunar sample collected by the Apollo 15 astronauts near the Apennine
Mountain front. The rock is a blocky, angular breccia. [From National
Aeronautics and Space Administration.]

from the earliest moments in lunar history, when the primitive
crust formed, to about 3.1 billion years ago, when the youngest
mare basalts poured out. From our experience on Earth (Chapter
20), we surmise that planetary magnetic fields are caused by fluid
convection in a molten-iron core, and many scientists have concluded that the Moon was hot enough to produce a small, molten-
iron core in its early history. Some time after 3.1 billion years
ago, the Moon supposedly cooled down enough for the iron to
freeze into a solid, and the magnetic field was suppressed. Thus,
the magnetism of its rocks may imply that the Moon had a hot
interior, at least until about 3 billion years ago.

The word **mascons** is a contraction for mass concentrations.
Scientists at the Jet Propulsion Laboratory discovered that spacecraft orbiting the Moon were mysteriously pulled toward the
surface when they passed over certain locations. These places
with higher gravitational attraction turned out to be the circular
maria (Fig. 23-21). Apparently, massive bodies or mascons lie
in or under the circular basins, and they tug gravitationally on a
satellite each time it flies over. Aside from the question of what
the mascons might be, the most fascinating question is how these

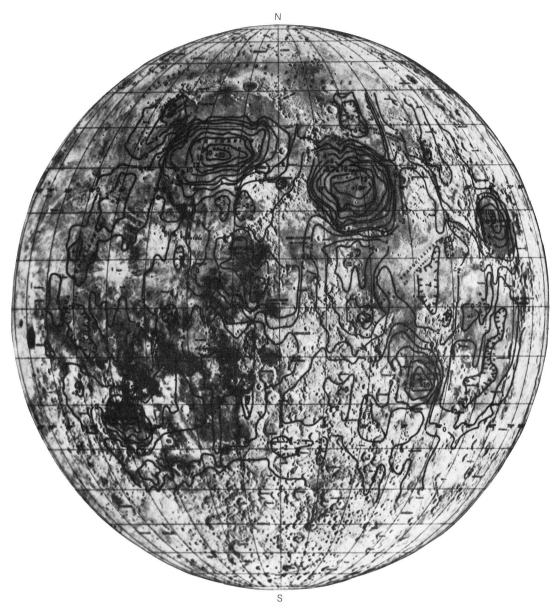

**Figure 23-21**
Gravity-anomaly contours on the moon. Note how the higher gravity values, shown by the concentration of circular contours, are centered on the circular maria—evidence that massive bodies, or mascons, lie in or under these basins. [From National Aeronautics and Space Administration.]

heavy objects could survive near the lunar surface without sinking into the interior in the 3–4 billion year period since the basins were formed. Recall that, on Earth, a surface load such as a continental glacier depresses the crust in only a few thousand years, effectively sinking until it is supported buoyantly (Chapter 20). The large positive gravity anomalies of the mascons show that this has not yet occurred on the Moon, although 3–4 billion years have elapsed. This means that an outer layer of the Moon, perhaps 1000 kilometers thick (see Box 23-1), has been strong, rigid, and, therefore, cold for 3–4 billion years, not warm, plastic, and weak, as we might suppose from the evidence of lunar volcanism and magnetized Moon rocks. As to the nature of mascons, some scientists believe them to be the buried remains of the massive meteorites whose impacts created the basins. Others speculate that, when the primitive crust was punctured, lava spurted out—more than enough to fill the excavation—and cooled to form a dense plug of mantle material where light crust formerly had been.

**What Does It All Mean?** It is a difficult task for the authors of a textbook to pull together this wealth of new and often contradictory material on the Moon when the experts themselves have not yet done so to everyone's satisfaction. Perhaps, in time, a concept with the simple elegance and synthesizing power of the concept of plate tectonics on Earth will emerge to unify the lunar data.

Whatever concept does emerge must begin with a Moon that formed 4.6 billion years ago from starting materials that were probably different chemically from those that formed the Earth. The Moon heated up rapidly, and extensive melting occurred very early in lunar history. A primitive crust of anorthosite floated to the surface, and iron probably went down to form a small molten core and start a magnetic field. The first 1.5 billion years was a time of intense bombardment by meteorites; the ancient crust was heavily cratered, as evidenced by the relics in the highlands. Great impacts excavated mare basins, which were later filled with basalts from the lunar interior, perhaps due to a second episode of melting of the outer portions of the Moon. The core solidified and the outer layers cooled rapidly about 3 billion years ago, forming a strong lithosphere about ten times thicker than Earth's lithosphere. What the scientists must come up with is a thermal mechanism whereby the Moon was hot enough to sustain a magnetic field and fill the mare basins with basalt, yet could cool fast enough to shut off volcanism abruptly and form a thick, strong lithosphere to support the mascons. The Moon has been essentially dead for 3 billion years. Its surface has been modified somewhat by a reduced amount of meteorite bombardment, but the large-scale features have remained unchanged.

Moon geologists do not have all the answers at this time. However, they are working on materials that date back to the early days of the solar system—in a sense, Rosetta Stones waiting to be deciphered.

The features of Mercury—the smallest, innermost, and densest planet—reflect its close proximity to the Sun. There is no atmosphere because, with Mercury's low gravity and high temperatures, gases would boil off into space from the daytime side, which receives 5–10 times more solar energy than the Earth does. Temperatures may reach 370°C (roughly 700°F) during the day and drop to −180°C (roughly −300°F) at night. Closeness to the Sun may also be the reason why the lighter, more volatile elements escaped from the matter that formed Mercury, leaving behind this small but heavy planet. The chances of finding life on Mercury are almost nil.

# VENUS

Venus is the nearest thing to being Earth's twin—in mass, diameter, density, gravity, and distance from the Sun (see Table 23-1). But here the similarity ends, according to the latest results using radar, optical telescopes, and automated space probes. From radar observations comes the information that Venus rotates at the extremely slow rate of once in 243 Earth days and the peculiar discovery that its rotation, though opposite to that of Earth, is synchronized with the movements of the Earth in that it presents the same face to us every time it swings by in its orbit.

The surface of Venus is completely masked by clouds, but radar waves bounced off the planet reveal crustal properties that are typical of a crust of silicate rocks and also that some mountains and craters are present.

Venus differs profoundly from Earth in the nature of its atmosphere and in its surface temperature. The most recent information comes from the United States' Mariner spacecraft, which flew past the planet, and the Soviet Union's Venera spacecraft, which performed the remarkable feat of sending a probe parachuting into the atmosphere. Venus is covered by a thick blanket of carbon dioxide. Perhaps 90–95 percent of the Venusian atmosphere is made up of this gas; the Earth's atmosphere by comparison, is only 0.03 percent carbon dioxide. Water vapor and oxygen are present, but in extremely small amounts compared to the great quantities present in the Earth's biosphere. At the surface of Venus, atmospheric pressure is 100 times more than on Earth—nearly one ton per square inch—and the temperature reaches 800°F—hot enough to melt such metals as lead and tin. This certainly means that no organic life could exist on the surface.

What makes Venus hot is its insulating blanket of carbon dioxide. Solar energy is trapped by the atmosphere, which warms up

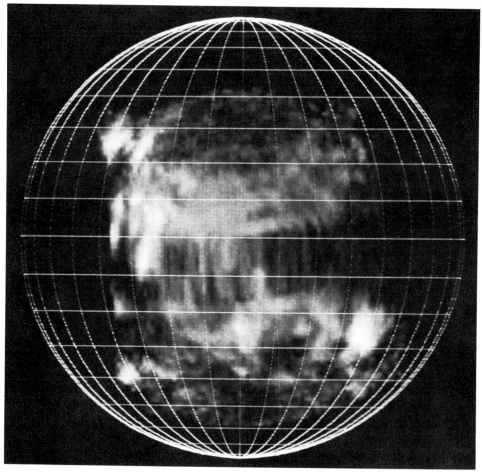

**Figure 23-22**
Radar waves bounced off the surface of Venus provide the first data on the topography of this cloud-covered planet. In the radar map above, the brightness of the radar echo is a measure of the roughness of the surface. From this and other maps, scientists conclude that Venus has mountains and basins comparable in scale to those on Earth. Mercury seems to have a smoother surface. [From A. E. E. Rogers and R. P. Ingalls, *Science,* v. 165, p. 797. Copyright 1969 by the American Association for the Advancement of Science.]

as a result. This is entirely analogous to the much milder greenhouse effect on Earth (Chapters 1 and 14).

Venus has no magnetic field—perhaps because it lacks a fluid core, or perhaps the rotation of a planet is also important in generating a magnetic field and Venus rotates too slowly, or maybe like the Moon, it once had one but lost it.

These profound differences between Venus and Earth pose difficult questions: How could two planets, so similar in many respects, evolve so differently? Could it be that Venus, though Earth's neighbor, was just enough closer to the Sun to have accreted at higher temperatures, lost water and other volatiles, and ended up with an entirely different atmosphere?

Mars has been a planet favored by scientific speculators and science-fiction writers. Some have thought that Mars was like Earth, and that it had canals, a seasonally varying cover of vegetation, and even intelligent life. Others have thought it to be a tectonically dead, cratered planet like the Moon. These and other speculations have been laid to rest as a result of the Mariner 9 mission, which reached Mars toward the end of 1971. The Mariner spacecraft orbited Mars twice a day for months. It transmitted back thousands of color photographs, taken from a height of 1500 kilometers (roughly 1000 miles), with unprecedented resolution: features only 0.1 kilometer across were clearly visible. Contrast this with the 80-kilometer resolving ability of Earth-based telescopes, upon which so many of the earlier speculations depended.

Mars, the outermost of the terrestrial planets, travels in the next orbit to Earth's, going out from the Sun. Its diameter is a little over half that of the Earth, its mass about one-eighth, and its gravity is about two-fifths as strong. Mars has a density somewhat less than Earth's, but, when allowance is made for its smaller size and reduced internal pressures, its "uncompressed density" turns out to be the same as the Earth would have. This means that its overall composition is probably not much different from Earth's.

By coincidence, the Martian day and the Earth day are of nearly the same length, and the angle that the axis of rotation forms with the orbital plane is nearly the same for both planets. The Mars year is 687 Earth days. Mars has no large-scale magnetic field today, which implies the absence of a fluid core.

The Mariner data confirm that Mars has an atmosphere about 200 times thinner than Earth's. It is composed mostly of carbon dioxide, with a trace of water vapor present. Water cannot exist freely on the surface; it would evaporate quickly in the near vacuum. The surface temperatures are more extreme than those on Earth, ranging from about 25°C (roughly 80°F) at the equator to about −125°C (roughly −190°F) at the poles. Wind velocities of 270 kilometers per hour (roughly 170 miles per hour) can blow up in gigantic storms that cover the planet with dust for weeks at a time. One such dust storm blurred the view of the Mariner 9 television cameras for many weeks and delayed the schedule of receiving pictures.

The portrait of Mars that has emerged from the thousands of Mariner 9 photographs is one of a dynamic planet with extensive tectonic activity, volcanism, cratering by impact, and erosion by wind and, possibly, water (Figs. 23-23 and 23-24). Although all of these processes occur on Earth, we will see that Mars has a unique character, vastly different from Earth's.

Geologists have divided Mars into four regions on the basis of the new data. The first is a volcanic province with giant volcanic

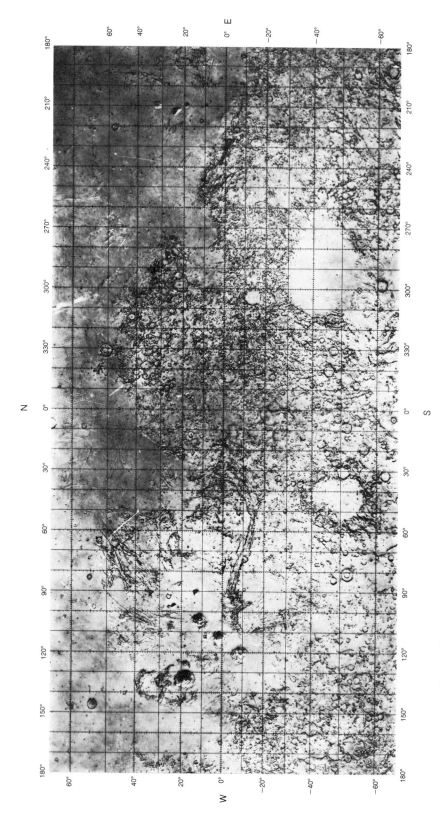

**Figure 23-23**
Photo composite of the first detailed, "shaded relief" map of the entire surface of Mars, based upon mosaics of thousands of photographs returned by the Mariner 9 spacecraft. Mountains, craters, plains, channels, and canyons are depicted over an area of about 140 million km². [Compiled by U.S. Geological Survey for the Jet Propulsion Laboratory; H. Masursky, leader of Mariner Television Team.]

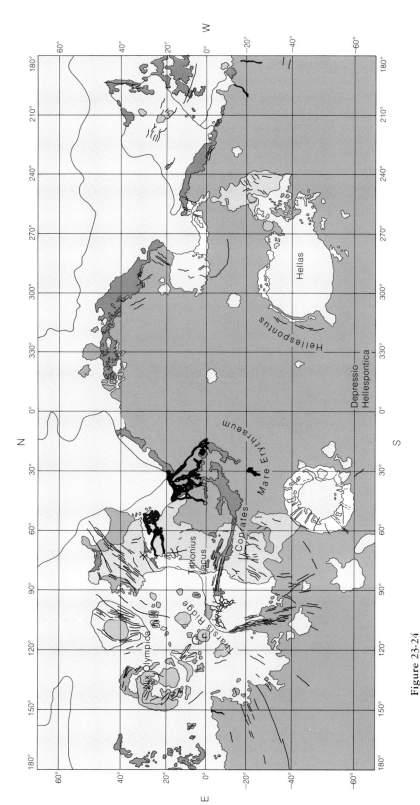

**Figure 23-24**

Geological map of Mars, showing a classification of surface features revealed by television photographs. Pale brown areas are smooth plains; medium-brown regions are cratered plains; dark-brown areas are old cratered terrains; white areas are mountainous terrain; light gray represents volcanic regions. Areas in dark gray identify terrain that has been modified by some process such as the series of huge east-west canyons. The black areas are channel deposits, and the irregular black lines are inferred faults. The map was prepared by M. Carr, J. F. McCauley, D. Milton, and D. Wilhelms of the U.S. Geological Survey. [After "Mars from Mariner 9" by Bruce C. Murray. Copyright © 1973 by Scientific American, Inc. All rights reserved.]

mountains. One of them, Olympus Mons—500 kilometers wide at the base, 8 kilometers high, and with a crater on top 70 kilometers in diameter (Fig. 23-25)—dwarfs anything found on Earth.

The second province is an equatorial plateau cut by deep canyons and great crack-like faults (see Figs. 23-23 and 23-24). One immense chasm, Vallis Marineris, makes the Grand Canyon of the Colorado look puny: in round figures, it is 120 kilometers (roughly 75 miles) wide, 6 kilometers (roughly 4 miles) deep, and sinuously winds along the Martian equator for some 5000 kilometers (about 3000 miles). Many of these rifts are clearly tectonic, which is evidence that the crust of Mars has cracked and slipped, in its recent history, due to internally generated forces. Networks of dendritic tributaries feed into many of the chasms, and delta-like

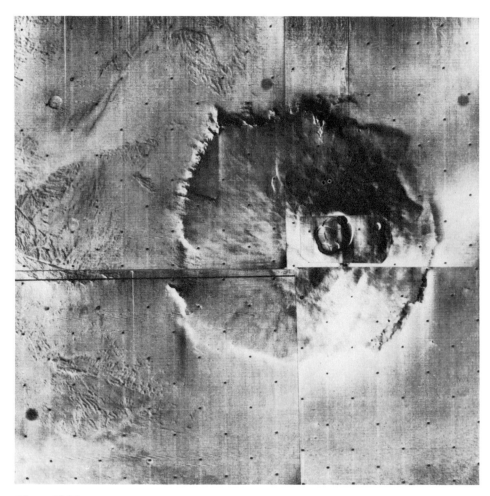

Figure 23-25
Olympus Mons, a gigantic volcanic mountain on Mars, photographed by Mariner 9. This feature—500 km across at its base, with a caldera 65 km in diameter—is twice as big as the Hawaiian Islands, Earth's most massive volcanic pile. [Jet Propulsion Laboratory and National Aeronautics and Space Administration.]

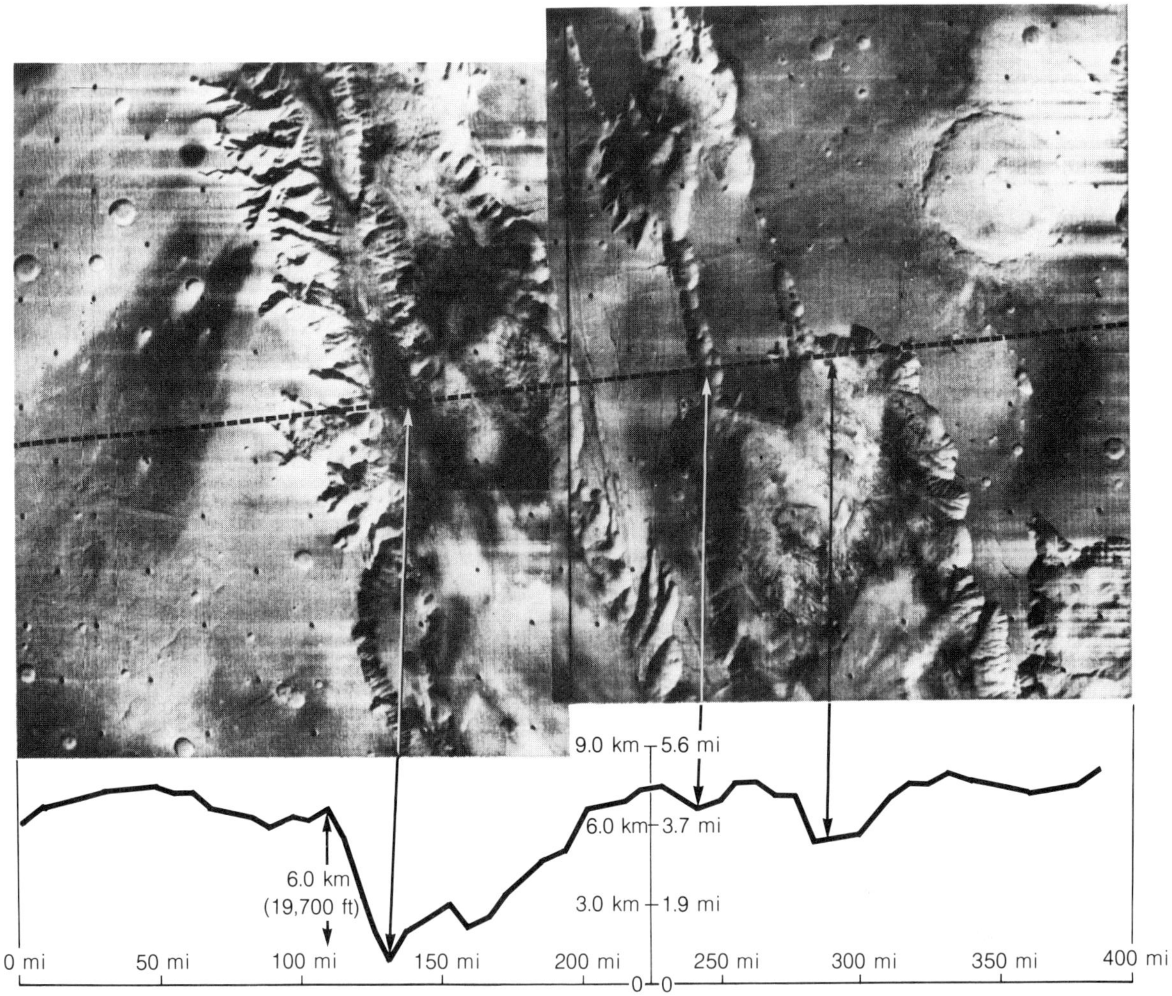

**Figure 23-26**
Deep canyon in Tithonius Lacus region of Mars, photographed by Mariner 9. This feature is four times deeper and six times wider than the Grand Canyon. The topographic profile below shows elevation changes. Note the craters, the tectonic rifts indicative of tension, and the branching canyons suggestive of some sort of erosional process. [Jet Propulsion Laboratory and National Aeronautics and Space Administration.]

regions are found at the ends (Fig. 23-26). Long meandering valleys can also be seen. All of this strongly suggests that erosion by water occurred in the past; some scientists, however, have proposed that sculpturing by winds or some combination of other processes is responsible.

The third province, which covers perhaps half of the Martian surface, includes both smooth and cratered terranes that are not unlike the mare basins of the Moon. They probably have the same impact origin. The most cratered plains are thought to be the oldest feature on Mars. One circular basin, Hellas, appears to be larger than any impact basin on the Moon. Great sand-dune fields are found in this area (Fig. 23-27). The earlier Mariner flights sent

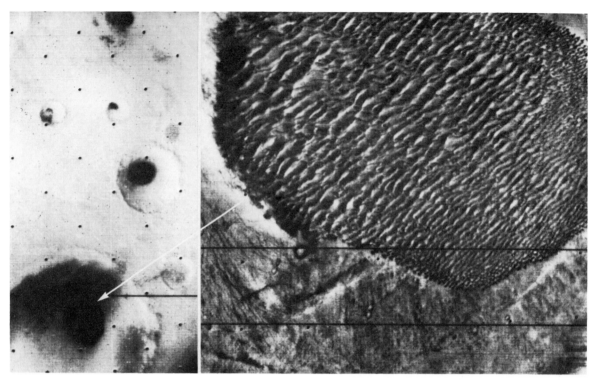

**Figure 23-27**
Dunes on Mars, photographed by Mariner 9. Strong winds blowing unconsolidated materials formed these long dunes, spaced about 1.5 km apart. This dune field is on the floor of a crater about 150 km wide in the Hellespontus region. [Jet Propulsion Laboratory and National Aeronautics and Space Administration.]

back pictures mostly from this province, and scientists were misled by this unrepresentative sample to conclude that Mars was a dead planet, much like the Moon.

The fourth province features what appears to be debris laid down in step-like terraces and deep grooves radiating from the south polar regions. There is no firm evidence of glacial movement, but these layered terranes are related to the ice caps.

A rough chronology, based on crater counting, enables us to say that the evidence of tectonic activity and volcanism on Mars indicates that it is geologically young. This means that tectonism and volcanism are still underway on Mars, just as they are on Earth, albeit on a different scale and with a different rate from what we know on Earth. The huge, long-lasting dust storms must make erosion and deposition by wind an important factor. However, water erosion (for which there is so much evidence) is indeed difficult to explain, for there is only a trace of water in the atmosphere and liquid water could not now exist on the surface. Harold Masursky of the U.S. Geological Survey offers two possible mechanisms for the release of water in large enough quantities to be an erosional agent: Mars has two polar ice caps, which grow and contract alternately on a seasonal basis. The ice caps seem to be composed of ice and frozen carbon dioxide (dry ice). Because Mars' axis wobbles or precesses, it happens that, periodically (every 25,000 years), both poles receive a larger dose of sunlight, perhaps enough to melt large quantities of water. By this mechanism, Mars could have a rainy season, with heavy erosion, perhaps every 25,000 years. The second possibility suggested by Masursky is that water is stored beneath the surface as ice trapped in a permafrost zone, and that every once in a while there is sufficient volcanic heat to thaw the ice and release large quantities of water into the atmosphere.

No sign of life has been detected on Mars, but the Viking mission, which will land a capsule on the surface in 1976, will be devoted principally to a search for living forms. The chances of finding life have always been thought to be slim. The thin Martian atmosphere cannot stop the sterilizing ultraviolet light from reach-

Figure 23-28
Sinuous valley in the Mare Erythraeum region of Mars, photographed by Mariner 9. This feature resembles water-cut gullies in the southwestern United States, but there is no evidence of water in the Martian atmosphere. [Jet Propulsion Laboratory and National Aeronautics and Space Administration.]

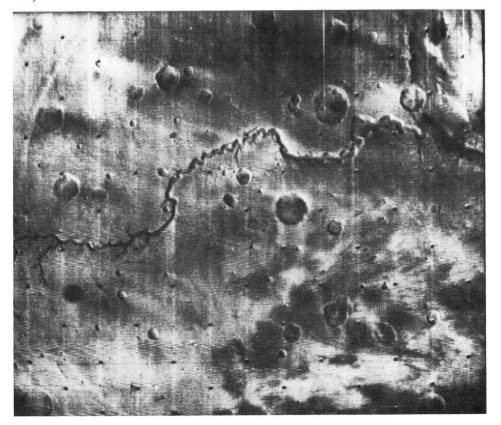

ing the surface, and the lack of a magnetic field means the absence of a shield against other forms of radiation. The absence of oxygen and the lack of water—except, perhaps, during occasional floods spaced thousands of years apart—does not help matters. The probability of finding simple forms of life on Mars is still thought to be small, but scientists are slightly more hopeful as a result of the new information from Mariner 9.

## JUPITER

The first of the giant gas planets, Jupiter, with a volume 1300 times greater than Earth's, contains more than twice the mass of all the other planets combined. Had it been somewhat more massive, Jupiter might have attained internal temperatures as high as the ignition point for nuclear reactions, and it would have flamed as a star in its own right. As mentioned earlier, Jupiter and the other

**Figure 23-29**
Jupiter, as photographed by the 200-inch telescope. The great red spot is visible. Also shown is the satellite Ganymede and its shadow. [From Hale Observatories.]

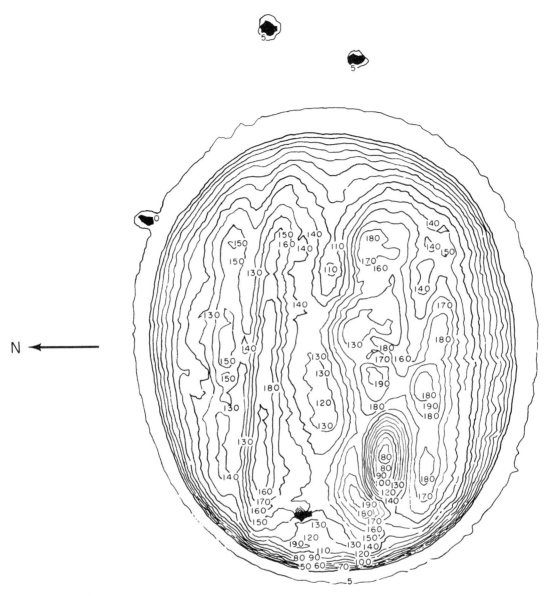

**Figure 23-30**
An example of modern techniques of planetary exploration. An image of Jupiter was made by an extremely sensitive television camera attached to a 60-inch telescope. Contours of brightness were drawn by computer; larger numbers represent brighter regions. The great red spot shows as circular, low-valued contours in the southwest quadrant. [From T. McCord, M.I.T.]

giant planets are of a low-density type quite distinct from the terrestrial planets: they are composed predominantly of such substances as hydrogen, helium, ammonia, and methane. Much of Jupiter's interior may be in the form of solid, metallic hydrogen. Normally, hydrogen is a gas; but under pressures of millions of kilograms per square centimeter, which exist in the deep interior of Jupiter, the hydrogen atoms may lock together to form a solid.

Some scientists believe that the innermost core of Jupiter may be rocky, or metallic like the core of the Earth.

Jupiter rotates very fast, once every 9.8 hours. As a result, its clouds, which are composed largely of frozen and liquid ammonia compounds, form turbulent, pastel-colored bands that circle the planet at different speeds in different latitudes (Fig. 23-29). The puzzling Great Red Spot changes size as it hovers in the southern hemisphere. It may be a gigantic atmospheric disturbance (Earth could easily fit inside it) caused by a bump on the surface of Jupiter's solid hydrogen core.

Jupiter has twelve satellites; one of them Ganymede, is larger than the planet Mercury.

Radio telescopes pointed at Jupiter pick up intense bursts of radio energy, apparently originating in a radiation belt quite similar to the Van Allen radiation belt, which surrounds the Earth. This means that Jupiter must have a powerful magnetic field that traps charged particles from the Sun, sending out radio waves in the process. Fascinating, and as yet unexplained, is the observation that the radio emanations are definitely influenced by the position of Io, the second closest of Jupiter's moons.

## SATURN

The second largest planet, Saturn, is also the lightest planet. It is much like Jupiter—it rotates rapidly, shows colored bands of clouds across its face, and has a large family of ten satellites. Saturn, however, has one feature not shared by any other planet—its four rings (Fig. 23-31). The rings are thought to be countless ice particles, or ice-covered dust and rock, whirling in orbit around the planet in a belt anywhere from a few inches to a few miles thick. The rings are thought to be the fragments of a satellite that broke up, because of tidal forces, when it spiraled too close to Saturn—or, alternatively, material that never accreted to form a satellite, because a single body of any size could not exist so close to the planet.

## URANUS, NEPTUNE, AND PLUTO

Not much is known about Uranus and Neptune, the outermost of the giant planets, nor of Pluto, the ninth and most distant planet, beyond the data given in Table 23-1. Following the discovery of Uranus by telescope in 1781, the existence of Neptune and Pluto was predicted mathematically, because Uranus seemed to deviate from its predicted orbit, indicating that it might be influenced by the gravitational attraction of unknown planets. With

**Figure 23-31**
Saturn and its ring system, photographed by the 100-inch telescope. [From Hale Observatories.]

the aid of these mathematical hints, Neptune was found in 1846 and Pluto in 1930.

## METEORITES

The prevailing view is that meteorites are broken fragments formed by collisions between asteroids orbiting the Sun in the belt between Mars and Jupiter. They are deflected into orbits that cross the path of Earth, which sweeps them up.

Meteorites range in size from fine dust particles to bodies many miles in diameter, which gouge out large craters or astroblemes when they impact the Moon, Mars, or Earth. Meteoritic dust falls almost continually on Earth. However, the larger, more spectacular meteorites that survive the fiery flight through the atmosphere were the only source of information about extraterrestrial materials prior to the collection of lunar samples. Their nature and origin are highly controversial, and hundreds of books and articles have been written about them.

Stony meteorites are by far the most abundant. They are made up of silicate minerals, mostly olivine and pyroxene. Iron meteorites consist essentially of an iron–nickel alloy. Nothing resembling sedimentary or metamorphic rocks has been found in meteoritic material.

**Figure 23-32**
Photograph of a piece of the Allende meteorite, thought to have formed
early in the history of the solar system. Black crust on top is due to
melting from frictional heating in flight through the atmosphere. [J. Wood,
Smithsonian Astrophysical Observatory.]

Perhaps the most significant result of meteorite studies is the
fact that their mineral inclusions crystallized 4.5 billion years ago.
This age, determined by radioactive dating, was used to fix the
time of formation of the solar system. Meteorites, and now the
Moon rocks, provide the best examples of the earliest nonvolatile
components of the solar system.

1. The planets of the solar system fall in two distinct groups: the small, stony, inner, Earth-like planets, which have been opened up for geological examination as a result of space exploration, and the gaseous, giant, outer planets, which are more Sun-like in their compositions.

2. The Moon was geologically active in its early history. Lunar rocks show the planet to have been differentiated and subjected to volcanism, but, for reasons not fully understood, the Moon became essentially inactive about 3 billion years ago. The absence of young rocks, the small number of moonquakes, and the great age of surface features confirm this conclusion. Without water or an atmosphere to foster erosion, and without the renewing forces of internal activity, the lunar surface has remained essentially unchanged for billions of years. Its features were mostly shaped during its earlier active phase, with meteorite impacts playing a major role. The major formations on the Moon are its original anorthositic crust, now found primarily in the highlands, basaltic lava flows primarily filling the maria, ejecta blankets of varying ages, and the lunar regolith. The lunar lithosphere is much thicker and stronger than Earth's and is able to support mascons.

3. Mercury is the smallest and densest planet. Perhaps closeness to the Sun accounts for its lack of an atmosphere and the absence of the lighter, more volatile elements. Not much else is known about our innermost planet.

4. Venus is similar to Earth in mass, size, density, and distance from the Sun, but there the resemblance ends. The surface is hidden by a thick and heavy blanket of carbon dioxide, which traps solar energy so that the surface temperature reaches 800°F.

5. The Mariner 9 flight has revealed that Mars has an extremely thin atmosphere, but that it is a dynamic planet with extensive tectonic activity, volcanism, erosion, and deposition by wind and (possibly) water. Cratering by impacts is also in evidence. Four provinces have been recognized: (1) a volcanic region with giant volcanic mountains; (2) an equatorial plateau cut by deep canyons and great crack-like faults that seem to be fed by networks of dendritic tributaries; (3) smooth and cratered terrane like the lunar maria; and (4) polar regions with surface features attributable to moving glaciers. The evidence for erosion by water is quite striking—sinuous valleys, branching canyons—yet no water has been found on the surface or in the atmosphere.

# EXERCISES

1. The surface features of the Moon are profoundly different from those of Earth. Describe these differences and explain why they occur.

2. The Earth and Moon are of comparable age, yet Earth's surface is youthful compared to the old surface features of the Moon. Most rocks on the surface of the Moon are about 3 billion years or older, whereas most rocks on Earth's surface are younger than about 200 million years. Why?

3. What is the implication of the fact that mascons are found on the Moon and not on Earth?

4. Moon rocks differ from Earth rocks. How? Why?

5. Contrast the soils found on the Earth and the Moon.

6. In size, mass, and distance from the Sun, Venus is nearly Earth's twin. Yet the two planets are profoundly different. Describe these differences and hazard a guess as to why they occur.

7. Mars is to be explored with unmanned spacecraft, which will land on the surface. The primary mission will be to search for life. If you were the chief scientist, to which of the four "geological" provinces of Mars would you send your spacecraft to maximize the chance of finding living forms? Why?

## BIBLIOGRAPHY

Bergamini, David, *The Universe.* New York: Time–Life Books, 1969.

Brandt, J. C., and S. P. Maran, *New Horizons in Astronomy.* San Francisco: W. H. Freeman and Company, 1972.

Hinners, N. W., "The New Moon: A View." *Reviews of Geophysics and Space Physics,* vol. 9, pp. 447–522, 1971.

Lewis, J. S., "The Atmosphere, Clouds and Surface of Venus." *American Scientist,* vol. 59, pp. 557–566, 1971.

Murray, B. C., "Mars from Mariner 9." *Scientific American,* vol. 228, no. 1, pp. 49–69, January 1973.

Simmons, Gene, *On the Moon with Apollo 17.* Washington, D.C.: U.S. Government Printing Office (Stock Number 3300-00470), 1972.

# 24

# Matter and Energy from the Earth

*In the course of the evolution of civilization, Man has steadily broadened his use of the mineral and energy resources of the Earth. Since the eighteenth century, the abundance of fossil fuels has provided the industrial nations of the world with cheap power, but they are faced today with a declining supply of these conventional sources of energy. In the next few decades, the geological hunt for new supplies of oil, gas, and coal will be coupled with exploration for new uranium ore reserves and a search for alternative energy supplies. High-grade sources of many vital metal ores and other useful minerals are becoming harder to find, and the many lower-grade deposits can be exploited only if abundant energy is available.*

Unlike other living things, Man has learned to use the materials of the Earth to modify his environment extensively. We mine metal ores, coal, and salt. We drill for oil and gas. We strip away surface deposits on land and evaporate sea water to get dissolved metals. In our increasingly systematic search of the globe for new sources of the materials we depend upon, we use our geological knowledge of the ways in which already discovered natural segregations are distributed in order to find more of them. At the same time, we rely on a plentiful supply of energy for concentrating, smelting, and refining the materials from their natural, impure forms to the final products that are machined or molded into various components of our technology. The source of the energy is itself in the Earth as oil, gas, coal, or uranium ore.

The metals that we use for machinery, the phosphate rock prepared for fertilizer, and the salt used in food never disappear. Some we return to Earth as refuse—one of several forms of man-made deposits. Some dissolve in sanitary drainage waters that find their way into groundwaters and into rivers and eventually into the ocean, just like the dissolved materials from weathering. In this sense, Man is a hyperactive agent of erosion. Even though the used material is not really destroyed, much of it is so widely dispersed that it is recoverable only with a huge expenditure of energy. For this reason, inordinate amounts of energy may be required to gather together and refine used materials for recycling. Energy, in contrast to the matter that we consume, is not returnable. It is permanently lost to Earth. Regardless of which process we use, electrical, kinetic, or chemical, energy is converted to heat and irretrievably radiated to space.

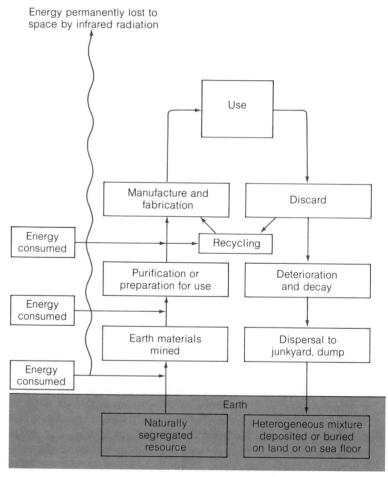

**Figure 24-1**
The cycle of mining, preparation, use, and discard of useful Earth materials is one in which a naturally segregated resource, such as an iron ore body, is mined, processed, and fabricated and then returned to Earth in a dispersed, heterogeneous mixture.

Because the prospect of maintaining or increasing our supply of energy underlies all resource problems, we will first review the existing and potential sources.

## ENERGY FROM FOSSIL FUELS

A wood fire, the extremely rapid oxidation of organic matter, is, in a superficial sense, deferred respiration. Throughout its lifetime, a tree respires, essentially as animals do, by slow controlled oxidation of organic substances in cells. The decay of a tree following its death is another slow oxidation process, depending, like respiration, on the organic matter produced by photosynthesis. Thus we can look upon a piece of wood or any piece of plant matter as a photosynthetic product that can be returned, by respiration during life, by slow decay, or by burning, to the carbon dioxide and water from which it was made. If the wood was buried and transformed into coal (see Chapter 13) 250 million years ago and we burn it today, we are utilizing the energy stored by photosynthesis from late Paleozoic sunlight. We are burning a "fossil," and we can refer in the same way to all natural organic materials, ranging from coal to oil and gas, including oil shales, tar sands, and other geologic formations, as **fossil fuels**.

The industrial revolution of the nineteenth century was powered by the energy from coal — in Britain, from the coal fields of England

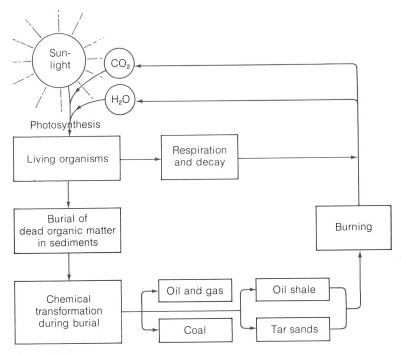

**Figure 24-2**
Photosynthesis produces organic matter that is buried, transformed, and so becomes a "fossilized" product of photosynthesis — a fossil fuel.

and Wales; in continental Europe, from the coal basins of western Germany and bordering countries; and in North America, from the Appalachian coal fields of Pennsylvania and West Virginia. As industrialization expanded, so did Man's hunger for coal, and geological exploration for this fuel spread over much of the world, from the Arctic deposits of Spitzbergen to Australia. Coal production climbed at an ever-accelerating pace.

Half a century after the first oil well was drilled in 1859, oil and gas were beginning to displace coal as the fuel of choice. Not only were they found to burn more cleanly, producing no ash, but they could be transported by pipeline as well as by rail or ship. When oil exploration entered its heyday, the application of geology and geophysics to the search for prospects was combined with a rapidly developing innovative technology for drilling ever more deeply and rapidly. The independent "wildcatter" who drilled in new territory was typically a self-trained geologist who learned how to recognize the permeable rocks and geologic structures in which oil might be found. Understanding of the geological conditions under which oil and gas are formed came quickly from the data of oil exploration and inferences from the chemical composition of petroleum and natural gas.

## PETROLEUM AND NATURAL GAS

Crude oil is made up of a great variety of **hydrocarbons** — compounds whose molecules are chains of carbon atoms with hydrogen atoms attached. Crude oil also contains some impurities, notably sulfur. The **paraffins**, a major subdivision of the hydrocarbons, are saturated compounds — that is, each carbon in the chain has attached to it all the hydrogens it can hold. The number of hydrogens in relation to the number of carbons is given by the general formula $C_nH_{2n+2}$. Methane ($CH_4$), or marsh gas, is the smallest and lightest member of the paraffin series. Methane and the other short-chain compounds are gases at room temperature and atmospheric pressure; examples are ethane ($C_2H_6$) and butane ($C_4H_{10}$). Natural gas is a mixture of methane and other light paraffins. The mixture heptane, octane, and nonane ($C_7H_{16}$ to $C_9H_{20}$) is what we know as gasoline. Lubricating oils are mixtures of still longer-chain paraffins.

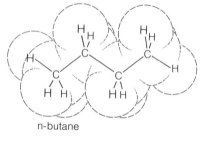

n-butane

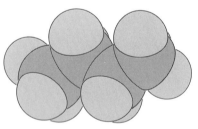

**Figure 24-3**
The structure of butane, a paraffin compound containing four carbon atoms and ten hydrogen atoms. [From *College Chemistry* (3rd ed.) by Linus Pauling. Copyright © 1964. W. H. Freeman and Company.]

**How Oil and Gas Form** Small amounts of hydrocarbons and related compounds occur in organisms of all kinds, from algae to elephants. Other biological compounds can be transformed into hydrocarbons by organic chemical reactions in sediments. Nowhere else on Earth, neither in igneous nor in metamorphic rocks, are hydrocarbons of similar nature formed. Petroleum, like coal, is a biological product — the organic debris of former life, buried, transformed, and preserved in sediments.

The conditions of preservation are a favorable balance between the production of organic matter and its destruction by the scavenging of other organisms or by inorganic oxidation. These conditions are met where productivity of organic matter is high, such as in the coastal waters of the sea, where large numbers of organisms thrive and where the supply of oxygen in bottom sediment layers is not enough to oxidize all organic matter. Many offshore sedimentary basins on continental shelves satisfy these conditions; in such environments, and to a lesser degree in many others, small amounts of organic matter are buried and protected from oxidation. During the millions of years of burial, chemical reactions slowly transform a fraction of this material into liquid and gaseous hydrocarbons. These fluids tend to be squeezed out of compacting muddy sediments into adjacent permeable beds, such as sandstones or porous limestones. The low density of oil and gas causes them to migrate to the highest place they can reach, where they float on top of the water in the pores of the formations.

The geologic environment that favors the large-scale accumulation of oil is a combination of structure and rock types that create an impermeable barrier to upward migration—an **oil trap**. In order for the accumulation to qualify as an **oil field**, the oil-bearing formation must be permeable enough to allow the crude oil to be pumped out. The typical trap-forming structure is an

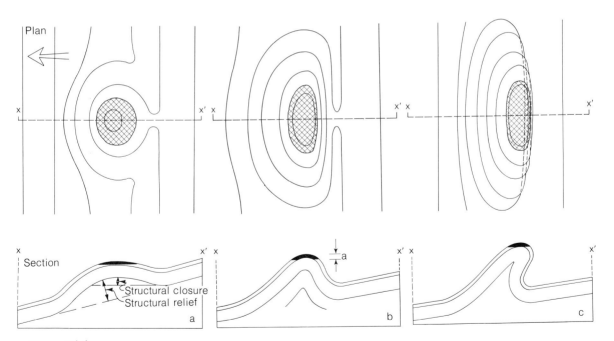

**Figure 24-4**
Idealized structural maps and sections of typical anticlinal dome folds; such folds are characteristic of many traps containing oil and gas pools (oil, shaded and black). (Arrow shows direction of dip.) Contour lines show elevation of top bed with respect to sea level. [From *Geology of Petroleum* (2nd ed.) by A. I. Levorsen. Copyright © 1967. W. H. Freeman and Company.]

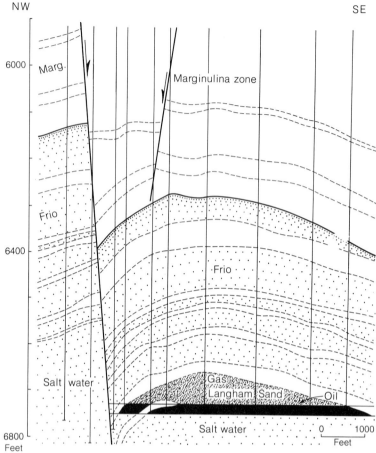

**Figure 24-5**
Section through the Amelia field, Texas, showing the relations between a fold and a fault and how gas and oil are trapped overlying a groundwater line. The dip into the down side of the fault is contrary to most faulting, except in the system of faults parallel to the Gulf Coast in Louisiana and Texas, of which this is an example. [From Hamner, *Bull. Amer. Assoc. Petrol. Geol.*, vol. 23, p. 1647.]

anticline in which a permeable sandstone is overlain by an impermeable shale. The oil and gas accumulate at the highest part of the fold, the gas highest, the oil next, both floating on the water that saturates all permeable formations. Other structures may trap oil in similar ways. For example, displacement at a fault may place a dipping permeable limestone bed opposite an impermeable shale, creating a trap for oil. A dipping sandstone bed may thin out against a shale and form a stratigraphic trap—one that is primarily the result of the original sedimentation pattern rather than of later structural deformation.

Geologists have mapped thousands of structural or stratigraphic traps, all conceivably favorable to oil accumulation, but only a fraction of them have proved to contain any oil or gas. The structures alone are not enough: the oil has to be available to migrate into them. There is a sequence of necessary geologic events that lead to oil accumulation, no one of which can be omitted, though the precise order of steps may vary.

1. Organic matter must be produced in some quantity in the environment.
2. Organic matter must be buried in the sediment before being destroyed by oxidation.
3. Some part of the organic matter must be transformed by slow chemical reactions to oil and gas.
4. The organic matter in the sediment must migrate into nearby permeable beds.
5. The permeable beds must be deformed either by folding or faulting to form structural traps or stratigraphic traps must be produced by conditions of sedimentation.
6. The oil-containing sediment must not be transformed into slate or higher-grade metamorphic rocks or the organic matter will be transformed by heat into a non-hydrocarbon black carbonaceous material of no value.
7. Structural deformation in general must not be extremely severe. Intensely fractured rocks may render traps ineffective by causing leakage.
8. The beds must remain buried. If exposed to erosion, the oil may be dissipated.
9. The pore spaces of the permeable formations must not be so filled with swelling clays or mineral cements deposited by groundwaters that the permeability will be reduced to the point that the oil cannot be pumped out of the formation.

With such a long list of requisites, one might wonder that there is as much oil as there is. Yet the probability of all those events happening in the right order is fairly high. A great quantity of organic matter *is* produced and buried in the sediments on and around continents, and the normal pattern *is* for that organic matter to be transformed slowly to oil. Stratigraphic and structural traps abound in most sedimentary basins. Although sediments are intensely deformed in mountain chains, the greater part of any continent consists of sediments of all ages that are only gently deformed and are subjected at most to mild warming by deep burial. Thus oil is not an unusual geologic product. Rather, we expect to find it where conditions are right, and the successful results of exploration during the past century have borne out those expectations.

**The World Distribution of Oil and Gas**  If you were to visit the exploration and research offices of any large oil company, you would be able to find maps and reports of all of the sedimentary

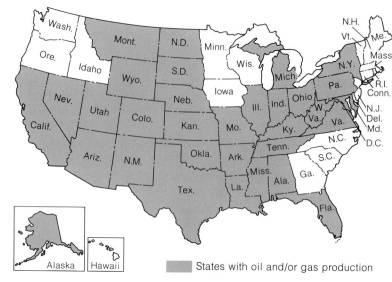

**Figure 24-6**
The oil producing industry in the United States, 1972. [From "The Oil
Producing Industry in Your State," Independent Petroleum Association of
America, 1972.]

rock terranes in which the company has operated, for where there
is sediment, there may be oil. For example, thirty-one states of
the fifty United States produce commercially marketable oil, and
some small occurrences are known in most of the others. Canada
produces oil from many of its provinces. Most of the countries
of South America and Europe produce oil. Once considered oil-
poor, Africa now produces a respectable quantity. This is not to
say, however, that the world distribution is uniform; rather, it is
most uneven. The two richest and most important oil-producing
regions are the Middle East and the area between the Gulf of
Mexico and the Caribbean. The Middle East includes about
two-thirds of the known reserves of the world in the oil fields of
Iran, Kuwait, Saudi Arabia, Iraq, and the Baku Region of the
U.S.S.R. The highly productive Gulf Coast-Caribbean area includes
the Louisiana-Texas province, Mexico, Colombia, Venezuela, and
Trinidad.

Why this uneven distribution? A political scientist once jokingly
asked one of us, "Why is it that oil is always found in underde-
veloped countries?" The distribution of oil, of course, is determined not by political boundaries but by geologic history. The age
range of sediments from which oil is produced gives some idea of
the reasons for the distribution. There is practically no oil in Precambrian rocks and very little in Cambrian rocks. Rocks of the
Triassic Period also contain very little. Oil produced from rocks
of these periods amounts to less than one percent of the world's
total production. For all practical purposes, we can say that no oil

is produced from Pleistocene rocks. But Cenozoic sediments, those deposited only in the last 70 million years of the Earth's 4.7 billion year history, account for about 60 percent of the total. Another large fraction, about 25 percent, comes from rocks of Mesozoic age, the next older era. Only about 15 percent comes from Paleozoic rocks.

The correlation of geologic age and oil production is made understandable in terms of the processes that destroy oil or allow it to disperse at the surface. The older the rock formation is, the more likely it will be eroded or involved in metamorphism and deformation that would eliminate the oil. The Cenozoic thus may represent the normal abundance of oil. Older oil-producing formations are those that have escaped the ravages of time—exposure to destructive geological events. If we were living in the Triassic period, we would probably find oil most abundant in late Paleozoic rocks.

Oil in the Middle East is concentrated largely in thick Cenozoic and Mesozoic sediments deposited in and adjacent to a major geosyncline situated along the southern margin of the Asian continent. Sedimentation has been accompanied by a variety of deformation structures and stratigraphic traps. There is no evidence of metamorphism, and relatively few of the sediments have been destroyed by erosion.

**Continental Shelf Oil** Continental shelves are good places to explore for oil, as has been known since the general nature of their structure was worked out after World War II. They easily meet the requisites for accumulation: the sediments are young, unmetamorphosed, mildly deformed, and little eroded. The shelves of the world have been geophysically combed for favorable structures by the major oil companies in anticipation of the large amounts of oil expected to be found there. Drilling on continental shelves, however, is not without its problems. A major one is the difficulty of drilling in deep water. At this time, wells in offshore areas of Louisiana, Texas, and California can be drilled in water as much as 100 meters deep. In the future, that depth will probably be increased at least two to four times. In order to drill in deeper water—on continental slopes and rises—new kinds of drilling technology will probably be needed. Perhaps the Glomar Challenger's equipment, which has been used to drill the holes for the deep-sea drilling program, will serve as a prototype.

Another major problem of offshore drilling is the prevention of oil pollution. The name "Santa Barbara" calls to mind the environmental damage that resulted from the accidental release of oil from an offshore drilling platform there in 1969, but many other accidents of that kind have not made the headlines. Though it is difficult if not impossible to guarantee a "100 percent safe" drilling well, it *is* within the realm of possibility to devise technological systems and controls that would greatly reduce the

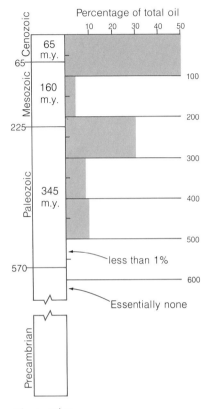

**Figure 24-7**
Percentage of the world's total past production and proven reserves of oil by age distribution in 100-million-year intervals. More than 80 percent of the world's known oil is found in rocks formed in the last 6 percent of geologic time.

Offshore areas

Active exploring
and producing
operations.

Potential oil- and gas-bearing
areas, only partially explored,
in water less than 600 feet deep.

**Figure 24-8**
The continental shelves of the southern part of North America, showing areas
being explored and investigated for future potential. [Courtesy Exxon
Corporation.]

chances of a serious accident. If there is adequate development
and enforcement of safety procedures, offshore drilling need not
pose an important threat to the oceans or nearby beaches. Flushing
of oil tanks by ships at sea probably poses a more serious current
danger. The important point is that there are large reserves of
oil and gas under the sea bed that will eventually have to be drilled
to satisfy the world's needs. We *can* tap those supplies without
severe cost to the environment if precautions are taken. There is
no doubt that we will be pushed to explore every conceivable oil
resource as our present reserves are depleted, whether our increasing demands are slowed or not.

**Oil: The Exhaustible Supply**   Inconceivable as it may seem to
the landowner who sees a "gusher" spurting oil from the derrick
on his property, that oil well will eventually run dry. What the
world wants to know is when all the wells will run dry. What is
behind this loaded question are the facts of oil occurrence on
Earth. Thirty years ago, many oil producers were optimistic; sup-

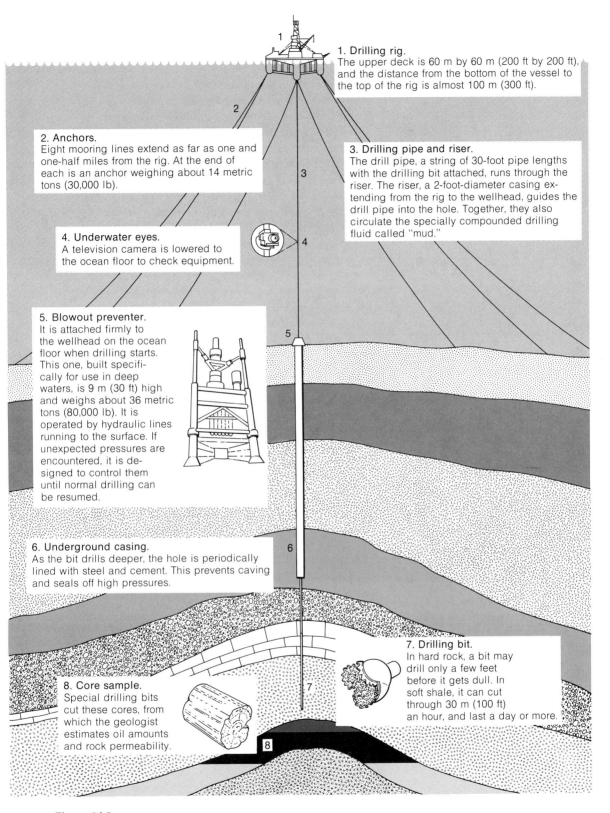

**1. Drilling rig.**
The upper deck is 60 m by 60 m (200 ft by 200 ft), and the distance from the bottom of the vessel to the top of the rig is almost 100 m (300 ft).

**2. Anchors.**
Eight mooring lines extend as far as one and one-half miles from the rig. At the end of each is an anchor weighing about 14 metric tons (30,000 lb).

**3. Drilling pipe and riser.**
The drill pipe, a string of 30-foot pipe lengths with the drilling bit attached, runs through the riser. The riser, a 2-foot-diameter casing extending from the rig to the wellhead, guides the drill pipe into the hole. Together, they also circulate the specially compounded drilling fluid called "mud."

**4. Underwater eyes.**
A television camera is lowered to the ocean floor to check equipment.

**5. Blowout preventer.**
It is attached firmly to the wellhead on the ocean floor when drilling starts. This one, built specifically for use in deep waters, is 9 m (30 ft) high and weighs about 36 metric tons (80,000 lb). It is operated by hydraulic lines running to the surface. If unexpected pressures are encountered, it is designed to control them until normal drilling can be resumed.

**6. Underground casing.**
As the bit drills deeper, the hole is periodically lined with steel and cement. This prevents caving and seals off high pressures.

**7. Drilling bit.**
In hard rock, a bit may drill only a few feet before it gets dull. In soft shale, it can cut through 30 m (100 ft) an hour, and last a day or more.

**8. Core sample.**
Special drilling bits cut these cores, from which the geologist estimates oil amounts and rock permeability.

**Figure 24-9**
Diagrammatic view of an offshore drilling rig. With the exception of the special equipment in the water, the drill hole is drilled in much the same way as in land-based drilling. [Courtesy Exxon Corporation.]

plies seemed so immense, and so many scientific and engineering innovations held promise of tapping new sources, that no one needed to worry about the future. Since then, the rapidly accelerating world demand and a more sober analysis of the total quantity of oil remaining on Earth has changed the picture.

The efforts of one man played an important role in changing attitudes about energy reserves. *M. King Hubbert,* a geologist-geophysicist working for the Shell Oil Company, started his analysis of the world oil situation about thirty years ago by estimating the total oil reserves. He argued that essentially all the oil in the Earth had formed during the 600 million years since the beginning of the Cambrian period and that new oil is being made at about the same rate. This means that it will take another million years for the natural formation of 1/600 of the oil that has so far been found. Thus it is clear that we cannot rely on natural processes to replenish the oil we pump out of the ground at so rapid a rate. We are dealing with a fixed supply that steadily dwindles as we use it. Three significant questions that need to be answered are: how great is the total supply, how much of it have we already used, and at what rate do we expect to continue depleting the supply?

How does a petroleum geologist go about estimating the total amount of recoverable oil in the Earth? The individual oil well that discovers a new oil field gives only a hint of the ultimate size of the field. But from it the geologist learns how thick the oil-saturated porous beds are and how permeable they are, the permeability being a good guide to the ease of pumping the oil. As more wells are drilled, the productive areas are outlined and the volume of oil in the oil field becomes better known. Experience with older wells in the same geologic province gives some idea of how much oil might eventually be produced, for not all of the oil in the ground can be recovered, though improvements in recovery engineering are constantly being made. The general abundance and distribution of already drilled oil fields in the various kinds of sedimentary basins are guides to estimating future reserves. For example, exploration has been so thorough in the older oil provinces, such as those of Pennsylvania, Illinois, Oklahoma, and Texas, that there is little doubt that more than 90 percent of the oil there has already been discovered (though not all of it has been pumped). These **proven reserves** are those that have been geologically outlined and at least partially drilled. Estimates of proven reserves are constantly being updated as the balance changes between the increase from new discoveries and the decrease by production at pumping wells.

The next step is the geological estimate of the future exploration of as-yet-untapped oil fields. With more than a hundred years worth of experience in drilling oil and the increase in the geological knowledge of the world, geologists can make a good guess about how much oil is left to find. Geologic mapping of the land

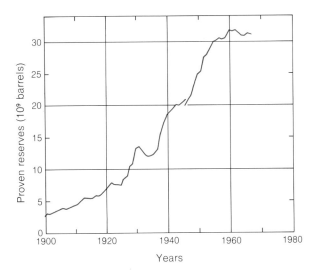

**Figure 24-10**
Proven reserves of crude oil in the United States, exclusive
of Alaska. [From *Resources and Man,* National Academy of
Sciences–National Research Council. Copyright © 1969.
W. H. Freeman and Company.]

areas of the world is essentially complete, at least on a general
scale, and we know where all of the sedimentary basins are. The
large discoveries of the past four years in Alaska were welcome
but not exactly a great surprise, for geologists had known for years
that the area was favorable for oil exploration. We now have a
good idea of the sedimentary basins under the permanent glacial
ice of Antarctica, and sooner or later exploration and production
methods may be developed that will allow oil drillers to go there
too.

The companion to geologic mapping of exposed formations on
land is geophysical exploration. Since the early 1920's the major
practical use of seismology has been in oil-field exploration (see
Box 19-1). Reflected seismic waves reveal deeper subsurface structures and formation changes that are not apparent from surface
mapping. This ability to "see" beneath the surface was utilized
fully in the exploration of the continental shelves for structures
favorable to oil accumulation. The combination of geology and
geophysics has made possible reliable estimates of all the oil that
may eventually be discovered and produced on land, in the ocean,
and under glacial ice—what we call **petroleum resources.** In
spite of all of the major new discoveries in Alaska and elsewhere,
the known petroleum resources of the world have not changed
much in the past decade, increasing confidence in the estimates.
Of course, we cannot completely rule out discoveries of a totally
unpredictable kind. But we would be unwise to count on them.

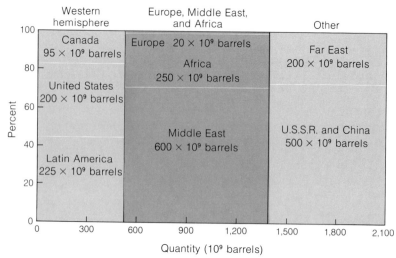

**Figure 24-11**
Petroleum resources of the world are depicted in an arrangement that can be
read horizontally for original total resources and vertically for percentage. Thus
the block on the left shows that the western hemisphere originally contained a
little more than 500 billion barrels, of which the United States held about 19
percent. The figures for petroleum are derived from estimates made in 1967 by
W. P. Ryman of the Standard Oil Company of New Jersey. They represent
ultimate crude-oil production, including oil from offshore areas, and consist of
oil already produced, proven and probable reserves, and future discoveries.
Estimates as low as $1350 \times 10^9$ barrels have also been made. [From "The Energy
Resources of the Earth" by M. K. Hubbert. Copyright © 1971 by Scientific
American, Inc. All rights reserved.]

The standard measure for oil
production is the U.S. barrel, equal
to 42 gallons or 159 liters. Because
the density of oils from different
fields varies appreciably, one can
only state an average weight per
barrel, about 140 kg or 310 lbs.

**How Much Oil Is Left?**    A recent estimate of world oil resources
is shown in Figure 24-11. Huge as the total, about 2100 billion
barrels, may seem, projections for the future must be based on
how much of that total we have already used, the rate we are now
using it, and the anticipated rate in the future. Since oil was first
discovered in Pennsylvania in 1859, the total production up to
1969 was 277 billion barrels, about 10 percent of the total resource.
From past production figures we can also see how the rate of use
has jumped with steadily expanding industrialization. It took a
little more than the first 100 years of oil production to pump the
first half of all oil produced since the beginning, but only another
ten years to pump the remaining half.

Though 90 percent of the world's oil is still in the ground, the
proportion left in the ground in the United States (exclusive of
Alaska), is much less than that, for the United States was one of the
earliest countries to explore and produce oil systematically and
efficiently. Of the approximately 165 billion barrels of ultimate
production estimated for the United States, about two-thirds, or
about 110 billion barrels, has already been produced and burned
up. Of the 165 billion barrels of petroleum resources, about 145
billion barrels have already been discovered (as of January 1, 1972)

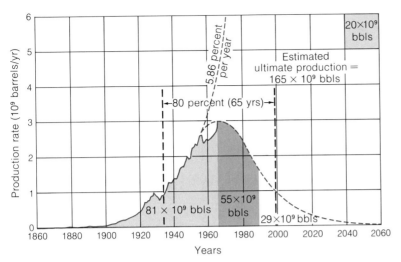

**Figure 24-12**
Complete cycle of crude-oil production in the United States and adjacent
continental shelves, exclusive of Alaska, assuming a total ultimate production of
$165 \times 10^9$ barrels. The broken line, showing a 5.86 percent increase in rate per
year, indicates what would have happened if the rate between 1945 and 1955
had been continued. [From *Resources and Man,* National Academy of Sciences–
National Research Council. Copyright © 1969. W. H. Freeman and Company.]

and thus are proven reserves. Even allowing for some error, the
picture is clear: we have already discovered at least 75 percent of
all that we may ultimately find. Much of the as-yet-undiscovered
resource lies beneath the continental shelves. The discovery of
Alaskan oil adds about another 35 billion barrels to the total resource (proven reserves of about 10 billion barrels at the Prudhoe
Bay oil field), but really doesn't change the picture greatly. One
must remember that 30 billion barrels is less than a 10-year supply
for the United States at current rates of use.

The picture for natural gas is much the same as that for oil.
Though it has been exploited a little less rapidly than oil, natural-
gas production in this country will probably peak in the 1970's
and then decline. The world political and economic consequences
of these facts of oil supply will continue to be a major factor in
the next decades. Every student of international politics should
study Figure 24-12 and ponder its implications. Yet oil and gas
do not tell all the story, even though we depend on them heavily
today. There are enormous supplies of energy from the other fossil
fuels, coal, oil shale and tar sands.

## COAL

Coal has been used as a fuel for millenia, perhaps since early inhabitants of Wales used it for funeral pyres about 3000 or 4000 years
ago. By the eighteenth century it was well established as a fuel for

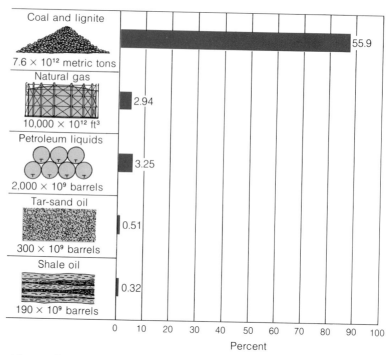

**Figure 24-13**
Energy content of the world's initial supply of recoverable fossil fuels is given in units of $10^{15}$ thermal kilowatt-hours. Coal and lignite, for example, contain $55.9 \times 10^{15}$ thermal kilowatt-hours of energy and represent 88.8 percent of the recoverable energy. [From "The Energy Resources of the Earth" by M. K. Hubbert. Copyright © 1971 by Scientific American, Inc. All rights reserved.]

heating houses and then became the mainstay of the industrial revolution in the nineteenth and early part of the twentieth centuries. Coal was discovered in North America by several early explorers; the first were Louis Joliet and Jacques Marquette, who found it in Illinois in 1673. Major production began in the 1800's, and increased steadily with demands brought on by the emerging steel industry, the development of the railroad steam engine, increased use for general heating, and then the growth of electrical power. As of 1970, about 133 billion metric tons of all kinds of coal had been mined. The increase in the rate of use has slackened in the twentieth century, largely because oil and gas supplanted coal for so many uses. There are now signs of a resurgence in coal production prompted by the oil reserve and price picture.

It is far easier to estimate coal reserves than oil and gas reserves. Like other sedimentary layers, coal beds extend over large areas, hence they can easily be mapped from surface exposures and their subsurface extent inferred. There has been much mining and exploration drilling for coal; in addition, many wells drilled for oil have penetrated coal-bearing formations. Because the coal-exploration

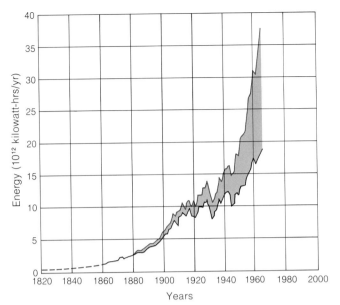

**Figure 24-14**
Energy contribution of coal and coal plus oil is portrayed in terms
of their heat of combustion. Before 1900 the energy contribution
from oil was barely significant. Since then the contribution from
oil has risen much more rapidly than that from coal. By 1968 oil
represented about 60 percent of the total. If the energy from
natural gas were included, petroleum would account for about 70
percent of the total. [From "The Energy Resources of the Earth"
by M. K. Hubbert. Copyright © 1971 by Scientific American, Inc.
All rights reserved.]

geologist has a good idea of what kinds of sedimentary rock sections include coal and where to look for them (see Chapter 13),
he is likely to be more concerned with thickness and quality of
coal than with its presence or absence. For all of these geological
reasons, accurate estimates of the world's coal reserves are easier
to make.

According to the best estimates made by the U.S. Geological
Survey, about 7.64 trillion metric tons of coal remain unexploited.
Thus we have so far used up only 1.7 percent. These figures include
only coal deposits minable by present technology. Trillions of
tons of additional coal exist, but the deposits are either so deep
or in such thin beds that we cannot easily mine them by current
methods. There are many new technologies that may possibly
allow the recovery of such coal. Some experiments have already
succeeded in chemically transforming the coal into gas while
it is underground, without using mines or miners. Other ideas
include underground combustion and conversion to steam or
electrical energy. In any event, there is enough coal to last for a
long time.

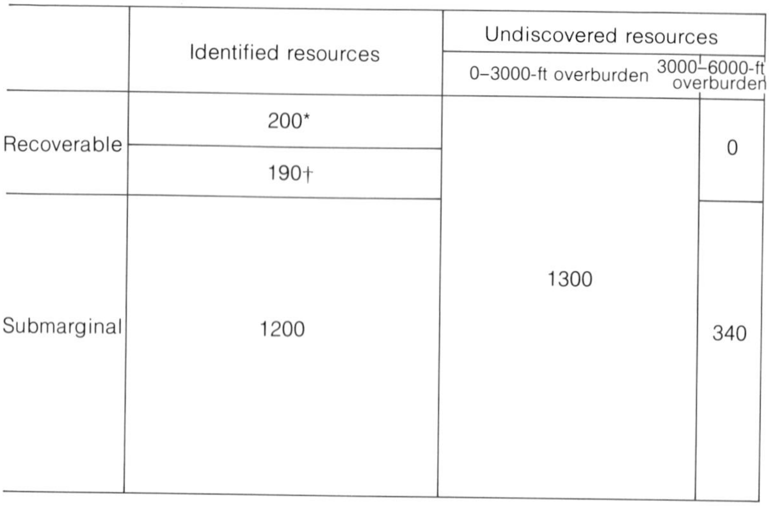

| | Identified resources | Undiscovered resources | |
|---|---|---|---|
| | | 0–3000-ft overburden | 3000–6000-ft overburden |
| Recoverable | 200* | 1300 | 0 |
| | 190† | | |
| Submarginal | 1200 | | 340 |

*In beds 42 in. or more thick for bituminous coal and anthracite, and beds 10 ft or more thick for subbituminous coal and lignite.

†In beds 28–42 in thick for bituminous coal and anthracite, and beds 3–5 ft thick for subbituminous and lignite.

**Figure 24-15**
Scaled diagram of coal resources in the United States (billions of short tons). The reliability of estimates decreases downward and to the right. [Compiled by P. Averitt, February 1972. U.S. Geological Survey.]

The problems involved in the recovery and use of coal—problems that make it less desirable than oil or gas—still remain. Much coal contains appreciable amounts of sulfur combined in the mineral pyrite (ferrous sulfide, $FeS_2$), which vaporizes during combustion to liberate noxious sulfur gas to the atmosphere. Coal ash amounts to several percent of the weight of the unburned coal, and poses a significant disposal problem. Though coal can be converted into oil in chemical engineering plants, the process is expensive. The smoke from coal combustion is a serious air pollution nuisance and must be controlled at the smokestack. Coal mining, particularly strip mining, can ravage the countryside. Underground mining accidents cost the lives of some miners each year, and many suffer from "black lung" and other pulmonary diseases. None of these drawbacks, however, is likely to prevent the increased use of this important fuel.

## OIL SHALE AND TAR SANDS

Known for many years but practically untouched are large reserves of oil that can be extracted from oil shale and tar sands. Oil shales are fine-grained sedimentary rocks containing a high proportion of solid organic matter mixed with the various minerals; some contain small amounts of liquid oil. This organic matter, called kerogen, has the same general origin as the other common organic materials, all ultimately derived from plants and animal matter. This complex and poorly understood solid mixture of carbon compounds has an extremely important property that is of practical value. If it is heated and the vapors are distilled, oil can be recovered from the shale. Some oil shales can produce up to 150 gallons of oil per ton but most yield about 25 to 50 gallons per ton.

Reserves of oil shales are known moderately well. Like coal, they occur as beds whose extent can be fairly reliably predicted. They are less abundant than coal, however, and have not been well explored in many parts of the world. Many oil shales are too low in amount of recoverable oil to make mining economical and so the total practical resource is far less than the total amount of oil shale. Geologists of the U.S. Geological Survey have estimated a total world resource of about 3100 billion barrels of oil contained in oil shales that would yield more than 10 gallons of oil per ton. That figure is half again as much as all liquid petroleum in the world, but it is deceptively high, for it must be reduced to only 190 billion barrels that are recoverable under mining technology known today. But this is still a respectable figure and clearly a significant future resource.

A little less than half of the world's supply of oil shale, about 80 billion barrels, is in the Green River Shale of Eocene age in

| | Identified resources | Undiscovered resources | |
| | | Extension of known resources | Undiscovered and unappraised |
|---|---|---|---|
| Recoverable | 0 | 0 | 0 |
| Paramarginal | 160–600 | 850 | 500 |
| Submarginal | 1600 | 2500 | 20,000 |

Figure 24-16
Scaled diagram of oil-shale resources in the United States (billions of barrels). The reliability of estimates decreases downward and to the right. [Compiled by D. S. Duncan, February 1972. U.S. Geological Survey.]

**Figure 24-17**
Typical cliffs in the principal oil-shale group of the Green River Formation. Garfield County, Colorado. [Photo by D. E. Winchester, U.S. Geological Survey.]

Wyoming, Utah, and Colorado. Though pilot plants have been in operation for years, little serious attempt has yet been made to exploit these rich reserves commercially, partly because up to now it has not been competitive in price with oil. In addition, disposal of immense tonnages of waste shale from the extraction process presents a serious problem.

A more unusual occurrence of organic matter useable as fuel is an asphalt or **tar sand**. Oil of a type similar to liquid petroleum can be recovered from these sands, and geologic evidence points to the tar as being a transformed product of a once-liquid petroleum. Some deposits of tar sands are oil pools that have "dried up" and become tarry by loss of the volatile hydrocarbon fractions; some are exposed at the surface by erosion. Others show signs of having been deposited as a combination of sand and oil, the oil having seeped from older formations. Some of the largest deposits, the Athabaska tar sands of Cretaceous age in Alberta, Canada, may be of this origin. The Athabaska deposit is a great resource that is now beginning to be exploited. Estimates are that 300 billion barrels of oil are recoverable from this one region. Tar sands are poorly known and there are no reliable figures or world reserves.

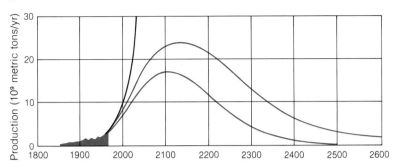

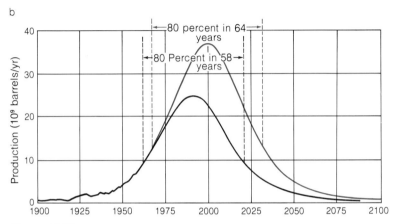

**Figure 24-18**
(a) Cycle of world coal production is plotted on the basis of estimated supplies
and rates of production. The top curve reflects Averitt's estimate of $7.6 \times 10^{12}$
metric tons as the initial supply of minable coal; the bottom curve reflects an
estimate of $4.3 \times 10^{12}$ metric tons. The curve that rises to the top of the graph
shows the trend if production were to continue rising at the present rate of
3.56 percent per year. The amount of coal mined and burned in the century
beginning in 1870 is shown by the black area at left. (b) Cycle of world oil
production is plotted on the basis of two estimates of the amount of oil that will
ultimately be produced. The colored curve reflects Ryman's estimate of
$2100 \times 10^{9}$ barrels, and the black curve represents an estimate of $1350 \times 10^{9}$
barrels. [From "The Energy Resources of the Earth" by M. K. Hubbert.
Copyright © 1971 by Scientific American, Inc. All rights reserved.]

## THE FUTURE OF FOSSIL FUELS

One of the currently popular occupations of geologists, economists,
and others concerned with energy sources is to estimate our future
dependence on fossil fuels. M. King Hubbert, the dean of such
predictors, has concluded that if oil and gas continue to be used
mainly for their energy, and as the major resource for satisfying the
world's continually climbing energy appetite, the great bulk of the
world's supply will be exhausted in a century. Under these conditions the world's coal supply will last 300 to 400 years, but not that

long if it is the main energy source. Oil shale and tar sands will add some years, but will not alter the picture greatly.

For the United States the energy problem is acute, for we will have used up the vast bulk of our oil and gas supplies sometime between 1990 and 2000. Already partially dependent on imports, we will then be wholly dependent on foreign supplies, on coal and oil shale, or sources of energy alternative to fossil fuels.

## ALTERNATIVE ENERGY SOURCES

Some alternative energy sources are rather familiar, such as hydro-electric power from turbines driven by falling water. Hydroelectric power is clean and cheap. If all the available sites in the world were utilized for hydroelectric plants, the world's energy needs at present levels, now largely supplied by fossil fuels, could be barely satisfied. But think of the cost. Much of our beautiful scenery — mountain canyons, gorges, and waterfalls — would be drowned in reservoirs behind dams containing power plants. Great areas would be flooded by reservoirs, which would have only a limited "lifetime"; within a few decades, or a century at most, they would become so silted up with sediment brought in by streams that the dams would become useless. It is doubtful that we can or would want to solve the world's energy problems by converting entirely to water power.

The movement of powerful tides can be used in some places to generate hydroelectric power. The French hydroelectric plant at

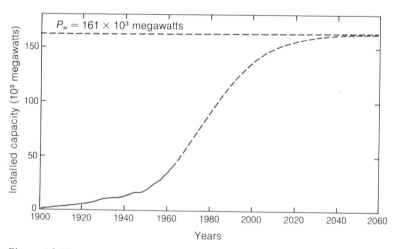

**Figure 24-19**
United States installed and potential water-power capacity. The ultimate maximum power capacity, $P_\infty$ has been estimated by the Federal Power Commission to be about 161,000 megawatts. [From *Resources and Man,* National Academy of Sciences–National Research Council. W. H. Freeman and Company. Copyright © 1969.]

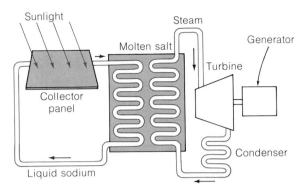

**Figure 24-20**
In a recently proposed solar power plant, sunlight falls on
specially coated collectors and raises the temperature of a
liquid metal to 1000°F. A heat exchanger transfers the heat
so collected to steam, which then turns a turbogenerator as
in a conventional power plant. A salt reservoir holds enough
heat to keep generating steam during the night and when
the sun is hidden by clouds. [From "The Conversion of
Energy" by C. M. Summers. Copyright © 1971 by
Scientific American, Inc. All rights reserved.]

La Rance uses strong tides surging back and forth across an estuary.
The Passamaquoddy tidal power project in Maine and New Bruns-
wick has been proposed to the U.S. Congress for years to do the
same. Yet if all the available tidal power sites in the world were
utilized, it would add only 1 percent to the total possible water
power of the Earth.

Solar energy is another source of power that scientists have
been investigating. In a very specialized way, it has shown its usefulness in powering artificial satellites and scientific experiments
on the Moon. Since all of our major natural energy sources come
from the Sun anyway, why not convert its rays to power? The
power of the Sun's rays, if all were converted, would amount to
about a hundred-thousand times the present electric-power capacity of the world, more than enough if it could be done. It is non-
depletable; the Sun will continue to shine at least for the next
several billion years. The major problem is that we must harvest
the Sun's rays over a great area if we are to produce much power.
The average solar power available is only about 0.024 of a watt for
each square centimeter of ground surface illuminated. It has been
calculated that to generate 1000 megawatts, a power plant would
need at least ten times that much solar energy, and that would
require a large collecting area, about 42 square kilometers. To
power a significant part of the North American continent would
require many thousands of square kilometers. One proposal en-
visions using a great area of desert in the southwestern United
States for a solar power plant. However and wherever solar energy
is utilized, it seems improbable that it will ever produce a significant

part of the world's power needs, for as population grows, need for space grows too. Solar energy may assume an important role, however, in heating and cooling buildings in regions where sunny days are frequent enough, as in most of the United States.

**Geothermal power** Because of the strong association of heat and steam with energy and power it was natural for power engineers to turn their attention to using the Earth's heat to drive electrical generators, especially in regions of active hot springs and geysers. In 1904, Italian engineers first put the natural steam vents at Larderello to work to make power. That plant can now produce 370 megawatts (million watts) of electricity. In the past few decades, generating plants have been built in the United States, Japan, the Soviet Union, New Zealand, and other countries, all taking advantage of abnormally high heat flow produced by volcanic activity. The United States plant, at the Geysers, 75 miles north of San Francisco, is in its second decade of production and is now producing almost 400 megawatts from 80 steam wells. In a few years that plant plans to expand to a capacity of 847 megawatts, enough to supply all the electricity needs of San Francisco. Interest is now great in the development of geothermal power in the Imperial Valley in southern California.

The United States now produces about 350,000 megawatts from all power sources. It is estimated, according to a National Science Foundation report, that by 1985 the Geysers and all other geothermal energy sources in the United States, if fully utilized at highest efficiency, have the potential of supplying a little less than half the power needed now. That is not quite as good as it sounds,

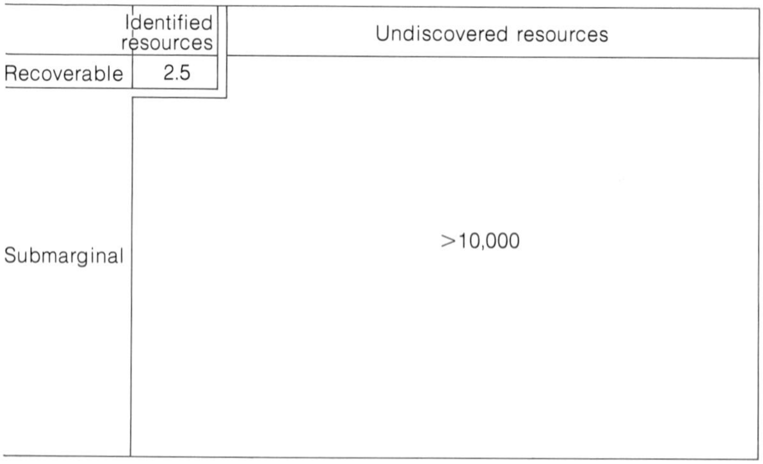

| | Identified resources | Undiscovered resources |
|---|---|---|
| Recoverable | 2.5 | |
| Submarginal | | >10,000 |

**Figure 24-21**
Diagram of geothermal resources in the United States ($10^{18}$ calories). The reliability of estimates decreases downward and to the right. [Compiled by L. J. P. Muffler and D. E. White, February 1972. U.S. Geological Survey.]

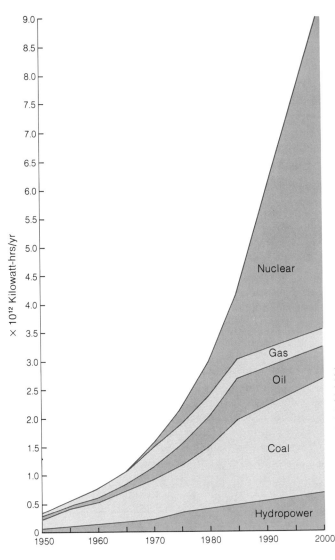

**Figure 24-22**
Net United States production of electricity by primary source, 1950–2000. This prediction assumes a steadily increasing role for nuclear power in the next decades. [From *Energy and the Environment*, U.S. Council on Environmental Quality, 1973.]

though, because energy demands in the United States are *doubling* every eight years. It is also a very optimistic estimate. A more realistic appraisal is given by Donald E. White of the U.S. Geological Survey. He calculated that if all the world's useable geothermal power were fully utilized, about 60,000 megawatts would be available each year for 50 years until total depletion, which is of the same general amount as would be available from tidal power if it were fully exploited. Geothermal power thus will not satisfy the world's demands by any means. In those regions of recent or active volcanism, where geothermal power can be produced, there are some serious disadvantages that temper its attractiveness. Tapping heat sources in the ground requires drilling wells and pumping steam or hot water from volcanically heated underground water reservoirs. That is why geothermal power is a depletable resource, for it is like the continued pumping of ordinary ground-water aquifers, which may result in using up the water (see Chapter 7). These hot waters cannot be recharged *and* heated rapidly enough to maintain the supply indefinitely. Experience at Lardarello, Italy, the oldest geothermal field, shows that wells maintain their heat and water output for only about 10 years. More and more wells have had to be drilled to maintain high power production, and it is apparent that the whole field is being depleted. Superheated water used in some plants dissolves much material from rock and pipes and thereby produces a significant pollution problem when discharged into streams. Thus in spite of some attractive properties, the lack of any pollution side-effects, the amount of power available, and costs may not make geothermal power the best answer for all geothermal regions. And it certainly does not answer the needs of the many millions of people living in the interior or along the eastern edge of North America, where volcanic heat sources have geologically long ago cooled off. For that population, the need for great power supplies is not likely to be met by any source but nuclear reactors.

## NUCLEAR POWER

Nuclear physicists foresaw the possibility of peaceful uses of the gigantic amounts of energy stored in the atomic nucleus even before they were able first to liberate it in the form of an atomic bomb in 1944. Their predictions have started to come true as countries all over the world are building new atomic reactors at an increasing rate. In its full potential, nuclear energy can supply virtually unlimited power to the world. In this book we cannot go into the complex of matters relating to the best engineering designs, reactor safety, radioactive leakage to the environment, or thermal pollution. Those are technological problems that will need to be solved before we can light up the world with all the power it wants, whether for the United States' highly advanced needs or for the growing demands of the newly developing countries. But one aspect of nuclear energy is most definitely in the geologic province: the question of reserves of nuclear fuel—in particular, uranium.

The older and still most widely used nuclear energy plants use the fission of Uranium-235 ($U^{235}$), the radioactive isotope whose nucleus spontaneously splits. When enough of this element is brought together in a carefully engineered assembly, called a pile, the controlled fission reaction proceeds steadily to produce a great quantity of energy, released as heat. Fission of one gram of $U^{235}$ liberates energy equivalent to about 2.7 metric tons of coal or 13.7 barrels of crude oil. The uranium is "used up" by this process, appearing as various fission products, other elements that cannot

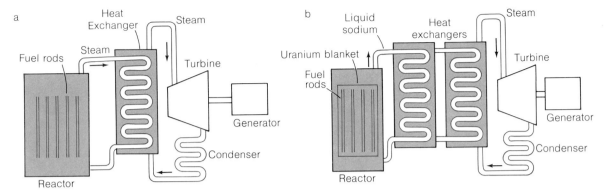

**Figure 24-23**
In a nuclear power plant (a) the fission of uranium 234 releases the energy to make steam, which then goes through the same cycle as in a fossil-fuel power plant. Under development are nuclear breeder reactors (b) in which surplus neutrons are captured by a blanket of nonfissile atoms of uranium 238 or thorium 232, which are transformed into fissile plutonium-239 or uranium-233. The heat of the reactor is removed by liquid sodium. [From "The Conversion of Energy" by C. M. Summers. Copyright © 1971 by Scientific American, Inc. All rights reserved.]

be used for energy production. An average nuclear power plant of this type, producing 1000 megawatts of electricity, will use up about 3 kilograms (6.6 pounds) of U-235 each day. Uranium-235 constitutes only one atom out of every 141 of the average mixture of uranium isotopes mined (usually referred to as $U_3O_8$), so the amount of $U_3O_8$ production needed to support a large number of reactors is large. The U.S. Atomic Energy Commission calculates that at projected rates of nuclear power plant production the United States will need a reserve of 650,000 tons of $U_3O_8$ by 1980.

The estimated reserves of $U_3O_8$ ore available in the United States that are of high enough quality to be mined under present economic conditions amount to just a little over that, about 660,000 tons. The world reserve picture is somewhat brighter, there being about twice as many tons in reserve as needed for projected increases, which are about double the needs of the United States. As with the fossil fuel reserve situation, there are still new discoveries of uranium being made. Nevertheless, geological estimates of uranium ore indicate that there will not be much more than a 25-year supply of high-grade ore before the supply runs out.

Uranium is present in very small amounts in the Earth's crust, constituting only 0.00016 percent of the average crustal rock. It is typically present as small quantities of the mineral uraninite ($UO_2$), frequently called by its common name, pitchblende, in granites and other silicic rocks and associated hydrothermal veins. Under near-surface groundwater conditions this uranium may become oxidized and dissolved, transported in the groundwater, and later reprecipitated as uraninite if it becomes reduced by organic matter, which it may encounter in sedimentary rocks. The richest ores in the United States occur in the Triassic and Jurassic

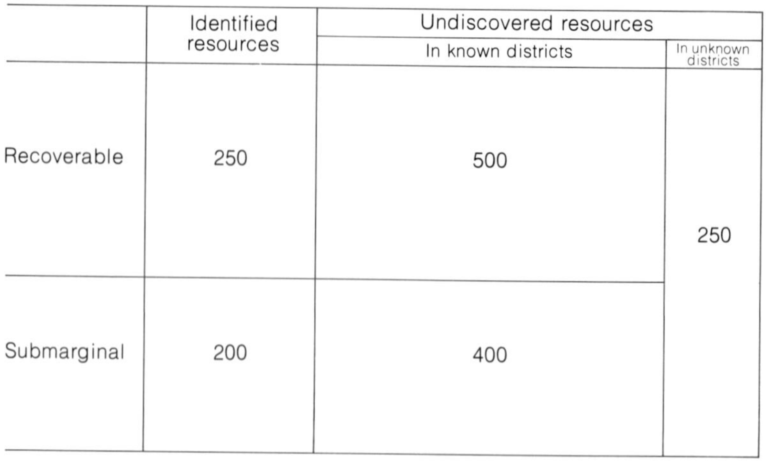

| | Identified resources | Undiscovered resources | |
|---|---|---|---|
| | | In known districts | In unknown districts |
| Recoverable | 250 | 500 | 250 |
| Submarginal | 200 | 400 | |

**Figure 24-24**
Scaled diagram of uranium resources of conventional deposits in the United States (thousands of short tons of $U_3O_8$). The reliability of estimates decreases downward and to the right. [Compiled by A. P. Butler, Jr., and W. I. Finch, February 1972. U.S. Geological Survey.]

sedimentary rocks of the Colorado Plateau in western Colorado and adjacent parts of Utah, Arizona, Wyoming, and New Mexico. Rich Canadian deposits are found both in hydrothermal veins in the Great Bear Lake region of the Northwest Territories and Precambrian conglomerate sediments north of Lake Huron in Ontario. It is these rich deposits that are estimated as reserves, and because of the great effort in uranium exploration after World War II, we know where most of them are.

The story does not end here, however, for there are much larger reserves of lower quality ore, many of which are in the Colorado Plateau region. In addition, there are abundant low-grade deposits in the Devonian Chattanooga Shale, a black, organic-rich sediment underlying large parts of Tennessee, Kentucky, Ohio, Indiana, and Illinois in the flat-lying region of sediments on the western border of the Appalachian Mountains and interior regions. This formation contains uranium equivalent in energy to 5000 barrels of oil for each square meter of its surface. These huge reserves can be exploited when new technology, the breeder reactor, is developed to the point that they can be economically mined. By going to lower-grade ores, we can plan on the basis of thirty or more times the present resources, more than enough to satisfy our needs for far into the future.

The breeder reactor will make the picture brighter because it does not simply use up the $U^{235}$, as do the fission reactors. In the breeder, $U^{235}$ fission is used to create new radioactive elements, such as plutonium-239 from uranium-238, and more nuclear fuel is produced than consumed. In addition, it increases the number of useable ores; the entire supply of natural uranium and thorium can be used instead of only $U^{235}$. Breeders are under active development in various countries, and some are already producing power. Technological problems remain, particularly in the area of reactor safety, but there is every reason to believe that the breeder can be made safe and efficient as a major supplier of power for the world.

Beyond the breeder, nuclear physicists are looking to the possibility of building a fusion reactor, utilizing the same nuclear processes as the Sun, in which the heavy hydrogen isotope deuterium can be made to react to produce helium with an immense production of energy. Small laboratory models have been built that show promise. Because deuterium is abundant as a minor constituent of the water of the oceans, reserves are not a problem. The fusion reactor of this type awaits the application of science and engineering to make a workable reactor that will make energy so abundant and cheap that the world will not need to worry about an energy crisis. If it cannot be made to work, and there is no guarantee that it can, we will have to wrestle with energy conservation in a far more serious way. Sooner or later, we will have to make difficult choices about what we spend our energy resources on.

## MINERALS AS ECONOMIC RESOURCES

Aside from agricultural products, just about everything we use in modern society comes from the ground, including all of the metals and their alloys and the thousands of products made from the chemicals refined from natural deposits. At various places in this book, we have mentioned the practical uses of different minerals and rocks. In this section, we will survey a broad range of useful materials and discuss them in an economic context, because, at the moment that commercial utility enters, matters of cost and price become important.

Just as fossil-fuel reserves are not limitless, neither are the economically recoverable deposits of many useful minerals. Thus, the study of mineral reserves is as important to our subject as the study of their geologic occurrence. The distribution of mineral resources, economic aspects of their recovery, and estimates of reserves are parts of the specialized field called **economic geology**. Inevitably, students of this subject must know modern economics in addition to geology. And, because it takes energy to mine and purify mineral deposits, the economics and geology of energy are important components of decision making in the mineral industry. One of the major decisions an economic geologist must make is whether a mineral deposit has economic significance, and that assessment depends on a knowledge of how mineral resources are distributed geologically.

**What Is a Mineral Deposit?** The chemical elements of the Earth's crust—the portion of the globe that is readily available to us for mining and drilling—are widely distributed in many different kinds of minerals, and those minerals are found in a great variety of rocks. From the beginning of this book, we have discussed examples of how nature homogenizes materials at some times and places and segregates them at others. In most places, a particular element will be found in amounts that tend to be close to its average abundance in the crust—that is, it is homogenized with the other elements. Segregation occurs in a smaller number of geologic situations. The occurrences of elements in much higher abundance—those in which some geologic process has operated to segregate much higher quantities of the element than normal—are the ones that interest us, because the richer the deposit, the cheaper it is to recover the resource, both in terms of energy and in terms of money. Rich deposits of metals are **ores**; the minerals

Table 24-1
Crustal abundance of economically important elements

| Name | Chemical symbol | Atomic number | Crustal abundance (% by weight) |
|---|---|---|---|
| Aluminum | Al | 13 | 8.00 |
| Iron | Fe | 26 | 5.8 |
| Magnesium | Mg | 12 | 2.77 |
| Potassium | K | 19 | 1.68 |
| Titanium | Ti | 22 | 0.86 |
| Hydrogen | H | 1 | 0.14 |
| Phosphorus | P | 15 | 0.101 |
| Manganese | Mn | 25 | 0.100 |
| Fluorine | F | 9 | 0.0460 |
| Sulfur | S | 16 | 0.030 |
| Chlorine | Cl | 17 | 0.019 |
| Vanadium | V | 23 | 0.017 |
| Chromium | Cr | 24 | 0.0096 |
| Zinc | Zn | 30 | 0.0082 |
| Nickel | Ni | 28 | 0.0072 |
| Copper | Cu | 29 | 0.0058 |
| Cobalt | Co | 27 | 0.0028 |
| Lead | Pb | 82 | 0.00010 |
| Boron | B | 5 | 0.0007 |
| Beryllium | Be | 4 | 0.00020 |
| Arsenic | As | 33 | 0.00020 |
| Tin | Sn | 50 | 0.00015 |
| Molybdenum | Mb | 42 | 0.00012 |
| Uranium | U | 92 | 0.00016 |
| Tungsten | W | 74 | 0.00010 |
| Silver | Ag | 47 | 0.000008 |
| Mercury | Hg | 80 | 0.000002 |
| Platinum | Pt | 78 | 0.0000005 |
| Gold | Au | 79 | 0.0000002 |

Table 24-2
Concentration factors of some economically important
elements needed for profitable mining

| Element | Crustal Abundance (% by weight) | Concentration factor* |
|---------|--------------------------------|----------------------|
| Aluminum | 8.00 | 3–4 |
| Iron | 5.8 | 5–10 |
| Copper | 0.0058 | 80–100 |
| Nickel | 0.0072 | 150 |
| Zinc | 0.0082 | 300 |
| Uranium | 0.00016 | 1,200 |
| Lead | 0.00010 | 2,000 |
| Gold | 0.0000002 | 4,000 |
| Mercury | 0.000002 | 100,000 |

Sources: Data from B. J. Skinner, Earth Resources, Prentice-Hall, 1969;
D. A. Brobst and W. P. Pratt, Mineral Resources of the U.S., U.S. Geological
Survey Prof. Paper 820, 1973.
*Concentration factor = abundance in deposit/crustal abundance.

containing these metals are **ore minerals**. Ore minerals include sulfides (the dominant group), oxides, and silicates. In addition, some metals, such as gold, are found in their **native** state—that is, uncombined with other elements.

The average crustal abundance of a number of economically important elements, determined by evaluation of a great many chemical analyses of all rock types, is shown in Table 24-1.

Some of the elements, such as aluminum, iron, and magnesium, are so abundant that any average crustal rock could conceivably be used as a raw material, though not necessarily economically. In contrast are the elements of low abundance, such as gold, platinum, and mercury, which are present in such small amounts in the average rock that enormous quantities of rock would have to be refined to recover even small amounts. *Mineral deposits of economic value are those in which an element occurs in much higher abundance than the average crustal rock, sufficiently high to make it economically worthwhile to mine.* Many of the most valuable mineral deposits are metal-ore deposits, and much of economic geology is concerned with them.

The **concentration factor**—that is, the ratio of the abundance of an element in a mineral deposit to its average abundance—is highly variable and depends on the particular element and its average crustal abundance (Table 24-2). Iron, one of the common elements of the crust, has an average abundance of 5.8%. A good iron ore contains 50% iron; thus, its concentration factor is about 10. A less abundant metal, such as copper, which has a crustal abundance of 0.0058%, is concentrated by factors from 60 to 100 in its economic ores. Even more spectacular are the rarer elements, such as mercury or gold, which have concentration factors in the thousands. The

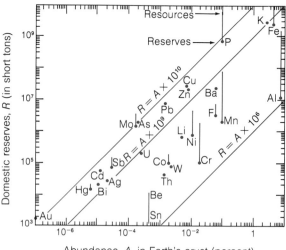

Figure 24-25
Domestic reserves of elements compared to their abundance
in the Earth's crust. Tonnage of ore minable now is shown
by a dot; tonnage of lower-grade ores whose exploitation
depends upon future technological advances or higher prices
is shown by a bar. [From McKelvey, 1973, "Mineral
Resource Estimates and Public Policy," *U.S. Mineral
Resources*, U.S. Geological Survey.]

crustal abundances of the elements are related to their atomic number and chemical affinities in a complex way, as has been described
in Chapter 1. But the concentration factors, though generally increasing with decreasing crustal abundance, depend largely on the
ways in which the metals are held in crystal structures, and their
solubilities in various geologic solutions, such as groundwaters,
hydrothermal solutions, or sea water.

Because the elements are so widely distributed in many common
rocks, the issue of resources and reserves is dominated by costs of
recovery. We could, theoretically, take almost any average rock
and extract both abundant and rare elements from it, given enough
money and energy. So, from that point of view, it is clear that we
will never "run out" of any vital elements. What is important is the
exhaustion of high-grade mineral deposits that are relatively inex-
pensive to mine and purify. Because the distribution patterns of
all elements are like those shown in Figure 24-24, once the deposits
of the highest grade are mined out, we proceed to mine those of
lower grades—that is, deposits whose concentration factors are
lower. This is not always an orderly retreat to a deposit that is only
a little lower in grade. We now mine some extraordinarily rich
deposits of certain metals, such as mercury, and, when they are
depleted, we will have to resort to deposits of ore that are much,
much poorer. At some point in the history of mining a certain

element, it may become economic to reuse (or "recycle") the element from worn-out manufactured materials that contain it. Gold and platinum, among other valuable metals, have been recycled for many years. Whether or not to recycle, of course, is not just a matter of comparing the economics of recycling versus that of mining new deposits. An important consideration is the disposal of used items. The practice of recycling automobiles is now increasing, not so much because we need the scrap metal but because of the desirability of keeping the junked cars from littering the landscape and choking the side streets of our cities.

All such questions can be resolved in a reasonable way only if all costs are calculated — not only the costs of mining, smelting, and transportation, but those of waste disposal and pollution control, both by the producer and the consumer. Because energy costs are rapidly rising in response to the belated recognition that energy sources are not unlimited, and therefore ought to be governed by the economics of scarcity rather than the economics of abundance, the cost pattern of mineral resources is bound to change drastically in the next decades. Smelting aluminum from its ore, for example, takes enormous quantities of electricity. It is estimated that the aluminum industry of the United States claimed 3 percent of the country's total power budget in 1965. For some materials, recovery from lower-grade ores will require such high expenditures of energy that it will be more economical to substitute other materials. In the future, we may well become familiar with costs of materials expressed in terms of units of energy, such as joules* per ton, as a measure on a par with dollars per ton.

The economic geologist, mindful of these costs, naturally seeks to minimize them, if he can, by finding new high-grade deposits rather than automatically resorting to the use of previously known lower-grade ones. The tools he uses for exploration are the subjects we have explored in this book. Understanding of Earth processes and Earth history, knowledge of how to use geophysics and geochemistry in exploration, and practical training in recognizing and mapping rocks and minerals in the field are brought together in the search. The prospector for mineral resources needs to be familiar with the geology of igneous, sedimentary, and metamorphic rocks, for mineral deposits are found in all three. But he becomes more specialized than most geologists, always looking for the special situation in which nature has segregated a purer product from the surrounding rock masses.

---

*Energy is defined as the capacity for doing work. A **joule** is the work done when one **newton** of force is applied through a distance of one meter. A newton is the force required to accelerate one kilogram of mass one meter per second per second. A joule is equal to 0.2390 **calories**, the calorie being the amount of heat necessary to raise the temperature of one gram of water from 14°C. to 15°C. In electrical terms, the unit of power (the time rate at which work is done) is the **watt**, which is the production of energy at a rate of one joule per second. See appendix I.

A mineral segregation is the product of so many different kinds of geologic processes that we would need to restate most of this book to catalogue them. To give some idea of the origins of mineral deposits, we will describe some of the main kinds of geological occurrence of some of the elements of major economic importance.

**Hydrothermal Deposits**  Many of the most useful ore deposits are found in hydrothermal veins — that is, in joints and fractures filled with minerals precipitated by heated waters, usually those emanating from an igneous intrusion (see Chapters 16 and 18). Some of the ores are found in the veins themselves; others are found in the rock adjacent to the veins that has been altered by heating and infiltration by the vein-forming solutions, a process similar, in some ways, to contact metamorphism.

Most hydrothermal deposits are thought to be magmatic in origin because most are associated with igneous intrusions; in some places, several deposits occur around a single intrusion, and the veins are more abundant nearer the rocks of the intrusive bodies. The minerals found in hydrothermal veins are often similar to those found in volcanic and contact-metamorphic rocks, especially the sulfides of various metals: iron ($FeS_2$, pyrite), Lead ($PbS$, galena), zinc ($ZnS$, sphalerite), mercury ($HgS$, cinnabar), and copper ($CuS$, covellite; $Cu_2S$, chalcocite; and $CuFeS_2$, chalcopyrite). Only some magmas appear to be closely associated with ore deposits; some of these liquids appear to have been particularly enriched in one or a group of metallic elements and so have generated hydrothermal veins rich in the sulfides of those metals. Some hydrothermal deposits do not seem to be clearly associated with any intrusion, and many economic geologists believe that these are the result of groundwater circulating through areas of abnormal heating that are related to an episode of regional metamorphism.

Where hydrothermal solutions reach the surface, they become hot springs and geysers, many of which precipitate different kinds of minerals — including lead, zinc, and mercury ores — as they cool. At depth, such fluids deposit their metallic ores as they traverse the surrounding country rock and cool. Geochemists studying the ore-forming process have concluded that the ability of water to dissolve and later precipitate many of the metallic elements is dependent on the high temperatures and pressures found at depth in the crust and on chemical interactions with surrounding rocks. The ability of hot waters to carry significant quantities of metals in solution is partly the result of changes in the properties of the water itself and partly because, when the solutions become relatively concentrated in dissolved substances at high temperatures and pressures, interactions between those substances in solution promote greater solubility.

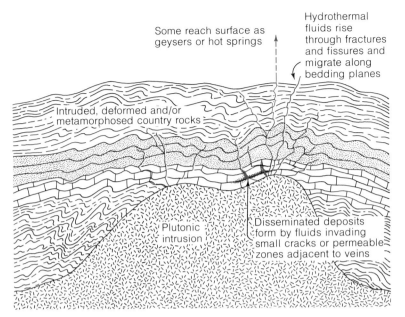

Some reach surface as
geysers or hot springs

Hydrothermal
fluids rise
through fractures
and fissures and
migrate along
bedding planes

Intruded, deformed and/or
metamorphosed country rocks

Plutonic
intrusion

Disseminated deposits
form by fluids invading
small cracks or permeable
zones adjacent to veins

**Figure 24-26**
Many ore deposits are found in hydrothermal veins formed by hot solutions
rising from magmas. As the solutions cool and react with surrounding rocks,
they may precipitate ore minerals together with quartz, calcite, or other common
vein-filling minerals.

Much excitement was generated among economic geologists in
1962 when a deep well drilled near the Salton Sea in southern
California (see Fig. 22-61) penetrated rocks with concentrated hydrothermal solutions carrying very large amounts of such metals
as copper and silver. The hydrothermal solutions from the Salton
Sea well apparently do not emanate from a deep magma, but are
surface waters that have percolated downward into an area of high
heat flow and have dissolved material from the rocks in their path.
In this area—which may be an extension of the East Pacific Rise
under the continent—the rocks through which the solutions have
passed are several kilometers in depth. The rocks there are apparently undergoing regional metamorphism, because the core samples
brought up, which have been stratigraphically correlated with undeformed sedimentary rocks elsewhere, are low-grade metamorphic rocks.

In 1963, one year after the Salton Sea well was drilled, a most
remarkable oceanographic find was made in the Red Sea: Several
deep basins were discovered at the bottom that were filled with
hot brines, dramatically hotter and much more concentrated in
dissolved salts than the overlying normal Red Sea water. The
bottom sediments of those basins are rich in precipitates of oxides
and sulfides of copper, zinc, and other metals. Apparently, these
hot brines are escaping from the sea floor after travelling upward

through subsurface formations below the Red Sea—which lies in a rift-valley area, a young zone of sea-floor spreading. Since then, coring by the deep-sea drilling program has led to the discovery of other areas whose sediments are abnormally rich in such metal ores; one core brought up from sediments of the continental rise about 550 kilometers (350 miles) southeast of New York City contained a small vein of native copper.

These thus far unusual occurrences may provide a clue to a new interpretation of the worldwide distribution of ore deposits, a subject of continuing interest to economic geologists, who use such theories to guide them in their exploration efforts. The central idea of such theories has been the concentration of a major part of the world's ore deposits along major structural zones of the crust in rocks of a relatively few narrow age ranges distributed from the Precambrian to the present. New interpretations are now being proposed that link plate tectonics to major ore deposits by hypothesizing that plate divergences and convergences are the sites of much hydrothermal activity. One recent view is that of Peter Rona, an oceanographer who has been involved in the evolution of the theories of plate tectonics. Some of the following material in this section has been developed from his suggestions. We include it here as an example—not an example of established theory, but one hypothesis typical of the ferment in the field.

A great many of the major deposits of sulfide ores of hydrothermal origin are found on modern or ancient plate-convergence boundaries, including those of the Coast Ranges, the Andes, the eastern Mediterranean to Pakistan, the Philippine Islands, and Japan. Many gold deposits, such as those of Alaska, Canada, and California, also seem to be associated with convergent plate boundaries of the present or past. The mechanism by which this gold was deposited may involve hydrothermal solutions associated with magmatic activity generated by a descending lithospheric plate in a subduction zone, the hydrothermal solutions having dissolved metals from the rocks they traversed in their circulation.

Divergent plate boundaries are the sites of the hot brines of the Red Sea and the Salton Sea, suggesting that hydrothermal fluids are generated by sea water or meteoric water invading fractured rocks in the high heat-flow zone along a spreading ridge, perhaps mixing with some magmatic water from upwelling basalt. Sulfide-ore deposits on the island of Cyprus in the Mediterranean are located along what may be a former mid-ocean ridge, a portion of which has since been uplifted. The stratigraphic sequence in this area extends downward from marine sediment, through extrusive volcanic rocks (including pillow lavas), to intrusive igneous rocks (mainly dikes), which overlie crystalline basement rocks. This sequence is typical of oceanic crustal sections. The hydrothermal ores of copper and iron are found in roughly elliptical basins— bodies comparable in size to the small hot-brine basins of the Red

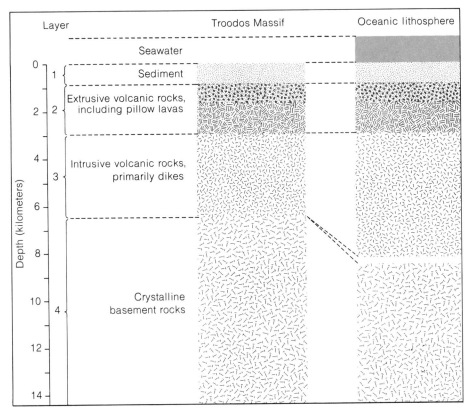

**Figure 24-27**
Close correspondence between the layered sequence of rocks in the Troodos Massif (left) and that of the oceanic lithosphere (right) is evident from this comparison. The geological structure of the Troodos Massif was determined directly from rock outcrops; the structure of the oceanic lithosphere was determined indirectly by seismic-refraction techniques. The sulfide ore bodies of the Troodos Massif are in the upper portion of layer made up of extrusive volcanic rocks. Pillow shapes form when volcanic lava cools on the sea floor. [From "Plate Tectonics and Mineral Resources" by P. A. Rona. Copyright © 1973 by Scientific American, Inc. All rights reserved.]

Sea—and the surrounding sedimentary rocks are chemically similar to metal-enriched sediments of active mid-ocean ridges.

How well these ideas will serve the economic geologist in his explorations for new high-grade ores and mineral deposits remains to be seen. Though it does provide a new general framework, the mountain belts so far suggested for exploration are, for the most part, those that have already been singled out for exploration on the basis of earlier theories. How much undersea exploration may be stimulated is an open question—depending, in part, on the development of an efficient marine technology for deep-sea mining. In any case, the most difficult part of mineral prospecting is pinpointing the generally small ore bodies, most of which seem to be distributed sparsely in an irregular pattern along structurally favorable zones. Zeroing in on them is still mostly a matter of geological

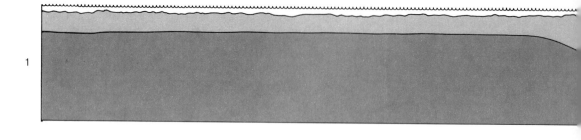

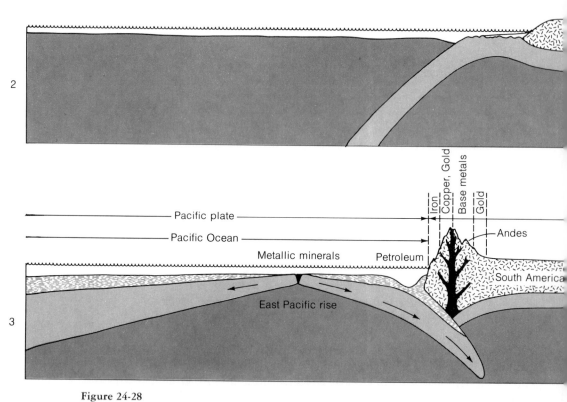

**Figure 24-28**
Role of plate boundaries in the accumulation of mineral deposits is exemplified in this sequence of cross-sectional views of the development of the South Atlantic Ocean, according to Peter A. Rona. In stage 1 a single ancestral continent, called Pangaea, is rifted into two continents (South America and Africa) about a divergent plate boundary. In stage 2 the oceanic crust created by the process of sea-floor spreading from the divergent plate boundary (a precursor of the mid-Atlantic ridge) rafts South America westward and is compensated for by the consumption of oceanic crust at a trench (a convergent plate boundary) that develops to the west of South America. Thick layers of rock salt, organic matter and metallic minerals accumulate in the Atlantic Sea during this early stage of continental drift. In stage 3 continued sea-floor spreading from the mid-Atlantic ridge widens the Atlantic into an ocean, rafts South America westward over the trench, reversing the inclination of

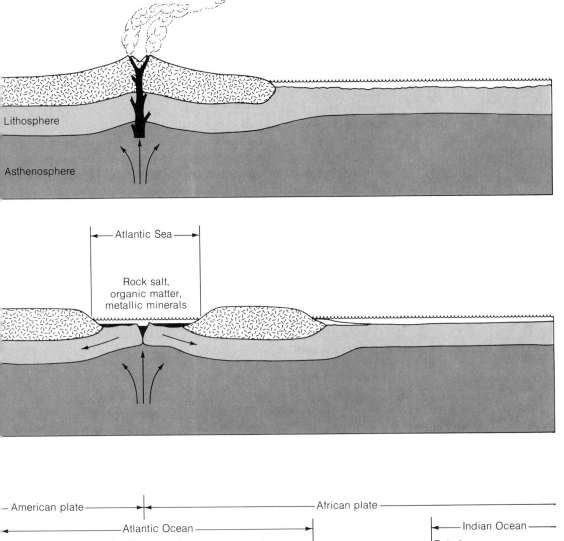

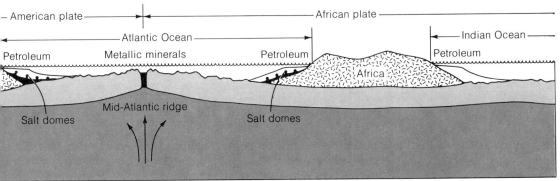

the trench and producing the Andes mountain chain as a consequence of the deformation that develops at the convergent plate boundary along the western margin of South America. Metallic minerals that are melted from the Pacific plate as it plunges under South America ascend through the overlying crustal layers and are deposited in them to form the metal-bearing provinces of the Andes. Meanwhile, in the Atlantic Ocean metallic minerals continue to accumulate about the mid-Atlantic ridge. Salt originating in the thick layers of rock salt that have been buried under the sediments of the continental margins rises in large, dome-shaped masses that act to trap the oil and gas that are generated from the organic matter that was preserved in the former Atlantic Sea. [From "Plate Tectonics and Mineral Resources" by P. A. Rona, Copyright © 1973 by Scientific American, Inc. All rights reserved.]

common sense and good luck, but newer methods of geophysical and geochemical prospecting show promise of being very helpful in this respect.

Once a hydrothermal deposit is found — perhaps from an outcrop, perhaps from a drill hole — the geologist and mining engineer must attempt to predict the extent of the deposit, both as a guide to future mining and as the basis for estimating reserves. Such predictions are based on the characteristic shapes of ore bodies: **Veins** are bodies of rock that have been deposited along joints, cracks, faults, dikes, bedding planes, or other zones of structural weakness that offered entry for the infiltrating hot waters. Like most igneous dikes, most veins are tabular — that is, thin in comparison to their length and width; others, sometimes called **lodes**, may be very thick, and some may even be pencil-like in shape. Some are uniform in thickness; others pinch and swell in an irregular manner. Many veins are easily distinguished from the surrounding rock because of their distinctly different mineral composition and texture; others may tend to merge imperceptibly with the surrounding rock in an extensive transition zone of rock alteration, and so may not be clearly distinguishable as veins in the field. Some veins contain cavities lined with crystals and crusts, remnants of gas pockets trapped in the fractures through which the hydrothermal solutions percolated.

Important hydrothermal deposits that are dispersed through much larger volumes of rock than vein deposits are called **disseminated deposits**. In igneous and sedimentary rocks alike, the dissemination takes place along abundant cracks and fractures; in sedimentary rocks, it may also take place along zones of higher permeability produced by variations in the lithology of the rocks. One important type of disseminated deposit is exemplified by the **copper porphyry** deposits of the southwestern United States and Chile. These deposits occur in a great number of tiny fractures in porphyritic felsic intrusives (granitic rocks with large feldspar or quartz crystals in a finer-grained matrix) and country rocks surrounding the higher parts of the plutons. Some unknown process associated with the intrusion or its aftermath broke the rocks into millions of pieces, and hydrothermal solutions penetrated and recemented the rocks with tiny veins and pore fillings. Such widespread dispersal of the ore deposit produces a low-grade but very large resource of many millions of tons of ore, which may be mined economically by large-scale mining methods (see photograph of an open pit mine in one of these deposits in Fig. 4-2). The most common copper mineral in the porphyry is chalcopyrite ($CuFeS_2$).

Sedimentary rocks are the hosts of disseminated hydrothermal deposits in the lead–zinc province of the upper Mississippi Valley, which extends from southwestern Wisconsin to Kansas and Oklahoma. These ores are not associated with any known intrusion, and the evidence from the chemical and isotopic compositions of remnants of ore-depositing fluid encased in tiny cavities in quartz crystals, called **fluid inclusions**, is that the solutions were relatively low in temperature, no more than 150°C. The origin of such solutions is not known, but speculation centers on possible localized heat sources in the lower part of the lithospheric plate. The major minerals of these deposits are lead sulfide (galena, PbS) and zinc sulfide (sphalerite, ZnS). They have been deposited as **replacements** of calcite and dolomite in the surrounding limestones—that is, the solutions have dissolved some carbonate and replaced it with an equal volume of new crystals of sulfide.

**Igneous and Contact-metamorphic Deposits**  The most important igneous deposits are found as segregations of ore minerals near the bases of layered intrusives, which are formed as minerals crystallizing from a melt settle to the floor of a magma chamber to form distinct layers (see Chapters 16 and 18). One of the largest ore bodies of this type ever found is at Sudbury, Ontario. It is a large mafic intrusive lopolith containing great quantities of layered nickel, copper, and iron sulfides near its base. These sulfide deposits are believed to have formed from crystallization of a dense sulfide-rich liquid that separated from the rest of the cooling magma and sank to the bottom of the chamber before congealing. Important platinum and chromium deposits have been found in layered intrusives in South Africa and Montana. One of the most valuable minerals, diamond, occurs in ultramafic rocks called **kimberlites** that extend to the surface from deep in the crust and upper mantle, where the extremely high pressures needed for their formation are found. These rocks are in the form of narrow pipes, and the mechanism of their eruption to the surface is a matter of controversy.

In Chapter 18, contact metamorphism is described as the transformation of the mineral composition and texture of country rock along zones of contact with igneous intrusives. In this process, minerals of commercial value—such as garnet and emery (corundum), which are used for abrasives—may be formed. Some of the most important deposits of this type are iron ores, such as the small but high-grade deposit of magnetite ($Fe_3O_4$) near Cornwall, Pennsylvania, where a diabase has intruded and metamorphosed shale. In this, as in most contact-metamorphic deposits, the ore is the result of an interaction of the invaded rock with permeating material from the hot intrusion, not just an alteration produced only by heat and pressure without exchange of material.

**Sedimentary Mineral Deposits**  Chemical and mechanical segregations of many economically important minerals are the ordinary result of sedimentary processes. Limestones, separated out as chemical precipitates mainly by organisms, are used for agricultural lime, cement, and building stone. Pure sands, abraded and winnowed by waves and currents so that all materials other than quartz are removed, are the raw materials for glassmaking. Coarse sand and gravel for construction purposes has been abundantly distributed in many areas of the northern United States and southern Canada by the Pleistocene glaciations, and it is also widely distributed in channels and former channels of many rivers. Clays of high purity produced by prolonged weathering are used for pottery and ceramics, both for home and for industrial use. Evaporite deposits of gypsum, separated from sea water by fractional crystallization, are used for plaster, and sodium and potassium salts from evaporites have varied uses from table salt to fertilizer. Phosphate rocks—marine shales and limestones enriched in phosphate by the chemical action of deep-sea waters—are the major raw materials of the world's fertilizer industry.

Sedimentary ore deposits are some of the world's most important sources of copper, iron, and other metals. They are chemical precipitates formed in sedimentary environments to which large quantities of the metals were transported in solution. Some of the important sedimentary copper ores, such as those of the Permian Kupferschiefer (German for copper shale) beds of Germany, are possibly the products of interaction between metal-rich sulfides derived from hot brines of hydrothermal origin with bottom sediments.

The major iron ores of the Earth have been found in Precambrian sedimentary rocks. It is now thought that the low oxygen content of the Earth's atmosphere at that time (see Chapter 14, p. 490) allowed great quantities of iron to be transported in its soluble, reduced (ferrous) form into broad, shallow marine environments where it was oxidized to its insoluble (ferric) form and precipitated. In many of these basins, it was deposited in thin layers alternating with layers of chert. Such iron ores of alternating silica and hematite have been called **banded iron ores**. Post-Cambrian iron ores were formed in relatively small, restricted marine basins at times when abundant iron was being leached from the land surface by deep weathering. One of many hypotheses is that the iron was transported in reduced form from land to these basins by marine seepage of groundwaters that were sufficiently acid and poor in oxygen to dissolve large quantities of iron from soils and subsurface formations. In the marine environment, the iron combines with clays or oxidizes and precipitates.

Possibly the most widely publicized (and romanticized) type of mineral prospecting is "panning" for gold, in which the gold seeker shakes a flat pan of river sediment in hopes of turning up the glint of a nugget. Many rich deposits of gold, diamonds, and such heavy-metal ore minerals as magnetite and chromite ($Fe_2CrO_4$) are found in **placers**, deposits concentrated by the mechanical sorting action of currents. Ore minerals are some of the most common heavy minerals—those much denser than the abundant quartz and feldspar of most sand—that are concentrated by current action in some places in streams or along beaches. Because the heavy minerals settle more quickly out of the current than the lighter quartz and

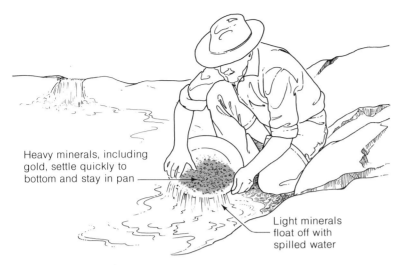

Heavy minerals, including
gold, settle quickly to
bottom and stay in pan

Light minerals
float off with
spilled water

**Figure 24-29**
The gold pan, an old prospecting tool that remains useful.

feldspar, they tend to concentrate in accumulations on river bottoms and bars, where the current is strong enough to keep the lighter minerals suspended and in transport. The same kind of concentration is induced by waves, which preferentially deposit heavy minerals on the beach or on shallow offshore bars. The gold panner does the same thing by shaking the water-filled pan and washing out the lighter minerals.

Some placers can be followed upstream to the location of the original mineral deposit from which they were eroded. Erosion of the Mother Lode, an extensive gold-bearing vein system lying along the western flanks of the Sierra Nevada batholith, produced the placers that were discovered in 1848 and led to the California gold rush. The placers were discovered first, then their source. That was also the story of the Kimberley diamond mines of South Africa two decades later.

This brief summary of the geology of mineral deposits barely touches on the great diversity of geologic situations in which various minerals of value are found. Some minerals or ores are found mainly or only in one kind of deposit; others are found in many different situations. Table 24-3 shows the geologic occurrence of some of the principal kinds of mineral deposits.

## FINDING NEW MINERAL DEPOSITS

As the Earth's human population grows larger at an ever increasing rate, and people all over the world demand higher standards of living (usually stated in terms of more food, goods, and materials), the requirements for mineral resources shoot upward. The total dollar value of all mineral resources, including fuels, produced in

Table 24-3
Principal types of economic mineral deposit

| Mineral deposit | Typical minerals | Geological occurrence | Uses | Major deposits Remarks |
|---|---|---|---|---|
| METALS PRESENT IN MAJOR AMOUNTS IN EARTH'S CRUST | | | | |
| Iron | Hematite, $Fe_2O_3$<br>Magnetite, $Fe_3O_4$<br>Limonite, $FeO(OH)$ | Sedimentary banded iron formation<br>Contact metamorphic<br>Magmatic segregation<br>Sedimentary bog iron ore | Manufactured materials, construction, etc. | Mesabi, Minn.; Cornwall, Pa.; Kiruna, Sweden Resources immense; economics determines exploitation. |
| Aluminum | Gibbsite, $Al(OH)_3$<br>Diaspore, $AlO(OH)$ | Bauxite: residual soils formed by deep chemical weathering | Lightweight manufactured materials | Jamaica Resources great, but expensive to smelt. |
| Magnesium | Dolomite, $CaMg(CO_3)_2$<br>Magnesite, $MgCO_3$ | Dissolved in sea water<br>Hydrothermal veins, limestones | Lightweight alloy metal, insulators, chemical raw material | Most extracted from sea water; unlimited supply. |
| Titanium | Ilmenite, $FeTiO_3$<br>Rutile, $TiO_2$ | Magmatic segregations<br>Placers | High-temperature alloys; paint pigment | Allard Lake, Quebec; Kerala, India Reserves large in relation to demand. |
| Chromium | Chromite, $(Mg, Fe)_2CrO_4$ | Magmatic segregations of mafic and ultramafic rocks | Steel alloys | Bushveldt, S. Africa Extensive reserves in a number of large deposits. |
| Manganese | Pyrolusite, $MnO_2$ | Chemical sedimentary deposits, residual weathering deposits, sea-floor nodules | Essential to steel making | Ukraine, U.S.S.R. World's land resources moderate, but seafloor deposits immerse. |
| METALS PRESENT IN MINOR AMOUNTS IN EARTH'S CRUST | | | | |
| Copper | Covellite, $CuS$<br>Chalcocite, $Cu_2S$<br>Digenite, $Cu_9S_5$<br>Chalcopyrite, $CuFeS_2$<br>Bornite, $Cu_5FeS_4$ | Porphyry copper deposits<br>Hydrothermal veins<br>Contact metamorphic<br>Sedimentary deposits in shales (Kupferschiefer type) | Electrical wire and other products | Bingham Canyon, Utah; Kuperschiefer: Germany; Poland |
| Lead | Galena, $PbS$ | Hydrothermal (replacement)<br>Contact metamorphic<br>Sedimentary deposits (Kupferschiefer type) | Storage batteries, gasoline additive (tetraethyl lead) | Mississippi Valley; Broken Hill, Australia Large resources, Many lower-grade deposits. |
| Zinc | Sphalerite, $ZnS$ | Same as lead | Alloy metal | Same as lead |
| Nickel | Pentlandite, $(Ni, Fe)_9S_8$<br>Garnierite, $Ni_3Si_2O_5(OH)_4$ | Magmatic segregations<br>Residual weathering deposits | Alloy metal | Sudbury, Ontario High grade ores limited; large resources of low-grade ores; also in sea-floor Mn nodules. |
| Silver | Argentite, $Ag_2S$<br>In solid solution in copper, lead, and zinc sulfides | Hydrothermal veins with lead, zinc, and copper | Photographic chemicals; electrical equipment | Most produced as by-product of copper, lead, and zinc recovery. |

| Mineral deposit | Typical minerals | Geological occurrence | Uses | Major deposits Remarks |
|---|---|---|---|---|
| Mercury | Cinnabar, HgS | Hydrothermal veins | Electrical equipment, pharmaceuticals | *Almadén, Spain* Few high-grade deposits with limited reserves. |
| Platinum | Native metal | Magmatic segregations (mafic rocks) Placers | Chemical industry; electrical; alloying metal | *Bushveldt, S. Africa* Large reserves in relation to demand. |
| Gold | Native metal | Hydrothermal veins Placers | Coinage; dentistry, jewelry | *Witwatersrand, S. Africa* Reserves concentrated in a few larger deposits. |
| NONMETALS | | | | |
| Salt | Halite, NaCl | Evaporite deposits Salt domes | Food; chemicals | Resources unlimited; economics determines exploitation |
| Phosphate rock | Apatite, $Ca_5(PO_4)_3OH$ | Marine phosphatic sedimentary rock Residual concentrations of nodules | Fertilizer | *Florida* High-grade deposits limited but extensive resources of low grade deposits. |
| Sulfur | Native sulfur Sulfide ore minerals | Caprock of salt domes (main source) Hydrothermal and sedimentary sulfides | Fertilizer manufacture; chemical industry | *Texas; Louisiana; Sicily.* Native sulfur reserves limited but immense resources of sulfides. |
| Potassium | Sylvite, KCl Carnallite, $KCl \cdot MgCL_2 \cdot 6H_2O$ | Evaporite deposits | Fertilizer | *Carlsbad, New Mexico* Great resources of rich deposits. |
| Diamond | Diamond, C | Kimberlite pipes Placers | Industrial abrasives | *Kimberly, S. Africa* Synthetic diamond now commercially available. |
| Gypsum | Gypsum, $CaSO_4 \cdot 2H_2O$ Anhydrite, $CaSO_4$ | Evaporite deposits | Plaster | Immense resources widely distributed. |
| Limestone | Calcite, $CaCO_3$ Dolomite, $CaMg(CO_3)_2$ | Sedimentary carbonate rocks | Building stone; Agricultural lime; cement | Widely distributed; transportation a major cost. |
| Clay | Kaolinite $Al_2Si_2O_5(OH)_4$ Montmorillonite* Illite* | Residual weathering deposits; sedimentary clays and shales | Ceramics: china, electrical; structural tile | Many large pure deposits; immense reserves of all grades. |
| Asbestos | Chrysotile, $Mg_3Si_2O_5(OH)_4$ | Ultramafic rocks altered and hydrated in near-surface crustal zones | Nonflammable fibers and products | *Southeastern Quebec* Limited high grade reserves but great low grade reserves. |

*Formula highly variable; a hydrous aluminum silicate with other cations, such as Na, K, Ca, Mg.

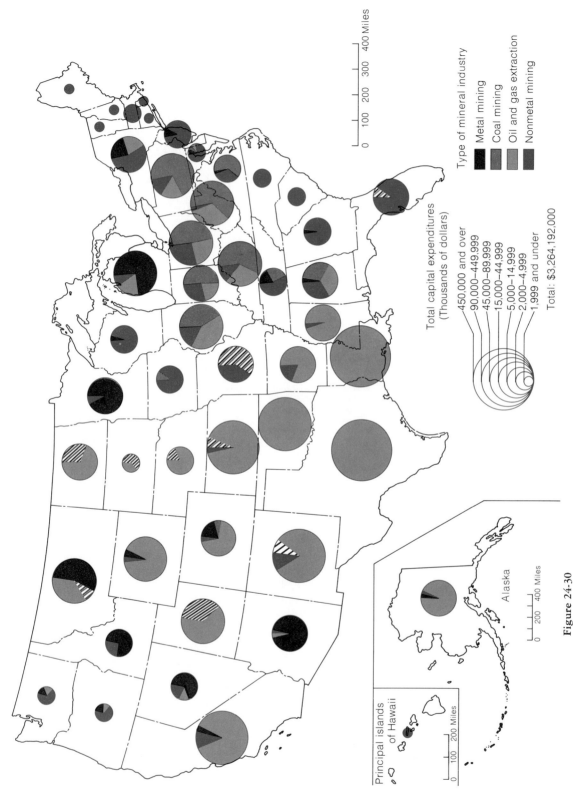

**Figure 24-30**
Capital expenditures in metal mining, oil and gas extraction, and nonmetal mining in the United States, 1963. [After *The National Atlas of the United States of America*, U.S. Geological Survey, 1970.]

Type of mineral industry

Metal mining
Coal mining
Oil and gas extraction
Nonmetal mining

Total capital expenditures
(Thousands of dollars)

450,000 and over
90,000–449,999
45,000–89,999
15,000–44,999
5,000–14,999
2,000–4,999
1,999 and under

Total: $3,264,192,000

0   100  200  300  400 Miles

Alaska

0   200   400 Miles

Principal islands
of Hawaii

0   100   200 Miles

the United States has grown from less than \$5,000,000,000 in 1925 to more than \$21,000,000,000 in 1965. This fourfold growth in forty years (which includes some inflationary increases because of the lowering of the value of the dollar in that time period) took place in a highly industrialized society that had already built a huge technological capability and whose population had grown by only 150 percent. The rate of increase in countries that are rapidly developing their industries is even faster, and all countries have aspirations to increase their rates further.

The rate at which mineral resources are being used is much greater than the rate of population growth in the United States. This is shown by the increase in per capita consumption of almost all metals and other rock and mineral products. Most startling is the jump in the per capita consumption of aluminum, which has multiplied more than twofold in the last fifty years so that now more than 16 kilograms (about 35 pounds) per year is produced for each person in the United States. Though other mineral products show smaller gains, they are all increasing at a rapid pace. These trends show no signs of diminishing: all segments of society demand more and new kinds of manufactured products and a volume of agricultural commodities highly dependent on the fertilizer industry, and those with lower incomes seek a greater share in the material wealth of their society.

Unequal sharing in the exploitation of mineral resources is an important fact on an international scale. North America, with less than a tenth of the world's population, uses almost three-quarters of the world's production of aluminum, while Asia and Africa, with about two-thirds of the world's population, use a little over 5 percent of the production. Again, the same extreme imbalance is true of other materials as well. We are now seeing how international relations can be deeply affected by struggles over the control of resources, as some nations have nationalized, or demanded a greater share in the profits of, oil or mining companies owned by corporations based in North America or Europe. In late 1973, some nations of the Middle East began to restrict exports of oil as a political response to Arab-Israeli hostilities.

A hard fact of life is that an equal per capita sharing of the world's resources would not bring everyone to a "satisfactory" level of consumption—if that could ever be defined—certainly not to a level of consumption anywhere near those of the affluent countries of Europe or North America. How can all of the world's peoples enjoy the benefits of the Earth's resources, benefits that require a highly technological society to utilize them in support of a great population? One of the needs is a great increase in the discovery and exploitation of mineral and energy resources. In a sense, we are all dependent on the skill of the economic geologist.

The prospect for future discovery of major resources is neither rosy nor bleak. It is not rosy because most of the Earth's surface

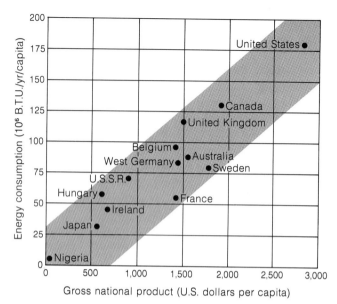

**Figure 24-31**
Commercial energy use and gross national product show a
reasonably close correlation. The same general relationship also
holds true for use of many metal ores. [From "Energy and Power"
by C. Starr. Copyright © 1971 by Scientific American, Inc. All
rights reserved.]

has been explored geologically to one extent or another, and it is
not very probable that major ore bodies of the magnitude of the
Sudbury nickel deposit will be discovered in outcrop. In addition,
most known major mineral deposits are at least moderately well
mapped, so there is little chance of finding appreciable new extensions of these deposits. Yet optimism remains a common quality
among economic geologists. Though the major provinces in which
abundant mineral deposits have been found are well known, the
investigation of new regions may be stimulated by further exploration of the relation between plate boundaries and mineral deposits.
In this search for new deposits, we will be searching below the
surface more than ever, using a combination of methods: inference
of subsurface geology from mapping of surface rocks, exploration
drilling, and a battery of exploration techniques from geophysics
and geochemistry.

The methods of experimental geophysics have been applied
with great success to many resource-exploration campaigns. Most
notable has been the use of shallow-explosion seismology in the
hunt for geologic structures favorable for oil and gas accumulation,
but other applications of seismology are also important in the
search for distinctive structures or rock properties of ore bodies
or other mineral deposits. Magnetic, electrical (mainly electrical-
conductivity), and gravity mapping have likewise been used in the

service of mineral exploration, both to reveal structures and to locate specific ore bodies. Airborne surveys of this type have been effective in locating good prospects, particularly in country difficult to cover on the ground, such as great areas of northern Canada and other polar terrain. Radioactivity is a property of great importance in prospecting for uranium ore. In the late 1940s and the 1950s, no prospector for uranium on the Colorado Plateau was well equipped without a radioactive-particle counter (usually a Geiger counter) to detect small quantities of radiation from uranium-235.

The application of geochemistry to mineral prospecting is a little less direct. The main object is to see if unusual concentrations of minor elements, which show up in some places in surface or subsurface waters, in soils, or in vegetation, are indicators of a buried mineral deposit. The amounts of an element present in a sample may be so small that only the most sensitive laboratory instruments can detect them, yet their variation may be sufficient to draw maps that can be interpreted to locate the source of the elements in a mineral deposit. The elements may have been distributed to surface materials by small leakages from the original vein-depositing fluids of hydrothermal deposits, by weathering processes and soil formation of near-surface rock masses, or by circulating ground waters. The trace-element content (and possibly even the appearance) of vegetation may be a guide to small amounts of particular elements, diagnostic of a nearby ore body, that have been absorbed by the plants.

The effect of trace elements on soils and vegetation is now being explored in a new way by a satellite. Launched in July, 1972, the Earth Resources Technology Satellite (ERTS) has been designed to map the Earth's surface photographically in ways that may be applied to the discovery of mineral resources. The one-ton satellite is equipped with a television-camera system that photographs the Earth as it orbits, each frame covering an area of about 34,250 square kilometers (about 13,225 square miles). The camera takes pictures of the terrain by reflected red, blue-green, and infrared light—that is, it acts as if it used film sensitive only to those colors. By manipulating these images, various composite photographic images can be made that bring out fires, flooding, volcanic eruptions, different types of vegetation, and many different rock and soil types. The satellite photographs the same area every 18 days, so it can detect changes efficiently. In 1973, an Earth Resources Experiment Package, in some ways similar to ERTS, was included on Skylab, the first NASA space station.

Through this type of surveying of the surface from a great distance, called **remote sensing**, geologists can detect the indirect effects on photographic imagery of small geochemical differences that might be related to mineral deposits. Of course, remote sensing does not need to be done exclusively by satellite. Black-and-white aerial photography from airplanes is the oldest form of

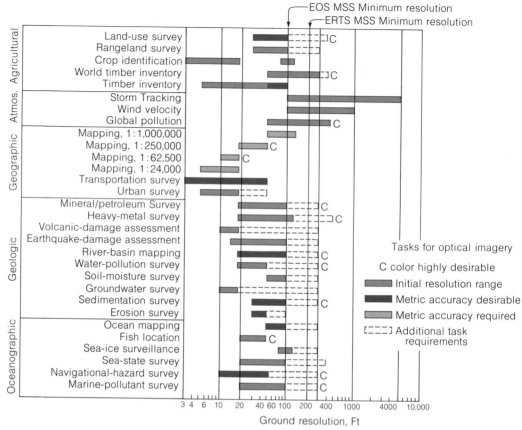

**Figure 24-32**
Ground resolution required in remote sensing photography or television imagery for various resource and environmental mapping needs. The vertical bar at 200 feet is for the present ERTS capability; that at 100 feet, for a proposed new satellite. [From "Remote Sensing of Earth Resources," Comm. on Science and Astronautics, U.S. House of Representatives.]

remote sensing, and, by the time of World War II, such photography had been extended to color and infrared. But the capacity for photographic resolution and separation of the different parts of the electromagnetic spectrum has increased greatly in the past decades. Newer developments include mapping of terrain by a kind of radar, called side-scanning airborne radar, in which small-wavelength radio waves, beamed at topographic features, are reflected back to sensing equipment that produces a photographic type of image. The modern exploration geologist uses all of these methods to produce a better picture of the geology of a target region and to pick up patterns suggestive of mineral deposits that he can later check out on the ground. To put it figuratively, these methods have given the geologist bigger and better eyes for more effective surveillance of the Earth and seven-league boots with which to cover more territory.

Remote sensing has also been applied to the study of ocean resources, primarily the detection of changing patterns of life in the sea. Most of the minerals that we may recover from the ocean will come either from the water or from the sea floor, which has been studied by a variety of remote-sensing devices, such as echo-sounding instruments. The ocean is an immense storehouse of some easily extractable elements. Almost all magnesium is now recovered from evaporated sea water, for example. It is the ease of recoverability from sea water that makes an element attractive rather than its abundance, because, as Tables 24-2 and 24-4 show, the concentration of almost every element is greater in crustal rocks than in sea water. Because silver is 900 times more concentrated in the average rock than in sea water, the cost of recovering it from the sea would have to be at least 900 times less than the cost of recovering it from rock in order for sea-water extraction to be economical. These figures suggest that, for most metals, it will be a long time before we "mine" sea water.

The element bromine, however, is a different story, as can be seen from Table 24-4: it is much more plentiful in sea water than in average crustal rocks. Bromine has been economically recovered from sea water for many years, and only recently has it been extracted more profitably from concentrated saline groundwaters, the so-called brines (see Chapter 7). These brines are a potential source for many other materials as well, for there are indications that some heavy trace metals may be concentrated in them.

Sodium chloride has been recovered from sea water probably since prehistoric times. In a great many coastal regions of the world, the separation of salt from sea water is an economic process. One of the best known processing plants is in San Francisco Bay, where a series of evaporating basins is used to extract the pure salt.

The sea floor, which is being explored now for economic mineral deposits, has been hailed by many as the answer to many problems of apparent resource depletion. We discussed earlier how the idea of plate tectonics may be a stimulus to mineral exploration in the sea. But long before this, there were mining operations near shores that extended under the ocean floor from land. Because the continental borderlands have the same range of rocks as the nearby continental areas, they may contain the same range of mineral deposits as the land. The metal that seems likely to be the first candidate for deep-sea mining is manganese. In Chapter 13, we pointed out the widespread occurrence on the sea floor of manganese nodules, spheroidal aggregates of manganese, iron, and other metal oxides that range in size from tiny encrustations weighing less than a gram to large masses of several hundred kilograms. The majority of the nodules are a few centimeters in diameter. Interest in the possibility of economic recovery of the deposits has been strong in the past few years, both because of the gradual depletion of high-grade manganese ores on land and because the nodules are enriched in many other valuable metals, such as copper,

Table 24-4

Concentrations of the elements in seawater and their values in
elemental or combined form. (The elements are listed in order of
abundance; those in italic type are in concentrations valued at $1
or more per million gallons of seawater.)

| Element | Concentration (lb/10⁶ gal)* | Value ($/10⁶ gal) | |
|---------|---------|---------|---------|
| | | as | |
| Chlorine | 166,000 | NaCl | 924 |
| Sodium | 92,000 | $Na_2CO_3$ | 378 |
| Magnesium | 11,800 | Mg | 4,130 |
| Sulfur | 7,750 | S | 101 |
| Calcium | 3,500 | $CaCl_2$ | 150 |
| Potassium | 3,300 | $K_2O$ (equiv) | 91 |
| Bromine | 570 | $Br_2$ | 190 |
| Carbon | 250 | Graphite | $8 \times 10^{-5}$ |
| Strontium | 70 | $SrCO_3$ | 2 |
| Boron | 40 | $H_3BO_3$ | 3 |
| Fluorine | 11 | $CaF_2$ | 0.35 |
| Nitrogen | 4 | $NH_4NO_3$ | 1 |
| Lithium | 1.5 | $Li_2CO_3$ | 36 |
| Rubidium | 1.0 | Rb | 125 |
| Phosphorus | 0.6 | $CaHPO_4$ | 0.08 |
| Iodine | 0.5 | $I_2$ | 1 |
| Zinc | 0.09 | Zn | 0.013 |
| Iron | 0.09 | $Fe_2O_3$ | 0.001 |
| Aluminum | 0.09 | Al | 0.04 |
| Molybdenum | 0.09 | Mo | 0.004 |
| Tin | 0.03 | Sn | 0.05 |
| Copper | 0.03 | Cu | 0.01 |
| Arsenic | 0.03 | $As_2O_3$ | 0.002 |
| Uranium | 0.03 | $U_3O_8$ | 0.3 |
| Nickel | 0.02 | Ni | 0.02 |
| Manganese | 0.02 | Mn | 0.006 |
| Titanium | 0.009 | $TiO_2$ | 0.003 |
| Silver | 0.003 | Ag | 0.02 |
| Chromium | 0.0004 | $Cr_2O_3$ | 0.00001 |
| Thorium | 0.0004 | $ThO_2$ | 0.0009 |
| Lead | 0.0003 | Pb | 0.00004 |
| Mercury | 0.0003 | Hg | 0.002 |
| Gold | 0.00004 | Au | 0.02 |
| Radium | $9 \times 10^{-10}$ | Ra (in salts) | 0.002 |

Source: From *Resources and Man*, National Academy of Sciences–National
Research Council. W. H. Freeman and Company, 1969.
*One lb/10⁶ gal is approximately 0.000012 percent.

nickel, and cobalt. The average nodule contains more than 20
percent manganese, 6 percent iron, and about 1 percent each of
copper and nickel—a highly enriched ore by any standard.

The economic recovery of manganese nodules awaits the development of a technology of sea-floor surface mining that will
be sufficiently cheap to match the current costs of recovery of

**Figure 24-33**
Manganese nodules dredged from the ocean floor helped to demonstrate the
feasibility of mining the seabed for them. Deposits are estimated to be in the
trillions of tons. [Photo by B. J. Nixon/Deepsea Ventures; Courtesy National
Science Foundation.]

known ores from land deposits. The location of rich fields of nodules
on the sea floor, known earlier from coring, dredging, and bottom
photography, is becoming better known by efficient scanning of
the bottom by towed television cameras. The resolution of these
cameras is good enough to read headlines in old newspapers that
make up part of man's slowly accumulating garbage on the sea
floor below shipping lanes. Though all of the oceans have not
been mapped in detail, good reconnaissance maps of the major
ocean basins have been prepared that show regions of high nodule
abundance, some of which are areas potentially favorable for
mining operations. Mining operations have been carried out on
an experimental basis since 1970 by American and Japanese companies. The prospects for mining the sea floor await — and make
more urgent — an international agreement regarding economic

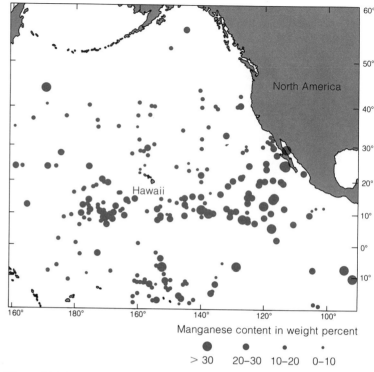

**Figure 24-34**
Distribution of manganese nodule deposits in the Northeast Pacific Ocean.
[From Horn, Delach, and Horn, 1972 "Metal Content of Ferromanganese
Deposits of the Oceans," Technical Report, Nat. Sci. Foundation Int'l. Decade
of Ocean Exploration.]

exploitation of the sea. In the summer of 1973, a major convoca-
tion of nations began to work on the thorny international legal
problems, all of which have political overtones. Ultimately, some
resolution of the question of ownership must be found before the
resources of the sea can be used.

There are other resources under the sea that can be mined,
including phosphate deposits, salt and sulfur from evaporite deposits and salt domes, and abundant sand and gravel for construction needs. Some of the regions that use the greatest amounts of
sand and gravel are populous urban areas along coasts in northern
latitudes, particularly in Europe and North America. In many of
these regions, Pleistocene glaciation has left many sand and gravel
deposits in kames, eskers, and outwash-plain deposits (see Chapter
11). Building has been so extensive in these areas for so many
years that many high grade sources are now depleted. Offshore of
these regions are continental-shelf deposits of much the same
nature, originally deposited when continental glaciers overrode
the offshore areas, like Cape Cod and Long Island, and also extensive glaciomarine deposits laid down by melting icebergs calved

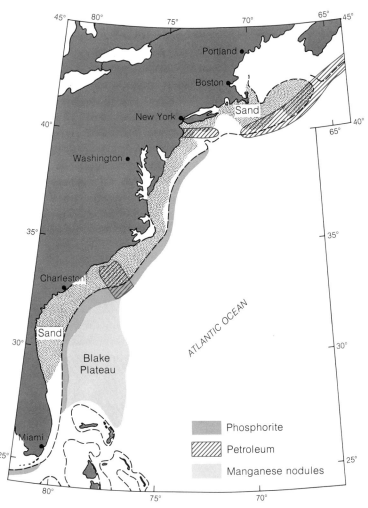

**Figure 24-35**
Distribution of the most favorable areas for some potential mineral resources
off the Atlantic coast of the United States. The broken line denotes the position
of the edge of the continental shelf, about 80 meters deep in the south and
about 140 meters deep in the north. The area indicated for manganese nodules
corresponds to the surface of the Blake Plateau. [After Emery, 1965, "Some
Potential Mineral Resources of the Atlantic Continental Margin." U.S. Geol.
Survey Prof. Paper 525-C.]

from the glaciers. Though glacial gravels are somewhat restricted
to these areas, sand is much more widely distributed, covering
much of the Atlantic continental shelf of North America, for
example. About $100 million worth of these deposits is mined
each year, and the expectations are that this figure will increase
greatly in the future. The large bulk and low unit price of sand and
gravel require efficient cheap recovery from the sea floor by
dredging in relatively shallow water. Operations carried on so far
have shown little or no ecological damage to the life of the sea in
the area mined.

Rich as the ocean may be in some mineral resources, its major value as a resource for human needs may be something entirely different. It is now an important source of food for many maritime regions, and it is potentially a major source of food for the expanding population of the Earth. Pollution of the oceans presents a serious threat to the "farming" of the oceans for food, as well as offending our senses and interfering with the ocean as a prime recreation site. Up to now, we have used the oceans as a garbage dump. We have thrown into the sea everything from city sewage and garbage to toxic wastes from industrial processes and radioactive materials. All of this has not yet made a great impact on its composition, but there is no doubt that it will if we continue to accelerate such dumping. In comparison to the pollution from existing sources, the threat to the ocean that will come from mining operations will be negligible. There may be local effects, and, to the extent to which they are unacceptable, mining companies will have to be responsible for limiting them. The matter of oil pollution has been discussed earlier, in connection with oil exploration on the continental shelves. As important as the matter of ownership is to future congresses of nations on the law of the sea, the matter of ocean pollution may yet overshadow it if world fisheries are threatened. But another question of conservation is now taking greater precedence: overfishing. Just as the bison was exterminated on the western plains in the nineteenth century, we are now exterminating the whale population of the sea. Many kinds of fish are slated for extinction, too, if some control is not exercised. Thus, we have to include in our definition of exhaustible resources the fish population of the seas. How well we use the sea will depend on how well we understand how it works and the extent to which the countries of the world will cooperate intelligently in its exploitation, regulation, and control.

In a larger sense, that is true of the whole Earth. We have only one. To live on it best, we must understand it better.

## SUMMARY

1. Matter of the Earth is theoretically recyclable, but energy is permanently consumed when we use its resources.

2. Energy from fossil fuels—coal, oil, and gas—has been the mainstay of the world's industrialization.

3. Oil and gas form by the preservation and alteration of organic matter deposited in sediments that chemically transform part of it into liquid and gaseous hydrocarbons, compounds of carbon and hydrogen.

4. Oil and gas accumulate in structural and stratigraphic traps that confine the fluids with impermeable barriers.

5. Oil and gas are found in many of the buried sedimentary rocks of the Earth that have not been too compacted, deformed or metamorphosed to allow preservation of the hydrocarbons.

6. The geographic distribution of oil and gas is related to geologic age of sedimentary rocks accumulated in major geosynclines. Middle East oil is accumulated in thick Cenozoic and Mesozoic sediments in a major geosyncline.

7. We may expect that continental shelves, which are favorable areas for oil accumulation, will be explored more heavily in the future. In addition to dealing with the problems of drilling in deep water, those who drill for oil on the continental shelves must safeguard against oil spills that pollute the ocean and adjacent shorelines.

8. Oil is a depletable resource whose reserves will be steadily exhausted at present accelerating rates of use. The United States, exclusive of Alaska, has already produced about two-thirds of its total reserve. World reserves can be estimated, and the exhaustion of all oil and gas resources can be expected within the next century at projected rates of use.

9. Coal, derived from the compaction and chemical alteration of swamp vegetation, is present in great reserves in sedimentary rocks. We have used only 1.7 percent of the world's minable coal resources. Pollution caused by coal burning is a problem, but increased use of coal in the next decades seems certain.

10. Reserves of oil shale and tar sands are great, and may be extensively exploited in the next few decades as costs go up and technology improves.

11. Alternative energy sources are water power, solar energy, geothermal power, and tidal power, none of which has any immediate prospects of being a major answer to world energy needs.

12. The major alternative now envisioned is nuclear power from controlled fission reactors or, ultimately, in fusion plants. Known high-grade reserves of U-235 ore will not be adequate to support major use of conventional nuclear power plants for much more than a few decades, thus making urgent the development of fast breeder reactors and basic research leading toward fusion reactors.

13. Mineral deposits of economic value are those in which an element occurs in much higher abundance than the average crustal rock, sufficiently high to make it economically worthwhile to mine.

14. When high grade ores or other economic deposits are depleted, we can move to recover elements from less concentrated ores, though at greater cost, until in many cases it may become profitable to recycle.

15. Hydrothermal veins, which are some of the most important ore deposits, are formed by hot water emanating from igneous intrusions or by circulating groundwater in areas of high heat flow. Many of these may be related to boundaries of lithospheric plates. They may occur in vein or lode deposits or in such disseminated deposits as the copper-porphyry type.

16. Igneous and contact-metamorphic deposits occur as segregations of layered intrusives, such as the copper and nickel deposits of lopoliths, or in contact-metamorphic aureoles.

17. Sedimentary deposits include many of the chemical and mechanical segregations of such ordinary rocks as limestone, sand and gravel, and evaporite salt deposits. Sedimentary ores of copper and iron have formed in special sedimentary environments, the iron ores chiefly in Precambrian times. Placers are current-laid deposits that are rich in gold or other heavy minerals.

18. Finding new mineral deposits is vitally necessary to support an increasingly industrialized world civilization. Prospects for finding new resources, based on geological, geophysical, and geochemical prospecting, are good. The sea represents a largely untapped resource.

## EXERCISES

1. What sedimentary environments favor the formation of sediments that contain much organic matter that would later be transformed to petroleum?

2. Do you think the mid-Atlantic ridge might be a good prospect for oil drilling, assuming that we can invent the technology to drill there? Why?

3. Given two permeable, anticlinal sandstones, one Ordovician, the other Miocene, which would you choose to drill first in the hope of finding large reserves of oil and gas?

4. Which of the following factors are most important in estimating the future supply of oil and gas: (1) rate of oil accumulation, (2) rate of natural seepage of oil, (3) rate of pumping of oil from known reserves, (4) rate of discovery of new reserves, (5) the total amount of oil now present in the Earth?

5. In terms of use and economic recovery, rank according to relative importance all of the different forms of fossil fuels today, and explain how their ranking might differ at the end of the next century.

6. How would you use knowledge of the distribution of plate boundaries to make a map showing the most likely areas of Earth to investigate for geothermal power?

7. In what important respects do uranium reserves differ from fossil fuel reserves?

8. What evidence might you marshal to show that a particular ore deposit was formed by hydrothermal solutions emanating from an igneous intrusion?

9. Why do you think that economic geologists might worry more about adequate future supplies of mercury ore than iron ore?

## BIBLIOGRAPHY

Committee on Resources and Man (National Academy of Sciences), *Resources and Man.* San Francisco: W. H. Freeman and Company, 1969.

Levorsen, A. I., *Geology of Petroleum* (2nd ed.). San Francisco: W. H. Freeman and Company, 1967.

Mero, J. L., *The Mineral Resources of the Sea.* Amsterdam: Elsevier Publishing Company, 1965.

Park, C. F., Jr., and R. A. MacDiarmid, *Ore Deposits.* San Francisco: W. H. Freeman and Company, 1964.

Scientific American, *Energy and Power,* San Francisco: W. H. Freeman and Company, 1971.

Skinner, B. J., *Earth Resources.* Englewood Cliffs, New Jersey: Prentice-Hall, Inc., 1969.

Skinner, B. J. and K. K. Turekian, *Man and the Ocean.* Englewood Cliffs, New Jersey: Prentice-Hall, Inc., 1973.

# APPENDIXES

# Conversion Factors: Metric–English

## LENGTH

| | |
|---|---|
| 1 centimeter | 0.3937 inch |
| 1 inch | 2.5400 centimeters |
| 1 meter | 3.2808 feet |
| 1 foot | 0.3048 meters |
| 1 meter | 1.0936 yards |
| 1 yard | 0.9144 meters |
| 1 kilometer | 0.6214 mile (statute) |
| 1 kilometer | 3281 feet |
| 1 mile (statute) | 1.6093 kilometers |
| 1 mile (nautical) | 1.8531 kilometers |
| 1 fathom | 6 feet |
| 1 fathom | 1.8288 meters |
| 1 angstrom | $10^{-8}$ centimeters |
| 1 micron | 0.0001 centimeters |

## VELOCITY

| | |
|---|---|
| 1 kilometer/hour | 27.78 centimeters/second |
| 1 mile/hour | 17.60 inches/second |

## AREA

| | |
|---|---|
| 1 square centimeter | 0.1550 square inch |
| 1 square inch | 6.452 square centimeters |
| 1 square meter | 10.764 square feet |
| 1 square meter | 1.1960 square yards |
| 1 square foot | 0.0929 square meter |
| 1 square kilometer | 0.3861 square mile |
| 1 square mile | 2.590 square kilometers |
| 1 acre (U.S.) | 4840 square yards |

## VOLUME

| | |
|---|---|
| 1 cubic centimeter | 0.0610 cubic inch |
| 1 cubic inch | 16.3872 cubic centimeters |
| 1 cubic meter | 35.314 cubic feet |
| 1 cubic foot | 0.02832 cubic meter |
| 1 cubic meter | 1.3079 cubic yards |
| 1 cubic yard | 0.7646 cubic meter |
| 1 liter | 1000 cubic centimeters |
| 1 liter | 1.0567 quarts (U.S. liquid) |
| 1 liter | 33.815 ounces (U.S. fluid) |
| 1 gallon (U.S. liquid) | 3.7853 liters |

## MASS

| | |
|---|---|
| 1 gram | 0.03527 ounce (Avoirdupois) |
| 1 ounce (Avoirdupois) | 28.3495 grams |
| 1 gram | 0.03215 ounce (Troy) |
| 1 kilogram | 2.20462 pounds (Avoirdupois) |
| 1 pound (Avoirdupois) | 0.45359 kilograms |

## DENSITY

| | |
|---|---|
| 1 gram/cubic centimeter | 62.4280 pounds/cubic foot |

## PRESSURE

| | |
|---|---|
| 1 kilogram/square centimeter | 0.96784 atmosphere |
| 1 kilogram/square centimeter | 0.98067 bar |
| 1 kilogram/square centimeter | 14.2233 pounds/square inch |
| 1 bar | 0.98692 atmosphere |
| 1 foot of water | 0.03048 kilogram/square centimeter |
| 1 kilometer of granite | 265 kilograms/square centimeter (approx.) |

## ENERGY

| | |
|---|---|
| 1 erg | $2.39006 \times 10^{-8}$ calories (gram) |
| 1 erg | $9.48451 \times 10^{-11}$ BTU |
| 1 erg | $10^{-7}$ joules |
| Explosion equivalent to 1000 tons of TNT | $4 \times 10^{19}$ ergs |

## POWER

| | |
|---|---|
| 1 watt | $10^7$ ergs/second |
| 1 watt | 0.001341 horsepower (U.S.) |
| 1 watt | 0.05688 BTU/minute |
| 1 watt | 0.73756 foot pound/second |

### *Prefix Names of Multiples and Submultiples of Units*

| *Prefix* | *Factor by which unit is multiplied* |
|---|---|
| giga | $10^9$ |
| mega | $10^6$ |
| kilo | $10^3$ |
| hecto | $10^2$ |
| deka | 10 |
| deci | $10^{-1}$ |
| centi | $10^{-2}$ |
| milli | $10^{-3}$ |
| micro | $10^{-6}$ |
| nano | $10^{-9}$ |
| pico | $10^{-12}$ |

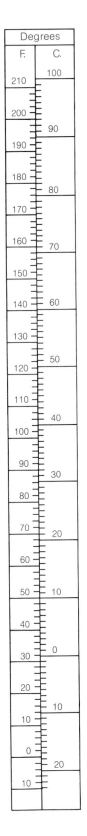

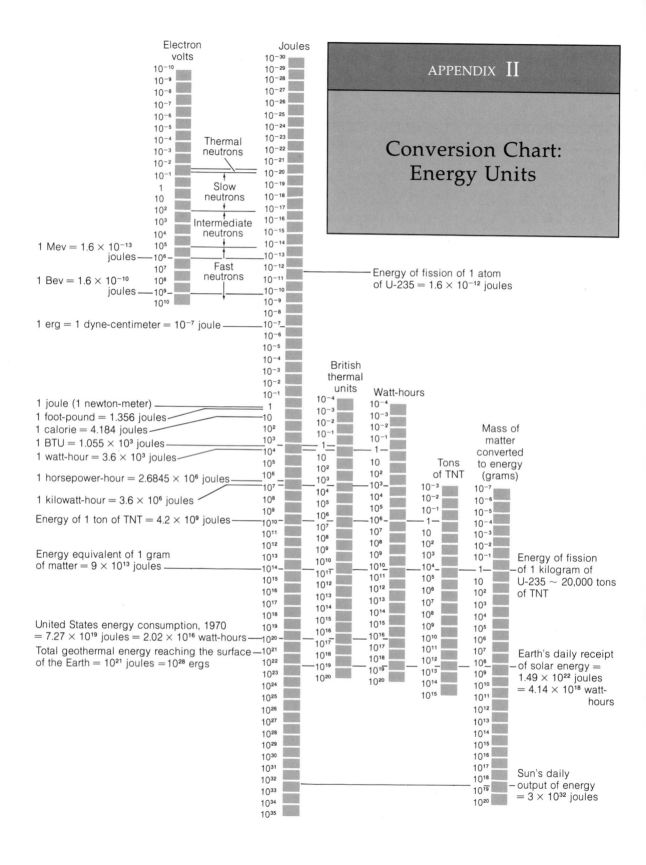

APPENDIX **II**

# Conversion Chart: Energy Units

Electron volts

$10^{-10}$
$10^{-9}$
$10^{-8}$
$10^{-7}$
$10^{-6}$
$10^{-5}$
$10^{-4}$
$10^{-3}$
$10^{-2}$
$10^{-1}$
1
10
$10^2$
$10^3$
$10^4$
$10^5$
$10^6$
$10^7$
$10^8$
$10^9$
$10^{10}$

Thermal neutrons

Slow neutrons

Intermediate neutrons

Fast neutrons

1 Mev = $1.6 \times 10^{-13}$ joules

1 Bev = $1.6 \times 10^{-10}$ joules

1 erg = 1 dyne-centimeter = $10^{-7}$ joule

Joules

$10^{-30}$
$10^{-29}$
$10^{-28}$
$10^{-27}$
$10^{-26}$
$10^{-25}$
$10^{-24}$
$10^{-23}$
$10^{-22}$
$10^{-21}$
$10^{-20}$
$10^{-19}$
$10^{-18}$
$10^{-17}$
$10^{-16}$
$10^{-15}$
$10^{-14}$
$10^{-13}$
$10^{-12}$
$10^{-11}$
$10^{-10}$
$10^{-9}$
$10^{-8}$
$10^{-7}$
$10^{-6}$
$10^{-5}$
$10^{-4}$
$10^{-3}$
$10^{-2}$
$10^{-1}$
1
10
$10^2$
$10^3$
$10^4$
$10^5$
$10^6$
$10^7$
$10^8$
$10^9$
$10^{10}$
$10^{11}$
$10^{12}$
$10^{13}$
$10^{14}$
$10^{15}$
$10^{16}$
$10^{17}$
$10^{18}$
$10^{19}$
$10^{20}$
$10^{21}$
$10^{22}$
$10^{23}$
$10^{24}$
$10^{25}$
$10^{26}$
$10^{27}$
$10^{28}$
$10^{29}$
$10^{30}$
$10^{31}$
$10^{32}$
$10^{33}$
$10^{34}$
$10^{35}$

Energy of fission of 1 atom of U-235 = $1.6 \times 10^{-12}$ joules

1 joule (1 newton-meter)
1 foot-pound = 1.356 joules
1 calorie = 4.184 joules
1 BTU = $1.055 \times 10^3$ joules
1 watt-hour = $3.6 \times 10^3$ joules

1 horsepower-hour = $2.6845 \times 10^6$ joules

1 kilowatt-hour = $3.6 \times 10^6$ joules

Energy of 1 ton of TNT = $4.2 \times 10^9$ joules

Energy equivalent of 1 gram of matter = $9 \times 10^{13}$ joules

United States energy consumption, 1970 = $7.27 \times 10^{19}$ joules = $2.02 \times 10^{16}$ watt-hours
Total geothermal energy reaching the surface of the Earth = $10^{21}$ joules = $10^{28}$ ergs

British thermal units

$10^{-4}$
$10^{-3}$
$10^{-2}$
$10^{-1}$
1
10
$10^2$
$10^3$
$10^4$
$10^5$
$10^6$
$10^7$
$10^8$
$10^9$
$10^{10}$
$10^{11}$
$10^{12}$
$10^{13}$
$10^{14}$
$10^{15}$
$10^{16}$
$10^{17}$
$10^{18}$
$10^{19}$
$10^{20}$

Watt-hours

$10^{-4}$
$10^{-3}$
$10^{-2}$
$10^{-1}$
1
10
$10^2$
$10^3$
$10^4$
$10^5$
$10^6$
$10^7$
$10^8$
$10^9$
$10^{10}$
$10^{11}$
$10^{12}$
$10^{13}$
$10^{14}$
$10^{15}$

Tons of TNT

$10^{-3}$
$10^{-2}$
$10^{-1}$
1
10
$10^2$
$10^3$
$10^4$
$10^5$
$10^6$
$10^7$
$10^8$
$10^9$
$10^{10}$
$10^{11}$
$10^{12}$
$10^{13}$
$10^{14}$
$10^{15}$

Mass of matter converted to energy (grams)

$10^{-7}$
$10^{-6}$
$10^{-5}$
$10^{-4}$
$10^{-3}$
$10^{-2}$
$10^{-1}$
1
10
$10^2$
$10^3$
$10^4$
$10^5$
$10^6$
$10^7$
$10^8$
$10^9$
$10^{10}$
$10^{11}$
$10^{12}$
$10^{13}$
$10^{14}$
$10^{15}$
$10^{16}$
$10^{17}$
$10^{18}$
$10^{19}$
$10^{20}$

Energy of fission of 1 kilogram of U-235 ~ 20,000 tons of TNT

Earth's daily receipt of solar energy = $1.49 \times 10^{22}$ joules = $4.14 \times 10^{18}$ watt-hours

Sun's daily output of energy = $3 \times 10^{32}$ joules

# Numerical Data Pertaining to Earth

| | |
|---|---|
| Equatorial radius: | 6378 kilometers |
| Polar radius: | 6357 kilometers |
| Radius of sphere with Earth's volume: | 6371 kilometers |
| Volume: | $1.083 \times 10^{27}$ cubic centimeters |
| Surface area: | $5.1 \times 10^{18}$ square centimeters |
| Percent surface area of oceans: | 71 |
| Percent surface area of land: | 29 |
| Average elevation of lands: | 623 meters |
| Average depth of oceans: | 3.8 kilometers |
| Mass: | $5.976 \times 10^{27}$ grams |
| Density: | 5.517 grams/cubic centimeter |
| Gravity at equator: | 978.032 centimeters/second/second |
| Mass of atmosphere: | $5.1 \times 10^{21}$ grams |
| Mass of ice: | $25\text{–}30 \times 10^{21}$ grams |
| Mass of oceans: | $1.4 \times 10^{24}$ grams |
| Mass of crust: | $2.5 \times 10^{25}$ grams |
| Mass of mantle: | $4.05 \times 10^{27}$ grams |
| Mass of core: | $1.90 \times 10^{27}$ grams |
| Mean distance to Sun: | $1.496 \times 10^{8}$ kilometers |
| Rotational velocity: | $7.292 \times 10^{-5}$ radians/second (40,000 km/day linear velocity at equator) |
| Average velocity around Sun: | 29.77 kilometers/second |
| Ratio—mass of Sun/mass of Earth: | $3.329 \times 10^{5}$ |
| Ratio—mass of Earth/mass of Moon: | 81.303 |

# APPENDIX IV

## Properties of the Most Common Minerals of the Earth's Crust

| | | Mineral or group name | Varieties and chemical composition | Form, diagnostic characters | Cleavage, fracture | Color | Hardness |
|---|---|---|---|---|---|---|---|
| **LIGHT** COLORED MINERALS, **VERY ABUNDANT** IN EARTH'S CRUST IN ALL **MAJOR** ROCK TYPES | FRAMEWORK SILICATES | **FELDSPAR** | *POTASSIUM FELDSPARS* Sanidine Orthoclase Microcline [$KAlSi_3O_8$] | Cleavable coarsely crystalline or finely granular masses. Isolated crystals or grains in rocks, most commonly not showing crystal faces. | Two at right angles, one perfect and one good; pearly luster on perfect cleavage. | White to gray, frequently pink or yellowish; some green. | 6 |
| | | | *PLAGIOCLASE FELDSPARS* Albite [$NaAlSi_3O_8$] Anorthite [$CaAl_2Si_2O_8$] | | Two at nearly right angles; both good. Fine parallel striations on perfect cleavage. | White to gray, less commonly greenish or yellowish. | |
| | | **QUARTZ** | $SiO_2$ | Single crystals or masses of 6-sided prismatic crystals. Also formless crystals and grains or finely granular or massive. | Very poor or nondetectable; conchoidal fracture. | Colorless, usually transparent; also slightly colored smoky gray, pink, yellow. | 7 |
| | SHEET SILICATES | **MICA** | *MUSCOVITE* [$KAl_3Si_3O_{10}(OH)_2$] | Thin, disc-shaped crystals, some with hexagonal outlines. Dispersed or aggregates. | One perfect; splittable into very thin, flexible, transparent sheets. | Colorless; slight gray or green to brown in thick pieces. | 2–2½ |
| **DARK** COLORED MINERALS **ABUNDANT** IN MANY KINDS OF **IGNEOUS** AND **METAMORPHIC** ROCKS | | | *BIOTITE* [$K(Mg,Fe)_3AlSi_3O_{10}(OH)_2$] | Irregular, foliated masses; scaly aggregates. | One perfect; splittable into thin, flexible sheets. | Black to dark brown. Translucent to opaque. | 2½–3 |
| | | | *CHLORITE* [$(Mg,Fe)_5(Al,Fe)_2Si_3O_{10}(OH)_8$] | Foliated masses or aggregates of small scales. | One perfect: thin sheets flexible but not elastic. | Various shades of green. | 2–2½ |
| | DOUBLE CHAINS | **AMPHIBOLE** | *TREMOLITE–ACTINOLITE* [$Ca_2(Mg,Fe)_5Si_8O_{22}(OH)_2$] | Long, prismatic crystals, usually 6-sided. | Two good cleavage directions at 56° and 124° angles. | Pale to deep green. Pure tremolite white, vitreous luster. | 5–6 |
| | | | *HORNBLENDE* [Complex Ca,Na,Mg,Fe, Al silicate] | Commonly in fibrous masses or irregular aggregates. | | Dark green to black. | |

| Group | Structure | Mineral / Formula | Form and occurrence | Cleavage / fracture | Color | Hardness |
|---|---|---|---|---|---|---|
| | SINGLE CHAINS | PYROXENE — ENSTATITE–HYPERSTHENE [Mg,Fe)$_2$Si$_2$O$_6$] | Prismatic crystals, either 4- or 8-sided. Granular masses and scattered grains. | Two good cleavage directions at about 90°. | grayish or greenish white. | |
| | | DIOPSIDE [Ca,Mg)Si$_2$O$_6$] | | | Light to dark green. | |
| | | AUGITE [Complex Ca,Na,Mg,Fe, Al silicate] | | | Very dark green to black. | |
| | ISOLATED TETRAHEDRA | OLIVINE [(Mg,Fe)$_2$SiO$_4$] | Granular masses and disseminated small grains. | Conchoidal fracture. | Olive to grayish green and brown. | 6½–7 |
| | | GARNET [Ca,Mg,Fe,Al silicate] | Isometric crystals, well-formed or rounded; high specific gravity, 3.5–4.3. | Conchoidal and irregular fracture. | Red and brown, less commonly pale colors. | 6½–7½ |
| | CARBONATES | CALCITE CaCO$_3$ | Coarsely to finely crystalline in beds, veins, and other aggregates. Cleavage faces may show in coarser masses. Calcite effervesces rapidly, dolomite slowly, only in powders. | Three perfect cleavages, at oblique angles; splits to rhombohedral cleavage pieces. | Colorless, transparent to translucent; variously colored by impurities. | 3 |
| | | DOLOMITE CaMg(CO$_3$)$_2$ | | | | 3½–4 |
| | HYDROUS ALUMINO-SILICATES | CLAY MINERALS — KAOLINITE [Al$_2$Si$_2$O$_5$(OH)$_4$], ILLITE [similar to Muscovite + Mg,Fe], MONTMORILLONITE [Complex Ca,Na,Mg,Fe Al silicate + H$_2$O] | Earthy masses in soils; bedded; in association with other clays, iron oxides, or carbonates. Plastic when wet; montmorillonite swells when wet. | Earthy, irregular. | White to light gray and buff; also gray to dark gray, greenish gray, and brownish depending on impurities and associated minerals. | 1½–2½ |
| | SULFATES | GYPSUM CaSO$_4$·2H$_2$O | Granular, earthy, or finely crystalline masses. Tabular crystals. | One perfect, splitting to fairly thin slabs or sheets. Two other good cleavages. | Colorless to white. Transparent to translucent. | 2 |
| LIGHT COLORED, TYPICALLY AS ABUNDANT CONSTITUENTS OF SEDIMENTS AND SEDIMENTARY ROCKS | | ANHYDRITE CaSO$_4$ | Massive or crystalline aggregates in beds and veins. | One perfect, one nearly perfect, one good; at right angles. | Colorless, some tinged with blue. | 3–3½ |
| | | HALITE NaCl | Granular masses in beds. Some cubic crystals. Salty taste. | Three excellent cleavages at right angles. | Colorless, transparent to translucent. | 2½ |
| | | OPAL–CHALCEDONY SiO$_2$ [Opal is an amorphous variety; chalcedony is a formless microcrystalline quartz.] | Beds in siliceous sediments and chert; in veins or banded aggregates. | Conchoidal fracture. | Colorless or white when pure, but tinged with various colors by impurities in bands, especially in agates. | 5–6½ |

| Category | Group | Mineral or group name | Varieties and chemical composition | Form, diagnostic characters | Cleavage, fracture | Color | Hardness |
|---|---|---|---|---|---|---|---|
| DARK MINERALS COMMON IN MANY ROCK TYPES | IRON OXIDES | MAGNETITE | $Fe_3O_4$ | Disseminated grains, granular masses; occasional octahedral isometric crystals. High specific gravity: 5.2. | Conchoidal or irregular fracture. | Black, metallic luster. | 6 |
| | | HEMATITE | $Fe_2O_3$ | Earthy to dense masses, some with rounded forms, some granular or foliated. High specific gravity: 4.9–5.3. | None; uneven, sometimes splintery fracture. | Reddish-brown to black. | $5\frac{1}{2}$–$6\frac{1}{2}$ |
| | | "LIMONITE" | $HFeO_2$ [GOETHITE is the major mineral of the mixture called "limonite," a field term.] | Earthy masses, massive bodies or encrustations, irregular layers. High specific gravity: 3.3–4.7. | One excellent in the rare crystals; usually an earthy fracture. | Yellowish-brown to dark brown and black. | 5–$5\frac{1}{2}$ |
| LIGHT COLORED MINERALS, MAINLY IN IGNEOUS AND METAMORPHIC ROCKS AS COMMON OR MINOR CONSTITUENTS | ALUMINOSILICATES | KYANITE | | Long, bladed or tabular crystals or aggregates. | One perfect and one poor, parallel to length of crystals. | White to light-colored or pale blue. | 5 parallel to crystal length  7 across crystals |
| | | SILLIMANITE | $Al_2SiO_5$ | Long, slender crystals or fibrous, felted masses. | One perfect parallel to length, not usually seen. | Colorless, gray to white. | 6–7 |
| | | ANDALUSITE | | Coarse, nearly square prismatic crystals, some with symmetrically arranged impurities. | One distinct; irregular fracture. | Red, reddish-brown, olive-green | $7\frac{1}{2}$ |
| | ALKALI SILICATES | FELDSPATHOIDS | NEPHELINE [(Na,K)AlSiO₄] | Compact masses or as embedded grains, rarely as small prismatic crystals. | One distinct. Irregular fracture. | Colorless, white, light gray. Gray-greenish in masses, with greasy luster. | $5\frac{1}{2}$–6 |
| | | | LEUCITE [KAlSi₂O₆] | Trapezohedral crystals embedded in volcanic rocks. | One very imperfect. | White to gray. | $5\frac{1}{2}$–6 |
| | MAGNESIUM SILICATES | SERPENTINE | $Mg_6Si_4O_{10}(OH)_8$ | Fibrous (asbestos) or platy masses. | Splintery fracture. | Green; some yellowish, brownish, or gray. Waxy or greasy luster in massive habit; silky luster in fibrous habit. | 4–6 |
| | | TALC | $Mg_3Si_4O_{10}(OH)_2$ | Foliated or compact masses or aggregates. | One perfect, making thin flakes or scales. Soapy feel | White to pale green. Pearly or greasy luster. | 1 |

| Category | Mineral | Class | Formula | Occurrence / Form | Cleavage / Fracture | Color | Hardness |
|---|---|---|---|---|---|---|---|
| **DARK** COLORED MINERALS **COMMON** IN **METAMORPHIC** ROCKS | CORUNDUM | OXIDE | $Al_2O_3$ | Some rounded, barrel-shaped crystals; most often as disseminated grains or granular (emery) masses. | ...gular fracture. | or blue. Emery black. Gem stone varieties: ruby, sapphire. | |
| | EPIDOTE | SILICATES | $Ca_2(Al,Fe)Al_2Si_3O_{12}(OH)$ | Aggregates of long prismatic crystals, granular or compact masses, embedded grains. | One good, one poor at greater than right angles. Conchoidal and irregular fracture. | Green, yellow-green, gray, some varieties dark brown to black. | 6–7 |
| | STAUROLITE | | $Fe_2Al_9Si_4O_{22}(O,OH)_2$ | Short prismatic crystals, some cross-shaped, usually coarser than matrix of rock. | One poor. | Brown, reddish, or dark brown to black. | 7 |
| **METALLIC** LUSTER, **COMMON** IN MANY ROCK TYPES, **ABUNDANT** IN **VEINS** | PYRITE | SULFIDES | $FeS_2$ | Granular masses or well-formed cubic crystals in veins and beds or disseminated. Specific gravity high: 4.9–5.2. | Uneven fracture. | Pale brass-yellow. | 6–6½ |
| | GALENA | | $PbS$ | Granular masses in veins and disseminated. Some cubic crystals. Specific gravity very high: 7.3–7.6. | Three perfect cleavages at mutual right angles, giving cubic cleavage fragments. | Silver-gray. | 2½ |
| | SPHALERITE | | $ZnS$ | Granular masses or compact crystalline aggregates. Specific gravity high: 3.9–4.1. | Six perfect cleavages at 60° to one another. | White to green, brown, and black. Resinous to submetallic luster. | 3½–4 |
| | CHALCOPYRITE | | $CuFeS_2$ | Granular or compact masses; disseminated crystals. Specific gravity: 4.1–4.3. | Uneven fracture. | Brassy to golden-yellow. | 3½–4 |
| | CHALCOCITE | | $Cu_2S$ | Fine-grained masses. Specific gravity: 5.5–5.8. | Conchoidal fracture. | Lead-gray to black. May tarnish green or blue. | 2½–3 |
| MINERALS, FOUND IN **MINOR** AMOUNTS IN A VARIETY OF ROCK TYPES AND IN **VEINS** OR **PLACERS** | RUTILE | TITANIUM OXIDES | $TiO_2$ | Slender to prismatic crystals; granular masses. Specific gravity: 4.25. | One distinct, one less distinct. Conchoidal fracture. | Reddish-brown, some yellowish, violet, or black. | 6–6½ |
| | ILMENITE | | $FeTiO_3$ | Compact masses, embedded grains, detrital grains in sand. Specific gravity: 4.79. | Conchoidal fracture | Iron-black metallic to submetallic luster. | 5–6 |
| | ZEOLITES | SILICATES | Complex hydrous silicates, many varieties of minerals, including analcime, natrolite, heulandite, and chabazite. | Well-formed radiating crystals in cavities in volcanics, in veins, and hot springs. Also as fine-grained and earthy bedded deposits. | One perfect for most. | Colorless, white, some pinkish. | 4–5 |

# Topographic and Geologic Maps

A map is a quantitative representation of the spatial distribution of some attribute or property of the Earth. It is a kind of graph in which the axes are lines of latitude and longitude and the positions of points on the surface (or beneath it) are plotted in relation to those axes or some other established reference. Geologists have frequent need of showing the configuration and nature of the geological materials at or near the surface in a meaningful way, so that they can construct a three-dimensional mental picture of the geology from this two-dimensional graph. Once the nature of maps becomes familiar, the map reader can become practiced at deducing much of the geologic structure and history of an area.

The use of topographic and geologic maps has spread widely throughout our culture. To the more traditional users of such maps — the geologists and surveyors — have been added city planners, industrial zoning commissions, and a large number of the public seeking recreational areas for hiking, camping, fishing, and other activities. Maps are a necessity for all kinds of geologic and mineral resource studies, as well as for studies of groundwater, flood control, soil management, and such environmental concerns as land-use planning, which involves the location of highways, industrial areas, oil and gas pipelines, and recreational areas. Each may use maps with different emphasis for their specific needs, but all follow the same general conventions.

## TOPOGRAPHIC MAPS

The beginning of a map is the choice of a suitable graphical framework to use for plotting the position of points on the globe onto a flat piece of paper. Since a curved surface cannot be made to lie flat without distortion, there will necessarily be some adjustments in order to approximate the true size and shape of an area, such as a lake, when it is mapped. Though there are a large number of ways of doing this, most topographic and geologic maps are made by means of **projections**, the term used for the method of transferring a three-dimensional surface onto a two-dimensional one, just as a camera transfers what it "sees" onto photographic film. The projections used for most maps strike a compromise between distortion of shape and maintaining true relative sizes of areas. The deviation from true size and shape is small for maps of relatively small areas of the globe, such as a county or a state, because the surface of a small area of this huge globe is for all practical purposes flat.

Because the size of the area covered, and thus the amount of detail that can be shown, is always important, we use the concept of **scale** — that is, the relation of a distance (or area) on the map to the true distance on the Earth. This is simply done by stating a ratio, such as 1:24,000, which indicates that a distance of one unit on the map represents a distance of 24,000 such units on the Earth. It does not matter what the units are: a map of scale 1:24,000 is the same whether we use metric or English systems. The scale can be thought of in any convenient units: 1 inch = 2000 feet, or 1 meter = 24 kilometers, or 10 centimeters = 2.4 kilometers. For convenience, maps have a graphic scale, usually at the bottom margin, in which distances such as one kilometer or one mile, usually with subdivisions, are shown as they would appear on the map. Common scales for detailed topographic and geologic maps are the 1:24,000, used by the U.S. Geological Survey for most modern maps, or a scale that is somewhat smaller (because the ratio is smaller) of 1:62,500, roughly an inch to a mile, which was used for many of the older maps. The scale used for regional maps covering much larger areas is 1:250,000. Scales of 1:1,000,000 are used for aeronautical charts.

On most maps, natural and man-made features of the surface are represented by conventional symbols. Those used by the U.S. Geological Survey are typical. Rivers, lakes, and oceans are shown in blue. Topography is shown in brown. Man's structures are shown in black, with main highways and urban areas shown in red. Green shaded areas show wooded land. Some special symbols may be shown on the explanation, or **legend**, of the map, which is usually displayed along the bottom margin. Most complex to represent are the topographic elevations of the surface, usually shown on North American maps by contours (see Chapter 5 for discussion of various ways of showing topography.) Special maps are sometimes prepared to show environmental variables, such as the distribution of slopes of various steepness.

Topographic maps used to be made entirely by geologists or surveyors in the field. They first established a major network of points accurately located by surveying instruments with respect to latitude, longitude, and elevation. This major network, still used as a framework for most maps, was gradually extended from coast to coast across the continent. Within specific areas to be mapped, smaller networks of surveyed points were established and tied to the major network. Between points the locations were drawn in by sight by practiced topographers: surveyors, topographic engineers, or geologists. One can still see such surveyors operating at new highway or construction sites where high precision is needed. Since the advent of the airplane and the development of aerial photography in the 1930's and 1940's, most topographic maps are made by **photogrammetry**, the science of making measurements of position and elevation from aerial photographs. Because maps so prepared are checked on the ground and tied to the major surveyed network, they are the most accurate and precise maps available.

## GEOLOGIC MAPS

Geologic maps are a representation of the distribution of rocks and other geologic materials of different lithologies and ages over the Earth's surface or below it. The geologist perceives the Earth not only in its surface expression of topography and patterns of land and water but in terms of its pattern of subsurface structures, stratigraphic sequences, igneous intrusions, unconformities, and other geometric relationships of rocks. Just as an anatomist can visualize the muscles and bones beneath the skin, so can a geologist visualize details of the Earth's subsurface. What the map looks like is very much a product of the geologic ideas — the concepts of origin of rocks and structure — that the geologist uses. An area mapped a hundred years ago might look somewhat different today as remapped. For example, some older maps of metamorphosed sedimentary rock terranes show relations of one rock type to another only in terms of lithology. The pattern is a complex and disorganized array of schists, gneisses, and other rock types. The same region, mapped later by a geologist who conceptualized the nature of the original sedimentary rocks by "looking through" the metamorphism, appears as a simpler, more organized pattern of a deformed sedimentary rock sequence with a superimposed pattern of different grades of metamorphism. So even though the rocks themselves have not changed, the way we look at them, and therefore the way we map them, has changed over the years.

Detailed geologic maps are normally constructed on a topographic map base. This serves the useful purpose of making it easy to locate geologic structures with respect to surface features of the Earth. It is also important because topography is so often related to the nature of the underlying rocks and their structures. Because it contains so much more information than a topographic map alone, a geologic map is the most valuable for many of the purposes for which maps are used.

Geologic maps are ordinarily made by a geologist who roams over the area and notes the kinds of rocks, sediments, and soils and their structural and stratigraphic relationships. In modern times this is supplemented or even entirely supplanted by remote sensing by aerial photography or geophysical instruments. Remote, inaccessible regions, such as those in some polar or desert regions, may be mapped almost entirely by this method, with the geologist ground checking in scattered places. The mapping of the Moon is an extreme example of this approach. Mars is being mapped with no ground check at all. The best and most accurate maps, however, are those made by traditional means, the geologist covering the ground on his own two legs to see most if not all of the outcrops; he is of course helped enormously by automobile, sometimes by helicopter, and, in some places, by horse or donkey.

The mapping proceeds by the following steps:

### Principal Observations

Description of outcrop location, lithology, age, fossil content, and structural attitude as measured by dip and strike, direction of fault movement, fold axes, etc. (see Chapter 22). Plotting observations on work map.

### First Integration

Conceptualizing the spatial relationship of one outcrop to another by stratigraphic correlation of rocks of the same age, facies, degree of metamorphism and deformation. Grouping of mappable rock units into formations. Drawing of lines on the primitive geologic map of inferred connections where formations are hidden. Compilation of the complete or composite stratigraphic sequence, ages of deformational or igneous intrusive events.

### Synthesizing the Map

Visualizing the larger pattern of geologic relationships and constructing the map, together with geologic cross sections made both to help the geologist in his thinking and to illustrate more detail and inference from the map.

In the later stages, the geologist incorporates analyses of rock composition, radioactive determinations of absolute age, and such geophysical information as seismic, gravity, and magnetic data. He further draws on the geologic literature or his own experience of the geology of nearby and similar kinds of regions. The final result is the finished geologic map, a codified mass of information displayed in a form in which anyone familiar with the geology can quickly read the nature of the Earth's crust in the area and a good deal of its geologic history.

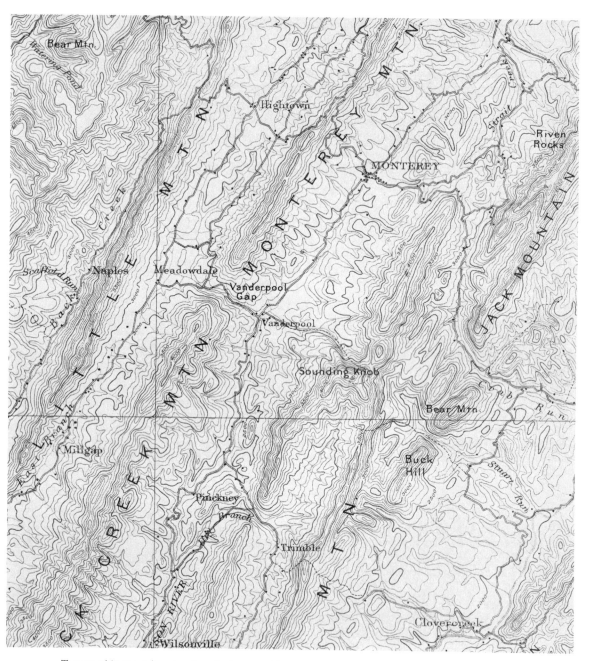

Topographic map (above) and geologic map with cross sections (facing page) of folded sedimentary rocks in the Valley and Ridge province of the Appalachian Mountains. Contours show the pronounced trends of valleys and ridges that reflect the parallel folds. The ridges have developed along the formations that are resistant to erosion, some at the crests of anticlines, such as Jack Mountain north of Crab Run, and others along the flanks of folds, such as Little Mountain. The valleys are in the easily eroded formations, some in synclines, such as Jackson River, some on anticlines, such as

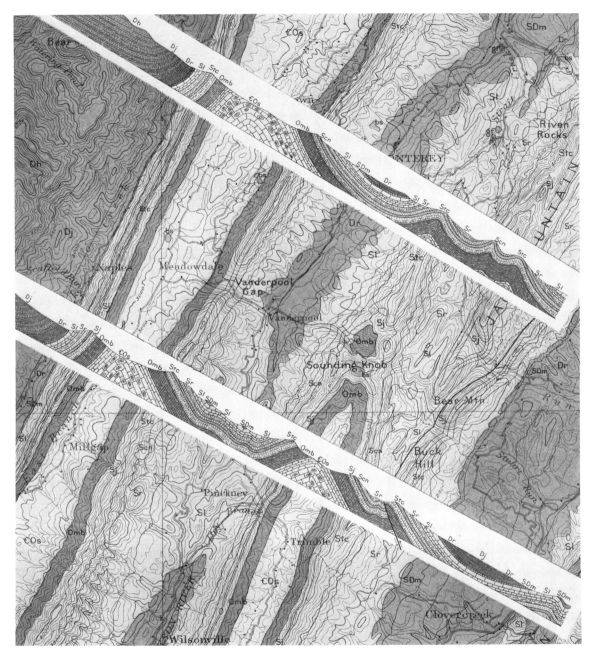

East Branch, and some in the flanks of folds, such as Back Creek. On the geologic map the pattern of anticlines and synclines can be read from the positions of formations of different age, such as at East Branch, where the oldest rocks, Cambrian and Ordovician formations, (EOs) are at the surface bordered on both sides by younger formations of Ordovician and Silurian age (Omb, Stc, Sj, and others). The cross sections make these relationships clearer and add some detail. [ From U.S.G.S.]

Geologic maps are of many kinds. The most common shows the bedrock geology and gives a picture of what the land would look like if all soil were stripped away. Surficial geological maps, on the other hand, emphasize the nature of soils, unconsolidated river sediment, sand dunes, and whatever other materials, including outcrops, that appear at the surface. Tectonic maps show the disposition of large groups of rocks and their structural relationships to each other. Paleogeologic maps show a geologic map of a former land surface now buried beneath an unconformity. Whatever the geologic purpose, there is a map that can be made to show the relevant data. There is no question that the map is at one time both the best device for geological research into the origin of the distribution of important geologic characteristics over the Earth and the best way to illustrate the patterns discovered from such research.

# Glossary

Two-word entries are arranged according to common usage
(for example, "Thrust fault" rather than "Fault, thrust").
Mineral names and some rock names are defined in Appendix
IV and omitted from the Glossary. Words that appear in
*italic* type within a definition are defined elsewhere in
the glossary.

**Aa:** A blocky and fragmented form of lava occurring in flows with fissured and angular surfaces.

**Ablation zone:** The lower part of a glacier, where annual water loss exceeds snow accumulation.

**Abyssal hill:** A low, rounded submarine hill with a relief of 100 to 200 meters, common in deep ocean basins.

**Abyssal plain:** A flat, sediment-covered province of the sea floor that slopes at less than 1:1000.

**Accumulation zone:** The upper part of a glacier, where annual snowfall exceeds melting and evaporation.

**Aeration:** The amount of interstitial space in a soil that contains air and gases; the process of increasing this volume.

**A-horizon:** The uppermost layer of a soil, containing organic material and leached minerals.

**Airy isostatic compensation:** The variation in thickness of a constant-density crust that serves to balance out the excess or deficient weight of topographic features: hence, mountains might be underlain by thick crustal "roots."

**Algal mat:** A layered communal growth of algae observed in fossils and in present-day tidal zones associated with carbonate sedimentation.

**Alkali metal:** A strongly basic metal like potassium or sodium.

**Alluvial fan:** A low, cone-shaped deposit of terrestrial sediment formed where a stream undergoes an abrupt reduction in slope.

**Alluvium:** An unconsolidated terrestrial sediment composed of sorted or unsorted sand, gravel, and clay that had been deposited by water.

**Amphibolite:** A metamorphic rock containing mostly amphibole and plagioclase feldspar.

**Amygdule:** A vesicle or gas bubble in an igneous rock that has been filled with another mineral after the solidification of the lava.

**Angle of repose:** The steepest slope angle in which a particular sediment will lie without cascading down.

**Angstrom:** A length of $10^{-10}$ meter; one hundred-millionth part of a centimeter.

**Angular momentum:** The product of a body's angular velocity, or rotation rate, and its moment of inertia.

**Angular unconformity:** An unconformity in which the bedding planes of the rocks above and below are not parallel.

**Anion:** Any negatively charged ion; the opposite of a *cation*.

**Anisotropic:** Any material in which physical properties (for example, light transmission or seismic wave velocity) vary quantitatively with the direction in which they are measured.

**Anthracite:** The most highly metamorphosed form of coal, containing 92 to 98 percent of fixed carbon. It is black, hard, and glassy.

**Anticline:** A fold, usually from 100 meters to 300 kilometers in width, that is convex upward with the oldest strata at the center.

**Antiroot:** An accumulation of higher-density material in the suboceanic crust that compensates for the low density of sea water.

**Aphanitic texture:** In igneous rocks, a grain size that is so uniformly small that crystals are invisible to the naked eye.

**Aquiclude:** An impermeable strata that acts as a barrier to the flow of groundwater.

**Aquifer:** A permeable formation that stores and transmits groundwater in sufficient quantity to supply wells.

**Arkose:** A variety of sandstone containing abundant feldspar and quartz, frequently in angular, poorly-sorted grains.

**Arroyo:** A steep-sided and flat-bottomed gulley in an arid region that is occupied by a stream only intermittently, after rains.

**Artesian well:** A well that penetrates an *aquiclude* to reach an aquifer containing water under pressure. Thus water in the well rises above the surrounding water table.

**Aseismic region:** One that is relatively free of earthquakes. (Actually, all areas show some seismicity over a sufficiently long interval.)

**Ash:** *See* Volcanic ash.

**Asthenosphere:** The worldwide layer below the lithosphere which is marked by low seismic wave velocities and high *seismic attenuation.* The asthenosphere is a soft layer, probably partially molten. It may be the site of convection.

**Astrobleme:** A circular erosional feature that has been ascribed to the impact of a meteorite or comet.

**Asymmetrical fold:** A fold that is inclined to one side. The dips of the two limbs are unequal.

**Asymmetrical ripple:** A ripple whose cross section is asymmetric, with a gentle slope on the upcurrent side and a steeper face on the downcurrent side.

**Atmosphere (unit):** A unit of pressure equal to 101,325 newtons per square meter, or about 14.7 pounds per square inch.

**Atoll:** A continuous or broken circle of coral reef and low coral islands surrounding a central lagoon.

**Atomic number:** The number of protons in the nucleus of an atom.

**Atomic weight:** The average weight of one atom of an element, relative to a standard weight of 12 for carbon.

**Augen gneiss:** A gneiss containing *phenocrysts* or *porphyroblasts* that have been deformed into eye-shaped grains or grain clusters.

**Aureole:** The area surrounding an intrusion that has been affected by contact metamorphism.

**Axial plane:** In folds, the plane that most nearly separates two symmetrical limbs. In a simple anticline it is vertical; in a complex folding it is perpendicular to the direction of compression.

**Axis (fold):** Within each stratum involved in a fold, there is an axis connecting all the points in the center of the fold, from which both limbs bend.

**Axis of symmetry:** An imaginary line about which an object (for example, a crystal) may be rotated $1/2$, $1/3$, $1/4$, $1/6$, or any other simple fraction of a turn without changing its appearance. The denominator corresponds to the order of the axis.

**Backwash:** The return flow of water down a beach after a wave has broken.

**Banded iron ore:** A sediment consisting of layers of chert alternating with bands of ferric iron oxides (hematite and limonite) in valuable concentrations.

**Bankfull stage:** The height of water in a stream that just corresponds to the level of the surrounding floodplain.

**Bar:** A unit of pressure equal to $10^6$ dynes/square centimeter; approximately one atmosphere.

**Bar (stream):** An accumulation of sediment, usually sandy, which forms at the borders or in the channels of streams or offshore from a beach.

**Barchan:** A crescent-shaped sand dune moving across a clean surface with its convex face upwind and its concave *slip face* downwind.

**Bar-finger sand:** An elongated lens of sand deposited during the growth of a distributary in a delta. The bar at the distributary mouth is the growing segment of the bar finger.

**Barrier island:** A long, narrow island parallel to the shore, composed of sand, and built by wave action.

**Basalt:** A fine-grained, dark, mafic igneous rock composed largely of plagioclase feldspar and pyroxene.

**Base level:** The level below which a stream cannot erode: usually sea level, sometimes locally the level of a lake or resistant formation.

**Basement:** The oldest rocks recognized in a given area, a complex of metamorphic and igneous rocks that underlies all the sedimentary formations. Usually Precambrian or Paleozoic in age.

**Basic rock:** Any igneous rock containing mafic minerals rich in iron and magnesium, but containing no quartz and little sodium-rich plagioclase feldspar (see preferred term, *mafic rock*).

**Basin:** In tectonics, a circular, syncline-like depression of strata. In sedimentology, the site of accumulation of a large thickness of sediments.

**Batholith:** A great irregular mass of coarse-grained igneous rock with an exposed surface of more than 100 square kilometers, which has either intruded the country rock or been derived from it through metamorphism.

**Bathymetry:** The study and mapping of sea-floor topography.

**Bauxite:** A rock composed primarily of hydrous aluminum oxides and formed by weathering in tropical areas with good drainage; a major ore of aluminum.

**Bedding:** A characteristic of sedimentary rocks in which parallel planar surfaces separating different grain sizes or compositions indicate successive depositional surfaces that existed at the time of sedimentation.

**Bed load:** The sediment that a stream moves along the bottom of its channel by rolling and bouncing (see *saltation*).

**Beta-particle:** An electron emitted with high energy and velocity from a nucleus undergoing *radioactive decay.*

**B-horizon:** The intermediate layer in a soil, situated below the A-horizon and consisting of clays and oxides. Also called the zone of accumulation.

**Bicarbonate ion:** The anion group $HCO_3^-$ with a charge of minus one.

**Biochemical precipitate:** A sediment, especially of limestone or iron, formed from elements extracted from sea water by living organisms.

**Bituminous coal:** A soft coal formed by an intermediate degree of metamorphism and containing 15 to 20 percent volatiles. The most common grade of coal.

**Block fault:** A structure formed when the crust is divided into blocks of different elevation by a set of normal faults.

**Blowout:** A shallow circular or elliptical depression in sand or dry soil formed by wind erosion (see also *deflation*).

**Blueschist facies:** A high-pressure (in excess of 5 kb) form of metamorphic rock containing the blue amphibole glaucophane.

**Bolson:** In arid regions, a basin filled with alluvium and intermittent playa lakes and having no outlet.

**Bond:** The force that holds together two atoms in a compound. It may be derived from the sharing of electrons (covalent) or from electrostatic attraction between ions.

**Bottomset bed:** A flat-lying bed of fine sediment deposited in front of a delta and then buried by continued delta growth.

**Bouguer correction:** The correction of a measured gravity value by an amount theoretically calculated to compensate for the mass of known topography around the station.

**Braided stream:** A stream so choked with sediment that it divides and recombines numerous times, forming many small and meandering channels.

**Breaker:** An ocean wave that becomes unstably steep on encountering shallow water and collapses turbulently.

**Breccia:** A clastic rock composed principally of large angular fragments. Usually the clasts are all derived from the same parent formation.

**Breeder reactor:** A planned-for nuclear reactor that would use the high-energy particles created in fission to create more fissionable fuel out of stable uranium or thorium isotopes.

**Brine:** Sea water whose salinity has been increased by evaporation; or groundwater with an unusual concentration of salts.

**Butte:** A steep-sided and flat-topped hill formed by erosion of flat-lying strata where remnants of a resistant layer protect softer rocks underneath.

**Calcium carbonate compensation depth:** The depth in the oceans below which the solution rate becomes so great that no carbonate organisms are preserved on the sea floor.

**Caldera:** A large, circular depression in a volcanic terrane, typically originating in collapse, explosion, or erosion.

**Canyon:** A very large, deep valley with precipitous walls formed principally by stream downcutting.

**Capacity (stream):** A measure of the amount of sediment and detritus a stream can transport past any point in a given time.

**Carbonate ion:** The anion group $CO_3^{-2}$ with a charge of minus two.

**Carbonate platform:** A submarine or intertidal shelf whose elevation is maintained by active shallow-water carbonate deposition.

**Carbonate rock:** A rock composed of carbonate minerals, especially limestone and dolomite.

**Carbon-14 activity:** The number of atoms of the isotope of carbon with atomic weight 14 that decay radioactively in a unit time.

**Carbonic acid:** The weak acid $H_2CO_3$, formed by the dissolution of $CO_2$ in water.

**Cataclastic rock:** A breccia or powdered rock formed by crushing and shearing during tectonic movements.

**Cation:** Any ion with a positive electric charge.

**Central vent:** The largest vent of a volcano, situated at the center of its cone.

**Chemical differentiation:** The formation of more than one igneous rock composition from a common magma as a result of crystals settling out as they form, thus removing certain elements from the melt.

**Chemical sediment:** One that is formed at or near its place of deposition by chemical precipitation, usually from sea water.

**Chemical weathering:** The total set of all chemical reactions that act on rock exposed to water and atmosphere and so change its minerals to more stable forms.

**Chert:** A sedimentary form of amorphous or extremely fine-grained silica, partially hydrous, found in concretions and beds.

**C-horizon:** The lowest layer of a soil, consisting of fragments of rock and their chemically weathered products.

**Cinder cone:** A steep, conical hill built up about a volcanic vent and composed of coarse pyroclasts expelled from the vent by escaping gases.

**Cirque:** The head of a glacial valley, usually with the form of one half of an inverted cone. The upper edges have the steepest slopes, approaching vertical, and the base may be flat or hollowed out and occupied by a small lake or pond.

**Clastic rock:** A sedimentary rock formed from mineral particles (clasts) that were mechanically transported.

**Clay:** Any of a number of hydrous aluminosilicate minerals formed by weathering and hydration of other silicates; also, any mineral fragment smaller than $1/256$ mm.

**Coal:** The metamorphic product of stratified plant remains. It contains more than 50 percent carbon compounds and burns readily.

**Coastal plain:** A low plain of little relief adjacent to the ocean and covered with gently dipping sediments.

**Coefficient of thermal expansion:** A measure of the increase in volume of a material relative to the original volume with increasing temperature.

**Collision hypothesis:** The theory that the material forming the planets was separated from the solar mass by an interstellar collision or near-miss.

**Columnar jointing:** The division of an igneous rock body into prismatic columns by cracks produced by thermal contraction on cooling.

**Compaction:** The decrease in volume and porosity of a sediment caused by burial.

**Compensation (gravity):** The mechanism by which segments of the crust rise or sink to equilibrium positions, depending upon the mass and density of the rocks above and below a certain depth, called the compensation depth.

**Compensation depth:** In relation to gravity, the critical depth in the above definition (see also *calcium carbonate compensation depth*).

**Competence (rock):** The ability of a strata to withstand deformation without fracturing or changing in thickness.

**Competence (stream):** A measure of the largest particle a stream is able to transport, not the total volume.

**Composite cone:** The volcanic cone of a *stratovolcano,* composed of both cinders and lava flows.

**Concordant contact:** The planar contact of an intrusion that follows the bedding of the country rock.

**Confined water reservoir:** A body of groundwater surrounded by impermeable strata.

**Conformable succession:** A sequence of sedimentary rocks that indicates continuous deposition with no erosion over a geologically long period.

**Conglomerate:** A sedimentary rock a significant fraction of which is composed of rounded pebbles and boulders; the lithified equivalent of gravel.

**Conservation of angular momentum:** The physical law that the total angular momentum of a system of isolated bodies cannot change without outside interference.

**Contact metamorphism:** Mineralogical and textural changes and deformation of rock resulting from the heat and pressure of an igneous intrusion in the near vicinity.

**Continental divide:** An imaginary line connecting high points across a continent and dividing regions whose streams drain into one ocean from regions that drain into another.

**Continental drift:** The horizontal displacement or rotation of continents relative to one another.

**Continental glacier:** A continuous, thick glacier covering more than 50,000 square kilometers and moving independently of minor topographic features.

**Continental rise:** A broad and gently sloping ramp that rises from an abyssal plain to the *continental slope* at a rate of less than 1:40.

**Continental shelf:** The gently sloping submerged edge of a continent, extending commonly to a depth of about 200 meters or the edge of the continental slope.

**Continental slope:** The region of steep slopes between the continental shelf and continental rise.

**Contour map:** A map that shows topography by means of contour lines. Each contour line connects points of equal elevation, and the elevation interval between lines is constant.

**Convection:** A mechanism of heat transfer through a liquid in which hot material from the bottom rises because of its lesser density, while cool surface material sinks.

**Convergence zone:** A band along which moving plates collide and area is lost either by shortening and crustal thickening or subduction and destruction of crust. The site of volcanism, earthquakes, trenches, and mountain-building.

**Coordination number:** In a mineral, the number of ions bonded to a given ion of opposite charge.

**Cordillera:** If capitalized: the continuous mountain system extending from Alaska to extreme South America and ranging up to 1500 kilometers in width. If not: any similar chain of parallel mountain ranges.

**Core:** The central part of the Earth below a depth of 2900 kilometers. It is thought to be composed of iron and nickel and to be molten on the outside with a central solid inner core.

**Coriolis effect:** An apparent force that a moving object feels, tending to the right in the Northern hemisphere and to the left in the Southern. It is due to the Earth's rotation, which requires that objects near the equator have a greater eastward velocity than those nearer the poles.

**Country rock:** The rock into which an igneous rock intrudes or a mineral deposit is emplaced.

**Covalent bond:** A bond between atoms in which outer electrons are shared between them.

**Crater:** An abrupt circular depression formed by extrusion of volcanic material and its deposition in a surrounding rim or by explosive ejection of matter upon meteorite impact.

**Craton:** A portion of a continent that has not been subjected to major deformation for a prolonged time, typically since Precambrian or early Paleozoic time.

**Crevasse:** Any large vertical crack in the surface of a glacier or snowfield.

**Cross-bedding:** Inclined beds of depositional origin in a sedimentary rock. Formed by currents of wind or water in the direction toward which the bed slopes downward.

**Cross section:** A drawing showing the features that would be exposed by a vertical cut through a structure.

**Crystal:** A form of matter in which a *unit cell* of atoms, having a fixed chemical formula, is arranged regularly in all directions to form a repeating network.

**Crystal face:** A planar growth surface of a crystal.

**Cuesta:** A ridge with one steep and one gentle face formed by the outcrop and slower erosion of a resistant, gently dipping bed.

**Curie point:** The temperature above which a given mineral cannot retain any permanent magnetization.

**Datum plane:** An artificially established, well-surveyed horizontal plane against which elevations, depths, tides, etc. are measured (for example, mean sea level).

**Daughter element:** Also "daughter product." An element that occurs in a rock as the end-product of the radioactive decay of another element.

**Debris avalanche:** A fast downhill mass movement of soil and rock.

**Declination:** At any place on Earth, the angle between the magnetic and rotational poles.

**Deflation:** The removal of clay and dust from dry soil by strong winds.

**Delta:** A body of sediment deposited in an ocean or lake at the mouth of a stream.

**Delta kame:** A deposit having the form of a steep, flat-topped hill, left at the front of a retreating continental glacier.

**Dendritic drainage:** A stream system that branches irregularly and resembles a branching tree in plan.

**Density:** The mass per unit volume of a substance, commonly expressed in grams/cubic centimeter.

**Density current:** A subaqueous current that flows on the bottom of a sea or lake because entering water is more dense due to temperature or suspended sediments.

**Deposition:** A general term for the accumulation of sediments by either physical or chemical sedimentation.

**Depositional remanent magnetization:** A weak magnetization created in sedimentary rocks by the rotation of magnetic crystals into line with the ambient field during settling.

**Desert pavement:** A residual deposit produced by continued deflation, which removes the fine grains of a soil and leaves a surface covered with close-packed cobbles.

**Detrital sediment:** A sediment deposited by a physical process.

**Diagenesis:** The physical and chemical changes undergone by a sediment during lithification and compaction, excluding erosion and metamorphism.

**Diatom:** A one-celled plant that has a siliceous framework and grows in oceans and lakes.

**Diatomite:** A siliceous chert-like sediment formed from the hard parts of diatoms.

**Diatom ooze:** A fine muddy sediment consisting of the hard parts of diatoms.

**Diatreme:** A volcanic vent filled with breccia by the explosive escape of gases.

**Differentiated planet:** One that is chemically zoned because heavy materials have sunk to the center and light materials have accumulated in a crust.

**Diffraction pattern:** The pattern formed by the reflection of X-rays from the internal planes of atoms when a crystal is exposed to a beam of X-radiation.

**Dike:** A roughly planar body of intrusive igneous rock that has *discordant contacts* with the surrounding rock.

**Dike swarm:** A group of dikes emanating from a common magma chamber.

**Dip:** The angle by which a stratum or other planar feature deviates from the horizontal. The angle is measured in a plane perpendicular to the strike.

**Dip needle:** A type of compass that measures the inclination of the Earth's magnetic field from the horizontal.

**Dip-slip fault:** A fault in which the relative displacement is along the direction of dip of the fault plane; either a normal or a reverse fault.

**Discharge:** The rate of water movement through a stream, measured in volume units per unit time.

**Discontinuity:** See *seismic discontinuity.*

**Discontinuous reaction series:** A *reaction series* in which the end members have different crystal structures.

**Discordant contact:** A contact that cuts across bedding or foliation planes, such as the contact between a dike and the surrounding rocks.

**Disintegration constant:** The time required for a radioactive substance to decay to $1/e$ of its original quantity, where $e$ is 2.718, the base of natural logarithms.

**Disseminated deposit:** A deposit of ore in which the metal is distributed in small amounts throughout the rock, and not concentrated in veins.

**Distributary:** A smaller branch of a large stream that receives water from the main channel; the opposite of a tributary.

**Divergence zone:** A belt along which plates move apart and new crust and lithosphere is created: the site of mid-ocean ridges, earthquakes, and volcanism.

**Divide:** A ridge of high ground separating two drainage basins emptied by different streams.

**Dome:** In structural geology, a round or elliptical upwarp of strata resembling a short anticline.

**Drainage basin:** A region of land surrounded by divides and crossed by streams that eventually converge to one river or lake.

**Drift (continental):** See *Continental drift.*

**Drift (glacial):** A collective term for all the rock, sand, and clay that is transported and deposited by a glacier either as till or as outwash.

**Drumlin:** A smooth, streamlined hill composed of till.

**Dry wash:** An intermittent stream bed in an arroyo or canyon that carries water only briefly after a rain.

**Ductile rock:** A rock that can withstand 5 to 10 percent *Strain* without fracturing.

**Dune:** An elongated mound of sand formed by wind or water.

**Earthflow:** A detachment of soil and broken rock and its subsequent downslope movement at slow or moderate rates in a stream- or tongue-like form.

**Earthquake:** The violent oscillatory motion of the ground caused by the passage of seismic waves radiating from a fault along which sudden movement has taken place.

**Ebb tide:** The part of the tide cycle during which the water level is falling.

**Echo-sounder:** An oceanographic instrument that emits sound pulses into the water and measures its depth by the time elapsed before they return.

**Ecliptic:** The plane that contains the Earth's orbit around the Sun.

**Eclogite:** An extremely high-pressure metamorphic rock containing garnet and pyroxene.

**Ecology:** The science of the life cycles, populations, and interactions of various biological species as controlled by their physical environment, including also the effect of life forms upon the environment.

**Economic geology:** The study of and location of ores, fossil fuels, and other useful materials that occur in sufficient concentration to be marketable in a particular economy.

**Elastic limit:** The maximum stress that can be applied to a body without resulting in permanent strain.

**Elastic rebound theory:** A theory of fault movement and earthquake generation that holds that faults remain locked while strain energy accumulates in the country rock, and then suddenly slip and release this energy.

**Electron:** A negatively charged particle with negligible mass orbiting around the nucleus of an atom. An electron has a mass of $9.1 \times 10^{-28}$ gram and a negative charge of $1.6 \times 10^{-19}$ coulomb.

**Electron microprobe:** An instrument that bombards a very minute sample with electrons to determine its chemical composition from the resulting X-radiation.

**Elevation:** The vertical height of one point on the Earth above a given *datum plane,* usually sea level.

**Elliptical orbit:** An orbit with the shape of a geometrical ellipse. All orbits are elliptical or hyperbolic, with the Sun occupying one focus.

**Emission spectroscopy:** The identification of elements by the characteristic wavelengths of light that they emit when heated. A prism separates the spectrum into characteristic bright lines.

**Environment of deposition:** A geographically limited area where sediments are preserved; characterized by its landforms, relative energy of currents, and chemical equilibria.

**Eolian:** Pertaining to or deposited by wind.

**Eon:** The largest division of geologic time, embracing

several Eras, (for example, the Phanerozoic, 600 m.y. ago to present); also any span of one billion years.

**Epeirogeny:** Large-scale, primarily vertical movement of the crust. It is characteristically so gradual that rocks are little folded and faulted.

**Epicenter:** The point on the Earth's surface directly above the focus or hypocenter of an Earthquake.

**Epoch:** One subdivision of a geologic period, often chosen to correspond to a stratgraphic series. Also used for a division of time corresponding to a paleomagnetic interval.

**Era:** A time period including several periods, but smaller than an eon. Commonly recognized eras are Precambrian, Paleozoic, Mesozoic, and Cenozoic.

**Erosion:** The set of all processes by which soil and rock are loosened and moved downhill or downwind.

**Esker:** A glacial deposit in the form of a continuous, winding ridge, formed from the deposits of a stream flowing beneath the ice.

**Eugeosyncline:** The seaward part of a geosyncline; characterized by clastic sediments and volcanism.

**Eustatic change:** Sea-level changes that affect the whole Earth.

**Eutrophication:** A superabundance of algal life in a body of water; caused by an unusual influx of nitrate, phosphate, or other nutrients.

**Evaporite:** A chemical sedimentary rock consisting of minerals precipitated by evaporating waters, especially salt and gypsum.

**Exfoliation:** A physical weathering process in which sheets of rock are fractured and detached from an outcrop.

**Exobiology:** The study of life outside the Earth.

**Extinction angle:** The angle between a crystallographic direction, such as a face or cleavage plane, and the direction in which all light is blocked by a pair of crossed polarizers.

**Facies:** The set of all characteristics of a sedimentary rock that indicate its particular environment of deposition and which distinguish it from other facies in the same rock unit.

**Fault:** A planar or gently curved fracture in the Earth's crust across which there has been relative displacement.

**Fault-block mountain:** A mountain or range formed as a horst when it was elevated (or as the surrounding region sank) between parallel normal faults.

**Fault plane:** The plane that best approximates the fracture surface of a fault.

**Faunal succession:** The evolutionary sequence of life forms, especially as recorded by the fossil remains in a stratigraphic sequence.

**Feedback:** Any system in which a process controls itself, either through negative feedback, which has the effect of damping out fluctuations in rate or process, or positive feedback, which has the effect of amplifying and preserving fluctuations.

**Felsic:** An adjective used to describe a light-colored igneous rock poor in iron and magnesium content, abundant in feldspars and quartz.

**Field relations:** The total pattern of contacts, faults, intrusions, unconformities, and other surfaces where rock formations meet, and from which the field geologist reconstructs the chronology and history of an area.

**Fiord:** A former glacial valley with steep walls and a U-shaped profile now occupied by the sea.

**First motion:** On a seismogram, the direction of ground motion at the beginning of the arrival of a P-wave. Upward ground motion indicates a compression; downward motion, a dilatation.

**Flood basalt:** A plateau basalt extending many kilometers in flat, layered flows originating in fissure eruptions.

**Floodplain:** A level plain of stratified alluvium on either side of a stream; submerged during floods and built up by silt and sand carried out of the main channel.

**Flood tide:** The part of the tide cycle during which the water is rising or leveling off at high water.

**Flow cleavage:** In a metamorphic rock, the parallel arrangement of all planar or linear crystals as a result of rock flowage during metamorphism.

**Fluid inclusion:** A small body of fluid that is entrapped in a crystal and has the same composition as the fluid from which the crystal formed.

**Flume:** A laboratory model of stream flow and sedimentation consisting of a rectangular channel filled with sediment and running water.

**Focus (earthquake):** The point at which the rupture occurs; synonymous with hypocenter.

**Fold:** A planar feature, such as a bedding plane, that has been strongly warped, presumably by deformation.

**Fold axis:** See *Axis.*

**Fold belt:** Synonym of *orogenic belt.*

**Foliation:** Any planar set of minerals or banding of mineral concentrations, including cleavage, found in a metamorphic rock.

**Foraminifer:** A class of oceanic protozoa most of which have shells composed of calcite.

**Foraminiferal ooze:** A calcareous sediment composed of the shells of dead Foraminifera.

**Foreset bed:** One of the inclined beds found in cross-bedding; also an inclined bed deposited on the outer front of a delta.

**Formation:** The basic unit for the naming of rocks in stratigraphy: a set of rocks that are or once were horizontally continuous, that share some distinctive feature of lithology, and are large enough to be mapped.

**Fossil:** An impression, cast, outline, or track of any animal or plant that is preserved in rock after the original organic material is transformed or removed.

**Fossil fuel:** A general term for combustible geologic deposits of carbon in reduced (organic) form and of biological origin, including coal, oil, natural gas, oil shales, and tar sands.

**Fractional crystallization:** The separation of a cooling magma into components by successive formation of crystals at progressively lower temperatures.

**Fracture (mineralogy):** The irregular breaking of a crystal along a surface not parallel to a crystal face. The nature of the surface texture is considered diagnostic.

**Fracture cleavage:** A set of closely spaced, parallel joints in a rock resulting from deformation and metamorphism.

**Free-air correction:** In gravity, the addition to a measured value of gravity of an amount theoretically calculated to correct for the effect of elevation only.

**Free oscillation:** The "ringing" or periodic deformation of the whole Earth at characteristic low frequencies after a major earthquake.

**Friction breccia:** A breccia formed in a fault zone or volcanic pipe by the relative motion of two rock bodies.

**Fringing reef:** A coral reef that is directly attached to a landmass not made of coral.

**Fumarole:** A small vent in the ground from which volcanic gases and heated groundwater emerge, but no lava.

**Gabbro:** A black, coarse-grained, intrusive igneous rock, composed of calcic feldspars and pryoxene. The intrusive equivalent of basalt.

**Geochronology:** The science of absolute dating and relative dating of geologic formations and events, primarily through the measurement of daughter elements produced by radioactive decay in minerals.

**Geologic cycle:** The sequence through which rock material passes in going from its sedimentary form, through diastrophism and deformation of sedimentary rock, then through metamorphism and eventual melting and magma formation, then through volcanism and plutonism to igneous rock formation, and finally through erosion to form new sediments.

**Geomorphic cycle:** An idealized model of erosion wherein a plain is uplifted epeirogenically, then dissected by rapid streams (youth), then rounded by downslope movements into a landscape of steep hills (maturity), and finally reduced to a new peneplain at sea level (old age).

**Geomorphology:** The science of surface landforms and their interpretation on the basis of geology and climate.

**Geosyncline:** A major downwarp in the Earth's crust, usually more than 1000 kilometers in length, in which sediments accumulate to thicknesses of many kilometers. The sediments may eventually be deformed and metamorphosed during a mountain-building episode.

**Geotherm:** Also geoisotherm. A curving surface within the Earth along which the temperature is constant.

**Geothermal power:** Power generated by utilizing the heat energy of the crust, especially in volcanic regions.

**Geyser:** A hot spring that throws hot water and steam into the air. The heat is thought to result from the contact of groundwater with magma bodies.

**Giant planet:** One of the large planets outside the orbit of Mars, as opposed to the terrestrial planets.

**Glacial drift:** See *Drift.*

**Glacial rebound:** Epeirogenic uplift of the crust that takes place after the retreat of a continental glacier, in response to earlier subsidence under the weight of the ice.

**Glacial striations:** Scratches left on bedrock and boulders by overriding ice, and showing the direction of motion.

**Glacial valley:** A valley occupied or formerly occupied by a glacier, typically with a U-shaped profile.

**Glacier:** A mass of ice and surficial snow that persists throughout the year and flows downhill under its own weight. The size range is from 100 meters to 10,000 kilometers.

**Glacier surge:** A period of unusually rapid movement of one glacier, sometimes lasting more than a year.

**Glass:** A rock formed when magma is too rapidly cooled (quenched) to allow crystal growth.

**Glassiness:** The content of extent of glass in an igneous rock.

**Gneiss:** A coarse-grained regional metamorphic rock that shows compositional banding and parallel alignment of minerals.

**Graben:** A downthrown block between two normal faults of parallel strike but converging dips; hence a tensional feature. See also *horst.*

**Graded bedding:** A bed in which the coarsest particles are concentrated at the bottom and grade gradually upward into fine silt, the whole bed having been deposited by a waning current.

**Graded stream:** A stream whose smooth profile is unbroken by resistant ledges, lakes, or waterfalls, and which maintains exactly the velocity required to carry the sediment provided to it.

**Granite:** A coarse-grained, intrusive igneous rock composed of quartz, orthoclase feldspar, sodic plagioclase feldspar, and micas. Also sometimes a metamorphic product.

**Granitization:** The formation of metamorphic granite from other rocks by recrystallization with or without complete melting.

**Granular snow:** Snow that has been metamorphosed into small granules of ice.

**Granulite:** A metamorphic rock with coarse interlocking grains and little or no foliation.

**Gravel:** The coarsest of alluvial sediments, containing mostly particles larger than 2 mm in size and including cobbles and boulders.

**Gravity anomaly:** The value of gravity left after subtracting from a gravity measurement the reference value based on latitude, and possibly the free-air and Bouguer corrections.

**Gravity survey:** The measurement of gravity at regularly-spaced grid points with repetitions to control instrument drift.

**Greenhouse effect:** The heating of the atmosphere by the absorption of infrared energy re-emitted by the Earth as it receives light energy in the visible band from the Sun.

**Greenschist:** A metamorphic schist containing chlorite and epidote (which are green) and formed by low-temperature, low-pressure metamorphism.

**Ground moraine:** A glacial deposit of till with no marked relief, interpreted as having been transported at the base of the ice.

**Groundwater:** The mass of water in the ground below the phreatic zone, occupying the total pore space in the rock and moving slowly downhill where permeability allows.

**Gully:** A small steep-sided valley or erosional channel from 1 meter to about 10 meters across.

**Guyot:** A flat-topped submerged mountain or seamount found in the ocean.

**Gyre:** The circular rotation of the waters of each major sea, driven by prevailing winds and the *Coriolis effect.*

Half-life: The time required for half of a homogeneous sample of radioactive material to decay.

Hanging valley: A former glacial tributary valley that enters a larger glacial valley above its base, high up on the valley wall.

Hard water: Water that contains sufficient dissolved calcium and magnesium to cause a carbonate scale to form when the water is boiled or to prevent the sudsing of soap.

Heat conduction: The transfer of the rapid vibrational energy of atoms and molecules, which constitutes heat energy, through the mechanism of atomic or molecular impact.

Heat engine: A device that transfers heat from a place of high temperature to a place of lower temperature and does mechanical work in the process.

Hill: A natural land elevation, usually less than 1000 feet above its surroundings, with a rounded outline. The distinction between hill and mountain depends on the locality.

Hogback: A formation similar to a Cuesta in that it is a ridge formed by slower erosion of hard strata, but having two steep, equally inclined slopes.

Hooke's Law: The principle that the stress within a solid is proportional to the strain. It holds only for strains of a few percent or less.

Hornfels: A high-temperature, low-pressure metamorphic rock of uniform grain size showing no foliation. Usually formed by contact metamorphism.

Horst: An elongate, elevated block of crust forming a ridge or plateau, typically bounded by parallel, outward-dipping normal faults.

Hot spring: A spring whose waters are above both human body and soil temperature as a result of *plutonism* at depth.

Humus: The decayed part of the organic matter in a soil.

Hydration: A chemical reaction, usually in weathering, which adds water or $OH^-$ to a mineral structure.

Hydraulic conductivity: A measure of the permeability of a rock or soil: the volume of flow through a unit surface in unit time with unit hydraulic pressure difference as the driving force.

Hydrocarbon: An organic chemical compound made up of carbon and hydrogen atoms arranged in chains or rings.

Hydrologic cycle: The cyclical movement of water from the ocean to the atmosphere, through rain to the surface, through runoff and groundwater to streams, and back to the sea.

Hydrology: The science of that part of the hydrologic cycle between rain and return to the sea; the study of water on and within the land.

Hydrothermal activity: Any process involving high-temperature groundwaters, especially the alteration and emplacement of minerals and the formation of hot springs and geysers.

Hydrothermal vein: A cluster of minerals precipitated by hydrothermal activity in a rock cavity.

Hypocenter: The point below the epicenter at which an earthquake actually begins; the focus.

Hypsometric diagram: A graph that shows in any way the relative amounts of the Earth's surface at different elevations with regard to sea level.

Igneous rock: A rock formed by congealing rapidly or slowly from a molten state.

Ignimbrite: An igneous rock formed by the lithification of volcanic ash and volcanic breccia.

Inclination: The angle between a line in the Earth's magnetic field and the horizontal plane; also a synonym for *dip*.

Index of refraction: The ratio of the speed of light in a vacuum to the speed in a material; this ratio determines the amount that light is refracted as it passes into a crystal.

Infiltration: The movement of groundwater or hydrothermal water into rock or soil through joints and pores.

Interfacial angle: The angle between two crystal faces of a crystal, characteristic of a mineral's symmetry.

Interior drainage: A system of streams that converge in a closed basin and evaporate without reaching the sea.

Intermontane basin: A basin between mountain ranges, often formed over a graben.

Intrusion: An igneous rock body that has forced its way in a molten state into surrounding country rock.

Intrusive rock: Igneous rock that is interpreted as a former intrusion from its cross-cutting contacts, chilled margins, or other field relations.

Ion: An atom or group of atoms that has gained or lost electrons and so has a net electric charge.

Ionic bond: A bond formed between atoms by electrostatic attraction between oppositely charged ions.

Iron formation: A sedimentary rock containing much iron, usually more than 15 percent as sulfide, oxide, hydroxide, or carbonate; a low-grade ore of iron.

Isograd: A line or curved surface connecting rocks that have undergone an equivalent degree of metamorphism.

Isostasy: The mechanism whereby areas of the crust rise or subside until the mass of their topography is buoyantly supported or compensated by the thickness of crust below, which "floats" on the denser mantle. The theory that continents and mountains are supported by low-density crustal "roots."

Isotope: One of several forms of one element, all having the same number of protons in the nucleus, but differing in their number of neutrons and thus atomic weight.

Isotope geology: The study of the relative abundances of isotopes in rocks to determine their ages (see *geochronology*) or conditions of formation.

Isotropic substance: One in which the magnitude of a physical property, such as transmission of light is independent of crystallographic direction.

Joint: A large and relatively planar fracture in a rock across which there is no relative displacement of the two sides.

Juvenile gas: Gases that come to the surface for the first time from the deep interior.

Kame: A ridge-like or hilly local glacial deposit of coarse alluvium formed as a delta at the glacier front by meltwater streams.

Karst topography: An irregular topography characterized by sinkholes, caverns, and lack of surface streams in humid regions because an underlying carbonate formation has been riddled with underground drainage channels that capture the surface streams.

**Kerogen:** A mixture of organic substances found in many fine-grained sedimentary rocks and a major constituent of oil shale.

**Kettle:** A small hollow or depression formed in glacial deposits when outwash was deposited around a residual block of ice that later melted.

**Kilobar:** A unit of pressure equal to 1000 *bars*.

**Kimberlite:** A peridotite containing garnet and olivine and found in volcanic pipes, through which it may come from the upper Mantle.

**Laccolith:** A sill-like igneous intrusion that forces apart two strata and forms a round, lens-shaped body many times wider than it is thick.

**Lahar:** A mudflow of unconsolidated volcanic ash, dust, breccia, and boulders mixed with rain or the water of a lake displaced by a lava flow.

**Laminar flow:** A flow regime in which particle paths are straight or gently curved and parallel.

**Landslide:** The rapid downslope movement of soil and rock material, often lubricated by groundwater, over a basal shear zone; also the tongue of stationary material deposited by such an event.

**Lapilli:** A fragment of volcanic rock formed when magma is ejected into the air by expanding gases. The size of the fragments ranges from sand- to cobble-size.

**Lateral moraine:** A moraine formed along the side of a valley glacier and composed of rock scraped off or fallen from the valley sides.

**Lava:** Magma or molten rock that has reached the surface.

**Lava tube:** A sinuous, hollow tunnel formed when the outside of a lava flow cools and solidifies and the molten material passing through it is drained away.

**Leaching:** The removal of elements from a soil by dissolution in water moving downward in the ground.

**Left-lateral fault:** A strike-slip fault on which the displacement of the far block is to the left when viewed from either side.

**Levee:** A low ridge along a stream bank, formed by deposits left when floodwater decelerates on leaving the channel; also an artificial barrier to floods built in the same form.

**Limb (fold):** The relatively planar part of a fold or of two adjacent folds (for example, the steeply dipping part of a stratum between an anticline and syncline).

**Limestone:** A sedimentary rock composed principally of calcium carbonate ($CaCO_3$), usually as the mineral calcite.

**Lineation:** Any linear arrangement of features found in a rock.

**Lithification:** The processes that convert a sediment into a sedimentary rock.

**Lithology:** The systematic description of rocks, in terms of mineral composition and texture.

**Lithosphere:** The outer, rigid shell of the Earth, situated above the asthenosphere and containing the crust, continents, and plates.

**Lode:** An unusually large vein or set of veins containing ore minerals.

**Longitudinal dune:** A long dune parallel to the direction of the prevailing wind.

**Longitudinal profile:** A cross section of a stream from its mouth to its head, showing elevation versus distance to the mouth.

**Longshore current:** A current that moves parallel to a shore and is formed from the momentum of breaking waves that approach the shore obliquely.

**Longshore drift:** The movement of sediment along a beach by swash and backwash of waves that approach the shore obliquely.

**Lopolith:** A large laccolith that is bowl-shaped and depressed in the center, possibly by subsidence of an emptied magma chamber beneath the intrusion.

**Lowland:** Land of general low relief at the lower levels of regional elevation.

**Low-velocity zone:** A region in the Earth, especially a planar layer, that has lower seismic-wave velocities than the region immediately above it.

**Luster:** The general textural impression of a mineral surface, given by the light reflected from it. Terms such as metallic, submetallic are standardized but subjective.

**Maar volcano:** A volcanic crater without a cone, believed to have been formed by an explosive eruption of trapped gases.

**Mafic mineral:** A dark-colored mineral rich in iron and magnesium, especially a pyroxene, amphibole, or olivine.

**Magma:** Molten rock material that forms igneous rocks upon cooling. Magma that reaches the surface is referred to as lava.

**Magma chamber:** A magma-filled cavity within the lithosphere.

**Magmatic water:** Water that is dissolved in a magma or that is derived from such water.

**Magnetic anomaly:** The value of the local magnetic field remaining after the subtraction of the dipole portion of the Earth's field.

**Magnetic coupling:** The transfer of momentum between celestial bodies, especially dust and gas clouds, through magnetic forces.

**Magnetic north pole:** (1) The point where the Earth's surface intersects the axis of the dipole that best approximates the Earth's field. (2) The point where the Earth's magnetic field dips vertically downward.

**Magnetic stratigraphy:** The study and correlation of polarity epochs and events in the history of the Earth's magnetic field as contained in magnetic rocks.

**Magnetometer:** An instrument for measuring either one orthogonal component or the entire intensity of the Earth's magnetic field at various points.

**Magnitude:** A measure of earthquake size, determined by taking the common logarithm (base 10) of the largest ground motion observed during the arrival of a *P*-wave or seismic surface wave and applying a standard correction for distance to the epicenter.

**Manganese nodule:** A small, rounded concretion found on the deep ocean floor that may contain as much as 20 percent manganese and smaller amounts of iron, copper, and nickel oxides and hydroxides.

**Mantle:** The main bulk of the Earth, between the crust and core, ranging from depths of about 40 to 3480 kilometers. It is composed of dense mafic silicates and divided into concentric layers by phase changes that are caused by the increase in pressure with depth.

**Mare:** A dark, low-lying lunar plain, filled to an undetermined depth with mafic volcanic rocks. Plural: maria.

**Mascon:** A *con*centration of *mass* below the lunar surface; detected as a positive gravity anomaly, cause unknown.

**Massive rock:** A rock that is little or not at all broken by joints, cracks, foliation, or bedding, tending to present a homogeneous appearance.

**Mass movement:** A downhill movement of soil or fractured rock under the force of gravity.

**Mass spectrometer:** An instrument for separating ions of different mass but equal charge (mainly isotopes in geology) and measuring their relative quantities.

**Maturity:** A stage in the *geomorphic cycle* in which maximum relief and well-developed drainage are both present.

**Meander:** Broad, semicircular curves in a stream that develop as the stream erodes the outer bank of a curve and deposits sediment against the inner bank.

**Mechanical weathering:** The set of all physical processes by which an outcrop is broken up into small particles.

**Medial moraine:** A long stripe of rock debris carried on or within a glacier resulting from the convergence of lateral moraines where two glaciers join.

**Medical geology:** The application of geologic science to problems of health, especially those relating to mineral sources of toxic or nutritious elements and natural dispersal of toxic pollutants.

**Mélange:** A formation consisting of a heterogenous mixture of rock materials on a mappable scale. Fragments of diverse composition, size, and texture have been mixed and consolidated by tremendous deformational pressure.

**Mesa:** A flat-topped, steep-sided upland topped by a resistant formation and larger than a butte.

**Mesosphere:** The lower mantle.

**Metamorphism:** The changes of mineralogy and texture imposed on a rock by pressure and temperature in the Earth's interior.

**Meteoric water:** Rainwater, snow, hail, and sleet.

**Meteorite:** A stoney or metallic object from interplanetary space that penetrates the atmosphere to impact on the surface.

**Micrometeorite:** A meteorite less than 1 millimeter in diameter.

**Microseism:** A weak vibration of the ground that can be detected by seismographs and which is caused by waves, wind, or human activity, but not by an earthquake.

**Migmatite:** A rock with both igneous and metamorphic characteristics that shows large crystals and laminar flow structures. Probably formed metamorphically in the presence of water and without melting.

**Mineral:** A naturally occurring element or compound with a precise chemical formula and a regular internal lattice structure. Organic products are usually not included.

**Mineralogy:** The study of mineral composition, structure, appearance, stability, occurrence, and associations.

**Miogeosyncline:** A Geosyncline that is situated near a craton and receives chemical and well-sorted clastic sediments from the continent.

**Mohorovičić discontinuity:** The boundary between crust and mantle, marked by a rapid increase in seismic wave velocity to more than 8 kilometers per second. Depth: 5 to 45 kilometers. Abbreviated "Moho" or "M-discontinuity."

**Mohs scale of hardness:** An empirical, ascending scale of mineral hardness with talc as 1, gypsum 2, calcite 3, fluorite 4, apatite 5, orthoclase 6, quartz 7, topaz 8, corundum 9, and diamond 10.

**Monadnock:** An isolated hill or mountain rising above a peneplain.

**Monocline:** The S-shaped fold connecting two horizontal parts of the same stratum at different elevations. Its central limb is usually not overturned.

**Moraine:** A glacial deposit of till left at the margin of an ice sheet. (See specifically by name, *ground moraine, longitudinal moraine, medial moraine,* and *terminal moraine.*

**Mountain:** A steep-sided topographic elevation larger than a *hill;* also a single prominence forming part of a ridge or mountain range.

**Mudflow:** A mass movement of material finer than sand, lubricated with large amounts of water.

**Mudstone:** The lithified equivalent of mud, a fine-grained sedimentary rock similar to shale but more massive.

**M.y.:** Abbreviation for "million years."

**Mylonite:** A very fine lithified fault breccia commonly found in major thrust faults and produced by shearing and rolling during fault movement.

**Native metal:** A natural deposit of a metallic element in pure metallic form, neither oxidized nor combined with sulfur or other elements.

**Neap tide:** A tide cycle of unusually small amplitude, which occurs twice monthly when the lunar and solar tides are opposed—that is, when the gravitational pull of the Sun is at right angles to that of the Moon.

**Nebula:** An immense, diffuse body of interstellar gas and dust that has not condensed into a star.

**Nebular hypothesis:** A theory of the formation of the planets that states that a rotating nebula contracted and was then torn into fragments by centrifugal forces, with planets condensing from the fragments.

**Neutron:** An electrically neutral elementary particle in the atomic nucleus having the mass of one proton.

**Neutron-activation analysis:** A method of identifying isotopes of an element by bombarding them with neutrons and observing the characteristic radioactive decay products emitted.

**Normal fault:** A dip-slip fault in which the block above the fault has moved downward relative to the block below.

**Nuclear geology:** See *Isotope geology.*

**Nuée ardente:** A "glowing cloud" of hot volcanic ash, dust, and gas that moves rapidly downhill as a density current in the atmosphere.

**Oblique-slip fault:** A fault that combines some strike-slip motion with some dip-slip motion.

**Obsidian:** Dark volcanic glass of felsic composition.

**Octahedral coordination:** The packing of six ions around an ion of opposite charge to form an octahedron.

**Oil field:** An underground accumulation of oil and gas concentrated beneath an impermeable trap, preventing its escape upward.

**Oil shale:** A dark-colored shale containing organic material that can be crushed and heated to liberate gaseous hydrocarbons.

**Oil trap:** See *Trap.*

**Old age:** A stage in the *geomorphic cycle,* characterized by formation of a peneplain near sea level.

**Oolite:** A sedimentary carbonate particle composed of spherical grains precipitated from warm ocean water on carbonate platforms. Also a rock composed of such particles.

**Opaque mineral:** A mineral which transmits no light through a thin section under a microscope. Usually a native metal, sulfide, or metallic oxide mineral.

**Ophiolite suite:** An assemblage of mafic and ultramafic igneous rocks with deep-sea sediments supposedly associated with divergence zones and the sea-floor environment.

**Orbit:** The elliptical or hyperbolic path traced by a planet or meteorite or satellite in the presence of a more massive body.

**Ore:** A natural deposit in which a valuable metallic element occurs in high enough concentration to make mining economically feasible.

**Ore mineral:** The mineral of an ore that contains the useful element.

**Original Horizontality, Principle of:** The proposition of Steno, that all sedimentary bedding is horizontal at the time of deposition.

**Orogenic belt:** A linear region, often a former geosyncline, that has been subjected to folding, and other deformation in a mountain-building episode.

**Orogeny:** The tectonic process in which large areas are folded, thrust-faulted, metamorphosed, and subjected to plutonism. The cycle ends with uplift and the formation of mountains.

**Oscillation ripple:** A ripple with a symmetrical cross section and a sharp peak formed by waves.

**Outcrop:** A segment of bedrock exposed to the atmosphere.

**Outgassing:** The release of juvenile gases to the atmosphere and oceans by volcanism.

**Outwash:** A glaciofluvial sediment that is deposited by meltwater streams emanating from a glacier.

**Overturned fold:** A fold in which a limb has tilted past vertical so that the older strata are uppermost.

**Oxbow lake:** A long, broad, crescent-shaped lake formed when a stream abandons a meander and takes a new course.

**Oxidation:** A chemical reaction in which electrons are lost from an atom and its charge becomes more positive.

**Oxidized element:** An element occurring in the more positively charged of two common ionic forms (for example, the $Fe^{+3}$ form of iron, as opposed to $Fe^{+2}$).

**Pahoehoe:** A basaltic lava flow with a glassy, smooth, and undulating, or ropy, surface.

**Paleoclimate:** The average state or typical conditions of climate during some past geologic period.

**Paleocurrent map:** A map of depositional currents that have been inferred from cross-bedding, ripples, or other sedimentary structures.

**Paleogeographic map:** A map showing the surface landforms and coastline of an area at some time in the geologic past.

**Paleomagnetism:** The science of the reconstruction of the Earth's ancient magnetic field and the positions of the continents from the evidence of remanent magnetization in ancient rocks.

**Paleontology:** The science of fossils, of ancient life-forms, and their evolution.

**Paleowind:** A prevailing wind direction in an area, inferred from dune structure or the distribution of volcanic ash for one particular time in geologic history.

**Pangaea:** According to some theories, a great protocontinent from which all present continents have broken off by the mechanism of sea-floor spreading and continental drift.

**Panthalassa:** A hypothetical primaeval ocean covering two-thirds of the world except for the continent of Pangaea.

**Paraffin:** Hydrocarbons that contain chains of carbon and hydrogen atoms with the general formula $C_nH_{2n+2}$.

**Parent element:** An element that is transformed by radioactive decay to a different (daughter) element.

**Peat:** A marsh or swamp deposit of water-soaked plant remains containing more than 50 percent carbon.

**Pedalfer:** A common soil type in humid regions, characterized by an abundance of iron oxides and clay minerals deposited in the B-horizon by leaching.

**Pediment:** A planar, sloping rock surface forming a ramp up to the front of a mountain range in an arid region. It may be covered locally by thin alluvium.

**Pedocal:** A common soil type of arid regions, characterized by accumulation of calcium carbonate in the A-horizon.

**Pegmatite:** An igneous rock with extremely large grains, more than a centimeter in diameter. It may be of any composition but most frequently is granitic.

**Pelagic sediment:** Deep-sea sediments composed of fine-grained detritus that slowly settles from surface waters. Common constituents are clay, radiolarian ooze, and foraminiferal ooze.

**Peléan eruption:** A volcanic eruption accompanied by great explosions and emanations of hot gas and Nuées ardentes. Named for Mont Pelée, Martinique.

**Peneplain:** A hypothetical extensive area of low elevation and relief reduced to near sea level by a long period of erosion and representing the end product of the ideal geomorphic cycle.

**Perched groundwater:** An isolated body of groundwater that is perched above and separated from the main water table by an aquiclude.

**Peridotite:** A coarse-grained mafic igneous rock composed of olivine with accessory amounts of pyroxene and amphibole but little or no feldspar.

**Period (geologic):** The most commonly used unit of geologic time, representing one subdivision of an era.

**Period (wave):** The time interval between the arrival of successive crests in a homogeneous wave train; the period is the inverse of the frequency of a cyclic event.

**Permeability:** The ability of a formation to transmit ground-water or other fluids through pores and cracks.

**Petrographic microscope:** An optical microscope designed for looking at thin sections of rocks and equipped with polarizing filters on opposite sides of the specimen.

**Petroleum resources:** The total quantity of natural oil in the Earth's crust that it may be possible to extract.

**Petrology:** The study of the composition, structure, and origin of rocks.

**Phaneritic texture:** A rock texture in which individual crystals are visible to the unaided eye.

**Phase change:** (1) The transformation of an element or compound from solid to liquid, gaseous to solid, etc. (2) The transformation of a solid mineral to a different solid form of different structure and density.

**Phenocryst:** A large crystal surrounded by a finer matrix in a porphyry.

**Phosphate sediment:** A sediment composed largely of calcium phosphate, usually as a variety of the mineral apatite and largely in the form of concretions and nodules.

**Photogeology:** The study of geologic features as exposed in aerial photographs of Earth or other planets.

**Photolysis:** The chemical breakdown of water into hydrogen and oxygen in the upper atmosphere by solar radiation.

**Photosynthesis:** The conversion of water, carbon dioxide, and the energy of sunlight into oxygen and organic compounds such as sugars by plants containing chlorophyll.

**Phreatic eruption:** A volcanic eruption of mud and debris caused by the expansion of steam formed when magma comes in contact with confined groundwater.

**Phreatic zone:** The zone of soil and rock in which pores are completely filled with groundwater. Also called the saturated zone.

**Physical sedimentation:** The deposition of clastic particles derived by erosion.

**Pillow lava:** A type of lava flow formed underwater, in which many small pillow-shaped tongues break through the chilled surface and quickly solidify, leading to a rock formation resembling a pile of sandbags.

**Piracy (stream):** The headward growth of one stream until it intersects the channel of another, whereupon all the water is diverted from the latter stream below that point.

**Placer:** A detrital sedimentary deposit of a valuable mineral or native metal in unusually high concentration, usually segregated because of its greater density.

**Plain:** An area of low relief, without major hills, valleys, or ridges; not necessarily at a low elevation.

**Plane of symmetry:** An imaginary plane passing through an object in such an orientation that each feature of the object on one side of the plane has a mirror image on the other side.

**Planetary evolution:** The process by which a differentiated planet is formed.

**Planetesimal:** A body of rock in space, smaller than a planet and attracted by gravity to other planetesimals in the formation of a planet.

**Planetology:** The science of the distribution, composition, and origin of matter in the planets of the solar system.

**Plastic deformation:** Deformation that proceeds to large strains at constant stress without fracturing.

**Plate:** One of the dozen or more segments of the lithosphere that are internally rigid and move independently over the interior, meeting in convergence zones and separating at divergence zones.

**Plateau:** An extensive upland region at high elevation with respect to its surroundings.

**Plate tectonics:** The theory and study of plate formation, movement, interaction, and destruction; the attempt to explain seismicity, volcanism, mountain-building, and paleomagnetic evidence in terms of plate motions.

**Playa:** The flat floor of a closed basin in an arid region. It may be occupied by an intermittent lake.

**Plunging fold:** A fold whose axis is not horizontal but dips. Thus progressively younger strata are found at the center of the fold as one travels along the direction of plunge, and the geologic map pattern is one of nested V-shaped outcrops of formations.

**Pluton:** A large igneous intrusion, formed at depth in the crust.

**Plutonic:** Pertaining to igneous activity at depth.

**Plutonism:** (1) The formation and worldwide distribution of plutons. (2) The eighteenth- and nineteenth-century theory that the Earth was at one time entirely molten.

**Point bar:** A deposit of sediment on the inner bank of a meander that forms because the stream velocity is lower against the inner bank.

**Polarity epoch:** In paleomagnetism, a segment of geologic time in which the Earth's magnetic field was either predominantly in the present direction or predominantly reversed.

**Polar wandering:** An interpretation given to the observation that the position of the magnetic north pole, as inferred from the thermoremanent magnetization of rocks, varies with time as a result of plate movements.

**Pole of spreading:** An imaginary point on the Earth's surface that represents the emergence of an imaginary axis passing through the Earth's center and about which one plate moves relative to another; thus, for each pair of plates, there is a unique pole.

**Polymorph:** A chemical compound that may show two or more structures (for example, calcite and aragonite are polymorphs of calcium carbonate).

**Porosity:** The percentage of the total volume of a rock that is pore space.

**Porphyroblast:** A large cyrstal in a finer-grained matrix of a metamorphic rock, resembling a phenocryst in an igneous formation.

**Porphyry:** An igneous rock containing abundant phenocrysts of one mineral but very fine grains of the other minerals.

**Porphyry copper deposit:** A porphyritic igneous rock that contains a few percent of copper sulfides in numerous veins and cracks and is of economical value.

**Potable water:** Water that is agreeable to the taste and not dangerous to the health.

**Pothole:** A semispherical hole in the bedrock of a stream bed, formed by abrasion of small pebbles and cobbles in a strong current.

**Ppm:** Abbreviation for "parts per million."

**Pratt isostatic compensation:** The mechanism in which variations in crustal density act to counterbalance the varying weight of topographic features. The crust is here assumed to be of approximately uniform thickness, thus a mountain range would be underlain by lighter rocks.

**Preferred orientation:** Any deviation from randomness in the distribution of the crystallographic or grain shape axes of minerals of a rock (including flow cleavage and foliation), produced by deformation and nonuniform stress during crystallization in metamorphic rocks or by depositional currents in sediments.

**Proton:** An elementary particle found in the atomic nucleus with a positive charge of $1.602 \times 10^{-19}$ coulomb and a mass of 1836 electrons; one $H^+$ Ion.

**Proto-sun:** A large cloud of dust and gas gradually coalescing into a star under the force of gravity.

**Proven reserves:** Deposits of fossil fuels whose location and extent are known, as opposed to potential but unproved ("discovered") deposits.

**Pumice:** A form of volcanic glass, usually of silicic composition, so filled with vesicles that it resembles a sponge and is very light.

**P-wave:** The primary or fastest wave traveling away from a seismic event through the solid rock, and consisting of a train of compressions and dilations of the material.

**Pyroclastic rock:** A rock formed by the accumulation of fragments of volcanic rock scattered by volcanic explosions.

**Pyroclastic texture:** The unsorted, angular, and unrounded texture of the fragments in a pyroclastic rock.

**Pyroxene granulite:** A coarse-grained contact metamorphic rock containing pyroxene, formed at high temperatures and low pressures.

**Quartz arenite:** A sandstone containing very little except pure quartz grains and cement.

**Quartzite:** (1) A very hard, clean, white metamorphic rock formed from a quartz arenite sandstone. (2) A quartz arenite containing so much cement that it resembles (1).

**Quartzose sandstone:** (1) A quartz arenite. (2) A clean quartz sandstone, less pure than a quartz arenite, that may contain a moderate amount of other detrital minerals and/or calcite cement.

**Radial drainage:** A system of streams running in a radial pattern away from the center of a circular elevation, such as a volcano or dome.

**Radiative transfer:** One mechanism for the movement of heat, in which it takes the form of long-wavelength infrared radiation.

**Radioactive decay:** The spontaneous breakdown of certain kinds of atomic nuclei into one or more nuclei of different elements, involving the release of energy and subatomic particles.

**Radiolarian:** A class of one-celled marine animals with siliceous skeletons that have existed in the ocean throughout the Phanerozoic Eon.

**Radiolarian ooze:** A siliceous deep-sea sediment composed largely of the skeletons of radiolaria.

**Radiolarite:** The lithified sedimentary rock formed from radiolarian ooze.

**Radius ratio:** The ratio of the radius of an anion to the radius of a cation, which ratio determines the coordination number.

**Ray:** A linear landform of the lunar surface emanating from a large crater and extending as much as 100 kilometers outward, probably consisting of fine ejecta thrown out by the impact of a meteorite.

**Reaction series:** A series of chemical reactions occurring in a cooling magma by which a mineral formed at high temperature becomes unstable in the melt and reacts to form another mineral (see also *Discontinuous reaction series*).

**Recharge:** In hydrology, the replenishment of groundwater by infiltration of meteoric water through the soil.

**Recrystallization:** The growth of new mineral grains in a rock at the expense of old grains, which supply the material.

**Rectangular drainage:** A system of streams in which each straight segment of each stream takes one of two characteristic perpendicular directions, with right-angle bends between. The streams are usually following two perpendicular sets of joints.

**Recumbent fold:** An overturned fold with both limbs nearly horizontal.

**Reduced element:** Of elements that have two different ions of different charge, the one with the less positive (more negative) charge is said to be reduced (for example, $Fe^{+2}$ as opposed to $Fe^{+3}$).

**Refraction (wave):** The departure of a wave from its original direction of travel at the interface with a material of different index of refraction (light) or seismic wave velocity (see also *Seismic refraction*).

**Regional metamorphism:** Metamorphism occurring over a wide area and caused by deep burial and high internal temperatures of the Earth.

**Regolith:** Any solid material lying on top of bedrock. Includes soil, alluvium, and rock fragments weathered from the bedrock.

**Regression:** A drop in sea level that causes an area of the Earth to be uncovered by sea water, ending marine deposition.

**Relief:** The maximum regional difference in elevation.

**Remote sensing:** The study of Earth surface conditions and materials from airplanes and satellites by means of photography, spectroscopy, or radar.

**Replacement deposit:** A deposit of ore minerals by hydrothermal solutions that have first dissolved the original mineral to form a small cavity.

**Reserves:** See proven reserves.

**Residence time:** The amount of a given element or compound in a given body (especially the sea or atmosphere) divided by the average removal rate, assuming the system is in a steady state.

**Respiration:** The chemical reaction by which carbohydrates are oxidized and by which all animals and plants convert their food into energy. Carbon dioxide is released and oxygen used up.

**Reverse fault:** See *Thrust fault*.

**Reversible reaction:** A chemical reaction which can proceed in either direction, depending on the concentration of reacting materials.

**Rheidity:** (1) The ability of a substance to yield to viscous flow under large strains. (2) One thousand times the time required for a substance to stop changing shape when stress is no longer applied.

**Rhyolite:** The fine-grained volcanic or extrusive equivalent of granite, light brown to gray and compact.

**Richter magnitude scale:** See *Magnitude*.

**Ridge (mid-ocean):** A major linear elevated landform of the ocean floor, from 200 to 20,000 kilometers in extent. It is not a single ridge, but resembles a mountain range and may have a central rift valley.

**Rift valley:** A fault trough formed in a divergence zone or other area of tension.

**Right-lateral fault:** A strike-slip fault on which the displacement of the far block is to the right when viewed from either side.

**Ring dike:** A dike in the form of a segment of a cone or cylinder, having an arcuate outcrop.

**Rip current:** A current that flows strongly away from the sea shore through gaps in the surf zone at intervals along the shoreline.

**Ripple:** A very small dune of sand or silt whose long dimension is formed at right angles to the current.

**River order:** See *Stream order*.

**Rock cycle:** The geologic cycle, with emphasis on the rocks produced; sedimentary rocks are metamorphosed to metamorphic rocks, or melted to create igneous rocks, and all rocks may be uplifted and eroded to make sediments, which lithify to sedimentary rocks.

**Rock flour:** A glacial sediment of extremely fine (silt- and clay-size) ground rock formed by abrasion of rocks at the base of the glacier.

**Rock glacier:** A glacier-like mass of rock fragments or talus with interstitial ice that moves downhill under the force of gravity.

**Rockslide:** A landslide involving mainly large blocks of detached bedrock with little or no soil or sand.

**Rounding:** The degree to which the edges and corners of a particle become worn and rounded as a result of abrasion during transportation. Expressed as angular, subrounded, well-rounded, etc.

**Runoff:** The amount of rain water directly leaving an area in surface drainage, as opposed to the amount that seeps out as groundwater.

**Rupture strength:** The greatest stress that a material can sustain without fracturing at one atmosphere pressure.

**Saltation:** The movement of sand or fine sediment by short jumps above the ground or stream bed under the influence of a current too weak to keep it permanently suspended.

**Sandblasting:** A physical weathering process in which rock is eroded by the impact of sand grains carried by the wind, frequently leading to ventifact formation of pebbles and cobbles.

**Sandstone:** A detrital sedimentary rock composed of grains from 1/16 to 2 millimeters in diameter, dominated in most sandstones by quartz, feldspar, and rock fragments, bound together by a cement of silica, carbonate, or other minerals or a matrix of clay minerals.

**Schist:** A metamorphic rock characterized by strong foliation or schistosity.

**Schistosity:** The parallel arrangement of sheety or prismatic minerals like micas and amphiboles resulting from nonhydrostatic stress in metamorphism.

**Scoria:** Congealed lava, usually of mafic composition, with a large number of vesicles formed by gases coming out of solution.

**Sea-floor spreading:** The mechanism by which new sea floor crust is created at ridges in divergence zones and adjacent plates are moved apart to make room. This process may continue at 0.5 to 10 centimeters/year through many geologic periods.

**Seamount:** An isolated tall mountain on the sea floor that may extend more than 1 kilometer from base to peak (see also *Guyot*).

**Secular variation:** Slow changes in the orientation of the Earth's magnetic field that appear to be long lasting and internal in origin as opposed to rapid fluctuations, which are external in origin.

**Sedimentary environment:** See *Environment of deposition*.

**Sedimentary rock:** A rock formed by the accumulation and cementation of mineral grains transported by wind, water, or ice to the site of deposition or chemically precipitated at the depositional site.

**Sedimentary structure:** Any structure of a sedimentary or weakly metamorphosed rock that was formed at the time of deposition; includes bedding, cross-bedding, graded bedding, ripples, scour marks, mudcracks.

**Sedimentation:** The process of deposition of mineral grains or precipitates in beds or other accumulations.

**Seif dune:** A longitudinal dune that shows the sculpturing effect of cross-winds not parallel to its axis.

**Seismic discontinuity:** A surface within the Earth across which *P*-wave or *S*-wave velocities change rapidly, usually by more than ±0.2 kilometer/second.

**Seismicity:** The world-wide or local distribution of earthquakes in space and time; a general term for the number of earthquakes in a unit of time.

**Seismic profile:** The data collected from a set of seismographs arranged in a straight line with an artificial seismic source, especially the times of *P*-wave arrivals.

**Seismic reflection:** A mode of seismic prospecting in which the seismic profile is examined for waves that have reflected from near-horizontal strata below the surface.

**Seismic refraction:** A mode of seismic prospecting in which the seismic profile is examined for waves that have been refracted upward from seismic discontinuities below the profile. Greater depths may be reached than through seismic reflection.

**Seismic surface wave:** A seismic wave that follows the earth's surface only, with a speed less than that of *S*-waves. There are Raleigh waves (forward and vertical vibrations) and Love waves (transverse vibrations).

**Seismic transition zone:** A seismic discontinuity, found in all parts of the Earth, at which the velocity increases rapidly with depth; especially the one at 300 to 600 kilometers.

**Seismic travel time:** See *Travel time curve*.

**Seismograph:** An instrument for magnifying and recording the motions of the Earth's surface that are caused by seismic waves.

**Seismology:** The study of earthquakes, seismic waves, and their propagation through the Earth.

**Self-exciting dynamo:** A system involving a rotating electrical conductor that creates and sustains its own magnetic field.

**Series:** A set of rocks formed in one area during a geologic epoch.

**Settling velocity:** The rate at which a sedimentary particle of a given size falls through water or air.

**Shale:** A very fine-grained detrital sedimentary rock composed of silt and clay which tends to part along bedding planes.

**Shard (volcanic):** An angular fragment of volcanic glass.

**Sheeting:** See *Exfoliation*.

**Sheetwash:** A flow of rainwater that covers the entire ground surface with a thin film and is not concentrated into streams.

**Shield volcano:** A large, broad volcanic cone with very gentle slopes built up by nonviscous basalt lavas.

**Shooting flow:** A very fast form of water flow with a high velocity induced by steep slopes in the flow bed, typically developed in rapids.

**Silicate:** Any of a vast class of minerals containing silicon and oxygen and constructed from the tetrahedral group $(SiO_4)^{-4}$. The bulk of the crust is composed of silicate minerals.

**Silicic rock:** An igneous rock containing more than two-thirds silicon-oxygen tetrahedra by weight, usually as quartz and feldspar (for example, granite).

**Sill:** A horizontal tabular intrusion with concordant contacts.

**Sill (topographic):** A small submarine ridge that nearly separates or restricts water flow between two adjacent bodies of water.

**Sinkhole:** A small, steep depression caused in Karst topography by the dissolution and collapse of subterranean caverns in carbonate formations.

**Sinking current:** A downward movement of sea or lake water that has become denser through cooling or increased salinity or that sinks because of an onshore wind piling up water against the shore.

**Sinuosity:** The length of the channel of a stream divided by the straight-line distance between its ends.

**Slate:** The metamorphic equivalent of shale; a hard, gray, red, green, or black fine-grained rock with slaty cleavage.

**Slaty cleavage:** A foliation consisting of the parallel arrangement of sheety metamorphic minerals (see *Schistosity*) at an angle to bedding planes, and related to deformational structures.

**Slickensides:** Parallel grooves, ramps, and scratches on one or both of the inside faces of a fault, showing the direction of slip.

**Slip (fault):** The relative motion of one face of a fault relative to the other.

**Slip face:** The steep downwind face of a dune on which sand is deposited in crossbeds at the angle of repose.

**Slope wash:** The motion of water and sediment down a slope by the mechanism of sheet wash.

**Slump:** A collapse structure in unlithified sediments, caused by gravity sliding or collapse under the weight of later sediments.

**Snowfield:** An area of permanent snow that is not moving or whose movement is not visible.

**Soft water:** Water that is free of calcium and magnesium carbonates and other dissolved materials of hard water.

**Soil:** The surface accumulation of sand, clay, and humus that compose the regolith, but excluding the larger fragments of unweathered rock.

**Soil creep:** The imperceptible downhill flow of soil under the force of gravity. It is a shear flow with velocity decreasing downward, and, occurs even on gentle slopes.

**Solid-solution series:** A series of minerals of identical structure that can contain a mixture of two elements over a range of ratios (for example, plagioclase feldspars).

**Solidus:** A curve on a pressure versus temperature graph representing the beginning of melting of a rock or mineral.

**Solifluction:** The soil creep of material saturated with water and/or ice; most common in polar regions.

**Solubility:** The mass of a substance that can be dissolved in a certain amount of solvent, if chemical equilibrium is attained.

**Solute:** The substance dissolved in a solvent.

**Solvent:** A medium, usually liquid, in which other substances can be dissolved.

**Sorting:** A measure of the homogeneity of the sizes of particles in a sediment or sedimentary rock.

**Spatter cone:** A conical deposit of cooled lava congealed around a volcanic vent that disgorges mostly gas with occasional globs of molten rock.

**Spheroidal weathering:** The formation of spherical residual inner cores by the weathering of boulders.

**Spit:** A long ridge of sand deposited by longshore current and drift where the coast takes an abrupt inward turn. It is attached to land at the upstream end.

**Splay deposit:** A small delta deposited on a floodplain when the stream breaches a levee during a flood.

**Spring tide:** A tide cycle of unusually large amplitude that occurs twice monthly when the lunar and solar tides are in phase (compare *Neap tide*).

**Stage (stream):** The elevation of the water level of a stream measured against some constant reference.

**Stalactite:** An icicle- or tooth-like deposit of calcite or aragonite hanging from the roof of a cave and deposited by evaporation and precipitation from solutions seeping through limestone.

**Stalagmite:** An inverted icicle-shaped deposit that builds up on a cave floor beneath a stalactite, and formed by the same process as a stalactite.

**Steady-state:** An adjective that implies that a system is in a stable dynamic state in which inputs balance outputs.

**Stock (volcanic):** An intrusion with the characteristics of a batholith but less than 100 square kilometers in area.

**Stoping:** The process by which an intruding igneous mass makes room for itself in the country rock — by breaking rock fragments off the walls and "absorbing" them when they sink.

**Strain:** A quantity describing the exact deformation of each point in a body. Roughly, the change in a dimension or volume divided by the original dimension or volume.

**Strain seismograph:** An instrument that measures

changes of strain in surface rocks to detect seismic waves.

**Stratification:** A structure of sedimentary rocks, which have recognizable parallel beds of considerable lateral extent.

**Stratigraphic sequence:** A set of beds deposited that reflects the geologic history of a region.

**Stratigraphy:** The science of the description, correlation, and classification of strata in sedimentary rocks, including the interpretation of the depositional environments of those strata.

**Stratovolcano:** A volcanic cone consisting of both lava and pyroclastic rocks, often conical.

**Streak:** The fine deposit of mineral dust left on an abrasive surface when a mineral is scraped across it; especially the characteristic color of the dust.

**Streak plate:** A ceramic abrasive surface for streak tests.

**Streaming flow:** A tranquil flow slower than shooting flow.

**Streamline:** A curved line representing the successive positions of a particle in a flow as time passes.

**Stream order:** The hierarchical number of a stream segment in dendritic drainage: the smallest tributary streams have order one and at each junction of streams of equal order the order of the subsequent segment is one higher.

**Stress:** A quantity describing the forces acting on each part of a body in units of force per unit area.

**Striation:** See *Glacial striation.*

**Strike:** The angle between true North and the horizontal line contained in any planar feature (inclined bed, dike, fault plane, etc.); also the geographic direction of this horizontal line.

**Strike-slip fault:** A fault whose relative displacement is purely horizontal (see also *Right-* and *left-lateral fault*).

**Stromatolite:** A fossil form representing the growth habit of an algal mat: concentric spherules, stacked hemispheres, or flat sheets of calcium carbonate and trapped silt encountered in limestones.

**Subduction zone:** A dipping planar zone descending away from a trench and defined by high seismicity, interpreted as the shear zone between a sinking oceanic plate and an overriding plate (see also *Convergence zone*).

**Sublimation:** A phase change from the solid to the gaseous state, without passing through the liquid state.

**Submarine canyon:** An underwater canyon in the continental shelf.

**Subsidence:** A gentle epeirogenic movement where a broad area of the crust sinks without appreciable deformation.

**Superposed stream:** A stream that flows through resistant formations because its course was established at a higher level on uniform rocks before down-cutting began.

**Superposition, Principle of:** The principle stated by Steno that, except in extremely deformed strata, a bed that overlies another bed is always the younger.

**Supersaturation:** The unstable state of a solution that contains more solute than its solubility allows.

**Surf:** The breaking or tumbling forward of water waves as they approach the shore.

**Surface wave:** See *Seismic surface wave.*

**Surf zone:** An offshore belt along which the waves collapse into breakers as they approach the shore.

**Suspended load:** The fine sediment kept suspended in a stream because the settling velocity is lower than the upward velocity of eddies.

**Swash:** The landward rush of water from a breaking wave up the slope of the beach.

**S-wave:** The secondary seismic wave, traveling slower than the *P*-wave, and consisting of elastic vibrations transverse to the direction of travel. It cannot penetrate a liquid.

**Swell:** An oceanic water wave with a wavelength on the order of 30 meters or more and a height of perhaps 2 meters or less that may travel great distances from its source.

**Symbiosis:** The interaction of two mutually supporting species who do not compete with or prey upon each other.

**Syncline:** A large fold whose limbs are higher than its center; a fold with the youngest strata in the center (compare *Anticline*).

**System (stratigraphy):** A stratigraphic unit larger than a series, consisting of all the rocks deposited in one period of an era.

**Tableland:** A large elevated region with a relatively low relief surface (see *Plateau*).

**Tar sand:** A sandstone containing the densest asphaltic components of petroleum—the end-product of evaporation of volatile components or of some thickening process.

**Talus:** A deposit of large angular fragments of physically weathered bedrock, usually at the base of a cliff or steep slope.

**Tectonics:** The study of the movements and deformation of the crust on a large scale, including epeirogeny, metamorphism, folding, faulting, and plate tectonics.

**Terminal moraine:** A sinuous ridge of unsorted glacial till deposited by a glacier at the line of its farthest advance.

**Terrestrial planet:** A planet similar in size and composition to the Earth; especially Mars, Earth, Venus, and Mercury.

**Terrestrial sediment:** A deposit of sediment that accumulated above sea level in lakes, alluvial fans, floodplains, moraines, etc., regardless of its present elevation.

**Tetrahedron:** A geometric form of four equal triangular sides, or an arrangement of four atoms around a fifth at the corners of such a form, as in $(SiO_4)^{-4}$.

**Tetrapod:** A vertebrate animal with four legs.

**Texture (rock):** The rock characteristics of grain or crystal size, size variability, rounding or angularity, and preferred orientation.

**Thalweg:** A sinuous imaginary line following the deepest part of a stream.

**Thermal conductivity:** A measure of a rock's capacity for heat conduction.

**Thermal expansion:** The property of increasing in volume as a result of an increase in internal temperature.

**Thermonuclear reaction:** A reaction in which atomic nuclei fuse into new elements with a large release of heat; especially a reaction that is self-sustaining. Occasionally used to include fission reactions as well.

**Thermoremanent magnetization:** A permanent magnetization acquired by igneous rocks in the presence of the Earth's magnetic field as they cool through the *Curie point.*

**Thrust fault:** A dip-slip fault in which the upper block above the fault plane moves up and over the lower block, so that older strata are placed over younger.

**Tidal current:** A horizontal displacement of ocean water under the gravitational influence of Sun and Moon, causing the water to pile up against the coast at high tide and move outward at low tide.

**Tidal flat:** A broad, flat region of muddy or sandy sediment, covered and uncovered in each tidal cycle.

**Till:** An unconsolidated sediment containing all sizes of fragments from clay to boulders deposited by glacial action, usually unbedded.

**Time scale:** The division of geologic history into eras, periods, and epochs accomplished through stratigraphy and paleontology.

**Topographic map:** See *Contour map*; also a schematic drawing of prominent landforms indicated by conventionalized symbols, such as hachures or contours.

**Topography:** The shape of the Earth's surface, above and below sea level; the set of landforms in a region; the distribution of elevations.

**Topset bed:** A horizontal sedimentary bed formed at the top of a delta and overlying the foreset beds.

**Trace element:** An element that appears in minerals in a concentration of less than 1 percent (often less than 0.001 percent).

**Tranquil flow:** See *Streaming flow.*

**Transform fault:** A strike-slip fault connecting the ends of an offset in a mid-ocean ridge. Some pairs of plates slide past each other along transform faults.

**Transgression:** A rise in sea level relative to the land which causes areas to be submerged and marine deposition to begin in that region.

**Transition element:** Elements of atomic number 21 to 29, 38 to 46, and 71 to 78, whose second outermost electron shell is only partially filled.

**Transition zone:** See *Seismic transition zone.*

**Transpiration:** The removal of water from the ground into plants, ultimately to be evaporated into the atmosphere by them.

**Transverse dune:** A dune that has its axis transverse to the prevailing winds or to a current. The upwind or upcurrent side has a gentle slope, and the downwind side lies at the angle of repose.

**Trap (oil):** A sedimentary or tectonic structure that impedes the upward movement of oil and gas and allows it to collect beneath the barrier.

**Travel-time curve:** A curve on a graph of travel time versus distance for the arrival of seismic waves from distant events. Each type of seismic wave has its own curve.

**Travertine:** A terrestrial deposit of limestone formed in caves and around hot springs where cooling, carbonate-saturated groundwater is exposed to the air.

**Trellis drainage:** A system of streams in which tributaries tend to lie in parallel valleys formed in steeply dipping beds in folded belts.

**Trench:** A long and narrow deep trough in the sea floor; interpreted as marking the line along which a plate bends down into a subduction zone.

**Triple junction:** A point that is common to three plates and which must also be the meeting place of three boundary features, such as divergence zones, convergence zones, or transform faults.

**Tsunami:** A large destructive wave caused by sea-floor movements in an earthquake.

**Tufa:** See *Travertine.* Not related to *Tuff.*

**Tuff:** A consolidated rock composed of pyroclastic fragments and fine ash. If particles are melted slightly together from their own heat, it is a "welded tuff."

**Turbidite:** The sedimentary deposit of a turbidity current, typically showing graded bedding and sedimentary structures on the undersides of the sandstones.

**Turbidity current:** A mass of mixed water and sediment that flows downhill along the bottom of an ocean or lake because it is denser than the surrounding water. It may reach high speeds and erode rapidly (see also *Density current*).

**Turbulent flow:** A high-velocity flow in which streamlines are neither parallel nor straight but curled into small tight eddies (compare *Laminar flow*).

**Ultramafic rock:** An igneous rock consisting dominantly of mafic minerals, containing less than 10 percent feldspar. Includes dunite, peridotite, amphibolite, and pyroxenite.

**Unconformity:** A surface that separates two strata. It represents an interval of time in which deposition stopped, erosion removed some sediments and rock, and then deposition resumed (see also *Angular unconformity*).

**Unconsolidated material:** Nonlithified sediment that has no mineral cement or matrix binding its grains.

**Uniformitarianism, Principle of:** The concept that the processes that have shaped the Earth through geologic time are the same as those observable today.

**Unit cell:** The smallest contiguous group of atomic structural units in a mineral that can be repeated in three directions to form a crystal.

**Uplift:** A broad and gentle epeirogenic increase in the elevation of a region without a eustatic change of sea level.

**Upwelling current:** The upward movement of cold bottom water in the sea, which occurs when wind or currents displace the lighter surface water.

**U-shaped valley:** A deep valley with steep upper walls that grade into a flat floor, usually eroded by a glacier.

**Vadose zone:** The region in the ground between the surface and the water table in which pores are not filled with water. Also called the unsaturated zone.

**Valence electron:** An electron of the outermost shell of an atom; one of those most active in bonding.

**Valley glacier:** A glacier that is smaller than a continental glacier or an icecap, and which flows mainly along well-defined valleys, many with tributaries.

**Van der Waals bond:** A bond much weaker than the ionic or covalent, which bonds atoms by small electrostatic attraction.

**Varve:** A thin layer of sediment grading upward from coarse to fine and light to dark, found in a lake bed and representing one year's deposition of glacial outwash.

**Vector:** A mathematical element that has a direction and magnitude, but no fixed position. Examples are force and gravity.

**Vein:** A deposit of foreign minerals within a rock fracture or joint.

**Ventifact:** A rock that exhibits the effects of sand-blasting or "snowblasting" on its surfaces, which become flat with sharp edges in between.

**Vertical exaggeration:** The ratio of the horizontal scale (for example, 100,000:1) to the vertical scale (for example, 500:1) in an illustration.

**Vesicle:** A cavity in an igneous rock that was formerly occupied by a bubble of escaping gas.

**Viscosity:** A measure of resistance to flow in a liquid.

**Volcanic ash:** A volcanic sediment of rock fragments, usually glass, less than 4 millimeters in diameter that is formed when escaping gases force out a fine spray of magma.

**Volcanic ash fall:** A deposit of volcanic ash resting where it was dropped by eruptions and winds.

**Volcanic ash flow:** A mixture of volcanic ash and gases that moves downhill as a density current in the atmosphere.

**Volcanic block:** A pyroclastic rock fragment ranging from about fist- to car-sized.

**Volcanic bomb:** A pyroclastic rock fragment that shows the effects of cooling in flight in its stream-lined or "bread-crust" surface.

**Volcanic breccia:** A pyroclastic rock in which all fragments are more than 2 millimeters in diameter.

**Volcanic cone:** The deposit of lava and pyroclastic materials that has settled close to the volcano's central vent.

**Volcanic dome:** A rounded accumulation around a volcanic vent of congealed lava too viscous to flow away quickly; hence usually rhyolite lava.

**Volcanic dust:** See *Volcanic ash.*

**Volcanic ejecta blanket:** A collective term for all the pyroclastic rocks deposited around a volcano, especially by a volcanic explosion.

**Volcanic emanations:** Gases, especially steam, emitted from a vent or released from lava.

**Volcanic pipe:** The vertical chamber along which magma and gas ascend to the surface; also, a formation of igneous rock that cooled in a pipe and remains after the erosion of the volcano.

**Volcano:** Any opening through the crust that has allowed magma to reach the surface, including the deposits immediately surrounding this vent.

**V-shaped valley:** A valley whose walls have a more-or-less uniform slope from top to bottom, usually formed by stream erosion.

**Wadi:** A steep-sided valley containing an intermittent stream in an arid region.

**Warping:** In tectonics, refers to the gentle, regional bending of the crust, which occurs in epeirogenic movements.

**Water mass:** A mass of water that fills part of an ocean or lake and is distinguished by its uniform physical and chemical properties, such as temperature and salinity.

**Water table:** A gently-curved surface below the ground at which the vadose zone ends and the phreatic zone begins; the level to which a well would fill with water.

**Wave-cut terrace:** A level surface formed by wave erosion of coastal bedrock to the bottom of the turbulent breaker zone. May appear above sea level if uplifted.

**Wavelength:** The distance between two successive peaks, or between troughs, of a cyclic propagating disturbance.

**Wave steepness:** The maximum height or amplitude of a wave divided by its wavelength.

**Weathering:** The set of all processes that decay and break up bedrock, by a combination of physically fracturing or chemical decomposition.

**Xenolith:** A piece of country rock found engulfed in an intrusion.

**X-ray diffraction:** In mineralogy, the process of identifying mineral structures by exposing crystals to X-rays and studying the resulting diffraction pattern.

**Youth (geomorphology):** A stage in the geomorphic cycle in which a landscape has just been uplifted and is beginning to be dissected by canyons cut by young streams.

**Zeolite:** A class of silicates containing $H_2O$ in cavities within the crystal structure. Formed by alteration at low temperature and pressure of other silicates, often volcanic glass.

**Zoned crystal:** A single crystal of one mineral that has a different chemical composition in its inner and outer parts. Formed from minerals belonging to a solid-solution series, and caused by the changing concentration of elements in a cooling magma that results from crystals settling out.

# Index

Page numbers in **boldface** refer to a definition.
Page numbers followed by an asterisk indicate
an illustration. The Appendixes and Glossary
are not covered by this index.

# The Plates of Earth's Lithosphere

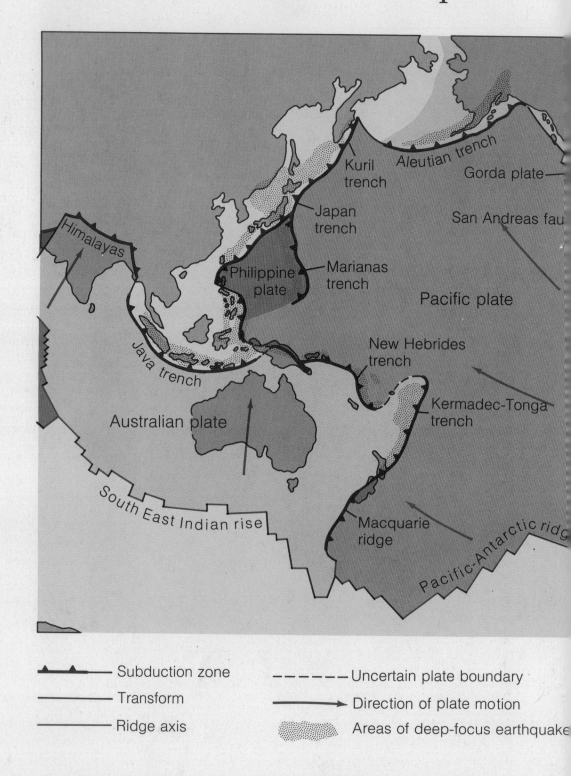

Kuril trench

Aleutian trench

Gorda plate

Japan trench

San Andreas fau

Marianas trench

Philippine plate

Pacific plate

Himalayas

New Hebrides trench

Java trench

Kermadec-Tonga trench

Australian plate

South East Indian rise

Macquarie ridge

Pacific-Antarctic ridg

Subduction zone  ———  Uncertain plate boundary

Transform  ———  Direction of plate motion

Ridge axis  ———  Areas of deep-focus earthquake